SI UNITS USED IN MECHANICS

Quantity	Unit	SI
(Base Units)		
Length	meter°	m
Mass	kilogram	kg
Time	second	s
(Derived Units)		
Acceleration, linear	meter/second^2	m/s^2
Acceleration, angular	radian/second^2	rad/s^2
Area	meter^2	m^2
Density	kilogram/meter^3	kg/m^3
Force	newton	N ($= $ kg$\cdot$m/s^2)
Frequency	hertz	Hz ($= $ 1/s)
Impulse, linear	newton-second	N$\cdot$s
Impulse, angular	newton-meter-second	N$\cdot$m$\cdot$s
Moment of force	newton-meter	N$\cdot$m
Moment of inertia, area	meter^4	m^4
Moment of inertia, mass	kilogram-meter^2	kg$\cdot$m^2
Momentum, linear	kilogram-meter/second	kg$\cdot$m/s ($= $ N$\cdot$s)
Momentum, angular	kilogram-meter^2/second	kg$\cdot$m^2/s ($= $ N$\cdot$m$\cdot$s)
Power	watt	W ($= $ J/s $= $ N$\cdot$m/s)
Pressure, stress	pascal	Pa ($= $ N/m^2)
Product of inertia, area	meter^4	m^4
Product of inertia, mass	kilogram-meter^2	kg$\cdot$m^2
Spring constant	newton/meter	N/m
Velocity, linear	meter/second	m/s
Velocity, angular	radian/second	rad/s
Volume	meter^3	m^3
Work, energy	joule	J ($= $ N$\cdot$m)
(Supplementary and Other Acceptable Units)		
Distance (navigation)	nautical mile	($= $ 1.852 km)
Mass	ton (metric)	t ($= $ 1000 kg)
Plane angle	degrees (decimal)	°
Plane angle	radian	—
Speed	knot	(1.852 km/h)
Time	day	d
Time	hour	h
Time	minute	min

° Also spelled *metre*.

SI UNIT PREFIXES

Multiplication Factor	Prefix	Symbol
1 000 000 000 000 $= 10^{12}$	terra	T
1 000 000 000 $= 10^{9}$	giga	G
1 000 000 $= 10^{6}$	mega	M
1 000 $= 10^{3}$	kilo	k
100 $= 10^{2}$	hecto	h
10 $= 10$	deka	da
0.1 $= 10^{-1}$	deci	d
0.01 $= 10^{-2}$	centi	c
0.001 $= 10^{-3}$	milli	m
0.000 001 $= 10^{-6}$	micro	μ
0.000 000 001 $= 10^{-9}$	nano	n
0.000 000 000 001 $= 10^{-12}$	pico	p

SELECTED RULES FOR WRITING METRIC QUANTITIES

1. (a) Use prefixes to keep numerical values generally between 0.1 and 1000.
   (b) Use of the prefixes hecto, decka, deci, and centi should be generally avoided except for certain areas or volumes where the numbers would be otherwise awkward.
   (c) Use prefixes only in the numerator of unit combinations. The one exception is the base unit kilogram. (*Example:* write kN/m not N/mm; J/kg not mJ/g)
   (d) Avoid double prefixes. (*Example:* write GN not kMN)
2. Unit designations
   (a) Use a dot for multiplication of units. (*Example:* write N$\cdot$m not Nm)
   (b) Avoid ambiguous double solidus. (*Example:* write N/m^2 not N/m/m)
   (c) Exponents refer to entire unit. (*Example:* mm^2 means (mm)^2)
3. Number grouping
   Use a space rather than a comma to separate numbers in groups of three, counting from the decimal point in both directions. (*Example:* 4 607 321.048 72)
   Space may be omitted for numbers of four digits. (*Example:* 4296 or 0.0476)

ENGINEERING MECHANICS

DYNAMICS

ENGINEERING MECHANICS

VOLUME 2

DYNAMICS

THIRD EDITION

J. L. MERIAM
University of California
Santa Barbara

L. G. KRAIGE
Virginia Polytechnic Institute and
State University

JOHN WILEY & SONS, Inc.
New York • Chichester • Brisbane • Toronto • Singapore

SI/ENGLISH VERSION

Acquisitions Editor	Charity Robey
Marketing Manager	Susan Elbe
Senior Production Supervisor	Nancy Prinz
Cover Designer	Pedro Noa
Manufacturing Manager	Lorraine Fumoso
Copy Editing Supervisor	Deborah Herbert
Photo Researcher	Jennifer Atkins
Illustration	John Balbalis

This book was set in 9.5/12 Century Schoolbook by Waldman Graphics and printed and bound by Von Hoffmann.

Recognizing the importance of preserving what has been written, it is a policy of John Wiley & Sons, Inc. to have books of enduring value published in the United States printed on acid-free paper, and we exert our best efforts to that end.

Library of Congress Cataloging in Publication Data:

Meriam, J. L. (James L.)
    Engineering mechanics / J. L. Meriam, L. G. Kraige. — 3rd ed.
      p.   cm.
    "SI/English version."
    Includes indexes.
    Contents: v. 2. Dynamics.
    ISBN 0-471-60293-0 (v. 2)
    1. Mechanics, Applied.   I. Kraige, L. G. (L. Glenn) II. Title.
TA350.M458   1992
620.1—dc20   91-42953
                                                   CIP

Printed in the United States of America

10 9 8 7 6 5 4 3 2 1

FOREWORD

The innovations and contributions of Dr. James L. Meriam to the field of engineering mechanics cannot be overstated. He has undoubtedly had as much influence on instruction in mechanics during the last forty years as any one individual. His first books on mechanics in 1951 literally reconstructed undergraduate mechanics and became the definitive textbooks for the decades that followed. His texts were logically organized, easy to read, directed to the average engineering undergraduate, and were packed with exciting examples of real-life engineering problems which were superbly illustrated. These books became the model for other engineering mechanics texts in the 1950s and beyond.

Dr. Meriam began his work in mechanical engineering at Yale University where he earned his B.E., M. Eng., and Ph.D. degrees. He had early industrial experience with Pratt and Whitney Aircraft and the General Electric Company, which stimulated his first contributions to mechanics in mathematical and experimental stress analysis. During the Second World War he served in the U.S. Coast Guard.

Dr. Meriam was a member of the faculty of the University of California, Berkeley, for twenty-one years where he served as Professor of Engineering Mechanics, Assistant Dean of Graduate Studies, and Chairman of the Division of Mechanics and Design. In 1963 he became Dean of Engineering at Duke University where he devoted his full energies to the development of its School of Engineering. In 1972 Professor Meriam followed his desire to return to full-time teaching and served as Professor of Mechanical Engineering at California Polytechnic State University, where he retired in 1980. During the following ten years he served as visiting professor at the University of California, Santa Barbara, retiring for the second time in 1990. Professor Meriam has always placed great emphasis on teaching, and this trait has been recognized by his students wherever he has taught. At Berkeley in 1963 he was the first recipient of the Outstanding Faculty Award of Tau Beta Pi, given primarily for excellence in teaching, and in 1978 he received the Distinguished Educator Award for Outstanding Service to Engineering Mechanics Education from the American Society for Engineering Education.

Professor Meriam was the first author to show clearly how the method of virtual work in statics can be employed to solve a class of problems largely neglected by previous authors. In dynamics, plane motion became understandable, and in his later editions, three-dimensional kinematics and kinetics received the same treatment. He is credited with original developments in the theory of variable-mass dynamics, which are contained in his *Dynamics, Second Edition* (1971). Professor Meriam has also been a leader in promoting the use of SI units, and his *SI Versions* of *Statics* and *Dynamics* (1975) were the first mechanics textbooks in SI units in this country.

Dr. L. Glenn Kraige, coauthor for the second time in the *Engineering Mechanics* series, has also made significant contributions to mechanics education. Dr. Kraige earned his B.S., M.S., and Ph.D. degrees at the University of Virginia, principally in aerospace engineering, and he currently serves as Professor of Engineering Science and Mechanics at Virginia Polytechnic Institute and State University.

In addition to his recognized research and publications in the field of spacecraft dynamics, Professor Kraige has devoted his attention to the teaching of mechanics at both introductory and advanced levels. His outstanding teaching has been widely recognized and has earned him several awards, including, in 1988, an AT&T award for outstanding teaching in the Southeastern Section of the American Society for Engineering Education and, also in 1988, the Outstanding Educator Award from the State Council of Higher Education for the Commonwealth of Virginia. In his teaching, Professor Kraige stresses the development of analytical capabilities along with the strengthening of physical insight and engineering judgment. In the mid 1980s, he was a leader in the development of motion simulation software for use on personal computers. More recently, he has begun a long-term effort in the area of multimedia approaches to the instruction and learning of statics and dynamics.

The third edition of *Engineering Mechanics* continues the same high standards set by previous editions and adds new features of help and interest to students. It contains a vast collection of interesting and instructive problems. Analysis and applications are the cornerstones of a successful learning experience in engineering mechanics, and J. L. Meriam and L. G. Kraige have shown again that they are the best at melding these essential characteristics.

Robert F. Steidel, Jr.
Professor Emeritus of Mechanical Engineering
University of California, Berkeley

PREFACE
To the Student

As you undertake the study of engineering mechanics, first statics and then dynamics, you will be building a foundation of analytical capability for the solution of a great variety of engineering problems. Modern engineering practice demands a high level of analytical capability, and you will find that your study of mechanics will help you immensely in developing this capacity.

In engineering mechanics we learn to construct and to solve mathematical models which describe the effects of force and motion on a variety of structures and machines that are of concern to engineers. In applying our principles of mechanics we formulate these models by incorporating appropriate physical assumptions and mathematical approximations. In both the formulation and solution of mechanics problems you will have frequent occasion to use your background in plane and solid geometry, scalar and vector algebra, trigonometry, analytic geometry, and calculus. Indeed, you are likely to discover new significance to these mathematical tools as you make them work for you in mechanics.

Your success in mechanics (and throughout engineering) will be highly contingent on developing a well-disciplined method of attack from hypothesis to conclusion in which the applicable principles are applied rigorously. Years of experience in teaching and engineering disclose the importance of developing the ability to represent one's work in a clear, logical, and concise manner. Mechanics is an excellent place in which to develop these habits of logical thinking and effective communication.

Engineering Mechanics contains a large number of sample problems in which the solutions are presented in detail. Also included in these examples are helpful observations that mention common errors and pitfalls to be avoided. In addition, the book contains a large selection of simple, introductory problems and problems of intermediate difficulty to help you gain initial confidence and understanding of each new topic. Also included are many problems that illustrate significant and contemporary engineering situations to stimulate your interest and help you to develop an appreciation for the many applications of mechanics in engineering.

We are pleased to extend our encouragement to you as a student of mechanics. We hope this book will provide both help and stimulation as you develop your background in engineering.

J. L. Meriam

Santa Barbara, California
February 1992

L. Glenn Kraige

Blacksburg, Virginia
February 1992

PREFACE
To the Instructor

The primary purpose of the study of engineering mechanics is to develop the capacity to predict the effects of force and motion in the course of carrying out the creative design function of engineering. Successful prediction requires more than a mere knowledge of the physical and mathematical principles of mechanics. Prediction also requires the ability to visualize physical configurations in terms of real materials, actual constraints, and the practical limitations that govern the behavior of machines and structures. One of our primary objectives in teaching mechanics is to help the student develop this ability to visualize, which is so vital to problem formulation. Indeed, the construction of a meaningful mathematical model is often a more important experience than its solution. Maximum progress is made when the principles and their limitations are learned together within the context of engineering application.

Courses in mechanics are often regarded by students as a difficult requirement and frequently as an uninteresting academic hurdle as well. The difficulty stems from the extent to which reasoning from fundamentals, as distinguished from rote learning, is required. The lack of interest that is frequently experienced is due primarily to the extent to which mechanics is presented as an academic discipline often lacking in engineering purpose and challenge. This attitude is traceable to the frequent tendency in the presentation of mechanics to use problems mainly as a vehicle to illustrate theory rather than to develop theory for the purpose of solving problems. When the first view is allowed to predominate, problems tend to become overly idealized and unrelated to engineering with the result that the exercise becomes dull, academic, and uninteresting. This approach deprives the student of much of the valuable experience in formulating problems and thus of discovering the need for and meaning of theory. The second view provides by far the stronger motive for learning theory and leads to a better balance between theory and application. The crucial role of interest and purpose in providing the strongest possible motive for learning cannot be overemphasized. Furthermore, we should stress the view that, at best, theory can only approximate the real world of mechanics rather than the view that the real world approximates the theory. This difference in philosophy is indeed basic and distinguishes the *engineering* of mechanics from the *science* of mechanics.

During the past thirty years there has been a strong trend in engineering education to increase the extent and level of theory in the engineering-science courses. Nowhere has this trend been more evident than in mechanics courses. To the extent that students are prepared to handle the accelerated treatment, the trend is beneficial. There is evidence and justifiable concern, however, that a significant disparity has more recently appeared between coverage and comprehension. Among the contributing factors we note three trends. First, emphasis on the geometric and physical meanings of prerequisite mathematics appears to have diminished. Second, there has been a significant reduction and even elimination of instruction in graphics, which in the past enhanced the visualization and representation of mechanics problems. Third, in advancing the mathematical level of our treatment of mechanics, there has been a tendency to allow the notational manipulation of vector operations to mask or replace geometric visualization. Mechanics is inherently a subject that depends on geometric and physical perception, and we should increase our efforts to develop this ability.

One of our responsibilities as teachers of mechanics is to use the mathematics that is most appropriate for the problem at hand. The use of vector notation for one-dimensional problems is usually trivial; for two-dimensional problems it is often optional; but for three-dimensional problems it is usually essential. As we introduce vector operations in two-dimensional problems, it is especially important that their geometric meaning be emphasized. A vector equation is brought to life by a sketch of the corresponding vector polygon, which often discloses through its geometry the shortest solution. There are, of course, many mechanics problems where the complexity of variable interdependence is beyond the normal powers of visualization and physical perception, and reliance on analysis is essential. Nevertheless, our students become better engineers when their abilities to perceive, visualize, and represent are developed to the fullest.

As teachers of engineering mechanics, we have the strongest obligation to the engineering profession to set reasonable standards of performance and to uphold them. In addition, we have a serious responsibility to encourage our students to think for themselves. Too much help with details that students should be reasonably able to handle from prerequisite subjects can be as bad as too little help and can easily condition them to become overly dependent on others rather than to exercise their own initiative and ability. Also, when mechanics is subdivided into an excessive number of small compartments, each with detailed and repetitious instructions, students can have difficulty seeing the "forest for the trees" and, consequently, will fail to perceive the unity of mechanics and the far-reaching applicability of its few basic principles and methods.

This third edition of *Engineering Mechanics,* as with previous editions, is written with the foregoing philosophy in mind. It is intended primarily for the first engineering course in mechanics, generally taught in the second year of study. *Engineering Mechanics* is written in a style that is both concise and friendly. The major emphasis is on basic principles and methods rather than on a multitude of special cases. Strong effort has been made to show both the cohesiveness of the relatively few fundamental ideas and the great variety of problems that these few ideas will solve. A major feature of the book is the extensive treatment of sample problems, which are presented in a single-page format for convenient self-study. In addition to presenting the solution in detail, each sample problem also contains comments and cautions keyed to salient points in the solution and printed in colored type.

Volume 2, Dynamics, contains 114 sample problems and 1500 unsolved problems from which a wide choice of assignments can be made. Of these problems more than 50 percent are new to the Third Edition. In recognition of the need for the predominant emphasis on SI units, there are approximately two problems in SI units for every one in U.S. customary units. This apportionment between the two sets of units permits anywhere from a 50-50 emphasis to a 100 percent SI treatment. Many practical problems and examples of interesting engineering situations drawn from a wide range of applications are represented in the problem collection.

In a feature new to the Third Edition, most problem sets are divided into two sections entitled *Introductory Problems* and *Representative Problems*. In the first section are simple, uncomplicated problems designed to help students gain confidence with the new topic, while most of the problems in the second section are of average difficulty and length. The problems are arranged generally in order of increasing difficulty; near the end of the *Representative Problems* are more difficult exercises, which are marked with the symbol ▶. Computer-oriented problems are in a special section at the conclusion of the Review Problems at the end of each chapter. The answers to all odd-numbered problems and to all difficult problems have been provided. Simple numerical values have been used throughout so as not to complicate the solutions and divert attention from the principles. All numerical solutions have been carried out and checked with an electronic calculator without rounding intermediate values. Consequently, the final answers should be correct to within the number of significant figures cited.

A special note on the use of computers is in order. We wish to emphasize that the experience of formulating problems, where reason and judgment are developed, is vastly more important for the student than is the manipulative exercise in carrying out the solution. For this reason, we believe that computer usage must be carefully controlled. At the present time, the processes of constructing free-body diagrams and formulating governing equations are best done with pencil and paper. On the other hand, there are instances in which the *solution* to the governing equations can best be carried out and displayed via the computer. Computer-oriented problems should be genuine in the sense that there is a condition of design or criticality to be found, rather than "makework" problems in which some parameter is varied for no apparent reason other than to force artificial use of the computer. These thoughts have been in mind as we have designed the computer-oriented problems in the Third Edition. To conserve adequate time for problem formulation, it is suggested that the student be assigned only a limited number of the computer problems.

The logical division between particle dynamics and rigid-body dynamics has been preserved, with each part treating the kinematics prior to the kinetics. This arrangement greatly facilitates a more thorough and rapid excursion in rigid-body dynamics with the prior benefit of a comprehensive introduction to particle dynamics in Chapters 2 and 3.

Chapter 3 on particle kinetics focuses on the three basic methods, force-mass-acceleration, work-energy, and impulse-momentum. The special topics of impact, central-force motion, and relative motion are grouped together in Section D of Chapter 3 on special applications and serve as optional material to be assigned according to instructor preference and available time. With this arrangement, the attention of the student is focused more strongly on the three basic approaches to kinetics, which are developed in Sections A, B, and C of the chapter.

Chapter 4 on systems of particles is an extension of the principles of motion for a single particle and develops the general relationships that are so basic to the modern comprehension of dynamics. The chapter also includes the topics of steady mass flow and variable mass, which may be considered as optional material depending on the time available.

In Chapter 5 on the kinematics of rigid bodies in plane motion where the equations of relative velocity and relative acceleration are encountered, emphasis is placed jointly on solution by vector geometry and solution by vector algebra. This dual approach serves the purpose of reinforcing the meaning of vector mathematics.

In Chapter 6 on the kinetics of rigid bodies, we place great emphasis on the basic equations which govern all categories of plane motion. Strong dependence is placed on the identification of knowns and unknowns and on the necessary and sufficient motion equations that guarantee solution. Special emphasis is also placed on forming the direct equivalence between the actual applied forces and couples and their $m\bar{a}$ and $\bar{I}\alpha$ resultants. In this way, the versatility of the moment principle is emphasized, and the student is encouraged to think directly in terms of resultant dynamics effects.

Chapter 7, which may be treated as optional, provides a basic introduction to three-dimensional dynamics which is sufficient to solve many of the more common space-motion problems. For students who later pursue more advanced work in dynamics, Chapter 7 will provide a solid foundation. Gyroscopic motion with steady precession is treated in two ways. The first approach makes use of the analogy between the relation of force and linear-momentum vectors and the relation of moment and angular-momentum vectors. With this treatment the student can understand the gyroscopic phenomenon of steady precession and can handle most of the engineering problems on gyros without a detailed study of three-dimensional dynamics. The second approach makes use of the more general momentum equations for three-dimensional rotation where all components of momentum are accounted for.

Chapter 8 is devoted to the topic of vibrations. This full-chapter coverage will be especially useful for engineering students whose only exposure to vibrations is acquired in the basic dynamics course.

Moments and products of inertia of mass are presented in Appendix B. Appendix C contains a summary review of selected topics of elementary mathematics as well as several numerical techniques that the student should be prepared to use in computer-solved problems.

We wish to specially cite the outstanding contributions to this series of mechanics books over a period of twenty-five years by illustrator John Balbalis, who died in October 1991. His dedication to high standards of illustrative achievement greatly increased the educational potential of the *Engineering Mechanics* series by providing clarity, reality, and interest to the many thousands of students who have been challenged by his efforts.

Special recognition is due Dr. A. L. Hale of the Bell Telephone Laboratories for his continuing contribution in the form of invaluable suggestions and accurate checking of the manuscript. Dr. Hale has rendered similar service to all previous versions in this entire series of mechanics books, and his input has been a great asset. Appreciation is expressed to Professor J. M. Henderson of the University of California, Davis, for helpful comments and suggestions of selected problems. Also acknowledged are the observations and constructive comments over a period of years of Professor Alfonso Diaz-Jiménez of Bogota, Colombia. A number of members of the Department of Engineering Science and Mechanics at Virginia Polytechnic Institute and State University have offered helpful suggestions, including Professors Norman E. Dowling, J. Wallace Grant, Scott L. Hendricks, Arpad A. Pap, Saad A. Ragab, and George W. Swift. The contribution by the staff of John Wiley &

Sons, Inc., including editor Charity Robey, reflects a high degree of professional competence and is duly recognized. The support, in the form of a study-research leave, of VPI & SU is acknowledged. Finally, we acknowledge the patience, support, and assistance of our wives, Julia and Dale, during the many hours required to prepare this manuscript.

J. L. Meriam

Santa Barbara, California
February 1992

L. Glenn Kraige

Blacksburg, Virginia
February 1992

CONTENTS

PART I
DYNAMICS OF PARTICLES

The engineering of space flight is one example of the extensive application of the basic principles of dynamics in our modern technical world. The generation of rocket propulsion, the precise prediction of required orbits, and the control and stability of orbital maneuvers are among the many challenges that require a thorough knowledge of dynamics.

INTRODUCTION TO DYNAMICS

1

1/1 *HISTORY AND MODERN APPLICATIONS*

Dynamics is that branch of mechanics which deals with the motion of bodies under the action of forces. The study of dynamics in engineering usually follows the study of statics, which deals with the action of forces on bodies at rest. Dynamics has two distinct parts: *kinematics*, which is the study of motion without reference to the forces which cause motion, and *kinetics*, which relates the action of forces on bodies to their resulting motions. The student of engineering will find that a thorough comprehension of dynamics will provide one of his or her most useful and powerful tools for analysis in engineering.

Historically, dynamics is a relatively recent subject compared with statics. The beginning of a rational understanding of dynamics is credited to Galileo (1564–1642), who made careful observations concerning bodies in free fall, motion on an inclined plane, and motion of the pendulum. He was largely responsible for bringing a scientific approach to the investigation of physical problems. Galileo was continually under severe criticism for refusing to accept the established beliefs of his day, such as the philosophies of Aristotle which held, for example, that heavy bodies fall more rapidly than light bodies. The lack of accurate means for the measurement of time was a severe handicap to Galileo, and further significant development in dynamics awaited the invention of the pendulum clock by Huygens in 1657. Newton (1642–1727), guided by Galileo's work, was able to make an accurate formulation of the laws of motion and, hence, to place dynamics on a sound basis. Newton's famous work was published in the first edition of his *Principia*,* which is generally recognized as one of the greatest of all recorded contributions to knowledge. In addition to stating the laws governing the motion of a particle, Newton was the first to correctly formulate the law of universal gravitation. Although his mathematical

---

*The original formulations of Sir Isaac Newton may be found in the translation of his *Principia* (1687), revised by F. Cajori, University of California Press, 1934.

description was accurate, he felt that the concept of remote transmission of gravitational force without a supporting medium was an absurd notion. Following Newton's time, important contributions to mechanics were made by Euler, D'Alembert, Lagrange, Laplace, Poinsot, Coriolis, Einstein, and others.

In terms of engineering application, dynamics is an even more recent science. Only since machines and structures have operated with high speeds and appreciable accelerations has it been necessary to make calculations based on the principles of dynamics rather than on the principles of statics. The rapid technological developments of the present day require increasing application of the principles of mechanics, particularly dynamics. These principles are basic to the analysis and design of moving structures, to fixed structures subject to shock loads, to robotic devices, to automatic control systems, to rockets, missiles, and spacecraft, to ground and air transportation vehicles, to electron ballistics of electrical devices, and to machinery of all types such as turbines, pumps, reciprocating engines, hoists, machine tools, etc. Students whose interests lead them into one or more of these and many other activities will find a constant need for applying the fundamentals of dynamics.

1/2 BASIC CONCEPTS

The concepts basic to mechanics were set forth in Art. 1/2 of *Vol. 1 Statics*. They are summarized here along with additional comments of special relevance to the study of dynamics.

Space is the geometric region occupied by bodies. Position in space is determined relative to some geometric reference system by means of linear and angular measurements. The basic frame of reference for the laws of Newtonian mechanics is the *primary inertial system* or *astronomical frame of reference*, which is an imaginary set of rectangular axes assumed to have no translation or rotation in space. Measurements show that the laws of Newtonian mechanics are valid for this reference system as long as any velocities involved are negligible compared with the speed of light, which is 300 000 km/s or 186,000 mi/sec. Measurements made with respect to this reference are said to be *absolute*, and this reference system may be considered "fixed" in space. A reference frame attached to the surface of the earth has a somewhat complicated motion in the primary system, and a correction to the basic equations of mechanics must be applied for measurements made relative to the earth's reference frame. In the calculation of rocket and space flight trajectories, for example, the absolute motion of the earth becomes an important parameter. For most engineering problems of machines and structures that remain on the earth's surface, the corrections are extremely small and may be neglected. For these problems the laws of mechanics may be applied directly for measurements made relative to the earth, and in a practical sense such measurements will be referred to as *absolute*.

Time is a measure of the succession of events and is considered an absolute quantity in Newtonian mechanics.

Mass is the quantitative measure of the inertia or resistance to change in motion of a body. Mass can also be considered as the quantity of matter in a body as well as the property that gives rise to gravitational attraction.

Force is the vector action of one body on another. The properties of forces have been thoroughly treated in *Vol. 1 Statics*.

A ***particle*** is a body of negligible dimensions. Also, when the dimensions of a body are irrelevant to the description of its motion or the action of forces on it, the body may be treated as a particle. An airplane, for example, may be treated as a particle for the description of its flight path.

A ***rigid body*** is a body whose changes in shape are negligible compared with the overall dimensions of the body or with the changes in position of the body as a whole. As an example of the assumption of rigidity, the small flexural movement of the wing tip of an airplane flying through turbulent air is clearly of no consequence to the description of the motion of the airplane as a whole along its flight path. For this purpose, then, the treatment of the airplane as a rigid body offers no complication. On the other hand, if the problem is one of examining the internal stresses in the wing structure due to changing dynamic loads, then the deformable characteristics of the structure would have to be examined, and for this purpose the airplane could no longer be considered a rigid body.

Vector and ***scalar*** quantities have been treated extensively in *Vol. 1 Statics*, and their distinction should be perfectly clear by now. Scalar quantities are printed in lightface italic type, and vectors are shown in boldface type. Thus, V is the scalar magnitude of the vector **V**. It is important that we use an identifying mark, such as an underline $\underline{V}$, for all handwritten vectors to take the place of the boldface designation in print. For two nonparallel vectors recall, for example, that $\mathbf{V}_1 + \mathbf{V}_2$ and $V_1 + V_2$ have two entirely different meanings.

It is assumed that the reader is familiar with the geometry and algebra of vectors through previous study of statics and mathematics. Students who need to review these topics will find a brief summary of them in Appendix C along with other mathematical relations that find frequent use in mechanics. Experience has shown that the geometry of mechanics is often a source of difficulty. Mechanics by its very nature is geometrical, and students should bear this in mind as they review their mathematics. In addition to vector algebra, dynamics requires the use of vector calculus, and the essentials of this topic will be developed in the text as they are needed.

Dynamics involves the frequent use of time derivatives of both vectors and scalars. As a notational shorthand, a dot over a quantity will frequently be used to indicate a derivative with respect to time. Thus, $\dot{x}$ means dx/dt and $\ddot{x}$ stands for d^2x/dt^2.

1/3 NEWTON'S LAWS

Newton's three laws of motion, stated in Art. 1/4 of *Vol. 1 Statics*, are restated here because of their special significance to dynamics. In modern terminology they are

Law I.   A particle remains at rest or continues to move in a straight line with a constant velocity if there is no unbalanced force acting on it.

Law II.   The acceleration of a particle is proportional to the resultant force acting on it and is in the direction of this force.*

Law III.   The forces of action and reaction between interacting bodies are equal in magnitude, opposite in direction, and collinear.

These laws have been verified by countless physical measurements. The first two laws hold for measurements made in an absolute frame of reference, but are subject to some correction when the motion is measured relative to a reference system having acceleration, such as one attached to the earth's surface.

Newton's second law forms the basis for most of the analysis in dynamics. For a particle of mass m subjected to a resultant force **F**, the law may be stated as

$$\boxed{\mathbf{F} = m\mathbf{a}} \tag{1/1}$$

where **a** is the resulting acceleration measured in a nonaccelerating frame of reference. Newton's first law is a consequence of the second law since there is no acceleration when the force is zero, and the particle is either at rest or is moving with constant velocity. The third law constitutes the principle of action and reaction with which we should be thoroughly familiar from our work in statics.

1/4 UNITS

Both the International System of metric units (SI) and the U.S. customary system of units are defined and used in *Vol. 2 Dynamics*, although a stronger emphasis is placed on the metric system since

---

*To some it is preferable to interpret Newton's second law as meaning that the resultant force acting on a particle is proportional to the time rate of change of momentum of the particle and that this change is in the direction of the force. Both formulations are equally correct when applied to a particle of constant mass.

it is replacing the U.S. customary system. To become familiar with each system, it is necessary to think directly in that system. Although in engineering practice numerical conversion from one system to the other must be accomplished for some years, it would be a mistake to assume that familiarity with the new system can be achieved simply by the conversion of numerical results from the old system. Tables defining the SI units and giving numerical conversions between U.S. customary and SI units are included inside the front cover of the book. Charts comparing selected quantities in SI and U.S. customary units are included inside the back cover of the book to facilitate conversion and to help establish a feel for the relative size of units in both systems.

The four fundamental quantities of mechanics and their units and symbols for the two systems are summarized in the following table:

QUANTITY	DIMENSIONAL SYMBOL	SI UNITS		U.S. CUSTOMARY UNITS	
		UNIT	SYMBOL	UNIT	SYMBOL
Mass	M	Base units { kilogram	kg	slug	—
Length	L	meter*	m	Base units { foot	ft
Time	T	second	s	second	sec
Force	F	newton	N	pound	lb

*Also spelled *metre*.

In SI the units for mass, length, and time are taken as base units and the units for force are derived from Newton's second law of motion, Eq. 1/1. In the U.S. customary system the units for force, length, and time are base units and the units for mass are derived from the second law. The SI system is termed an *absolute* system since mass is taken to be an absolute or base quantity. The U.S. customary system is termed a *gravitational* system since force (as measured from gravitational pull) is taken as a base quantity. This distinction is a fundamental difference between the two systems.

In SI units by definition one newton is that force which will give a one-kilogram mass an acceleration of one meter per second squared. In the U.S. customary system a 32.1740-pound mass (1 slug) will have an acceleration of one foot per second squared when acted on by a force of one pound. Thus, for each system we have from Eq. 1/1

SI UNITS	U.S. CUSTOMARY UNITS
$(1 \text{ N}) = (1 \text{ kg})(1 \text{ m/s}^2)$	$(1 \text{ lb}) = (1 \text{ slug})(1 \text{ ft/sec}^2)$
$\text{N} = \text{kg}\cdot\text{m/s}^2$	$\text{slug} = \text{lb-sec}^2/\text{ft}$

In SI units each unit and symbol stands for only one quantity. The kilogram is to be used *only* as a unit of mass and *never* as a unit of force. On the other hand, in U.S. customary units the *pound* is used both as a unit of force (lbf) and a unit of mass (lbm). The distinction between the meaning as force or mass is usually clear from the situation at hand. The symbol for pound force is most frequently written merely lb.

Additional quantities used in mechanics and their equivalent base units will be defined as they are introduced in the chapters that follow. However, for convenient reference these quantities along with their SI base units are listed in one place in the first table inside the front cover of the book.

Detailed guidelines have been established for the consistent use of SI units, and these guidelines have been followed throughout this book. The most essential ones are summarized inside the front cover, and the student should observe these rules carefully.

1/5 GRAVITATION

Newton's law of gravitation that governs the mutual attraction between bodies is

$$F = G \frac{m_1 m_2}{r^2} \tag{1/2}$$

where

F = the mutual force of attraction between two particles
G = a universal constant called the constant of gravitation
m_1, m_2 = the masses of the two particles
r = the distance between the centers of the particles

The value of the gravitational constant from experimental data is $G = 6.673(10^{-11})$ m^3/(kg·s^2). The only gravitational force of appreciable magnitude is the force due to the attraction of the earth. It was shown in *Vol. 1 Statics*, for example, that each of two iron spheres 100 mm in diameter is attracted to the earth with a gravitational force of 37.1 N, which is called its *weight*, but the force of mutual attraction between them if they are just touching is only 0.000 000 095 1 N.

Since the gravitational attraction or weight of a body is a force, it should always be expressed in force units, newtons (N) in SI units and pounds force (lb) in U.S. customary units. Unfortunately, the mass unit kilogram (kg) has been used widely in common practice as a measure of weight. When expressed in kilograms, the word "weight" technically means mass. To avoid confusion, the word "weight" in this book shall be restricted to mean the force of gravitational attraction, and it will always be expressed in newtons or pounds force.

The force of gravitational attraction of the earth on a body depends on the position of the body relative to the earth. If the earth were a perfect sphere of the same volume, a body with a mass of exactly 1 kg would be attracted to the earth by a force of 9.825 N on the earth's surface, 9.822 N at an altitude of 1 km, 9.523 N at an altitude of 100 km, 7.340 N at an altitude of 1000 km, and 2.456 N at an altitude equal to the mean radius 6371 km of the earth. It is at once apparent that the variation in gravitational attraction of high-altitude rockets and spacecraft becomes a major consideration.

Every object that is allowed to fall in a vacuum at a given position near the earth's surface will have the same acceleration g, as can be seen by combining Eqs. 1/1 and 1/2 and canceling the term representing the mass of the falling object. This combination gives

$$g = \frac{Gm_e}{R^2}$$

where m_e is the mass of the earth and R is the radius of the earth.* The mass m_e and the mean radius R of the earth have been found through experimental measurements to be $5.976(10^{24})$ kg and $6.371(10^6)$ m, respectively. These values, together with the value of G already cited, when substituted into the expression for g, give a mean value of $g = 9.825$ m/s^2.

The acceleration due to gravity as determined from the gravitational law is the acceleration that would be measured from a set of axes whose origin is at the center of the earth but which does not rotate with the earth. With respect to these "fixed" axes, then, this value may be termed the absolute value of g. Because of the fact that the earth rotates, the acceleration of a freely falling body as measured from a position attached to the earth's surface is slightly less than the absolute value. Accurate values of the gravitational acceleration as measured relative to the earth's surface account for the fact that the earth is a rotating oblate spheroid with flattening at the poles. These values may be calculated to a high degree of accuracy from the 1980 International Gravity Formula, which is

$$g = 9.780\ 327(1 + 0.005\ 279 \sin^2 \gamma + 0.000\ 023 \sin^4 \gamma + \cdots)$$

where γ is the latitude and g is expressed in meters per second squared. The formula is based on an ellipsoidal model of the earth and also accounts for the effect of the earth's rotation. The absolute acceleration due to gravity as determined for a nonrotating earth may be computed from the relative values to a close approximation

---

*It can be proved that the earth, when taken as a sphere with a symmetrical distribution of mass about its center, may be considered a particle with its entire mass concentrated at its center.

by adding $3.382(10^{-2}) \cos^2 \gamma$ m/s^2, which removes the effect of the earth's rotation. The variation of both the absolute and the relative values of g with latitude is shown in Fig. 1/1 for sea-level conditions.*

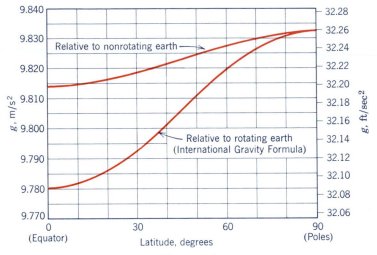

Figure 1/1

The standard value that has been adopted internationally for the gravitational acceleration relative to the rotating earth at sea level and at a latitude of 45° is 9.806 65 m/s^2 or 32.1740 ft/sec^2. This value differs very slightly from that obtained by evaluating the International Gravity Formula for $\gamma = 45°$. The reason for the small difference is that the earth is not exactly ellipsoidal, as assumed in the formulation of the International Gravity Formula.

The proximity of large land masses and the variations in the density of the earth's crust also influence the local value of g by a small but detectable amount. In almost all engineering problems where measurements are made on the surface of the earth, the difference between the absolute and relative values of the gravitational acceleration and the effect of local variations are neglected, and 9.81 m/s^2 in SI units and 32.2 ft/sec^2 in U.S. customary units are used for the sea-level value of g.

The variation of g with altitude is easily determined by the gravitational law. If g_0 represents the absolute acceleration due to gravity at sea level, the absolute value at an altitude h is

$$g = g_0 \frac{R^2}{(R + h)^2}$$

*Students will be able to derive these relations for a spherical earth following their study of relative motion in Chapter 3.

where R is the radius of the earth.

The earth's gravitational attraction on a body may be calculated from the results of the simple gravitational experiment. If the gravitational force of attraction or true weight of a body is **W**, then, since the body will fall with an absolute acceleration **g** in a vacuum, Eq. 1/1 gives

$$\boxed{\mathbf{W} = m\mathbf{g}} \qquad\qquad (1/3)$$

The apparent weight of a body as determined by a spring balance, calibrated to read the correct force and attached to the surface of the earth, will be slightly less than its true weight. The difference is due to the rotation of the earth. The ratio of the apparent weight to the apparent or relative acceleration due to gravity still gives the correct value of mass. The apparent weight and the relative acceleration due to gravity are, of course, the quantities that are measured in experiments conducted on the surface of the earth.

1/6 DIMENSIONS

A given dimension such as length can be expressed in a number of different units such as meters, millimeters, or kilometers. Thus, the word *dimension* is distinguished from the word *unit*. Physical relations must always be dimensionally homogeneous, that is, the dimensions of all terms in an equation must be the same. It is customary to use the symbols L, M, T, and F to stand for length, mass, time, and force, respectively. In SI units force is a derived quantity and from Eq. 1/1 has the dimensions of mass times acceleration or

$$F = ML/T^2$$

One important use of the theory of dimensions is found in checking the dimensional correctness of some derived physical relation. The following expression for the velocity v of a body of mass m that is moved from rest a horizontal distance x by a force F may be derived:

$$Fx = \tfrac{1}{2}mv^2$$

where the $\tfrac{1}{2}$ is a dimensionless coefficient resulting from integration. This equation is dimensionally correct since substitution of L, M, and T gives

$$[MLT^{-2}][L] = [M][LT^{-1}]^2$$

Dimensional homogeneity is a necessary condition for correctness, but it is not sufficient since the correctness of dimensionless coefficients cannot be checked in this way.

1/7 *FORMULATION AND SOLUTION OF DYNAMICS PROBLEMS*

The study of dynamics is directed toward the understanding and description of the various quantities involved in the motions of bodies. This description, which is largely mathematical, enables predictions of dynamical behavior to be made. A dual thought process is necessary in formulating this description. It is necessary to think both in terms of the physical situation and in terms of the corresponding mathematical description. Analysis of every problem will require this repeated transition of thought between the physical and the mathematical. Without question, one of the greatest difficulties encountered by students is the inability to make this transition of thought freely. They should recognize that the mathematical formulation of a physical problem represents an ideal and limiting description, or model, that approximates but never quite matches the actual physical situation.

In the course of constructing the idealized mathematical model for any given engineering problem, certain approximations will always be involved. Some of these approximations may be mathematical, whereas others will be physical. For instance, it is often necessary to neglect small distances, angles, or forces compared with large distances, angles, or forces. If the change in velocity of a body with time is nearly uniform, then an assumption of constant acceleration may be justified. An interval of motion that cannot be easily described in its entirety is often divided into small increments, each of which can be approximated. The retarding effect of bearing friction on the motion acquired by a machine as the result of applied forces or moments may often be neglected if the friction forces are small. However, these same friction forces cannot be neglected if the purpose of the inquiry is a determination of the drop in efficiency of the machine due to the friction process. Thus, the degree of assumption involved depends on what information is desired and on the accuracy required. The student should be constantly alert to the various assumptions called for in the formulation of real problems. The ability to understand and make use of the appropriate assumptions in the course of the formulation and solution of engineering problems is certainly one of the most important characteristics of a successful engineer. Along with the development of the principles and analytical tools needed for modern dynamics, one of the major aims of this book is to provide a maximum of opportunity to develop ability in formulating good mathematical models. Strong emphasis is placed on a wide range of practical problems that not only require the full exercise of theory but also force consideration of the decisions that must be made concerning relevant assumptions.

An effective method of attack on dynamics problems, as in all engineering problems, is essential. The development of good habits in formulating problems and in representing their solutions will prove to be an invaluable asset. Each solution should proceed with a logical sequence of steps from hypothesis to conclusion, and its representation should include a clear statement of the following parts, each clearly identified:

1. Given data
2. Results desired
3. Necessary diagrams
4. Calculations
5. Answers and conclusions

In addition, it is well to incorporate a series of checks on the calculations at intermediate points in the solution. The reasonableness of numerical magnitudes should be observed, and the accuracy and dimensional homogeneity of terms should be checked frequently. It is also important that the arrangement of work be neat and orderly. Careless solutions that cannot be read easily by others are of little or no value. It will be found that the discipline involved in adherence to good form will in itself be an invaluable aid to the development of the abilities for formulation and analysis. Many problems that at first may seem difficult and complicated become clear and straightforward once they are begun with a logical and disciplined method of attack.

The subject of dynamics is based on a surprisingly few fundamental concepts and principles that, however, are extended and applied over an exceedingly wide range of conditions. One of the most valuable aspects of the study of dynamics is the experience afforded in reasoning from fundamentals. This experience cannot be obtained merely by memorizing the kinematical and dynamical equations that describe various motions. It must be obtained through exposure to a wide variety of problem situations that force the choice, use, and extension of basic principles to meet the given conditions.

In describing the relations between forces and the motions they produce, it is essential that the system to which a principle is applied be clearly defined. At times a single particle or a rigid body is the system to be isolated, whereas at other times two or more bodies taken together constitute the system. The definition of the system to be analyzed is made clear by constructing its *free-body diagram*. This diagram consists of a closed outline of the external boundary of the system defined. All bodies that contact and exert forces on the system but are not a part of it are removed and replaced by vectors representing the forces they exert *on* the system isolated.

In this way, we make a clear distinction between the action and reaction of each force, and account is taken of *all* forces on and external to the system. It is assumed that all students are familiar with the technique of drawing free-body diagrams from their prior work in statics.

In applying the laws of dynamics, numerical values of the quantities may be used directly in proceeding toward the solution, or algebraic symbols may be used to represent the quantities involved and the answer left as a formula. With numerical substitution the magnitudes of all quantities expressed in their particular units are evident at each stage of the calculation. This approach offers advantage when the practical significance of the magnitude of each term is important. The symbolic solution, however, has several advantages over the numerical solution. First, the abbreviation achieved by the use of symbols aids in focusing attention on the interconnection between the physical situation and its related mathematical description. Second, a symbolic solution permits a dimensional check to be made at every step, whereas dimensional homogeneity may not be checked when only numerical values are used. Third, a symbolic solution may be used repeatedly for obtaining answers to the same problem when different sets and sizes of units are used. Facility with both forms of solution is essential, and ample practice with each should be sought in the problem work.

Students will find that solutions to the various equations of dynamics may be obtained in one of three ways. First, a direct mathematical solution by hand calculation may be carried out, with answers appearing either as algebraic symbols or as numerical results. The large majority of the problems come under this category. Second, certain problems may be readily handled by graphical solutions, such as with the determination of velocities and accelerations in two-dimensional relative motion of rigid bodies. Third, there are a number of problems in *Vol. 2 Dynamics* that are designated as *computer-oriented problems*. They appear at the end of the Review Problem sets and are selected to illustrate the type of problem for which solution by computer offers a distinct advantage. The choice of the most expedient method of solution is an important aspect of the experience to be gained from the problem work. We emphasize, however, that the most important experience in learning mechanics lies in the formulation of problems, as distinct from their solution per se.

PROBLEMS

(Refer to Table D/2 in Appendix D for relevant solar-system values.)

1/1 Determine the weight in newtons of a person whose mass is 80 kg. Convert the person's mass to slugs and calculate the corresponding weight in pounds.
  Ans. $W = 785$ N, $m = 5.48$ slugs, $W = 176.4$ lb

1/2 Determine your mass in slugs. Convert your weight to newtons and calculate the corresponding mass in kilograms.

1/3 At what altitude h above the north pole is the weight of an object reduced to 10% of its earth-surface value? Assume a spherical earth of radius R and express h in terms of R.   *Ans.* $h = 2.16R$

1/4 A space shuttle is in a circular orbit at an altitude of 150 mi. Calculate the absolute value of g at this altitude and determine the corresponding weight of a shuttle passenger who weighs 200 lb when standing on the surface of the earth at a latitude of 45°. Are the terms "zero-g" and "weightless," which are sometimes used to describe conditions aboard orbiting spacecraft, correct in the absolute sense?

1/5 Calculate the force F_s exerted by the sun on a 90-kg man as he stands on the surface of the moon. Compare F_s to the force F_m exerted on him by the moon.
  Ans. $F_s = 0.534$ N, $F_m = 146$ N

1/6 Consider two iron spheres, each of diameter 100 mm, that are just touching. At what distance r from the center of the earth will the force of mutual attraction between the contacting spheres be equal to the force exerted by the earth on one of the spheres?

1/7 Calculate the acceleration due to gravity relative to the rotating earth and the absolute value if the earth were not rotating for a sea-level position at a north or south latitude of 45°. Compare your results with the values of Fig. 1/1.   *Ans.* $g_{\text{rel}} = 9.806$ m/s^2
$g_{\text{abs}} = 9.823$ m/s^2

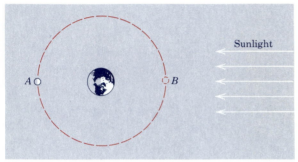

1/8 Determine the ratio R_A of the force exerted by the sun on the moon to that exerted by the earth on the moon for position A of the moon. Repeat for moon position B.

Problem 1/8

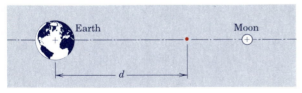

1/9 Calculate the distance d from the center of the earth at which a particle experiences equal attractions from the earth and moon. The particle is restricted to the line that joins the centers of the earth and moon. Justify the two solutions physically.
Ans. $d = 346\ 022$ km or $432\ 348$ km

Problem 1/9

1/10 Check the following equation for dimensional homogeneity:

$$mv = \int_{t_1}^{t_2} (F \cos \theta)\, dt$$

where m is mass, v is velocity, F is force, θ is an angle, and t is time.

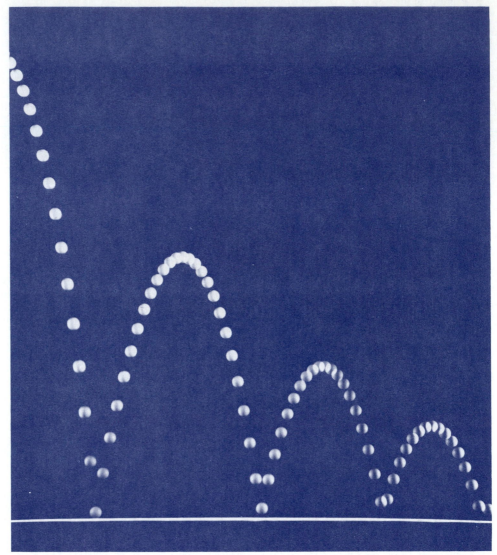

The description of the path of a particle in terms of geometry and time is a basic part of dynamics. This time-lapse photograph of a bouncing ball records its trajectory and agrees very closely with the path that we are able to predict through calculation accounting for both gravity and impact.

KINEMATICS OF PARTICLES

2

2/1 INTRODUCTION

Kinematics is that branch of dynamics which describes the motion of bodies without reference to the forces that either cause the motion or are generated as a result of the motion. Kinematics is often referred to as the "geometry of motion." The design of cams, gears, linkages, and other machine elements to control or produce certain desired motions and the calculation of flight trajectories for aircraft, rockets, and spacecraft are a few examples of kinematic problems that engage the attention of engineers. A thorough working knowledge of kinematics is an absolute prerequisite to kinetics, which is the study of the relationships between motion and the corresponding forces that cause or accompany the motion.

We start our study of kinematics by first discussing in this chapter the motions of points or particles. A particle is a body whose physical dimensions are so small compared with the radius of curvature of its path that we can treat the motion of the particle as that of a point. For example, the wingspan of a jet transport flying between Los Angeles and New York is of no consequence compared with the radius of curvature of its flight path, and the treatment of the airplane as a particle or point should raise no question.

There are a number of ways in which the motion of a particle can be described, and the choice of the most convenient or appropriate method depends a great deal on experience and on how the data are given. Let us get an overview of the several methods developed in this chapter by referring to Fig. 2/1 which shows a particle P moving along some general path in space. If the particle is confined to a specified path, as with a bead sliding along a fixed wire, its motion is said to be *constrained*. If there are no physical guides, the motion is said to be *unconstrained*. A small rock tied to the end of a string and whirled in a circle undergoes constrained motion until the string breaks, after which instant its motion is unconstrained.

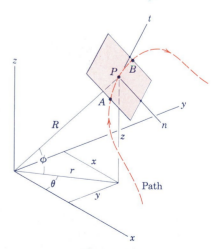

Figure 2/1

The position of particle P at any time t can be described by specifying its rectangular coordinates* x, y, z, its cylindrical coordinates r, θ, z, or its spherical coordinates R, θ, ϕ. The motion of P may also be described by measurements along the tangent t and normal n to the curve. The direction of n lies in the local plane of the curve.† These last measurements are known as *path variables*.

The motion of particles (or rigid bodies) may be described by using coordinates measured from fixed reference axes (*absolute-motion* analysis) or by using coordinates measured from moving reference axes (*relative-motion* analysis). Both descriptions will be developed and applied in the articles that follow.

With this conceptual picture of the description of particle motion in mind, we will now restrict our attention in the first part of this chapter to the case of *plane motion* where all movement occurs in or can be represented as occurring in a single plane. A large proportion of the motions of machines and structures in engineering can be represented as plane motion. Later, in Chapter 7 an introduction to three-dimensional motion is presented. We will begin our discussion of plane motion with *rectilinear motion*, which is motion along a straight line, and follow it with a description of motion along a plane curve.

2/2 RECTILINEAR MOTION

Consider a particle P moving along a straight line, Fig. 2/2. The position of P at any instant of time t may be specified by its distance s measured from some convenient reference point O fixed on the line. At time $t + \Delta t$ the particle has moved to P' and its coordinate becomes $s + \Delta s$. The change in the position coordinate during the interval Δt is called the *displacement* Δs of the particle. The displacement would be negative if the particle moved in the negative s-direction.

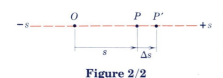

Figure 2/2

The average velocity of the particle during the interval Δt is the displacement divided by the time interval or $v_{\text{av}} = \Delta s/\Delta t$. As Δt becomes smaller and approaches zero in the limit, the average velocity approaches the *instantaneous velocity* of the particle, which is $v = \lim\limits_{\Delta t \to 0} \dfrac{\Delta s}{\Delta t}$ or

$$\boxed{v = \frac{ds}{dt} = \dot{s}} \qquad (2/1)$$

---

*Often called *Cartesian* coordinates, named after René Descartes (1596–1650), a French mathematician who was one of the inventors of analytic geometry.

†This plane is known as the *osculating* plane, which comes from the Latin word *osculari* meaning "to kiss." The plane that contains P and the two points A and B, one on either side of P, becomes the osculating plane as the distances between the points approach zero.

Thus, the velocity is the time rate of change of the position coordinate s. The velocity is positive or negative depending on whether the corresponding displacement is positive or negative.

The average acceleration of the particle during the interval Δt is the change in its velocity divided by the time interval or $a_{av} = \Delta v/\Delta t$. As Δt becomes smaller and approaches zero in the limit, the average acceleration approaches the *instantaneous acceleration* of the particle, which is $a = \lim\limits_{\Delta t \to 0} \dfrac{\Delta v}{\Delta t}$ or

$$a = \frac{dv}{dt} = \dot{v} \qquad \text{or} \qquad a = \frac{d^2s}{dt^2} = \ddot{s} \qquad\qquad (2/2)$$

The acceleration is positive or negative depending on whether the velocity is increasing or decreasing. Note that the acceleration would be positive if the particle had a negative velocity that was becoming less negative. If the particle is slowing down, the particle is said to be *decelerating*.

Velocity and acceleration are actually vector quantities, as we will see for curvilinear motion beginning with Art. 2/3. For rectilinear motion in the present article, where the direction of the motion is that of the given straight-line path, the sense of the vector along the path is described by a plus or minus sign. In our treatment of curvilinear motion, we will account for the change in direction of the velocity and acceleration vectors as well as their change in magnitude.

By eliminating the time dt between Eq. 2/1 and the first of Eqs. 2/2, a differential equation relating displacement, velocity, and acceleration results* that is

$$v\,dv = a\,ds \qquad \text{or} \qquad \dot{s}\,d\dot{s} = \ddot{s}\,ds \qquad\qquad (2/3)$$

Equations 2/1, 2/2, and 2/3 are the differential equations for the rectilinear motion of a particle. Problems in rectilinear motion involving finite changes in the motion variables are solved by integration of these basic differential relations. The position coordinate s, the velocity v, and the acceleration a are algebraic quantities, so that their signs, positive or negative, must be carefully observed. Note that the positive directions for v and a are the same as the positive direction for s.

Interpretation of the differential equations governing rectilinear motion is considerably clarified by representing the relationships among s, v, a, and t graphically. Figure 2/3a represents a

---

*Differential quantities can be multiplied and divided in exactly the same way as other algebraic quantities.

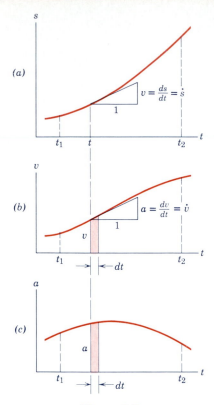

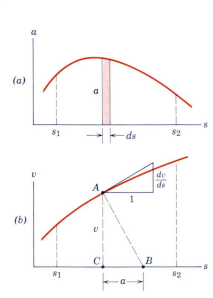

Figure 2/3

Figure 2/4

schematic plot of the variation of s with t from time t_1 to time t_2 for some given rectilinear motion. By constructing the tangent to the curve at any time t, we obtain the slope, which is the velocity $v = ds/dt$. Thus, the velocity may be determined at all points on the curve and plotted against the corresponding time as shown in Fig. 2/3b. Similarly, the slope dv/dt of the v-t curve at any instant gives the acceleration at that instant, and the a-t curve can therefore be plotted as in Fig. 2/3c.

We now see from Fig. 2/3b that the area under the v-t curve during time dt is $v\,dt$ which from Eq. 2/1 is the displacement ds. Consequently, the net displacement of the particle during the interval from t_1 to t_2 is the corresponding area under the curve, which is

$$\int_{s_1}^{s_2} ds = \int_{t_1}^{t_2} v\,dt \qquad \text{or} \qquad s_2 - s_1 = \text{(area under } v\text{-}t \text{ curve)}$$

Similarly, from Fig. 2/3c we see that the area under the a-t curve during time dt is $a\,dt$ which, from the first of Eqs. 2/2, is dv. Thus, the net change in velocity between t_1 and t_2 is the corresponding area under the curve, which is

$$\int_{v_1}^{v_2} dv = \int_{t_1}^{t_2} a\,dt \qquad \text{or} \qquad v_2 - v_1 = \text{(area under } a\text{-}t \text{ curve)}$$

Two additional graphical relations are noted. When the acceleration a is plotted as a function of the position coordinate s, Fig. 2/4a, the area under the curve during a displacement ds is $a\,ds$ which, from Eq. 2/3, is $v\,dv = d(v^2/2)$. Thus, the net area under the curve between position coordinates s_1 and s_2 is

$$\int_{v_1}^{v_2} v\,dv = \int_{s_1}^{s_2} a\,ds \quad \text{or} \quad \tfrac{1}{2}(v_2{}^2 - v_1{}^2) = \text{(area under } a\text{-}s \text{ curve)}$$

When the velocity v is plotted as a function of the position coordinate s, Fig. 2/4b, the slope of the curve at any point A is dv/ds. By constructing the normal AB to the curve at this point, we see from the similar triangles that $\overline{CB}/v = dv/ds$. Thus, from Eq. 2/3, $\overline{CB} = v(dv/ds) = a$, the acceleration. It is necessary that the velocity and position coordinate axes have the same numerical scales so that the acceleration read on the position coordinate scale in meters (or feet), say, will represent the actual acceleration in the units meters (or feet) per second squared.

The graphical representations described are useful not only in visualizing the relationships among the several motion quantities but also in approximating results by graphical integration or differentiation when a lack of knowledge of the mathematical relationship prevents its expression as an explicit mathematical function. Experimental data and motions that involve discontinuous relationships between the variables are frequently analyzed graphically.

If the position coordinate s is known for all values of the time t, then successive mathematical or graphical differentiation with respect to t gives the velocity v and acceleration a. In many problems, however, the functional relationship between position coordinate and time is unknown, and we must determine it by successive integration from the acceleration. Acceleration is determined by the forces that act on moving bodies and is computed from the equations of kinetics discussed in subsequent chapters. Depending on the nature of the forces, the acceleration may be specified as a function of time, velocity, position coordinate, or as a combined function of these quantities. The procedure for integrating the differential equation in each case is indicated as follows.

(a) Constant acceleration.　When a is constant, the first of Eqs. 2/2 and 2/3 may be integrated directly. For simplicity with $s = s_0$, $v = v_0$, and $t = 0$ designated at the beginning of the interval, then for a lapse of time t the integrated equations become

$$\int_{v_0}^{v} dv = a \int_{0}^{t} dt \qquad \text{or} \qquad v = v_0 + at$$

$$\int_{v_0}^{v} v\, dv = a \int_{s_0}^{s} ds \qquad \text{or} \qquad v^2 = v_0^{\,2} + 2a(s - s_0)$$

Substitution of the integrated expression for v into Eq. 2/1 and integration with respect to t give

$$\int_{s_0}^{s} ds = \int_{0}^{t} (v_0 + at)\, dt \qquad \text{or} \qquad s = s_0 + v_0 t + \tfrac{1}{2}at^2$$

These relations are necessarily restricted to the special case where the acceleration is constant. The integration limits depend on the initial and final conditions and for a given problem may be different from those used here. It may be more convenient, for instance, to begin the integration at some specified time t_1 rather than at time $t = 0$.

Caution: One of the most frequent mistakes made by students is the attempt to use the foregoing equations for problems of variable acceleration where they do not apply since they have been integrated for constant acceleration only.

(b) Acceleration given as a function of time, a = f(t). Substitution of the function into the first of Eqs. 2/2 gives $f(t) = dv/dt$. Multiplying by dt separates the variables and permits integration. Thus,

$$\int_{v_0}^{v} dv = \int_{0}^{t} f(t)\, dt \qquad \text{or} \qquad v = v_0 + \int_{0}^{t} f(t)\, dt$$

From this integrated expression for v as a function of t, the position coordinate s is obtained by integrating Eq. 2/1 which, in form, would be

$$\int_{s_0}^{s} ds = \int_{0}^{t} v \, dt \qquad \text{or} \qquad s = s_0 + \int_{0}^{t} v \, dt$$

If the indefinite integral is employed, the end conditions are used to establish the constants of integration with results that are identical with those obtained by using the definite integral.

If desired, the displacement s may be obtained by a direct solution of the second-order differential equation $\ddot{s} = f(t)$ obtained by substitution of $f(t)$ into the second of Eqs. 2/2.

(c) Acceleration given as a function of velocity, a = f(v). Substitution of the function into the first of Eqs. 2/2 gives $f(v) = dv/dt$, which permits separating the variables and integrating. Thus,

$$t = \int_{0}^{t} dt = \int_{v_0}^{v} \frac{dv}{f(v)}$$

This result gives t as a function of v. Then it would be necessary to solve for v as a function of t so that Eq. 2/1 can be integrated to obtain the position coordinate s as a function of t.

Alternatively, the function $a = f(v)$ can be substituted into the first of Eqs. 2/3, giving $v \, dv = f(v) \, ds$. The variables are now separated and the equation integrated in the form

$$\int_{v_0}^{v} \frac{v \, dv}{f(v)} = \int_{s_0}^{s} ds \qquad \text{or} \qquad s = s_0 + \int_{v_0}^{v} \frac{v \, dv}{f(v)}$$

Note that this equation gives s in terms of v without explicit reference to t.

(d) Acceleration given as a function of displacement, a = f(s). Substitution of the function into Eq. 2/3 and integrating give the form

$$\int_{v_0}^{v} v \, dv = \int_{s_0}^{s} f(s) \, ds \qquad \text{or} \qquad v^2 = v_0{}^2 + 2 \int_{s_0}^{s} f(s) \, ds$$

Next we solve for v to give $v = g(s)$, a function of s. Now we can substitute ds/dt for v, separate variables, and integrate in the form

$$\int_{s_0}^{s} \frac{ds}{g(s)} = \int_{0}^{t} dt \qquad \text{or} \qquad t = \int_{s_0}^{s} \frac{ds}{g(s)}$$

which gives t as a function of s. Lastly we can rearrange to obtain s as a function of t.

In each of the foregoing cases when the acceleration varies according to some functional relationship, the ability to solve the equations by direct mathematical integration will depend on the form of the function. In cases where the integration is excessively awkward or difficult, integration by graphical, numerical, or computer methods may be utilized.

Sample Problem 2/1

The position coordinate of a particle that is confined to move along a straight line is given by $s = 2t^3 - 24t + 6$, where s is measured in meters from a convenient origin and t is in seconds. Determine (a) the time required for the particle to reach a velocity of 72 m/s from its initial condition at $t = 0$, (b) the acceleration of the particle when $v = 30$ m/s, and (c) the net displacement of the particle during the interval from $t = 1$ s to $t = 4$ s.

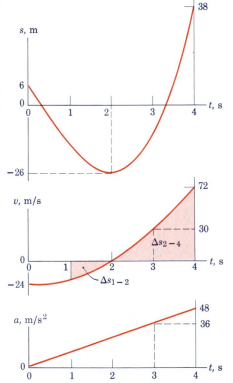

Solution. The velocity and acceleration are obtained by successive differentiation of s with respect to the time. Thus,

$$[v = \dot{s}] \qquad v = 6t^2 - 24 \text{ m/s}$$

$$[a = \dot{v}] \qquad a = 12t \text{ m/s}^2$$

(a) Substituting $v = 72$ m/s into the expression for v gives us $72 = 6t^2 - 24$, from which $t = \pm 4$ s. The negative root describes a mathematical solution for t before the initiation of motion, so this root ① is of no physical interest. Thus, the desired result is

$$t = 4 \text{ s} \qquad\qquad Ans.$$

(b) Substituting $v = 30$ m/s into the expression for v gives $30 = 6t^2 - 24$, from which the positive root is $t = 3$ s, and the corresponding acceleration is

$$a = 12(3) = 36 \text{ m/s}^2 \qquad\qquad Ans.$$

(c) The net displacement during the specified interval is

$$\Delta s = s_4 - s_1 \qquad \text{or}$$

$$\Delta s = [2(4^3) - 24(4) + 6] - [2(1^3) - 24(1) + 6]$$

$$= 54 \text{ m} \qquad\qquad Ans.$$

② which represents the net advancement of the particle along the s-axis from the position it occupied at $t = 1$ s to its position at $t = 4$ s.

To help visualize the motion, the values of s, v, and a are plotted against the time t as shown. Because the area under the v-t curve rep- ③ resents displacement, we see that the net displacement from $t = 1$ s to $t = 4$ s is the positive area Δs_{2-4} less the negative area Δs_{1-2}.

① Be alert to the proper choice of sign when taking a square root. When the situation calls for only one answer, the positive root is not always the one you may need.

② Note carefully the distinction between italic s for the position coordinate and the vertical s for seconds.

③ Note from the graphs that the values for v are the slopes ($\dot{s}$) of the s-t curve and that the values for a are the slopes ($\dot{v}$) of the v-t curve. *Suggestion:* Integrate $v\,dt$ for each of the two intervals and check the answer for Δs. Show that the total distance traveled during the interval $t = 1$ s to $t = 4$ s is 74 m.

Sample Problem 2/2

A particle moves along the x-axis with an initial velocity $v_x = 50$ ft/sec at the origin when $t = 0$. For the first 4 seconds it has no acceleration, and thereafter it is acted on by a retarding force that gives it a constant acceleration $a_x = -10$ ft/sec². Calculate the velocity and the x-coordinate of the particle for the conditions of $t = 8$ sec and $t = 12$ sec and find the maximum positive x-coordinate reached by the particle.

① Learn to be flexible with symbols. The position coordinate x is just as valid as s.

Solution. The velocity of the particle after $t = 4$ sec is computed from

$$\left[\int dv = \int a \, dt\right] \qquad \int_{50}^{v_x} dv_x = -10\int_{4}^{t} dt \qquad v_x = 90 - 10t \text{ ft/sec}$$

② Note that we integrate to a general time t and then substitute specific values.

and is plotted as shown. At the specified times, the velocities are

$$t = 8 \text{ sec}, \qquad v_x = 90 - 10(8) = 10 \text{ ft/sec}$$

$$t = 12 \text{ sec}, \qquad v_x = 90 - 10(12) = -30 \text{ ft/sec} \qquad Ans.$$

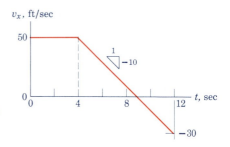

The x-coordinate of the particle at any time greater than 4 seconds is the distance traveled during the first 4 seconds plus the distance traveled after the discontinuity in acceleration occurred. Thus,

$$\left[\int ds = \int v \, dt\right] \quad x = 50(4) + \int_{4}^{t} (90 - 10t) \, dt = -5t^2 + 90t - 80 \text{ ft}$$

For the two specified times,

$$t = 8 \text{ sec}, \qquad x = -5(8^2) + 90(8) - 80 = 320 \text{ ft}$$

$$t = 12 \text{ sec}, \qquad x = -5(12^2) + 90(12) - 80 = 280 \text{ ft} \qquad Ans.$$

The x-coordinate for $t = 12$ sec is less than that for $t = 8$ sec since the motion is in the negative x-direction after $t = 9$ sec. The maximum positive x-coordinate is, then, the value of x for $t = 9$ sec which is

$$x_{\max} = -5(9^2) + 90(9) - 80 = 325 \text{ ft} \qquad Ans.$$

③ These displacements are seen to be the net positive areas under the v-t graph up to the value of t in question.

③ Show that the total distance traveled by the particle in the 12 seconds is 370 ft.

Sample Problem 2/3

The spring-mounted slider moves in the horizontal guide with negligible friction and has a velocity v_0 in the s-direction as it crosses the mid-position where $s = 0$ and $t = 0$. The two springs together exert a retarding force to the motion of the slider, which gives it an acceleration proportional to the displacement but oppositely directed and equal to $a = -k^2 s$, where k is constant. (The constant is arbitrarily squared for later convenience in the form of the expressions.) Determine the expressions for the displacement s and velocity v as functions of the time t.

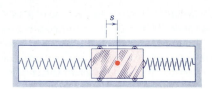

Solution I. Since the acceleration is specified in terms of the displacement, the differential relation $v\,dv = a\,ds$ may be integrated. Thus,

① $$\int v\,dv = \int -k^2 s\,ds + C_1 \text{ a constant,} \qquad \text{or} \qquad \frac{v^2}{2} = -\frac{k^2 s^2}{2} + C_1$$

① We have used an indefinite integral here and evaluated the constant of integration. For practice, obtain the same results by using the definite integral with the appropriate limits.

When $s = 0$, $v = v_0$, so that $C_1 = v_0^2/2$, and the velocity becomes

$$v = +\sqrt{v_0^2 - k^2 s^2}$$

The plus sign of the radical is taken when v is positive (in the plus s-direction). This last expression may be integrated by substituting $v = ds/dt$. Thus,

② $$\int \frac{ds}{\sqrt{v_0^2 - k^2 s^2}} = \int dt + C_2 \text{ a constant,} \qquad \text{or} \qquad \frac{1}{k}\sin^{-1}\frac{ks}{v_0} = t + C_2$$

② Again try the definite integral here as above.

With the requirement of $t = 0$ when $s = 0$, the constant of integration becomes $C_2 = 0$, and we may solve the equation for s so that

$$s = \frac{v_0}{k}\sin kt \qquad\qquad Ans.$$

The velocity is $v = \dot{s}$ which gives

$$v = v_0 \cos kt \qquad\qquad Ans.$$

Solution II. Since $a = \ddot{s}$, the given relation may be written at once as

$$\ddot{s} + k^2 s = 0$$

This is an ordinary linear differential equation of second order for which the solution is well known and is

$$s = A \sin Kt + B \cos Kt$$

where A, B, and K are constants. Substitution of this expression into the differential equation shows that it satisfies the equation, provided that $K = k$. The velocity is $v = \dot{s}$ which becomes

$$v = Ak \cos kt - Bk \sin kt$$

The initial condition $v = v_0$ when $t = 0$ requires that $A = v_0/k$, and the condition $s = 0$ when $t = 0$ gives $B = 0$. Thus, the solution is

③ $$s = \frac{v_0}{k}\sin kt \qquad \text{and} \qquad v = v_0 \cos kt \qquad\qquad Ans.$$

③ This motion is called *simple harmonic motion* and is characteristic of all oscillations where the restoring force, and hence the acceleration, is proportional to the displacement but opposite in sign.

Sample Problem 2/4

A freighter is moving at a speed of 8 knots when its engines are suddenly stopped. If it takes 10 minutes for the freighter to reduce its speed to 4 knots, determine and plot the distance s in nautical miles moved by the ship and its speed v in knots as functions of the time t during this interval. The deceleration of the ship is proportional to the square of its speed, so that $a = -kv^2$.

① Recall that one knot is the speed of one nautical mile (6076 ft) per hour. Work directly in the units of nautical miles and hours.

Solution. The speeds and the time are given, so we may substitute the expression for acceleration directly into the basic definition $a = dv/dt$ and integrate. Thus,

$$-kv^2 = \frac{dv}{dt} \qquad \frac{dv}{v^2} = -k\,dt \qquad \int_8^v \frac{dv}{v^2} = -k \int_0^t dt$$

$$-\frac{1}{v} + \frac{1}{8} = -kt \qquad v = \frac{8}{1 + 8kt}$$

② We choose to integrate to a general value of v and its corresponding time t so that we may obtain the variation of v with t.

Now we substitute the end limits of $v = 4$ knots and $t = \frac{10}{60} = \frac{1}{6}$ hour and get

$$4 = \frac{8}{1 + 8k(1/6)} \qquad k = \frac{3}{4}\ \mathrm{mi}^{-1} \qquad v = \frac{8}{1 + 6t} \qquad Ans.$$

The speed is plotted against the time as shown.

The distance is obtained by substituting the expression for v into the definition $v = ds/dt$ and integrating. Thus,

$$\frac{8}{1 + 6t} = \frac{ds}{dt} \qquad \int_0^t \frac{8\,dt}{1 + 6t} = \int_0^s ds \qquad s = \frac{4}{3} \ln\,(1 + 6t) \qquad Ans.$$

The distance s is also plotted against the time as shown, and we see that the ship has moved through a distance $s = \frac{4}{3} \ln\,(1 + \frac{6}{6}) = \frac{4}{3} \ln 2 = 0.924$ mi (nautical) during the 10 minutes.

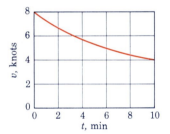

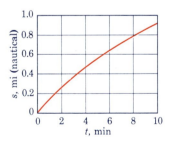

PROBLEMS

Introductory problems

Problems 2/1 through 2/5 treat the motion of a particle that moves along the *s*-axis shown in the figure.

2/1 The velocity of a particle is given by $v = 20t^2 - 100t + 50$, where v is in meters per second and t is in seconds. Plot the velocity v and acceleration a versus time for the first 6 seconds of motion and evaluate the velocity when a is zero.

Ans. $v = -75$ m/s

Problems 2/1–2/5

2/2 The displacement of a particle is given by $s = 2t^3 - 30t^2 + 100t - 50$, where s is in feet and t is in seconds. Plot the displacement, velocity, and acceleration as functions of time for the first 12 seconds of motion. Determine the time at which the velocity is zero.

2/3 The displacement of a particle is given by $s = (-2 + 3t)e^{-0.5t}$, where s is in meters and t is in seconds. Plot the displacement, velocity, and acceleration versus time for the first 20 seconds of motion. Determine the time at which the acceleration is zero.

Ans. $v = (4 - 1.5t)e^{-0.5t}$
$a = (-3.5 + 0.75t)e^{-0.5t}$, $t = 4.67$ s

2/4 The velocity of a particle that moves along the *s*-axis is given by $v = 2 + 5t^{3/2}$, where t is in seconds and v is in meters per second. Evaluate the displacement s, velocity v, and acceleration a when $t = 4$ s. The particle is at the origin $s = 0$ when $t = 0$.

2/5 The acceleration of a particle is given by $a = 4t - 30$, where a is in meters per second squared and t is in seconds. Determine the velocity and displacement as functions of time. The initial displacement at $t = 0$ is $s_0 = -5$ m, and the initial velocity is $v_0 = 3$ m/s.

Ans. $v = 3 - 30t + 2t^2$
$s = -5 + 3t - 15t^2 + \frac{2}{3}t^3$

2/6 Calculate the constant acceleration a in g's which the catapult of an aircraft carrier must provide to produce a launch velocity of 180 mi/hr in a distance of 300 ft. Assume that the carrier is at anchor.

2/7 In the final stages of a moon landing, the lunar module descends under retrothrust of its descent engine to within $h = 5$ m of the lunar surface where it has a downward velocity of 2 m/s. If the descent engine is cut off abruptly at this point, compute the impact velocity of the landing gear with the moon. Lunar gravity is $\frac{1}{6}$ of the earth's gravity.

Ans. $v = 4.51$ m/s

Problem 2/7

2/8 A car comes to a complete stop from an initial speed of 50 mi/hr in a distance of 100 ft. With the same constant acceleration, what would be the stopping distance s from an initial speed of 70 mi/hr?

2/9 A projectile is fired vertically with an initial velocity of 200 m/s. Calculate the maximum altitude h reached by the projectile and the time t after firing for it to return to the ground. Neglect air resistance and take the gravitational acceleration to be constant at 9.81 m/s^2.

Ans. $h = 2040$ m, $t = 40.8$ s

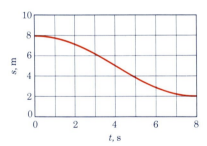

Problem 2/10

2/10 The graph shows the displacement-time history for the rectilinear motion of a particle during an 8-second interval. Determine the average velocity v_{av} during the interval and, to within reasonable limits of accuracy, find the instantaneous velocity v when $t = 4$ s.

2/11 During an 8-second interval the velocity of a particle moving in a straight line varies with the time as shown. Within reasonable limits of accuracy, determine the amount Δa by which the acceleration at $t = 4$ s exceeds the average acceleration during the interval. What is the displacement Δs during the interval?

Ans. $\Delta a = 0.50$ m/s^2, $\Delta s = 64$ m

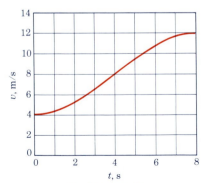

Problem 2/11

2/12 Experimental data for the motion of a particle along a straight line yield measured values of the velocity v for various position coordinates s. A smooth curve is drawn through the points as shown in the graph. Determine the acceleration of the particle when $s = 20$ ft.

Representative problems

2/13 The pilot of a jet transport brings the engines to full takeoff power before releasing the brakes as the aircraft is standing on the runway. The jet thrust remains constant, and the aircraft has a near-constant acceleration of $0.4g$. If the takeoff speed is 200 km/h, calculate the distance s and time t from rest to takeoff.   *Ans.* $s = 393$ m, $t = 14.16$ s

2/14 Determine the vertical velocity v_0 with which a ball must be thrown at A in order for the time of flight to the bottom of the cliff to be (a) 4 s and (b) 2 s.

2/15 A girl rolls a ball up an incline and allows it to return to her. For the particular inclination angle θ and ball involved, the acceleration of the ball along the incline is constant at $0.25g$, directed down the incline. If the ball is released with a speed of 4 m/s, determine the distance s it moves up the incline before reversing its direction and the total time t required for the ball to return to the child's hand.
Ans. $s = 3.26$ m, $t = 3.26$ s

2/16 To test the effects of "weightlessness" for short periods of time, a test facility is designed which accelerates a test package vertically up from A to B by means of a gas-activated piston and allows it to ascend and descend from B to C to B under freefall conditions. The test chamber consists of a deep well and is evacuated to eliminate any appreciable air resistance. If a constant acceleration of $40g$ from A to B is provided by the piston and if the total test time for the "weightless" condition from B to C to B is 10 s, calculate the required working height h of the chamber. Upon returning to B, the test package is recovered in a basket filled with polystyrene pellets inserted in the line of fall.

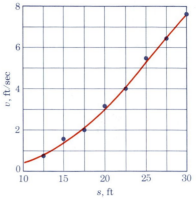

Problem 2/12

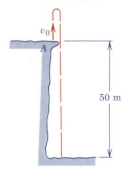

Problem 2/14

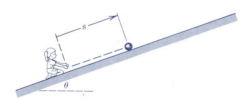

Problem 2/15

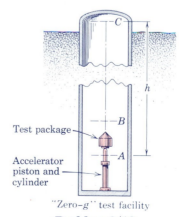

Problem 2/16

350 m

A

Problem 2/19

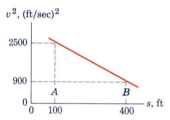

v^2, (ft/sec)^2

2500

900

0

0   100   400

A   B

s, ft

Problem 2/20

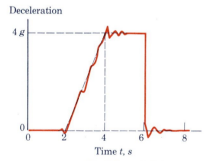

Deceleration

$4g$

0

0   2   4   6   8

Time t, s

Problem 2/22

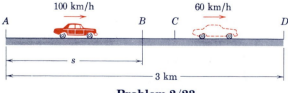

100 km/h   60 km/h

A   B   C   D

s

3 km

Problem 2/23

2/17 The velocity of a particle along the s-axis is given by $v = 5s^{3/2}$, where s is in millimeters and v is in millimeters per second. Determine the acceleration when s is 2 millimeters.   *Ans.* $a = 150$ mm/s^2

2/18 A sprinter reaches his top speed of 35 ft/sec in t seconds from rest with essentially constant acceleration. If he maintains his speed and covers the 100-yd distance in 9.6 sec, find the acceleration interval t and his average starting acceleration a.

2/19 The main elevator A of the CN Tower in Toronto rises about 350 m and for most of its run has a constant speed of 22 km/h. Assume that both the acceleration and deceleration have a constant magnitude of $\frac{1}{4}g$ and determine the time duration t of the elevator run.   *Ans.* $t = 59.8$ s

2/20 A body moves in a straight line with a velocity whose square decreases linearly with the displacement between two points A and B, which are 300 ft apart as shown. Determine the displacement Δs of the body during the last 2 sec before arrival at B.

2/21 A jet aircraft with a landing speed of 200 km/h has a maximum of 600 m of available runway after touchdown in which to reduce its ground speed to 30 km/h. Compute the average acceleration a required of the aircraft during braking.
Ans. $a = -2.51$ m/s^2

2/22 A retarding force acts on a particle moving initially with a velocity of 100 m/s and gives it a deceleration as recorded by the oscilloscope record shown. Approximate the velocity of the particle at $t = 4$ s and at $t = 8$ s.

2/23 In traveling a distance of 3 km between points A and D, a car is driven at 100 km/h from A to B for t seconds and at 60 km/h from C to D also for t seconds. If the brakes are applied for 4 s between B and C to give the car a uniform deceleration, calculate t and the distance s between A and B.
Ans. $t = 65.5$ s, $s = 1.819$ km

2/24 The position of a particle in millimeters is given by $s = 27 - 12t + t^2$, where t is in seconds. Plot the s-t and v-t relationships for the first 9 seconds. Determine the net displacement Δs during that interval and the total distance D traveled. By inspection of the s-t relationship, what conclusion can you reach regarding the acceleration?

2/25 A motorcycle patrolman starts from rest at A two seconds after a car, speeding at the constant rate of 120 km/h, passes point A. If the patrolman accelerates at the rate of 6 m/s² until he reaches his maximum permissible speed of 150 km/h, which he maintains, calculate the distance s from point A to the point at which he overtakes the car.

Ans. s = 912 m

Problem 2/25

2/26 The following table gives the acceleration a_x of the piston of a reciprocating internal combustion engine in terms of its displacement x from the top of its stroke. From a plot of the data, determine the maximum velocity v_{max} of the piston.

x, m	a_x, m/s²	x, m	a_x, m/s²
0	3300	0.10	−300
0.01	2890	0.12	−844
0.02	2490	0.14	−1306
0.04	1723	0.16	−1676
0.06	994	0.18	−1943
0.08	317	0.20	−2100

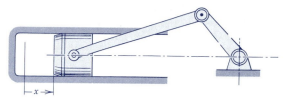

Problem 2/26

2/27 A particle moves along the positive x-axis with an acceleration a_x in m/s² which increases linearly with x expressed in millimeters, as shown on the graph for an interval of its motion. If the velocity of the particle at $x = 40$ mm is 0.4 m/s, determine the velocity at $x = 120$ mm.   *Ans. v = 0.8 m/s*

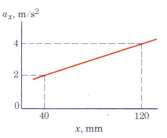

Problem 2/27

2/28 The 14-in. spring is compressed to an 8-in. length, where it is released from rest and accelerates the sliding block A. The acceleration has an initial value of 400 ft/sec² and then decreases linearly with the x-movement of the block, reaching zero when the spring regains its original 14-in. length. Calculate the time t for the block to go (*a*) 3 in. and (*b*) 6 in.

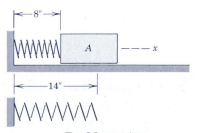

Problem 2/28

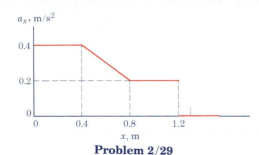

Problem 2/29

2/29 The acceleration of a particle that moves in the positive x-direction varies with its position as shown. If the velocity of the particle is 0.8 m/s when $x = 0$, determine the velocity v of the particle when $x = 1.4$ m.   *Ans.* $v = 1.166$ m/s

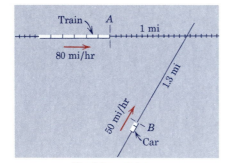

Problem 2/30

2/30 The deceleration of the mass center G of a car during a crash test is measured by an accelerometer with the results shown, where the distance x moved by G after impact is 0.8 m. Obtain a close approximation to the impact velocity v from the data given.

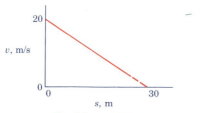

Problem 2/31

2/31 A train that is traveling at 80 mi/hr applies its brakes as it reaches point A and slows down with a constant deceleration. Its decreased velocity is observed to be 60 mi/hr as it passes a point 1/2 mi beyond A. A car moving at 50 mi/hr passes point B at the same instant that the train reaches point A. In an unwise effort to beat the train to the crossing, the driver "steps on the gas." Calculate the constant acceleration a that the car must have in order to beat the train to the crossing by 4 sec and find the velocity v of the car as it reaches the crossing.
Ans. $a = 1.168$ ft/sec^2, $v = 99.8$ mi/hr

2/32 If the velocity v of a particle moving along a straight line decreases linearly with its displacement s from 20 m/s to a value approaching zero at $s = 30$ m, determine the acceleration a of the particle when $s = 15$ m and show that the particle never reaches the 30-m displacement.

v, m/s

Problem 2/32

2/33 A ship reduces speed as it approaches the entrance to a harbor in still water. The speed is reduced from 10 knots as the ship passes the first entrance buoy to 4 knots 6 minutes later as it passes the second buoy. Approximate the distance Δs between the buoys in nautical miles by using the graph directly. (Recall that 1 knot equals one nautical mile per hour.) *Ans.* $\Delta s = 0.60$ mi (nautical)

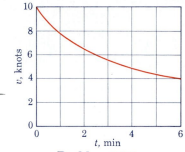

Problem 2/33

2/34 A particle moving along the positive x-direction with an initial velocity of 12 m/s is subjected to a retarding force that gives it a negative acceleration which varies linearly with time for the first 4 seconds as shown. For the next 5 seconds the force is constant and the acceleration remains constant. Plot the velocity of the particle during the 9 seconds and specify its value at $t = 4$ s. Also find the distance Δx traveled by the particle from its position at $t = 0$ to the point where it reverses its direction.

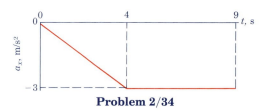

Problem 2/34

2/35 A particle starts from rest at $x = -2$ m and moves along the x-axis with the velocity history shown. Plot the corresponding acceleration and displacement histories for the two seconds. Find the time t when the particle crosses the origin. *Ans.* $t = 0.917$ s

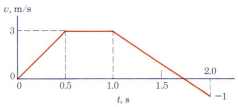

Problem 2/35

2/36 A car starts from rest with an acceleration of 6 m/s^2 which decreases linearly with time to zero in 10 seconds, after which the car continues at a constant speed. Determine the time t required for the car to travel 400 m from the start.

2/37 A retarding force is applied to a body moving in a straight line so that, during an interval of its motion, its speed v decreases with increased position coordinate s according to the relation $v^2 = k/s$, where k is a constant. If the body has a forward speed of 2 in./sec and its position coordinate is 9 in. at time $t = 0$, determine the speed v at $t = 3$ sec. *Ans.* $v = 1.587$ in./sec

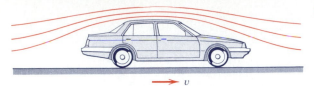

Problem 2/38

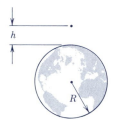

Problem 2/39

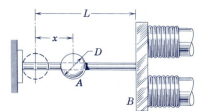

Problem 2/40

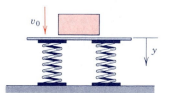

Problem 2/41

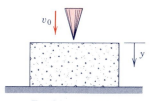

Problem 2/42

2/38 The aerodynamic resistance to motion of a car is nearly proportional to the square of its velocity. Additional frictional resistance is constant, so that the acceleration of the car when coasting may be written $a = -C_1 - C_2 v^2$, where C_1 and C_2 are constants that depend on the mechanical configuration of the car. If the car has an initial velocity v_0 when the engine is disengaged, derive an expression for the distance D required for the car to coast to a stop.

2/39 Compute the impact speed of a body released from rest at an altitude $h = 500$ mi. (*a*) Assume a constant gravitational acceleration $g_0 = 32.2$ ft/sec^2 and (*b*) account for the variation of g with altitude (refer to Art. 1/5). Neglect the effects of atmospheric drag.

$\qquad$ *Ans.* (*a*) $v = 13,040$ ft/sec, (*b*) $v = 12,290$ ft/sec

2/40 The steel ball A of diameter D slides freely on the horizontal rod that leads to the pole face of the electromagnet. The force of attraction obeys an inverse-square law, and the resulting acceleration of the ball is $a = K/(L - x)^2$, where K is a measure of the strength of the magnetic field. If the ball is released from rest at $x = 0$, determine the velocity v with which it strikes the pole face.

2/41 The body falling with speed v_0 strikes and maintains contact with the platform supported by a nest of springs. The acceleration of the body after impact is $a = g - cy$, where c is a positive constant and y is measured from the original platform position. If the maximum compression of the springs is observed to be y_m, determine the constant c.

$$Ans.\ c = \frac{v_0{}^2 + 2gy_m}{y_m{}^2}$$

2/42 The cone falling with a speed v_0 strikes and penetrates the block of packing material. The acceleration of the cone after impact is $a = g - cy^2$, where c is a positive constant and y is the penetration distance. If the maximum penetration depth is observed to be y_m, determine the constant c.

2/43 A certain lake is proposed as a landing area for large jet aircraft. The touchdown speed of 100 mi/hr upon contact with the water is to be reduced to 20 mi/hr in a distance of 1500 ft. If the deceleration is proportional to the square of the velocity of the aircraft through the water, $a = -Kv^2$, find the value of the design parameter K, which would be a measure of the size and shape of the landing gear vanes that plow through the water. Also find the time t elapsed during the specified interval.

Ans. $K = 1.073(10^{-3})$ ft^{-1}, $t = 25.4$ sec

2/44 The electronic throttle control of a model train is programmed so that the train speed varies with position as shown in the plot. Determine the time t required for the train to complete one lap.

2/45 A test projectile is fired horizontally into a viscous liquid with a velocity v_0. The retarding force is proportional to the square of the velocity, so that the acceleration becomes $a = -kv^2$. Derive expressions for the distance D traveled in the liquid and the corresponding time t required to reduce the velocity to $v_0/2$. Neglect any vertical motion.

Ans. $D = 0.693/k$, $t = 1/(kv_0)$

2/46 Approximate the v-t data for the decelerating ship of Prob. 2/33 by using the exponential relation $v = v_0 e^{-kt}$, where v_0 and k are constants determined by the initial and final conditions. Then calculate the distance s between the buoys.

2/47 The acceleration of the drag racer is modeled as $a = c_1 - c_2 v^2$, where the v^2-term accounts for aerodynamic drag and where c_1 and c_2 are positive constants. If c_2 is known (from wind-tunnel tests) to be $5(10^{-5})$ ft^{-1}, determine c_1 if the final speed is 190 mi/hr. A drag race is a 1/4-mile straight run from a standing start.

Ans. $c_1 = 31.4$ ft/sec^2

2/48 Use the value for c_1 cited in the answer to Prob. 2/47 and determine the time t required for the drag racer described in that problem to complete the 1/4-mile run.

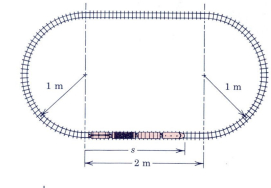

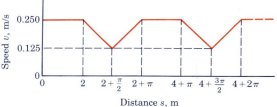

Problem 2/44

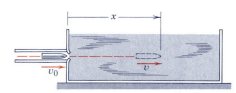

Problem 2/45

Problem 2/47

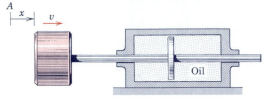

Problem 2/49

2/49 The horizontal motion of the plunger and shaft is arrested by the resistance of the attached disk that moves through the oil bath. If the velocity of the plunger is v_0 in the position A where $x = 0$ and $t = 0$, and if the deceleration is proportional to v so that $a = -kv$, derive expressions for the velocity v and position coordinate x in terms of the time t. Also express v in terms of x.

$$Ans. \quad v = v_0 e^{-kt}, \; x = \frac{v_0}{k}[1 - e^{-kt}]$$
$$v = v_0 - kx$$

2/50 A small object is released from rest in a tank of oil. The downward acceleration of the object is $g - kv$, where g is the constant acceleration due to gravity, k is a constant that depends on the viscosity of the oil and shape of the object, and v is the downward velocity of the object. Derive expressions for the velocity v and vertical drop y as functions of the time t after release.

2/51 In a test drop in still air, a small steel ball is released from rest at a considerable height. Its initial acceleration g is reduced by the decrement kv^2, where k is a constant and v is the downward velocity. Determine the maximum velocity attainable by the ball and find the vertical drop y in terms of the time t from release.

$$Ans. \quad v_{\max} = \sqrt{\frac{g}{k}}, \; y = \frac{1}{k} \ln \cosh \sqrt{kg}\, t$$

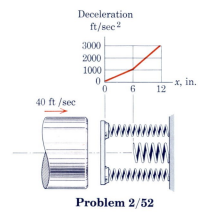

Problem 2/52

2/52 A bumper, consisting of a nest of three springs, is used to arrest the horizontal motion of a large mass that is traveling at 40 ft/sec as it contacts the bumper. The two outer springs cause a deceleration proportional to the spring deformation. The center spring increases the deceleration rate when the compression exceeds 6 in. as shown on the graph. Determine the maximum compression x of the outer springs.

2/53 When the effect of aerodynamic drag is included, the y-acceleration of a baseball when moving vertically upward is $a_u = -g - kv^2$, while the acceleration when moving downward is $a_d = -g + kv^2$, where k is a positive constant and v is the speed in ft/sec. If the ball is thrown upward at 100 ft/sec from essentially ground level, compute its maximum height h and its speed v_f upon impact with the ground. Take k to be 0.002 ft^{-1} and assume that g is constant.

$$Ans. \quad h = 120.8 \text{ ft}, \; v_f = 78.5 \text{ ft/sec}$$

2/54 For the baseball of Prob. 2/53 thrown upward with an initial speed of 100 ft/sec, determine the time t_u from ground to apex and the time t_d from apex to ground.

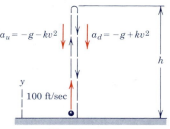

$a_u = -g - kv^2$　　　$a_d = -g + kv^2$

y

100 ft/sec

Problem 2/53

2/55 The fuel of a model rocket is burned so quickly that one may assume that the rocket acquires its burnout velocity of 120 m/s while essentially still at ground level. The rocket then coasts vertically upward to the trajectory apex. With the inclusion of aerodynamic drag, the y acceleration is $a_u = -g - 0.0005v^2$ during this motion, where the units are meters and seconds. At apex a parachute pops out of the nose cone, and the rocket quickly acquires a constant downward speed of 4 m/s. Estimate the flight time t_f.　　　*Ans.* $t_f = 147.7$ s

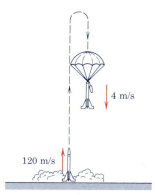

4 m/s

120 m/s

Problem 2/55

▶**2/56** A particle that is constrained to move in a straight line is subjected to an accelerating force that increases with time and a retarding force that increases directly with the position coordinate x. The resulting acceleration is $a = Kt - k^2 x$, where K and k are positive constants and where both x and $\dot{x}$ are zero when $t = 0$. Determine x as a function of t.

$$\text{Ans. } x = \frac{K}{k^3} (kt - \sin kt)$$

▶**2/57** The vertical acceleration of a certain solid-fuel rocket is given by $a = ke^{-bt} - cv - g$, where k, b, and c are constants, v is the vertical velocity acquired, and g is the gravitational acceleration, essentially constant for atmospheric flight. The exponential term represents the effect of a decaying thrust as fuel is burned, and the term $-cv$ approximates the retardation due to atmospheric resistance. Determine the expression for the vertical velocity of the rocket t seconds after firing.

$$\text{Ans. } v = \frac{g}{c} (e^{-ct} - 1) + \frac{k}{c - b} (e^{-bt} - e^{-ct})$$

▶**2/58** The preliminary design for a rapid transit system calls for the train velocity to vary with time as shown in the plot as the train runs the two miles between stations A and B. The slopes of the cubic transition curves (which are of form $a + bt + ct^2 + dt^3$) are zero at the end points. Determine the total run time t between the stations and the maximum acceleration.

　　Ans. $t = 105$ sec, $a_{max} = 11.73$ ft/sec^2

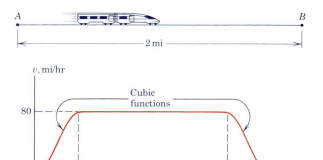

A　　　　　　　　　　　　　　　　　　　　　　B

2 mi

v, mi/hr

Cubic functions

80

0

15　　　　　Δt　　　　　15

A　　　　　　　　　　　　　　　　　　　B

t, sec

Problem 2/58

2/3 *PLANE CURVILINEAR MOTION*

We now begin our description of the motion of a particle along a curved path that lies in a single plane. This motion is seen to be a special case of the more general three-dimensional motion introduced in Art. 2/1 and illustrated in Fig. 2/1. If we let the plane of motion be the *x-y* plane, for instance, then the coordinates *z* and ϕ of Fig. 2/1 are both zero and *R* becomes the same as *r*. As mentioned previously, the vast majority of the motions of points or particles encountered in engineering practice may be represented as plane motion.

Before pursuing the description of plane curvilinear motion in any specific set of coordinates, we will first describe the motion with the aid of vector analysis, the results of which are independent of any particular coordinate system. What follows in this article will constitute one of the most basic of all treatments in dynamics, namely, development of the *time derivative of a vector*. A large fraction of the entire subject of dynamics hinges directly on the time rates of change of vector quantities, and all students of dynamics are well advised to master this topic at the outset because they will have frequent occasion to use it.

Consider now the continuous motion of a particle along a plane curve as represented in Fig. 2/5. At time *t* the particle is at position *A*, which is located by the *position vector* **r** measured from some convenient fixed origin *O*. If both the magnitude and direction of **r** are known at time *t*, then the position of the particle is completely specified. At time $t + \Delta t$, the particle is at *A'*, located by the position vector $\mathbf{r} + \Delta \mathbf{r}$. We note, of course, that this combination is vector addition and not scalar addition. The *displacement* of the particle during time Δt is the vector $\Delta \mathbf{r}$ that represents the vector change of position and is clearly independent of the choice of origin. If an origin were chosen at some different location, the position vector **r** would be changed, but $\Delta \mathbf{r}$ would be unchanged. The *distance* actually traveled by the particle as it moves along the path from *A* to *A'* is the scalar length Δs measured along the path. Thus, we

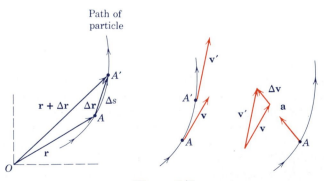

Figure 2/5

distinguish between the vector displacement $\Delta\mathbf{r}$ and the scalar distance Δs.

The *average velocity* of the particle between A and A' is defined as $\mathbf{v}_{av} = \Delta\mathbf{r}/\Delta t$, which is a vector whose direction is that of $\Delta\mathbf{r}$ and whose magnitude is the magnitude of $\Delta\mathbf{r}$ divided by Δt. The average *speed* of the particle between A and A' is the scalar quotient $\Delta s/\Delta t$. Clearly, the magnitude of the average velocity and the speed approach one another as the interval Δt decreases and A and A' become closer together.

The *instantaneous velocity* $\mathbf{v}$ of the particle is defined as the limiting value of the average velocity as the time interval approaches zero. Thus,

$$\mathbf{v} = \lim_{\Delta t \to 0} \frac{\Delta\mathbf{r}}{\Delta t}$$

We observe that the direction of $\Delta\mathbf{r}$ approaches that of the tangent to the path as Δt approaches zero and, hence, the velocity $\mathbf{v}$ is always a vector tangent to the path. We now extend the basic definition of the derivative of a scalar quantity to include a vector quantity and write

$$\mathbf{v} = \frac{d\mathbf{r}}{dt} = \dot{\mathbf{r}} \qquad\qquad \textbf{(2/4)}$$

The derivative of a vector is itself a vector having both a magnitude and a direction. The magnitude of $\mathbf{v}$ is called the *speed* and is the scalar

$$v = |\mathbf{v}| = \frac{ds}{dt} = \dot{s}$$

At this point we make a careful distinction between the *magnitude of the derivative* and the *derivative of the magnitude*. The magnitude of the derivative may be written in any one of the several ways $|d\mathbf{r}/dt| = |\dot{\mathbf{r}}| = \dot{s} = |\mathbf{v}| = v$ and represents the magnitude of the velocity or the speed of the particle. On the other hand, the derivative of the magnitude is written $d|\mathbf{r}|/dt = dr/dt = \dot{r}$, and represents the rate at which the length of the position vector $\mathbf{r}$ is changing. Thus, these two derivatives have two entirely different meanings, and we must be extremely careful to distinguish between them in our thinking and in our notation. For this and other reasons, students are urged to adopt a consistent notation for their handwritten work for all vector quantities to distinguish them from scalar quantities. For simplicity the underline $\underline{v}$ is recommended. Other handwritten symbols such as $\bar{v}$, $\underset{\sim}{v}$, and $\hat{v}$ are sometimes used.

With the concept of velocity as a vector established, we return to Fig. 2/5 and denote the velocity of the particle at A by the tangent vector **v** and the velocity at A' by the tangent vector **v**'. Clearly, there is a vector change in the velocity during the time Δt. The velocity **v** at A plus (vectorially) the change $\Delta\mathbf{v}$ must equal the velocity at A', so we may write $\mathbf{v}' - \mathbf{v} = \Delta\mathbf{v}$. Inspection of the vector diagram shows that $\Delta\mathbf{v}$ depends both on the change in magnitude (length) of **v** and on the change in direction of **v**. These two changes are fundamental characteristics of the derivative of a vector.

The *average acceleration* of the particle between A and A' is defined as $\Delta\mathbf{v}/\Delta t$, which is a vector whose direction is that of $\Delta\mathbf{v}$. The magnitude of this average acceleration is the magnitude of $\Delta\mathbf{v}$ divided by Δt.

The *instantaneous acceleration* **a** of the particle is defined as the limiting value of the average acceleration as the time interval approaches zero. Thus,

$$\mathbf{a} = \lim_{\Delta t \to 0} \frac{\Delta\mathbf{v}}{\Delta t}$$

By definition of the derivative, then, we write

$$\mathbf{a} = \frac{d\mathbf{v}}{dt} = \dot{\mathbf{v}} \qquad\qquad (2/5)$$

As the interval Δt becomes smaller and approaches zero, the direction of the change $\Delta\mathbf{v}$ approaches that of the differential change $d\mathbf{v}$ and, hence, **a**. The acceleration **a**, then, includes the effect of change in magnitude of **v** and the change of direction of **v**. It is apparent, in general, that the direction of the acceleration of a particle in curvilinear motion is neither tangent to the path nor normal to the path. We do observe, however, that the acceleration component that is normal to the path points toward the center of curvature of the path.

A further approach to the visualization of acceleration is shown in Fig. 2/6, where the position vectors to three arbitrary positions on the path of the particle are shown for illustrative purpose. There is a velocity vector tangent to the path corresponding to each posi-

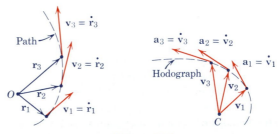

Figure 2/6

tion vector, and the relation is $\mathbf{v} = \dot{\mathbf{r}}$. Now if these velocity vectors are plotted from some arbitrary point C, then a curve, known as the *hodograph*, is formed. The derivatives of these velocity vectors will be the acceleration vectors $\mathbf{a} = \dot{\mathbf{v}}$ that are tangent to the hodograph. We see that the acceleration bears the same relation to the velocity as the velocity bears to the position vector.

     The geometric portrayal of the derivatives of the position vector $\mathbf{r}$ and velocity vector $\mathbf{v}$ in Fig. 2/5 can be used to describe the derivative of any vector quantity with respect to t or with respect to any other scalar variable. With the introduction of the derivative of a vector in the definitions of velocity and acceleration, it is important at this point to establish the rules under which the differentiation of vector quantities may be carried out. These rules are the same as for the differentiation of scalar quantities, except for the case of the cross product where the order of the terms must be preserved. These rules are covered in Art. C/7 of Appendix C and should be reviewed at this point.

     For curvilinear motion of a particle in a plane, there are three different coordinate systems that are in common use to describe the vector relationships developed in this article. An important lesson to be learned from the study of these coordinate systems is the proper choice of a reference system for a given problem. This choice is usually revealed by the manner in which the motion is generated or by the form in which the data are specified. Each of the three coordinate systems will now be developed and illustrated.

2/4   RECTANGULAR COORDINATES (*x-y*)

     This system of coordinates is particularly useful for describing motions where the x- and y-components of acceleration are independently generated or determined. The resulting curvilinear motion is, then, obtained by a vector combination of the x- and y-components of the position vector, the velocity, and the acceleration.

     The particle path of Fig. 2/5 is shown again in Fig. 2/7 along with x- and y-axes. The position vector $\mathbf{r}$, the velocity $\mathbf{v}$, and the acceleration $\mathbf{a}$ of the particle as developed in Art. 2/3 are represented in Fig. 2/7 together with their x- and y-components. With the aid of the unit vectors $\mathbf{i}$ and $\mathbf{j}$, we may write the vectors $\mathbf{r}$, $\mathbf{v}$, and $\mathbf{a}$ in terms of their x- and y-components. Thus,

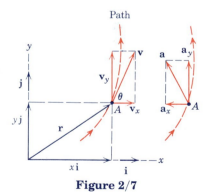

Figure 2/7

$$\mathbf{r} = x\mathbf{i} + y\mathbf{j}$$
$$\mathbf{v} = \dot{\mathbf{r}} = \dot{x}\mathbf{i} + \dot{y}\mathbf{j} \qquad \textbf{(2/6)}$$
$$\mathbf{a} = \dot{\mathbf{v}} = \ddot{\mathbf{r}} = \ddot{x}\mathbf{i} + \ddot{y}\mathbf{j}$$

As we differentiate with respect to time, we observe that the unit vectors have no time derivatives since their magnitudes and directions remain constant. The scalar magnitudes of the components of **v** and **a** are merely $v_x = \dot{x}$, $v_y = \dot{y}$ and $a_x = \dot{v}_x = \ddot{x}$, $a_y = \dot{v}_y = \ddot{y}$. (As drawn in Fig. 2/7, a_x is in the negative x-direction, so that $\ddot{x}$ would be a negative number.)

As observed previously, the direction of the velocity is always tangent to the path, and from the figure it is clear that

$$v^2 = v_x^2 + v_y^2 \qquad v = \sqrt{v_x^2 + v_y^2} \qquad \tan\theta = \frac{v_y}{v_x}$$

$$a^2 = a_x^2 + a_y^2 \qquad a = \sqrt{a_x^2 + a_y^2}$$

If the angle θ is measured counterclockwise from the x-axis to **v** for the configuration of axes shown, then we may also observe that $dy/dx = \tan\theta = v_y/v_x$.

If the coordinates x and y are each known independently as a function of time, $x = f_1(t)$ and $y = f_2(t)$, then for any value of the time we may combine them to obtain **r**. Similarly, we combine their first derivatives $\dot{x}$ and $\dot{y}$ to obtain **v** and their second derivatives $\ddot{x}$ and $\ddot{y}$ to obtain **a**. Or, if the acceleration components a_x and a_y are given as functions of the time, we may integrate each one separately with respect to time, once to obtain v_x and v_y and again to obtain $x = f_1(t)$ and $y = f_2(t)$. Elimination of the time t between these last two parametric equations gives the equation of the curved path $y = f(x)$.

From the foregoing discussion it should be recognized that the rectangular-coordinate representation of curvilinear motion is merely the superposition of the components of two simultaneous rectilinear motions in the x- and y-directions. Therefore, everything covered in Art. 2/2 on rectilinear motion may be applied separately to the x-motion and to the y-motion.

Projectile motion. An important application of two-dimensional kinematic theory is the problem of projectile motion. For a first treatment of the subject, we neglect aerodynamic drag and the earth's curvature and rotation, and we assume that the altitude range is small enough so that the acceleration due to gravity can be considered constant. With these assumptions, rectangular coordinates are useful for the trajectory analysis. For the axes shown in Fig. 2/8, the acceleration components are

$$a_x = 0 \qquad a_y = -g$$

Integration of these accelerations follows the results obtained previously in Art. 2/2a for constant acceleration and yields

$$v_x = (v_x)_0 \qquad\qquad v_y = (v_y)_0 - gt$$

$$x = x_0 + (v_x)_0 t \qquad y = y_0 + (v_y)_0 t - \tfrac{1}{2}gt^2$$

$$v_y{}^2 = (v_y)_0{}^2 - 2g(y - y_0)$$

In all these expressions, the subscript zero denotes initial conditions, frequently taken as those at launch where, for the case illustrated, $x_0 = y_0 = 0$. Note that the quantity g is taken to be positive throughout this text.

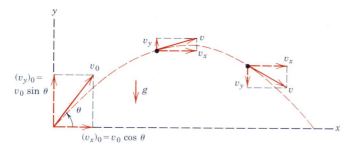

Figure 2/8

We can see that the x- and y-motions are independent for the simple projectile conditions under consideration. Elimination of the time t between the x- and y-displacement equations shows the path to be parabolic (see Sample Problem 2/6). If we were to introduce a drag force that depends on the speed squared (for example), then the x- and y-motions would be coupled (interdependent), and the trajectory would be nonparabolic.

When extreme accuracy is to be achieved in describing projectile motion when large velocities and high altitudes are involved, the shape of the projectile, the variation of g with altitude, the variation of the air density with altitude, and the rotation of the earth must all be accounted for. These factors introduce considerable complexity in the motion equations, and numerical integration of the acceleration equations is usually necessary.

Sample Problem 2/5

The curvilinear motion of a particle is defined by $v_x = 50 - 16t$ and $y = 100 - 4t^2$, where v_x is in meters per second, y is in meters, and t is in seconds. It is also known that $x = 0$ when $t = 0$. Plot the path of the particle and determine its velocity and acceleration when the position $y = 0$ is reached.

Solution. The x-coordinate is obtained by integrating the expression for v_x, and the x-component of the acceleration is obtained by differentiating v_x. Thus,

$$\left[\int dx = \int v_x \, dt\right] \qquad \int_0^x dx = \int_0^t (50 - 16t) \, dt \qquad x = 50t - 8t^2 \text{ m}$$

$$[a_x = \dot{v}_x] \qquad a_x = \frac{d}{dt}(50 - 16t) \qquad a_x = -16 \text{ m/s}^2$$

The y-components of velocity and acceleration are

$$[v_y = \dot{y}] \qquad v_y = \frac{d}{dt}(100 - 4t^2) \qquad v_y = -8t \text{ m/s}$$

$$[a_y = \dot{v}_y] \qquad a_y = \frac{d}{dt}(-8t) \qquad a_y = -8 \text{ m/s}^2$$

We now calculate corresponding values of x and y for various values of t and plot x against y to obtain the path as shown.

When $y = 0$, $0 = 100 - 4t^2$, so $t = 5$ s. For this value of the time, we have

$$v_x = 50 - 16(5) = -30 \text{ m/s}$$

$$v_y = -8(5) = -40 \text{ m/s}$$

$$v = \sqrt{(-30)^2 + (-40)^2} = 50 \text{ m/s}$$

$$a = \sqrt{(-16)^2 + (-8)^2} = 17.9 \text{ m/s}^2$$

The velocity and acceleration components and their resultants are shown on the separate diagrams for point A, where $y = 0$. Thus, for this condition we may write

$$\mathbf{v} = -30\mathbf{i} - 40\mathbf{j} \text{ m/s} \qquad\qquad Ans.$$

$$\mathbf{a} = -16\mathbf{i} - 8\mathbf{j} \text{ m/s}^2 \qquad\qquad Ans.$$

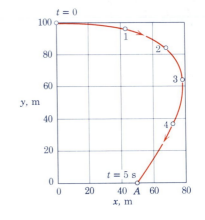

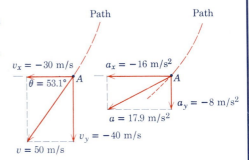

We observe that the velocity vector lies along the tangent to the path as it should, but that the acceleration vector is not tangent to the path. Note especially that the acceleration vector has a component that points toward the inside of the curved path. We concluded from our diagram in Fig. 2/5 that it is impossible for the acceleration to have a component that points toward the outside of the curve.

Sample Problem 2/6

A rocket has expended all its fuel when it reaches position A, where it has a velocity $\mathbf{u}$ at an angle θ with respect to the horizontal. It then begins unpowered flight and attains a maximum added height h at position B after traveling a horizontal distance s from A. Determine the expressions for h and s, the time t of flight from A to B, and the equation of the path. For the interval concerned, assume a flat earth with a constant gravitational acceleration g and neglect any atmospheric resistance.

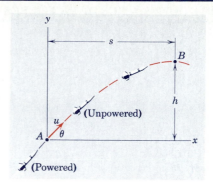

Solution. Since all motion components are directly expressible in terms of horizontal and vertical coordinates, a rectangular set of axes x-y will be employed. With the neglect of atmospheric resistance, $a_x = 0$ and $a_y = -g$, and the resulting motion is a direct superposition of two rectilinear motions with constant acceleration. Thus,

$$[dx = v_x \, dt] \qquad x = \int_0^t u \cos \theta \, dt \qquad x = ut \cos \theta$$

$$[dv_y = a_y \, dt] \qquad \int_{u \sin \theta}^{v_y} dv_y = \int_0^t (-g) \, dt \qquad v_y = u \sin \theta - gt$$

$$[dy = v_y \, dt] \qquad y = \int_0^t (u \sin \theta - gt) \, dt \qquad y = ut \sin \theta - \tfrac{1}{2}gt^2$$

Position B is reached when $v_y = 0$, which occurs for $0 = u \sin \theta - gt$ or

$$t = (u \sin \theta)/g \qquad\qquad Ans.$$

Substitution of this value for the time into the expression for y gives the maximum added altitude

$$h = u \left(\frac{u \sin \theta}{g} \right) \sin \theta - \frac{1}{2} g \left(\frac{u \sin \theta}{g} \right)^2 \qquad h = \frac{u^2 \sin^2 \theta}{2g} \qquad Ans.$$

The horizontal distance is seen to be

$$s = u \left(\frac{u \sin \theta}{g} \right) \cos \theta \qquad s = \frac{u^2 \sin 2\theta}{2g} \qquad Ans.$$

which is clearly a maximum when $\theta = 45°$. The equation of the path is obtained by eliminating t from the expressions for x and y, which gives

$$y = x \tan \theta - \frac{gx^2}{2u^2} \sec^2 \theta \qquad Ans.$$

③ This equation describes a vertical parabola as indicated in the figure.

① Note that this problem is simply the description of projectile motion neglecting atmospheric resistance.

② We see that the total range and time of flight for a projectile fired above a horizontal plane would be twice the respective values of s and t given here.

③ If atmospheric resistance were to be accounted for, the dependency of the acceleration components on the velocity would have to be established before an integration of the equations could be carried out. This becomes a much more difficult problem.

PROBLEMS

(In the following problems where motion as a projectile in air is involved, neglect air resistance unless otherwise stated and use $g = 9.81$ m/s^2 or $g = 32.2$ ft/sec^2.)

Introductory problems

2/59 The position vector of a particle moving in the x-y plane at time $t = 3.60$ s is $2.76\mathbf{i} - 3.28\mathbf{j}$ m. At $t = 3.62$ s its position vector has become $2.79\mathbf{i} - 3.33\mathbf{j}$ m. Determine the magnitude v of its average velocity during this interval and the angle θ made by the average velocity with the x-axis.

Ans. $v = 2.92$ m/s, $\theta = -59.0°$

2/60 A particle moving in the x-y plane has a velocity at time $t = 6$ s given by $4\mathbf{i} + 5\mathbf{j}$ m/s, and at $t = 6.1$ s its velocity has become $4.3\mathbf{i} + 5.4\mathbf{j}$ m/s. Calculate the magnitude a of its average acceleration during the 0.1-s interval and the angle θ it makes with the x-axis.

2/61 The velocity of a particle moving in the x-y plane is given by $6.12\mathbf{i} + 3.24\mathbf{j}$ m/s at time $t = 3.65$ s. Its average acceleration during the next 0.02 s is $4\mathbf{i} + 6\mathbf{j}$ m/s^2. Determine the velocity v of the particle at $t = 3.67$ s and the angle θ between the average acceleration vector and the velocity vector at $t = 3.67$ s.

Ans. $\mathbf{v} = 6.20\mathbf{i} + 3.36\mathbf{j}$ m/s, $\theta = 27.9°$

2/62 A particle that moves with curvilinear motion has coordinates in millimeters that vary with the time t in seconds according to $x = 2t^2 - 4t$ and $y = 3t^2 - \frac{1}{3}t^3$. Determine the magnitudes of the velocity $\mathbf{v}$ and acceleration $\mathbf{a}$ and the angles that these vectors make with the x-axis when $t = 2$ s.

2/63 A roofer tosses a small tool toward a coworker on the ground. What is the minimum horizontal velocity v_0 necessary so that the tool clears point B? Locate the point of impact by specifying the distance s shown in the figure.

Ans. $v_0 = 6.64$ m/s, $s = 2.49$ m

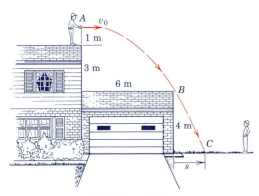

Problem 2/63

2/64 The x- and y-motions of guides A and B with right-angle slots control the curvilinear motion of the connecting pin P, which slides in both slots. For a short interval, the motions are governed by $x = 20 + \frac{1}{4}t^2$ and $y = 15 - \frac{1}{6}t^3$, where x and y are in millimeters and t is in seconds. Calculate the magnitudes of the velocity $\mathbf{v}$ and acceleration $\mathbf{a}$ of the pin for $t = 2$ s. Sketch the direction of the path and indicate its curvature for this instant.

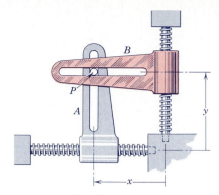

Problem 2/64

2/65 A broadjumper approaches his takeoff board A with a horizontal velocity of 30 ft/sec. Determine the vertical component v_y of the velocity of his center of gravity at takeoff for him to make the jump shown. What is the vertical rise h of his center of gravity?

$Ans.\ v_y = 11.81$ ft/sec, $h = 2.16$ ft

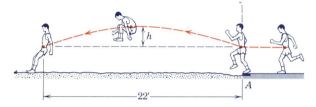

Problem 2/65

2/66 The y-coordinate of a particle in curvilinear motion is given by $y = 4t^3 - 3t$, where y is in inches and t is in seconds. Also, the particle has an acceleration in the x-direction given by $a_x = 12t$ in./sec^2. If the velocity of the particle in the x-direction is 4 in./sec when $t = 0$, calculate the magnitudes of the velocity $\mathbf{v}$ and acceleration $\mathbf{a}$ of the particle when $t = 1$ sec. Construct $\mathbf{v}$ and $\mathbf{a}$ in your solution.

2/67 A rocket runs out of fuel in the position shown and continues in unpowered flight above the atmosphere. If its velocity in this position was 600 mi/hr, calculate the maximum additional altitude h acquired and the corresponding time t to reach it. The gravitational acceleration during this phase of its flight is 30.8 ft/sec^2.

$Ans.\ h = 1.786$ mi, $t = 24.7$ sec

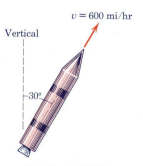

Problem 2/67

Representative problems

2/68 The position vector of a point that moves in the x-y plane is given by

$$\mathbf{r} = \left(\frac{2}{3}t^3 - \frac{3}{2}t^2\right)\mathbf{i} + \frac{t^4}{12}\mathbf{j}$$

where $\mathbf{r}$ is in meters and t is in seconds. Determine the angle between the velocity $\mathbf{v}$ and the acceleration $\mathbf{a}$ when $(a)\ t = 2$ s and $(b)\ t = 3$ s.

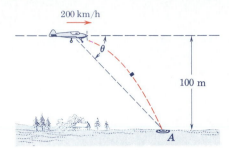

Problem 2/69

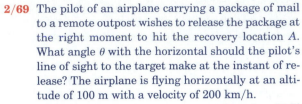

2/69 The pilot of an airplane carrying a package of mail to a remote outpost wishes to release the package at the right moment to hit the recovery location A. What angle θ with the horizontal should the pilot's line of sight to the target make at the instant of release? The airplane is flying horizontally at an altitude of 100 m with a velocity of 200 km/h.

$Ans.$ $\theta = 21.7°$

2/70 Prove the well-known result that, for a given launch speed v_0, the launch angle $\theta = 45°$ yields the maximum range R. Determine the maximum range. (Note that this result does not hold when aerodynamic drag is included in the analysis.)

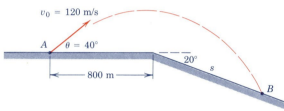

Problem 2/71

2/71 A projectile is launched from point A with the initial conditions shown in the figure. Determine the slant distance s that locates the point B of impact. Calculate the time of flight t.

$Ans.$ $s = 1057$ m, $t = 19.50$ s

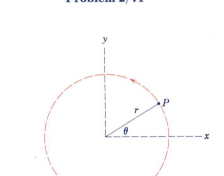

Problem 2/72

2/72 A particle P moves on the circular path of radius r as shown. If $\dot{\theta} = \omega$, a constant, prove that the acceleration of P is always directed toward the center.

2/73 A particle moves in the x-y plane with a y-component of velocity in feet per second given by $v_y = 8t$ with t in seconds. The acceleration of the particle in the x-direction in feet per second squared is given by $a_x = 4t$ with t in seconds. When $t = 0$, $y = 2$ ft, $x = 0$, and $v_x = 0$. Find the equation of the path of the particle and calculate the magnitude of the velocity $\mathbf{v}$ of the particle for the instant when its x-coordinate reaches 18 ft.

$Ans.$ $(y - 2)^3 = 144x^2$, $v = 30$ ft/sec

Problem 2/74

2/74 A projectile is launched with a speed $v_0 = 25$ m/s from the floor of a 5-m-high tunnel as shown. Determine the maximum range R of the projectile and the corresponding launch angle θ.

2/75 For a certain interval of motion, the pin P is forced to move in the fixed parabolic slot by the vertical slotted guide, which moves in the x-direction at the constant rate of 20 mm/s. All measurements are in millimeters and seconds. Calculate the magnitudes of the velocity $\mathbf{v}$ and acceleration $\mathbf{a}$ of pin P when $x = 60$ mm. *Ans.* $v = 25$ mm/s, $a = 5$ mm/s^2

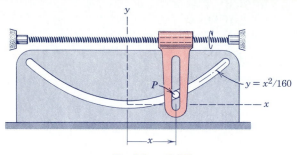

Problem 2/75

2/76 If the tennis player serves the ball horizontally ($\theta = 0$), calculate its velocity v if the center of the ball clears the 36-in. net by 6 in. Also find the distance s from the net to where the ball hits the court surface. Neglect air resistance and the effect of ball spin.

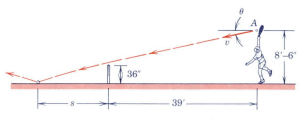

Problem 2/76

2/77 If the tennis player shown in Prob. **2/76** serves the ball with a velocity v of 80 mi/hr at the angle $\theta = 5°$, calculate the vertical clearance h of the center of the ball above the net and the distance s from the net where the ball hits the court surface. Neglect air resistance and the effect of ball spin. *Ans.* $h = 3.55$ in., $s = 16.57$ ft

2/78 The muzzle velocity of a long-range rifle at A is $u = 400$ m/s. Determine the two angles of elevation θ that will permit the projectile to hit the mountain target B.

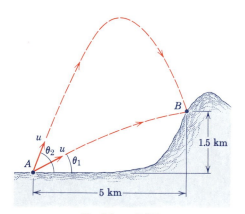

Problem 2/78

2/79 A projectile is launched with an initial speed of 200 m/s at an angle of 60° with respect to the horizontal. Compute the range R as measured up the incline. *Ans.* $R = 2970$ m

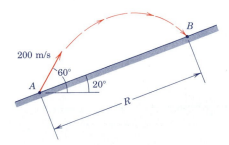

Problem 2/79

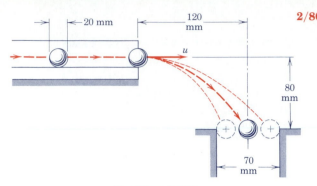

2/80 Ball bearings leave the horizontal trough with a velocity of magnitude u and fall through the 70-mm-diameter hole as shown. Calculate the permissible range of u that will enable the balls to enter the hole. Take the dotted positions to represent the limiting conditions.

Problem 2/80

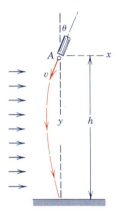

Problem 2/81

2/81 A particle is ejected from the tube at A with a velocity v at an angle θ with the vertical y-axis. A strong horizontal wind gives the particle a constant horizontal acceleration a in the x-direction. If the particle strikes the ground at a point directly under its released position, determine the height h of point A. The downward y-acceleration may be taken as the constant g.

$$Ans. \quad h = \frac{2v^2}{a} \sin \theta \left(\cos \theta + \frac{g}{a} \sin \theta \right)$$

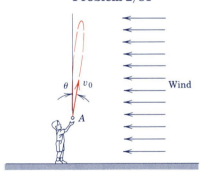

Problem 2/82

2/82 A boy throws a ball upward with a speed $v_0 = 12$ m/s. The wind imparts a horizontal acceleration of 0.4 m/s^2 to the left. At what angle θ must the ball be thrown so that it returns to the point of release? Assume that the wind does not affect the vertical motion.

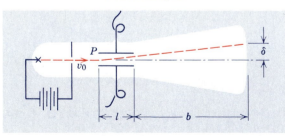

Problem 2/83

2/83 In the cathode-ray tube, electrons traveling horizontally from their source with the velocity v_0 are deflected by an electric field E due to the voltage gradient across the plates P. The deflecting force causes an acceleration in the vertical direction on the sketch equal to eE/m, where e is the electron charge and m is its mass. When clear of the plates, the electrons travel in straight lines. Determine the expression for the deflection δ for the tube and plate dimensions shown.

$$Ans. \quad \delta = \frac{eEl}{mv_0^2} \left(\frac{l}{2} + b \right)$$

2/84 Electrons are emitted at A with a velocity v at the angle θ into the space between two charged plates. The electric field between the plates is in the direction E and repels the electrons approaching the upper plate. The field produces an acceleration of the electrons in the E-direction of eE/m, where e is the electron charge and m is its mass. Determine the field strength E that will permit the electrons to cross one-half of the gap between the plates. Also find the distance s.

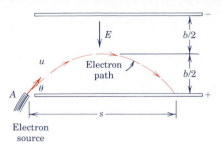

Problem 2/84

2/85 To pass inspection, small ball bearings must bounce through an opening of limited size at the top of their trajectory when rebounding from a heavy plate as shown. Calculate the angle θ made by the rebound velocity with the horizontal and the velocity v of the balls as they pass through the opening.

Ans. $\theta = 68.2°$, $v = 1.253$ m/s

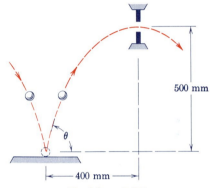

Problem 2/85

2/86 The water nozzle ejects water at a speed $v_0 = 45$ ft/sec at the angle $\theta = 40°$. Determine where, relative to the wall base point B, the water lands. Neglect the effects of the thickness of the wall.

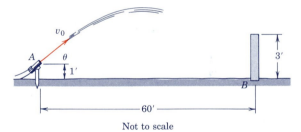

Not to scale

Problem 2/86

2/87 Water is ejected from the water nozzle of Prob. 2/86 with a speed $v_0 = 45$ ft/sec. For what value of the angle θ will the water land closest to the wall after clearing the top? Neglect the effects of wall thickness and air resistance. Where does the water land?

Ans. $\theta = 50.6°$, 2.50 ft to the right of B

2/88 Determine the location h of the spot toward which the pitcher must throw if the ball is to hit the catcher's mitt. The ball is released with a speed of 40 m/s.

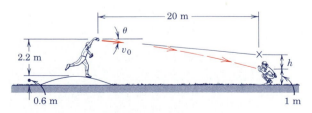

Problem 2/88

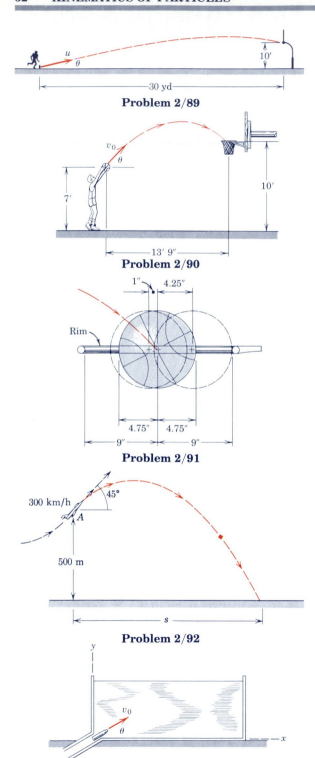

Problem 2/89

Problem 2/90

Problem 2/91

Problem 2/92

Problem 2/93

2/89 A football player attempts a 30-yd field goal. If he is able to impart a velocity u of 100 ft/sec to the ball, compute the minimum angle θ for which the ball will clear the crossbar of the goal. (*Hint:* Let $m = \tan\theta$.)   *Ans.* $\theta = 14.91°$

2/90 The basketball player likes to release his foul shots at an angle $\theta = 50°$ to the horizontal as shown. What initial speed v_0 will cause the ball to pass through the center of the rim?

2/91 A detail of the basketball and rim of Prob. 2/90 is shown in the figure. Assume that the player always releases the ball with a speed $v_0 = 24$ ft/sec. For the range of horizontal positions at the rim level indicated in the figure, determine the corresponding range for the launching angle θ. (Note that the actual range of horizontal positions would not be symmetrical about the rim center due to clearance considerations as the ball goes over the front edge of the rim.)   *Ans.* $43.1° \leq \theta \leq 45.2°$
$56.8° \leq \theta \leq 59.3°$

2/92 The pilot of an airplane pulls into a steep 45° climb at 300 km/h and releases a package at position A. Calculate the horizontal distance s and the time t from the point of release to the point at which the package strikes the ground.

2/93 A projectile is ejected into an experimental fluid at time $t = 0$. The initial speed is v_0 and the angle to the horizontal is θ. The drag on the projectile results in an acceleration term $\mathbf{a}_D = -k\mathbf{v}$, where k is a constant and $\mathbf{v}$ is the velocity of the projectile. Determine the x- and y-components of both the velocity and displacement as functions of time. What is the terminal velocity? Include the effects of gravitational acceleration.

Ans. $v_x = (v_0 \cos\theta)e^{-kt}, x = \dfrac{v_0 \cos\theta}{k}(1 - e^{-kt})$

$v_y = \left(v_0 \sin\theta + \dfrac{g}{k}\right)e^{-kt} - \dfrac{g}{k}$

$y = \dfrac{1}{k}\left(v_0 \sin\theta + \dfrac{g}{k}\right)(1 - e^{-kt}) - \dfrac{g}{k}t$

$v_x \to 0, v_y \to -\dfrac{g}{k}$

2/94 A point P is located by its position vector $\mathbf{r} = (b_1 \cos \theta)\mathbf{i} + (b_2 \sin \theta)\mathbf{j}$, where b_1 and b_2 are constants and θ is the angle between $\mathbf{r}$ and the x-axis. If θ increases at the constant rate $\dot{\theta}$, show that P moves on an elliptical path with an acceleration proportional to r and directed along $\mathbf{r}$ toward the origin.

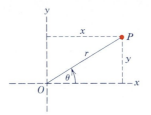

Problem 2/94

2/95 Determine the equation for the envelope a of the parabolic trajectories of a projectile fired at any angle but with a fixed muzzle velocity u. (*Hint:* Substitute $m = \tan \theta$, where θ is the firing angle with the horizontal, into the equation of the trajectory. The two roots m_1 and m_2 of the equation written as a quadratic in m give the two firing angles for the two trajectories shown such that the shells pass through the same point A. Point A will approach the envelope a as the two roots approach equality.) Neglect air resistance and assume g is constant.

$$Ans. \ y = \frac{u^2}{2g} - \frac{gx^2}{2u^2}$$

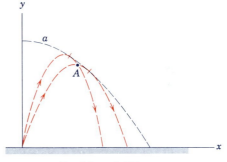

Problem 2/95

2/96 A projectile is launched with speed v_0 from point A. Determine the launch angle θ that results in the maximum range R up the incline of angle α (where $0 \le \alpha \le 90°$). Evaluate your results for $\alpha = 0$, $30°$, and $45°$.

$$Ans. \ \theta = \frac{90° + \alpha}{2}, \ \theta = 45°, 60°, 67.5°$$

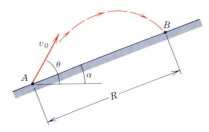

Problem 2/96

2/5 *NORMAL AND TANGENTIAL COORDINATES (n-t)*

In the introduction to this chapter, Art. 2/1, it was mentioned that one of the common descriptions of curvilinear motion involves the use of path variables, which are measurements made along the tangent t and normal n to the path of the particle. These coordinates provide a very natural description for curvilinear motion and are frequently the most direct and convenient coordinates to use. The n- and t-coordinates are considered to move along the path with the particle, as seen in Fig. 2/9 where the particle advances from A to B to C. The positive direction for n at any position is always taken toward the center of curvature of the path. As seen from Fig. 2/9, the positive n-direction will shift from one side of the curve to the other side if the curvature changes direction.

The coordinates n and t will now be used to describe the velocity $\mathbf{v}$ and acceleration $\mathbf{a}$ that were introduced in Art. 2/3 for the curvilinear motion of a particle. For this purpose, we introduce unit vectors $\mathbf{e}_n$ in the n-direction and $\mathbf{e}_t$ in the t-direction, as shown in Fig. 2/10a for the position of the particle at point A on its path. During a differential increment of time dt, the particle moves a differential distance ds along the curve from A to A'. With the radius of curvature of the path at this position designated by ρ, we see that $ds = \rho \, d\beta$, where β is in radians. We see that it is unnecessary to consider the differential change in ρ between A and A' since a higher-order term would be introduced that disappears in the limit. Thus, the magnitude of the velocity may be written $v = ds/dt = \rho \, d\beta/dt$, and we may write the velocity as the vector

$$\mathbf{v} = v\mathbf{e}_t = \rho\dot{\beta}e_t \qquad (2/7)$$

The acceleration $\mathbf{a}$ of the particle was defined in Art. 2/3 as $\mathbf{a} = d\mathbf{v}/dt$, and we observed from Fig. 2/5 that the acceleration is a vector that reflects both the change in magnitude and the change in direction of $\mathbf{v}$. We now differentiate $\mathbf{v}$ in Eq. 2/7 by applying the ordinary rule for the differentiation of the product of a scalar and a vector* and get

$$\mathbf{a} = \frac{d\mathbf{v}}{dt} = \frac{d(v\mathbf{e}_t)}{dt} = v\dot{\mathbf{e}}_t + \dot{v}\mathbf{e}_t \qquad (2/8)$$

where the unit vector $\mathbf{e}_t$ now has a derivative because its direction changes. To find $\dot{\mathbf{e}}_t$ we analyze the change in $\mathbf{e}_t$ during a differential increment of motion as the particle moves from A to A' in Fig. 2/10a. The unit vector $\mathbf{e}_t$ correspondingly changes to $\mathbf{e}_t'$, and the vector difference $d\mathbf{e}_t$ is shown in the b-part of the figure. The vector $d\mathbf{e}_t$ in the limit has a magnitude equal to the length of the arc

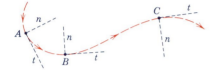

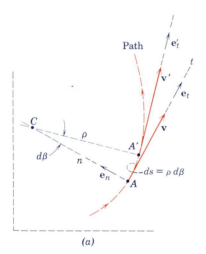

Figure 2/9

(a)

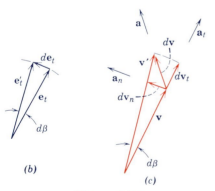

(b)

(c)

Figure 2/10

---

*See Art. C/7 of Appendix C.

$|\mathbf{e}_t|\,d\beta = d\beta$ obtained by swinging the unit vector $\mathbf{e}_t$ through the angle $d\beta$ expressed in radians. The direction of $d\mathbf{e}_t$ is given by $\mathbf{e}_n$. Thus, we may write $d\mathbf{e}_t = \mathbf{e}_n\,d\beta$. Dividing by $d\beta$ gives

$$\frac{d\mathbf{e}_t}{d\beta} = \mathbf{e}_n$$

or dividing by dt gives $d\mathbf{e}_t/dt = (d\beta/dt)\mathbf{e}_n$ which may be written

$$\dot{\mathbf{e}}_t = \dot{\beta}e_n \qquad\qquad (2/9)$$

With the substitution of Eq. 2/9 and $\dot{\beta}$ from the relation $v = \rho\dot{\beta}$, Eq. 2/8 for the acceleration becomes

$$\mathbf{a} = \frac{v^2}{\rho}\,\mathbf{e}_n + \dot{v}\mathbf{e}_t \qquad\qquad (2/10)$$

where

$$a_n = \frac{v^2}{\rho} = \rho\dot{\beta}^2 = v\dot{\beta}$$

$$a_t = \dot{v} = \ddot{s}$$

$$a = \sqrt{a_n^{\,2} + a_t^{\,2}}$$

We may also note that $a_t = \dot{v} = d(\rho\dot{\beta})/dt = \rho\ddot{\beta} + \dot{\rho}\dot{\beta}$. However, this relation finds little use since we seldom have reason to compute $\dot{\rho}$.

A clear understanding of Eq. 2/10 comes only when the geometry of the physical changes that it describes can be clearly seen. Figure 2/10c shows the velocity vector $\mathbf{v}$ when the particle is at A and $\mathbf{v}'$ when it is at A'. The vector change in the velocity is $d\mathbf{v}$, which establishes the direction of the acceleration $\mathbf{a}$. The n-component of $d\mathbf{v}$ is labeled $d\mathbf{v}_n$, and in the limit its magnitude equals the length of the arc generated by swinging the vector $\mathbf{v}$ as a radius through the angle $d\beta$. Thus, $|d\mathbf{v}_n| = v\,d\beta$ and the n-component of acceleration is $a_n = |d\mathbf{v}_n|/dt = v(d\beta/dt) = v\dot{\beta}$ as before. The t-component of $d\mathbf{v}$ is labeled $d\mathbf{v}_t$, and its magnitude is simply the change dv in the magnitude or length of the velocity vector. Therefore, the t-component of acceleration is $a_t = dv/dt = \dot{v} = \ddot{s}$ as before. The acceleration vectors resulting from the corresponding vector changes in velocity are shown in Fig. 2/10c.

It is especially important to observe that the normal component of acceleration a_n is *always directed toward the center of curvature C.* The tangential component of acceleration, on the other hand, will be in the positive t-direction of motion if the speed v is increasing and in the negative t-direction if the speed is decreasing.

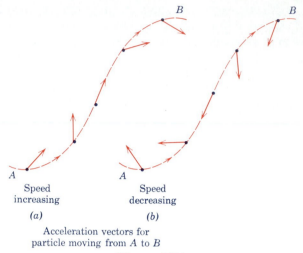

A

Speed
increasing

(a)

A

Speed
decreasing

(b)

Acceleration vectors for
particle moving from *A* to *B*

Figure 2/11

In Fig. 2/11 are shown schematic representations of the variation in the acceleration vector for a particle moving from *A* to *B* with (*a*) increasing speed and (*b*) decreasing speed. At an inflection point on the curve, the normal acceleration v^2/ρ goes to zero since ρ becomes infinite.

Circular motion is an important special case of plane curvilinear motion where the radius of curvature ρ becomes the constant radius r of the circle and the angle β is replaced by the angle θ measured from any convenient radial reference to *OP*, Fig. 2/12. The velocity and the acceleration components for the circular motion of the particle *P* become

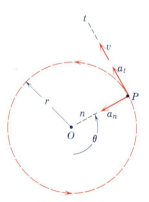

Figure 2/12

$$
\begin{aligned}
v &= r\dot{\theta} \\
a_n &= v^2/r = r\dot{\theta}^2 = v\dot{\theta} \\
a_t &= \dot{v} = r\ddot{\theta}
\end{aligned}
\qquad \textbf{(2/11)}
$$

We find repeated use for Eqs. 2/10 and 2/11 in dynamics, and these relations and the principles behind them should be *mastered thoroughly.*

Sample Problem 2/7

To anticipate the dip and hump in the road, the driver of a car applies her brakes to produce a uniform deceleration. Her speed is 100 km/h at the bottom A of the dip and 50 km/h at the top C of the hump, which is 120 m along the road from A. If the passengers experience a total acceleration of 3 m/s^2 at A and if the radius of curvature of the hump at C is 150 m, calculate (a) the radius of curvature ρ at A, (b) the acceleration at the inflection point B, and (c) the total acceleration at C.

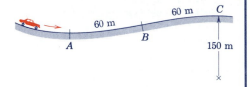

① **Solution.** The dimensions of the car are small compared with those of the path, so we will treat the car as a particle. The velocities are

$$v_A = \left(100 \frac{km}{h}\right)\left(\frac{1 \ h}{3600 \ s}\right)\left(1000 \frac{m}{km}\right) = 27.78 \text{ m/s}$$

$$v_C = 50 \frac{1000}{3600} = 13.89 \text{ m/s}$$

We find the constant deceleration along the path from

$$\left[\int v \ dv = \int a_t \ ds\right] \qquad \int_{v_A}^{v_C} v \ dv = a_t \int_0^s ds$$

$$a_t = \frac{1}{2s}(v_C{}^2 - v_A{}^2) = \frac{(13.89)^2 - (27.78)^2}{2(120)} = -2.41 \text{ m/s}^2$$

(a) Condition at A. With the total acceleration given and a_t determined, we can easily compute a_n and hence ρ from

$[a^2 = a_n{}^2 + a_t{}^2] \qquad a_n{}^2 = 3^2 - (2.41)^2 = 3.19 \qquad a_n = 1.78 \text{ m/s}^2$

$[a_n = v^2/\rho] \qquad \rho = v^2/a_n = (27.78)^2/1.78 = 432 \text{ m}$ *Ans.*

(b) Condition at B. Since the radius of curvature is infinite at the inflection point, $a_n = 0$ and

$$a = a_t = -2.41 \text{ m/s}^2 \qquad \qquad Ans.$$

(c) Condition at C. The normal and total acceleration become

$[a_n = v^2/\rho] \qquad a_n = (13.89)^2/150 = 1.29 \text{ m/s}^2$

With unit vectors $\mathbf{e}_n$ and $\mathbf{e}_t$ in the n- and t-directions, the acceleration may be written

$$\mathbf{a} = 1.29\mathbf{e}_n - 2.41\mathbf{e}_t \text{ m/s}^2$$

where the magnitude of $\mathbf{a}$ is

$[a = \sqrt{a_n{}^2 + a_t{}^2}] \qquad a = \sqrt{(1.29)^2 + (-2.41)^2} = 2.73 \text{ m/s}^2$ *Ans.*

The acceleration vectors representing the conditions at each of the three points are shown for clarification.

① Actually, the radius of curvature to the road differs by about 1 m from that to the path followed by the center of mass of the passengers, but we have neglected this relatively small difference.

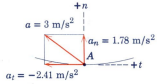

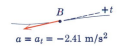

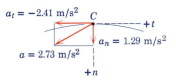

Sample Problem 2/8

A certain rocket maintains a horizontal attitude of its axis during the powered phase of its flight at high altitude. The thrust imparts a horizontal component of acceleration of 20 ft/sec^2, and the downward acceleration component is the acceleration due to gravity at that altitude, which is $g = 30$ ft/sec^2. At the instant represented, the velocity of the mass center G of the rocket along the 15° direction of its trajectory is 12,000 mi/hr. For this position determine (a) the radius of curvature of the flight trajectory, (b) the rate at which the speed v is increasing, (c) the angular rate $\dot{\beta}$ of the radial line from G to the center of curvature C, and (d) the vector expression for the total acceleration $\mathbf{a}$ of the rocket.

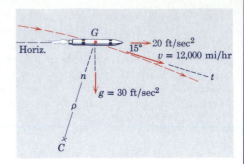

Solution. We observe that the radius of curvature appears in the expression for the normal component of acceleration, so we use n- and t-coordinates to describe the motion of G. The n- and t-components of the total acceleration are obtained by resolving the given horizontal and ① vertical accelerations into their n- and t-components and then combining. From the figure we get

$$a_n = 30 \cos 15° - 20 \sin 15° = 23.8 \text{ ft/sec}^2$$

$$a_t = 30 \sin 15° + 20 \cos 15° = 27.1 \text{ ft/sec}^2$$

(a) We may now compute the radius of curvature from

② $[a_n = v^2/\rho]$ $\rho = \dfrac{v^2}{a_n} = \dfrac{[(12,000)(44/30)]^2}{23.8} = 13.01(10^6) \text{ ft}$ *Ans.*

(b) The rate at which v is increasing is simply the t-component of acceleration.

$[\dot{v} = a_t]$ $\dot{v} = 27.1 \text{ ft/sec}^2$ *Ans.*

(c) The angular rate $\dot{\beta}$ of line GC depends on v and ρ and is given by

$[v = \rho\dot{\beta}]$ $\dot{\beta} = v/\rho = \dfrac{12,000(44/30)}{13.01(10^6)} = 13.53(10^{-4}) \text{ rad/sec}$ *Ans.*

(d) With unit vectors $\mathbf{e}_n$ and $\mathbf{e}_t$ for the n- and t-directions, respectively, the total acceleration becomes

$$\mathbf{a} = 23.8\mathbf{e}_n + 27.1\mathbf{e}_t \text{ ft/sec}^2$$ *Ans.*

① Alternatively, we could find the resultant acceleration and then resolve it into n- and t-components.

② To convert from mi/hr to ft/sec, multiply by $\dfrac{5280 \text{ ft/mi}}{3600 \text{ sec/hr}} = \dfrac{44 \text{ ft/sec}}{30 \text{ mi/hr}}$ which is easily remembered, as 30 mi/hr is the same as 44 ft/sec.

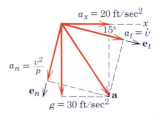

PROBLEMS

Introductory problems

2/97 A particle moves in a circular path of 0.4-m radius. Calculate the magnitude a of the acceleration of the particle (a) if its speed is constant at 0.6 m/s and (b) if its speed is 0.6 m/s but is increasing at the rate of 1.2 m/s each second.

     Ans. (a) $a = 0.9$ m/s^2, (b) $a = 1.5$ m/s^2

2/98 Six acceleration vectors are shown for the car whose velocity vector is directed forward. For each acceleration vector describe in words the instantaneous motion of the car.

2/99 A particle P moves in a circular path of 3-m radius. At the instant considered, the speed of the particle is increasing at the rate of 6 m/s^2, and the magnitude of its total acceleration is 10 m/s^2. Determine the speed v of the particle at this instant.

     Ans. $v = 4.90$ m/s

2/100 A particle moves along the curved path shown. If the particle has a speed of 40 ft/sec at A at time t_A and a speed of 44 ft/sec at B at time t_B, determine the average values of the acceleration of the particle between A and B, both normal and tangent to the path.

2/101 The car passes through a dip in the road at A with a constant speed that gives its mass center G an acceleration equal to $0.5g$. If the radius of curvature of the road at A is 100 m, and if the distance from the road to the mass center G of the car is 0.6 m, determine the speed v of the car.

     Ans. $v = 79.5$ km/h

2/102 The driver of the truck has an acceleration of $0.4g$ as the truck passes over the top A of the hump in the road at constant speed. The radius of curvature of the road at the top of the hump is 98 m, and the center of mass G of the driver (considered a particle) is 2 m above the road. Calculate the speed v of the truck.

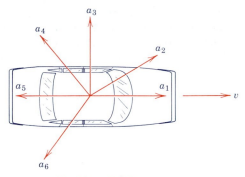

Problem 2/98

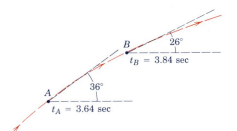

Problem 2/100

Problem 2/101

Problem 2/102

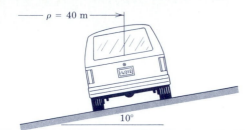

Problem 2/103

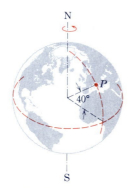

Problem 2/104

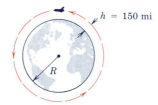

Problem 2/106

Problem 2/107

2/103 A minivan starts from rest on the road whose constant radius of curvature is 40 m and whose bank angle is 10°. The motion occurs in a horizontal plane. If the constant forward acceleration of the minivan is 1.8 m/s², determine the magnitude a of its total acceleration 5 s after starting.

> *Ans.* $a = 2.71$ m/s²

Representative problems

2/104 A space shuttle that moves in a circular orbit around the earth at a height $h = 150$ mi above its surface must have a speed of 17,369 mi/hr. Calculate the gravitational acceleration g for this altitude. The mean radius of the earth is 3959 mi. (Check your answer by computing g from the gravitational law $g = g_0 \left(\dfrac{R}{R + h} \right)^2$, where $g_0 = 32.22$ ft/sec² from Table D/2 in Appendix D.)

2/105 A satellite travels with constant speed v in a circular orbit 320 km above the earth's surface. Calculate v knowing that the acceleration of the satellite is the gravitational acceleration at its altitude. (*Note:* Review Art. 1/5 as necessary and use the mean value of g and the mean value of the earth's radius. Also recognize that v is the magnitude of the velocity of the satellite with respect to the center of the earth.)

> *Ans.* $v = 27.8(10^3)$ km/h

2/106 Consider the polar axis of the earth to be fixed in space and compute the magnitude of the acceleration **a** of a point P on the earth's surface at latitude 40° north. The mean diameter of the earth is 12 742 km and its angular velocity is $0.729(10^{-4})$ rad/s.

2/107 To simulate a condition of "weightlessness" in its cabin, a jet transport traveling at 800 km/h moves on a sustained vertical curve as shown. At what rate $\dot{\beta}$ in degrees per second should the pilot drop his longitudinal line of sight to effect the desired condition? The maneuver takes place at a mean altitude of 8 km, and the gravitational acceleration may be taken as 9.79 m/s². *Ans.* $\dot{\beta} = 2.52$ deg/s

2/108 The preliminary plan for a "small" space station to orbit the earth in a circular path consists of a ring (torus) with a circular cross section as shown. The living space within the torus is shown in section A, where the "ground level" is 20 ft from the center of the section. Calculate the angular speed N in revolutions per minute required to simulate standard gravity at the surface of the earth (32.17 ft/sec^2). Recall that you would be unaware of a gravitational field if you were in a nonrotating spacecraft in a circular orbit around the earth.

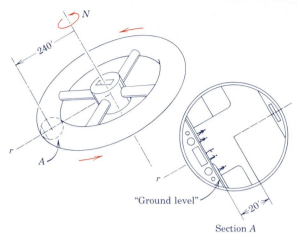

"Ground level"

Section A

Problem 2/108

2/109 The camshaft drive system of a four-cylinder automobile engine is shown. As the engine is revved up, the belt speed v changes uniformly from 3 m/s to 6 m/s over a two-second interval. Calculate the magnitudes of the accelerations of points P_1 and P_2 halfway through this time interval.

Ans. $a_{P_1} = 338$ m/s^2, $a_{P_2} = 1.5$ m/s^2

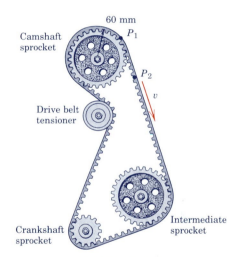

60 mm

Camshaft sprocket

P_1

P_2

v

Drive belt tensioner

Crankshaft sprocket

Intermediate sprocket

Problem 2/109

2/110 Write the vector expression for the acceleration **a** of the mass center G of the simple pendulum in both $n\text{-}t$ and $x\text{-}y$ coordinates for the instant when $\theta = 60°$ if $\dot{\theta} = 2.00$ rad/sec and $\ddot{\theta} = 4.025$ rad/sec^2.

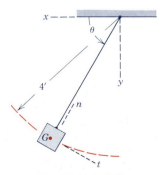

Problem 2/110

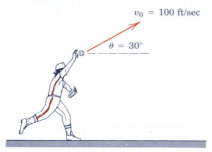

Problem 2/111

2/111 A baseball player releases a ball with the initial conditions shown in the figure. Determine the radius of curvature of the trajectory (*a*) just after release and (*b*) at the apex. For each case, compute the time rate of change of the speed.

$$\text{Ans. } (a) \ \rho = 359 \text{ ft}, \ \dot{v} = -16.1 \text{ ft/sec}^2$$
$$(b) \ \rho = 233 \text{ ft}, \ \dot{v} = 0$$

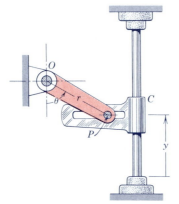

Problem 2/112

2/112 Pin *P* in the crank *PO* engages the horizontal slot in the guide *C* and controls its motion on the fixed vertical rod. Determine the velocity $\dot{y}$ and the acceleration $\ddot{y}$ of guide *C* for a given value of the angle θ if (*a*) $\dot{\theta} = \omega$ and $\ddot{\theta} = 0$ and (*b*) if $\dot{\theta} = 0$ and $\ddot{\theta} = \alpha$.

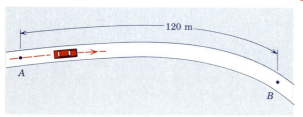

Problem 2/113

2/113 A car travels along the level curved road with a speed that is decreasing at the constant rate of 0.6 m/s each second. The speed of the car as it passes point *A* is 16 m/s. Calculate the magnitude of the total acceleration of the car as it passes point *B* which is 120 m along the road from *A*. The radius of curvature of the road at *B* is 60 m.

$$\text{Ans. } a = 1.961 \text{ m/s}^2$$

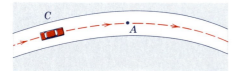

Problem 2/114

2/114 The car *C* increases its speed at the constant rate of 1.5 m/s^2 as it rounds the curve shown. If the magnitude of the total acceleration of the car is 2.5 m/s^2 at point *A* where the radius of curvature is 200 m, compute the speed *v* of the car at this point.

2/115 The motion of the pin A in the fixed circular slot is controlled by the guide B, which is being elevated by its lead screw with a constant upward velocity $v_0 = 2$ m/s for an interval of its motion. Calculate both the normal and tangential components of acceleration of pin A as it passes the position for which $\theta = 30°$.

$$\textit{Ans. } a_n = 21.3 \text{ m/s}^2, a_t = -12.32 \text{ m/s}^2$$

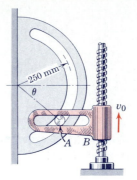

Problem 2/115

2/116 At the bottom A of the vertical inside loop, the magnitude of the total acceleration of the airplane is $3g$. If the airspeed is 800 km/h and is increasing at the rate of 20 km/h per second, calculate the radius of curvature ρ of the path at A.

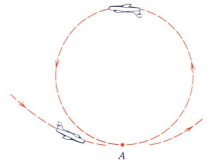

Problem 2/116

2/117 At a certain point in the reentry of the space shuttle into the earth's atmosphere, the total acceleration of the shuttle may be represented by two components. One component is the gravitational acceleration of $g = 9.66$ m/s² at this altitude. The second component equals 12.90 m/s² due to atmospheric resistance and is directed opposite to the velocity. The shuttle is at an altitude of 48.2 km and has reduced its orbital velocity of 28 300 km/h to 15 450 km/h in the direction $\theta = 1.50°$. For this instant calculate the radius of curvature ρ of the path and the rate $\dot{v}$ at which the speed is changing.

$$\textit{Ans. } \dot{v} = -12.65 \text{ m/s}^2, \rho = 1907 \text{ km}$$

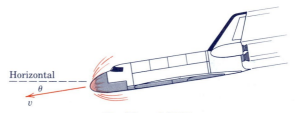

Problem 2/117

2/118 A particle moves on a circular path of radius $r = 2$ ft with a constant speed of 10 ft/sec. During the interval from A to B, the velocity undergoes a vector change $\Delta\mathbf{v}$. Divide this change by the time interval between the two points to obtain the average normal acceleration for (a) $\Delta\theta = 30°$, (b) $\Delta\theta = 15°$, and (c) $\Delta\theta = 5°$. Compare the results with the instantaneous normal acceleration.

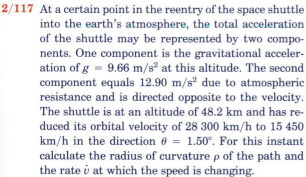
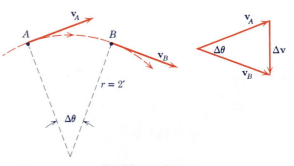

Problem 2/118

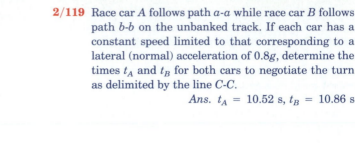

2/119 Race car A follows path a-a while race car B follows path b-b on the unbanked track. If each car has a constant speed limited to that corresponding to a lateral (normal) acceleration of $0.8g$, determine the times t_A and t_B for both cars to negotiate the turn as delimited by the line C-C.

Ans. $t_A = 10.52$ s, $t_B = 10.86$ s

Problem 2/119

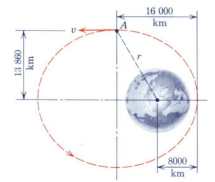

Problem 2/120

2/120 An earth satellite that moves in the elliptical equatorial orbit shown has a velocity v in space of 17 970 km/h when it passes the end of the semiminor axis at A. The earth has an absolute surface value of g of 9.821 m/s^2 and has a radius of 6371 km. Determine the radius of curvature ρ of the orbit at A.

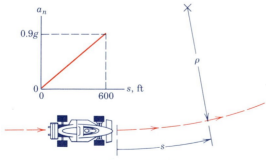

Problem 2/121

2/121 For a race car entering a turn, it is desirable that the normal acceleration not be acquired abruptly; otherwise, unstable behavior may develop. For a car traveling at a constant speed of 180 mi/hr, determine the relation between the radius of curvature ρ of the turn and the distance s along the turn for the first 600 ft if a_n is to change as shown in the graph.

Ans. $\rho = 1.443(10^6)/s$ (ρ and s in ft)

2/122 A rocket traveling above the atmosphere at an altitude of 500 km would have a free-fall acceleration $g = 8.43$ m/s^2 in the absence of forces other than gravitational attraction. Because of thrust, however, the rocket has an additional acceleration component a_1 of 8.80 m/s^2 tangent to its trajectory, which makes an angle of 30° with the vertical at the instant considered. If the velocity v of the rocket is 30 000 km/h at this position, compute the radius of curvature ρ of the trajectory and the rate at which v is changing with time.

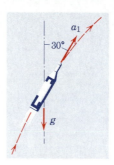

Problem 2/122

2/123 Magnetic tape runs over the idler pulley in a computer as shown. If the total acceleration of a point P on the tape in contact with the pulley makes an angle of 4° with the tangent to the tape at time $t = 0$ when the velocity v of the tape is 4 m/s, determine the time t required to bring the pulley to a stop with constant deceleration. Assume no slipping between the pulley and the tape.   *Ans.* $t = 2.10(10^{-3})$ s

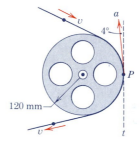

Problem 2/123

2/124 A small particle P starts from point O with a negligible speed and increases its speed to a value $v = \sqrt{2gy}$, where y is the vertical drop from O. When $x = 50$ ft, determine the n-component of acceleration of the particle. (See Art. C/10 for the radius of curvature.)

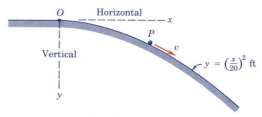

Problem 2/124

2/125 A particle moving in the x-y plane has a position vector given by $\mathbf{r} = \frac{3}{2}t^2\mathbf{i} + \frac{2}{3}t^3\mathbf{j}$, where $\mathbf{r}$ is in inches and t is in seconds. Calculate the radius of curvature ρ of the path for the position of the particle when $t = 2$ sec. Sketch the velocity $\mathbf{v}$ and the curvature of the path for this particular instant.
                                                Ans. $\rho = 41.7$ in.

2/126 As a handling test, a car is driven through the slalom course shown. It is assumed that the car path is sinusoidal and that the maximum lateral acceleration is 0.7g. If the testers wish to design a slalom through which the maximum speed is 80 km/h, what cone spacing L should be used?

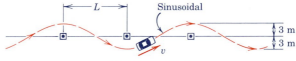

Problem 2/126

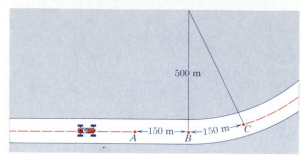

Problem 2/127

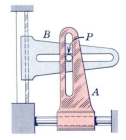

Problem 2/129

2/127 A race driver traveling at a speed of 250 km/h on the straightaway applies his brakes at point A and reduces his speed at a uniform rate to 200 km/h at C in a distance of $150 + 150 = 300$ m. Calculate the magnitude of the total acceleration of the race car an instant after it passes point B.

Ans. $a = 8.42$ m/s^2

2/128 At a given instant, a particle has the following position, velocity, and acceleration components relative to a fixed coordinate system: $x = -3$ m, $\dot{x} = 10$ m/s, $\ddot{x} = 2$ m/s^2, $y = 3$ m, $\dot{y} = 10$ m/s, $\ddot{y} = 9$ m/s^2. Determine and sketch the unit vectors $\mathbf{e}_n$ and $\mathbf{e}_t$ in terms of the unit vectors $\mathbf{i}$ and $\mathbf{j}$. Calculate v, a_t, a_n, and ρ.

▶**2/129** The pin P is constrained to move in the slotted guides that move at right angles to one another. At the instant represented, A has a velocity to the right of 0.2 m/s which is decreasing at the rate of 0.75 m/s each second. At the same time, B is moving down with a velocity of 0.15 m/s which is decreasing at the rate of 0.5 m/s each second. For this instant determine the radius of curvature ρ of the path followed by P. Is it possible to determine also the time rate of change of ρ?   *Ans. $\rho = 1.25$ m*

▶**2/130** A particle starts from rest at the origin and moves along the positive branch of the curve $y = 2x^{3/2}$ so that the distance s measured from the origin along the curve varies with the time t according to $s = 2t^3$, where x, y, and s are in inches and t is in seconds. Find the magnitude of the total acceleration $\mathbf{a}$ of the particle when $t = 1$ sec. (Refer to Art. C/10 of Appendix C for an expression for ρ.)

Ans. $a = 12.17$ in./sec^2

2/6 POLAR COORDINATES (r-θ)

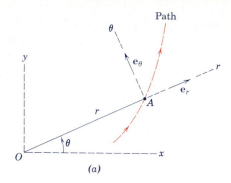

We now consider the third description of plane curvilinear motion, namely, polar coordinates where the particle is located by the radial distance r from a fixed pole and by an angular measurement θ to the radial line. Polar coordinates are particularly useful when a motion is constrained through the control of a radial distance and an angular position or when an unconstrained motion is observed by measurements of a radial distance and an angular position.

Figure 2/13*a* shows the polar coordinates r and θ that locate a particle traveling on a curved path. An arbitrary fixed line, such as the *x*-axis, is used as a reference for the measurement of θ. Unit vectors $\mathbf{e}_r$ and $\mathbf{e}_\theta$ are established in the positive *r*- and *θ*-directions, respectively. The position vector $\mathbf{r}$ to the particle at *A* has a magnitude equal to the radial distance r and a direction specified by the unit vector $\mathbf{e}_r$. Thus, we express the location of the particle at *A* by the vector

$$\mathbf{r} = r\mathbf{e}_r$$

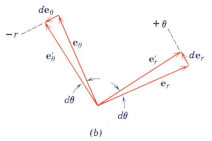

Figure 2/13

As we differentiate this relation with respect to time to obtain $\mathbf{v} = \dot{\mathbf{r}}$ and $\mathbf{a} = \dot{\mathbf{v}}$, we will need expressions for the time derivatives of both unit vectors $\mathbf{e}_r$ and $\mathbf{e}_\theta$. We obtain $\dot{\mathbf{e}}_r$ and $\dot{\mathbf{e}}_\theta$ in exactly the same way that we derived $\dot{\mathbf{e}}_t$ in the preceding article. During time dt the coordinate directions rotate through the angle $d\theta$, and the unit vectors also rotate through the same angle from $\mathbf{e}_r$ and $\mathbf{e}_\theta$ to $\mathbf{e}_r{}'$ and $\mathbf{e}_\theta{}'$, as shown in Fig. 2/13*b*. We note that the vector change $d\mathbf{e}_r$ is in the plus θ-direction and that $d\mathbf{e}_\theta$ is in the minus *r*-direction. Since their magnitudes in the limit are equal to the unit vector as radius times the angle $d\theta$ in radians, we may write them as $d\mathbf{e}_r = \mathbf{e}_\theta\,d\theta$ and $d\mathbf{e}_\theta = -\mathbf{e}_r\,d\theta$. If we divide these equations by $d\theta$, we have

$$\frac{d\mathbf{e}_r}{d\theta} = \mathbf{e}_\theta \qquad \text{and} \qquad \frac{d\mathbf{e}_\theta}{d\theta} = -\mathbf{e}_r$$

Or, if we divide them by dt, we have $d\mathbf{e}_r/dt = (d\theta/dt)\mathbf{e}_\theta$ and $d\mathbf{e}_\theta/dt = -(d\theta/dt)\mathbf{e}_r$ or simply

$$\boxed{\dot{\mathbf{e}}_r = \dot{\theta}\mathbf{e}_\theta \qquad \text{and} \qquad \dot{\mathbf{e}}_\theta = -\dot{\theta}\mathbf{e}_r} \qquad (2/12)$$

We are now ready to differentiate $\mathbf{r} = r\mathbf{e}_r$ with respect to time. Using the rule for differentiating the product of a scalar and a vector gives

$$\mathbf{v} = \dot{\mathbf{r}} = \dot{r}\mathbf{e}_r + r\dot{\mathbf{e}}_r$$

With the substitution of $\dot{\mathbf{e}}_r$ from Eq. 2/12, the vector expression for the velocity becomes

$$\mathbf{v} = \dot{r}\mathbf{e}_r + r\dot{\theta}\mathbf{e}_\theta \qquad (2/13)$$

where
$$v_r = \dot{r}$$
$$v_\theta = r\dot{\theta}$$
$$v = \sqrt{v_r^2 + v_\theta^2}$$

The r-component of $\mathbf{v}$ is merely the rate at which the vector $\mathbf{r}$ stretches. The θ-component of $\mathbf{v}$ is due to the rotation of $\mathbf{r}$.

We now differentiate the expression for $\mathbf{v}$ to obtain the acceleration $\mathbf{a} = \dot{\mathbf{v}}$. Note that the derivative of $r\dot{\theta}\mathbf{e}_\theta$ will produce three terms since all three factors are variable. Thus,

$$\mathbf{a} = \dot{\mathbf{v}} = (\ddot{r}\mathbf{e}_r + \dot{r}\dot{\mathbf{e}}_r) + (\dot{r}\dot{\theta}\mathbf{e}_\theta + r\ddot{\theta}\mathbf{e}_\theta + r\dot{\theta}\dot{\mathbf{e}}_\theta)$$

Substitution of $\dot{\mathbf{e}}_r$ and $\dot{\mathbf{e}}_\theta$ from Eq. 2/12 and collecting terms give

$$\mathbf{a} = (\ddot{r} - r\dot{\theta}^2)\mathbf{e}_r + (r\ddot{\theta} + 2\dot{r}\dot{\theta})\mathbf{e}_\theta \qquad (2/14)$$

where
$$a_r = \ddot{r} - r\dot{\theta}^2$$
$$a_\theta = r\ddot{\theta} + 2\dot{r}\dot{\theta}$$
$$a = \sqrt{a_r^2 + a_\theta^2}$$

We may write the θ-component alternatively as

$$a_\theta = \frac{1}{r}\frac{d}{dt}(r^2\dot{\theta})$$

which can be verified easily by carrying out the differentiation. This form for a_θ will be found useful when we treat the angular momentum of particles in the next chapter.

An adequate appreciation of the terms in Eq. 2/14 comes only when the geometry of the physical changes can be clearly seen. For this purpose, Fig. 2/14*a* is developed to show the velocity vectors and their r- and θ-components at position A and at position A' after an infinitesimal movement. Each of these components undergoes a change in magnitude and direction as shown in Fig. 2/14*b*. In this figure we see the following changes:

Magnitude change of $\mathbf{v}_r$. This change is simply the increase in length of v_r or $dv_r = d\dot{r}$, and the corresponding acceleration term in $d\dot{r}/dt = \ddot{r}$ in the positive r-direction.

Direction change of $\mathbf{v}_r$. The magnitude of this change is seen from the figure to be $v_r\,d\theta = \dot{r}\,d\theta$, and its contribution to the acceleration becomes $\dot{r}\,d\theta/dt = \dot{r}\dot{\theta}$ which is in the positive θ-direction.

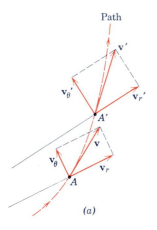

Path

(a)

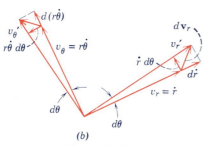

(b)

Figure 2/14

  Magnitude change of $\mathbf{v}_\theta$.　This term is the change in length of $\mathbf{v}_\theta$ or $d(r\dot\theta)$, and its contribution to the acceleration is $d(r\dot\theta)/dt = r\ddot\theta + \dot r\dot\theta$ and is in the positive θ-direction.

  Direction change of $\mathbf{v}_\theta$.　The magnitude of this change is $v_\theta\,d\theta = r\dot\theta\,d\theta$, and the corresponding acceleration term is observed to be $r\dot\theta(d\theta/dt) = r\dot\theta^2$ in the negative r-direction.

  Collecting terms gives $a_r = \ddot r - r\dot\theta^2$ and $a_\theta = r\ddot\theta + 2\dot r\dot\theta$ as obtained previously. We see that the term $\ddot r$ is the acceleration that the particle would have along the radius in the absence of a change in θ. The term $-r\dot\theta^2$ is the normal component of acceleration if r were constant, as in circular motion. The term $r\ddot\theta$ is the tangential acceleration that the particle would have if r were constant, but is only a part of the acceleration due to the change in magnitude of $\mathbf{v}_\theta$ when r is variable. Lastly, the term $2\dot r\dot\theta$ is composed of two effects. The first effect comes from that portion of the change in magnitude $d(r\dot\theta)$ of v_θ due to the change in r, and the second effect comes from the change in direction of $\mathbf{v}_r$. The term $2\dot r\dot\theta$ represents, therefore, a combination of changes and is not so easily perceived as are the other acceleration terms.

  We note carefully the difference between the vector change $d\mathbf{v}_r$ in $\mathbf{v}_r$ and the change dv_r in the magnitude of v_r. Similarly, the vector change $d\mathbf{v}_\theta$ is not the same as the change dv_θ in the magnitude of v_θ. When we divide these changes by dt to obtain expressions for the derivatives, we see clearly that the magnitude of the derivative $|d\mathbf{v}_r/dt|$ and the derivative of the magnitude dv_r/dt are *not* the same. We also note carefully that a_r is not $\dot v_r$ and that a_θ is not $\dot v_\theta$.

  The total acceleration $\mathbf{a}$ and its components are represented in Fig. 2/15. If $\mathbf{a}$ has a component normal to the path, we know from our analysis of n- and t-components in Art. 2/5 that the sense of the n-component *must* be toward the center of curvature.

  For motion in a circular path with r constant, the components of Eqs. 2/13 and 2/14 become simply

$$v_r = 0 \qquad\qquad v_\theta = r\dot\theta$$

$$a_r = -r\dot\theta^2 \qquad\qquad a_\theta = r\ddot\theta$$

This description is the same as that obtained with n- and t-components, where the θ- and t-directions coincide but the positive r-direction is in the negative n-direction. Hence, $a_r = -a_n$ for circular motion centered at the origin of the polar coordinates.

  The expressions for a_r and a_θ in scalar form may also be obtained by direct differentiation of the coordinate relations $x = r\cos\theta$ and $y = r\sin\theta$ to get $a_x = \ddot x$ and $a_y = \ddot y$. Each of these rectangular components of acceleration may then be resolved into r- and θ-components that, when combined, will yield the expressions of Eq. 2/14.

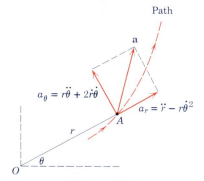

Figure 2/15

Sample Problem   2/9

Rotation of the radially slotted arm is governed by $\theta = 0.2t + 0.02t^3$, where θ is in radians and t is in seconds. Simultaneously, the power screw in the arm engages the slider B and controls its distance from O according to $r = 0.2 + 0.04t^2$, where r is in meters and t is in seconds. Calculate the magnitudes of the velocity and acceleration of the slider for the instant when $t = 3$ s.

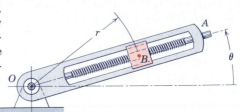

Solution. The coordinates and their time derivatives that appear in the expressions for velocity and acceleration in polar coordinates are obtained first and evaluated for $t = 3$ s.

$$r = 0.2 + 0.04t^2 \qquad r_3 = 0.2 + 0.04(3^2) = 0.56 \text{ m}$$

$$\dot{r} = 0.08t \qquad \dot{r}_3 = 0.08(3) = 0.24 \text{ m/s}$$

$$\ddot{r} = 0.08 \qquad \ddot{r}_3 = 0.08 \text{ m/s}^2$$

$$\theta = 0.2t + 0.02t^3 \qquad \theta_3 = 0.2(3) + 0.02(3^3) = 1.14 \text{ rad}$$
$$\text{or } \theta_3 = 1.14(180/\pi) = 65.3°$$

$$\dot{\theta} = 0.2 + 0.06t^2 \qquad \dot{\theta}_3 = 0.2 + 0.06(3^2) = 0.74 \text{ rad/s}$$

$$\ddot{\theta} = 0.12t \qquad \ddot{\theta}_3 = 0.12(3) = 0.36 \text{ rad/s}^2$$

The velocity components are obtained from Eq. 2/13 and for $t = 3$ s are

$$[v_r = \dot{r}] \qquad v_r = 0.24 \text{ m/s}$$

$$[v_\theta = r\dot{\theta}] \qquad v_\theta = 0.56(0.74) = 0.414 \text{ m/s}$$

$$[v = \sqrt{v_r^2 + v_\theta^2}] \qquad v = \sqrt{(0.24)^2 + (0.414)^2} = 0.479 \text{ m/s} \qquad \textit{Ans.}$$

The velocity and its components are shown for the specified position of the arm.

The acceleration components are obtained from Eq. 2/14 and for $t = 3$ s are

$$[a_r = \ddot{r} - r\dot{\theta}^2] \qquad a_r = 0.08 - 0.56(0.74)^2 = -0.227 \text{ m/s}^2$$

$$[a_\theta = r\ddot{\theta} + 2\dot{r}\dot{\theta}] \qquad a_\theta = 0.56(0.36) + 2(0.24)(0.74) = 0.557 \text{ m/s}^2$$

$$[a = \sqrt{a_r^2 + a_\theta^2}] \qquad a = \sqrt{(-0.227)^2 + (0.557)^2} = 0.601 \text{ m/s}^2 \qquad \textit{Ans.}$$

The acceleration and its components are also shown for the 65.3° position of the arm.

① We see that this problem is an example of constrained motion where the center B of the slider is mechanically constrained by the rotation of the slotted arm and by engagement with the turning screw.

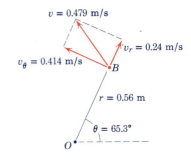

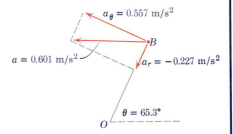

Sample Problem   2/10

A tracking radar lies in the vertical plane of the path of a rocket that is coasting in unpowered flight above the atmosphere. For the instant when $\theta = 30°$, the tracking data give $r = 25(10^4)$ ft, $\dot{r} = 4000$ ft/sec, and $\dot{\theta} = 0.80$ deg/sec. The acceleration of the rocket is due only to gravitational attraction and for its particular altitude is 31.4 ft/sec^2 vertically down. For these conditions determine the velocity v of the rocket and the values of $\ddot{r}$ and $\ddot{\theta}$.

Solution. The components of velocity from Eq. 2/13 are

$[v_r = \dot{r}]$ $\qquad v_r = 4000$ ft/sec

① $[v_\theta = r\dot{\theta}]$ $\qquad v_\theta = 25(10^4)(0.80)\left(\dfrac{\pi}{180}\right) = 3490$ ft/sec

$[v = \sqrt{v_r^2 + v_\theta^2}]$ $\qquad v = \sqrt{(4000)^2 + (3490)^2} = 5310$ ft/sec $\qquad$ *Ans.*

Since the total acceleration of the rocket is $g = 31.4$ ft/sec^2 down, we can easily find its r- and θ-components for the given position. As shown in the figure, they are

② $\qquad a_r = -31.4 \cos 30° = -27.2$ ft/sec^2

$\qquad a_\theta = 31.4 \sin 30° = 15.7$ ft/sec^2

We now equate these values to the polar-coordinate expressions for a_r and a_θ that contain the unknowns $\ddot{r}$ and $\ddot{\theta}$. Thus, from Eq. 2/14

③ $[a_r = \ddot{r} - r\dot{\theta}^2]$ $\qquad -27.2 = \ddot{r} - 25(10^4)\left(0.80\dfrac{\pi}{180}\right)^2$

$\qquad\qquad \ddot{r} = 21.5$ ft/sec^2 $\qquad$ *Ans.*

$[a_\theta = r\ddot{\theta} + 2\dot{r}\dot{\theta}]$ $\qquad 15.7 = 25(10^4)\ddot{\theta} + 2(4000)\left(0.80\dfrac{\pi}{180}\right)$

$\qquad\qquad \ddot{\theta} = -3.84(10^{-4})$ rad/sec^2 $\qquad$ *Ans.*

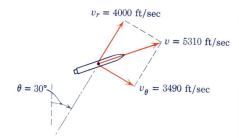

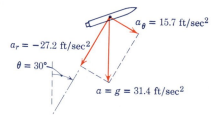

① We observe that the angle θ in polar coordinates need not always be taken positive in a counterclockwise sense.

② Note that the r-component of acceleration is in the negative r-direction, so it carries a minus sign.

③ We must be careful to convert $\dot{\theta}$ from deg/sec to rad/sec.

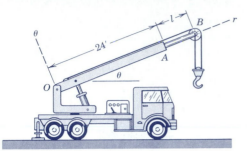

Problem 2/131

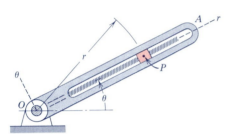

Problem 2/132

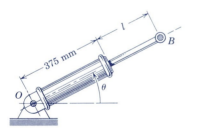

Problem 2/133

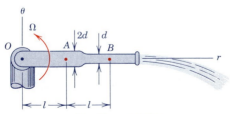

Problem 2/134

PROBLEMS

Introductory problems

2/131 The boom OAB pivots about point O, while section AB simultaneously extends from within section OA. Determine the velocity and acceleration of the center B of the pulley for the following conditions: $\theta = 20°$, $\dot{\theta} = 5$ deg/sec, $\ddot{\theta} = 2$ deg/sec^2, $l = 7$ ft, $\dot{l} = 1.5$ ft/sec, $\ddot{l} = -4$ ft/sec^2. The quantities $\dot{l}$ and $\ddot{l}$ are the first and second time derivatives, respectively, of the length l of section AB.

> *Ans.* $\mathbf{v} = 1.5\mathbf{e}_r + 2.71\mathbf{e}_\theta$ ft/sec
> $\mathbf{a} = -4.24\mathbf{e}_r + 1.344\mathbf{e}_\theta$ ft/sec^2

2/132 The position of the slider P in the rotating slotted arm OA is controlled by a power screw as shown. At the instant represented, $\dot{\theta} = 8$ rad/s and $\ddot{\theta} = -20$ rad/s^2. Also at this same instant, $r = 200$ mm, $\dot{r} = -300$ mm/s, and $\ddot{r} = 0$. For this instant determine the r- and θ-components of the acceleration of P.

2/133 As the hydraulic cylinder rotates around O, the exposed length l of the piston rod P is controlled by the action of oil pressure in the cylinder. If the cylinder rotates at the constant rate $\dot{\theta} = 60$ deg/s and l is decreasing at the constant rate of 150 mm/s, calculate the magnitudes of the velocity $\mathbf{v}$ and acceleration $\mathbf{a}$ of end B when $l = 125$ mm.

> *Ans.* $v = 545$ mm/s, $a = 632$ mm/s^2

2/134 The nozzle shown rotates with constant angular speed Ω about a fixed horizontal axis through point O. Because of the change in diameter by a factor of 2, the water speed relative to the nozzle at A is v, while that at B is $4v$. The water speeds at both A and B are constant. Determine the velocity and acceleration of a water particle as it passes (a) point A and (b) point B.

2/135 An internal mechanism is used to maintain a constant angular rate $\Omega = 0.05$ rad/s about the z-axis of the spacecraft as the telescopic booms are extended at a constant rate. The length l is varied from essentially zero to 3 m. The maximum acceleration to which the sensitive experiment modules P may be subjected is 0.011 m/s². Determine the maximum allowable boom extension rate $\dot{l}$.

Ans. $\dot{l} = 32.8$ mm/s

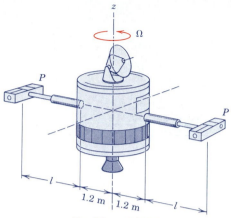

Problem 2/135

2/136 The curvilinear motion of a particle is governed by the polar coordinates $r = t^3/3$ and $\theta = 2\cos(\pi t/6)$, where r is in meters, θ is in radians, and t is in seconds. Specify the velocity **v** and acceleration **a** of the particle when $t = 2$ s.

2/137 The rocket is fired vertically and tracked by the radar shown. When θ reaches 60°, other corresponding measurements give the values $r = 30,000$ ft, $\ddot{r} = 70$ ft/sec², and $\dot{\theta} = 0.02$ rad/sec. Calculate the magnitudes of the velocity and acceleration of the rocket at this position.

Ans. $v = 1200$ ft/sec
$a = 67.0$ ft/sec²

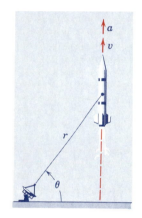

Problem 2/137

Representative problems

2/138 For a certain curvilinear motion of a particle expressed in polar coordinates, the product $r^2\dot{\theta}$ in m²/s varies with the time t in seconds measured over a short period as shown. Approximate the θ-component of the acceleration of the particle at the instant when $t = 5$ s, at which time $r = 1/3$ m.

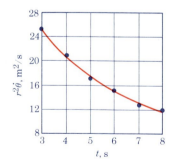

Problem 2/138

2/139 Car A is moving with constant speed v on the straight and level highway. The police officer in the stationary car P attempts to measure the speed v with radar. If the radar measures "line-of-sight" velocity, what velocity v' will the officer observe? Evaluate your general expression for the values $v = 70$ mi/hr, $L = 500$ ft, and $D = 20$ ft, and draw any appropriate conclusions.

Ans. $v' = v\dfrac{L}{\sqrt{L^2 + D^2}}$, $v' = 69.9$ mi/hr

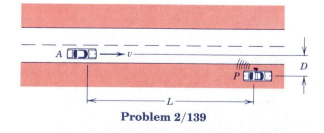

Problem 2/139

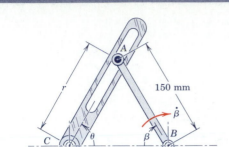

Problem 2/140

2/140 Link AB rotates through a limited range of the angle β, and its end A causes the slotted link AC to rotate also. For the instant represented where $\beta = 60°$ and $\dot{\beta} = 0.6$ rad/s constant, determine the corresponding values of $\dot{r}$, $\ddot{r}$, $\dot{\theta}$, and $\ddot{\theta}$. Make use of Eqs. 2/13 and 2/14.

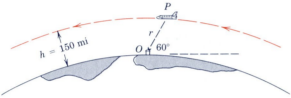

Problem 2/141

2/141 At the instant depicted in the figure, the radar station at O measures the range rate of the space shuttle P to be $\dot{r} = -12,272$ ft/sec, with O considered fixed. If it is known that the shuttle is in a circular orbit at an altitude $h = 150$ mi, determine the orbital speed of the shuttle from this information.

Ans. $v = 25,474$ ft/sec

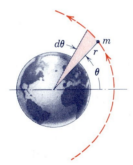

Problem 2/142

2/142 A satellite m moves in an elliptical orbit around the earth. There is no force on the satellite in the θ-direction, so that $a_\theta = 0$. Prove Kepler's second law of planetary motion which says that the radial line r sweeps through equal areas in equal times. The area dA swept by the radial line during time dt is shaded in the figure.

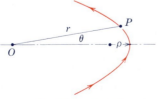

Problem 2/143

2/143 A particle P moves along a path, given by $r = f(\theta)$, which is symmetrical about the line $\theta = 0$. As the particle passes the position $\theta = 0$ where the radius of curvature of the path is ρ, the velocity of P is v. Derive an expression for $\ddot{r}$ in terms of v, r, and ρ for the motion of the particle at this point.

Ans. $\ddot{r} = -v^2 \left(\dfrac{1}{\rho} - \dfrac{1}{r} \right)$

2/144 The slider P can be moved inward by means of the string S, while the slotted arm rotates about point O. The angular position of the arm is given by $\theta = 0.8t - \dfrac{t^2}{20}$, where θ is in radians and t is in seconds. The slider is at $r = 1.6$ m when $t = 0$ and thereafter is drawn inward at the constant rate of 0.2 m/s. Determine the magnitude and direction (expressed by the angle α relative to the x-axis) of the velocity and acceleration of the slider when $t = 4$ s.

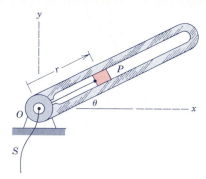

Problem 2/144

2/145 The cam has a shape such that the center of the roller A that follows the contour moves on a limaçon defined by $r = b - c \cos \theta$, where $b > c$. If the cam does not rotate, determine the magnitude of the total acceleration of A in terms of θ if the slotted arm revolves with a constant counterclockwise angular rate $\dot{\theta} = \omega$.

$$\text{Ans. } a = \omega^2 \sqrt{4c^2 - 4bc \cos \theta + b^2}$$

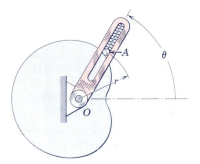

Problem 2/145

/146 The robot arm is elevating and extending simultaneously. At a given instant, $\theta = 30°$, $\dot{\theta} = 10$ deg/s $=$ constant, $l = 0.5$ m, $\dot{l} = 0.2$ m/s, and $\ddot{l} = -0.3$ m/s². Compute the magnitudes of the velocity $\mathbf{v}$ and acceleration $\mathbf{a}$ of the gripped part P. In addition, express $\mathbf{v}$ and $\mathbf{a}$ in terms of the unit vectors $\mathbf{i}$ and $\mathbf{j}$.

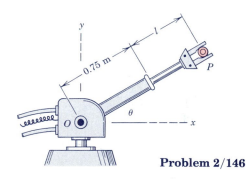

Problem 2/146

/147 The fixed horizontal guide carries a slider and pin P whose motion is controlled by the rotating slotted arm OA. If the arm is revolving about O at the constant rate $\dot{\theta} = 2$ rad/s for an interval of its motion, determine the magnitudes of the velocity and acceleration of the slider in the slot for the instant when $\theta = 60°$. Also find the r-components of the velocity and acceleration.

$$\text{Ans. } v = 533 \text{ mm/s}, \ v_r = -267 \text{ mm/s}$$
$$a = 1232 \text{ mm/s}^2, \ a_r = 616 \text{ mm/s}^2$$

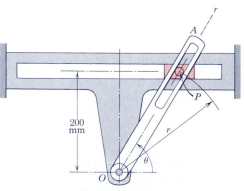

Problem 2/147

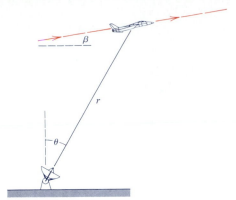

Problem 2/148

2/148 An aircraft flying in a straight line at a climb angle β to the horizontal is tracked by radar located directly below the line of flight. At a certain instant, the following data are recorded: $r = 12{,}000$ ft, $\dot{r} = 360$ ft/sec, $\ddot{r} = 19.60$ ft/sec^2, $\theta = 30°$, and $\dot{\theta} = 2.20$ deg/sec. For this instant determine the aircraft altitude h, velocity v, angle of climb β, $\ddot{\theta}$, and acceleration a.

2/149 At a given instant, a particle has the following position, velocity, and acceleration components relative to a fixed x-y coordinate system: $x = 4$ m, $y = 2$ m, $\dot{x} = 2\sqrt{3}$ m/s, $\dot{y} = -2$ m/s, $\ddot{x} = -5$ m/s^2, $\ddot{y} = 5$ m/s^2. Determine the following properties associated with polar coordinates: θ, $\dot{\theta}$, $\ddot{\theta}$, r, $\dot{r}$, $\ddot{r}$. Sketch the geometry of your solution as you proceed.

Ans. $\theta = 26.6°$, $\dot{\theta} = -0.746$ rad/s
$\ddot{\theta} = 2.24$ rad/s^2
$r = 2\sqrt{5}$ m, $\dot{r} = 2.20$ m/s, $\ddot{r} = 0.255$ m/s^2

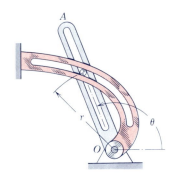

Problem 2/150

2/150 The slotted arm OA forces the small pin to move in the fixed spiral guide defined by $r = K\theta$. Arm OA starts from rest at $\theta = \pi/4$ and has a constant counterclockwise angular acceleration $\ddot{\theta} = \alpha$. Determine the magnitude of the acceleration of the pin when $\theta = 3\pi/4$.

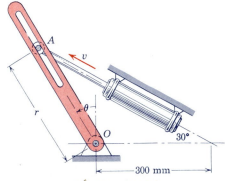

Problem 2/151

2/151 The hydraulic cylinder gives pin A a constant velocity $v = 2$ m/s along its axis for an interval of motion and, in turn, causes the slotted arm to rotate about O. Determine the values of $\dot{r}$, $\ddot{r}$, and $\ddot{\theta}$ for the instant when $\theta = 30°$. (*Hint:* Recognize that all acceleration components are zero when the velocity is constant.)

Ans. $\dot{r} = 1.732$ m/s
$\ddot{r} = 3.33$ m/s^2, $\ddot{\theta} = -38.5$ rad/s^2

2/152 The circular disk rotates about its center O with a constant angular velocity $\omega = \dot{\theta}$ and carries the two spring-loaded plungers shown. The distance b that each plunger protrudes from the rim of the disk varies according to $b = b_0 \sin 2\pi nt$, where b_0 is the maximum protrusion, n is the constant frequency of oscillation of the plungers in the radial slots, and t is the time. Determine the maximum magnitudes of the r- and θ-components of the acceleration of the ends A of the plungers during their motion.

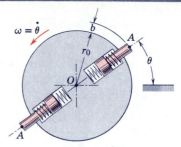

Problem 2/152

2/153 The earth satellite of Prob. 2/120, shown again here, has a velocity $v = 17\,970$ km/h as it passes the end of the semiminor axis at A. Gravitational attraction produces an acceleration $a = a_r = -1.556$ m/s² as calculated from the gravitational law. For this position calculate the rate $\dot{v}$ at which the speed of the satellite is changing and the quantity $\ddot{r}$.

$$\text{Ans. } \dot{v} = -0.778 \text{ m/s}^2, \ddot{r} = -0.388 \text{ m/s}^2$$

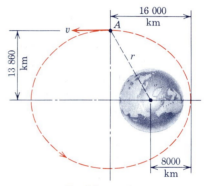

Problem 2/153

2/154 The small block P starts from rest at time $t = 0$ at point A and moves up the incline with constant acceleration a. Determine $\dot{r}$ as a function of time.

2/155 For the conditions of Prob. 2/154, determine $\dot{\theta}$ as a function of time.

$$\text{Ans. } \dot{\theta} = \frac{Rat \sin \alpha}{R^2 + Rat^2 \cos \alpha + \frac{1}{4}a^2t^4}$$

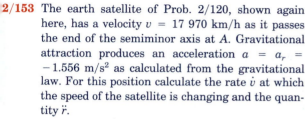

Problem 2/154

2/156 A rocket follows a trajectory in the vertical plane and is tracked by radar from point A. At a certain instant, the radar measurements give $r = 35,000$ ft, $\dot{r} = 1600$ ft/sec, $\dot{\theta} = 0$, and $\ddot{\theta} = -0.00720$ rad/sec². Sketch the position of the rocket for this instant and determine the radius of curvature ρ of the trajectory at this position of the rocket.

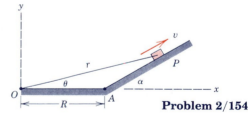

Problem 2/156

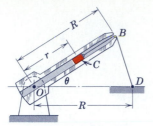

Problem 2/157

2/157 The slotted arm is pivoted at O and carries the slider C. The position of C in the slot is governed by the cord that is fastened at D and remains taut. The arm turns counterclockwise with a constant angular rate $\dot{\theta} = 4$ rad/sec during an interval of its motion. The length DBC of the cord equals R, which makes $r = 0$ when $\theta = 0$. Determine the magnitude a of the acceleration of the slider at the position for which $\theta = 30°$. The distance R is 15 in.

Ans. $a = 489$ in./sec^2

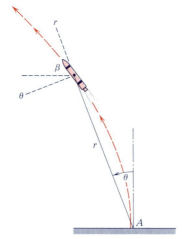

Problem 2/158

2/158 A rocket is tracked by radar from its launching point A. When it is 10 seconds into its flight, the following radar measurements are recorded: $r = 2200$ m, $\dot{r} = 500$ m/s, $\ddot{r} = 4.66$ m/s^2, $\theta = 22°$, $\dot{\theta} = 0.0788$ rad/s, and $\ddot{\theta} = -0.0341$ rad/s^2. For this instant determine the angle β between the horizontal and the direction of the trajectory of the rocket and find the magnitudes of its velocity $\mathbf{v}$ and acceleration $\mathbf{a}$.

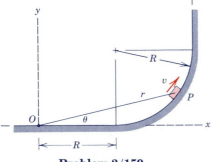

Problem 2/159

2/159 The block P slides on the surface shown with constant speed $v = 0.6$ m/s and passes point O at time $t = 0$. If $R = 1.2$ m, determine the following quantities at time $t = 2\left(1 + \dfrac{\pi}{3}\right)$: $r, \theta, \dot{r}, \dot{\theta}, \ddot{r},$ and $\ddot{\theta}$.

Ans. $r = 2.32$ m, $\dot{r} = 0.424$ m/s
$\ddot{r} = -0.1345$ m/s^2
$\theta = 15°, \dot{\theta} = 0.1830$ rad/s, $\ddot{\theta} = 0.025$ rad/s^2

2/160 The slotted arm *OA* oscillates about *O* within the limits shown and drives the crank *CP* through the pin *P*. For an interval of the motion, $\dot{\theta} = K$, a constant. Determine the magnitude of the corresponding total acceleration of *P* for any value of θ within the range for which $\dot{\theta} = K$. Use polar coordinates *r* and θ. Show that the magnitudes of the velocity and acceleration of *P* in its circular path are constant.

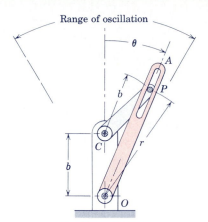

Range of oscillation

Problem 2/160

2/161 During a portion of a vertical loop, an airplane flies in an arc of radius $\rho = 600$ m with a constant speed $v = 400$ km/h. When the airplane is at *A*, the angle made by **v** with the horizontal is $\beta = 30°$, and radar tracking gives $r = 800$ m and $\theta = 30°$. Calculate v_r, v_θ, a_r, and $\ddot{\theta}$ for this instant.

$\quad$ *Ans.* $v_r = 96.2$ m/s, $v_\theta = 55.6$ m/s

$\qquad a_r = 10.29$ m/s^2, $\ddot{\theta} = -0.0390$ rad/s^2

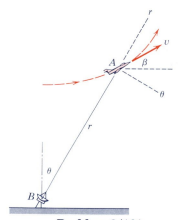

Problem 2/161

2/162 A cable attached to the vehicle at *A* passes over the small fixed pulley at *B* and around the drum at *C*. If the vehicle moves with a constant speed $v_0 = \dot{x}$, determine an expression for the acceleration of a point *P* on the cable between *B* and *C* in terms of θ. Also express $\ddot{\theta}$ in terms of θ. (*Hint:* Observe that the *r*- and θ-components of the acceleration of *A* are both zero.)

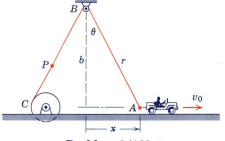

Problem 2/162

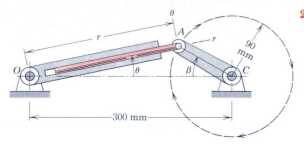

Problem 2/163

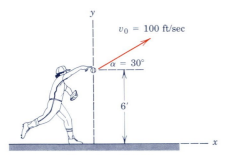

Problem 2/164

Problem 2/165

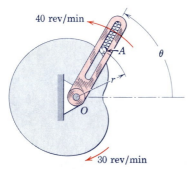

Problem 2/166

2/163 Pin *A* moves in a circle of 90-mm radius as crank *AC* revolves at the constant rate $\dot{\beta} = 60$ rad/s. The slotted link rotates about point *O* as the rod attached to *A* moves in and out of the slot. For the position $\beta = 30°$, determine $\dot{r}$, $\ddot{r}$, $\dot{\theta}$, and $\ddot{\theta}$.

$$Ans.\ \dot{r} = 3.58\ \text{m/s},\ \ddot{r} = 315\ \text{m/s}^2$$
$$\dot{\theta} = 17.86\ \text{rad/s}$$
$$\ddot{\theta} = -1510\ \text{rad/s}^2$$

2/164 The baseball player of Prob. 2/111 is repeated here with additional information supplied. At time $t = 0$, the ball is thrown with an initial speed of 100 ft/sec at an angle of 30° to the horizontal. Determine the quantities r, $\dot{r}$, $\ddot{r}$, θ, $\dot{\theta}$, and $\ddot{\theta}$, all relative to the *x*-*y* coordinate system shown, at time $t = 0.5$ sec.

2/165 During reentry a space capsule *A* is tracked by radar station *B* located in the vertical plane of the trajectory. The values of *r* and θ are read against time and are recorded in the following table. Determine the velocity *v* of the capsule when $t = 40$ s. Explain how the acceleration of the capsule could be found from the given data.

t, s	r, km	θ, deg	t, s	r, km	θ, deg
0	36.4	110.5	60	19.0	45.0
5	29.9	100.0	70	18.8	38.5
10	26.2	91.0	80	18.7	32.8
15	24.1	83.7	90	18.7	27.0
20	22.7	77.7	100	18.7	21.6
30	20.9	67.7	110	18.8	16.8
40	20.1	58.6	120	19.0	12.0
50	19.3	52.0			

$$Ans.\ v = 1020\ \text{km/h}$$

▶**2/166** If the slotted arm (Prob. 2/145) is revolving counterclockwise at the constant rate of 40 rev/min and the cam is revolving clockwise at the constant rate of 30 rev/min, determine the magnitude of the acceleration of the center of the roller *A* when the cam and arm are in the relative position for which $\theta = 30°$. The limaçon has the dimensions $b = 100$ mm and $c = 75$ mm. (*Caution:* Redefine the coordinates as necessary after noting that the θ in the expression $r = b - c \cos \theta$ is not the absolute angle appearing in Eq. 2/14.) $Ans.\ a = 3.68\ \text{m/s}^2$

2/7 SPACE CURVILINEAR MOTION

The general case of three-dimensional motion of a particle along a space curve was introduced in Art. 2/1 and illustrated in Fig. 2/1. Mention was made of the three coordinate systems, rectangular $(x\text{-}y\text{-}z)$, cylindrical $(r\text{-}\theta\text{-}z)$, and spherical $(R\text{-}\theta\text{-}\phi)$, which are commonly used to describe this motion. These systems are indicated in Fig. 2/16, which also shows the unit vectors for the three coordinate systems.*

Before describing the use of these coordinate systems, we may observe that a path-variable description, using n- and t-coordinates, which we developed in Art. 2/5, may be applied in the osculating plane shown in Fig. 2/1. We defined this plane earlier as the plane that contains the curve at the location in question. We see that the velocity **v**, which is along the tangent t to the curve, lies in the osculating plane. The acceleration **a** also lies in the osculating plane and, as in the case of plane motion, has a component $a_t = \dot{v}$ tangent to the path due to the change in magnitude of the velocity and a component $a_n = v^2/\rho$ normal to the curve due to the change in direction of the velocity. As before, ρ is the radius of curvature of the path at the point in question and would be measured in the osculating plane. This description of motion, which we found to be natural and direct for many plane-motion problems, finds little use for space motion because the osculating plane continually shifts its orientation, thus making its use as a reference awkward. We shall confine our attention, therefore, to the three fixed coordinate systems shown in Fig. 2/16.

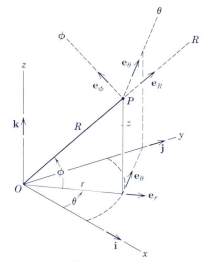

Figure 2/16

(a) Rectangular coordinates (x-y-z). The extension from two to three dimensions offers no particular difficulty. We merely add the z-coordinate and its two time derivatives to the two-dimensional expressions of Eqs. 2/6 so that the position vector **R**, the velocity **v**, and the acceleration **a** become

$$
\begin{aligned}
\mathbf{R} &= x\mathbf{i} + y\mathbf{j} + z\mathbf{k} \\
\mathbf{v} = \dot{\mathbf{R}} &= \dot{x}\mathbf{i} + \dot{y}\mathbf{j} + \dot{z}\mathbf{k} \\
\mathbf{a} = \dot{\mathbf{v}} = \ddot{\mathbf{R}} &= \ddot{x}\mathbf{i} + \ddot{y}\mathbf{j} + \ddot{z}\mathbf{k}
\end{aligned}
\tag{2/15}
$$

Note that in three dimensions we are using **R** in place of **r** for the position vector.

(b) Cylindrical coordinates (r-θ-z). If we understand the polar-coordinate description of plane motion, then there should be

---

*In a variation of spherical coordinates commonly used, angle ϕ is replaced by its complement.

no difficulty with cylindrical coordinates because all that is required is the addition of the *z*-coordinate and its two time derivatives. The position vector **R** to the particle for cylindrical coordinates is simply

$$\mathbf{R} = r\mathbf{e}_r + z\mathbf{k}$$

In place of Eq. 2/13 for plane motion, we may write the velocity as

$$\mathbf{v} = \dot{r}\mathbf{e}_r + r\dot{\theta}\mathbf{e}_\theta + \dot{z}\mathbf{k} \tag{2/16}$$

where

$$v_r = \dot{r}$$

$$v_\theta = r\dot{\theta}$$

$$v_z = \dot{z}$$

$$v = \sqrt{v_r{}^2 + v_\theta{}^2 + v_z{}^2}$$

Similarly, the acceleration is written by adding the *z*-component to Eq. 2/14, which gives us

$$\mathbf{a} = (\ddot{r} - r\dot{\theta}^2)\mathbf{e}_r + (r\ddot{\theta} + 2\dot{r}\dot{\theta})\mathbf{e}_\theta + \ddot{z}\mathbf{k} \tag{2/17}$$

where

$$a_r = \ddot{r} - r\dot{\theta}^2$$

$$a_\theta = r\ddot{\theta} + 2\dot{r}\dot{\theta} = \frac{1}{r}\frac{d}{dt}(r^2\dot{\theta})$$

$$a_z = \ddot{z}$$

$$a = \sqrt{a_r{}^2 + a_\theta{}^2 + a_z{}^2}$$

Whereas the unit vectors $\mathbf{e}_r$ and $\mathbf{e}_\theta$ have time derivatives due to the changes in their directions, we note that the unit vector **k** in the *z*-direction remains fixed in direction and therefore has no time derivative.

(c) Spherical coordinates (R-θ-φ). When a radial distance and two angles are used to locate the position of a particle, as in the case of radar measurements, for example, spherical coordinates R, θ, ϕ are used. Derivation of the expression for the velocity **v** is easily obtained, but the expression for the acceleration **a** is more complex because of the added geometry. Consequently, only the results will be cited here.* First we designate unit vectors $\mathbf{e}_R$, $\mathbf{e}_\theta$, $\mathbf{e}_\phi$ as shown

*For a complete derivation of **v** and **a** in spherical coordinates, see the senior author's book *Dynamics*, 2nd Edition, 1971 or SI Version, 1975 (John Wiley & Sons, Inc.).

in Fig. 2/16. Note that the unit vector $\mathbf{e}_R$ is in the direction that the particle P would move if R increases but θ and ϕ are held constant. Also the unit vector $\mathbf{e}_\theta$ is in the direction that P would move if θ increases while R and ϕ are held constant. Lastly, the unit vector $\mathbf{e}_\phi$ is in the direction in which P would move if ϕ increases while R and θ are held constant. The resulting expressions for $\mathbf{v}$ and $\mathbf{a}$ are

$$\mathbf{v} = v_R \mathbf{e}_R + v_\theta \mathbf{e}_\theta + v_\phi \mathbf{e}_\phi \qquad (2/18)$$

where

$$v_R = \dot{R}$$
$$v_\theta = R\dot{\theta} \cos \phi$$
$$v_\phi = R\dot{\phi}$$

and

$$\mathbf{a} = a_R \mathbf{e}_R + a_\theta \mathbf{e}_\theta + a_\phi \mathbf{e}_\phi \qquad (2/19)$$

where

$$a_R = \ddot{R} - R\dot{\phi}^2 - R\dot{\theta}^2 \cos^2 \phi$$

$$a_\theta = \frac{\cos \phi}{R} \frac{d}{dt}(R^2\dot{\theta}) - 2R\dot{\theta}\dot{\phi} \sin \phi$$

$$a_\phi = \frac{1}{R} \frac{d}{dt}(R^2\dot{\phi}) + R\dot{\theta}^2 \sin \phi \cos \phi$$

We should mention that linear algebraic transformations between any two of the three coordinate-system expressions for velocity or acceleration may be developed. These transformations make it possible to express the motion components in rectangular coordinates, for example, if the components are known in spherical coordinates, or vice versa.* These transformations are easily handled with the aid of matrix algebra and a simple computer program.

---

*These coordinate transformations are developed and illustrated in the senior author's book *Dynamics*, 2nd Edition, 1971 or SI Version, 1975 (John Wiley & Sons, Inc.).

Sample Problem 2/11

The power screw starts from rest and is given a rotational speed $\dot{\theta}$ that increases uniformly with time t according to $\dot{\theta} = kt$, where k is a constant. Determine the expressions for the velocity v and acceleration a of the center of ball A when the screw has turned through one complete revolution from rest. The lead of the screw (advancement per revolution) is L.

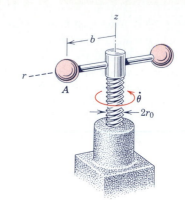

Solution. The center of ball A moves in a helix on the cylindrical surface of radius b, and the cylindrical coordinates r, θ, z are clearly indicated.

Integrating the given relation for $\dot{\theta}$ gives $\theta = \Delta\theta = \int \dot{\theta}\, dt = \frac{1}{2}kt^2$. For one revolution from rest we have

$$2\pi = \tfrac{1}{2}kt^2$$

giving

$$t = 2\sqrt{\pi/k}$$

Thus, the angular rate at one revolution is

$$\dot{\theta} = kt = k(2\sqrt{\pi/k}) = 2\sqrt{\pi k}$$

① The helix angle γ of the path followed by the center of the ball governs the relation between the θ- and z-components of velocity and
② is given by $\tan \gamma = L/(2\pi b)$. Now from the figure we see that $v_\theta = v \cos \gamma$. Substituting $v_\theta = r\dot{\theta} = b\dot{\theta}$ from Eq. 2/16 gives $v = v_\theta/\cos \gamma = b\dot{\theta}/\cos \gamma$. With $\cos \gamma$ obtained from $\tan \gamma$ and with $\dot{\theta} = 2\sqrt{\pi k}$, we have for the one-revolution position

$$v = 2b\sqrt{\pi k}\,\frac{\sqrt{L^2 + 4\pi^2 b^2}}{2\pi b} = \sqrt{\frac{k}{\pi}}\sqrt{L^2 + 4\pi^2 b^2} \qquad Ans.$$

The acceleration components from Eq. 2/17 become

③ $[a_r = \ddot{r} - r\dot{\theta}^2]$ $a_r = 0 - b(2\sqrt{\pi k})^2 = -4b\pi k$

$[a_\theta = r\ddot{\theta} + 2\dot{r}\dot{\theta}]$ $a_\theta = bk + 2(0)(2\sqrt{\pi k}) = bk$

$[a_z = \ddot{z} = \dot{v}_z]$ $a_z = \dfrac{d}{dt}(v_z) = \dfrac{d}{dt}(v_\theta \tan \gamma) = \dfrac{d}{dt}(b\dot{\theta}\tan \gamma)$

$$= (b\tan\gamma)\ddot{\theta} = b\,\frac{L}{2\pi b}\,k = \frac{kL}{2\pi}$$

Now we combine the components to give the magnitude of the total acceleration, which becomes

$$a = \sqrt{(-4b\pi k)^2 + (bk)^2 + \left(\frac{kL}{2\pi}\right)^2}$$

$$= bk\sqrt{(1 + 16\pi^2) + L^2/(4\pi^2 b^2)} \qquad Ans.$$

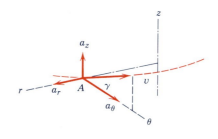

① Unless we recognize that the helix angle of the path governs the z-motion in relation to the rotation of the screw, we cannot complete the problem. We must also be careful to divide the lead L by the circumference $2\pi b$ and not the diameter $2b$ to obtain $\tan \gamma$. If in doubt, unwrap one turn of the helix traced by the center of the ball.

② Sketch a right triangle and recall that for $\tan \beta = a/b$ the cosine of β becomes $b/\sqrt{a^2 + b^2}$.

③ The negative sign for a_r is consistent with our previous knowledge that the normal component of acceleration is directed toward the center of curvature.

PROBLEMS

Introductory problems

2/167 Consider the power screw of Sample Problem 2/11 with a lead $L = 0.5$ in. If $b = 6$ in. and if the screw turns at a constant rate of 2 rev/sec, calculate the magnitudes of the velocity and acceleration of the center of ball A.

> *Ans.* $v = 75.4$ in./sec, $a = 947$ in./sec^2

2/168 The velocity and acceleration of a particle are given for a certain instant by $\mathbf{v} = 6\mathbf{i} - 3\mathbf{j} + 2\mathbf{k}$ m/s and $\mathbf{a} = 3\mathbf{i} - \mathbf{j} - 5\mathbf{k}$ m/s^2. Determine the angle β between $\mathbf{v}$ and $\mathbf{a}$, $\dot{v}$, and the radius of curvature ρ in the osculating plane.

2/169 The rectangular coordinates of a particle are: $x = r \cos \dfrac{v_0}{r} t$, $y = r \sin \dfrac{v_0}{r} t$, and $z = v_0 t + \dfrac{1}{2} bt^2$. Sketch the path of the particle and determine the magnitudes of the velocity $\mathbf{v}$ and acceleration $\mathbf{a}$ as a function of time t. The quantities r, v_0, and b are constants.

> *Ans.* $v = \sqrt{2v_0^2 + 2v_0 bt + b^2 t^2}$, $a = \sqrt{\dfrac{v_0^4}{r^2} + b^2}$

2/170 An amusement ride called the "corkscrew" takes the passengers through the upside-down curve of a horizontal cylindrical helix. The velocity of the cars as they pass position A is 15 m/s, and the component of their acceleration measured along the tangent to the path is $g \cos \gamma$ at this point. The effective radius of the cylindrical helix is 5 m, and the helix angle is $\gamma = 40°$. Compute the magnitude of the acceleration of the passengers as they pass position A.

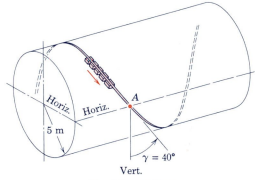

Problem 2/170

2/171 The rotating element in a mixing chamber is given a periodic axial movement $z = z_0 \sin 2\pi nt$ while it is rotating at the constant angular velocity $\dot{\theta} = \omega$. Determine the expression for the maximum magnitude of the acceleration of a point A on the rim of radius r. The frequency n of vertical oscillation is constant.

> *Ans.* $a_{max} = \sqrt{r^2 \omega^4 + 16 n^4 \pi^4 z_0^2}$

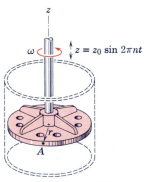

Problem 2/171

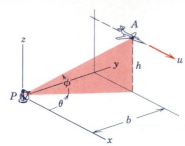

Problem 2/172

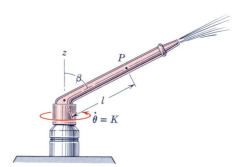

Problem 2/173

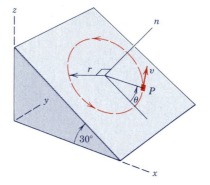

Problem 2/174

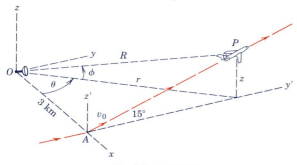

Problem 2/175

Representative problems

2/172 The radar antenna at P tracks the jet aircraft A that is flying horizontally at a speed u and an altitude h above the level of P. Determine the expressions for the components of the velocity in the spherical coordinates of the antenna motion.

2/173 The rotating nozzle sprays a large circular area and turns with the constant angular rate $\dot{\theta} = K$. Particles of water move along the tube at the constant rate $\dot{l} = c$ relative to the tube. Write expressions for the magnitudes of the velocity and acceleration of a water particle P for a given position l in the rotating tube.
Ans. $v = \sqrt{c^2 + K^2 l^2 \sin^2 \beta}$
$a = K \sin \beta \sqrt{K^2 l^2 + 4c^2}$

2/174 The small block P travels with constant speed v in the circular path of radius r on the inclined surface. If $\theta = 0$ at time $t = 0$, determine the x-, y-, and z-components of velocity and acceleration as functions of time.

2/175 An aircraft P takes off at A with a velocity v_0 of 250 km/h and climbs in the vertical y'-z' plane at the constant 15° angle with an acceleration along its flight path of 0.8 m/s². Flight progress is monitored by radar at point O. Resolve the velocity of P into cylindrical-coordinate components 60 s after takeoff and find $\dot{r}$, $\dot{\theta}$, and $\dot{z}$ for that instant. (*Suggestion:* Draw the related x-y and x-z projections of the velocity components.)
Ans. $\dot{r} = 99.2$ m/s
$\dot{\theta} = 8.88(10^{-3})$ rad/s, $\dot{z} = 30.4$ m/s

2/176 The robotic device rotates about a fixed vertical axis while its arm extends and elevates. At a given instant, $\phi = 30°$, $\dot\phi = 10$ deg/s = constant, $l = 0.5$ m, $\dot l = 0.2$ m/s, $\ddot l = -0.3$ m/s², and $\Omega = 20$ deg/s = constant. Determine the magnitudes of the velocity **v** and the acceleration **a** of the gripped part P.

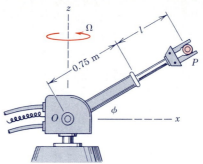

Problem 2/176

2/177 The base structure of the firetruck ladder rotates about a vertical axis through O with a constant angular velocity $\Omega = 10$ deg/s. At the same time, the ladder unit OB elevates at a constant rate $\dot\phi = 7$ deg/s, and section AB of the ladder extends from within section OA at the constant rate of 0.5 m/s. At the instant under consideration, $\phi = 30°$, $\overline{OA} = 9$ m, and $\overline{AB} = 6$ m. Determine the magnitudes of the velocity and acceleration of the end B of the ladder.   *Ans.* $v = 2.96$ m/s, $a = 0.672$ m/s²

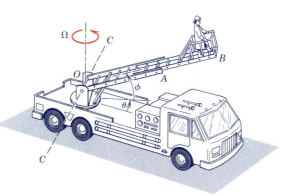

Problem 2/177

2/178 In a test of the actuating mechanism for a telescoping antenna on a spacecraft, the supporting shaft rotates about the fixed z-axis with an angular rate $\dot\theta$. Determine the R-, θ-, and ϕ-components of the acceleration **a** of the end of the antenna at the instant when $L = 1.2$ m and $\beta = 45°$ if the rates $\dot\theta = 2$ rad/s, $\dot\beta = \frac{3}{2}$ rad/s, and $\dot L = 0.9$ m/s are constant during the motion.

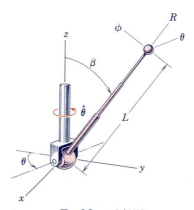

Problem 2/178

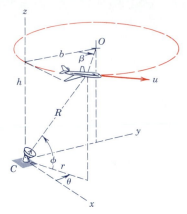

Problem 2/179

2/179 An aircraft is flying in a horizontal circle of radius b with a constant speed u at an altitude h. A radar tracking unit is located at C. Write expressions for the components of the velocity of the aircraft in the spherical coordinates of the radar station for a given position β.

$$Ans. \quad v_R = \frac{bu \sin \beta}{\sqrt{4b^2 \sin^2 \frac{\beta}{2} + h^2}}, \quad v_\theta = u \sin \frac{\beta}{2}$$

$$v_\phi = \frac{-hu \cos \frac{\beta}{2}}{\sqrt{4b^2 \sin^2 \frac{\beta}{2} + h^2}}$$

2/180 For the conditions of Prob. 2/175, resolve the velocity of the aircraft P into spherical-coordinate components 60 s after takeoff and find $\dot{R}$, $\dot{\theta}$, and $\dot{\phi}$ for that instant. (*Suggestion:* Draw the x-y projections of the velocity and the related projections in the vertical plane containing r and R.)

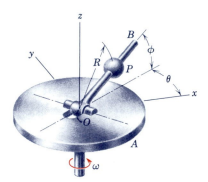

Problem 2/181

2/181 The disk A rotates about the vertical z-axis with a constant speed $\omega = \dot{\theta} = \pi/3$ rad/s. Simultaneously, the hinged arm OB is elevated at the constant rate $\dot{\phi} = 2\pi/3$ rad/s. At time $t = 0$, both $\theta = 0$ and $\phi = 0$. The angle θ is measured from the fixed reference x-axis. The small sphere P slides out along the rod according to $R = 50 + 200t^2$, where R is in millimeters and t is in seconds. Determine the magnitude of the total acceleration $\mathbf{a}$ of P when $t = \frac{1}{2}$ s.

$$Ans. \quad a = 0.904 \text{ m/s}^2$$

▶**2/182** Assign unit vectors $\mathbf{e}_R$, $\mathbf{e}_\theta$, and $\mathbf{e}_\phi$ along the spherical coordinate directions, Fig. 2/16, and determine $\dot{\mathbf{e}}_R$, $\dot{\mathbf{e}}_\theta$, and $\dot{\mathbf{e}}_\phi$.

$$Ans. \quad \dot{\mathbf{e}}_R = \dot{\phi}\mathbf{e}_\phi + (\dot{\theta} \cos \phi)\mathbf{e}_\theta$$
$$\dot{\mathbf{e}}_\theta = -(\dot{\theta} \cos \phi)\mathbf{e}_R + (\dot{\theta} \sin \phi)\mathbf{e}_\phi$$
$$\dot{\mathbf{e}}_\phi = -\dot{\phi}\mathbf{e}_R - (\dot{\theta} \sin \phi)\mathbf{e}_\theta$$

2/183 Use the results of Prob. 2/182 to obtain the acceleration components in spherical coordinates, Eqs. 2/19, by a direct differentiation of the position vector $\mathbf{R} = R\mathbf{e}_R$.

2/184 The cars of an amusement-park ride are attached to arms of length R that are hinged to a central rotating collar that drives the assembly about the vertical axis with a constant angular rate $\omega = \dot{\theta}$. The cars rise and fall with the track according to the relation $z = (h/2)(1 - \cos 2\theta)$. Determine the expressions for the R-, θ-, and ϕ-components of the velocity $\mathbf{v}$ of each car as it passes the position $\theta = \pi/4$ rad.

$$Ans. \ v_R = 0, \ v_\theta = R\omega\sqrt{1 - \left(\frac{h}{2R}\right)^2}$$

$$v_\phi = \frac{h\omega}{\sqrt{1 - \left(\frac{h}{2R}\right)^2}}$$

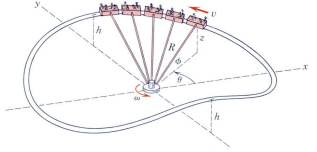

Problem 2/184

2/8 *RELATIVE MOTION (TRANSLATING AXES)*

In the previous articles of this chapter, we have described particle motion using coordinates that were referred to fixed reference axes. The displacements, velocities, and accelerations so determined are termed *absolute*. But it is not always possible or convenient to use a fixed set of axes for the observation of motion, and there are many engineering problems for which the analysis of motion is simplified by using measurements made with respect to a moving reference system. These measurements, when combined with the absolute motion of the moving coordinate system, permit us to determine the absolute motion in question. This approach is known as a *relative-motion* analysis.

The motion of the moving coordinate system is specified with respect to a fixed coordinate system. Strictly speaking, this fixed system in Newtonian mechanics is the primary inertial system that is assumed to have no motion in space. For engineering purposes, the fixed system may be taken as any system whose absolute motion is negligible for the problem at hand. For most earthbound engineering problems, it is sufficiently precise to take for the fixed reference system a set of axes attached to the earth, in which case we neglect the motion of the earth. For the motion of satellites around the earth, a nonrotating coordinate system with origin on the earth's axis of rotation is chosen. For interplanetary travel, a nonrotating coordinate system fixed to the sun would be used. Hence, the choice of the fixed system depends on the type of problem involved.

We will confine our attention in this article to moving reference systems that translate but do not rotate. Motion measured in rotating systems will be discussed in Art. 5/7 of Chapter 5 on rigid-body kinematics, where this approach finds special but important application. We will also confine our attention here to relative motion analysis for plane motion.

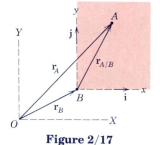

Figure 2/17

Consider now two particles A and B that may have separate curvilinear motions in a given plane or in parallel planes, Fig. 2/17. We will arbitrarily attach the origin of a set of translating (nonrotating) axes x-y to particle B and observe the motion of A from our moving position on B. The position vector of A as measured relative to the frame x-y is $\mathbf{r}_{A/B} = x\mathbf{i} + y\mathbf{j}$, where the subscript notation "A/B" means "A relative to B" or "A with respect to B." The unit vectors along the x- and y-axes are $\mathbf{i}$ and $\mathbf{j}$, and x and y are the coordinates of A measured in x-y. The absolute position of B is defined by the vector $\mathbf{r}_B$ measured from the origin of the fixed axes X-Y. The absolute position of A is seen, therefore, to be determined by the vector

$$\mathbf{r}_A = \mathbf{r}_B + \mathbf{r}_{A/B}$$

We now differentiate this vector equation once with respect to time to obtain velocities and twice to obtain accelerations. Thus,

$$\dot{\mathbf{r}}_A = \dot{\mathbf{r}}_B + \dot{\mathbf{r}}_{A/B} \qquad \text{or} \qquad \boxed{\mathbf{v}_A = \mathbf{v}_B + \mathbf{v}_{A/B}} \qquad (2/20)$$

$$\ddot{\mathbf{r}}_A = \ddot{\mathbf{r}}_B + \ddot{\mathbf{r}}_{A/B} \qquad \text{or} \qquad \boxed{\mathbf{a}_A = \mathbf{a}_B + \mathbf{a}_{A/B}} \qquad (2/21)$$

In Eq. 2/20 the velocity that we observe A to have from our position at B attached to the moving axes x-y is $\dot{\mathbf{r}}_{A/B} = \mathbf{v}_{A/B} = \dot{x}\mathbf{i} + \dot{y}\mathbf{j}$. This term is the velocity of A with respect to B. Similarly, in Eq. 2/21 the acceleration that we observe A to have from our nonrotating position on B is $\ddot{\mathbf{r}}_{A/B} = \dot{\mathbf{v}}_{A/B} = \ddot{x}\mathbf{i} + \ddot{y}\mathbf{j}$. This term is the acceleration of A with respect to B. We note carefully that the unit vectors $\mathbf{i}$ and $\mathbf{j}$ have no derivatives because their directions as well as their magnitudes remain unchanged. (Later when we discuss rotating reference axes, we will be obliged to account for the derivatives of the unit vectors when they change direction.)

In words, Eq. 2/20 (or 2/21) states that the absolute velocity (or acceleration) of A equals the absolute velocity (or acceleration) of B plus, vectorially, the velocity (or acceleration) of A relative to B. The relative term is the velocity (or acceleration) measurement that an observer attached to the moving coordinate system x-y would make. We may express the relative motion terms in whatever coordinate system is convenient—rectangular, normal and tangential, or polar—and the formulations in the preceding articles may be used for this purpose. The appropriate fixed system of the previous articles becomes the moving system in the present article.

The selection of the moving point B for attachment of the reference coordinate system is arbitrary. As shown in Fig. 2.18, point A could be used just as well for the attachment of the moving system, in which case the three corresponding relative motion equations for position, velocity, and acceleration are

$$\mathbf{r}_B = \mathbf{r}_A + \mathbf{r}_{B/A} \qquad \mathbf{v}_B = \mathbf{v}_A + \mathbf{v}_{B/A} \qquad \mathbf{a}_B = \mathbf{a}_A + \mathbf{a}_{B/A}$$

It is seen, therefore, that $\mathbf{r}_{B/A} = -\mathbf{r}_{A/B}$, $\mathbf{v}_{B/A} = -\mathbf{v}_{A/B}$, and $\mathbf{a}_{B/A} = -\mathbf{a}_{A/B}$.

An important observation to be made in relative motion analysis is that the acceleration of a particle as observed in a translating system x-y is the same as that observed in a fixed system X-Y if the moving system has a constant velocity. This conclusion broadens the application of Newton's second law of motion, treated in Chapter 3. We conclude, consequently, that a set of axes that has a constant absolute velocity may be used in place of a "fixed" system for the determination of accelerations. A translating reference system that has no acceleration is known as an *inertial system*.

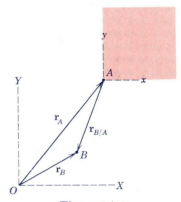

Figure 2/18

Sample Problem 2/12

Passengers in the jet transport A flying east at a speed of 800 km/h observe a second jet plane B that passes under the transport in horizontal flight. Although the nose of B is pointed in the 45° northeast direction, plane B appears to the passengers in A to be moving away from the transport at the 60° angle as shown. Determine the true velocity of B.

Solution. The moving reference axes x-y are attached to A from which the relative observations are made. We write, therefore,

① $$\mathbf{v}_B = \mathbf{v}_A + \mathbf{v}_{B/A}$$

Next we identify the knowns and unknowns. The velocity $\mathbf{v}_A$ is given both in magnitude and direction. The 60° direction of $\mathbf{v}_{B/A}$, the velocity that B appears to have to the moving observers in A, is known, and the true velocity of B is in the 45° direction in which it is heading. The two remaining unknowns are the magnitudes of $\mathbf{v}_B$ and $\mathbf{v}_{B/A}$. We may solve the vector equation in any one of three ways.

② ③

(I) Graphical. We start the vector sum at some point P by drawing $\mathbf{v}_A$ to a convenient scale and then construct a line through the tip of $\mathbf{v}_A$ with the known direction of $\mathbf{v}_{B/A}$. The known direction of $\mathbf{v}_B$ is then drawn through P, and the intersection C yields the unique solution enabling us to complete the vector triangle and scale off the unknown magnitudes, which are found to be

$$v_{B/A} = 586 \text{ km/h} \quad \text{and} \quad v_B = 717 \text{ km/h} \qquad \text{Ans.}$$

(II) Trigonometric. A sketch of the vector triangle is made to reveal the trigonometry, which gives

④ $$\frac{v_B}{\sin 60°} = \frac{v_A}{\sin 75°} \qquad v_B = 800 \frac{\sin 60°}{\sin 75°} = 717 \text{ km/h} \qquad \text{Ans.}$$

(III) Vector algebra. Using unit vectors $\mathbf{i}$ and $\mathbf{j}$, we express each of the velocities in vector form as

$$\mathbf{v}_A = 800\mathbf{i} \text{ km/h} \qquad \mathbf{v}_B = (v_B \cos 45°)\mathbf{i} + (v_B \sin 45°)\mathbf{j}$$

$$\mathbf{v}_{B/A} = (v_{B/A} \cos 60°)(-\mathbf{i}) + (v_{B/A} \sin 60°)\mathbf{j}$$

Substituting these relations into the relative-velocity equation and solving separately for the $\mathbf{i}$ and $\mathbf{j}$ terms give

$$\text{(i-terms} \qquad v_B \cos 45° = 800 - v_{B/A} \cos 60°$$

$$\text{(j-terms)} \qquad v_B \sin 45° = v_{B/A} \sin 60°$$

⑤ Solving simultaneously yields the unknown velocity magnitudes

$$v_{B/A} = 586 \text{ km/h} \quad \text{and} \quad v_B = 717 \text{ km/h} \qquad \text{Ans.}$$

It is worth noting the solution of this problem from the viewpoint of an observer in B. With reference axes attached to B, we would write $\mathbf{v}_A = \mathbf{v}_B + \mathbf{v}_{A/B}$. The apparent velocity of A as observed by B is then $\mathbf{v}_{A/B}$, which is the negative of $\mathbf{v}_{B/A}$.

① We treat each airplane as a particle.

② We assume no side slip due to cross wind.

③ Students should become familiar with all three solutions.

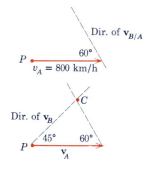

④ We must be prepared to recognize the appropriate trigonometric relation, which here is the law of sines.

⑤ We can see that the graphical or trigonometric solution is shorter than the vector algebra solution in this particular problem.

Sample Problem   2/13

Car A is accelerating in the direction of its motion at the rate of 3 ft/sec^2. Car B is rounding a curve of 440-ft radius at a constant speed of 30 mi/hr. Determine the velocity and acceleration that car B appears to have to an observer in car A if car A has reached a speed of 45 mi/hr for the positions represented.

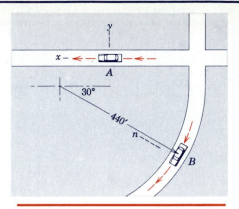

① Alternatively, we could use either a graphical or a vector algebraic solution.

Solution.   We choose nonrotating reference axes attached to car A since the motion of B with respect to A is desired.

Velocity.   The relative-velocity equation is

$$\mathbf{v}_B = \mathbf{v}_A + \mathbf{v}_{B/A}$$

and the velocities of A and B for the position considered have the magnitudes

$$v_A = 45\,\frac{5280}{60^2} = 45\,\frac{44}{30} = 66 \text{ ft/sec} \qquad v_B = 30\,\frac{44}{30} = 44 \text{ ft/sec}$$

The triangle of velocity vectors is drawn in the sequence required by the equation, and application of the law of cosines and the law of sines gives

$$v_{B/A} = 58.2 \text{ ft/sec} \qquad \theta = 40.9° \qquad \textit{Ans.}$$

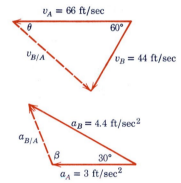

② Be careful to choose between the two values 69.8° and $180 - 69.8 = 110.2°$.

Acceleration.   The relative-acceleration equation is

$$\mathbf{a}_B = \mathbf{a}_A + \mathbf{a}_{B/A}$$

The acceleration of A is given, and the acceleration of B is normal to the curve in the n-direction and has the magnitude

$$[a_n = v^2/\rho] \qquad a_B = (44)^2/440 = 4.40 \text{ ft/sec}^2$$

The triangle of acceleration vectors is drawn in the sequence required by the equation as illustrated. Solving for the x- and y-components of $\mathbf{a}_{B/A}$ gives us

$$(a_{B/A})_x = 4.4 \cos 30° - 3 = 0.810 \text{ ft/sec}^2$$

$$(a_{B/A})_y = 4.4 \sin 30° = 2.2 \text{ ft/sec}^2$$

from which $a_{B/A} = \sqrt{(0.810)^2 + (2.2)^2} = 2.344 \text{ ft/sec}^2$ *Ans.*

The direction of $\mathbf{a}_{B/A}$ may be specified by the angle β which, by the law of sines, becomes

$$\frac{4.4}{\sin \beta} = \frac{2.344}{\sin 30°} \qquad \beta = \sin^{-1}\left(\frac{4.4}{2.344}\,0.5\right) = 110.2° \qquad \textit{Ans.}$$

Suggestion: To gain familiarity with the manipulation of vector equations, it is suggested that the student rewrite the relative-motion equations in the form $\mathbf{v}_{B/A} = \mathbf{v}_B - \mathbf{v}_A$ and $\mathbf{a}_{B/A} = \mathbf{a}_B - \mathbf{a}_A$ and redraw the vector polygons to conform with these alternative relations.

Caution: So far we are only prepared to handle motion relative to *nonrotating* axes. If we had attached the reference axes rigidly to car B, they would rotate with the car, and we would find that the velocity and acceleration terms relative to the rotating axes are *not* the negative of those measured from the nonrotating axes moving with A. Rotating axes are treated in Art. 5/7.

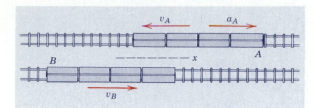

Problem 2/185

PROBLEMS

Introductory problems

2/185 Rapid-transit trains A and B travel on parallel tracks. Train A has a speed of 80 km/h and is slowing at the rate of 2 m/s^2, while train B has a constant speed of 40 km/h. Determine the velocity and acceleration of train B relative to train A.

> *Ans.* $\mathbf{v}_{B/A} = 120\mathbf{i}$ km/h, $\mathbf{a}_{B/A} = -2\mathbf{i}$ m/s^2

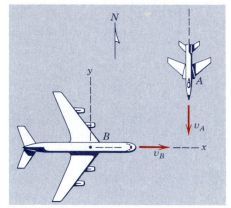

Problem 2/186

2/186 The passenger aircraft B is flying east with a velocity $v_B = 800$ km/h. A military jet traveling south with a velocity $v_A = 1200$ km/h passes under B at a slightly lower altitude. What velocity does A appear to have to a passenger in B, and what is the direction of that apparent velocity?

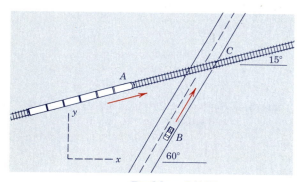

Problem 2/187

2/187 Train A travels with a constant speed $v_A = 120$ km/h along the straight and level track. The driver of car B, anticipating the railway grade crossing C, decreases the car speed of 90 km/h at the rate of 3 m/s^2. Determine the velocity and acceleration of the train relative to the car.

> *Ans.* $\mathbf{v}_{A/B} = 70.9\mathbf{i} - 46.9\mathbf{j}$ km/h
> $\mathbf{a}_{A/B} = 1.5\mathbf{i} + 2.60\mathbf{j}$ m/s^2

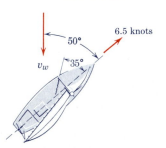

Problem 2/188

2/188 A sailboat moving in the direction shown is tacking to windward against a north wind. The log registers a hull speed of 6.5 knots. A "telltale" (light string tied to the rigging) indicates that the direction of the apparent wind is 35° from the centerline of the boat. What is the true wind velocity v_w?

2/189 To increase his speed, the water skier A cuts across the wake of the tow boat B, which has a velocity of 60 km/h. At the instant when $\theta = 30°$, the actual path of the skier makes an angle $\beta = 50°$ with the tow rope. For this position determine the velocity v_A of the skier and the value of $\dot{\theta}$.

Ans. $v_A = 80.8$ km/h, $\dot{\theta} = 0.887$ rad/s

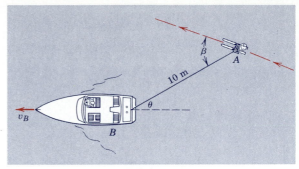

Problem 2/189

Representative problems

2/190 Two cars A and B are traveling along straight roads as shown. If the time rate of increase of the distance $\overline{AB}$ separating the two cars equals the magnitude of the relative velocity between the cars, what can be said concerning the velocities of the two cars?

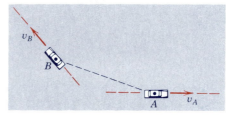

Problem 2/190

2/191 The car A has a forward speed of 18 km/h and is accelerating at 3 m/s². Determine the velocity and acceleration of the car relative to observer B, who rides in a nonrotating chair on the ferris wheel. The angular rate $\Omega = 3$ rev/min of the ferris wheel is constant.

Ans. $\mathbf{v}_{A/B} = 3.00\mathbf{i} + 2.00\mathbf{j}$ m/s
$\mathbf{a}_{A/B} = 3.63\mathbf{i} + 0.628\mathbf{j}$ m/s²

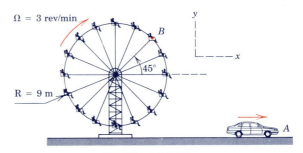

Problem 2/191

2/192 A small ship capable of making a speed of 6 knots through still water maintains a heading due east while being set to the south by an ocean current. The actual course of the boat is from A to B, a distance of 10 nautical miles that requires exactly 2 hours. Determine the speed v_w of the current and its direction measured clockwise from the north.

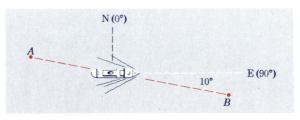

Problem 2/192

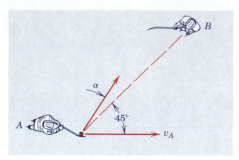

Problem 2/193

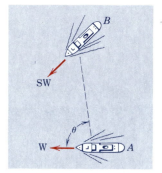

Problem 2/194

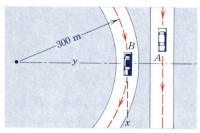

Problem 2/195

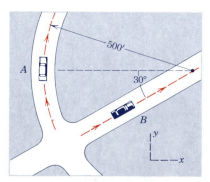

Problem 2/196

2/193 Hockey player A carries the puck on his stick and moves in the direction shown with a speed $v_A = 4$ m/s. In passing the puck to his stationary teammate B, by what shot angle α should the direction of his shot trail the line of sight if he launches the puck with a speed of 7 m/s relative to himself?

Ans. $\alpha = 23.8°$

2/194 Ship A is headed west at a speed of 15 knots, and ship B is headed southwest. The relative bearing θ of B with respect to A is 80° and is unchanging. If the distance between A and B is 8 nautical miles at 3:00 P.M., when would collision occur if neither ship altered course or speed? At what speed is ship B traveling?

2/195 For the instant represented, car A has a speed of 100 km/h, which is increasing at the rate of 8 km/h each second. Simultaneously, car B also has a speed of 100 km/h as it rounds the turn and is slowing down at the rate of 8 km/h each second. Determine the acceleration that car B appears to have to an observer in car A.

Ans. $\mathbf{a}_{B/A} = -4.44\mathbf{i} + 2.57\mathbf{j}$ m/s²

2/196 For the instant represented, car A is rounding the circular curve at a constant speed of 30 mi/hr, while car B is slowing down at the rate of 5 mi/hr per second. Determine the acceleration that car A appears to have to an observer in car B.

2/197 Satellites A and B are in a circular orbit of altitude $h = 1500$ km. Determine the magnitude of the acceleration of satellite B relative to a nonrotating observer in satellite A. Use $g_0 = 9.825$ m/s^2 for the surface-level gravitational acceleration and $R = 6371$ km for the radius of the earth.

Ans. $a_{B/A} = 9.10$ m/s^2

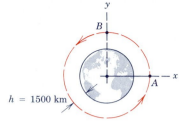

Problem 2/197

2/198 For the coplanar or parallel-plane motion of three particles A, B, and C, prove that the relative velocities and relative accelerations measured from nonrotating axes obey the equations $\mathbf{v}_{A/B} = \mathbf{v}_{A/C} + \mathbf{v}_{C/B}$ and $\mathbf{a}_{A/B} = \mathbf{a}_{A/C} + \mathbf{a}_{C/B}$.

2/199 For the instant represented, car A has an acceleration in the direction of its motion and car B has a speed of 45 mi/hr which is increasing. If the acceleration of B as observed from A is zero for this instant, determine the acceleration of A and the rate at which the speed of B is changing.

Ans. $a_A = 10.27$ ft/sec^2, $\dot{v}_B = 7.26$ ft/sec^2

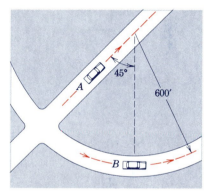

Problem 2/199

2/200 Two aircraft, A and B, are flying horizontally in the same vertical plane at the respective altitudes of 2.40 km and 1.80 km and at the respective velocities of 800 km/h and 600 km/h. When the line-of-sight angle θ from B to A reaches 30°, determine the rate at which the distance $r_{A/B}$ from B to A is increasing and the rate at which θ is increasing.

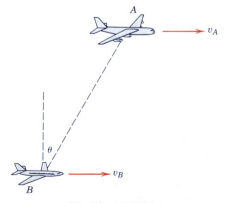

Problem 2/200

2/201 An earth satellite is put into a circular polar orbit at an altitude of 240 km, which requires an orbital velocity of 27 940 km/h with respect to the center of the earth considered fixed in space. In going from south to north, when the satellite passes over an observer on the equator, in which direction does the satellite appear to be moving? The equatorial radius of the earth is 6378 km, and the angular velocity of the earth is $0.729(10^{-4})$ rad/s.

Ans. Apparent direction 3.43° west of north

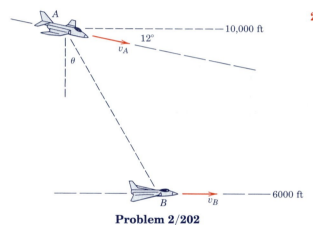

2/202 Aircraft B is flying at a constant speed $v_B = 400$ mi/hr at an altitude of 6000 ft. When aircraft A is at an altitude of 10,000 ft, its line of sight to B is in the vertical plane containing the flight path of B and makes an angle $\theta = 30°$ with the vertical. If the velocity of A is constant, determine its magnitude that would result in an air collision. How long after the aircraft reach the positions described would the collision occur if no evasive action were taken?

Problem 2/202

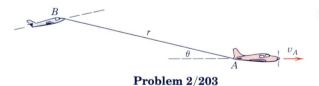

Problem 2/203

2/203 Airplane A is flying horizontally with a constant speed of 200 km/h and is towing the glider B, which is gaining altitude. If the tow cable has a length $r = 60$ m and θ is increasing at the constant rate of 5 degrees per second, determine the magnitudes of the velocity $\mathbf{v}$ and acceleration $\mathbf{a}$ of the glider for the instant when $\theta = 15°$.

Ans. $v_B = 206$ km/h

$a_B = 0.457$ m/s^2 from B to A

2/204 After starting from the position marked with the "x", a football receiver B runs the slant-in pattern shown, making a cut at P and thereafter running with a constant speed $v_B = 7$ yd/sec in the direction shown. The quarterback releases the ball with a horizontal velocity of 100 ft/sec at the instant the receiver passes point P. Determine the angle α at which the quarterback must throw the ball, and the velocity of the ball relative to the receiver when the ball is caught. Neglect any vertical motion of the ball.

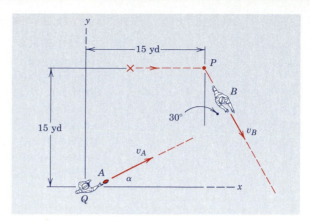

Problem 2/204

2/205 A batter hits the baseball A with an initial velocity of $v_0 = 100$ ft/sec directly toward fielder B at an angle of 30° to the horizontal; the initial position of the ball is 3 ft above ground level. Fielder B requires $\frac{1}{4}$ sec to judge where the ball should be caught and begins moving to that position with constant speed. Because of great experience, fielder B chooses his running speed so that he arrives at the "catch position" simultaneously with the baseball. The catch position is the field location at which the ball altitude is 7 ft. Determine the velocity of the ball relative to the fielder at the instant the catch is made.

Ans. $\mathbf{v}_{A/B} = 71.5\mathbf{i} - 47.4\mathbf{j}$ ft/sec

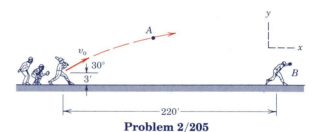

Problem 2/205

2/206 The aircraft A with radar detection equipment is flying horizontally at 40,000 ft and is increasing its speed at the rate of 4 ft/sec each second. Its radar locks onto an aircraft flying in the same direction and in the same vertical plane at an altitude of 60,000 ft. If A has a speed of 600 mi/hr at the instant that $\theta = 30°$, determine the values of $\ddot{r}$ and $\ddot{\theta}$ at this same instant if B has a constant speed of 900 mi/hr.

Ans. $\ddot{r} = -2.25$ ft/sec^2
$\ddot{\theta} = 1.548(10^{-4})$ rad/sec^2

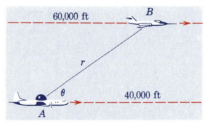

Problem 2/206

2/9 *CONSTRAINED MOTION OF CONNECTED PARTICLES*

There are occasions when the motions of particles are inter-related by virtue of the constraints of interconnecting members, and it becomes necessary to account for these constraints in order to determine the respective motions of the particles.

Consider first the very simple system of two interconnected particles A and B shown in Fig. 2/19. Although it should be quite evident by inspection that the horizontal motion of A is twice the vertical motion of B, we will use this example to illustrate the method of analysis that will carry over to more complex situations

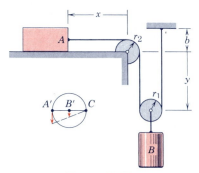

Figure 2/19

where the results cannot be easily perceived by inspection. Clearly, the motion of B is the same as that of the center of its pulley, so we establish position coordinates y and x measured from a convenient fixed datum. The total length of the cable is

$$L = x + \frac{\pi r_2}{2} + 2y + \pi r_1 + b$$

With L, r_2, r_1, and b all constant, the first and second time derivatives of the equation give

$$0 = \dot{x} + 2\dot{y} \quad \text{or} \quad 0 = v_A + 2v_B$$

$$0 = \ddot{x} + 2\ddot{y} \quad \text{or} \quad 0 = a_A + 2a_B$$

The velocity and acceleration constraint equations indicate that, for the coordinates selected, the velocity of A must have a sign which is opposite to that of the velocity of B and similarly for the accelerations. The constraint equations are valid for the motion of the system in either direction.

Since the results do not depend on the lengths or pulley radii, we should be able to analyze the motion without considering them. In the lower-left portion of Fig. 2/19 is shown an enlarged view of the horizontal diameter $A'B'C$ of the lower pulley at an instant of time. Clearly, A' and A have the same motion magnitudes, as do B

and B'. During an infinitesimal motion of A', it is easy to see that B' moves half as far since point C as a point on the fixed portion of the cable momentarily has no motion. Thus, with differentiation by time in mind, we can see the velocity and acceleration magnitude relationships by inspection. The pulley, in effect, is a wheel that rolls on the fixed vertical cable. (The kinematics of a rolling wheel will be treated more extensively in Chapter 5 on rigid-body motion.) The system of Fig. 2/19 is said to have *one degree of freedom* since only one variable, either x or y, is needed to specify the positions of all parts of the system.

A system with *two degrees of freedom* is shown in Fig. 2/20. Here the positions of the lower cylinder and pulley C depend on the separate specifications of the two coordinates y_A and y_B. The lengths of the cables attached to cylinders A and B may be written, respectively, as

$$L_A = y_A + 2y_D + \text{constant}$$

$$L_B = y_B + y_C + (y_C - y_D) + \text{constant}$$

and their time derivatives are

$$0 = \dot{y}_A + 2\dot{y}_D \quad \text{and} \quad 0 = \dot{y}_B + 2\dot{y}_C - \dot{y}_D$$

$$0 = \ddot{y}_A + 2\ddot{y}_D \quad \text{and} \quad 0 = \ddot{y}_B + 2\ddot{y}_C - \ddot{y}_D$$

Eliminating the terms in $\dot{y}_D$ and $\ddot{y}_D$ gives

$$\dot{y}_A + 2\dot{y}_B + 4\dot{y}_C = 0 \quad \text{or} \quad v_A + 2v_B + 4v_C = 0$$

$$\ddot{y}_A + 2\ddot{y}_B + 4\ddot{y}_C = 0 \quad \text{or} \quad a_A + 2a_B + 4a_C = 0$$

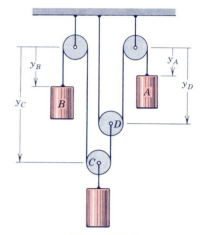

Figure 2/20

Clearly, it is impossible for the signs of all three terms to be positive simultaneously. So, for example, if both A and B have downward (positive) velocities, then C will have an upward (negative) velocity.

These results may also be found for inspection of the motions of the two pulleys at C and D. For an increment dy_A (with y_B held fixed), the center of D moves up an amount $dy_A/2$, which causes an upward movement $dy_A/4$ of the center of C. Due to an increment dy_B (with y_A held fixed), the center of C moves up a distance $dy_B/2$. A combination of the two movements gives an upward movement

$$-dy_C = \frac{dy_A}{4} + \frac{dy_B}{2}$$

so that $-v_C = v_A/4 + v_B/2$ as before. Visualization of the actual geometry of the motion is an important ability.

A second type of constraint where the direction of the connecting member changes with the motion is illustrated in the second of the two sample problems that follow.

Sample Problem 2/14

In the pulley configuration shown, cylinder A has a downward velocity of 0.3 m/s. Determine the velocity of B. Solve in two ways.

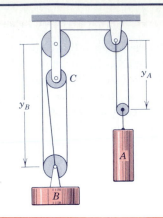

Solution (I). The centers of the pulleys at A and B are located by the coordinates y_A and y_B measured from fixed positions. The total constant length of cable in the pulley system is

$$L = 3y_B + 2y_A + \text{constants}$$

where the constants account for the fixed lengths of cable in contact with
① the circumferences of the pulleys and the constant vertical separation between the two upper left-hand pulleys. Differentiation with time gives

$$0 = 3\dot{y}_B + 2\dot{y}_A$$

Substitution of $v_A = \dot{y}_A = 0.3$ m/s and $v_B = \dot{y}_B$ gives

$$0 = 3(v_B) + 2(0.3) \qquad \text{or} \qquad v_B = -0.2 \text{ m/s} \qquad \textit{Ans.}$$

Solution (II). An enlarged diagram of the pulleys at A, B, and C is shown. During a differential movement ds_A of the center of pulley A, the left end of its horizontal diameter has no motion since it is attached to the fixed part of the cable. Therefore, the right-hand end has a movement of $2ds_A$ as shown. This movement is transmitted to the left-hand end of the horizontal diameter of the pulley at B. Further, from pulley C with its fixed center, we see that the displacements on each side are equal and opposite. Thus, for pulley B, the right-hand end of the diam-
② eter has a downward displacement equal to the upward displacement ds_B of its center. By inspection of the geometry, we conclude that

$$2ds_A = 3ds_B \qquad \text{or} \qquad ds_B = \tfrac{2}{3}ds_A$$

Dividing by dt gives

$$|v_B| = \tfrac{2}{3}v_A = \tfrac{2}{3}(0.3) = 0.2 \text{ m/s (upward)} \qquad \textit{Ans.}$$

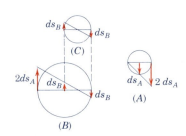

① We neglect the small angularity of the cables between B and C.

② The negative sign indicates that the velocity of B is *upward*.

Sample Problem 2/15

The tractor A is used to hoist the bale B with the pulley arrangement shown. If A has a forward velocity v_A, determine an expression for the upward velocity v_B of the bale in terms of x.

Solution. We designate the position of the tractor by the coordinate x and the position of the bale by the coordinate y, both measured from a fixed reference. The total constant length of the cable is

$$L = 2(h - y) + l = 2(h - y) + \sqrt{h^2 + x^2}$$

① Differentiation with time yields

$$0 = -2\dot{y} + \frac{x\dot{x}}{\sqrt{h^2 + x^2}}$$

Substituting $v_A = \dot{x}$ and $v_B = \dot{y}$ gives

$$v_B = \frac{1}{2}\frac{xv_A}{\sqrt{h^2 + x^2}} \qquad \textit{Ans.}$$

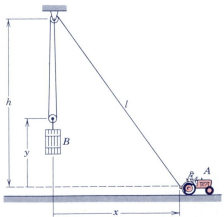

① Differentiation of the relation for a right triangle occurs frequently in mechanics.

PROBLEMS

Introductory problems

2/207 If block *B* has a leftward velocity of 1.2 m/s, determine the velocity of cylinder *A*.

Ans. $v_A = 0.4$ m/s down

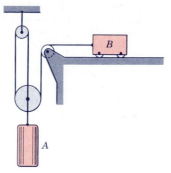

Problem 2/207

2/208 A truck equipped with a power winch on its front end pulls itself up a steep incline with the cable and pulley arrangement shown. If the cable is wound up on the drum at the constant rate of 40 mm/s, how long does it take for the truck to move 4 m up the incline?

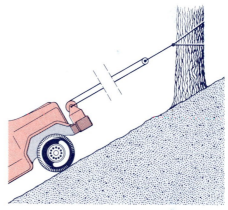

Problem 2/208

2/209 Cylinder *B* has a downward velocity of 2 ft/sec and an upward acceleration of 0.5 ft/sec². Calculate the velocity and acceleration of block *A*.

Ans. $v_A = 3$ ft/sec up, $a_A = 0.75$ ft/sec² down

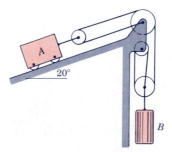

Problem 2/209

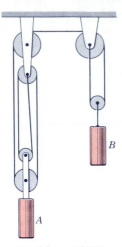

Problem 2/210

2/210 Cylinder B has a downward velocity in feet per second given by $v_B = t^2/2 + t^3/6$, where t is in seconds. Calculate the acceleration of A when $t = 2$ sec.

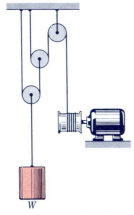

Problem 2/211

2/111 Determine the vertical rise h of the load W during 5 seconds if the hoisting drum wraps cable around it at the constant rate of 320 mm/s.

Ans. $h = 400$ mm

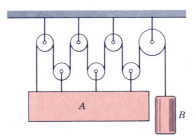

Problem 2/212

2/212 Determine the constraint equation that relates the accelerations of bodies A and B. Assume that the upper surface of A remains horizontal.

Representative problems

2/213 Determine the relationship that governs the velocities of the four cylinders. Express all velocities as positive down. How many degrees of freedom are there?
  Ans. $4v_A + 8v_B + 4v_C + v_D = 0$
  three degrees of freedom

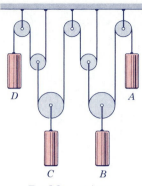

Problem 2/213

2/214 The power winches on the industrial scaffold enable it to be raised or lowered. For rotation in the sense indicated, the scaffold is being raised. If each drum has a diameter of 200 mm and turns at the rate of 40 rev/min, determine the upward velocity v of the scaffold.

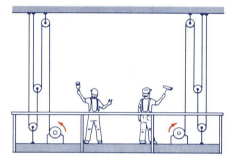

Problem 2/214

2/215 The scaffold of Prob. 2/214 is modified here by placing the power winches on the ground instead of on the scaffold. Other conditions remain the same. Calculate the upward velocity v of the scaffold.
  Ans. $v = 104.7$ mm/s

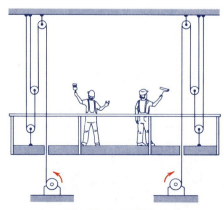

Problem 2/215

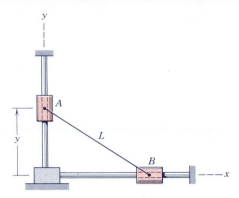

Problem 2/216

2/216 Collars A and B slide along the fixed right-angled rods and are connected by a cord of length L. Determine the acceleration a_x of collar B as a function of y if collar A is given a constant upward velocity v_A.

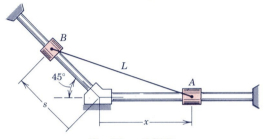

Problem 2/217

2/217 Collars A and B slide along the fixed rods and are connected by a cord of length L. If collar A has a velocity $v_A = \dot{x}$ to the right, express the velocity $v_B = -\dot{s}$ of B in terms of x, v_A, and s.

$$Ans. \quad v_B = \frac{s + \sqrt{2}\,x}{x + \sqrt{2}\,s}\,v_A$$

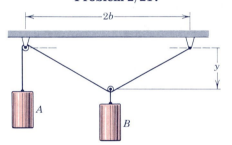

Problem 2/218

2/218 For a given value of y, determine the upward velocity of A in terms of the downward velocity of B. Neglect the diameters of the pulleys

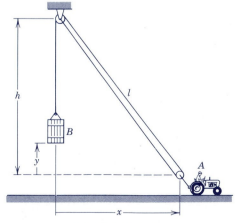

Problem 2/219

2/219 In order to speed up the hoisting of bales depicted in Sample Problem 2/15, the pulley arrangement is altered as shown here. If the tractor A has a forward velocity v_A, determine an expression for the upward velocity v_B of the bale in terms of x. Neglect the small distance between the tractor and its pulley so that both have essentially the same motion. Compare your results with those for Sample Problem 2/15.

$$Ans. \quad v_B = \frac{2xv_A}{\sqrt{x^2 + h^2}}$$

2/220 The particle A is mounted on a light rod pivoted at O and therefore is constrained to move in a circular arc of radius r. Determine the velocity of A in terms of the downward velocity v_B of the counterweight for any angle θ.

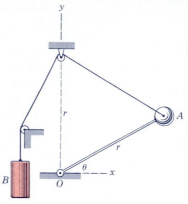

Problem 2/220

2/221 Under the action of force P, the constant acceleration of block B is 3 m/s^2 to the right. At the instant when the velocity of B is 2 m/s to the right, determine the velocity of B relative to A, the acceleration of B relative to A, and the absolute velocity of point C of the cable.

Ans. $v_{B/A} = 0.5$ m/s, $a_{B/A} = 0.75$ m/s^2
$v_C = 1$ m/s, all to the right

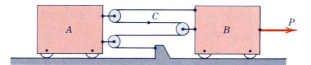

Problem 2/221

2/222 Neglect the diameter of the small pulley attached to body A and determine the magnitude of the total velocity of B in terms of the velocity v_A that body A has to the right. Assume that the cable between B and the pulley remains vertical and solve for a given value of x.

Ans. $v_B = v_A \sqrt{\dfrac{2x^2 + h^2}{x^2 + h^2}}$

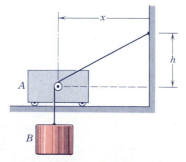

Problem 2/222

2/10 PROBLEM FORMULATION AND REVIEW

In Chapter 2 we have developed and illustrated the basic elements of theory and their use for the description of particle motion. It is important for the student to review this material before proceeding to consolidate and enlarge his or her perspective of particle kinematics. The concepts and procedures developed and illustrated in this chapter form the basis for building much of the entire subject of dynamics, and satisfactory progress in the topics ahead will be contingent on a firm grasp of Chapter 2.

By far the most important concept in Chapter 2 is the time derivative of a vector. The independent contributions of magnitude change and direction change to the time derivative are, indeed, basic. As we proceed in our study of dynamics, we will have further occasion to examine the time derivatives of vectors other than position and velocity vectors, and the principles and procedures developed in Chapter 2 will be utilized directly.

When one concentrates primarily on the individual parts of a subject, the relationships among its parts are not always clearly grasped. When guided by a specific topical assignment, the student automatically recognizes the problem category and the indicated method of solution. But without this guide, he must choose his own method of solution. The ability to choose the most appropriate method is certainly a key to the successful formulation and solution of engineering problems. One of the best ways to acquire this ability is to compare alternative methods after each is studied.

To facilitate this comparison, it will be helpful if we recognize the following breakdowns:

(a) Type of motion.   The three conspicuous categories are

Rectilinear motion (one coordinate)
Plane curvilinear motion (two coordinates)
Space curvilinear motion (three coordinates)

Generally, the geometry of a given problem permits this identification to be made readily. One exception to this categorization is encountered when only the magnitudes of the motion quantities measured along the path are of interest. In this event, we may use the single distance coordinate measured along the curved path, together with its scalar time derivatives giving the speed $|\dot{s}|$ and the tangential acceleration $\ddot{s}$.

Plane motion is easier to generate and control, particularly in machinery, than space motion, so that a large fraction of our motion problems come under the plane-curvilinear or rectilinear categories.

(b) Reference fixity.   We commonly make motion measurements with respect to fixed reference axes (absolute motion) and moving axes (relative motion). The acceptable choice of the fixed

axes depends on the problem. Axes attached to the surface of the earth are sufficiently "fixed" for most engineering problems, although important exceptions include earth–satellite and interplanetary motion, accurate projectile trajectories, navigation, and other problems. The equations of relative motion discussed in Chapter 2 are restricted to translating reference axes.

(c) Coordinates.   Of prime importance is the choice of coordinates. We have developed the description of motion using

Rectangular (Cartesian) coordinates (x-y) and (x-y-z)
Normal and tangential coordinates (n-t)
Polar coordinates (r-θ)
Cylindrical coordinates (r-θ-z)
Spherical coordinates (R-θ-ϕ)

When the coordinates are not specified, the choice usually depends on how the motion is generated or measured. Thus, for a particle that slides radially along a rotating rod, polar coordinates are the natural ones to use. Radar tracking calls for polar or spherical coordinates. When measurements are made along a curved path, normal and tangential coordinates are indicated. An x-y plotter clearly involves rectangular coordinates.

Figure 2/21 is a composite representation of the x-y, n-t, and r-θ coordinate descriptions of the velocity **v** and acceleration **a** for curvilinear motion. It is frequently essential to transfer motion description from one set of coordinates to another, and Fig. 2/21 contains the information necessary for that transition.

(d) Approximations.   One of the most important abilities to acquire is that of making appropriate approximations. The assumption of constant acceleration is valid when the forces that cause the acceleration do not vary appreciably. When motion data are acquired experimentally, we must utilize the nonexact data to acquire the best possible description, often with the aid of graphical or numerical approximations.

(e) Mathematical method.   We frequently have a choice of solution using scalar algebra, vector algebra, trigonometric geometry, or graphical geometry. All of these methods have been illustrated and all are important to learn. The choice of method will depend on the geometry of the problem, how the motion data are given, and the accuracy desired. Inasmuch as mechanics by its very nature is geometrical, the student is encouraged to develop facility in sketching vector relationships, both as an aid to the disclosure of appropriate geometrical and trigonometrical relations and as a means of solving vector equations graphically. Geometric portrayal is the most direct representation of the vast majority of mechanics problems.

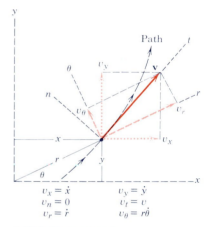

$$v_x = \dot{x} \qquad v_y = \dot{y}$$
$$v_n = 0 \qquad v_t = v$$
$$v_r = \dot{r} \qquad v_\theta = r\dot{\theta}$$

(a) Velocity components

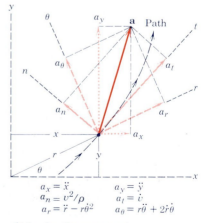

$$a_x = \ddot{x} \qquad a_y = \ddot{y}$$
$$a_n = v^2/\rho \qquad a_t = \dot{v}$$
$$a_r = \ddot{r} - r\dot{\theta}^2 \qquad a_\theta = r\ddot{\theta} + 2\dot{r}\dot{\theta}$$

(b) Acceleration components

Figure 2/21

REVIEW PROBLEMS

2/223 The position s of a particle along a straight line is given by $s = 8e^{-0.4t} - 6t + t^2$, where s is in meters and t is the time in seconds. Determine the velocity v when the acceleration is 3 m/s^2.

Ans. $v = -7.27$ m/s

2/224 The polar coordinates of a particle are given by $r = 30$ mm, constant, and $\theta = 5t$ radians, where t is in seconds. Determine $\ddot{y}$ when $\theta = 60°$.

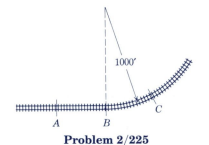

Problem 2/225

2/225 An inexperienced designer of a roadbed for a new high-speed train proposes to join a straight section of track to a circular section of 1000-ft radius as shown. For a train that would travel at a constant speed of 90 mi/hr, plot the magnitude of its acceleration as a function of distance along the track between points A and C and explain why this design is unacceptable.

Ans. $a_n = 17.42$ ft/sec^2 (from B to C)

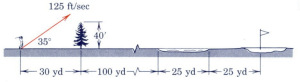

Problem 2/226

2/226 Just after being struck by the club, a golf ball has a velocity of 125 ft/sec directed at 35° to the horizontal as shown. Determine the location of the point of impact.

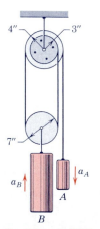

Problem 2/227

2/227 In the differential pulley hoist shown, the two upper pulleys are fastened together to form an integral unit. The cable is wrapped around the smaller pulley with its end secured to the pulley so that it cannot slip. Determine the upward acceleration a_B of cylinder B if cylinder A has a downward acceleration of 2 ft/sec^2. (*Suggestion:* Analyze geometrically the consequences of a differential movement of cylinder A.)

Ans. $a_B = 0.25$ ft/sec^2

2/228 The small cylinder is made to move along the rotating rod with a motion between $r = r_0 + b$ and $r = r_0 - b$ given by $r = r_0 + b \sin \dfrac{2\pi t}{\tau}$, where t is the time counted from the instant the cylinder passes the $r = r_0$ position and τ is the period of the oscillation (time for one complete oscillation). Simultaneously, the rod rotates about the vertical at the constant angular rate $\dot{\theta}$. Determine the value of r for which the radial (r-direction) acceleration is zero.

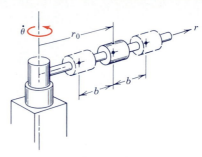

Problem 2/228

2/229 A small projectile is fired from point O with an initial velocity $u = 500$ m/s at the angle of 60° from the horizontal as shown. Neglect atmospheric resistance and any change in g and compute the radius of curvature ρ of the path of the projectile 30 s after the firing. *Ans.* $\rho = 9.53$ km

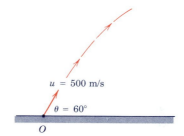

Problem 2/229

2/230 A stone is thrown down the slope as shown. Determine the magnitude u and direction θ of its initial velocity so that the stone will rise 12 m and still have a range of 50 m down the slope.

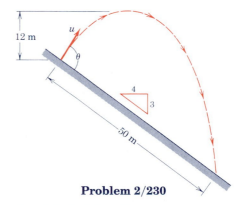

Problem 2/230

2/231 The third stage of a rocket is injected by its booster with a velocity u of 15 000 km/h at A into an unpowered coasting flight to B. At B its rocket motor is ignited when the trajectory makes an angle of 20° with the horizontal. Operation is effectively above the atmosphere, and the gravitational acceleration during this interval may be taken as 9 m/s², constant in magnitude and direction. Determine the time t to go from A to B. (This quantity is needed in the design of the ignition control system.) Also determine the corresponding increase h in altitude.
Ans. $t = 3$ min 28 s, $h = 418$ km

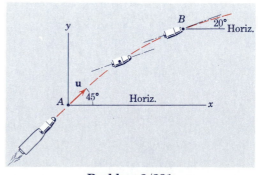

Problem 2/231

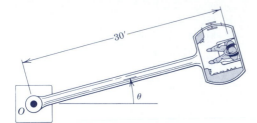

Problem 2/232

2/232 The angular displacement of the centrifuge is given by $\theta = 4[t + 30e^{-0.03t} - 30]$ rad, where t is in seconds and $t = 0$ is the startup time. If the person loses consciousness at an acceleration level of $10g$, determine the time t at which this would occur. Verify that the tangential acceleration is negligible as the normal acceleration approaches $10g$.

Problem 2/233

2/233 A rocket is fired vertically and tracked by the radar shown with Prob. 2/137 and repeated here. At the instant when $\theta = 60°$, measurements give $\dot{\theta} = 0.03$ rad/sec and $r = 25{,}000$ ft, and the vertical acceleration of the rocket is found to be $a = 64$ ft/sec^2. For this instant determine the values of $\ddot{r}$ and $\ddot{\theta}$.

$$Ans. \quad \ddot{r} = 77.9 \text{ ft/sec}^2$$
$$\ddot{\theta} = -1.838(10^{-3}) \text{ rad/sec}^2$$

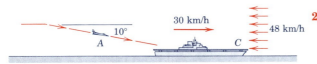

Problem 2/234

2/234 As part of a training exercise, the pilot of aircraft A adjusts her airspeed (speed relative to the wind) to 220 km/h while in the level portion of the approach path and thereafter holds her absolute speed constant as she negotiates the 10° glide path. The absolute speed of the aircraft carrier is 30 km/h and that of the wind is 48 km/h. What will be the angle β of the glide path with respect to the horizontal as seen by an observer on the ship?

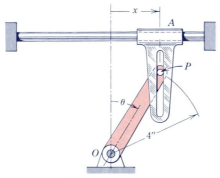

Problem 2/235

2/235 Rotation of the arm PO is controlled by the horizontal motion of the vertical slotted link. If $\dot{x} = 4$ ft/sec and $\ddot{x} = 30$ ft/sec^2 when $x = 2$ in., determine $\dot{\theta}$ and $\ddot{\theta}$ for this instant.

$$Ans. \quad \dot{\theta} = 13.86 \text{ rad/sec}, \quad \ddot{\theta} = 215 \text{ rad/sec}^2$$

2/236 A particle P moving in plane curvilinear motion is located by the polar coordinates shown. At a particular instant $r = 2$ m, $\theta = 60°$, $v_r = 3$ m/s, $v_\theta = 4$ m/s, $a_r = -10$ m/s^2, and $a_\theta = -5$ m/s^2. For this instant calculate the radius of curvature ρ of the path of the particle. Also locate the center of curvature C graphically.

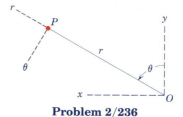

Problem 2/236

2/237 Cylinder A has a constant downward speed of 1 m/s. Compute the velocity of cylinder B for (a) $\theta = 45°$, (b) $\theta = 30°$, and (c) $\theta = 15°$. The spring is in tension throughout the motion range of interest, and the pulleys are connected by the cable of fixed length.

   Ans. (a) $v_B = 0.293$ m/s to the right
          (b) $v_B = 0$, (c) $v_B = 0.250$ m/s to the left

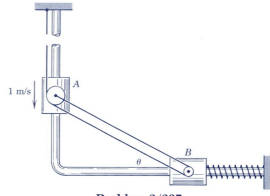

Problem 2/237

2/238 A particle has the following position, velocity, and acceleration components: $x = 50$ ft, $y = 25$ ft, $\dot{x} = -10$ ft/sec, $\dot{y} = 10$ ft/sec, $\ddot{x} = -10$ ft/sec^2, and $\ddot{y} = 5$ ft/sec^2. Determine the following quantities: v, a, $\mathbf{e}_t$, $\mathbf{e}_n$, a_t, $\mathbf{a}_t$, a_n, $\mathbf{a}_n$, ρ, $\mathbf{e}_r$, $\mathbf{e}_\theta$, v_r, $\mathbf{v}_r$, v_θ, $\mathbf{v}_\theta$, a_r, $\mathbf{a}_r$, a_θ, $\mathbf{a}_\theta$, r, $\dot{r}$, $\ddot{r}$, θ, $\dot{\theta}$, and $\ddot{\theta}$. Express all vectors in terms of $\mathbf{i}$ and $\mathbf{j}$, and graph all vectors on one set of x-y axes as you proceed.

2/239 The rod OA is held at the constant angle $\beta = 30°$ while it rotates about the vertical with a constant angular rate of 120 rev/min. Simultaneously, the slider P oscillates along the rod with a variable distance from the fixed pivot O given in inches by $R = 8 + 2 \sin 2\pi nt$, where the frequency n of oscillation along the rod is a constant 2 cycles per second and where the time t is in seconds. Calculate the magnitude of the acceleration of the slider for an instant when its velocity $\dot{R}$ along the rod is a maximum.

   Ans. $a = 706$ in./sec^2

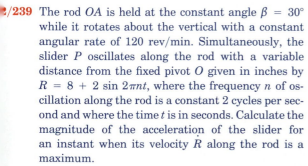

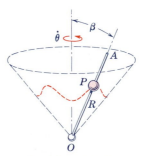

Problem 2/239

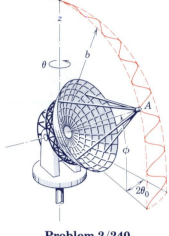

Problem 2/240

▶ **2/240** The radar tracking antenna oscillates about its vertical axis according to $\theta = \theta_0 \cos \omega t$, where ω is the constant circular frequency and $2\theta_0$ is the double amplitude of oscillation. Simultaneously, the angle of elevation ϕ is increasing at the constant rate $\dot{\phi} = K$. Determine the expression for the magnitude a of the acceleration of the signal horn (a) as it passes position A and (b) as it passes the top position B, assuming that $\theta = 0$ at this instant.

$$Ans. \ (a) \ a = b\sqrt{K^4 + \omega^4\theta_0^2 \cos^2\phi}$$
$$(b) \ a = bK\sqrt{K^2 + 4\omega^2\theta_0^2}$$

Computer-oriented problems

*2/241** Two particles A and B start from rest at $x = 0$ and move along parallel paths according to $x_A = 0.16 \sin \frac{\pi t}{2}$ and $x_B = 0.08t$, where x_A and x_B are in meters and t is in seconds counted from the start. Determine the time t (where $t > 0$) when both particles have the same displacement and calculate this displacement x.   *Ans.* $t = 1.473$ s, $x = 0.1178$ m

*2/242** The fixed cam and rotating arm of Prob. 2/166 are repeated here. The shape of the cam causes the follower A to move along a limaçon defined by $r = b - c \cos\theta$, where $b > c$. If $b = 100$ mm, $c = 50$ mm, and the arm revolves at the constant rate $\dot{\theta} = 2$ rad/s, plot the magnitudes of the velocity **v** and acceleration **a** of pin A from $\theta = 0$ to $\theta = 180°$. Justify your result for the acceleration at $\theta = 0$.

Problem 2/242

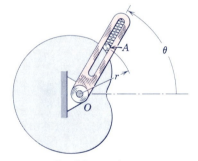

Problem 2/243

*2/243** The drag racer of Prob. 2/47 is repeated here. The acceleration is given by $a = c_1 - c_2v^2$, where c_1 is known to be 30 ft/sec² but c_2 is unknown. Determine c_2 if the racer completes the $\frac{1}{4}$-mi run in 9.4 sec. Then plot the velocity and displacement as functions of time.

*2/244 The guide with the vertical slot is given a horizontal oscillatory motion according to $x = 4 \sin 2t$, where x is in inches and t is in seconds. The oscillation causes the pin P to move in the fixed parabolic slot whose shape is given by $y = x^2/4$, with y also in inches. Plot the magnitude v of the velocity of the pin as a function of time during the interval required for pin P to go from the center to the extremity $x = 4$ in. Find and locate the maximum value of v and verify your results analytically.

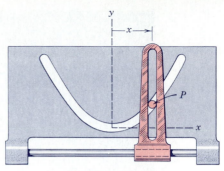

Problem 2/244

2/245 A bullet with a muzzle velocity of 2000 ft/sec is fired vertically upward and reaches a maximum height of 1 mi. Air resistance causes an additional component of downward acceleration kv^2 proportional to the square of the velocity v. Take g to be constant at 32.2 ft/sec^2 and calculate the coefficient k.

$$Ans. \quad k = 3.63(10^{-4}) \text{ ft}^{-1}$$

2/246 A ship with a total displacement of 16 000 metric tons (1 metric ton = 1000 kg) starts from rest in still water under a constant propeller thrust $T = 250$ kN. The ship develops a total resistance to motion through the water given by $R = 4.50v^2$, where R is in kilonewtons and v is in meters per second. The acceleration of the ship is $a = (T - R)/m$, where m equals the mass of the ship in metric tons. Plot the speed v of the ship in knots as a function of the distance s in nautical miles that the ship goes for the first 5 nautical miles from rest. Find the speed after the ship has gone 1 nautical mile. What is the maximum speed that the ship can reach?

Curved motion of a particle is always accompanied by acceleration which, in turn, requires force to support the motion. The design of the supporting structure for this amusement-park loop must account for the force that accompanies the resulting acceleration.

KINETICS
OF PARTICLES

3

3/1 INTRODUCTION

According to Newton's second law, a particle will accelerate when it is subjected to unbalanced forces. Kinetics is the study of the relations between unbalanced forces and the changes in motion that they produce. In Chapter 3 we shall study the kinetics of particles. This topic requires that we combine our knowledge of two previously learned parts of mechanics, namely, the properties of forces that we developed in our earlier study of statics, and the kinematics of particle motion that we have just covered in Chapter 2. With the aid of Newton's second law of motion, we are now ready to combine these two topics and prepare for the solution of engineering problems involving force, mass, and motion.

There are three general approaches to the solution of kinetics problems: (A) direct application of Newton's second law (called the force-mass-acceleration method), (B) use of work and energy principles, and (C) solution by impulse and momentum methods. Each approach has its special characteristics and advantages, and Chapter 3 is subdivided into sections according to these three methods of solution. In addition, a fourth section treats special applications and combinations of the three basic approaches. Before proceeding, the student is strongly advised to review carefully the definitions and concepts of Chapter 1, as they are fundamental to the developments that follow.

SECTION A. FORCE, MASS, AND ACCELERATION

3/2 NEWTON'S SECOND LAW

The basic relation between force and acceleration is found in Newton's second law of motion, Eq. 1/1, the verification of which is entirely experimental. We will describe the fundamental meaning of this law by an ideal experiment in which force and acceleration are assumed to be measured without error. A mass particle is isolated in the primary inertial system* and is subjected to the action of a single force $\mathbf{F}_1$. The acceleration $\mathbf{a}_1$ of the particle is measured, and the ratio F_1/a_1 of the magnitudes of the force and the acceleration will be some number C_1 whose value depends on the units used for measurement of force and acceleration. The experiment is now repeated by subjecting the same particle to a different force $\mathbf{F}_2$ and measuring the corresponding acceleration $\mathbf{a}_2$. Again the ratio of the magnitudes F_2/a_2 will produce a number C_2. The experiment is repeated as many times as desired. We draw two important conclusions from the results. First, the ratios of applied force to corresponding acceleration all equal the *same* number, provided the units used for measurement are not changed in the experiments. Thus,

$$\frac{F_1}{a_1} = \frac{F_2}{a_2} = \cdots = \frac{F}{a} = C, \qquad \text{a constant}$$

We conclude that the constant C is a measure of some property of the particle that does not change. This property is the *inertia* of the particle which is its *resistance to rate of change of velocity*. For a particle of high inertia (large C), the acceleration will be small for a given force F. On the other hand, if the inertia is small, the acceleration will be large. The mass m is used as a quantitative measure of inertia, and therefore the expression $C = km$ may be written, where k is a constant to account for the units used. Thus, the experimental relation becomes

$$F = kma \tag{3/1}$$

where F is the magnitude of the resultant force acting on the particle of mass m, and a is the magnitude of the resulting acceleration of the particle.

The second conclusion we draw from the ideal experiment is that the acceleration is always in the direction of the applied force. Thus, Eq. 3/1 becomes a *vector* relation and may be written

$$\mathbf{F} = km\mathbf{a} \tag{3/2}$$

Although an actual experiment cannot be performed in the ideal manner described, the conclusions are inferred from the measurements of countless accurately performed experiments where the results are correctly predicted from the hypothesis of the ideal experiment. One of the most accurate checks lies in the precise prediction of the motions of planets based on Eq. 3/2.

---

*The primary inertial system or astronomical frame of reference is an imaginary set of reference axes that are assumed to have no translation or rotation in space. See Art. 1/2, Chapter 1.

(a) Inertial system. Whereas the results of the ideal experiment are obtained for measurements made relative to the "fixed" primary inertial system, they are equally valid for measurements made with respect to any nonrotating reference system that translates with a constant velocity with respect to the primary system. From our study of relative motion in Art. 2/8, we saw that the acceleration measured in a system translating with no acceleration is the same as that measured in the primary system. Thus, Newton's second law holds equally well in a nonaccelerating system, so that we may define an *inertial system* as any system in which Eq. 3/2 is valid.

If the ideal experiment described were performed on the surface of the earth and all measurements were made relative to a reference system attached to the earth, the measured results would show a slight discrepancy upon substitution into Eq. 3/2. This discrepancy would be due to the fact that the measured acceleration would not be the correct absolute acceleration. The discrepancy would disappear when we introduced the corrections due to the acceleration components of the earth. These corrections are negligible for most engineering problems that involve the motions of structures and machines on the surface of the earth. In such cases, the accelerations measured with respect to reference axes attached to the surface of the earth may be treated as "absolute," and Eq. 3/2 may be applied with negligible error to experimental measurements made on the surface of the earth.*

There is an increasing number of problems, particularly in the fields of rocket and spacecraft design, where the acceleration components of the earth are of primary concern. For this work it is essential that the fundamental basis of Newton's law be thoroughly understood and that the appropriate absolute acceleration components be employed.

Before 1905 the laws of Newtonian mechanics had been verified by innumerable physical experiments and were considered the final description of the motion of bodies. The concept of *time*, considered an absolute quantity in the Newtonian theory, received a basically

---

*An example of the magnitude of the error introduced by neglect of the motion of the earth may be cited for the case of a particle that is allowed to fall from rest (relative to the earth) at a height h above the ground. We can show that the rotation of the earth gives rise to an eastward acceleration (Coriolis acceleration) relative to the earth and, neglecting air resistance, that the particle falls to the ground a distance

$$x = \frac{2}{3}\,\omega\,\sqrt{\frac{2h^3}{g}}\,\cos\gamma$$

east of the point on the ground directly under that from which it was dropped. The angular velocity of the earth is $\omega = 0.729(10^{-4})$ rad/s, and the latitude, north or south, is γ. At a latitude of $45°$ and from a height of 200 m, this eastward deflection would be $x = 43.9$ mm.

different interpretation in the theory of relativity announced by
Einstein in 1905. The new concept called for a complete reformu-
lation of the accepted laws of mechanics. The theory of relativity
was subjected to early ridicule, but has had experimental check
and is now universally accepted by scientists the world over. Although the difference between the mechanics of Newton and that
of Einstein is basic, there is a practical difference in the results
given by the two theories only when velocities of the order of the
speed of light (300 $\times$ 10^6 m/s) are encountered.* Important problems dealing with atomic and nuclear particles, for example, involve
calculations based on the theory of relativity and are of basic concern to both scientists and engineers.

(b) Units. It is customary to take k equal to unity in Eq. 3/2,
thus putting the relation in the usual form of Newton's second law

$$\boxed{\mathbf{F} = m\mathbf{a}} \tag{1/1}$$

A system of units for which k is unity is known as a *kinetic* system.
Thus, for a kinetic system the units of force, mass, and acceleration
are not independent. In SI units, as explained in Art. 1/4, the units
of force (newtons, N) are derived by Newton's second law from the
base units of mass (kilograms, kg) times acceleration (meters per
second squared, m/s^2). Thus, N = kg$\cdot$m/s^2. This system is known
as an *absolute* system since the unit for force is dependent on the
absolute value of mass.

In U.S. customary units, on the other hand, the units of mass
(slugs) are derived from the units of force (pounds force, lb) divided
by acceleration (feet per second squared, ft/sec^2). Thus, the mass
units are slugs = lb-sec^2/ft. This system is known as a *gravitational*
system since mass is derived from force as determined from gravitational attraction.

For measurements made relative to the rotating earth, the relative value of g should be used. The internationally accepted value
of g relative to the earth at sea level and at a latitude of 45° is
9.806 65 m/s^2. Except where greater precision is required, the value
of 9.81 m/s^2 will be used for g. For measurements relative to a
nonrotating earth, the absolute value of g should be used. At a latitude of 45° and at sea level, the absolute value is 9.8236 m/s^2. The

*The theory of relativity demonstrates that there is no such thing as a preferred
primary inertial system and that measurements of time which are made in two coordinate systems that have a velocity relative to one another are different. On this
basis, for example, the principles of relativity show that a clock carried by the pilot
of a spacecraft traveling around the earth in a circular polar orbit of 644 km altitude
at a velocity of 27 080 km/h would be slow compared with a clock at the pole by
0.000 001 85 s for each orbit.

sea-level variation in both the absolute and relative values of g with latitude is shown in Fig. 1/1 of Art. 1/5.

In the U.S. customary system, the standard value of g relative to the rotating earth at sea level and at a latitude of 45° is 32.1740 ft/sec². The corresponding value relative to a nonrotating earth is 32.2230 ft/sec².

By virtue of our need to use both SI units and U.S. customary units, we must be absolutely sure that we have a clear understanding of the correct force and mass units in each system. These units were explained in Art. 1/4, but it will be helpful to illustrate them here using simple numbers before moving ahead to the applications of Newton's second law. Consider, first, the free-fall experiment as depicted in Fig. 3/1a where we release an object from rest near the earth's surface and allow it to fall freely under the influence of the force of gravitational attraction W on the body, which we call its *weight*. In SI units for a mass $m = 1$ kg, the weight is $W = 9.81$ N, and the corresponding downward acceleration a is $g = 9.81$ m/s². In U.S. customary units for a mass $m = 1$ lbm (1/32.2 slug), the weight is $W = 1$ lbf and the resulting gravitational acceleration is $g = 32.2$ ft/sec². For a mass $m = 1$ slug (32.2 lbm), the weight is $W = 32.2$ lbf and the acceleration, of course, is also $g = 32.2$ ft/sec².

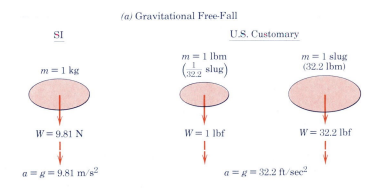

(a) Gravitational Free-Fall

SI   U.S. Customary

$m = 1$ kg $m = 1$ lbm $\left(\frac{1}{32.2}$ slug$\right)$ $m = 1$ slug (32.2 lbm)

$W = 9.81$ N $W = 1$ lbf $W = 32.2$ lbf

$a = g = 9.81$ m/s² $a = g = 32.2$ ft/sec²

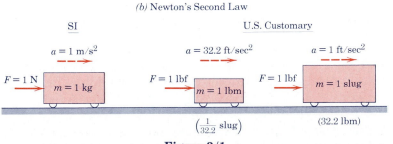

(b) Newton's Second Law

SI   U.S. Customary

$a = 1$ m/s² $a = 32.2$ ft/sec² $a = 1$ ft/sec²

$F = 1$ N $m = 1$ kg $F = 1$ lbf $m = 1$ lbm $F = 1$ lbf $m = 1$ slug

$\left(\frac{1}{32.2}$ slug$\right)$ (32.2 lbm)

Figure 3/1

In Fig. 3/1*b* we illustrate the proper units with the simplest example where we accelerate an object of mass m along the horizontal with a force F. In the coherent system of SI units, a force $F = 1$ N causes a mass $m = 1$ kg to accelerate at the rate $a = 1$ m/s^2. Thus, 1 N $= 1$ kg·m/s^2. In the U.S. customary system (which is a gravitational, noncoherent system), a force $F = 1$ lbf causes a mass $m = 1$ lbm ($1/32.2$ slug) to accelerate at the rate $a = 32.2$ ft/sec^2, whereas a force $F = 1$ lbf causes a mass $m = 1$ slug (32.2 lbm) to accelerate at the rate $a = 1$ ft/sec^2.

We note that in SI units where the mass is expressed in kilograms (kg), the weight W of the body in newtons (N) is given by $W = mg$, where $g = 9.81$ m/s^2. In U.S. customary units, the weight W of a body is expressed in pounds force (lbf), and the mass in slugs (lbf-sec^2/ft) is given by $m = W/g$, where $g = 32.2$ ft/sec^2. In U.S. customary units, we frequently speak of the weight of a body when we really mean mass. It is entirely proper to specify the mass of a body in pounds (lbm) which must be converted to mass in slugs before substituting into Newton's second law. Unless otherwise stated, the pound (lb) is normally used as the unit of force (lbf).

3/3   EQUATION OF MOTION AND SOLUTION OF PROBLEMS

When a particle of mass m is subjected to the action of concurrent forces $\mathbf{F}_1$, $\mathbf{F}_2$, $\mathbf{F}_3$, ... whose vector sum is $\Sigma\mathbf{F}$, Eq. 1/1 becomes

$$\boxed{\Sigma\mathbf{F} = m\mathbf{a}} \qquad (3/3)$$

In the solution of problems, Eq. 3/3 is usually expressed in scalar component form with the use of one of the coordinate systems developed in Chapter 2. The choice of the appropriate coordinate system is dictated by the type of motion involved and is a vital step in the formulation of any problem. Equation 3/3, or any one of the component forms of the force-mass-acceleration equation, is usually referred to as the *equation of motion*. The equation of motion gives the instantaneous value of the acceleration corresponding to the instantaneous values of the forces that are acting.

(a) The two problems of dynamics.   We encounter two types of problems when applying Eq. 3/3. In the first type, the acceleration is either specified or can be determined directly from known kinematic conditions. The corresponding forces that act on the particle whose motion is specified are then determined by direct substitution into Eq. 3/3. This problem is generally quite straightforward.

In the second type of problem, the forces are specified and the resulting motion is to be determined. If the forces are constant, the acceleration is constant and is easily found from Eq. 3/3. When the forces are functions of time, position, velocity, or acceleration, Eq. 3/3 becomes a differential equation that must be integrated to determine the velocity and displacement. Problems of this second type are often more formidable, as the integration may be difficult to carry out, particularly when the force is a mixed function of two or more motion variables. In practice, it is frequently necessary to resort to approximate integration techniques, either numerical or graphical, particularly when experimental data are involved. The procedures for a mathematical integration of the acceleration when it is a function of the motion variables were developed in Art. 2/2, and these same procedures apply when the force is a specified function of these same parameters, since force and acceleration differ only by the constant factor of the mass.

 (b) Constrained and unconstrained motion.   There are two physically distinct types of motion, both described by Eq. 3/3. The first type is *unconstrained* motion where the particle is free of mechanical guides and follows a path determined by its initial motion and by the forces that are applied to it from external sources. An airplane or rocket in flight and an electron moving in a charged field are examples of unconstrained motion. The second type is *constrained* motion where the path of the particle is partially or totally determined by restraining guides. An ice-hockey puck moves with the partial constraint of the ice. A train moving along its track and a collar sliding along a fixed shaft are examples of more fully constrained motion. Some of the forces acting on a particle during constrained motion may be applied from outside sources and others may be the reactions on the particle from the constraining guides. *All forces*, both applied and reactive, that act *on* the particle must be accounted for in applying Eq. 3/3.

 The choice of a coordinate system is frequently indicated by the number and geometry of the constraints. Thus, if a particle is free to move in space, as is the center of mass of the airplane or rocket in free flight, the particle is said to have *three degrees of freedom* since three independent coordinates are required to specify the position of the particle at any instant. All three of the scalar components of the equation of motion would have to be applied and integrated to obtain the space coordinates as a function of time. If a particle is constrained to move along a surface, as is the hockey puck or a marble sliding on the curved surface of a bowl, only two coordinates are needed to specify its position, and in this case it is said to have *two degrees of freedom*. If a particle is constrained to move along a fixed linear path, as is the collar sliding along a fixed shaft, its position may be specified by the coordinate measured along the shaft. In this case, the particle would have only *one degree of freedom*.

(c) Free-body diagram.   In the application of any of the force-mass-acceleration equations of motion, it is absolutely necessary to account correctly for *all* forces acting on the particle. The only forces that we may neglect are those whose magnitudes are negligible compared with other forces acting, such as the forces of mutual attraction between two particles compared with their attraction to a celestial body such as the earth. The vector sum $\Sigma \mathbf{F}$ of Eq. 3/3 means the vector sum of *all* forces acting *on* the particle in question. Likewise, the corresponding scalar force summation in any one of the component directions means the sum of the components of *all* forces acting *on* the particle in that particular direction. The only reliable way to account accurately and consistently for every force is to *isolate* the particle under consideration from *all* contacting and influencing bodies and replace the bodies removed by the forces they exert on the particle isolated. The resulting *free-body diagram* is the means by which every force, known and unknown, that acts on the particle is represented and hence accounted for. Only after this vital step has been completed should the appropriate equation or equations of motion be written. The free-body diagram serves the same key purpose in dynamics as it does in statics. This purpose is simply to establish a *thoroughly reliable method* for the correct evaluation of the resultant of all actual forces acting on the particle or body in question. In statics this resultant equals zero, whereas in dynamics it is equated to the product of mass and acceleration. If we recognize that the equations of motion must be interpreted literally and exactly, and if in so doing, we respect the full scalar and vector meaning of the equals sign in the motion equation, then a minimum of difficulty will be experienced. Every experienced student of engineering mechanics recognizes that careful and consistent observance of the *free-body method* is the *most important single lesson* to be learned in the study of engineering mechanics. As a part of the drawing of a free-body diagram, the coordinate axes and their positive directions should be clearly indicated. When the equations of motion are written, all force summations should be consistent with the choice of these positive directions. Also, as an aid to the identification of external forces that act on the body in question, these forces are shown as heavy red vectors in the illustrations throughout the remainder of the book. Sample Problems 3/1 through 3/5 in the next article contain five examples of free-body diagrams that may be easily reviewed as a reminder of their construction.

In solving problems, the student often wonders how to get started and what sequence of steps to follow in arriving at the solution. This difficulty may be minimized by forming the habit of first recognizing some relationship between the desired unknown quantity in the problem and other quantities, known and unknown. Additional relationships between these unknowns and other quantities, known and unknown, are then perceived. Finally, the dependence on the original data is established, and the procedure for the analysis and computation is indicated. A few minutes spent organizing the plan of attack through recognition of the dependence of one quantity on another will be time well spent and will usually prevent groping for the answer with irrelevant calculations.

3/4   RECTILINEAR MOTION

We now apply the concepts discussed in Arts. 3/2 and 3/3 to problems in particle motion, starting with rectilinear motion in this article and treating curvilinear motion in Art. 3/5. In both articles, we shall analyze the motions of bodies that can be treated as particles. This simplication is permitted as long as we are interested only in the motion of the mass center of the body, in which case, the forces may be treated as concurrent through the mass center. When we discuss the kinetics of rigid bodies in Chapter 6, we shall account for the action of nonconcurrent forces on the motions of bodies.

If we choose the x-direction, for example, as the direction of the rectilinear motion of a particle of mass m, the acceleration in the y- and z-directions will be zero and the scalar components of Eq. 3/3 become

$$\Sigma F_x = ma_x$$
$$\Sigma F_y = 0 \qquad\qquad (3/4)$$
$$\Sigma F_z = 0$$

For cases where we are not free to choose a coordinate direction along the motion, we would have in the general case all three component equations

$$\Sigma F_x = ma_x$$
$$\Sigma F_y = ma_y \qquad\qquad (3/5)$$
$$\Sigma F_z = ma_z$$

where the acceleration and resultant force are given by

$$\mathbf{a} = a_x\mathbf{i} + a_y\mathbf{j} + a_z\mathbf{k}$$
$$a = \sqrt{a_x{}^2 + a_y{}^2 + a_z{}^2}$$
$$\Sigma\mathbf{F} = \Sigma F_x\mathbf{i} + \Sigma F_y\mathbf{j} + \Sigma F_z\mathbf{k}$$
$$|\Sigma\mathbf{F}| = \sqrt{(\Sigma F_x)^2 + (\Sigma F_y)^2 + (\Sigma F_z)^2}$$

Sample Problem 3/1

A 75-kg man stands on a spring scale in an elevator. During the first 3 seconds of motion from rest, the tension T in the hoisting cable is 8300 N. Find the reading R of the scale in newtons during this interval and the upward velocity v of the elevator at the end of the 3 seconds. The total mass of the elevator, man, and scale is 750 kg.

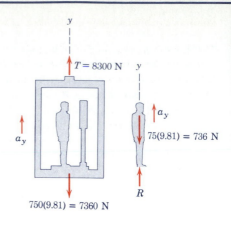

Solution. The force registered by the scale and the velocity both depend on the acceleration of the elevator, which is constant during the interval for which the forces are constant. From the free-body diagram of the elevator, scale, and man taken together, the acceleration is found to be

$$[\Sigma F_y = ma_y] \qquad 8300 - 7360 = 750a_y \qquad a_y = 1.257 \text{ m/s}^2$$

The scale reads the downward force exerted on it by the man's feet. The equal and opposite reaction R to this action is shown on the free-body diagram of the man alone together with his weight, and the equation of motion for him gives

① $[\Sigma F_y = ma_y] \qquad R - 736 = 75(1.257) \qquad R = 830 \text{ N}$ *Ans.*

The velocity reached at the end of the 3 seconds is

$$\left[\Delta v = \int a \, dt\right] \qquad v - 0 = \int_0^3 1.257 \, dt \qquad v = 3.77 \text{ m/s} \qquad \textit{Ans.}$$

① If the scale is calibrated in kilograms it would read $830/9.81 = 84.6$ kg which, of course, is not his true mass since the measurement was made in a noninertial (accelerating) system. *Suggestion:* Rework this problem in U.S. customary units.

Sample Problem 3/2

A small inspection car with a mass of 200 kg runs along the fixed overhead cable and is controlled by the attached cable at A. Determine the acceleration of the car when the control cable is horizontal and under a tension $T = 2.4$ kN. Also find the total force P exerted by the supporting cable on the wheels.

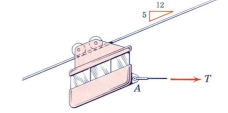

Solution. The free-body diagram of the car and wheels taken together and treated as a particle discloses the 2.4-kN tension T, the weight $W = mg = 200(9.81) = 1962$ N, and the force P exerted on the wheel assembly by the cable.

The car is in equilibrium in the y-direction since there is no acceleration in this direction. Thus,

$$[\Sigma F_y = 0] \qquad P - 2.4(\tfrac{5}{13}) - 1.962(\tfrac{12}{13}) = 0 \qquad P = 2.73 \text{ kN} \qquad \textit{Ans.}$$

① In the x-direction the equation of motion gives

$$[\Sigma F_x = ma_x] \qquad 2400(\tfrac{12}{13}) - 1962(\tfrac{5}{13}) = 200a \qquad a = 7.30 \text{ m/s}^2 \qquad \textit{Ans.}$$

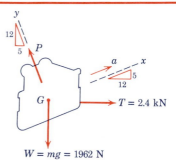

① By choosing our coordinate axes along and normal to the direction of the acceleration, we are able to solve the two equations independently. Would this be so if x and y were chosen as horizontal and vertical?

Sample Problem 3/3

The 250-lb concrete block A is released from rest in the position shown and pulls the 400-lb log up the 30° ramp. If the coefficient of kinetic friction between the log and the ramp is 0.5, determine the velocity of the block as it hits the ground at B.

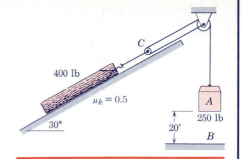

400 lb
$\mu_k = 0.5$
30°
A
250 lb
20'
B
C

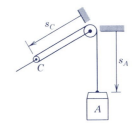

s_C
C
s_A
A

Solution. The motions of the log and the block A are clearly dependent. Although by now it should be evident that the acceleration of the log up the incline is half the downward acceleration of A, we may prove it formally. The constant total length of the cable is $L = 2s_C + s_A + $ constant, where the constant accounts for the cable portions

① wrapped around the pulleys. Differentiating twice with respect to time gives $0 = 2\ddot{s}_C + \ddot{s}_A$, or

$$0 = 2a_C + a_A$$

We assume here that the masses of the pulleys are negligible and that they turn with negligible friction. With these assumptions the free-body diagram of the pulley C discloses force and moment equilibrium. Hence, the tension in the cable attached to the log is twice that applied to the block. Note that the accelerations of the log and pulley C are identical.

The free-body diagram of the log shows the friction force $\mu_k N$ for motion up the plane. Equilibrium of the log in the y-direction gives

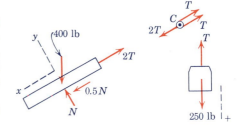

y
400 lb
$2T$
$0.5N$
N
x
T
C
$2T$
T
T
250 lb

② $[\Sigma F_y = 0]$ $N - 400 \cos 30° = 0$ $N = 346$ lb

and its equation of motion in the x-direction gives

$[\Sigma F_x = ma_x]$ $0.5(346) - 2T + 400 \sin 30° = \dfrac{400}{32.2} a_C$

For the block in the positive downward direction, we have

③ $[+\downarrow \Sigma F = ma]$ $250 - T = \dfrac{250}{32.2} a_A$

Solving the three equations in a_C, a_A, and T gives us

$a_A = 5.83$ ft/sec^2 $a_C = -2.92$ ft/sec^2 $T = 205$ lb

④ For the 20-ft drop with constant acceleration, the block acquires a velocity

$[v^2 = 2ax]$ $v_A = \sqrt{2(5.83)(20)} = 15.3$ ft/sec   *Ans.*

① The coordinates used in expressing the final kinematic constraint relationship must be consistent with those used for the kinetic equations of motion.

② We can verify that the log will indeed move up the ramp by calculating the force in the cable necessary to initiate motion from the equilibrium condition. This force is $2T = 0.5N + 400 \sin 30° = 373$ lb or $T = 186.5$ lb, which is less than the 250-lb weight of block A. Hence, the log will move up.

③ Note the serious error in assuming that $T = 250$ lb, in which case, block A would not accelerate.

④ Because the forces on this system remain constant, the resulting accelerations also remain constant.

Sample Problem 3/4

The design model for a new ship has a mass of 10 kg and is tested in an experimental towing tank to determine its resistance to motion through the water at various speeds. The test results are plotted on the accompanying graph, and the resistance R may be closely approximated by the dotted parabolic curve shown. If the model is released when it has a speed of 2 m/s, determine the time t required for it to reduce its speed to 1 m/s and the corresponding travel distance x.

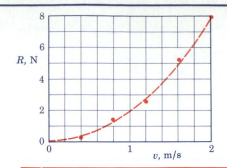

Solution. We approximate the resistance-velocity relation by $R = kv^2$ and find k by substituting $R = 8$ N and $v = 2$ m/s into the equation, which gives $k = 8/2^2 = 2$ N·s²/m². Thus, $R = 2v^2$.

The only horizontal force on the model is R, so that

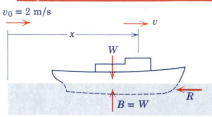

① $[\Sigma F_x = ma_x]$ $\qquad -R = ma_x \quad$ or $\quad -2v^2 = 10\dfrac{dv}{dt}$

We separate the variables and integrate to obtain

$$\int_0^t dt = -5 \int_2^v \frac{dv}{v^2} \qquad t = 5\left(\frac{1}{v} - \frac{1}{2}\right) \text{ sec}$$

Thus, when $v = v_0/2 = 1$ m/s, the time is $t = 5(\frac{1}{1} - \frac{1}{2}) = 2.5$ s $\qquad$ *Ans.*

The distance traveled during the 2.5 seconds is obtained by integrating $v = dx/dt$. Thus, $v = 10/(5 + 2t)$ so that

② $$\int_0^x dx = \int_0^{2.5} \frac{10}{5 + 2t}\, dt \qquad x = \frac{10}{2}\ln(5 + 2t)\Big]_0^{2.5} = 3.47 \text{ m} \qquad \textit{Ans.}$$

① Be careful to observe the minus sign for R.

② *Suggestion:* Express the distance x after release in terms of the velocity v and see if you agree with the resulting relation $x = 5\ln(v_0/v)$.

Sample Problem 3/5

The collar of mass m slides up the vertical shaft under the action of a force F of constant magnitude but variable direction. If $\theta = kt$ where k is a constant and if the collar starts from rest with $\theta = 0$, determine the magnitude F of the force that will result in the collar coming to rest as θ reaches $\pi/2$. The coefficient of kinetic friction between the collar and shaft is μ_k.

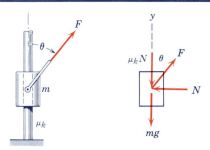

Solution. After drawing the free-body diagram, we apply the equation of motion in the y-direction to get

① $[\Sigma F_y = ma_y]$ $\qquad F\cos\theta = \mu_k N - mg = m\dfrac{dv}{dt}$

where equilibrium in the horizontal direction requires $N = F\sin\theta$. Substituting $\theta = kt$ and integrating first between general limits give

$$\int_0^t (F\cos kt - \mu_k F \sin kt - mg)\, dt = m\int_0^v dv$$

which becomes

$$\frac{F}{k}[\sin kt + \mu_k(\cos kt - 1)] - mgt = mv$$

For $\theta = \pi/2$ the time becomes $t = \pi/2k$ and $v = 0$ so that

② $$\frac{F}{k}[1 + \mu_k(0 - 1)] - \frac{mg\pi}{2k} = 0 \quad \text{and} \quad F = \frac{mg\pi}{2(1 - \mu_k)} \qquad \textit{Ans.}$$

① If θ were expressed as a function of the vertical displacement y instead of the time t, the acceleration would become a function of the displacement and we would use $v\,dv = a\,dy$.

② We see that the results do not depend on k, the rate at which the force changes direction.

PROBLEMS

Introductory problems

3/1 During a brake test, the front-wheel-drive car is stopped from an initial speed of 60 mi/hr in 175 ft. If it is known that the front wheels produce 90% of the braking force, determine the braking force F_f at each front wheel and the braking force F_r at each rear wheel. Assume a constant deceleration for the 2600-lb car. *Ans.* $F_f = 804$ lb, $F_r = 89.3$ lb

Problem 3/1

3/2 The 100-lb crate is carefully placed with zero velocity on the incline. Describe what happens if (a) $\theta = 15°$ and (b) $\theta = 10°$.

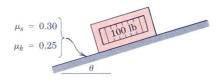

Problem 3/2

3/3 What fraction n of the weight of the jet airplane is the net thrust (nozzle thrust T minus air resistance R) required for the airplane to climb at an angle θ with the horizontal with an acceleration a in the direction of flight?
$$Ans. \quad n = \sin \theta + \frac{a}{g}$$

Problem 3/3

3/4 The 50-kg crate is projected along the floor with an initial speed of 7 m/s at $x = 0$. The coefficient of kinetic friction is 0.4. Calculate the time required for the crate to come to rest and the corresponding distance x traveled.

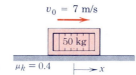

Problem 3/4

3/5 The 50-kg crate of Prob. 3/4 is now projected down an incline as shown with an initial speed of 7 m/s. Investigate the time t required for the crate to come to rest and the corresponding distance x traveled if (a) $\theta = 15°$ and (b) $\theta = 30°$.
Ans. (a) $t = 5.59$ s, $x = 19.58$ m
(b) Crate does not come to rest

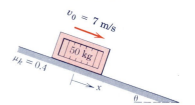

Problem 3/5

3/6 The 300-Mg jet airliner has four engines, each of which produces a nearly constant thrust of 180 kN during the takeoff roll. Determine the length s of runway required if the takeoff speed is 200 km/h. Compute s first for an uphill takeoff direction from A to B and second for a downhill takeoff from B to A on the slightly inclined runway. Neglect air and rolling resistance.

Problem 3/6

Problem 3/7

3/7 As part of the design process for a restraint chair, an engineer considers the following possible set of conditions: A 25-lb child is riding in the chair that, in turn, is securely fastened to the seat of an automobile. The automobile is involved in a head-on collision with another vehicle. The initial speed v_0 of the car is 30 mi/hr and this speed is reduced to zero during the collision of 0.2 sec duration. Assume a constant car deceleration during the collision and estimate the net horizontal force F that the straps of the restraint chair must exert on the child in order to keep her fixed to the chair. Treat the child as a particle and state any additional assumptions made during your analysis.   *Ans. F = 170.8 lb*

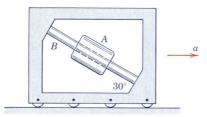

Problem 3/8

3/8 The collar A is free to slide along the smooth shaft B mounted in the frame. The plane of the frame is vertical. Determine the horizontal acceleration a of the frame necessary to maintain the collar in a fixed position on the shaft.

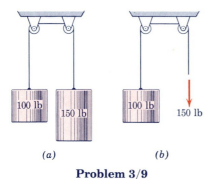

(a)   *(b)*

Problem 3/9

3/9 Calculate the vertical acceleration a of the 100-lb cylinder for each of the two cases illustrated. Neglect friction and the mass of the pulleys.
    Ans. (a) a = 6.44 ft/sec^2
    (b) a = 16.10 ft/sec^2

Representative problems

3/10 A car is descending the hill of slope θ_1 with the brakes slightly applied so that the speed v is constant. The slope decreases abruptly to θ_2 at point A. If the driver does not change the braking force, determine the acceleration a of the car after it passes point A. Evaluate your expression for θ_1 = 6° and θ_2 = 2°.

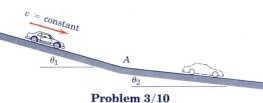

Problem 3/10

3/11 The coefficient of static friction between the flat bed of the truck and the crate it carries is 0.30. Determine the minimum stopping distance s that the truck can have from a speed of 70 km/h with constant deceleration if the crate is not to slip forward.

Ans. $s = 64.3$ m

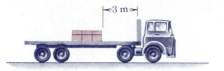

Problem 3/11

3/12 If the truck of Prob. 3/11 comes to a stop from an initial forward speed of 70 km/h in a distance of 50 m with uniform deceleration, determine whether or not the crate strikes the wall at the forward end of the flat bed. If the crate does strike the wall, calculate its speed relative to the truck as the impact occurs. Use the friction coefficients $\mu_s = 0.3$ and $\mu_k = 0.25$.

3/13 A train consists of a 400,000-lb locomotive and one hundred 200,000-lb hopper cars. If the locomotive exerts a friction force of 40,000 lb on the rails in starting the train from rest, compute the forces in couplers 1 and 100. Assume no slack in the couplers and neglect friction.

Ans. $T_1 = 39,200$ lb, $T_{100} = 392$ lb

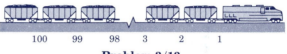

Problem 3/13

3/14 Small objects are delivered to the 72-in. inclined chute by a conveyor belt A that moves at a speed $v_1 = 1.2$ ft/sec. If the conveyor belt B has a speed $v_2 = 3.0$ ft/sec and the objects are delivered to this belt with no slipping, calculate the coefficient of friction μ_k between the objects and the chute.

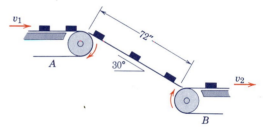

Problem 3/14

3/15 The block shown is observed to have a velocity $v_1 = 20$ ft/sec as it passes point A and a velocity $v_2 = 10$ ft/sec as it passes point B on the incline. Calculate the coefficient of kinetic friction μ_k between the block and the incline if $x = 30$ ft and $\theta = 15°$.

Ans. $\mu_k = 0.429$

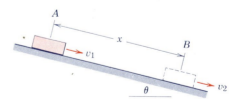

Problem 3/15

3/16 A cesium-ion engine for deep-space propulsion is designed to produce a constant thrust of 2.5 N for long periods of time. If the engine is to propel a 70-Mg spacecraft on an interplanetary mission, compute the time t required for a speed increase from 40 000 km/h to 65 000 km/h. Also find the distance s traveled during this interval. Assume that the spacecraft is moving in a remote region of space where the thrust from its ion engine is the only force acting on the spacecraft in the direction of its motion.

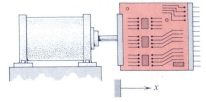

Problem 3/17

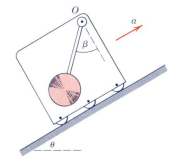

Problem 3/18

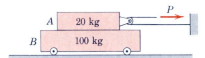

Problem 3/19

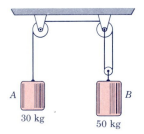

Problem 3/20

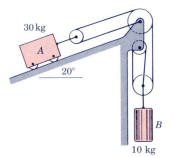

Problem 3/21

3/17 During a reliability test, a circuit board of mass m is attached to an electromagnetic shaker and subjected to a harmonic displacement $x = X \sin \omega t$, where X is the motion amplitude, ω is the motion frequency in radians per second, and t is time. Determine the magnitude F_{max} of the maximum horizontal force that the shaker exerts on the circuit board.   *Ans.* $F_{max} = mX\omega^2$

3/18 A simple pendulum is pivoted at O and is free to swing in the vertical plane of the plate. If the plate is given a constant acceleration a up the incline θ, write an expression for the steady angle β assumed by the pendulum after all initial start-up oscillations have ceased. Neglect the mass of the slender supporting rod.

3/19 If the coefficients of static and kinetic friction between the 20-kg block A and the 100-kg cart B are both essentially the same value of 0.50, determine the acceleration of each part for (a) $P = 60$ N and (b) $P = 40$ N.
Ans. (a) $a_A = 1.095$ m/s^2, $a_B = 0.981$ m/s^2
(b) $a_A = a_B = 0.667$ m/s^2

3/20 Determine the acceleration of each cylinder and the tension in the upper cable when the system is released from rest. Neglect friction and the mass of the pulleys.

3/21 The system of Prob. 2/209 is repeated here with additional mass information specified. Neglect all friction and the mass of the pulleys and determine the accelerations of bodies A and B upon release from rest.   *Ans.* $a_A = 1.024$ m/s^2 down the incline
$a_B = 0.682$ m/s^2 up

3/22 The system is released from rest with the cable taut. Neglect the small mass and friction of the pulley and calculate the acceleration of each body and the cable tension T upon release if (*a*) $\mu_s = 0.25$, $\mu_k = 0.20$ and (*b*) $\mu_s = 0.15$, $\mu_k = 0.10$.

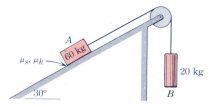

Problem 3/22

3/23 The pulley arrangement of Prob. 3/22 is modified as shown in the figure. For the friction coefficients $\mu_s = 0.25$ and $\mu_k = 0.20$, calculate the acceleration of each body and the tension T in the cable.

$$\text{Ans. } a_A = 1.450 \text{ m/s}^2 \text{ down}$$
$$a_B = 0.725 \text{ m/s}^2 \text{ up}$$
$$T = 105.4 \text{ N}$$

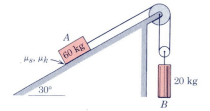

Problem 3/23

3/24 The device shown is used as an accelerometer and consists of a 4-oz plunger A that deflects the spring as the housing of the unit is given an upward acceleration a. Specify the necessary spring stiffness k that will permit the plunger to deflect 1/4 in. beyond the equilibrium position and touch the electrical contact when the steadily but slowly increasing upward acceleration reaches $5g$. Friction may be neglected.

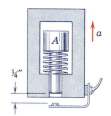

Problem 3/24

3/25 A cylinder of mass m rests in a supporting carriage as shown. If $\beta = 45°$ and $\theta = 30°$, calculate the maximum acceleration a that the carriage may be given up the incline so that the cylinder does not lose contact at B *Ans.* $a = 0.366g$

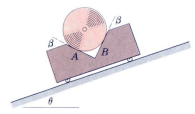

Problem 3/25

3/26 The acceleration of the 50-kg carriage A in its smooth vertical guides is controlled by the tension T exerted on the control cable that passes around the two circular pegs fixed to the carriage. Determine the value of T required to limit the downward acceleration of the carriage to 1.2 m/s^2 if the coefficient of friction between the cable and the pegs is 0.20. (Recall the relation between the tensions in a flexible cable that is slipping on a fixed peg: $T_2 = T_1 e^{\mu \beta}$.)

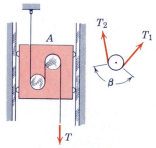

Problem 3/26

Problem 3/27

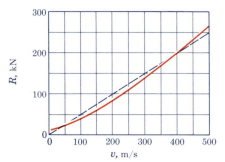

Problem 3/28

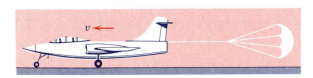

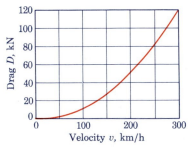

Problem 3/29

3/27 A player pitches a baseball horizontally toward a speed-sensing radar gun. The baseball weighs $5\frac{1}{8}$ oz and has a circumference of $9\frac{1}{8}$ in. If the speed at $x = 0$ is $v_0 = 90$ mi/hr, estimate the speed as a function of x. Assume that the horizontal aerodynamic drag on the baseball is given by $D = C_D(\frac{1}{2}\rho v^2)S$, where C_D is the drag coefficient, ρ is the air density, v is the speed, and S is the cross-sectional area of the baseball. Use a value of 0.3 for C_D. Neglect the y-component of the motion but comment on the validity of this assumption. Evaluate your answer for $x = 60$ ft, which is the approximate distance between a pitcher's hand and home plate.

$$Ans. \quad v = v_0 e^{(-1/2C_D\rho Sx/m)} = v_0 e^{-1.623(10^{-3})x}$$

$$v_{60} = 81.7 \text{ mi/hr}$$

3/28 The total resistance R to horizontal motion of a rocket test sled is shown by the solid line in the graph. Approximate R by the dotted line and determine the distance x that the sled must travel along the track from rest to reach a velocity of 400 m/s. The forward thrust T of the rocket motors is essentially constant at 300 kN, and the total mass of the sled is also nearly constant at 2 Mg.

3/29 A jet airplane with a mass of 5 Mg has a touchdown speed of 300 km/h, at which instant the braking parachute is deployed and the power shut off. If the total drag on the aircraft varies with velocity as shown in the accompanying graph, calculate the distance x along the runway required to reduce the speed to 150 km/h. Approximate the variation of the drag by an equation of the form $D = kv^2$, where k is a constant. $Ans. \quad x = 201$ m

3/30 In a test to determine the crushing characteristics of polystyrene packing material, a steel cone of mass m is dropped so that it falls a distance h and then penetrates the material. The resistance R of polystyrene to penetration depends on the cross-sectional area of the penetrating object and thus is proportional to the square of the cone penetration distance x, or $R = kx^2$. If the cone comes to rest at a distance $x = d$, determine the constant k in terms of the test conditions and results.

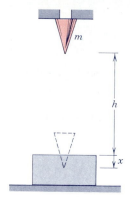

Problem 3/30

3/31 The 4-lb collar is released from rest against the light elastic spring, which has a stiffness of 10 lb/in. and has been compressed a distance of 6 in. Determine the acceleration a of the collar as a function of the vertical displacement x of the collar measured in feet from the point of release. Find the velocity v of the collar when $x = 0.5$ ft. Friction is negligible.

       Ans. $a_x = 32.2(14 - 30x)$, $v = 14.47$ ft/sec

Problem 3/31

3/32 A heavy chain with a mass ρ per unit length is pulled along a horizontal surface consisting of a smooth section and a rough section by the constant force P. If the chain is initially at rest on the smooth surface with $x = 0$ and if the coefficient of kinetic friction between the chain and the rough surface is μ_k, determine the velocity v of the chain when $x = L$. Assume that the chain remains taut and thus moves as a unit throughout the motion. What is the minimum value of P that will permit the chain to remain taut? (*Hint:* The acceleration must not become negative.)

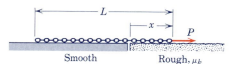

Problem 3/32

3/33 A chain of length $2l$ with a mass ρ per unit length is hanging in the equilibrium position shown. If end B is given a slight downward displacement, the imbalance causes an acceleration. Determine the acceleration of the chain in terms of the upward displacement x of end A and find the velocity v of end A as it reaches the top. Neglect the mass and diameter of the pulley.

       Ans. $a = \dfrac{x}{l} g$, $v = \sqrt{gl}$

Problem 3/33

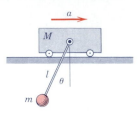

Problem 3/34

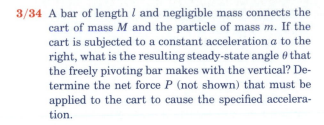

3/34 A bar of length l and negligible mass connects the cart of mass M and the particle of mass m. If the cart is subjected to a constant acceleration a to the right, what is the resulting steady-state angle θ that the freely pivoting bar makes with the vertical? Determine the net force P (not shown) that must be applied to the cart to cause the specified acceleration.

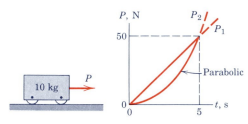

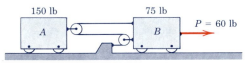

Problem 3/35

3/35 A force P is applied to the initially stationary cart. Determine the velocity and displacement at time $t = 5$ s for each of the force histories P_1 and P_2. Neglect friction.

$$\textit{Ans.}\quad \text{For } P_1: v = 12.5 \text{ m/s}, \ s = 20.8 \text{ m}$$
$$\text{For } P_2: v = 8.33 \text{ m/s}, \ s = 10.42 \text{ m}$$

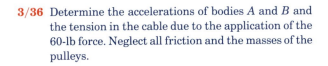

Problem 3/36

3/36 Determine the accelerations of bodies A and B and the tension in the cable due to the application of the 60-lb force. Neglect all friction and the masses of the pulleys.

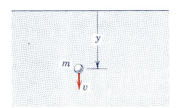

Problem 3/37

3/37 In a test of resistance to motion in an oil bath, a small steel ball of mass m is released from rest at the surface ($y = 0$). If the resistance to motion is given by $R = kv$ where k is a constant, derive an expression for the depth h required for the ball to reach a velocity v.

$$\textit{Ans.}\quad h = \frac{m^2 g}{k^2} \ln \left(\frac{1}{1 - kv/(mg)} \right) - \frac{mv}{k}$$

3/38 If the steel ball of Prob. 3/37 is released from rest at the surface of a liquid in which the resistance to motion is $R = cv^2$, where c is a constant and v is the downward velocity of the ball, determine the depth h required to reach a velocity v.

3/39 Two configurations for raising an elevator are shown. Elevator *A* with attached hoisting motor and drum has a total mass of 900 kg. Elevator *B* without motor and drum also has a mass of 900 kg. If the motor supplies a constant torque of 600 N·m to its 250-mm-diameter drum for 2 s in each case, select the configuration that results in the greater upward acceleration and determine the corresponding velocity v of the elevator 1.2 s after starting from rest. The mass of the motorized drum is small, thus permitting it to be analyzed as though it were in equilibrium. Neglect the mass of cables and pulleys and all friction.

> *Ans.* Case (*a*) has the higher acceleration
>
> $v = 7.43$ m/s

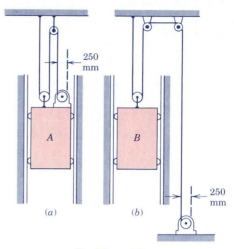

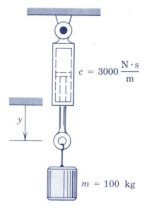

250 mm

250 mm

(*a*)   (*b*)

Problem 3/39

3/40 The spring of constant $k = 200$ N/m is attached to both the support and the 2-kg cylinder, which slides freely on the horizontal guide. If a constant 10-N force is applied to the cylinder at time $t = 0$ when the spring is undeformed and the system is at rest, determine the velocity of the cylinder when $x = 40$ mm and the maximum displacement of the cylinder.

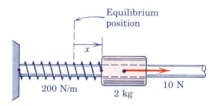

Equilibrium position

x

200 N/m   2 kg   10 N

Problem 3/40

3/41 A shock absorber is a mechanical device that provides resistance to compression or extension given by $R = cv$, where c is a constant and v is the time rate of change of the length of the absorber. An absorber of constant $c = 3000$ N·s/m is shown being tested with a 100-kg cylinder suspended from it. The system is released with the cable taut at $y = 0$ and allowed to extend. Determine (*a*) the steady-state velocity v_s of the lower end of the absorber and (*b*) the time t and displacement y of the lower end when the cylinder has reached 90 percent of its steady-state speed. Neglect the mass of the piston and attached rod.

> *Ans.* (*a*) $v_s = 0.327$ m/s
>
> (*b*) $t = 0.0768$ s, $y = 15.29$ mm

$c = 3000 \dfrac{\text{N} \cdot \text{s}}{\text{m}}$

y

$m = 100$ kg

Problem 3/41

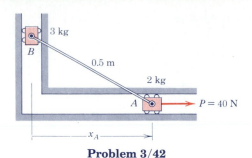

Problem 3/42

3/42 The sliders A and B are connected by a light rigid bar of length $l = 0.5$ m and move with negligible friction in the horizontal slots shown. For the position when $x_A = 0.4$ m, the velocity of A is $v_A = 0.9$ m/s to the right. Determine the acceleration of each slider and the force in the bar at this instant.

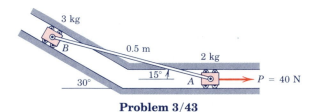

Problem 3/43

3/43 The sliders A and B are connected by a light rigid bar and move with negligible friction in the slots, both of which lie in a horizontal plane. For the position shown, the velocity of A is 0.4 m/s to the right. Determine the acceleration of each slider and the force in the bar at this instant.

Ans. $a_A = 7.95$ m/s^2 right, $a_B = 8.04$ m/s^2 down

$T = 25.0$ N

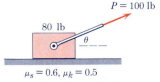

Problem 3/44

3/44 For what value(s) of the angle θ will the acceleration of the 80-lb block be 26 ft/sec^2 to the right?

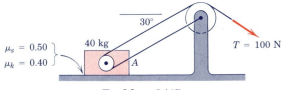

Problem 3/45

3/45 Compute the acceleration of block A for the instant depicted. Neglect the mass of the pulley.

Ans. $a_x = 1.406$ m/s^2

▶ **3/46** With the blocks initially at rest, the force P is increased slowly from zero to 60 lb. Plot the accelerations of both masses as functions of P.

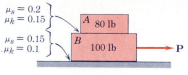

$\mu_s = 0.2$
$\mu_k = 0.15$

$\mu_s = 0.15$
$\mu_k = 0.1$

A　80 lb

B　100 lb　　→ **P**

Problem 3/46

▶ **3/47** The system is released from rest in the position shown. Calculate the tension T in the cord and the acceleration a of the 30-kg block. The small pulley attached to the block has negligible mass and friction. (*Suggestion:* First establish the kinematic relationship between the accelerations of the two bodies.)　　*Ans.* $T = 138.0$ N, $a = 0.766$ m/s^2

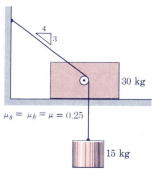

4 / 3

30 kg

$\mu_s = \mu_k = \mu = 0.25$

15 kg

Problem 3/47

3/48 Two iron spheres, each of which is 100 mm in diameter, are released from rest with a center-to-center separation of 1 m. Assume an environment in space with no forces other than the force of mutual gravitational attraction and calculate the time t required for the spheres to contact each other and the absolute speed v of each sphere upon contact.
　　　　　Ans. $t = 13$ h 33 min
　　　　　　　$v = 4.76(10^{-5})$ m/s

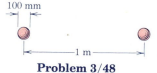

100 mm

⟵ 1 m ⟶

Problem 3/48

3/5 CURVILINEAR MOTION

We turn our attention now to the kinetics of particles that move along plane curvilinear paths. In applying Newton's second law, Eq. 3/3, we shall make use of the three coordinate descriptions of acceleration in curvilinear motion that we developed and applied in Arts. 2/4, 2/5, and 2/6.

The choice of coordinate system depends on the conditions of the problem and is one of the basic decisions to be made in solving curvilinear-motion problems. We now rewrite Eq. 3/3 in three ways, the choice of which depends on which coordinate system is most appropriate.

Rectangular coordinates (Art. 2/4, Fig. 2/7)

$$\Sigma F_x = ma_x$$
$$\Sigma F_y = ma_y$$

(3/6)

where $\qquad a_x = \ddot{x} \qquad$ and $\qquad a_y = \ddot{y}$

Normal and tangential coordinates (Art. 2/5, Fig. 2/10)

$$\Sigma F_n = ma_n$$
$$\Sigma F_t = ma_t$$

(3/7)

where

$$a_n = \rho\dot{\beta}^2 = v^2/\rho = v\dot{\beta}, \qquad a_t = \dot{v}, \qquad \text{and} \qquad v = \rho\dot{\beta}$$

Polar coordinates (Art. 2/6, Fig. 2/15)

$$\Sigma F_r = ma_r$$
$$\Sigma F_\theta = ma_\theta$$

(3/8)

where $\qquad a_r = \ddot{r} - r\dot{\theta}^2 \qquad$ and $\qquad a_\theta = r\ddot{\theta} + 2\dot{r}\dot{\theta}$

In the application of these motion equations, the general procedure established in the previous article on rectilinear motion should be followed. After the motion is identified and the coordinate system is designated, the free-body diagram of the body to be treated as a particle should be drawn. The appropriate force summations are then obtained from this diagram in the usual way. The free-body diagram should be complete so that no mistakes are made in figuring the force summations. Once an assignment of reference axes is made, the expressions for both the forces and the acceleration must be consistent with that assignment. In the first of Eqs. 3/7, for example, the positive sense of the n-coordinate axis is *toward* the center of curvature, and so the positive sense of our force summation ΣF_n must also be *toward* the center of curvature to agree with the positive sense of the acceleration $a_n = v^2/\rho$.

Sample Problem 3/6

Determine the maximum speed v that the sliding block may have as it passes point A without losing contact with the surface.

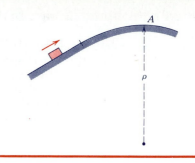

Solution. The condition for loss of contact is that the normal force N that the surface exerts on the block goes to zero. Summing forces in the normal direction gives

$$[\Sigma F_n = ma_n] \qquad mg = m\frac{v^2}{\rho} \qquad v = \sqrt{g\rho} \qquad Ans.$$

If the speed at A were less than $\sqrt{g\rho}$, then an upward normal force exerted by the surface on the block would exist. In order for the block to have a speed at A that is greater than $\sqrt{g\rho}$, some type of constraint, such as a second curved surface above the block, would have to be introduced to provide additional downward force.

Sample Problem 3/7

Small objects are released from rest at A and slide down the smooth circular surface of radius R to a conveyor B. Determine the expression for the normal contact force N between the guide and each object in terms of θ and specify the correct angular velocity ω of the conveyor pulley of radius r to prevent any sliding on the belt as the objects transfer to the conveyor.

Solution. The free-body diagram of the object is shown together with the coordinate directions n and t. The normal force N depends on the n-component of the acceleration that, in turn, depends on the velocity. The velocity will be cumulative according to the tangential acceleration a_t. Hence, we will find a_t first for any general position.

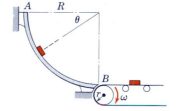

$$[\Sigma F_t = ma_t] \qquad mg\cos\theta = ma_t \qquad a_t = g\cos\theta$$

① Now we can find the velocity by integrating

$$[v\,dv = a_t\,ds] \qquad \int_0^v v\,dv = \int_0^\theta g\cos\theta\,d(R\theta) \qquad v^2 = 2gR\sin\theta$$

We obtain the normal force by summing forces in the positive n-direction, which is the direction of the n-component of acceleration.

$$[\Sigma F_n = ma_n] \qquad N - mg\sin\theta = m\frac{v^2}{R} \qquad N = 3mg\sin\theta \qquad Ans.$$

The conveyor pulley must turn at the rate $v = r\omega$ for $\theta = \pi/2$, so that

$$\omega = \sqrt{2gR}/r \qquad Ans.$$

① It is essential here that we recognize the need to express the tangential acceleration as a function of position so that v may be found by integrating the kinematical relation $v\,dv = a_t\,ds$, in which all quantities are measured along the path.

Sample Problem 3/8

A 1500-kg car enters a section of curved road in the horizontal plane and slows down at a uniform rate from a speed of 100 km/h at A to a speed of 50 km/h as it passes C. The radius of curvature ρ of the road at A is 400 m and at C is 80 m. Determine the total horizontal force exerted by the road on the tires at positions A, B, and C. Point B is the inflection point where the curvature changes direction.

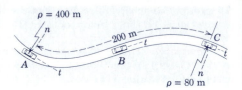

Solution. The car will be treated as a particle so that the effect of all forces exerted by the road on the tires will be treated as a single force. Since the motion is described along the direction of the road, normal and tangential coordinates will be used to specify the acceleration of the car. We will then determine the forces from the accelerations.

The constant tangential acceleration is in the negative t-direction and its magnitude is given by

① $[v_C{}^2 = v_A{}^2 + 2a_t\,\Delta s]$ $\quad a_t = \left| \dfrac{(50/3.6)^2 - (100/3.6)^2}{2(200)} \right| = 1.45 \text{ m/s}^2$

The normal components of acceleration at A, B, and C are

② $[a_n = v^2/\rho]$ At A, $a_n = \dfrac{(100/3.6)^2}{400} = 1.93 \text{ m/s}^2$

At B, $a_n = 0$

At C, $a_n = \dfrac{(50/3.6)^2}{80} = 2.41 \text{ m/s}^2$

Application of Newton's second law in both the n- and t-directions to the free-body diagrams of the car gives

$[\Sigma F_t = ma_t]$ $\qquad\qquad F_t = 1500(1.45) = 2170 \text{ N}$

③ $[\Sigma F_n = ma_n]$ At A, $F_n = 1500(1.93) = 2894 \text{ N}$

At B, $F_n = 0$

At C, $F_n = 1500(2.41) = 3617 \text{ N}$

Thus, the total horizontal force acting on the tires becomes

At A, $F = \sqrt{F_n{}^2 + F_t{}^2} = \sqrt{(2894)^2 + (2170)^2} = 3617 \text{ N}$ *Ans.*

At B, $F = F_t = 2170 \text{ N}$ *Ans.*

④ At C, $F = \sqrt{F_n{}^2 + F_t{}^2} = \sqrt{(3617)^2 + (2170)^2} = 4218 \text{ N}$ *Ans.*

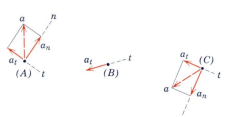

① Recognize the numerical value of the conversion factor from km/h to m/s as 1000/3600 or 1/3.6.

② Note that a_n is always directed toward the center of curvature.

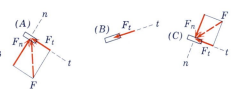

③ Note that the direction of F_n must agree with that of a_n.

④ The angle made by **a** and **F** with the direction of the path can be computed if desired.

Sample Problem 3/9

Compute the magnitude v of the velocity required for the spacecraft S to maintain a circular orbit of altitude 200 mi above the surface of the earth.

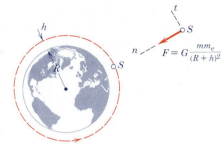

① Solution. The only external force acting on the spacecraft is the force of gravitational attraction to the earth (i.e., its weight), as shown in the free-body diagram. Summing forces in the normal direction yields

$$[\Sigma F_n = ma_n] \quad G\frac{mm_e}{(R+h)^2} = m\frac{v^2}{(R+h)}, \quad v = \sqrt{\frac{Gm_e}{(R+h)}} = R\sqrt{\frac{g}{(R+h)}}$$

where the substitution $gR^2 = Gm_e$ has been made. Substitution of numbers gives

$$v = (3959)(5280)\sqrt{\frac{32.234}{(3959+200)(5280)}} = 25{,}326 \text{ ft/sec} \quad Ans.$$

① Note that, for observations made within an inertial frame of reference, there is no such quantity as "centrifugal force" acting in the minus n-direction. Note also that neither the spacecraft nor its occupants are "weightless," because the weight in each case is given by Newton's law of gravitation. For this altitude, the weights are only about 10 percent less than the earth-surface values. Finally, the term "zero-g" is also misleading. It is only when we make our observations with respect to a coordinate system that has an acceleration equal to the gravitational acceleration (such as in an orbiting spacecraft) that we appear to be in a "zero-g" environment. The quantity that does go to zero aboard orbiting spacecraft is the familiar normal force associated with, for example, an object in contact with a horizontal surface within the spacecraft.

Sample Problem 3/10

Tube A rotates about the vertical O-axis with a constant angular rate $\dot{\theta} = \omega$ and contains a small cylindrical plug B of mass m whose radial position is controlled by the cord that passes freely through the tube and shaft and is wound around the drum of radius b. Determine the tension T in the cord and the horizontal component F_θ of force exerted by the tube on the plug if the constant angular rate of rotation of the drum is ω_0 first in the direction for case (a) and second in the direction for case (b). Neglect friction.

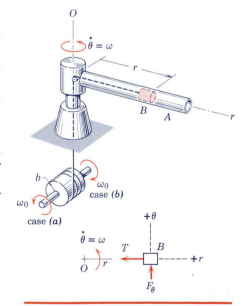

Solution. With r a variable, we use the polar-coordinate form of the equations of motion, Eqs. 3/8. The free-body diagram of B is shown in the horizontal plane and discloses only T and F_θ. The equations of motion are

$$[\Sigma F_r = ma_r] \qquad -T = m(\ddot{r} - r\dot{\theta}^2)$$
$$[\Sigma F_\theta = ma_\theta] \qquad F_\theta = m(r\ddot{\theta} + 2\dot{r}\dot{\theta})$$

Case (a). With $\dot{r} = +b\omega_0$, $\ddot{r} = 0$, and $\ddot{\theta} = 0$, the forces become

$$T = mr\omega^2 \qquad F_\theta = 2mb\omega_0\omega \qquad Ans.$$

① **Case (b).** With $\dot{r} = -b\omega_0$, $\ddot{r} = 0$, and $\ddot{\theta} = 0$, the forces become

$$T = mr\omega^2 \qquad F_\theta = -2mb\omega_0\omega \qquad Ans.$$

① The minus sign shows that F_θ is in the direction opposite to that shown on the free-body diagram.

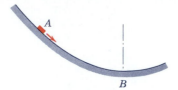

Problem 3/49

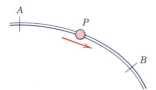

Problem 3/50

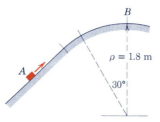

Problem 3/51

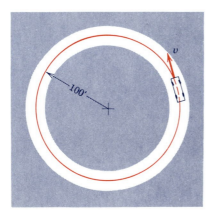

Problem 3/52

PROBLEMS

Introductory problems

3/49 The small 2-kg block A slides down the curved path and passes the lowest point B with a speed of 4 m/s. If the radius of curvature of the path at B is 1.5 m, determine the normal force N exerted on the block by the path at this point. Is knowledge of the friction properties necessary?   *Ans.* $N = 41.0$ N up, no

3/50 The 2-oz bead P is given an initial speed of 5 ft/sec at point A of the smooth guide, which is curved in the horizontal plane. If the horizontal force between the bead and the guide has a magnitude of 3 oz at point B, determine the radius of curvature ρ of the path at this point.

3/51 If the 2-kg block A passes over the top B of the circular portion of the path with a speed of 3 m/s, calculate the magnitude N of the normal force exerted by the path on the block at B. Determine the maximum speed v that the block can have at B without losing contact with the path.
    Ans. $N = 9.62$ N, $v = 4.20$ m/s

3/52 The standard test to determine the maximum lateral acceleration of a car is to drive it around a 200-ft-diameter circle painted on a level asphalt surface. The driver slowly increases the vehicle speed until he is no longer able to keep both wheel pairs straddling the line. If this maximum speed is 35 mi/hr for a 3000-lb car, determine its lateral acceleration capability a_n in g's and compute the magnitude F of the total friction force exerted by the pavement on the car tires.

3/53 The car of Prob. 3/52 is traveling at 25 mi/hr when the driver applies the brakes, and the car continues to move along the circular path. What is the maximum deceleration possible if the tires are limited to a total horizontal friction force of 2400 lb?
    Ans. $a_t = -22.0$ ft/sec^2

3/54 A 2-lb slider is propelled upward at A along the fixed curved bar that lies in a vertical plane. If the slider is observed to have a speed of 10 ft/sec as it passes position B, determine (*a*) the magnitude N of the force exerted by the fixed rod on the slider and (*b*) the rate at which the speed of the slider is decreasing. Assume that friction is negligible.

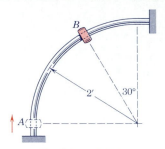

Problem 3/54

3/55 The small spheres are free to move on the inner surface of the rotating spherical chambers shown in section with radius R = 200 mm. If the spheres reach a steady-state angular position β = 45°, determine the angular velocity Ω of the device.

Ans. Ω = 3.64 rad/s

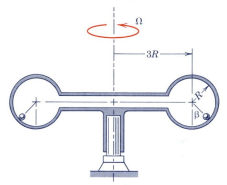

Problem 3/55

3/56 A child twirls a small 50-g ball attached to the end of a 1-m string so that the ball traces a circle in a vertical plane as shown. What is the minimum speed v that the ball must have when in position 1? If this speed is maintained throughout the circle, calculate the tension T in the string when the ball is in position 2. Neglect any small motion of the child's hand.

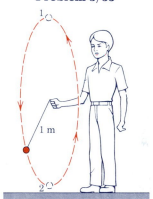

Problem 3/56

3/57 A pilot flies an airplane at a constant speed of 600 km/h in the vertical circle of radius 1000 m. Calculate the force exerted by the seat on the 90-kg pilot at point A and at point B.

Ans. N_A = 3380 N, N_B = 1617 N

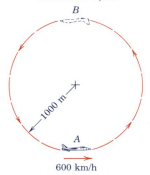

Problem 3/57

Problem 3/58

3/58 A jet transport plane flies in the trajectory shown in order to allow astronauts to experience the "weightless" condition similar to that aboard orbiting spacecraft. If the speed at the highest point is 600 mi/hr, what is the radius of curvature ρ necessary to exactly simulate the orbital "free-fall" environment?

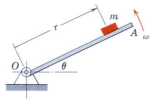

Problem 3/59

3/59 The member OA rotates about a horizontal axis through O with a constant counterclockwise velocity $\omega = 3$ rad/sec. As it passes the position $\theta = 0$, a small block of mass m is placed on it at a radial distance $r = 18$ in. If the block is observed to slip at $\theta = 50°$, determine the coefficient of static friction μ_s between the block and the member.

Ans. $\mu_s = 0.540$

Representative problems

3/60 Determine the altitude h (in kilometers) above the surface of the earth at which a satellite in a circular orbit has the same period, 23.9344 h, as the earth's absolute rotation. If such an orbit lies in the equatorial plane of the earth, it is said to be geosynchronous, because the satellite does not appear to move relative to an earth-fixed observer.

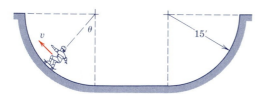

Problem 3/61

3/61 As the skateboarder negotiates the surface shown, his mass-center speeds at $\theta = 0$, 45°, and 90° are 28 ft/sec, 20 ft/sec, and 0, respectively. Determine the normal force between the surface and the skateboard wheels if the combined weight of the person and the skateboard is 150 lb and his center of mass is 30 in. from the surface.

Ans. $N_0 = 442$ lb, $N_{45°} = 255$ lb, $N_{90°} = 0$

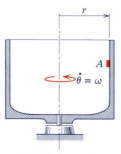

Problem 3/62

3/62 A small object A is held against the vertical side of the rotating cylindrical container of radius r due to centrifugal action. If the coefficient of static friction between the object and the container is μ_s, determine the expression for the minimum rotational rate $\dot{\theta} = \omega$ of the container that will keep the object from slipping down the vertical side.

3/63 The 30-Mg aircraft is climbing at the angle $\theta = 15°$ under a jet thrust T of 180 kN. At the instant shown, its speed is 300 km/h and is increasing at the rate of 1.96 m/s². Also θ is decreasing as the aircraft begins to level off. If the radius of curvature of the path at this instant is 20 km, compute the lift L and drag D. (Lift L and drag D are the aerodynamic forces normal to and opposite to the flight direction, respectively.)   *Ans. $D = 45.0$ kN, $L = 274$ kN*

Problem 3/63

3/64 At what angle θ should the racetrack turn of 1500-ft radius be banked in order that a racecar traveling at 120 mi/hr will have no tendency to slip sideways when rounding the turn?

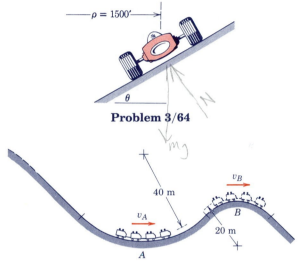

Problem 3/64

3/65 The cars of an amusement park ride have a speed $v_A = 22$ m/s at A and a speed $v_B = 12$ m/s at B. If a 75-kg rider sits on a spring scale (which registers the normal force exerted on it), determine the scale readings as the car passes points A and B. Assume that the person's arms and legs do not support appreciable force.
    Ans. $N_A = 1643$ N, $N_B = 195.8$ N

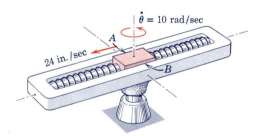

Problem 3/65

3/66 The slotted arm rotates about its center in a horizontal plane at the constant angular rate $\dot\theta = 10$ rad/sec and carries a 3.22-lb spring-mounted slider that oscillates freely in the slot. If the slider has a speed of 24 in./sec relative to the slot as it crosses the center, calculate the horizontal side thrust P exerted by the slotted arm on the slider at this instant. Determine which side, A or B, of the slot is in contact with the slider.

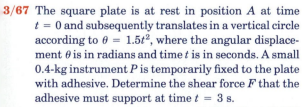

Problem 3/66

3/67 The square plate is at rest in position A at time $t = 0$ and subsequently translates in a vertical circle according to $\theta = 1.5t^2$, where the angular displacement θ is in radians and time t is in seconds. A small 0.4-kg instrument P is temporarily fixed to the plate with adhesive. Determine the shear force F that the adhesive must support at time $t = 3$ s.
    Ans. $F = 45.6$ N

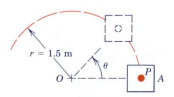

Problem 3/67

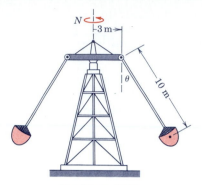

Problem 3/68

3/68 Calculate the necessary rotational speed N for the aerial ride in an amusement park in order that the arms of the gondolas will assume an angle $\theta = 60°$ with the vertical. Neglect the mass of the arms to which the gondolas are attached and treat each gondola as a particle.

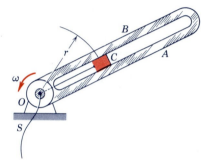

Problem 3/69

3/69 The slotted arm revolves in the horizontal plane about the fixed vertical axis through point O. The 3-lb slider C is drawn toward O at the constant rate of 2 in./sec by pulling the cord S. At the instant for which $r = 9$ in., the arm has a counterclockwise angular velocity $\omega = 6$ rad/sec and is slowing down at the rate of 2 rad/sec². For this instant, determine the tension T in the cord and the magnitude N of the force exerted on the slider by the sides of the smooth radial slot. Indicate which side, A or B, of the slot contacts the slider.

Ans. $T = 2.52$ lb, $N = 0.326$ lb, side B

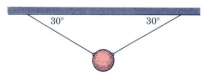

Problem 3/70

3/70 The small sphere of mass m is suspended initially at rest by the two wires. If one wire is suddenly cut, determine the ratio k of the tension in the remaining wire immediately after the other wire is cut to the initial equilibrium tension.

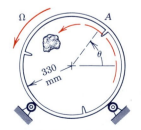

Problem 3/71

3/71 The rotating drum of a clothes dryer is shown in the figure. Determine the angular velocity Ω of the drum that results in loss of contact between the clothes and the drum at $\theta = 50°$. Assume that the small vanes prevent slipping until loss of contact.

Ans. $\Omega = 4.77$ rad/s

3/72 The amusement-park ride pivots about the fixed point O. A mechanism (not shown) drives the unit according to $\theta = (\pi/3) \sin 0.950t$, where θ is in radians and t is in seconds. Determine the maximum normal force N exerted by the seat on a rider of mass m, and state which riders are subjected to the maximum force.

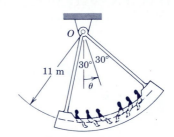

Problem 3/72

3/73 The small object is placed on the inner surface of the conical dish at the radius shown. If the coefficient of static friction between the object and the conical surface is 0.30, for what range of angular velocities ω about the vertical axis will the block remain on the dish without slipping? Assume that speed changes are made slowly so that any angular acceleration may be neglected.

Ans. $3.41 < \omega < 7.21$ rad/s

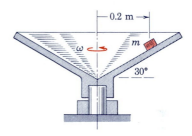

Problem 3/73

3/74 The small object of mass m is placed on the rotating conical surface at the radius shown. If the coefficient of static friction between the object and the rotating surface is 0.8, calculate the maximum angular velocity ω of the cone about the vertical axis for which the object will not slip. Assume very gradual angular velocity changes.

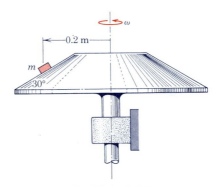

Problem 3/74

3/75 A 3220-lb car enters an S-curve at A with a speed of 60 mi/hr with brakes applied to reduce the speed to 45 mi/hr at a uniform rate in a distance of 300 ft measured along the curve from A to B. The radius of curvature of the path of the car at B is 600 ft. Calculate the total friction force exerted by the road on the tires at B. The road at B lies in a horizontal plane.   Ans. $F = 920$ lb

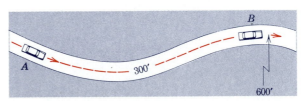

Problem 3/75

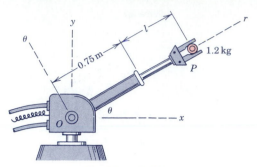

Problem 3/76

3/76 The robot arm of Prob. 2/146 is repeated here, with additional information supplied. At a given instant, $\theta = 30°$, $\dot{\theta} = 40$ deg/s, $\ddot{\theta} = 120$ deg/s², $l = 0.5$ m, $\dot{l} = 0.4$ m/s, and $\ddot{l} = -0.3$ m/s². Compute the radial and transverse forces F_r and F_θ that the arm must exert on the gripped part P, which has a mass of 1.2 kg. Compare to the case of static equilibrium in the same position.

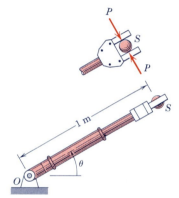

Problem 3/77

3/77 A 2-kg sphere S is being moved in a vertical plane by a robotic arm. When the arm angle θ is 30°, its angular velocity about a horizontal axis through O is 50 deg/s clockwise and its angular acceleration is 200 deg/s² counterclockwise. In addition, the hydraulic element is being shortened at the constant rate of 500 mm/s. Determine the necessary minimum gripping force P if the coefficient of static friction between the sphere and the gripping surfaces is 0.5. Compare P to the minimum gripping force P_s required to hold the sphere in static equilibrium in the 30° position.

Ans. $P = 27.0$ N, $P_s = 19.62$ N

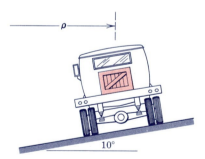

Problem 3/78

3/78 A flatbed truck going 100 km/h rounds a horizontal curve of 300-m radius inwardly banked at 10°. The coefficient of static friction between the truck bed and the 200-kg crate it carries is 0.70. Calculate the friction force F acting on the crate.

3/79 The flatbed truck of Prob. 3/78 starts from rest on a road whose constant radius of curvature is 30 m and whose bank angle is 10°. If the constant forward acceleration of the truck is 2 m/s², determine the time t after the start of motion at which the crate on the bed begins to slide. The coefficient of static friction between the crate and truck bed is $\mu_s = 0.3$, and the truck motion occurs in a horizontal plane.

Ans. $t = 5.58$ s

3/80 The spring-mounted 0.8-kg collar *A* oscillates along the horizontal rod that is rotating at the constant angular rate $\dot\theta$ = 6 rad/s. At a certain instant, *r* is increasing at the rate of 800 mm/s. If the coefficient of kinetic friction between the collar and the rod is 0.40, calculate the friction force *F* exerted by the rod on the collar at this instant.

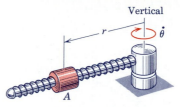
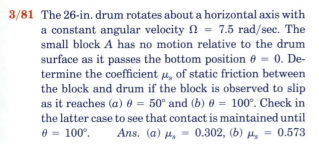

Problem 3/80

3/81 The 26-in. drum rotates about a horizontal axis with a constant angular velocity Ω = 7.5 rad/sec. The small block *A* has no motion relative to the drum surface as it passes the bottom position θ = 0. Determine the coefficient μ_s of static friction between the block and drum if the block is observed to slip as it reaches (*a*) θ = 50° and (*b*) θ = 100°. Check in the latter case to see that contact is maintained until θ = 100°.   *Ans.* (*a*) μ_s = 0.302, (*b*) μ_s = 0.573

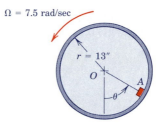

Problem 3/81

3/82 The 2-kg slider fits loosely in the smooth slot of the disk, which rotates about a vertical axis through point *O*. The slider is free to move slightly along the slot before one of the wires becomes taut. If the disk starts from rest at time *t* = 0 and has a constant clockwise angular acceleration of 0.5 rad/s², plot the tensions in wires 1 and 2 and the magnitude *N* of the force normal to the slot as functions of time *t* for the interval $0 \le t \le 5$ s.

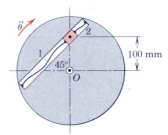

Problem 3/82

3/83 The rocket moves in a vertical plane and is being propelled by a thrust *T* of 32 kN. It is also subjected to an atmospheric resistance *R* of 9.6 kN. If the rocket has a velocity of 3 km/s and if the gravitational acceleration is 6 m/s² at the altitude of the rocket, calculate the radius of curvature ρ of its path for the position described and the time-rate-of-change of the magnitude *v* of the velocity of the rocket. The mass of the rocket at the instant considered is 2000 kg.

Ans. ρ = 3000 km, $\dot v$ = 6.00 m/s²

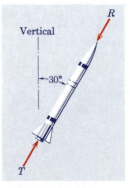

Problem 3/83

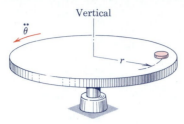

Vertical

$\ddot{\theta}$

r

Problem 3/84

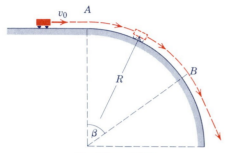

v_0 A

R

B

β

Problem 3/85

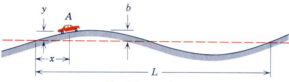

y

A b

x

L

Problem 3/86

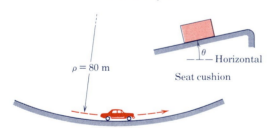

$\rho = 80$ m

θ — Horizontal

Seat cushion

Problem 3/87

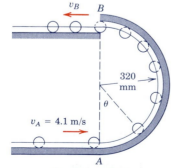

v_B B

320 mm

θ

$v_A = 4.1$ m/s

A

Problem 3/88

3/84 A small coin is placed on the horizontal surface of the rotating disk. If the disk starts from rest and is given a constant angular acceleration $\ddot{\theta} = \alpha$, determine an expression for the number of revolutions N through which the disk turns before the coin slips. The coefficient of static friction between the coin and the disk is μ_s.

3/85 A small vehicle enters the top A of the circular path with a horizontal velocity v_0 and gathers speed as it moves down the path. Determine an expression for the angle β to the position where the vehicle leaves the path and becomes a projectile. Evaluate your expression for $v_0 = 0$. Neglect friction and treat the vehicle as a particle.

$$Ans. \ \ \beta = \cos^{-1}\left(\frac{2}{3} + \frac{v_0{}^2}{3gR}\right), \ \beta = 48.2°$$

3/86 A stretch of highway includes a succession of evenly spaced dips and humps, the contour of which may be represented by the relation $y = b \sin (2\pi x/L)$. What is the maximum speed at which the car A can go over a hump and still maintain contact with the road? If the car maintains this critical speed, what is the total reaction N under its wheels at the bottom of a dip? The mass of the car is m.

3/87 The car has a speed of 70 km/h at the bottom of the dip when the driver applies the brakes, causing a deceleration of $0.5g$. What is the minimum seat cushion angle θ for which a package will not slide forward? The coefficient of static friction between the package and seat cushion is (a) 0.2 and (b) 0.4.

$$Ans. \ \ (a) \ \theta = 7.34°$$
$$(b) \ \theta = -3.16°$$

3/88 Small steel balls, each with a mass of 65 g, enter the semicircular trough in the vertical plane with a horizontal velocity of 4.1 m/s at A. Find the force R exerted by the trough on each ball in terms of θ and the velocity v_B of the balls at B. Friction is negligible.

3/89 A small rocket-propelled vehicle of mass m travels down the circular path of effective radius r under the action of its weight and a constant thrust T from its rocket motor. If the vehicle starts from rest at A, determine its speed v when it reaches B and the magnitude N of the force exerted by the guide on the wheels just prior to reaching B. Neglect any friction and any loss of mass of the rocket.

$$\text{Ans. } v = \sqrt{r\left(\frac{\pi T}{m} + 2g\right)}$$

$$N = T\pi + 3mg$$

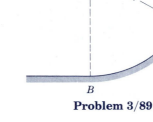

Problem 3/89

3/90 A satellite S, which weighs 322 lb when launched, is put into an elliptical orbit around the earth. At the instant represented, the satellite is a distance $r = 6667$ mi from the center of the earth and has a velocity $v = 16,610$ mi/hr making an angle $\beta = 60°$ with the radial line through S. The only force acting on the satellite is the gravitational attraction of the earth, which amounts to 114 lb for these conditions. Calculate the value of $\ddot{r}$ for the condition described.

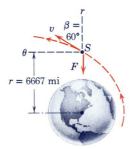

Problem 3/90

3/91 A right-handed baseball pitcher throws a curve ball initially aimed at the right edge of homeplate B. It curves so as to "break" 6 inches as shown. Assume that the horizontal speed is constant at $v = 120$ ft/sec, neglect vertical motion, and estimate (a) the average radius of curvature ρ of the baseball path and (b) the normal force R acting on the $5\frac{1}{8}$-oz baseball.   *Ans.* $\rho = 3600$ ft, $R = 0.0398$ lb

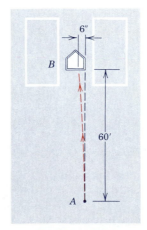

Problem 3/91

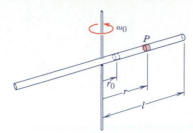

Problem 3/92

3/92 The particle P is released at time $t = 0$ from the position $r = r_0$ inside the smooth tube with no velocity relative to the tube, which is driven at the constant angular velocity ω_0 about a vertical axis. Determine the radial velocity v_r, the radial position r, and the transverse velocity v_θ as functions of time t. Explain why the radial velocity increases with time in the absence of radial forces. Plot the absolute path of the particle during the time it is inside the tube for $r_0 = 0.1$ m, $l = 1$ m, and $\omega_0 = 1$ rad/s.

3/93 Remove the assumption of smooth surfaces as stated in Prob. 3/92 and assume a coefficient of kinetic friction μ_k between the particle and rotating tube. Determine the radial position r of the particle as a function of time t if it is released with no relative velocity at $r = r_0$ when $t = 0$. Assume that static friction is overcome.
Ans.

$$r = \frac{r_0}{2\sqrt{\mu_k{}^2 + 1}}\left[(\mu_k + \sqrt{\mu_k{}^2 + 1})e^{\omega_0(-\mu_k + \sqrt{\mu_k{}^2 + 1})t}\right.$$
$$\left. + (-\mu_k + \sqrt{\mu_k{}^2 + 1})e^{\omega_0(-\mu_k - \sqrt{\mu_k{}^2 + 1})t}\right]$$

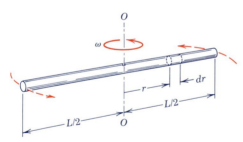

Problem 3/94

3/94 The uniform slender rod of length L, mass m, and cross-sectional area A is rotating in a horizontal plane about the vertical central axis O-O at a constant high angular velocity ω. By analyzing the horizontal forces on the accelerating differential element shown, derive an expression for the tensile stress σ in the rod as a function of r. The stress, commonly referred to as centrifugal stress, equals the tensile force divided by the cross-sectional area A.

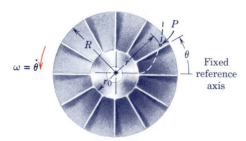

Problem 3/95

3/95 The centrifugal pump with smooth radial vanes rotates about its vertical axis with a constant angular velocity $\dot\theta = \omega$. Find the magnitude N of the force exerted by a vane on a particle of mass m as it moves out along the vane. The particle is introduced at $r = r_0$ without radial velocity. Assume that the particle contacts the side of the vane only.
Ans. $N = 2m\omega^2\sqrt{r^2 - r_0{}^2}$

▶ **3/96** Each tire on the 1350-kg car can support a maximum friction force parallel to the road surface of 2500 N. This force limit is nearly constant over all possible rectilinear and curvilinear car motions and is attainable only if the car does not skid. Under this maximum braking, determine the total stopping distance s if the brakes are first applied at point A when the car speed is 25 m/s and if the car follows the centerline of the road. *Ans.* $s = 47.4$ m

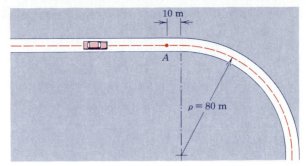

Problem 3/96

▶ **3/97** A hollow tube rotates about the horizontal axis through point O with constant angular velocity ω_0. A particle of mass m is introduced with zero relative velocity at $r = 0$ when $\theta = 0$ and slides outward through the smooth tube. Determine r as a function of θ.

$$Ans. \ r = \frac{g}{2\omega_0^2} (\sinh \theta - \sin \theta)$$

Problem 3/97

▶ **3/98** The small pendulum of mass m is suspended from a trolley that runs on a horizontal rail. The trolley and pendulum are initially at rest with $\theta = 0$. If the trolley is given a constant acceleration $a = g$, determine the maximum angle $\theta_{\max}$ through which the pendulum swings. Also find the tension T in the cord in terms of θ.

Ans. $\theta_{\max} = \pi/2$, $T = mg(3 \sin \theta + 3 \cos \theta - 2)$

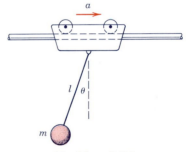

Problem 3/98

▶ **3/99** A small collar of mass m is given an initial velocity of magnitude v_0 on the horizontal circular track fabricated from a slender rod. If the coefficient of kinetic friction is μ_k, determine the distance traveled before the collar comes to rest. (*Hint:* Recognize that the friction force depends on the net normal force.)

$$Ans. \ s = \frac{r}{2\mu_k} \ln \left[\frac{v_0^2 + \sqrt{v_0^4 + r^2 g^2}}{rg} \right]$$

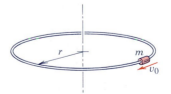

Problem 3/99

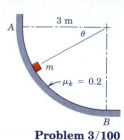

Problem 3/100

▶ **3/100** A small object is released from rest at A and slides with friction down the circular path. If the coefficient of friction is 0.2, determine the velocity of the object as it passes B. (*Hint:* Write the equations of motion in the n- and t-directions, eliminate N, and substitute $v \, dv = a_t r \, d\theta$. The resulting equation is a linear nonhomogeneous differential equation of the form $dy/dx + f(x)y = g(x)$, the solution of which is well known.)   *Ans. $v = 5.52$ m/s*

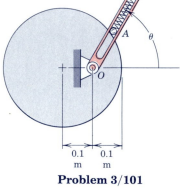

Problem 3/101

▶ **3/101** The slotted arm OB rotates in a horizontal plane about point O of the fixed circular cam with constant angular velocity $\dot{\theta} = 15$ rad/s. The spring has a stiffness of 5 kN/m and is uncompressed when $\theta = 0$. The smooth roller A has a mass of 0.5 kg. Determine the normal force N that the cam exerts on A and also the force R exerted on A by the sides of the slot when $\theta = 45°$. All surfaces are smooth. Neglect the small diameter of the roller.

Ans. $N = 81.6$ N, $R = 38.7$ N

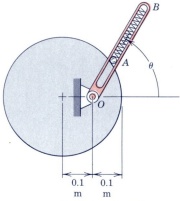

Problem 3/102

▶ **3/102** The small cart is nudged with negligible velocity from its horizontal position at A onto the parabolic path that lies in a vertical plane. Neglect friction and show that the cart maintains contact with the path for all values of k.

$$\text{Ans. } N = \frac{mg}{(1 + 4k^2x^2)^{3/2}} > 0$$

SECTION B. WORK AND ENERGY

3/6 *WORK AND KINETIC ENERGY*

In the previous two articles, we applied Newton's second law $\mathbf{F} = m\mathbf{a}$ to various problems of particle motion to establish the instantaneous relationship between the net force acting on a particle and the resulting acceleration of the particle. When intervals of motion were involved where the change in velocity or the corresponding displacement of the particle was required, we integrated the computed acceleration over the interval by using the appropriate kinematical equations.

There are two general classes of problems in which the cumulative effects of unbalanced forces acting on a particle over an interval of motion are of interest to us. These cases involve, respectively, integration of the forces with respect to the displacement of the particle and integration of the forces with respect to the time they are applied. We may incorporate the results of these integrations directly into the governing equations of motion so that it becomes unnecessary to solve directly for the acceleration. Integration with respect to displacement leads to the equations of work and energy, which are the subject of this article. Integration with respect to time leads to the equations of impulse and momentum, discussed in Section C.

(a) Work.   The quantitative meaning of the term work will now be developed.* Figure 3/2a shows a force $\mathbf{F}$ acting on a particle at A that moves along the path shown. The position vector $\mathbf{r}$ measured from some convenient origin O locates the particle as it passes point A, and $d\mathbf{r}$ is the differential displacement associated with an infinitesimal movement from A to A'. The work done by the force $\mathbf{F}$ during the displacement $d\mathbf{r}$ is defined as

$$dU = \mathbf{F} \cdot d\mathbf{r}$$

The magnitude of this dot product is $dU = F\,ds\cos\alpha$, where α is the angle between $\mathbf{F}$ and $d\mathbf{r}$ and where ds is the magnitude of $d\mathbf{r}$. This expression may be interpreted as the displacement multiplied by the force component $F_t = F\cos\alpha$ in the direction of the displacement, as represented by the dotted lines in Fig. 3/2b. Alternatively, the work dU may be interpreted as the force multiplied by the displacement component $ds\cos\alpha$ in the direction of the force, as represented by the full lines in Fig. 3/2b. With this definition of work, it should be noted that the component $F_n = F\sin\alpha$ normal to the

(a)

(b)

Figure 3/2

*The concept of work was also developed in the study of virtual work in Chapter 7 of *Vol. 1 Statics.*

displacement does no work. Hence, the work dU may be written as

$$dU = F_t \, ds$$

Work is positive if the working component F_t is in the direction of the displacement and negative if it is in the opposite direction. Forces that do work are termed *active forces*. Constraint forces that do no work are termed *reactive forces*.

In SI units work has the units of force (N) times displacement (m) or $N \cdot m$. This unit is given the special name *joule* (J), which is defined as the work done by a force of 1 N moving through a distance of 1 m in the direction of the force. Consistent use of the joule for work (and energy) rather than the units $N \cdot m$ will avoid possible ambiguity with the units of moment of a force or torque, which are also written $N \cdot m$.

In the U.S. customary system, work has the units of ft-lb. Dimensionally, work and moment are the same. In order to distinguish between the two quantities, it is recommended that work be expressed as foot pounds (ft-lb) and moment as pound feet (lb-ft). It should be noted that work is a scalar as given by the dot product and involves the product of a force and a distance, both measured along the same line. Moment, on the other hand, is a vector as given by the cross product and involves the product of force and distance measured at right angles to the force.

During a finite movement of the point of application of a force, the force does an amount of work equal to

$$U = \int \mathbf{F} \cdot d\mathbf{r} = \int (F_x \, dx + F_y \, dy + F_z \, dz)$$

or

$$U = \int F_t \, ds$$

In order to carry out this integration, it is necessary to know the relation between the force components and their respective coordinates or the relations between F_t and s. If the functional relationship is not known as a mathematical expression that can be integrated but is specified in the form of approximate or experimental data, then the work may be evaluated by carrying out a numerical or graphical integration that would be represented by the area under the curve of F_t versus s, as shown in Fig. 3/3.

A common example of the work done on a body by a variable force is found in the action of a spring on a movable body to which it is attached. We consider here the common linear spring of stiffness k, where the force F in the spring, tension or compression, is proportional to its deformation x, so that $F = kx$. Figure 3/4 shows the two cases where the body is moved by a force P so as to stretch

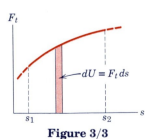

Figure 3/3

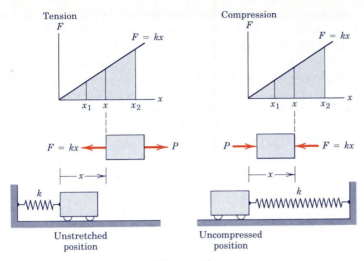

Figure 3/4

the spring a distance x or to compress the spring a distance x. The force exerted by the spring on the body in each case is in the sense *opposite* to the displacement so it does *negative* work on the body. Thus, for both stretching and compressing the spring, the work done *on* the body is *negative* and is given by

$$U_{1\text{-}2} = -\int_{x_1}^{x_2} F\,dx = -\int_{x_1}^{x_2} kx\,dx = -\tfrac{1}{2}k(x_2{}^2 - x_1{}^2)$$

When a spring under tension is being relaxed or when a spring under compression is being relaxed, then we see from both examples in Fig. 3/4 that the deformation changes from x_2 to a lesser deformation x_1. For this condition the force exerted on the body by the spring in both cases is in the *same* sense as the displacement, and, therefore, the work done *on* the body is *positive*.

The magnitude of the work, positive or negative, is seen to be equal to the shaded trapezoidal area shown for both cases in Fig. 3/4. In calculating the work done by a spring force, care must be taken to see that the units for k and x are consistent. Thus, if x is in meters (or feet), k must be in N/m (or lb/ft).

The expression $F = kx$ is actually a static relationship that is true only when elements of the spring have no acceleration. Dynamic behavior of a spring when its mass is accounted for is a fairly complex problem that will not be treated here. We shall assume that the mass of the spring is small compared with the masses of other accelerating parts of the system, in which case, the linear static relationship will not involve appreciable error.

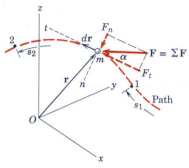

Figure 3/5

We now consider the work done on a particle of mass m, Fig. 3/5, moving along a curved path under the action of the force $\mathbf{F}$, which stands for the resultant $\Sigma\mathbf{F}$ of all forces acting on the particle. The position of m is established by the position vector $\mathbf{r}$, and its displacement along its path during time dt is represented by the change $d\mathbf{r}$ in its position vector. The work done by $\mathbf{F}$ during a finite movement of the particle from point 1 to point 2 is

$$U_{1\text{-}2} = \int_1^2 \mathbf{F}\cdot d\mathbf{r} = \int_{s_1}^{s_2} F_t \, ds$$

where the limits specify the initial and final end points of the interval of motion involved. When we substitute Newton's second law $\mathbf{F} = m\mathbf{a}$, the expression for the work of all forces becomes

$$U_{1\text{-}2} = \int_1^2 \mathbf{F}\cdot d\mathbf{r} = \int_1^2 m\mathbf{a}\cdot d\mathbf{r}$$

But $\mathbf{a}\cdot d\mathbf{r} = a_t \, ds$, where a_t is the tangential component of the acceleration of m. In terms of the velocity v of the particle, Eq. 2/3 gives $a_t \, ds = v \, dv$. Thus, the expression for the work of $\mathbf{F}$ becomes

$$U_{1\text{-}2} = \int_1^2 \mathbf{F}\cdot d\mathbf{r} = \int_{v_1}^{v_2} mv \, dv = \tfrac{1}{2}m(v_2{}^2 - v_1{}^2) \qquad (3/9)$$

where the integration is carried out between points 1 and 2 along the curve, at which points the velocities have the magnitudes v_1 and v_2, respectively.

(b) Kinetic energy.   The *kinetic energy* T of the particle is defined as

$$\boxed{T = \tfrac{1}{2}mv^2} \qquad \textbf{(3/10)}$$

and is the total work that must be done on the particle to bring it from a state of rest to a velocity v. Kinetic energy T is a scalar quantity with the units of $\text{N}\cdot\text{m}$ or *joules* (J) in SI units and ft-lb in U.S. customary units. Kinetic energy is *always* positive, regardless of the direction of the velocity. Equation 3/9 may be restated as

$$\boxed{U_{1\text{-}2} = T_2 - T_1 = \Delta T} \qquad \textbf{(3/11)}$$

which is the *work-energy equation* for a particle. The equation states that the *total work done* by all forces acting on a particle during an interval of its motion from condition 1 to condition 2 equals the corresponding *change in kinetic energy* of the particle. Although T is always positive, the change ΔT may be positive, negative, or zero.

When written in this concise form, Eq. 3/11 tells us that the work always results in a *change* of kinetic energy.

Alternatively, the work-energy relation may be expressed as the initial kinetic energy T_1 plus the work done $U_{1\text{-}2}$ equals the final kinetic energy T_2 or

$$\boxed{T_1 \ + \ U_{1\text{-}2} \ = \ T_2} \qquad\qquad \textbf{(3/11a)}$$

When written in this form, the terms correspond to the natural sequence of events. Clearly, the two forms 3/11 and 3/11a are equivalent.

We now see from Eq. 3/11 that a major advantage of the method of work and energy is that it avoids the necessity of computing the acceleration and leads directly to the velocity changes as functions of the forces that do work. Further, the work-energy equation involves only those forces that do work and thus give rise to changes in the magnitude of the velocities.

We consider now two particles joined together by a connection that is frictionless and incapable of any deformation. The forces in the connection constitute a pair of equal and opposite forces, and the points of application of these forces necessarily have identical displacement components in the direction of the forces. Hence, the net work done by these internal forces is zero during any movement of the system of the two connected particles. Thus, Eq. 3/11 is applicable to the entire system, where $U_{1\text{-}2}$ is the total or net work done on the system by forces external to it and ΔT is the change, $T_2 - T_1$, in the total kinetic energy of the system. The total kinetic energy is the sum of the kinetic energies of both elements of the system. It may now be observed that a further advantage of the work-energy method is that it permits the analysis of a system of particles joined in the manner described without dismembering the system.

Application of the work-energy method calls for an isolation of the particle or system under consideration. For a single particle a *free-body diagram* showing all externally applied forces should be drawn. For a system of particles rigidly connected without springs, an *active-force diagram* that shows only those external forces which do work (active forces) on the entire system may be drawn.*

(c) Power. The capacity of a machine is measured by the time rate at which it can do work or deliver energy. The total work or energy output is not a measure of this capacity since a motor, no matter how small, can deliver a large amount of energy if given

---

*The active-force diagram was introduced in the method of virtual work in statics. See Chapter 7 of *Vol. 1 Statics.*

sufficient time. On the other hand, a large and powerful machine is required to deliver a large amount of energy in a short period of time. Thus, the capacity of a machine is rated by its *power*, which is defined as the *time rate of doing work*.

Accordingly, the power P developed by a force $\mathbf{F}$ that does an amount of work U is $P = dU/dt = \mathbf{F} \cdot d\mathbf{r}/dt$. Since $d\mathbf{r}/dt$ is the velocity $\mathbf{v}$ of the point of application of the force, we have

$$\boxed{P = \mathbf{F} \cdot \mathbf{v}} \qquad (3/12)$$

Power is clearly a scalar quantity, and in SI units it has the units of $N \cdot m/s = J/s$. The special unit for power is the *watt* (W), which equals one joule per second (J/s). In U.S. customary units, the unit for mechanical power is the *horsepower* (hp). These units and their numerical equivalences are

$$1 \text{ W} = 1 \text{ J/s}$$

$$1 \text{ hp} = 550 \text{ ft-lb/sec} = 33,000 \text{ ft-lb/min}$$

$$1 \text{ hp} = 746 \text{ W} = 0.746 \text{ kW}$$

(d) Efficiency.   The ratio of the work done *by* a machine to the work done *on* the machine during the same time interval is called the *mechanical efficiency* e_m of the machine. This definition assumes that the machine operates uniformly so that there is no accumulation or depletion of energy within it. Efficiency is always less than unity since every device operates with some loss of energy and since energy cannot be created within the machine. In mechanical devices that involve moving parts, there will always be some loss of energy due to the negative work of kinetic friction forces. This work is converted to heat energy that, in turn, is dissipated to the surroundings. The mechanical efficiency at any instant of time may be expressed in terms of mechanical power P by

$$e_m = \frac{P_{\text{output}}}{P_{\text{input}}} \qquad (3/13)$$

In addition to energy loss by mechanical friction, there may also be electrical and thermal energy loss, in which case, the *electrical efficiency* e_e and *thermal efficiency* e_t are also involved. The *overall efficiency e* in such instances would be

$$e = e_m e_e e_t$$

Sample Problem 3/11

Calculate the velocity v of the 50-kg crate when it reaches the bottom of the chute at B if it is given an initial velocity of 4 m/s down the chute at A. The coefficient of kinetic friction is 0.30.

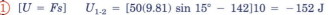

Solution. The free-body diagram of the crate is drawn and includes the normal force R and the kinetic friction force F calculated in the usual manner. The work done by the component of the weight down the plane is positive, whereas that done by the friction force is negative. The total work done on the crate during the motion is

① $[U = Fs]$ $U_{1\text{-}2} = [50(9.81) \sin 15° - 142]10 = -152$ J

The change in kinetic energy is $T_2 - T_1 = \Delta T$

$[T = \frac{1}{2}mv^2]$ $\Delta T = \frac{1}{2}(50)(v^2 - 4^2)$

The work-energy equation gives

$[U_{1\text{-}2} = \Delta T]$ $-152 = 25(v^2 - 16)$

$v^2 = 9.93$ (m/s)^2 $v = 3.15$ m/s   *Ans.*

① Since the net work done is negative, we obtain a decrease in the kinetic energy.

Sample Problem 3/12

The flatbed truck, which carries an 80-kg crate, starts from rest and attains a speed of 72 km/h in a distance of 75 m on a level road with constant acceleration. Calculate the work done by the friction force acting on the crate during this interval if the static and kinetic coefficients of friction between the crate and the truck bed are (*a*) 0.30 and 0.28, respectively, or (*b*) 0.25 and 0.20, respectively.

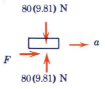

Solution. If the crate does not slip on the bed, its acceleration will be that of the truck, which is

$[v^2 = 2as]$ $a = \dfrac{v^2}{2s} = \dfrac{(72/3.6)^2}{2(75)} = 2.67$ m/s^2

Case (a). This acceleration requires a friction force on the block of

$[F = ma]$ $F = 80(2.67) = 213$ N

which is less than the maximum possible value of $\mu_s N = 0.30(80)(9.81) = 235$ N. Therefore, the crate does not slip and the work done by the actual static friction force of 213 N is

① $[U = Fs]$ $U_{1\text{-}2} = 213(75) = 16\ 000$ J   or   16.0 kJ

Case (b). For $\mu_s = 0.25$, the maximum possible friction force is $0.25(80)(9.81) = 196$ N, which is slightly less than the value of 213 N required for no slipping. Therefore, we conclude that the crate slips, and the friction force is governed by the kinetic coefficient and is $F = 0.20(80)(9.81) = 157$ N. The acceleration becomes

$[F = ma]$ $a = F/m = 157/80 = 1.96$ m/s^2

The distances traveled by the crate and the truck are in proportion to their accelerations. Thus, the crate has a displacement of $(1.96/2.67)75 = 55.2$ m, and the work done by kinetic friction is

② $[U = Fs]$ $U_{1\text{-}2} = 157(55.2) = 8660$ J   or   8.66 kJ   *Ans.*

① We note that static friction forces do no work when the contacting surfaces are both at rest. When they are in motion, however, as in this problem, the static friction force acting on the crate does positive work and that acting on the truck bed does negative work.

② This problem shows that a kinetic friction force can do positive work when the surface that supports the object and generates the friction force is in motion. If the supporting surface is at rest, then the kinetic friction force acting on the moving part always does negative work.

Sample Problem 3/13

The 50-kg block at A is mounted on rollers so that it moves along the fixed horizontal rail with negligible friction under the action of the constant 300-N force in the cable. The block is released from rest at A, with the spring to which it is attached extended an initial amount $x_1 = 0.233$ m. The spring has a stiffness $k = 80$ N/m. Calculate the velocity v of the block as it reaches position B.

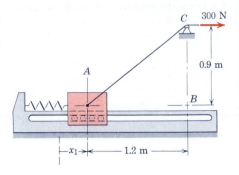

Solution. It will be assumed initially that the stiffness of the spring is small enough to allow the block to reach position B. The active-force diagram for the system composed of both block and cable is shown for a general position. The spring force $80x$ and the 300-N tension are the only forces external to this system that do work on the system. The force exerted on the block by the rail, the weight of the block, and the reaction of the small pulley on the cable do no work on the system and are not included on the active-force diagram.

As the block moves from $x = 0.233$ m to $x = 0.233 + 1.2 = 1.433$ m, the work done by the spring force acting on the block is negative and equals

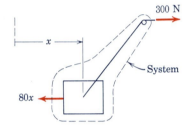

① $[U = \int F\,dx]$ $\quad U_{1\text{-}2} = -\int_{0.233}^{1.433} 80x\,dx = -40x^2\Big]_{0.233}^{1.433} = -80.0$ J

① If the variable x had been measured from the starting position A, the spring force would be $80(0.233 + x)$, and the limits of integration would be 0 and 1.2 m.

The work done on the system by the constant 300-N force in the cable is the force times the net horizontal movement of the cable over pulley C, which is $\sqrt{(1.2)^2 + (0.9)^2} - 0.9 = 0.6$ m. Thus, the work done is $300(0.6) = 180$ J. We now apply the work-energy equation to the system and get

$[U_{1\text{-}2} = \Delta T]$ $\quad -80.0 + 180 = \frac{1}{2}(50)(v^2 - 0)$ $\quad v = 2.0$ m/s $\quad$ *Ans.*

We take special note of the advantage to our choice of system. If the block alone had constituted the system, the horizontal component of the 300-N cable tension acting on the block would have to be integrated over the 1.2-m displacement. This step would require considerably more effort than was needed in the solution as presented. If there had been appreciable friction between the block and its guiding rail, we would have found it necessary to isolate the block alone in order to compute the variable normal force and, hence, the variable friction force. Integration of the friction force over the displacement would then be required to evaluate the negative work that it would do.

Sample Problem 3/14

The power winch A hoists the 800-lb log up the 30° incline at a constant speed of 4 ft/sec. If the power output of the winch is 6 hp, compute the coefficient of kinetic friction μ_k between the log and the incline. If the power is suddenly increased to 8 hp, what is the corresponding instantaneous acceleration a of the log?

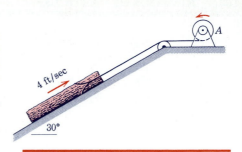

Solution. From the free-body diagram of the log, we get $N = 800 \cos 30° = 693$ lb, and the kinetic friction force becomes $693\mu_k$. For constant speed, the forces are in equilibrium so that

$$[\Sigma F_x = 0] \quad T - 693\mu_k - 800 \sin 30° = 0 \quad T = 693\mu_k + 400$$

The power output of the winch gives the tension in the cable

① $[P = Tv] \quad\quad T = P/v = 6(550)/4 = 825$ lb

Substituting T gives

$$825 = 693\mu_k + 400 \quad\quad \mu_k = 0.613 \quad\quad Ans.$$

When the power is increased, the tension momentarily becomes

$[P = Tv] \quad\quad T = P/v = 8(550)/4 = 1100$ lb

and the corresponding acceleration is given by

② $[\Sigma F_x = ma_x] \quad 1100 - 693(0.613) - 800 \sin 30° = \dfrac{800}{32.2} a$

$$a = 11.07 \text{ ft/sec}^2 \quad\quad Ans.$$

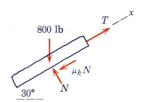

① Note the conversion from horsepower to ft-lb/sec.

② As the speed increases, the acceleration will drop until the speed stabilizes at a value higher than 4 ft/sec.

Sample Problem 3/15

A satellite of mass m is put into an elliptical orbit around the earth. At point A, its distance from the earth is $h_1 = 500$ km and it has a velocity $v_1 = 30\,000$ km/h. Determine the velocity v_2 of the satellite as it reaches point B, a distance $h_2 = 1200$ km from the earth.

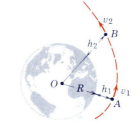

Solution. The satellite is moving outside of the earth's atmosphere so that the only force acting on it is the gravitational attraction of the earth. With the mass and radius of the earth expressed by m_e and R, respectively, the gravitational law of Eq. 1/2 gives $F = Gmm_e/r^2 = gR^2m/r^2$ when the substitution $Gm_e = gR^2$ is made for the surface values $F = mg$ and $r = R$. The work done by F is due only to the radial component of motion along the line of action of F and is negative for increasing r.

$$U_{1\text{-}2} = -\int_{r_1}^{r_2} F \, dr = -mgR^2 \int_{r_1}^{r_2} \frac{dr}{r^2} = mgR^2 \left(\frac{1}{r_2} - \frac{1}{r_1} \right)$$

The work-energy equation $U_{1\text{-}2} = \Delta T$ gives

① $mgR^2 \left(\dfrac{1}{r_2} - \dfrac{1}{r_1} \right) = \tfrac{1}{2}m(v_2{}^2 - v_1{}^2) \quad v_2{}^2 = v_1{}^2 + 2gR^2 \left(\dfrac{1}{r_2} - \dfrac{1}{r_1} \right)$

② Substituting the numerical values gives

$$v_2{}^2 = \left(\frac{30\,000}{3.6} \right)^2 + 2(9.81)[(6371)(10^3)]^2 \left(\frac{10^{-3}}{6371 + 1200} - \frac{10^{-3}}{6371 + 500} \right)$$

$$= 69.44(10^6) - 10.72(10^6) = 58.73(10^6) \text{ (m/s)}^2$$

$$v_2 = 7663 \text{ m/s} \quad\text{or}\quad v_2 = 7663(3.6) = 27\,590 \text{ km/h} \quad\quad Ans.$$

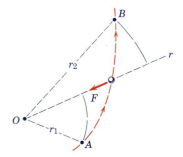

① Note that the result is independent of the mass of the satellite.

② Consult Table D/2, Appendix D, to find the radius R of the earth.

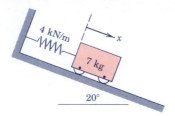

Problem 3/103

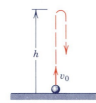

Problem 3/104

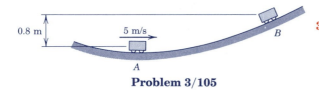

Problem 3/105

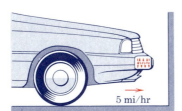

Problem 3/106

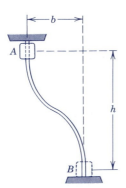

Problem 3/107

PROBLEMS

Introductory problems

3/103   The spring is unstretched when $x = 0$. If the body moves from the initial position $x_1 = 100$ mm to the final position $x_2 = 200$ mm, (*a*) determine the work done by the spring on the body and (*b*) determine the work done on the body by its weight.

Ans. (*a*) $U_{1\text{-}2} = -60$ J, (*b*) $U_{1\text{-}2} = 2.35$ J

3/104   Use the work-energy method to develop an expression for the maximum height attained by a projectile that is launched with initial speed v_0 from ground level. Evaluate your expression for $v_0 = 50$ m/s. Assume a constant gravitational acceleration and neglect air resistance.

3/105   The small body has a speed $v_A = 5$ m/s at point A. Neglecting friction, determine its speed v_B at point B after it has risen 0.8 m. Is knowledge of the shape of the track necessary?   *Ans.* $v_B = 3.05$ m/s, no

3/106   In the design of a spring bumper for a 3500-lb car, it is desired to bring the car to a stop from a speed of 5 mi/hr in a distance equal to 6 in. of spring deformation. Specify the required stiffness k for each of the two springs behind the bumper. The springs are undeformed at the start of impact.

3/107   The small collar of mass m is released from rest at A and slides down the curved rod in the vertical plane with negligible friction. Express the velocity v of the collar as it strikes the base at B in terms of the given conditions.   *Ans.* $v = \sqrt{2gh}$

3/108   For the sliding collar of Prob. 3/107, if $m = 0.5$ kg, $b = 0.8$ m, and $h = 1.5$ m, and if the velocity of the collar as it strikes the base B is 4.70 m/s after release of the collar from rest, calculate the work Q of friction. What happens to the energy that is lost?

3/109 The 1.5-lb collar slides with negligible friction on the fixed rod in the vertical plane. If the collar starts from rest at A under the action of the constant 2-lb horizontal force, calculate its velocity v as it hits the stop at B.   *Ans.* $v = 17.18$ ft/sec

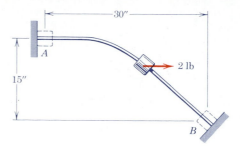

Problem 3/109

3/110 The 2-kg collar is released from rest at A and slides down the inclined fixed rod in the vertical plane. The coefficient of kinetic friction is 0.4. Calculate (*a*) the velocity v of the collar as it strikes the spring and (*b*) the maximum deflection x of the spring.

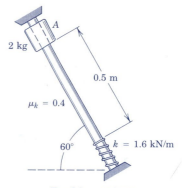

Problem 3/110

3/111 The car is moving with a speed $v_0 = 65$ mi/hr up the 6-percent grade and the driver applies the brakes at point A, causing all wheels to skid. The coefficient of kinetic friction for the rain-slicked road is $\mu_k = 0.6$. Determine the stopping distance s_{AB}. Repeat your calculations for the case when the car is moving downhill from B to A.

$Ans.$ $s_{AB} = 214$ ft, $s_{BA} = 262$ ft

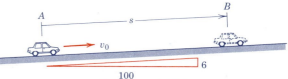

Problem 3/111

3/112 The 120-lb woman jogs up the flight of stairs in 5 seconds. Determine her average power output. Convert all given information to the SI system and repeat your calculation.

Representative problems

3/113 The position vector of a particle is given by $\mathbf{r} = 8t\mathbf{i} + 1.2t^2\mathbf{j} - 0.5(t^3 - 1)\mathbf{k}$, where t is the time in seconds from the start of the motion and where $\mathbf{r}$ is expressed in meters. For the condition when $t = 4$ s, determine the power P developed by the force $\mathbf{F} = 40\mathbf{i} - 20\mathbf{j} - 36\mathbf{k}$ N which acts on the particle.   *Ans.* $P = 0.992$ kW

Problem 3/112

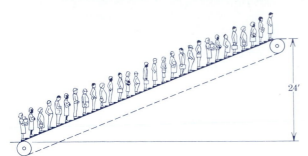

Problem 3/114

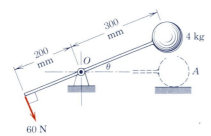

Problem 3/115

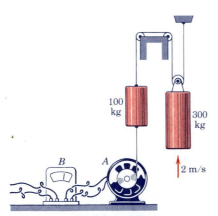

Problem 3/118

3/114 A department store escalator handles a steady load of 30 people per minute in elevating them from the first to the second floor through a vertical rise of 24 ft. The average person weighs 140 lb. If the motor that drives the unit delivers 4 hp, calculate the mechanical efficiency e of the system.

3/115 The 4-kg ball and the attached light rod rotate in the vertical plane about the fixed axis at O. If the assembly is released from rest at $\theta = 0$ and moves under the action of the 60-N force, which is maintained normal to the rod, determine the velocity v of the ball as θ approaches 90°. Treat the ball as a particle.   *Ans.* $v = 1.881$ m/s

3/116 A car with a mass of 1500 kg starts from rest at the bottom of a 10-percent grade and acquires a speed of 50 km/h in a distance of 100 m with constant acceleration up the grade. What is the power P delivered to the drive wheels by the engine when the car reaches this speed?

3/117 A 1200-kg car enters an 8-percent downhill grade at a speed of 100 km/h. The driver applies her brakes to bring the car to a speed of 25 km/h in a distance of 0.5 km measured along the road. Calculate the energy loss Q dissipated from the brakes in the form of heat. Neglect any friction losses from other causes such as air resistance.   *Ans.* $Q = 903$ kJ

3/118 The motor unit A is used to elevate the 300-kg cylinder at a constant rate of 2 m/s. If the power meter B registers an electrical input of 2.20 kW, calculate the combined electrical and mechanical efficiency e of the system.

3/119 The 15-lb cylindrical collar is released from rest in the position shown and drops onto the spring. Calculate the velocity v of the cylinder when the spring has been compressed 2 in.   *Ans.* $v = 7.08$ ft/sec

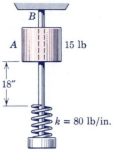

Problem 3/119

3/120 In the structural design of the upper floors of an industrial building, allowance must be made for the accidental dropping of heavy machinery through a small distance. For a machine of mass m dropped through a very small distance onto a floor that acts elastically, determine the maximum force F supported by the floor. (The problem is modeled by the mass m mounted on supports a negligible distance above a spring of stiffness k, with the action occurring when the supports are suddenly removed.)

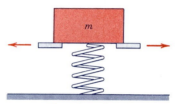

Problem 3/120

3/121 The pressure p in a certain rifle barrel varies with the position of the bullet as shown in the graph. If the diameter of the bore is 7.5 mm and the mass of the bullet is 14 g, determine the muzzle velocity v of the bullet. Neglect the effect of friction in the barrel compared with the force of the gases on the bullet. Observe that one megapascal (MPa) equals 10^6 newtons per square meter (N/m^2).

     Ans. $v = 940$ m/s

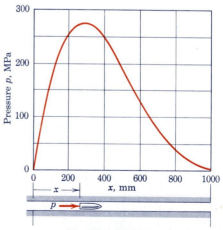

Problem 3/121

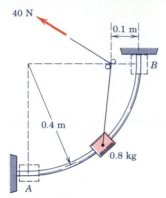

Problem 3/122

3/122 The 0.8-kg collar slides freely on the fixed circular rod. Calculate the velocity v of the collar as it hits the stop at B if it is elevated from rest at A by the action of the constant 40-N force in the cord. The cord is guided by the small fixed pulleys.

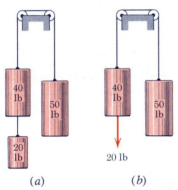

Problem 3/123

3/123 Each of the two systems is released from rest. Calculate the velocity v of each 50-lb cylinder after the 40-lb cylinder has dropped 6 ft. The 20-lb cylinder of case (a) is replaced by a 20-lb force in case (b).

Ans. (a) v = 5.93 ft/sec, (b) v = 6.55 ft/sec

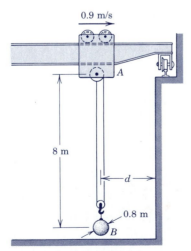

Problem 3/124

3/124 The carriage A of the overhead crane is traveling with a speed of 0.9 m/s when it is suddenly brought to a stop. (a) For the case when the distance d is large at the time of stopping, compute the maximum angle θ through which the supporting cables swing. (b) For the case when d is 0.75 m at the time of stopping, determine the speed v with which the heavy ball B strikes the wall. Neglect the masses of the cables, pulley, and hook, but take the 0.8-m diameter of ball B into account.

3/125 The ball is released from position *A* with a velocity of 3 m/s and swings in a vertical plane. At the bottom position, the cord strikes the fixed bar at *B*, and the ball continues to swing in the dotted arc. Calculate the velocity v_C of the ball as it passes position *C*.   *Ans.* $v_C = 3.59$ m/s

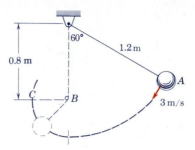

Problem 3/125

3/126 In a test to determine the crushing characteristics of a packing material, a steel cone of mass *m* is released, falls a distance *h*, and then penetrates the material. The radius of the cone depends on the square of the distance from its tip. The resistance *R* of the material to penetration depends on the cross-sectional area of the penetrating object and thus is proportional to the fourth power of the cone penetration distance *x*, or $R = kx^4$. If the cone comes to rest at a distance $x = d$, determine the constant *k* in terms of the test conditions and results. Utilize a single application of the work-energy equation.

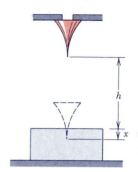

Problem 3/126

3/127 A constant horizontal force $P = 150$ lb is applied to the linkage as shown. With the 30-lb ball initially at rest on its support with $\theta = 60°$, calculate the velocity *v* of the ball as θ approaches zero where the ball reaches its highest position.
          Ans. $v = 12.68$ ft/sec

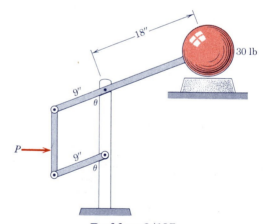

Problem 3/127

3/128 The 300-lb carriage has an initial velocity of 9 ft/sec down the incline at *A*, when a constant force of 110 lb is applied to the hoisting cable as shown. Calculate the velocity of the carriage when it reaches *B*. Show that in the absence of friction this velocity is independent of whether the initial velocity of the carriage at *A* was up or down the incline.

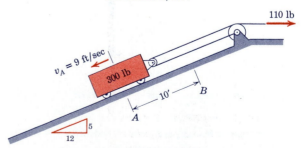

Problem 3/128

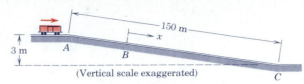

Problem 3/129

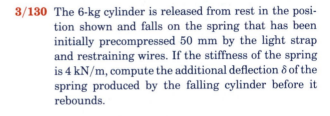

3/129 In a railroad classification yard, a 68-Mg freight car moving at 0.5 m/s at *A* encounters a retarder section of track at *B* that exerts a retarding force of 32 kN on the car in the direction opposite to motion. Over what distance *x* should the retarder be activated in order to limit the speed of the car to 3 m/s at *C*?

Ans. $x = 53.2$ m

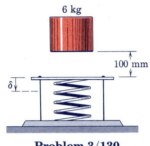

Problem 3/130

3/130 The 6-kg cylinder is released from rest in the position shown and falls on the spring that has been initially precompressed 50 mm by the light strap and restraining wires. If the stiffness of the spring is 4 kN/m, compute the additional deflection δ of the spring produced by the falling cylinder before it rebounds.

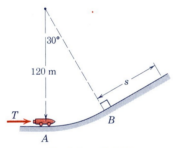

Problem 3/131

3/131 A small rocket-propelled test vehicle with a total mass of 100 kg starts from rest at *A* and moves with negligible friction along the track in the vertical plane as shown. If the propelling rocket exerts a constant thrust *T* of 1.5 kN from *A* to position *B* where it is shut off, determine the distance *s* that the vehicle rolls up the incline before stopping. The loss of mass due to the expulsion of gases by the rocket is small and may be neglected.   *Ans.* $s = 160.0$ m

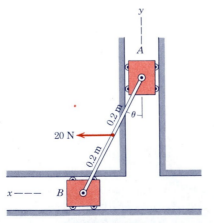

Problem 3/132

3/132 Each of the sliders *A* and *B* has a mass of 2 kg and moves with negligible friction in its respective guide, with *y* being in the vertical direction. A 20-N horizontal force is applied to the midpoint of the connecting link of negligible mass, and the assembly is released from rest with $\theta = 0$. Calculate the velocity v_A with which *A* strikes the horizontal guide when $\theta = 90°$.

3/133 The nest of two springs is used to bring the 0.5-kg plunger A to a stop from a speed of 5 m/s and reverse its direction of motion. The inner spring increases the deceleration, and the adjustment of its position is used to control the exact point at which the reversal takes place. If this point is to correspond to a maximum deflection δ = 200 mm for the outer spring, specify the adjustment of the inner spring by determining the distance s. The outer spring has a stiffness of 300 N/m and the inner one a stiffness of 150 N/m.   *Ans.* s = 142.3 mm

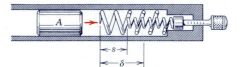

Problem 3/133

3/134 The 7-kg collar A slides with negligible friction on the fixed vertical shaft. When the collar is released from rest at the bottom position shown, it moves up the shaft under the action of the constant force F = 200 N applied to the cable. Calculate the stiffness k that the spring must have if its maximum compression is to be limited to 75 mm. The position of the small pulley at B is fixed.

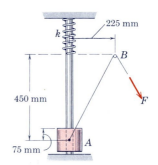

Problem 3/134

3/135 Extensive testing of an experimental 2000-lb automobile reveals the aerodynamic drag force F_D and the total nonaerodynamic rolling-resistance force F_R to be as shown in the plot. Determine (*a*) the power required for steady speeds of 30 and 60 mi/hr on a level road, (*b*) the power required for a steady speed of 60 mi/hr both up and down a 6-percent incline, and (*c*) the steady speed at which no power is required going down the 6-percent incline.

   Ans. (*a*) P_{30} = 5 hp, P_{60} = 16 hp
   (*b*) P_{up} = 35.2 hp, P_{down} = −3.17 hp
   (*c*) v = 70.9 mi/hr

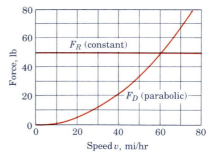

Problem 3/135

3/136 The vertical motion of the 50-lb block is controlled by the two forces P applied to the ends A and B of the linkage, where A and B are constrained to move in the horizontal guide. If forces P = 250 lb are applied with the linkage initially at rest with θ = 60°, determine the upward velocity v of the block as θ approaches 180°. Neglect friction and the weight of the links and note that P is greater than its equilibrium value of $(5W/2)$ cot 30° = 217 lb.

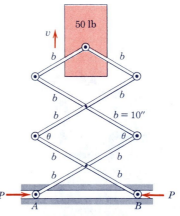

Problem 3/136

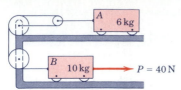

Problem 3/137

3/137 The force $P = 40$ N is applied to the system, which is initially at rest. Determine the speeds of A and B after A has moved 0.4 m.

Ans. $v_A = 1.180$ m/s, $v_B = 2.36$ m/s

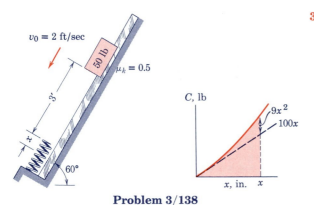

Problem 3/138

3/138 The 50-lb slider in the position shown has an initial velocity $v_0 = 2$ ft/sec on the inclined rail and slides under the influence of gravity and friction. The coefficient of kinetic friction between the slider and the rail is 0.5. Calculate the velocity of the slider as it passes the position for which the spring is compressed a distance $x = 4$ in. The spring offers a compressive resistance C and is known as a "hardening" spring, since its stiffness increases with deflection as shown in the accompanying graph.

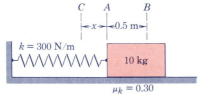

Problem 3/139

3/139 The 10-kg block is released from rest on the horizontal surface at point B, where the spring has been stretched a distance of 0.5 m from its neutral position A. The coefficient of kinetic friction between the block and the plane is 0.30. Calculate (a) the velocity v of the block as it passes point A and (b) the maximum distance x to the left of A that the block goes.

Ans. $v = 2.13$ m/s, $x = 0.304$ m

3/140 The three springs of equal moduli are unstretched when the cart is released from rest in the position $x = 0$. If $k = 120$ N/m and $m = 10$ kg, determine (a) the speed v of the cart when $x = 50$ mm, (b) the maximum displacement x_{max} of the cart, and (c) the steady-state displacement x_{ss} that would exist after all oscillations cease.

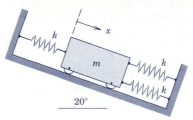

Problem 3/140

3/141 The car of mass m accelerates on a level road under the action of the driving force F from a speed v_1 to a higher speed v_2 in a distance s. If the engine develops a constant power output P, determine v_2. Treat the car as a particle under the action of the single horizontal force F.

Problem 3/141

$$\text{Ans. } v_2 = \left(\frac{3Ps}{m} + v_1{}^3\right)^{1/3}$$

3/142 The 10-kg cylinder is released from rest with $x = 1$ m, where the spring is unstretched. Determine (a) the maximum velocity v of the cylinder and the corresponding value of x and (b) the maximum value of x during the motion. The stiffness of the spring is 450 N/m.

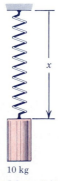

10 kg

Problem 3/142

3/7 POTENTIAL ENERGY

In the previous article on work and kinetic energy, a particle or a combination of joined particles was isolated, and the work done by gravity forces, spring forces, and other externally applied forces acting on the particle or system was determined to evaluate U in the work-energy equation. In the present article we shall treat the work done by gravity forces and by spring forces by introducing the concept of *potential energy*. This concept will simplify the analysis of many problems.

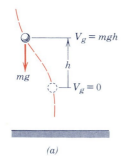

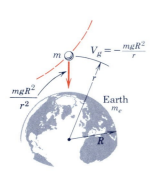

Figure 3/6

(a) Gravitational potential energy. We consider first the motion of a particle of mass m in close proximity to the earth's surface, where the gravitational attraction (weight) mg is essentially constant, Fig. 3/6a. The *gravitational potential energy* V_g of the particle is defined as the work mgh done *against* the gravitational field to elevate the particle a distance h above some arbitrary reference plane, where V_g is taken to be zero. Thus, we write the potential energy as

$$\boxed{V_g = mgh} \tag{3/14}$$

This work is called potential energy because it may be converted into energy if the particle is allowed to do work on a supporting body while it returns to its lower original datum plane. In going from one level at $h = h_1$ to a higher level at $h = h_2$, the *change* in potential energy becomes

$$\Delta V_g = mg(h_2 - h_1) = mg\,\Delta h$$

The corresponding work done *by* the gravitational force on the particle is $-mg\,\Delta h$. Thus, the work done by the gravitational force is the negative of the change in potential energy.

When large changes in altitude in the earth's field are encountered, Fig. 3/6b, the gravitational force $Gmm_e/r^2 = mgR^2/r^2$ is no longer constant. The work done *against* this force to change the radial position of the particle from r to r' is the change $V_g' - V_g$ in gravitational potential energy which is

$$\int_r^{r'} mgR^2 \frac{dr}{r^2} = mgR^2 \left(\frac{1}{r} - \frac{1}{r'} \right) = V_g' - V_g$$

It is customary to take $V_g' = 0$ when $r' = \infty$, so that with this datum we have

$$\boxed{V_g = -\frac{mgR^2}{r}} \tag{3/15}$$

In going from r_1 to r_2, the corresponding change in potential energy is

$$\Delta V_g = mgR^2 \left(\frac{1}{r_1} - \frac{1}{r_2} \right)$$

which, again, is the *negative* of the work done *by* the gravitational force. We note that the potential energy of a given particle depends only on its position, h or r, and not on the particular path it followed in reaching that position.

   (b) Elastic potential energy.   The second example of potential energy is found in the deformation of an elastic body, such as a spring. The work that is done on the spring to deform it is stored in the spring and is called its *elastic potential energy* V_e. This energy is recoverable in the form of work done by the spring on the body attached to its movable end during the release of the spring's deformation. For the one-dimensional linear spring of stiffness k, which we discussed in Art. 3/6 and illustrated in Fig. 3/4, the force supported by the spring at any deformation x, tensile or compressive, from its undeformed position is $F = kx$. Thus, we define the elastic potential energy of the spring as the work done on it to deform it an amount x, and we have

$$V_e = \int_0^x kx \, dx = \tfrac{1}{2}kx^2 \qquad\qquad (3/16)$$

   If the deformation, either tensile or compressive, of a spring increases from x_1 to x_2 during the motion interval, then the change in potential energy of the spring is its final value minus its initial value or

$$\Delta V_e = \tfrac{1}{2}k(x_2{}^2 - x_1{}^2)$$

which is positive. Conversely, if the deformation of a spring decreases from x_2 to x_1 during the motion interval, then the change in potential energy of the spring becomes negative. The magnitude of these changes is represented by the shaded trapezoidal area in the F-x diagram of Fig. 3/4.
   Since the force exerted *on* the spring *by* the moving body is equal and opposite to the force F exerted *on* the body *by* the spring (Fig. 3/4), it follows that the work done on the spring is the negative of the work done on the body. Therefore, we may replace the work U done by the spring on the body by $-\Delta V_e$, the negative of the potential energy change for the spring, provided the spring is now included within the system.

   (c) Work-energy equation.   With the elastic member included in the system, we now modify the work-energy equation to account for the potential energy terms. If $U'_{1\text{-}2}$ stands for the work of all external forces *other than* gravitational forces and spring forces, we may write Eq. 3/11 as $U'_{1\text{-}2} + (-\Delta V_g) + (-\Delta V_e) = \Delta T$ or

$$U'_{1\text{-}2} = \Delta T + \Delta V_g + \Delta V_e \qquad (3/17)$$

This alternative form of the work-energy equation is often far more convenient to use than Eq. 3/11, since the work of both gravity and spring forces is accounted for by focusing attention on the end-point positions of the particle and on the end-point lengths of the elastic spring. The path followed between these end-point positions is of no consequence in the evaluation of ΔV_g and ΔV_e.

Note that Eq. 3/17 may be rewritten in the equivalent form

$$T_1 + V_{g_1} + V_{e_1} + U'_{1\text{-}2} = T_2 + V_{g_2} + V_{e_2} \qquad (3/17a)$$

To help clarify the difference between the use of Eqs. 3/11 and 3/17, Fig. 3/7 shows schematically a particle of mass m constrained to move along a fixed path under the action of forces F_1 and F_2, the gravitational force $W = mg$, the spring force F, and the normal reaction N. In the *b*-part of the figure, the particle is isolated with its free-body diagram, and the work done by each of the forces F_1,

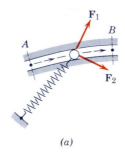

(a)

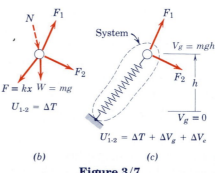

(b)   (c)

Figure 3/7

F_2, W, and the spring force $F = kx$ is evaluated for the interval of motion in question, say, from A to B, and equated to the change ΔT in kinetic energy using Eq. 3/11. The constraint reaction N, if normal to the path, will do no work. In the c-part of the figure for the alternative approach, the spring is included as a part of the isolated system. The work done during the interval by F_1 and F_2 constitutes the $U'_{1\text{-}2}$-term of Eq. 3/17 with the changes in elastic and gravitational potential energies included on the energy side of the equation. We note with the first approach that the work done by $F = kx$ could require a somewhat awkward integration to account for the changes in magnitude and direction of F as the particle moved from A to B. With the second approach, however, only the initial and final lengths of the spring would be required to evaluate ΔV_e, which would greatly simplify the calculation.

We may rewrite the alternative work-energy relation, Eq. 3/17, for a particle-and-spring system as

$$U'_{1\text{-}2} = \Delta(T + V_g + V_e) = \Delta E \qquad (3/17b)$$

where $E = T + V_g + V_e$ is the total mechanical energy of the particle and its attached linear spring. Equation 3/17b states that the net work done on the system by all forces other than gravitational forces and elastic forces equals the change in the total mechanical energy of the system. For problems where the only forces are gravitational, elastic, and nonworking constraint forces, the U'-term is zero, and the energy equation becomes merely

$$\Delta E = 0 \qquad \text{or} \qquad E = \text{constant} \qquad (3/18)$$

When E is constant, we see that transfers of energy between kinetic and potential may take place as long as the total mechanical energy $T + V_g + V_e$ does not change. Equation 3/18 expresses the *law of conservation of dynamical energy.*

(d) Conservative force fields. We have observed that the work done against a gravitational or an elastic force depends only on the net change of position and not on the particular path followed in reaching the new position. Forces with this characteristic are associated with *conservative force fields* that possess an important mathematical property. Consider a force field where the force $\mathbf{F}$ is a function of the coordinates, Fig. 3/8. The work done by $\mathbf{F}$ during a displacement $d\mathbf{r}$ of its point of application is $dU = \mathbf{F} \cdot d\mathbf{r}$. The total work done along its path from 1 to 2 is

$$U = \int \mathbf{F} \cdot d\mathbf{r} = \int (F_x \, dx + F_y \, dy + F_z \, dz)$$

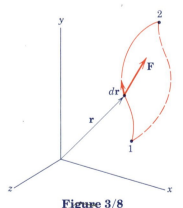

Figure 3/8

*Optional.

The integral $\int \mathbf{F} \cdot d\mathbf{r}$ is a line integral dependent, in general, on the particular path followed between any two points 1 and 2 in space. If, however, $\mathbf{F} \cdot d\mathbf{r}$ is an *exact differential*[†] $-dV$ of some scalar function V of the coordinates, then

$$U_{1\text{-}2} = \int_{V_1}^{V_2} -dV = -(V_2 - V_1) \qquad (3/19)$$

which depends only on the end points of the motion and which is thus *independent* of the path followed. The minus sign before dV is arbitrary but is chosen to agree with the customary designation of the sign of potential energy change in the earth's gravity field. If V exists, the differential change in V becomes

$$dV = \frac{\partial V}{\partial x} dx + \frac{\partial V}{\partial y} dy + \frac{\partial V}{\partial z} dz$$

Comparison with $-dV = \mathbf{F} \cdot d\mathbf{r} = F_x \, dx + F_y \, dy + F_z \, dz$ gives us

$$F_x = -\frac{\partial V}{\partial x} \qquad F_y = -\frac{\partial V}{\partial y} \qquad F_z = -\frac{\partial V}{\partial z}$$

The force may also be written as the vector

$$\mathbf{F} = -\boldsymbol{\nabla} V \qquad (3/20)$$

where the symbol $\boldsymbol{\nabla}$ stands for the vector operator "del" that is

$$\boldsymbol{\nabla} = \mathbf{i}\,\frac{\partial}{\partial x} + \mathbf{j}\,\frac{\partial}{\partial y} + \mathbf{k}\,\frac{\partial}{\partial z}$$

The quantity V is known as the *potential function*, and the expression $\boldsymbol{\nabla} V$ is known as the *gradient of the potential function*.

When force components are derivable from a potential as described, the force is said to be *conservative*, and the work done by $\mathbf{F}$ between any two points is independent of the path followed.

[†]Recall that a function $d\phi = P \, dx + Q \, dy + R \, dz$ is an exact differential in the coordinates x-y-z if

$$\frac{\partial P}{\partial y} = \frac{\partial Q}{\partial x} \qquad \frac{\partial P}{\partial z} = \frac{\partial R}{\partial x} \qquad \frac{\partial Q}{\partial z} = \frac{\partial R}{\partial y}$$

Sample Problem 3/16

The 10-kg slider A moves with negligible friction up the inclined guide. The attached spring has a stiffness of 60 N/m and is stretched 0.6 m in position A, where the slider is released from rest. The 250-N force is constant and the pulley offers negligible resistance to the motion of the cord. Calculate the velocity v of the slider as it passes point C.

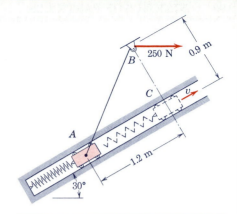

Solution. The slider and inextensible cord together with the attached spring will be analyzed as a system, which permits the use of Eq. 3/17. The only nonpotential force doing work on this system is the 250-N tension applied to the cord. While the slider moves from A to C, the point of application of the 250-N force moves a distance of $\overline{AB} - \overline{BC}$ or $1.5 - 0.9 = 0.6$ m.

① $$U'_{1\text{-}2} = 250(0.6) = 150 \text{ J}$$

The change in kinetic energy of the slider is

$$\Delta T = \tfrac{1}{2}m(v^2 - v_0^2) = \tfrac{1}{2}(10)v^2$$

where the initial velocity v_0 is zero. The change in gravitational potential energy is

② $$\Delta V_g = mg(\Delta h) = 10(9.81)(1.2 \sin 30°) = 58.9 \text{ J}$$

The change in elastic potential energy is

③ $$\Delta V_e = \tfrac{1}{2}k(x_2^2 - x_1^2) = \tfrac{1}{2}(60)([1.2 + 0.6]^2 - [0.6]^2) = 86.4 \text{ J}$$

Substitution into the alternative work-energy equation gives

$$[U'_{1\text{-}2} = \Delta T + \Delta V_g + \Delta V_e] \quad 150 = \tfrac{1}{2}(10)v^2 + 58.9 + 86.4$$

$$v = 0.974 \text{ m/s} \qquad Ans.$$

① The reactions of the guides on the slider are normal to the direction of motion and do no work.

② Since the center of mass of the slider has an upward component of displacement, ΔV_g is positive.

③ Be very careful not to make the mistake of using $\tfrac{1}{2}k(x_2 - x_1)^2$ for ΔV_e. We want the difference of the squares and not the square of the difference.

Sample Problem 3/17

The 6-lb slider is released from rest at point A and slides with negligible friction in a vertical plane along the circular rod. The attached spring has a stiffness of 2 lb/in. and has an unstretched length of 24 in. Determine the velocity of the slider as it passes position B.

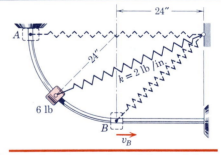

Solution. The work done by the weight and the spring force on the slider will be treated as changes in the potential energies, and the reaction of the rod on the slider is normal to the motion and does no work. Hence, $U'_{1\text{-}2} = 0$. The changes in the potential and kinetic energies for the system of slider and spring are

① $$\Delta V_e = \tfrac{1}{2}k(x_B^2 - x_A^2) = \tfrac{1}{2}(2)\{(24[\sqrt{2} - 1])^2 - (24)^2\} = -477 \text{ in.-lb}$$

$$\Delta V_g = W\Delta h = 6(-24) = -144 \text{ in.-lb}$$

$$\Delta T = \frac{1}{2}m(v_B^2 - v_A^2) = \frac{1}{2}\frac{6}{(32.2)(12)}(v_B^2 - 0) = \frac{v_B^2}{128.8}$$

$$[\Delta T + \Delta V_g + \Delta V_e = 0] \quad \frac{v_B^2}{128.8} - 144 - 477 = 0$$

$$v_B = 283 \text{ in./sec} \qquad Ans.$$

① Note that if we evaluated the work done by the spring force acting on the slider by means of the integral $\int \mathbf{F} \cdot d\mathbf{r}$, it would necessitate a lengthy computation to account for the change in the magnitude of the force, along with the change in the angle between the force and the tangent to the path. Note further that v_B depends only on the end conditions of the motion and does not require knowledge of the shape of the path.

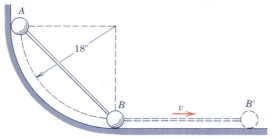

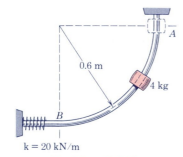

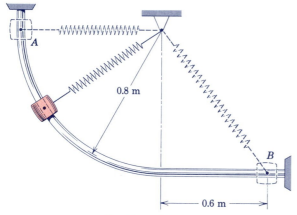

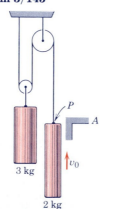

Problem 3/143

Problem 3/144

Problem 3/145

Problem 3/146

PROBLEMS

Introductory problems

3/143 The two particles of equal mass are joined by a rod of negligible mass. If they are released from rest in the position shown and slide on the smooth guide in the vertical plane, calculate their velocity v when A reaches B's position and B is at B'.

Ans. $v = 6.95$ ft/sec

3/144 The 4-kg slider is released from rest at A and slides with negligible friction down the circular rod in the vertical plane. Determine (a) the velocity v of the slider as it reaches the bottom at B and (b) the maximum deformation x of the spring.

3/145 The spring has an unstretched length of 0.4 m and a stiffness of 200 N/m. The 3-kg slider and attached spring are released from rest at A and move in the vertical plane. Calculate the velocity v of the slider as it reaches B in the absence of friction.

Ans. $v = 1.537$ m/s

3/146 Point P on the 2-kg cylinder has an initial velocity $v_0 = 0.8$ m/s as it passes position A. Neglect the mass of the pulleys and cable and determine the distance y of point P below A when the 3-kg cylinder has acquired an upward velocity of 0.6 m/s.

3/147 A bead with a mass of 0.25 kg is released from rest at *A* and slides down and around the fixed smooth wire. Determine the force *N* between the wire and the bead as it passes point *B*. *Ans. N* = 14.42 N

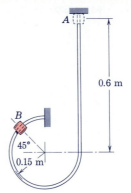

Problem 3/147

3/148 The spring of constant *k* is unstretched when the slider of mass *m* passes position *B*. If the slider is released from rest in position *A*, determine its speed as it passes points *B* and *C*. What is the normal force exerted by the guide on the slider at position *C*? Neglect friction between the mass and the circular guide, which lies in a vertical plane.

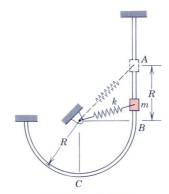

Problem 3/148

3/149 The light rod is pivoted at *O* and carries the 5- and 10-lb particles. If the rod is released from rest at $\theta = 60°$ and swings in the vertical plane, calculate (*a*) the velocity *v* of the 5-lb particle just before it hits the spring in the dotted position and (*b*) the maximum compression *x* of the spring. Assume that *x* is small so that the position of the rod when the spring is compressed is essentially horizontal.
Ans. (*a*) $v = 3.84$ ft/sec, (*b*) $x = 0.510$ in.

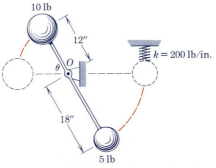

Problem 3/149

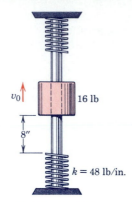

Problem 3/150

Representative problems

3/150 The 16-lb collar slides freely on the fixed vertical rod and is given an upward velocity $v_0 = 8$ ft/sec in the position shown. The collar compresses the upper spring and is then projected downward. Calculate the maximum resulting deformation x of the lower spring.

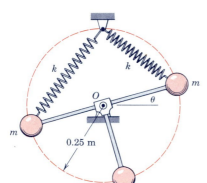

Problem 3/151

3/151 The two springs, each of stiffness $k = 1.2$ kN/m, are of equal length and undeformed when $\theta = 0$. If the mechanism is released from rest in the position $\theta = 20°$, determine its angular velocity $\dot{\theta}$ when $\theta = 0$. The mass m of each sphere is 3 kg. Treat the spheres as particles and neglect the masses of the light rods and springs. *Ans.* $\dot{\theta} = 4.22$ rad/s

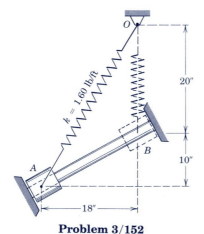

Problem 3/152

3/152 The 2-lb collar is released from rest at A and slides freely up the inclined rod, striking the stop at B with a velocity v. The spring of stiffness $k = 1.60$ lb/ft has an unstretched length of 15 in. Calculate v.

3/153 The identical links of negligible mass are released simultaneously from rest at $\theta = 30°$ and rotate in the vertical plane. Determine the velocity v of each 2-lb sphere when θ reaches 90°. The spring is unstretched when $\theta = 90°$. *Ans. $v = 9.86$ ft/sec*

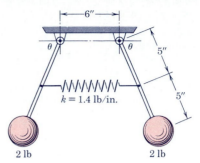

Problem 3/153

3/154 It is desired that the 100-lb container, when released from rest in the position shown, shall have no velocity after dropping 7 ft to the platform below. Specify the proper weight W of the counterbalancing cylinder.

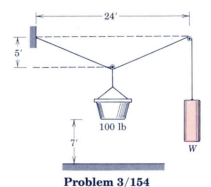

Problem 3/154

3/155 If the system is released from rest, determine the speeds of both masses after B has moved 1 m. Neglect friction and the masses of the pulleys. *Ans. $v_A = 0.616$ m/s, $v_B = 0.924$ m/s*

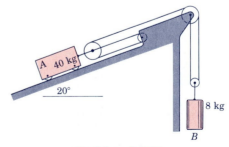

Problem 3/155

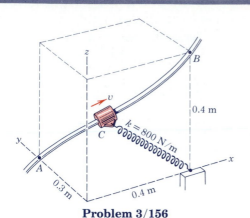

Problem 3/156

3/156 The 1.5-kg slider C moves along the fixed rod under the action of the spring whose unstretched length is 0.3 m. If the velocity of the slider is 2 m/s at point A and 3 m/s at point B, calculate the work U_f done by friction between these two points. Also, determine the average friction force acting on the slider between A and B if the length of the path is 0.70 m. The x-y plane is horizontal.

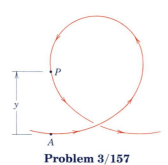

Problem 3/157

3/157 In the design of an inside loop for an amusement-park ride, it is desired to maintain the same centripetal acceleration throughout the loop. Assume negligible loss of energy during the motion and determine the radius of curvature ρ of the path as a function of the height y above the low point A, where the velocity and radius of curvature are v_0 and ρ_0, respectively. For a given value of ρ_0, what is the minimum value of v_0 for which the vehicle will not leave the track at the top of the loop?

$$Ans. \quad \rho = \rho_0 \left(1 - \frac{2gy}{v_0{}^2} \right), \quad v_{0_{min}} = \sqrt{\rho_0 g}$$

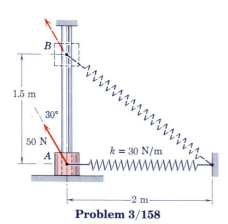

Problem 3/158

3/158 The collar has a mass of 2 kg and is attached to the light spring that has a stiffness of 30 N/m and an unstretched length of 1.5 m. The collar is released from rest at A and slides up the smooth rod under the action of the constant 50-N force. Calculate the velocity v of the collar as it passes position B.

3/159 A spacecraft m is heading toward the center of the moon with a velocity of 2000 mi/hr at a distance from the moon's surface equal to the radius R of the moon. Compute the impact velocity v with the surface of the moon if the spacecraft is unable to fire its rectro-rockets. Consider the moon to be fixed in space. The radius R of the moon is 1080 mi, and the acceleration due to gravity at its surface is 5.32 ft/sec². *Ans.* $v = 4250$ mi/hr

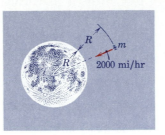

Problem 3/159

3/160 The two right-angled rods with attached spheres are released from rest in the position $\theta = 0$. If the system is observed to momentarily come to rest when $\theta = 45°$, determine the spring constant k. The spring is unstretched when $\theta = 0$. Treat the spheres as particles and neglect friction.

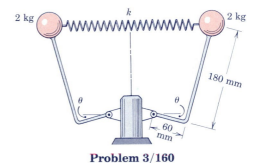

Problem 3/160

3/161 The 3-lb ball is given an initial velocity $v_A = 8$ ft/sec in the vertical plane at position A, where the two horizontal attached springs are unstretched. The ball follows the dotted path shown and crosses point B which is 5 in. directly below A. Calculate the velocity v_B of the ball at B. Each spring has a stiffness of 10 lb/in. *Ans.* $v_B = 8.54$ ft/sec

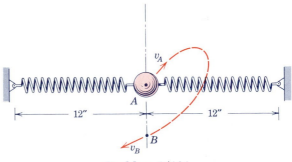

Problem 3/161

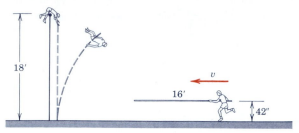

Problem 3/162

3/162 A 175-lb pole vaulter carrying a uniform 16-ft, 10-lb pole approaches the jump with a velocity v and manages to barely clear the bar set at a height of 18 ft. As he clears the bar, his velocity and that of the pole are essentially zero. Calculate the minimum possible value of v required for him to make the jump. Both the horizontal pole and the center of gravity of the vaulter are 42 in. above the ground during the approach.

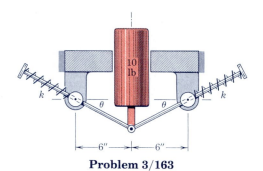

Problem 3/163

3/163 When the 10-lb plunger is released from rest in its vertical guide at $\theta = 0$, each spring of stiffness $k = 20$ lb/in. is uncompressed. The links are free to slide through their pivoted collars and compress their springs. Calculate the velocity v of the plunger when the position $\theta = 30°$ is passed.

Ans. $v = 3.06$ ft/sec

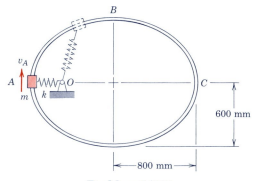

Problem 3/164

3/164 The fixed point O is located at one of the two foci of the elliptical guide. The spring has a stiffness of 3 N/m and is unstretched when the slider is at A. If the speed v_A is such that the speed of the 0.4-kg slider approaches zero at C, determine its speed at point B. The smooth guide lies in a horizontal plane. (If necessary, refer to Eqs. 3/39 for elliptical geometry.)

3/165 The mechanism is released from rest with $\theta = 180°$, where the uncompressed spring of stiffness $k = 900$ N/m is just touching the underside of the 4-kg collar. Determine the angle θ corresponding to the maximum compression of the spring. Motion is in the vertical plane, and the mass of the links may be neglected.   *Ans.* $\theta = 43.8°$

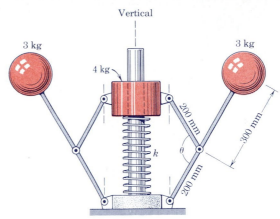

Problem 3/165

3/166 The cylinder of mass m is attached to the collar bracket at A by a spring of stiffness k. The collar fits loosely on the vertical shaft, which is lowering both the collar and the suspended cylinder with a constant velocity v. When the collar strikes the base B, it stops abruptly with essentially no rebound. Determine the maximum additional deflection δ of the spring after the impact.

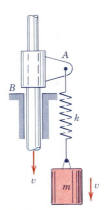

Problem 3/166

3/167 The 3-kg sphere is carried by the parallelogram linkage where the spring is unstretched when $\theta = 90°$. If the mechanism is released from rest at $\theta = 90°$, calculate the velocity v of the sphere when the position $\theta = 135°$ is passed. The links are in the vertical plane, and their mass is small and may be neglected.   *Ans.* $v = 1.143$ m/s

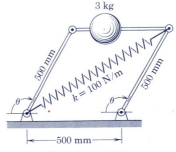

Problem 3/167

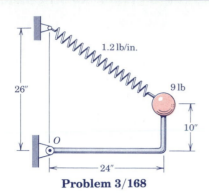

Problem 3/168

3/168 The spring has an unstretched length of 25 in. If the system is released from rest in the position shown, determine the speed v of the ball (*a*) when it has dropped a vertical distance of 10 in. and (*b*) when the rod has rotated 35°.

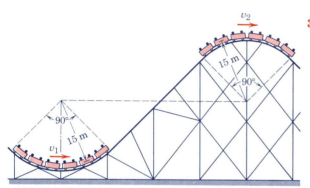

Problem 3/169

3/169 The cars of an amusement-park ride have a speed $v_1 = 90$ km/h at the lowest part of the track. Determine their speed v_2 at the highest part of the track. Neglect energy loss due to friction. (*Caution:* Give careful thought to the change in potential energy of the system of cars.)   *Ans.* $v_2 = 35.1$ km/h

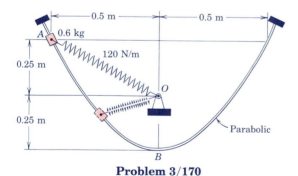

Problem 3/170

3/170 The 0.6-kg slider is released from rest at A and slides down the smooth parabolic guide (which lies in a vertical plane) under the influence of its own weight and of the spring of constant 120 N/m. Determine the speed of the slider as it passes point B and the corresponding normal force exerted on it by the guide. The unstretched length of the spring is 200 mm.

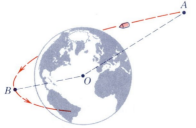

Problem 3/171

3/171 Upon its return voyage from a space mission, the spacecraft has a velocity of 24 000 km/h at point A, which is 7000 km from the center of the earth. Determine the velocity of the spacecraft when it reaches point B, which is 6500 km from the center of the earth. The trajectory between these two points is outside the effect of the earth's atmosphere.   *Ans.* $v_B = 26\ 300$ km/h

3/172 A satellite is put into an elliptical orbit around the earth and has a velocity v_P at the perigee position P. Determine the expression for the velocity v_A at the apogee position A. The radii to A and P are, respectively, r_A and r_P. Note that the total energy remains constant.

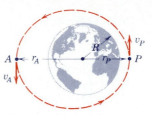

Problem 3/172

3/173 The shank of the 5-lb vertical plunger occupies the dotted position when resting in equilibrium against the spring of stiffness $k = 10$ lb/in. The upper end of the spring is welded to the plunger, and the lower end is welded to the base plate. If the plunger is lifted $1\frac{1}{2}$ in. above its equilibrium position and released from rest, calculate its velocity v as it strikes the button A. Friction is negligible.

Ans. $v = 3.43$ ft/sec

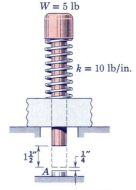

Problem 3/173

3/174 The rope shown in the a-part of the figure is released from rest in the smooth semicircular guide and acquires a velocity v when all of it is clear of the guide as seen in the b-part of the figure. Find v. (*Hint:* In finding the potential energy change, imagine that the semicircular portion of the rope in the initial position becomes the bottom part of the rope of equal length in the final position.)

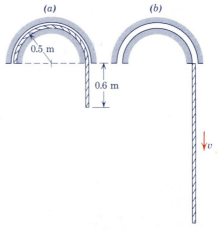

Problem 3/174

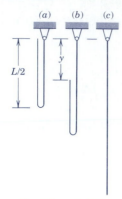

Problem 3/175

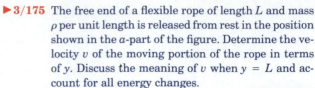

▶**3/175** The free end of a flexible rope of length L and mass ρ per unit length is released from rest in the position shown in the a-part of the figure. Determine the velocity v of the moving portion of the rope in terms of y. Discuss the meaning of v when $y = L$ and account for all energy changes.

$$\text{Ans. } v = \sqrt{2gy\,\frac{L - y/2}{L - y}}$$

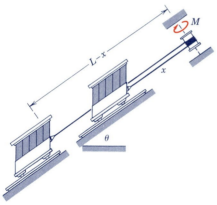

Problem 3/176

▶**3/176** The cable railway consists of two passenger gondolas, each of mass m, one on each end of the cable of total length L and mass ρ per unit length. The system is operated by applying a torque M to the drum of radius r at the top of the railway. Several turns of cable around the drum prevent slipping, and the length of cable around the drum may be neglected compared with L. If the gondolas start from rest at $x = 0$ with M constant, derive an expression for the velocity v of each gondola for a given value of x. Neglect the mass of the drum and all friction.

$$\text{Ans. } v = \sqrt{\frac{2}{2m + \rho L}}\sqrt{\frac{Mx}{r} - \rho gx(L - x)\sin\theta}$$

SECTION C. IMPULSE AND MOMENTUM

3/8 INTRODUCTION

In the previous two articles, we focused attention on the equations of work and energy that are obtained by integrating the equation of motion $\mathbf{F} = m\mathbf{a}$ with respect to the displacement of the particle. In consequence, we found that the velocity changes could be expressed directly in terms of the work done or in terms of the overall changes in energy. In the next two articles, we will direct our attention to the integration of the equation of motion with respect to time rather than displacement. This approach leads to the equations of impulse and momentum. We will find that these equations greatly facilitate the solution of many problems in which the applied forces act during extremely short periods of time (as in impact problems) or over specified intervals of time.

3/9 LINEAR IMPULSE AND LINEAR MOMENTUM

Consider again the general curvilinear motion in space of a particle of mass m, Fig. 3/9, where the particle is located by its position vector $\mathbf{r}$ measured from a fixed origin O. The velocity of the particle is $\mathbf{v} = \dot{\mathbf{r}}$ and is tangent to its path (shown as a dashed line). The resultant $\Sigma\mathbf{F}$ of all forces on m is in the direction of its acceleration $\dot{\mathbf{v}}$. We may now write the basic equation of motion for the particle, Eq. 3/3, as

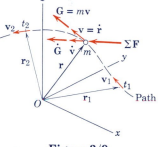

Figure 3/9

$$\Sigma\mathbf{F} = m\dot{\mathbf{v}} = \frac{d}{dt}(m\mathbf{v}) \qquad \text{or} \qquad \boxed{\Sigma\mathbf{F} = \dot{\mathbf{G}}} \qquad (3/21)$$

where the product of the mass and velocity is defined as the *linear momentum* $\mathbf{G} = m\mathbf{v}$ of the particle. Equation 3/21 states that *the resultant of all forces acting on a particle equals its time rate of change of linear momentum*. In SI the units of linear momentum $m\mathbf{v}$ are seen to be kg·m/s, which also equals N·s. In U.S. customary units, the units of linear momentum $m\mathbf{v}$ are [lb/(ft/sec²)][ft/sec] = lb-sec.

Since Eq. 3/21 is a vector equation, we recognize that, in addition to the equality of the magnitudes of $\Sigma\mathbf{F}$ and $\dot{\mathbf{G}}$, the direction of the resultant force coincides with the direction of the rate of change in linear momentum, which is the direction of the rate of change in velocity. Equation 3/21 is one of the most useful and important relationships in dynamics, and it is valid as long as the mass m of the particle is not changing with time. The case where m changes with time is discussed in Art. 4/7 of Chapter 4. We now write the three scalar components of Eq. 3/21 as

$$\Sigma F_x = \dot{G}_x \qquad \Sigma F_y = \dot{G}_y \qquad \Sigma F_z = \dot{G}_z \qquad (3/22)$$

These equations may be applied independently of one another.

All that we have done so far in this article is to rewrite Newton's second law in an alternative form in terms of momentum. But we are now able to describe the effect of the resultant force $\Sigma \mathbf{F}$ on the linear momentum of the particle over a finite period of time simply by integrating Eq. 3/21 with respect to the time t. Multiplying the equation by dt gives $\Sigma \mathbf{F}\, dt = d\mathbf{G}$, which we integrate from time t_1 to time t_2 to obtain

$$\int_{t_1}^{t_2} \Sigma \mathbf{F}\, dt = \mathbf{G}_2 - \mathbf{G}_1 = \Delta \mathbf{G} \qquad (3/23)$$

Here the linear momentum at time t_2 is $\mathbf{G}_2 = m\mathbf{v}_2$ and the linear momentum at time t_1 is $\mathbf{G}_1 = m\mathbf{v}_1$. The product of force and time is defined as *linear impulse* of the force, and Eq. 3/23 states that *the total linear impulse on m equals the corresponding change in linear momentum of m*.

Alternatively, we may write Eq. 3/23 as

$$\mathbf{G}_1 + \int_{t_1}^{t_2} \Sigma \mathbf{F}\, dt = \mathbf{G}_2 \qquad (3/23a)$$

which says that the initial linear momentum of the body plus the linear impulse applied to it equals its final linear momentum.

The impulse integral is a vector that, in general, may involve changes in both magnitude and direction during the time interval. Under these conditions, it will be necessary to express $\Sigma \mathbf{F}$ and $\mathbf{G}$ in component form and then combine the integrated components. The components of Eq. 3/23 become the scalar equations

$$\int_{t_1}^{t_2} \Sigma F_x\, dt = (mv_x)_2 - (mv_x)_1$$

$$\int_{t_1}^{t_2} \Sigma F_y\, dt = (mv_y)_2 - (mv_y)_1$$

$$\int_{t_1}^{t_2} \Sigma F_z\, dt = (mv_z)_2 - (mv_z)_1$$

These three scalar impulse-momentum equations are completely independent. The scalar expressions corresponding to the vector equations 3/23a are simply rearrangements of these equations.

There are times when a force acting on a particle varies with the time in a manner determined by experimental measurements or by other approximate means. In this case, a graphical or numerical integration must be accomplished. If, for example, a force F acting on a particle in a given direction varies with the time t as indicated in Fig. 3/10, then the impulse, $\int_{t_1}^{t_2} F\,dt$, of this force from t_1 to t_2 becomes the shaded area under the curve.

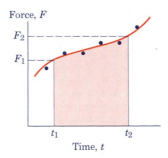

Force, F

F_2

F_1

t_1 t_2

Time, t

Figure 3/10

In evaluating the resultant impulse, it is necessary to include the effect of *all* forces acting on m, except those whose magnitudes are negligible. The student should be fully aware at this point that the only reliable method of accounting for the effects of *all* forces is to isolate the particle in question by drawing its *free-body diagram*.

Conservation of linear momentum. If the resultant force on a particle is zero during an interval of time, we see that Eq. 3/21 requires that its linear momentum $\mathbf{G}$ remain constant. In this case, the linear momentum of the particle is said to be *conserved*. Linear momentum may be conserved in one coordinate direction, such as x, but not necessarily in the y- or z-direction. A careful examination of the free-body diagram of the particle will disclose whether the total linear impulse on the particle in a particular direction is zero. If it is, the corresponding linear momentum is unchanged (conserved) in that direction.

Consider now the motion of two particles a and b that interact during an interval of time. If the interactive forces $\mathbf{F}$ and $-\mathbf{F}$ between them are the only unbalanced forces acting on the particles during the interval, it follows that the linear impulse on particle a is the negative of the linear impulse on particle b. Therefore, from Eq. 3/23 the change in linear momentum $\Delta\mathbf{G}_a$ of particle a is the negative of the change $\Delta\mathbf{G}_b$ in linear momentum of particle b. So we have $\Delta\mathbf{G}_a = -\Delta\mathbf{G}_b$ or $\Delta(\mathbf{G}_a + \mathbf{G}_b) = \mathbf{0}$. Thus, the total linear momentum $\mathbf{G} = \mathbf{G}_a + \mathbf{G}_b$ for the system of the two particles remains constant during the interval, and we write

$$\Delta\mathbf{G} = \mathbf{0} \qquad \text{or} \qquad \mathbf{G}_1 = \mathbf{G}_2 \qquad\qquad (3/24)$$

Equation 3/24 expresses the *principle of conservation of linear momentum*.

Sample Problem 3/18

A 2-lb particle moves in the vertical y-z plane (z up, y horizontal) under the action of its weight and a force $\mathbf{F}$ that varies with the time. The linear momentum of the particle in lb-sec is given by the expression $\mathbf{G} = \frac{3}{2}(t^2 + 3)\mathbf{j} - \frac{2}{3}(t^3 - 4)\mathbf{k}$, where t is the time in seconds. Determine $\mathbf{F}$ for the instant when $t = 2$ sec.

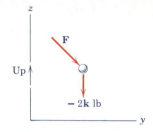

Solution. The weight expressed as a vector is $-2\mathbf{k}$ lb. Thus, the force-momentum equation becomes

① $[\Sigma\mathbf{F} = \dot{\mathbf{G}}]$ $\quad \mathbf{F} - 2\mathbf{k} = \dfrac{d}{dt}[\frac{3}{2}(t^2 + 3)\mathbf{j} - \frac{2}{3}(t^3 - 4)\mathbf{k}]$

$$= 3t\mathbf{j} - 2t^2\mathbf{k}$$

For $t = 2$ sec, $\quad \mathbf{F} = 2\mathbf{k} + 3(2)\mathbf{j} - 2(2^2)\mathbf{k} = 6\mathbf{j} - 6\mathbf{k}$ lb

Thus, $\quad F = \sqrt{6^2 + 6^2} = 6\sqrt{2}$ lb $\qquad$ *Ans.*

① Don't forget that $\Sigma\mathbf{F}$ includes *all* forces on the particle, of which the weight is one.

Sample Problem 3/19

A particle with a mass of 0.5 kg has a velocity $u = 10$ m/s in the x-direction at time $t = 0$. Forces $\mathbf{F}_1$ and $\mathbf{F}_2$ act on the particle, and their magnitudes change with time according to the graphical schedule shown. Determine the velocity $\mathbf{v}$ of the particle at the end of the 3 s.

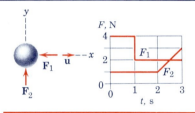

Solution. The impulse-momentum equation is applied in component form and gives for the x- and y-directions, respectively,

① $\left[\int \Sigma F_x \, dt = m\Delta v_x\right] \qquad -[4(1) + 2(3 - 1)] = 0.5(v_x - 10)$

$$v_x = -6 \text{ m/s}$$

$\left[\int \Sigma F_y \, dt = m\Delta v_y\right] \qquad [1(2) + 2(3 - 2)] = 0.5(v_y - 0)$

$$v_y = 8 \text{ m/s}$$

Thus,

$$\mathbf{v} = -6\mathbf{i} + 8\mathbf{j} \text{ m/s} \quad \text{and} \quad v = \sqrt{6^2 + 8^2} = 10 \text{ m/s}$$

$$\theta_x = \tan^{-1}\frac{8}{-6} = 126.9° \qquad \textit{Ans.}$$

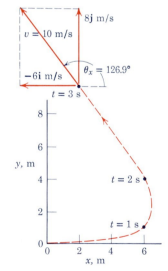

② Although not called for, the path of the particle for the first 3 s is plotted in the figure. The velocity at $t = 3$ s is shown together with its components.

① The impulse in each direction is the corresponding area under the force-time graph. Note that F_1 is in the negative x-direction so its impulse is negative.

② It is important to note that the algebraic signs must be carefully respected in applying the momentum equations. Also we must recognize that impulse and momentum are vector quantities in contrast to work and energy, which are scalars.

Sample Problem 3/20

The loaded 150-kg skip is rolling down the incline at 4 m/s when a force P is applied to the cable as shown at time $t = 0$. The force P is increased uniformly with the time until it reaches 600 N at $t = 4$ s, after which time it remains constant at this value. Calculate (a) the time t_1 at which the skip reverses its direction and (b) the velocity v of the skip at $t = 8$ s. Treat the skip as a particle.

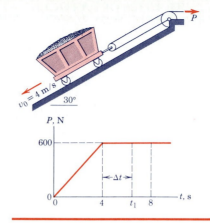

Solution. The stated variation of P with the time is plotted, and the free-body diagram of the skip is drawn.

Part (a). The skip reverses direction when its velocity becomes zero. We will assume that this condition occurs at $t = 4 + \Delta t$ s. The impulse-momentum equation applied consistently in the positive x-direction gives

$$\left[\int \Sigma F_x \, dt = m \Delta v_x \right]$$

① $\frac{1}{2}(4)(2)(600) + 2(600)\Delta t - 150(9.81) \sin 30°(4 + \Delta t) = 150(0 - [-4])$

$$464 \Delta t = 1143 \qquad \Delta t = 2.46 \text{ s} \qquad t = 4 + 2.46 = 6.46 \text{ s} \qquad Ans.$$

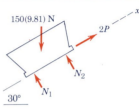

Part (b). Applying the impulse-momentum equation to the entire interval gives

$$\left[\int \Sigma F_x \, dt = m \Delta v_x \right]$$

$$\frac{1}{2}(4)(2)(600) + 4(2)(600) - 150(9.81) \sin 30°(8) = 150(v - [-4])$$

$$150v = 714 \qquad v = 4.76 \text{ m/s} \qquad Ans.$$

The same result is obtained by analyzing the interval from t_1 to 8 s.

① The free-body diagram keeps us from making the error of using the impulse of P rather than $2P$ or of forgetting the impulse of the component of the weight. The first term in the equation is the triangular area of the P-t relation for the first 4 s, doubled for the force of $2P$.

Sample Problem 3/21

The 50-g bullet traveling at 600 m/s strikes the 4-kg block centrally and is embedded within it. If the block slides on a smooth horizontal plane with a velocity of 12 m/s in the direction shown prior to impact, determine the velocity $\mathbf{v}$ of the block and embedded bullet immediately after impact.

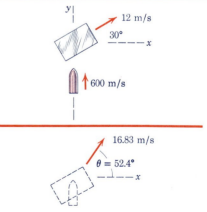

Solution. Since the force of impact is internal to the system composed of the block and bullet and since there are no other external forces acting on the system in the plane of motion, it follows that the linear momentum of the system is conserved. Thus,

① $[\mathbf{G}_1 = \mathbf{G}_2] \qquad 0.050(600\mathbf{j}) + 4(12)(\cos 30°\mathbf{i} + \sin 30°\mathbf{j}) = (4 + 0.050)\mathbf{v}$

$$\mathbf{v} = 10.26\mathbf{i} + 13.33\mathbf{j} \text{ m/s} \qquad Ans.$$

The final velocity and its direction are given by

$$[v = \sqrt{v_x^2 + v_y^2}] \qquad v = \sqrt{(10.26)^2 + (13.33)^2} = 16.83 \text{ m/s} \qquad Ans.$$

$$[\tan \theta = v_y/v_x] \qquad \tan \theta = \frac{13.33}{10.26} = 1.30 \qquad \theta = 52.4° \qquad Ans.$$

① Working with the vector form of the principle of conservation of linear momentum is clearly equivalent to working with the component form.

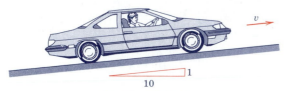

Problem 3/177

PROBLEMS

Introductory problems

3/177 The 1500-kg car has a velocity of 30 km/h up the 10-percent grade when the driver applies more power for 8 s to bring the car up to a speed of 60 km/h. Calculate the time average F of the total force tangent to the road exerted on the tires during the 8 s. Treat the car as a particle and neglect air resistance. *Ans.* $F = 3.03$ kN

3/178 The velocity of a 1.2-kg particle is given by $\mathbf{v} = 1.5t^3\mathbf{i} + (2.4 - 3t^2)\mathbf{j} + 5\mathbf{k}$, where $\mathbf{v}$ is in meters per second and the time t is in seconds. Determine the linear momentum $\mathbf{G}$ of the particle, its magnitude G, and the net force $\mathbf{R}$ that acts on the particle when $t = 2$ s.

3/179 The two orbital maneuvering engines of the space shuttle develop 26 kN of thrust each. If the shuttle is traveling in orbit at a speed of 26 000 km/h, how long would it take to reach a speed of 26 200 km/h after the two engines are fired? The mass of the shuttle is 90 Mg. *Ans.* $t = 1$ min 36 s

Problem 3/180

3/180 A jet-propelled airplane with a mass of 10 Mg is flying horizontally at a constant speed of 1000 km/h under the action of the engine thrust T and the equal and opposite air resistance R. The plot ignites two rocket-assist units, each of which develops a forward thrust T_0 of 8 kN for 9 s. If the velocity of the airplane in its horizontal flight is 1050 km/h at the end of the 9 s, calculate the time-average increase ΔR in air resistance. The mass of the rocket fuel used is negligible compared with that of the airplane.

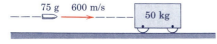

Problem 3/181

3/181 A 75-g projectile traveling at 600 m/s strikes and becomes embedded in the 50-kg block, which is initially stationary. Compute the energy lost during the impact. Express your answer as an absolute value $|\Delta E|$ and as a percentage n of the original system energy E.
Ans. $|\Delta E| = 13\ 480$ J, $n = 99.9\%$

3/182 Freight car A with a gross weight of 150,000 lb is moving along the horizontal track in a switching yard at 2 mi/hr. Freight car B with a gross weight of 120,000 lb and moving at 3 mi/hr overtakes car A and is coupled to it. Determine (a) the common velocity v of the two cars as they move together after being coupled and (b) the loss of energy $|\Delta E|$ due to the impact.

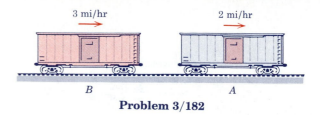

3 mi/hr   2 mi/hr

B   *A*

Problem 3/182

3/183 A boy weighing 100 lb runs and jumps on his 20-lb sled with a horizontal velocity of 15 ft/sec. If the sled and boy coast 80 ft on the level snow before coming to rest, compute the coefficient of kinetic friction μ_k between the snow and the runners of the sled.

Ans. $\mu_k = 0.030$

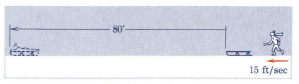

80'

15 ft/sec

Problem 3/183

3/184 A railroad car of mass m and initial speed v collides with and becomes coupled with the two identical cars. Compute the final speed v' of the group of three cars and the fractional loss n of energy if (a) the initial separation distance $d = 0$ (that is, the two stationary cars are initially coupled together with no slack in the coupling) and (b) the distance $d \neq 0$ so that the cars are uncoupled and slightly separated. Neglect rolling resistance.

v d

Problem 3/184

3/185 The 200-kg lunar lander is descending onto the moon's surface with a velocity of 6 m/s when its retro-engine is fired. If the engine produces a thrust T for 4 s that varies with the time as shown and then cuts off, calculate the velocity of the lander when $t = 5$ s, assuming that it has not yet landed. Gravitational acceleration at the moon's surface is 1.62 m/s².

Ans. $v = 2.10$ m/s

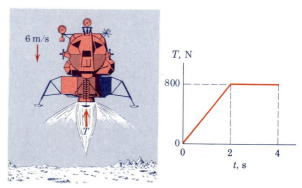

6 m/s

T, N

800

T

0   2   4

t, s

Problem 3/185

3/186 The 20-lb block is moving to the right with a velocity of 2 ft/sec on a horizontal surface when a force P is applied to it at time $t = 0$. Calculate the velocity v of the block when $t = 0.4$ sec. The kinetic coefficient of friction is $\mu_k = 0.3$.

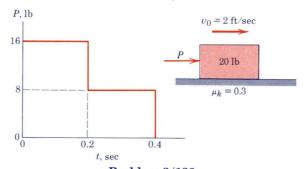

P, lb

16

8

0   0.2   0.4

t, sec

$v_0 = 2$ ft/sec

P 20 lb

$\mu_k = 0.3$

Problem 3/186

Problem 3/187

Representative problems

3/187 The supertanker has a total displacement (mass) of 150,000 long tons (one long ton equals 2240 lb) and is lying still in the water when the tug commences a tow. If a constant tension of 50,000 lb is developed in the tow cable, compute the time required to bring the tanker to a speed of 1 knot from rest. At this low speed, hull resistance to motion through the water is very small and may be neglected. (1 knot = 1.152 mi/hr)   *Ans.* $t = 6.25$ min

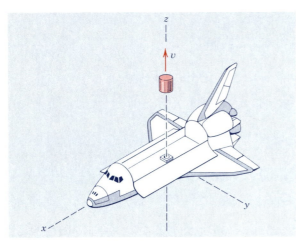

Problem 3/188

3/188 The space shuttle launches an 800-kg satellite by ejecting it from the cargo bay as shown. The ejection mechanism is activated and is in contact with the satellite for 4 s to give it a velocity of 0.3 m/s in the z-direction relative to the shuttle. The mass of the shuttle is 90 Mg. Determine the component of velocity v_f of the shuttle in the minus z-direction resulting from the ejection. Also find the time average F of the ejection force.

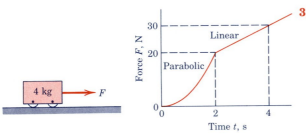

Problem 3/189

3/189 The 4-kg cart, at rest at time $t = 0$, is acted on by a horizontal force that varies with time t as shown. Neglect friction and determine the velocity of the cart at $t = 1$ s and at $t = 3$ s.
            Ans. $v_1 = 0.417$ m/s, $v_3 = 8.96$ m/s

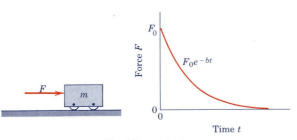

Problem 3/190

3/190 The cart of mass m is subjected to the exponentially decreasing force F, which represents a shock or blast loading. If the cart is stationary at time $t = 0$, determine its velocity v and displacement s as functions of time. What is the value of v for large values of t?

3/191 The 140-g projectile is fired with a velocity of 600 m/s and picks up three washers, each with a mass of 100 g. Find the common velocity v of the projectile and washers. Determine also the loss $|\Delta E|$ of energy during the interaction.

$Ans.$ $v = 190.9$ m/s, $|\Delta E| = 17.18(10)^3$ J

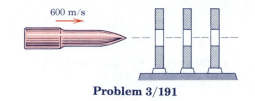

600 m/s

Problem 3/191

3/192 Car B is initially stationary and is struck by car A, which is moving with speed v. The cars become entangled and move with speed v' after the collision. The mass of car A is m and that of car B is pm, where p is the ratio of the mass of car B to that of car A. If the duration of the collision is Δt, express the common speed v' after the collision and the average acceleration of each car during the collision in terms of v, p, and Δt. Evaluate your expressions for $p = 0.5$.

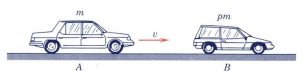

m pm

v

A B

Problem 3/192

3/193 The 450-kg ram of a pile driver falls 1.4 m from rest and strikes the top of a 240-kg pile embedded 0.9 m in the ground. Upon impact the ram is seen to move with the pile with no noticeable rebound. Determine the velocity v of the pile and ram immediately after impact. Can you justify using the principle of conservation of momentum even though the weights act during the impact?

$Ans.$ $v = 3.42$ m/s

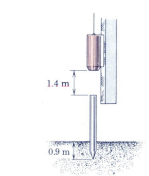

1.4 m

0.9 m

Problem 3/193

3/194 An emergency evacuation system at the launch tower for astronauts consists of a long slide-wire cable, down which the escape cage travels to a safe distance from the tower. The cage, together with its two occupants, has a mass of 320 kg and approaches the netting horizontally at a speed of 28 m/s. The netting is held to the cable by a breakaway lashing and is attached to 20 m of heavy chain with a mass of 18 kg/m. The coefficient of kinetic friction between the chain and the ground is 0.70. Determine the initial velocity v of the chain when the cage has engaged the net, and find the time t to bring the cage to a stop after engagement. Assume all links of the chain remain in contact with the ground.

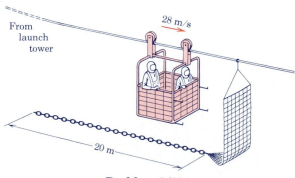

From launch tower

28 m/s

20 m

Problem 3/194

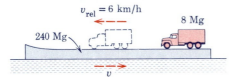

Problem 3/195

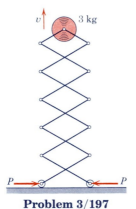

Problem 3/196

3/195 The tow truck with attached 1200-kg car accelerates uniformly from 30 km/h to 70 km/h over a 15-s interval. The average rolling resistance for the car over this speed interval is 500 N..Assume that the 60° angle shown represents the time average configuration and determine the average tension in the tow cable. *Ans.* $T = 2780$ N

3/196 An 8-Mg truck is resting on the deck of a barge that displaces 240 Mg and is at rest in still water. If the truck starts and drives toward the bow at a speed relative to the barge $v_{rel} = 6$ km/h, calculate the speed v of the barge. The resistance to the motion of the barge through the water is negligible at low speeds.

3/197 The vertical motion of the 3-kg cylinder is controlled by the forces P applied to the end rollers of the extensible framework shown. If the upward velocity v of the cylinder is increased from 2 m/s to 4 m/s in 2 seconds, calculate the average force R_{av} under each of the two rollers during the 2-s interval. Analyze the entire system as a unit and neglect the small mass of the framework. *Ans.* $R_{av} = 16.22$ N

3/198 Determine the time required by a diesel-electric locomotive, which produces a constant drawbar pull of 60,000 lb, to increase the speed of an 1800-ton freight train from 20 mi/hr to 30 mi/hr up a 1-percent grade. Train resistance is 10 lb per ton.

Problem 3/197

3/199 A spacecraft in deep space is programmed to increase its speed by a desired amount Δv by burning its engine for a specified time duration t. Twenty-five percent of the way through the burn, the engine suddenly malfunctions and thereafter produces only half of its normal thrust. What percent n of Δv is achieved if the rocket motor is fired for the planned time t? How much extra time t' would the rocket need to operate in order to compensate for the failure? *Ans.* $n = 62.5\%$, $t' = \frac{3}{4}t$

3/200 Careful measurements made during the impact of the 200-g metal cylinder with the spring-loaded plate reveal a semielliptical relation between the contact force F and the time t of impact as shown. Determine the rebound velocity v of the cylinder if it strikes the plate with a velocity of 6 m/s.

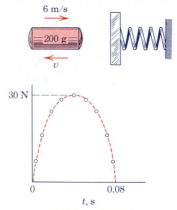

Problem 3/200

3/201 The motion of the 9-kg carriage in the smooth vertical guides is controlled by the tensions T_1 and T_2 in the cables. If the carriage has a downward velocity of 1.2 m/s at time $t = 0$, determine its velocity v when $t = 10$ s.   *Ans.* $v = 0.700$ m/s up

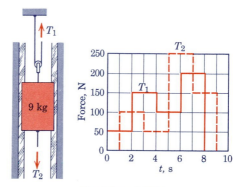

Problem 3/201

3/202 The third and fourth stages of a rocket are coasting in space with a velocity of 15 000 km/h when the fourth-stage engine ignites, causing separation of the stages under a thrust T and its reaction. If the velocity v of the fourth stage is 10 m/s greater than the velocity v_1 of the third stage at the end of the $\frac{1}{2}$-s separation interval, calculate the average thrust T during this period. The masses of the third and fourth stages at separation are 40 kg and 60 kg, respectively. Assume that the entire blast of the fourth-stage engine impinges against the third stage with a force equal and opposite to T.

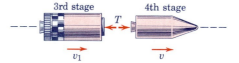

Problem 3/202

3/203 Car B weighing 3200 lb and traveling west at 30 mi/hr collides with car A weighing 3400 lb and traveling north at 20 mi/hr as shown. If the two cars become entangled and move together as a unit after the crash, compute the magnitude v of their common velocity immediately after the impact and the angle θ made by the velocity vector with the north direction.   *Ans.* $v = 17.82$ mi/hr, $\theta = 54.7°$

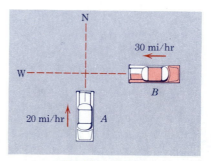

Problem 3/203

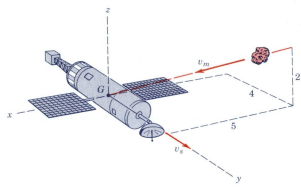

Problem 3/204

3/204 A 1000-kg spacecraft is traveling in deep space with a speed of v_s = 2000 m/s when a 10-kg meteor moving with a velocity $\mathbf{v}_m$ of magnitude 5000 m/s in the direction shown strikes and becomes embedded in the spacecraft. Determine the final velocity $\mathbf{v}$ of the mass center G of the spacecraft. Calculate the angle β between $\mathbf{v}$ and the initial velocity $\mathbf{v}_s$ of the spacecraft.

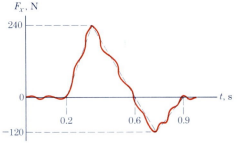

Problem 3/205

3/205 A 4-kg object, which is moving on a smooth horizontal surface with a velocity of 10 m/s in the $-x$-direction, is subjected to the force F_x that varies with the time as shown. Approximate the experimental data by the dotted line and determine the velocity of the object (a) at t = 0.6 s and (b) at t = 0.9 s. *Ans.* (a) v = 2 m/s, (b) v = -2.5 m/s

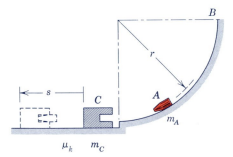

Problem 3/206

3/206 The cylindrical plug A of mass m_A is released from rest at B and slides down the smooth circular guide. The plug strikes the block C and becomes embedded in it. Write the expression for the distance s that the block and plug slide before coming to rest. The coefficient of kinetic friction between the block and the horizontal surface is μ_k.

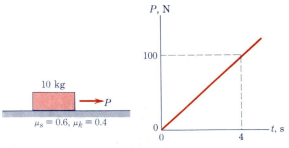

Problem 3/207

3/207 The force P, which is applied to the 10-kg block initially at rest, varies linearly with the time as indicated. If the coefficients of static and kinetic friction between the block and the horizontal surface are 0.6 and 0.4, respectively, determine the velocity of the block when t = 4 s. *Ans.* v = 6.61 m/s

3/208 The carriage weighs 64.4 lb and moves with negligible friction along the horizontal rail. A sinusoidal force $F = 10 \sin \pi t$ in pounds measured positive to the left acts on the carriage as shown. If the carriage has a velocity of 5 ft/sec to the right at $t = 0$, determine the velocity v of the carriage when $t = 1$ sec. By inspection what is the velocity of the carriage when $t = 2$ sec?

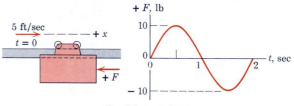

Problem 3/208

3/209 The 580-ton tug is towing the 1200-ton coal barge at a steady speed of 6 knots. For a short period of time, the stern winch takes in the towing cable at the rate of 2 ft/sec. Calculate the reduced speed v of the tug during this interval. Assume the tow cable to be horizontal. (Recall 1 knot = 1.690 ft/sec)

Ans. $v = 5.20$ knots

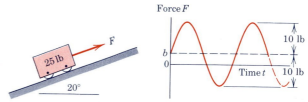

Problem 3/209

3/210 The 25-lb cart is stationary at time $t = 0$ and thereafter is subjected to the sinusoidal force $F = b + 10 \sin 6t$, where F and b are in pounds and time t is in seconds. (*a*) If $b = 5$ lb, determine the velocity v of the cart at $t = 1.5$ sec. (*b*) Determine the value of b for which the velocity of the cart would be zero after the first complete cycle of force application. Neglect friction.

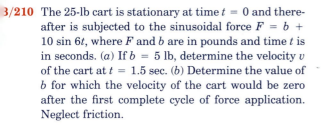

Problem 3/210

3/211 A 16.1-lb body is traveling in a horizontal straight line with a velocity of 12 ft/sec when a horizontal force P is applied to it at right angles to the initial direction of motion. If P varies according to the accompanying graph, remains constant in direction, and is the only force acting on the body in its plane of motion, find the magnitude of the velocity of the body when $t = 2$ sec and the angle θ it makes with the direction of P.

Ans. $v = 15.62$ ft/sec, $\theta = 50.2°$

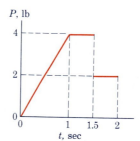

Problem 3/211

3/212 A 1-lb body oscillates along the horizontal x-axis under the sole action of an alternating force in the x-direction that decreases in amplitude with time t as shown and is given by $F_x = 4e^{-t} \cos 2\pi t$, where F_x is in pounds and t is in seconds. If the body has a velocity of 3 ft/sec in the negative x-direction at time $t = 0$, determine its velocity v_x at time $t = 2$ sec.

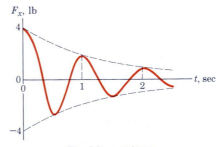

Problem 3/212

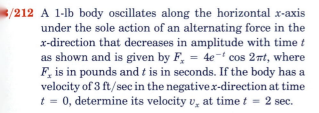

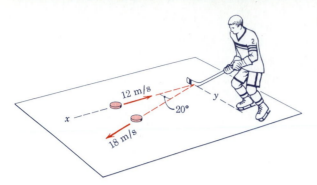

Problem 3/213

3/213 The ice-hockey puck with a mass of 0.20 kg has a velocity of 12 m/s before being struck by the hockey stick. After the impact the puck moves in the new direction shown with a velocity of 18 m/s. If the stick is in contact with the puck for 0.04 s, compute the magnitude of the average force $\mathbf{F}$ exerted by the stick on the puck during contact, and find the angle β made by $\mathbf{F}$ with the x-direction.

Ans. $F = 147.8$ N, $\beta = 12.02°$

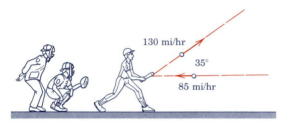

Problem 3/214

3/214 The baseball is traveling with a horizontal velocity of 85 mi/hr just before impact with the bat. Just after the impact, the velocity of the $5\frac{1}{8}$-oz ball is 130 mi/hr directed at 35° to the horizontal as shown. Determine the x- and y-components of the average force $\mathbf{R}$ exerted by the bat on the baseball during the 0.02-sec impact. Comment on treatment of the weight of the baseball (*a*) during the impact and (*b*) over the first few seconds after impact.

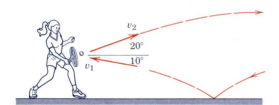

Problem 3/215

3/215 A tennis player strikes the tennis ball with her racket while the ball is still rising. The ball speed before impact with the racket is $v_1 = 15$ m/s and after impact its speed is $v_2 = 22$ m/s, with directions as shown in the figure. If the 60-g ball is in contact with the racket for 0.05 s, determine the magnitude of the average force $\mathbf{R}$ exerted by the racket on the ball. Find the angle β made by $\mathbf{R}$ with the horizontal. Comment on the treatment of the ball weight during impact.   *Ans.* $R = 43.0$ N, $\beta = 8.68°$

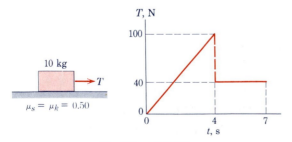

Problem 3/216

3/216 The 10-kg block is resting on the horizontal surface when the force T is applied to it for 7 seconds. The variation of T with time is shown. Calculate the maximum velocity reached by the block and the total time Δt during which the block is in motion. The coefficients of static and kinetic friction are both 0.50.

3/217 The loaded mine skip has a mass of 3 Mg. The hoisting drum produces a tension T in the cable according to the time schedule shown. If the skip is at rest against A when the drum is activated, determine the speed v of the skip when $t = 6$ s. Friction loss may be neglected.   *Ans.* $v = 9.13$ m/s

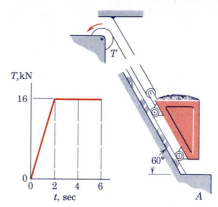

Problem 3/217

3/218 The simple pendulum A of mass m_A and length l is suspended from the trolley B of mass m_B. If the system is released from rest at $\theta = 0$, determine the velocity v_B of the trolley when $\theta = 90°$. Friction is negligible.

$$Ans. \quad v_B = \frac{m_A}{m_B}\sqrt{\frac{2gl}{1 + m_A/m_B}}$$

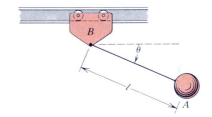

Problem 3/218

3/219 Two barges, each with a displacement (mass) of 500 metric tons, are loosely moored in calm water. A stunt driver starts his 1500-kg car from rest at A, drives along the deck, and leaves the end of the 15° ramp at a speed of 50 km/h relative to the barge and ramp. The driver successfully jumps the gap and brings his car to rest relative to barge 2 at B. Calculate the velocity v_2 imparted to barge 2 just after the car has come to rest on the barge. Neglect the resistance of the water to motion at the low velocities involved.   *Ans.* $v_2 = 40.0$ mm/s

Problem 3/219

3/220 A torpedo boat with a displacement of 60 long tons is moving at 18 knots when it fires a 300-lb torpedo horizontally with the launch tube at the 30° angle shown. If the torpedo has a velocity relative to the boat of 20 ft/sec as it leaves the tube, determine the momentary reduction Δv in forward speed of the boat. Recall that 1 knot = 1.689 ft/sec and 1 long ton = 2240 lb. Solve the problem also by referring the motion to a coordinate system moving with the initial speed of the boat.   *Ans.* $\Delta v = 0.04$ ft/sec

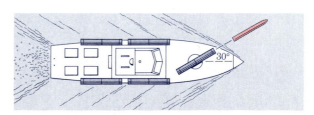

Problem 3/220

3/10 ANGULAR IMPULSE AND ANGULAR MOMENTUM

In addition to the equations of linear impulse and linear momentum, there exists a parallel set of equations for angular impulse and angular momentum. First, we define the term angular momentum. In Fig. 3/11a is shown a particle P of mass m moving along a curve in space. The particle is located by its position vector $\mathbf{r}$ with respect to a convenient origin O of fixed coordinates x-y-z. The velocity of the particle is $\mathbf{v} = \dot{\mathbf{r}}$, and its linear momentum is $\mathbf{G} = m\mathbf{v}$. The *moment* of the *linear momentum vector* $m\mathbf{v}$ about the origin O is defined as the *angular momentum* $\mathbf{H}_O$ of P about O and is given by the cross-product relation for the moment of a vector

$$\boxed{\mathbf{H}_O = \mathbf{r} \times m\mathbf{v}} \tag{3/25}$$

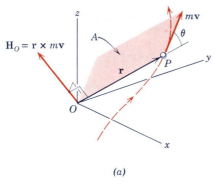

$$\mathbf{H}_O = \mathbf{r} \times m\mathbf{v}$$

The angular momentum then is a vector perpendicular to plane A defined by $\mathbf{r}$ and $\mathbf{v}$. The sense of $\mathbf{H}_O$ is clearly defined by the right-hand rule for cross products.

The scalar components of angular momentum may be obtained from the expansion

$$\mathbf{H}_O = \mathbf{r} \times m\mathbf{v} = m(v_z y - v_y z)\mathbf{i} + m(v_x z - v_z x)\mathbf{j} + m(v_y x - v_x y)\mathbf{k}$$

$$\mathbf{H}_O = m \begin{vmatrix} \mathbf{i} & \mathbf{j} & \mathbf{k} \\ x & y & z \\ v_x & v_y & v_z \end{vmatrix} \tag{3/26}$$

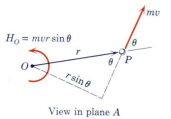

$$H_O = mvr\sin\theta$$

View in plane A

(b)

Figure 3/11

so that

$$H_x = m(v_z y - v_y z) \qquad H_y = m(v_x z - v_z x) \qquad H_z = m(v_y x - v_x y)$$

Each of these expressions for angular momentum may be checked easily from Fig. 3/12, which shows the three linear-momentum components, by taking the moments of these components about the respective axes.

To help visualize angular momentum, we show in Fig. 3/11b a two-dimensional representation in plane A of the vectors shown in the a-part of the figure. The motion is viewed in plane A defined by $\mathbf{r}$ and $\mathbf{v}$. The magnitude of the moment of $m\mathbf{v}$ about O is simply the linear momentum mv times the moment arm $r \sin \theta$ or $mvr \sin \theta$, which is the magnitude of the cross product $\mathbf{H}_O = \mathbf{r} \times m\mathbf{v}$.

Angular momentum is the moment of linear momentum and must not be confused with linear momentum. In SI units, angular momentum has the units $\text{kg} \cdot (\text{m/s}) \cdot \text{m} = \text{kg} \cdot \text{m}^2/\text{s} = \text{N} \cdot \text{m} \cdot \text{s}$. In the U.S. customary system, angular momentum has the units $[\text{lb}/(\text{ft/sec}^2)][\text{ft/sec}][\text{ft}] = \text{ft-lb-sec}$.

We are now ready to relate the moment of the forces acting on the particle P to its angular momentum. If $\Sigma \mathbf{F}$ represents the re-

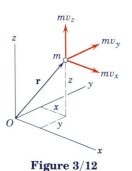

Figure 3/12

sultant of *all* forces acting on the particle P of Fig. 3/11, the moment $\mathbf{M}_O$ about the origin O is the vector cross product

$$\Sigma \mathbf{M}_O = \mathbf{r} \times \Sigma \mathbf{F} = \mathbf{r} \times m\dot{\mathbf{v}}$$

where Newton's second law $\Sigma \mathbf{F} = m\dot{\mathbf{v}}$ has been substituted. We now differentiate Eq. 3/25 with time, using the rule for the differentiation of a cross product (see item 9, Art. C/7, Appendix C) and get

$$\dot{\mathbf{H}}_O = \dot{\mathbf{r}} \times m\mathbf{v} + \mathbf{r} \times m\dot{\mathbf{v}} = \mathbf{v} \times m\mathbf{v} + \mathbf{r} \times m\dot{\mathbf{v}}$$

The term $\mathbf{v} \times m\mathbf{v}$ is zero since the cross product of parallel vectors is identically zero. Substitution into the expression for $\Sigma \mathbf{M}_O$ gives

$$\boxed{\Sigma \mathbf{M}_O = \dot{\mathbf{H}}_O} \tag{3/27}$$

Equation 3/27 states that the *moment about the fixed point O of all forces acting on m equals the time rate of change of angular momentum of m about O*. This relation, particularly when extended to a system of particles, rigid or nonrigid, constitutes one of the most powerful tools of analysis in all of dynamics. Equation 3/27 is a vector equation with scalar components

$$\Sigma M_{O_x} = \dot{H}_{O_x} \qquad \Sigma M_{O_y} = \dot{H}_{O_y} \qquad \Sigma M_{O_z} = \dot{H}_{O_z} \tag{3/28}$$

Equation 3/27 gives the instantaneous relation between the moment and the time rate of change of angular momentum. To obtain the effect of the moment $\Sigma \mathbf{M}_O$ on the angular momentum of the particle over a finite period of time, Eq. 3/27 may be integrated from time t_1 to time t_2. Multiplying the equation by dt gives $\Sigma \mathbf{M}_O \, dt = d\mathbf{H}_O$, which we integrate to obtain

$$\boxed{\int_{t_1}^{t_2} \Sigma \mathbf{M}_O \, dt = \mathbf{H}_{O_2} - \mathbf{H}_{O_1} = \Delta \mathbf{H}_O} \tag{3/29}$$

where $\mathbf{H}_{O_2} = \mathbf{r}_2 \times m\mathbf{v}_2$ and $\mathbf{H}_{O_1} = \mathbf{r}_1 \times m\mathbf{v}_1$. The product of moment and time is defined as *angular impulse*, and Eq. 3/29 states that the *total angular impulse on m about the fixed point O equals the corresponding change in angular momentum of m about O*.

Alternatively, we may write Eq. 3/29 as

$$\boxed{\mathbf{H}_{O_1} + \int_{t_1}^{t_2} \Sigma \mathbf{M}_O \, dt = \mathbf{H}_{O_2}} \tag{3/29a}$$

which states that the initial angular momentum of the particle plus the angular impulse applied to it equals its final angular momentum. The units of angular impulse are clearly those of angular momentum, which are N·m·s or kg·m²/s in SI units and ft-lb-sec in U.S. customary units.

As in the case of linear impulse and linear momentum, the equation of angular impulse and angular momentum is a vector equation where changes in direction as well as magnitude may occur during the interval of integration. Under these conditions, it is necessary to express $\Sigma \mathbf{M}_O$ and $\mathbf{H}_O$ in component form and then combine the integrated components. Thus, the x-component of Eq. 3/29 becomes

$$\int_{t_1}^{t_2} \Sigma M_{O_x}\, dt = (H_{O_x})_2 - (H_{O_x})_1$$

$$= m[(v_z y - v_y z)_2 - (v_z y - v_y z)_1]$$

where the subscripts 1 and 2 refer to the values of the respective quantities at times t_1 and t_2. Similar expressions exist for the y- and z-components of the angular momentum integral.

The foregoing angular-impulse and angular-momentum relations have been developed in their general three-dimensional sense. Most of the applications with which we shall be concerned, on the other hand, can be analyzed as plane-motion problems where moments are taken about a single axis normal to the plane of motion. In this event, the angular momentum may change magnitude and sense, but the direction of the vector remains unaltered. Thus, for a particle of mass m moving along a curved path in the x-y plane, Fig. 3/13, the angular momenta about O at points 1 and 2 have the magnitudes $H_{O_1} = |\mathbf{r}_1 \times m\mathbf{v}_1| = mv_1d_1$ and $H_{O_2} = |\mathbf{r}_2 \times m\mathbf{v}_2| = mv_2d_2$, respectively. In the illustration both H_{O_1} and H_{O_2} are represented in the counterclockwise sense according to the direction of the moment of the linear momentum. The scalar form of Eq. 3/29

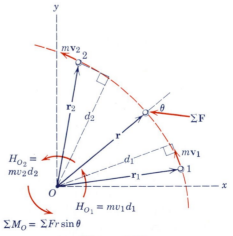

$$\Sigma M_O = \Sigma F r \sin \theta$$

Figure 3/13

applied to the motion between points 1 and 2 during the time interval t_1 to t_2 becomes

$$\int_{t_1}^{t_2} \Sigma M_O \, dt = H_{O_2} - H_{O_1} \quad \text{or} \quad \int_{t_1}^{t_2} \Sigma F r \sin \theta \, dt = mv_2 d_2 - mv_1 d_1$$

This example should help clarify the relation between the scalar and vector forms of the angular impulse-momentum relations.

Equations 3/21 and 3/27 add no new basic information since they are merely alternative forms of Newton's second law. However, we will discover in subsequent chapters that the motion equations expressed in terms of the time rate of change of momentum are applicable to the motion of rigid and nonrigid bodies and constitute a very general and powerful approach to many problems. The full generality of Eq. 3/27 is usually not required to describe the motion of a single particle or the plane motion of rigid bodies, but it does find important use in the analysis of the space motion of rigid bodies introduced in Chapter 7.

Conservation of angular momentum.   If the resultant moment about a fixed point O of all forces acting on a particle is zero during an interval of time, Eq. 3/27 requires that its angular momentum $\mathbf{H}_O$ about that point remain constant. In this case, the angular momentum of the particle is said to be *conserved*. Angular momentum may be conserved about one axis but not about another axis. A careful examination of the free-body diagram of the particle will disclose whether the moment of the resultant force on the particle about a fixed point is zero, in which case, the angular momentum about that point is unchanged (conserved).

Consider now the motion of two particles a and b that interact during an interval of time. If the interactive forces $\mathbf{F}$ and $-\mathbf{F}$ between them are the only unbalanced forces acting on the particles during the interval, it follows that the moments of the equal and opposite forces about any fixed point O not on their line of action are equal and opposite. If we apply Eq. 3/29 to particle a and then to particle b and add the two equations, we obtain $\Delta \mathbf{H}_a + \Delta \mathbf{H}_b = \mathbf{0}$ (where all angular momenta are referred to point O). Thus, the total angular momentum for the system of the two particles remains constant during the interval, and we write

$$\boxed{\Delta \mathbf{H}_O = \mathbf{0} \qquad \text{or} \qquad \mathbf{H}_{O_1} = \mathbf{H}_{O_2}} \qquad \textbf{(3/30)}$$

which expresses the *principle of conservation of angular momentum*.

Sample Problem 3/22

The small 3-lb block slides on a smooth horizontal surface under the action of the force in the spring and a force F. The angular momentum of the block about O varies with time as shown in the graph. When $t = 6.5$ sec, it is known that $r = 6$ in. and $\beta = 60°$. Determine F for this instant.

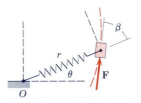

Solution. The only moment of the forces about O is due to F since the spring force passes through O. Thus, $\Sigma M_O = Fr \sin \beta$. From the graph the time rate of change of H_O for $t = 6.5$ sec is very nearly $(8 - 4)/(7 - 6)$ or $\dot{H}_O = 4$ ft-lb. The moment-angular momentum relation gives

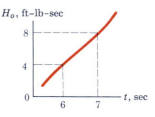

H_O, ft-lb-sec

① $[\Sigma M_O = \dot{H}_O]$ $\qquad F\left(\dfrac{6}{12}\right) \sin 60° = 4$ $\qquad F = 9.24$ lb $\qquad$ *Ans.*

① We do not need vector notation here since we have plane motion where the direction of the vector $\mathbf{H}_O$ does not change.

Sample Problem 3/23

A small mass particle is given an initial velocity $\mathbf{v}_0$ tangent to the horizontal rim of a smooth hemispherical bowl at a radius r_0 from the vertical centerline, as shown at point A. As the particle slides past point B, a distance h below A and a distance r from the vertical centerline, its velocity $\mathbf{v}$ makes an angle θ with the horizontal tangent to the bowl through B. Determine θ.

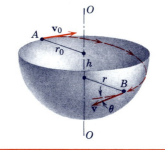

Solution. The forces on the particle are its weight and the normal reaction exerted by the smooth surface of the bowl. Neither force exerts a moment about the axis O-O, so that angular momentum is conserved about that axis. Thus,

① $[(H_O)_1 = (H_O)_2]$ $\qquad mv_0 r_0 = mvr \cos \theta$

Also, energy is conserved so that $E_1 = E_2$. Thus

$[T_1 + V_{g_1} = T_2 + V_{g_2}]$ $\qquad \tfrac{1}{2}mv_0^2 + mgh = \tfrac{1}{2}mv^2 + 0$

$$v = \sqrt{v_0^2 + 2gh}$$

Eliminating v and substituting $r^2 = r_0^2 - h^2$ give

$$v_0 r_0 = \sqrt{v_0^2 + 2gh}\sqrt{r_0^2 - h^2} \cos \theta$$

$$\theta = \cos^{-1} \frac{1}{\sqrt{1 + \dfrac{2gh}{v_0^2}}\sqrt{1 - \dfrac{h^2}{r_0^2}}} \qquad \textit{Ans.}$$

① The angle θ is measured in the plane tangent to the hemispherical surface at B.

PROBLEMS

Introductory problems

3/221 Determine the magnitude H_O of the angular momentum of the 2-kg sphere about point O (*a*) by using the vector definition of angular momentum and (*b*) by using an equivalent scalar approach. The center of the sphere lies in the *x*-*y* plane.

Ans. $H_O = 69.3 \text{ kg} \cdot \text{m}^2/\text{s}$

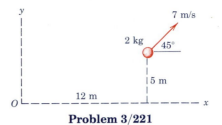

Problem 3/221

3/222 At a certain instant, the linear momentum of a particle is given by $\mathbf{G} = -3\mathbf{i} - 2\mathbf{j} + 3\mathbf{k}$ lb-sec and its position vector is $\mathbf{r} = 3\mathbf{i} + 4\mathbf{j} - 3\mathbf{k}$ ft. Determine the magnitude H_O of the angular momentum of the particle about the origin of coordinates.

3/223 A particle with a mass of 4 kg has a position vector in meters given by $\mathbf{r} = 3t^2\mathbf{i} - 2t\mathbf{j} - 3t\mathbf{k}$, where t is the time in seconds. For $t = 3$ s determine the magnitude of the angular momentum of the particle and the magnitude of the moment of all forces on the particle, both about the origin of coordinates.

Ans. $H = 389 \text{ N} \cdot \text{m} \cdot \text{s}, M = 260 \text{ N} \cdot \text{m}$

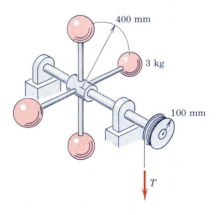

Problem 3/224

3/224 The assembly starts from rest and reaches an angular speed of 150 rev/min under the action of a 20-N force T applied to the string for t seconds. Determine t. Neglect friction and all masses except those of the four 3-kg spheres, which may be treated as particles.

3/225 A particle of mass m moves with negligible friction on a horizontal surface and is connected to a light spring fastened at O. At position A the particle has the velocity $v_A = 4$ m/s. Determine the velocity v_B of the particle as it passes position B.

Ans. $v_B = 5.43$ m/s

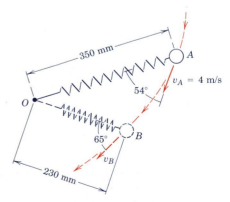

Problem 3/225

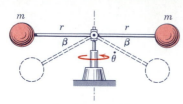

Problem 3/226

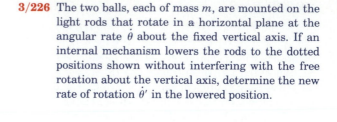

3/226 The two balls, each of mass m, are mounted on the light rods that rotate in a horizontal plane at the angular rate $\dot{\theta}$ about the fixed vertical axis. If an internal mechanism lowers the rods to the dotted positions shown without interfering with the free rotation about the vertical axis, determine the new rate of rotation $\dot{\theta}'$ in the lowered position.

Representative problems

3/227 The small sphere of mass m traveling with speed v strikes and becomes attached to the end of the stationary assembly that pivots freely about a vertical axis at O. Determine the angular velocity ω of the assembly after impact and calculate the change ΔE in the system energy. Neglect the mass of the rod compared with m.

$$Ans. \quad \omega = \frac{v}{2L}, \; \Delta E = -\frac{1}{4}mv^2$$

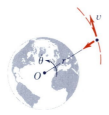

Problem 3/227

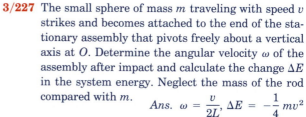

3/228 The only force acting on an earth satellite traveling outside of the earth's atmosphere is the radial gravitational attraction. The moment of this force is zero about the earth's center taken as a fixed point. Prove that $r^2\dot{\theta}$ remains constant for the motion of the satellite.

Problem 3/228

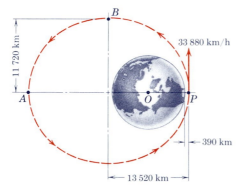

3/229 The central attractive force F on an earth satellite can exert no moment about the center O of the earth. For the particular elliptical orbit with major and minor axes as shown, a satellite will have a velocity of 33 880 km/h at the perigee altitude of 390 km. Determine the velocity of the satellite at point B and at apogee A. The radius of the earth is 6371 km.

$$Ans. \quad v_B = 19\,540 \text{ km/h}$$
$$v_A = 11\,290 \text{ km/h}$$

Problem 3/229

3/230 Use the relationship for time rate of change of angular momentum to express the angular acceleration $\ddot{\theta}$ of the simple pendulum in terms of θ.

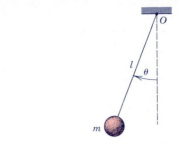

Problem 3/230

3/231 A small 0.1-kg particle is given a velocity of 2 m/s on the horizontal x-y plane and is guided by the fixed curved rail. Friction is negligible. As the particle crosses the y-axis at A, its velocity is in the x-direction, and as it crosses the x-axis at B, its velocity makes a 60° angle with the x-axis. The radius of curvature of the path at B is 500 mm. Determine the time rate of change of angular momentum H_O of the particle about the z-axis through O at both A and B.

$$Ans. \text{ At } A, \dot{H}_O = 0$$
$$\text{At } B, \dot{H}_O = -0.120 \text{ kg} \cdot \text{m}^2/\text{s}^2$$

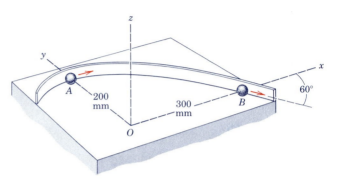

Problem 3/231

3/232 The two spheres of equal mass m are able to slide along the horizontal rotating rod. If they are initially latched in position a distance r from the rotating axis with the assembly rotating freely with an angular velocity ω_0, determine the new angular velocity ω after the spheres are released and finally assume positions at the ends of the rod at a radial distance of $2r$. Also find the fraction n of the initial kinetic energy of the system that is lost. Neglect the small mass of the rod and shaft.

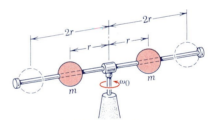

Problem 3/232

3/233 Estimate the magnitudes of the angular momentum of the moon about the center of the earth and that of the earth about the center of the sun. Assume a circular orbit in each case and refer to Table D/2 as needed.

$$Ans. \ H_E = 2.66(10^{40}) \text{ kg} \cdot \text{m}^2/\text{s}$$
$$H_M = 2.89(10^{34}) \text{ kg} \cdot \text{m}^2/\text{s}$$

3/234 The 6-kg sphere and 4-kg block (shown in section) are secured to the arm of negligible mass that rotates in the vertical plane about a horizontal axis at O. The 2-kg plug is released from rest at A and falls into the recess in the block when the arm has reached the horizontal position. An instant before engagement, the arm has an angular velocity $\omega_0 = 2$ rad/s. Determine the angular velocity ω of the arm immediately after the plug has wedged itself in the block.

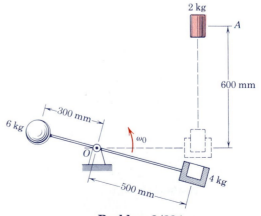

Problem 3/234

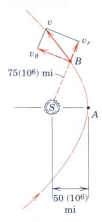

Problem 3/235

3/235 At the point A of closest approach to the sun, a comet has a velocity $v_A = 188,500$ ft/sec. Determine the radial and transverse components of its velocity v_B at point B, where the radial distance from the sun is $75(10^6)$ mi.

Ans. $v_r = 88,870$ ft/sec, $v_\theta = 125,700$ ft/sec

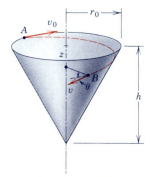

Problem 3/236

3/236 A particle moves on the inside surface of a smooth conical shell and is given an initial velocity $\mathbf{v_0}$ tangent to the horizontal rim of the surface at A. As the particle slides past point B, a distance z below A, its velocity $\mathbf{v}$ makes an angle θ with the horizontal tangent to the surface through B. Determine expressions for θ and the speed v.

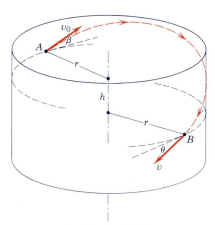

Problem 3/237

3/237 A particle is released on the smooth inside wall of a cylindrical tank at A with a velocity v_0 that makes an angle β with the horizontal tangent. When the particle reaches a point B a distance h below A, determine the expression for the angle θ made by its velocity with the horizontal tangent at B.

$$Ans. \quad \theta = \cos^{-1} \frac{\cos \beta}{\sqrt{1 + \dfrac{2gh}{v_0{}^2}}}$$

3/238 For Prob. 3/237 with given values of v_0, β, and r, find the maximum height b above point A reached by the particle and the magnitude v of its velocity at this point.

3/239 Determine the angular momentum H_O about the launch point O of the projectile of mass m that is launched with speed v_0 at the angle θ as shown (a) at the instant of launch and (b) at the instant of impact. Qualitatively account for the two results. Neglect atmospheric resistance.

$$\text{Ans. } (a)\ H_O = 0, \ (b)\ H_O = \frac{2mv_0^3 \sin^2 \theta \cos \theta}{g}$$

Problem 3/239

3/240 A pendulum consists of two 3.2-kg concentrated masses positioned as shown on a light but rigid bar. The pendulum is swinging through the vertical position with a clockwise angular velocity $\omega = 6$ rad/s when a 50-g bullet traveling with velocity $v = 300$ m/s in the direction shown strikes the lower mass and becomes embedded on it. Calculate the angular velocity ω' that the pendulum has immediately after impact and find the maximum angular deflection θ of the pendulum.

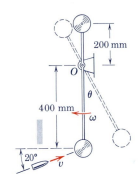

Problem 3/240

3/241 The small particle of mass m and its restraining cord are spinning with an angular velocity ω on the horizontal surface of a smooth disk, shown in section. As the force F is slightly relaxed, r increases and ω changes. Determine the rate of change of ω with respect to r and show that the work done by F during a movement dr equals the change in kinetic energy of the particle.

$$\text{Ans. } \frac{d\omega}{dr} = -\frac{2\omega}{r}$$

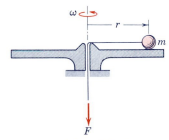

Problem 3/241

3/242 The assembly of two 5-kg spheres is rotating freely about the vertical axis at 40 rev/min with $\theta = 90°$. If the force F necessary to maintain the given position is increased to raise the base collar and reduce θ to $60°$, determine the new angular velocity ω. Also determine the work U done by F in changing the configuration of the system. Assume that the mass of the arms and collars is negligible.

$$\text{Ans. } \omega = 3.00 \text{ rad/s}, U = 5.34 \text{ J}$$

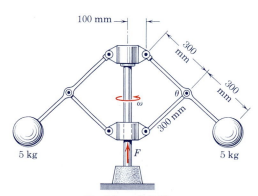

Problem 3/242

SECTION D. SPECIAL APPLICATIONS

3/11   *INTRODUCTION*

The basic principles and methods of particle kinetics were developed and illustrated in the first three sections of this chapter. This treatment included the direct use of Newton's second law, the equations of work and energy, and the equations of impulse and momentum. Special attention was paid to the kind of problem for which each of the approaches was most appropriate.

There are several topics of specialized interest in particle kinetics that will be briefly treated in Section D. These topics involve further extension and application of the fundamental principles of dynamics, and their study will help to broaden one's background in mechanics.

3/12   *IMPACT*

The principles of impulse and momentum find important use in describing the behavior of colliding bodies. *Impact* refers to the collision between two bodies and is characterized by the generation of relatively large contact forces that act over a very short interval of time. Before we begin treatment of the subject, it should be stated that an impact is a very complex event involving material deformation and recovery and the generation of heat and sound. Small changes in the impact conditions may cause large changes in the impact process and thus in the conditions immediately following the impact. Hence, we must be careful not to place excessive reliance on the results of impact calculations.

(a) *Direct central impact.* As an introduction to impact, we consider the collinear motion of two spheres of masses m_1 and m_2, Fig. 3/14a, traveling with velocities v_1 and v_2. If v_1 is greater than v_2, collision occurs with the contact forces directed along the line of centers. This condition is called *direct central impact*. Following initial contact, a short period of increasing deformation takes place

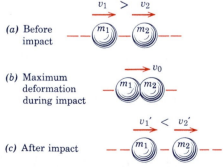

(a) Before impact

$$v_1 \; > \; v_2$$

(b) Maximum deformation during impact

$$v_0$$

(c) After impact

$$v_1' \; < \; v_2'$$

Figure 3/14

until the contact area between the spheres ceases to increase. At this instant, both spheres, Fig. 3/14*b*, are moving with the same velocity v_0. During the remainder of contact, a period of restoration occurs during which the contact area decreases to zero. The final condition is shown in the *c*-part of the figure where the spheres now have new velocities v_1' and v_2', where of necessity v_1' is less than v_2'. All velocities are arbitrarily assumed positive to the right, so that with this scalar notation a velocity to the left would carry a negative sign. If the impact is not overly severe and if the spheres are highly elastic, they will regain their original shape following the restoration. With a more severe impact and with less elastic bodies, a permanent deformation may result.

Inasmuch as the contact forces are equal and opposite during impact, the linear momentum of the system remains unchanged, as discussed in Art. 3/9. Thus, we apply the law of conservation of linear momentum and write

$$m_1 v_1 + m_2 v_2 = m_1 v_1' + m_2 v_2' \qquad (3/31)$$

It is assumed that any forces other than the large internal forces of contact that may act on the spheres during impact are relatively small and produce negligible impulses compared with the impulse associated with each internal impact force. In addition, we assume that no appreciable position change occurs during the short duration of the impact.

For given masses and initial conditions, the momentum equation contains two unknowns, v_1' and v_2'. Clearly, an additional relationship is required before the final velocities can be found. This relationship must reflect the capacity of the contacting bodies to recover from the impact and can be expressed by the ratio e of the magnitude of the restoration impulse to the magnitude of the deformation impulse. This ratio is called the *coefficient of restitution*. If F_r and F_d represent the magnitudes of the contact forces during the restoration and deformation periods, respectively, as shown in Fig. 3/15, for particle 1 the definition of e together with the impulse-momentum equation give us

$$e = \frac{\int_{t_0}^{t} F_r \, dt}{\int_{0}^{t_0} F_d \, dt} = \frac{m_1[-v_1' - (-v_0)]}{m_1[-v_0 - (-v_1)]} = \frac{v_0 - v_1'}{v_1 - v_0}$$

Similarly, for particle 2 we have

$$e = \frac{\int_{t_0}^{t} F_r \, dt}{\int_{0}^{t_0} F_d \, dt} = \frac{m_2(v_2' - v_0)}{m_2(v_0 - v_2)} = \frac{v_2' - v_0}{v_0 - v_2}$$

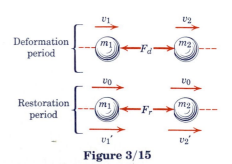

Figure 3/15

We are careful in these equations to express the change of momentum (and hence Δv) in the same direction as the impulse (and hence the force). The time for the deformation is taken as t_0 and the total time of contact is t. Eliminating v_0 between the two expressions for e gives us

$$e = \frac{v_2' - v_1'}{v_1 - v_2} = \frac{|\text{relative velocity of separation}|}{|\text{relative velocity of approach}|} \qquad (3/32)$$

In addition to the initial conditions, if e is known for the impact condition at hand, then Eqs. 3/31 and 3/32 give us two equations in the two unknown final velocities v_1' and v_2'.

Impact phenomena are almost always accompanied by energy loss, which may be calculated by subtracting the kinetic energy of the system just after impact from that just before impact. Energy is lost through the generation of heat during the localized inelastic deformation of the material, through the generation and dissipation of elastic stress waves within the bodies, and through the generation of sound energy.

According to this classical theory of impact, the value $e = 1$ means that the capacity of the two particles to recover equals their tendency to deform. This condition is one of *elastic impact* with no energy loss. The value $e = 0$, on the other hand, describes *inelastic* or *plastic impact* where the particles cling together after collision and the loss of energy is a maximum. All impact conditions lie somewhere between these two extremes. Also it should be noted that a coefficient of restitution must be associated with a *pair* of contacting bodies.

The coefficient of restitution is frequently considered a constant for given geometries and a given combination of contacting materials. Actually, it depends on the impact velocity and approaches unity as the impact velocity approaches zero as shown schematically in Fig. 3/16. A handbook value for e is generally unreliable.

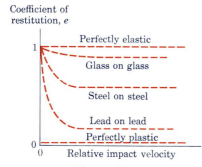

Figure 3/16

(b) Oblique central impact.   We now extend the relationships developed for direct central impact to the case where the initial and final velocities are not parallel, Fig. 3/17. Here spherical particles of mass m_1 and m_2 have initial velocities $\mathbf{v}_1$ and $\mathbf{v}_2$ in the same plane and approach each other on a collision course, as shown in the a-part of the figure. The directions of the velocity vectors are measured from the direction tangent to the contacting surfaces, Fig. 3/17b. Thus, the initial velocity components along the t- and n-axes are $(v_1)_n = -v_1 \sin \theta_1$, $(v_1)_t = v_1 \cos \theta_1$, $(v_2)_n = v_2 \sin \theta_2$, and $(v_2)_t = v_2 \cos \theta_2$. Note that $(v_1)_n$ is a negative quantity for the particular coordinate system and initial velocities shown. The final rebound conditions are shown in the c-part of the figure. The impact forces are $\mathbf{F}$ and $-\mathbf{F}$, as seen in the d-part of the figure. They vary

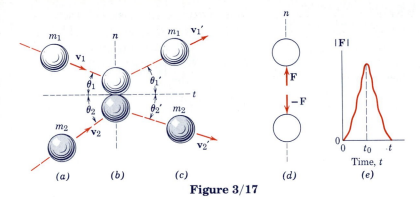

Figure 3/17

from zero to their peak value during the deformation portion of the impact and back again to zero during the restoration period, as indicated in the *e*-part of the figure where t is the duration of the impact interval.

For given initial conditions of m_1, m_2, $(v_1)_n$, $(v_1)_t$, $(v_2)_n$, and $(v_2)_t$, there will be four unknowns, namely, $(v_1')_n$, $(v_1')_t$, $(v_2')_n$, and $(v_2')_t$. The four needed equations are:

(1) Momentum of the system is conserved in the *n*-direction, giving

$$m_1(v_1)_n + m_2(v_2)_n = m_1(v_1')_n + m_2(v_2')_n$$

(2) and (3) The momentum for each particle is conserved in the *t*-direction since there is no impulse on either particle in the *t*-direction. Thus,

$$m_1(v_1)_t = m_1(v_1')_t$$

$$m_2(v_2)_t = m_2(v_2')_t$$

(4) The coefficient of restitution, as in the case of direct central impact, is the positive ratio of the recovery impulse to the deformation impulse. Equation 3/32 applies, then, to the velocity components in the *n*-direction, and for the notation adopted with Fig. 3/17, we have

$$e = \frac{(v_2')_n - (v_1')_n}{(v_1)_n - (v_2)_n}$$

Once the four final velocity components are found, the angles θ_1' and θ_2' of Fig. 3/17 may be easily determined.

Sample Problem 3/24

The ram of a pile driver has a mass of 800 kg and is released from rest 2 m above the top of the 2400-kg pile. If the ram rebounds to a height of 0.1 m after impact with the pile, calculate (a) the velocity v_p' of the pile immediately after impact, (b) the coefficient of restitution e, and (c) the percentage loss of energy due to the impact.

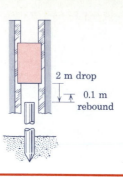

2 m drop

0.1 m rebound

Solution. Conservation of energy during free fall gives the initial and final velocities of the ram from $v = \sqrt{2gh}$. Thus,

$$v_r = \sqrt{2(9.81)(2)} = 6.26 \text{ m/s} \qquad v_r' = \sqrt{2(9.81)(0.1)} = 1.40 \text{ m/s}$$

(a) Conservation of momentum ($G_1 = G_2$) for the system of the ram and pile gives

① $\qquad 800(6.26) + 0 = 800(-1.40) + 2400v_p' \qquad v_p' = 2.55 \text{ m/s} \qquad Ans.$

(b) The coefficient of restitution yields

$$e = \frac{|\text{rel. vel. separation}|}{|\text{rel. vel. approach}|} \qquad e = \frac{2.55 + 1.40}{6.26 + 0} = 0.63 \qquad Ans.$$

(c) The kinetic energy of the system just before impact is the same as the potential energy of the ram above the pile and is

$$T = V_g = mgh = 800(9.81)(2) = 15\ 700 \text{ J}$$

The kinetic energy T' just after impact is

$$T' = \tfrac{1}{2}(800)(1.40)^2 + \tfrac{1}{2}(2400)(2.55)^2 = 8620 \text{ J}$$

The percentage loss of energy is, therefore,

$$\frac{15\ 700 - 8620}{15\ 700}(100) = 45.1\% \qquad Ans.$$

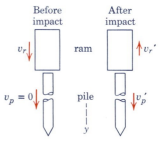

Before impact | After impact

① The impulses of the weights of the ram and pile are very small compared with the impulses of the impact forces and thus are neglected during the impact.

Sample Problem 3/25

A ball is projected onto the heavy plate with a velocity of 50 ft/sec at the 30° angle shown. If the effective coefficient of restitution is 0.5, compute the rebound velocity v' and its angle θ'.

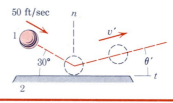

50 ft/sec

Solution. Let the ball be denoted body 1 and the plate body 2. The mass of the heavy plate may be considered infinite and its corresponding velocity zero after impact. The coefficient of restitution is applied to the velocity components normal to the plate in the direction of the impact force and gives

① $e = \dfrac{(v_2')_n - (v_1')_n}{(v_1)_n - (v_2)_n} \qquad 0.5 = \dfrac{0 - (v_1')_n}{-50 \sin 30° - 0} \qquad (v_1')_n = 12.5 \text{ ft/sec}$

Momentum of the ball in the t-direction is unchanged since, with assumed smooth surfaces, there is no force acting on the ball in that direction. Thus,

$$m(v_1)_t = m(v_1')_t \qquad (v_1')_t = (v_1)_t = 50 \cos 30° = 43.3 \text{ ft/sec}$$

The rebound velocity v' and its angle θ' are then

$$v' = \sqrt{(v_1')_n{}^2 + (v_1')_t{}^2} = \sqrt{12.5^2 + 43.3^2} = 45.1 \text{ ft/sec} \qquad Ans.$$

$$\theta' = \tan^{-1}\left(\frac{(v_1')_n}{(v_1')_t}\right) = \tan^{-1}\left(\frac{12.5}{43.3}\right) = 16.1° \qquad Ans.$$

$W \ll F_{impact}$

F_{impact}

① We observe here that for infinite mass there is no way of applying the principle of conservation of momentum for the system in the n-direction. From the free-body diagram of the ball during impact, we note that the impulse of the weight W is neglected since W is very small compared with the impact force.

Sample Problem 3/26

Spherical particle 1 has a velocity $v_1 = 6$ m/s in the direction shown and collides with spherical particle 2 of equal mass and diameter and initially at rest. If the coefficient of restitution for these conditions is $e = 0.6$, determine the resulting motion of each particle following impact. Also calculate the percentage loss of energy due to the impact.

Solution. The geometry at impact indicates that the normal n to the contacting surfaces makes an angle $\theta = 30°$ with the direction of $\mathbf{v}_1$, as indicated in the figure. Thus, the initial velocity components are $(v_1)_n = v_1 \cos 30° = 6 \cos 30° = 5.196$ m/s, $(v_1)_t = v_1 \sin 30° = 6 \sin 30° = 3$ m/s, and $(v_2)_n = (v_2)_t = 0$.

Momentum conservation for the two-particle system in the n-direction gives

$$m_1(v_1)_n + m_2(v_2)_n = m_1(v_1')_n + m_2(v_2')_n$$

or, with $m_1 = m_2$,

$$5.196 + 0 = (v_1')_n + (v_2')_n \qquad (a)$$

The coefficient of restitution relationship is

$$e = \frac{(v_2')_n - (v_1')_n}{(v_1)_n - (v_2)_n} \qquad 0.6 = \frac{(v_2')_n - (v_1')_n}{5.196 - 0} \qquad (b)$$

Simultaneous solution of Eqs. a and b yields

$$(v_1')_n = 1.039 \text{ m/s} \qquad (v_2')_n = 4.16 \text{ m/s}$$

Conservation of momentum for each particle holds in the t-direction because, with assumed smooth surfaces, there is no force in the t-direction. Thus for particles 1 and 2, we have

$$m_1(v_1)_t = m_1(v_1')_t \qquad (v_1')_t = (v_1)_t = 3 \text{ m/s}$$
$$m_2(v_2)_t = m_2(v_2')_t \qquad (v_2')_t = (v_2)_t = 0$$

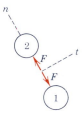

The final speeds of the particles are

$$v_1' = \sqrt{(v_1')_n^2 + (v_1')_t^2} = \sqrt{(1.039)^2 + 3^2} = 3.17 \text{ m/s} \qquad Ans.$$
$$v_2' = \sqrt{(v_2')_n^2 + (v_2')_t^2} = \sqrt{(4.16)^2 + 0^2} = 4.16 \text{ m/s} \qquad Ans.$$

The angle θ' that $\mathbf{v}_1'$ makes with the t-direction is

$$\theta' = \tan^{-1}\left(\frac{(v_1')_n}{(v_1')_t}\right) = \tan^{-1}\left(\frac{1.039}{3}\right) = 19.1° \qquad Ans.$$

The kinetic energies just before and just after impact, with $m = m_1 = m_2$, are

$$T = \tfrac{1}{2}m_1v_1^2 + \tfrac{1}{2}m_2v_2^2 = \tfrac{1}{2}m(6)^2 + 0 = 18m$$

$$T' = \tfrac{1}{2}m_1v_1'^2 + \tfrac{1}{2}m_2v_2'^2 = \tfrac{1}{2}m(3.17)^2 + \tfrac{1}{2}m(4.16)^2 = 13.68m$$

The percentage energy loss is then

$$\frac{|\Delta E|}{E}(100) = \frac{T - T'}{T}(100) = \frac{18m - 13.68m}{18m}(100) = 24\% \qquad Ans.$$

① Be sure to set up n- and t-coordinates that are, respectively, normal to and tangent to the contacting surfaces. Calculation of the 30° angle is critical to all that follows.

② Note that, even though there are four equations in four unknowns for the standard problem of oblique central impact, only one pair of the equations is coupled.

③ We note that particle 2 has no initial or final velocity component in the t-direction. Hence, its final velocity $\mathbf{v}_2'$ is restricted to the n-direction.

PROBLEMS

Introductory problems

$v_1 = 7$ m/s $v_2 = 5$ m/s

$m_1 = 2$ kg $m_2 = 3$ kg

Problem 3/243

3/243 Compute the final velocities v_1' and v_2' after colli-sion of the two cylinders that slide on the smooth horizontal shaft. The coefficient of restitution is $e = 0.6$.

> *Ans.* $v_1' = 4.52$ m/s to the left
> $v_2' = 2.68$ m/s to the right

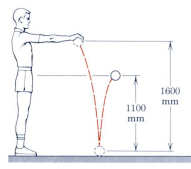

1600 mm

1100 mm

Problem 3/244

3/244 Tennis balls are usually rejected if they fail to re-bound to waist level when dropped from shoulder level. If a ball just passes the test as indicated in the figure, determine the coefficient of restitution e and the percentage n of the original energy lost during the impact.

3/245 If the tennis ball of Prob. 3/244 has a coefficient of restitution $e = 0.8$ during impact with the court sur-face, determine the velocity v_0 with which the ball must be thrown downward from the 1600-mm shoulder level if it is to return to the same level after bouncing once on the court surface.

> *Ans.* $v_0 = 4.20$ m/s

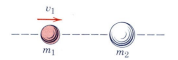

m pm

v

A B

Problem 3/246

3/246 The two cars of Prob. 3/192 are repeated here. Car B is initially stationary and is struck by car A, which is moving with speed v. The mass of car B is pm, where m is the mass of car A and p is a positive constant. If the coefficient of restitution is $e = 0.1$, express the speeds v_A' and v_B' of the two cars at the end of the impact in terms of p and v. Evaluate your expressions for $p = 0.5$.

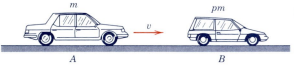

v_1

m_1 m_2

Problem 3/247

3/247 The sphere of mass m_1 travels with an initial veloc-ity v_1 directed as shown and strikes the sphere of mass m_2. For a given coefficient of restitution e, de-termine the mass ratio m_1/m_2 that results in m_1 being motionless after the impact.   *Ans.* $\dfrac{m_1}{m_2} = e$

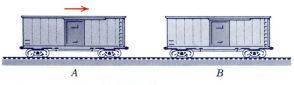

A B

Problem 3/248

3/248 Freight car A of mass m_A is rolling to the right when it collides with freight car B of mass m_B initially at rest. If the two cars are coupled together at impact, show that the fractional loss of energy equals $m_B/(m_A + m_B)$.

Representative problems

3/249 Three identical steel cylinders are free to slide horizontally on the fixed horizontal shaft. Cylinders 2 and 3 are at rest and are approached by cylinder 1 at a speed u. Express the final speed v of cylinder 3 in terms of u and the coefficient of restitution e.

$$Ans. \ v = \frac{u}{4}(1 + e)^2$$

Problem 3/249

3/250 Cylinder 1 of mass m moving with a velocity v_1 strikes cylinder 2 of mass $2m$ initially at rest. The impact force F varies with time as shown, where t_d is the duration of the deformation period and t_r is the duration of the restoration period. Determine the velocity v_2' of cylinder 2 immediately after impact in terms of the initial velocity v_1 of cylinder 1 for (a) $t_r = t_d$, (b) $t_r = 0.5t_d$, and (c) $t_r = 0$.

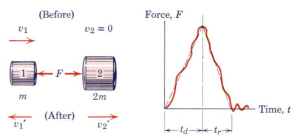

Problem 3/250

3/251 Determine the coefficient of restitution e for a steel ball dropped from rest at a height h above a heavy horizontal steel plate if the height of the second rebound is h_2.

$$Ans. \ e = \left(\frac{h_2}{h}\right)^{1/4}$$

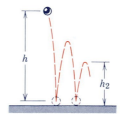

Problem 3/251

3/252 In the selection of the ram of a pile driver, it is desired that the ram lose all of its kinetic energy at each blow. Hence, the velocity of the ram is zero immediately after impact. The mass of each pile to be driven is 300 kg, and experience has shown that a coefficient of restitution of 0.3 can be expected. What should be the mass m of the ram? Compute the velocity v of the pile immediately after impact if the ram is dropped from a height of 4 m onto the pile. Also compute the energy loss ΔE due to impact at each blow.

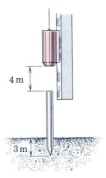

Problem 3/252

3/253 If the center of the ping-pong ball is to clear the net as shown, at what height h should the ball be horizontally served? Also determine h_2. The coefficient of restitution for the impacts between ball and table is $e = 0.9$, and the radius of the ball is $r = 0.75$ in.

$$Ans. \ h = 10.94 \ in., \ h_2 = 7.43 \ in.$$

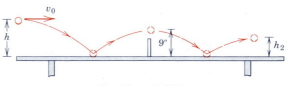

Problem 3/253

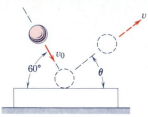

Problem 3/254

3/254 The steel ball strikes the heavy steel plate with a velocity $v_0 = 24$ m/s at an angle of 60° with the horizontal. If the coefficient of restitution is $e = 0.8$, compute the velocity v and its direction θ with which the ball rebounds from the plate.

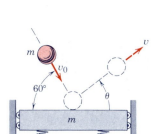

Problem 3/255

3/255 The previous problem is modified in that the plate struck by the ball now has a mass equal to that of the ball and is supported as shown. Compute the final velocities of both masses immediately after impact if the plate is initially stationary and all other conditions are the same as stated in the previous problem.

> *Ans.* Ball, $v_1' = 12.20$ m/s, $\theta = -9.83°$
> Plate, $v_2' = 18.71$ m/s down

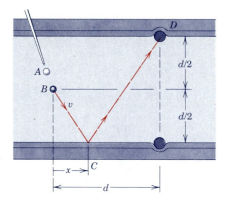

Problem 3/256

3/256 Pool ball B is to be shot into the side pocket D by banking it off the cushion at C. Specify the location x of the cushion impact for coefficients of restitution (*a*) $e = 1$ and (*b*) $e = 0.8$.

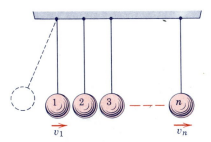

Problem 3/257

3/257 The figure shows n spheres of equal mass m suspended in a line by wires of equal length so that the spheres are almost touching each other. If sphere 1 is released from the dotted position and strikes sphere 2 with a velocity v_1, write an expression for the velocity v_n of the nth sphere immediately after being struck by the one adjacent to it. The common coefficient of restitution is e.

> *Ans.* $v_n = \left(\dfrac{1 + e}{2}\right)^{n-1} v_1$

3/258 The ball is released from position A and drops 0.75 m to the incline. If the coefficient of restitution in the impact is $e = 0.85$, determine the slant range R.

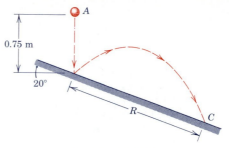

Problem 3/258

3/259 In a pool game the cue ball A must strike the eight ball in the position shown in order to send it to the pocket P with a velocity v_2'. The cue ball has a velocity v_1 before impact and a velocity v_1' after impact. The coefficient of restitution is 0.9. Both balls have the same mass and diameter. Calculate the rebound angle θ and the fraction n of the kinetic energy that is lost at impact.

Ans. $\theta = 2.86°$, $n = 0.0475$

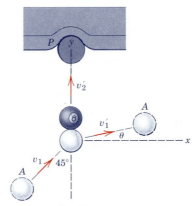

Problem 3/259

3/260 The two identical steel balls moving with initial velocities v_A and v_B collide as shown. If the coefficient of restitution is $e = 0.7$, determine the velocity of each ball just after impact and the percentage loss n of system kinetic energy.

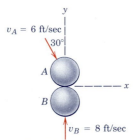

Problem 3/260

3/261 Sphere A has a mass of 23 kg and a radius of 75 mm, while sphere B has a mass of 4 kg and a radius of 50 mm. If the spheres are traveling initially along the parallel paths with the speeds shown, determine the velocities of the spheres immediately after impact. Specify the angles θ_A and θ_B with respect to the x-axis made by the rebound velocity vectors. The coefficient of restitution is 0.4 and friction is neglected.   *Ans.* $v_A' = 2.46$ m/s, $\theta_A = 40.3°$

$v_B' = 9.16$ m/s, $\theta_B = -88.7°$

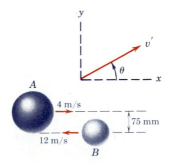

Problem 3/261

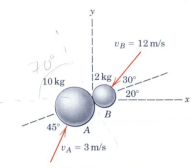

Problem 3/262

3/262 Sphere A collides with sphere B as shown in the figure. If the coefficient of restitution is $e = 0.5$, determine the x- and y-components of the velocity of each sphere immediately after impact.

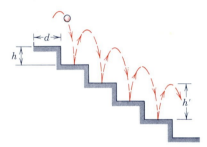

Problem 3/263

3/263 Determine the coefficient of restitution e that will allow the ball to bounce down the steps as shown. The tread and riser dimensions, d and h, respectively, are the same for every step, and the ball bounces the same distance h' above each step. What horizontal velocity v_x is required so that the ball lands in the center of each tread?

$$\text{Ans. } e = \sqrt{\frac{h'}{h' + h}}, \quad v_x = \frac{\sqrt{\frac{g}{2}}\,d}{\sqrt{h'} + \sqrt{h' + h}}$$

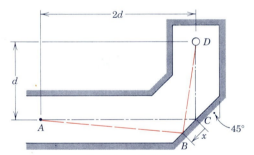

Problem 3/264

3/264 A miniature-golf shot from position A to the hole D is to be accomplished by "banking off" the 45° wall. Using the theory of this article, determine the location x for which the shot can be made. The coefficient of restitution associated with the wall collision is $e = 0.8$.

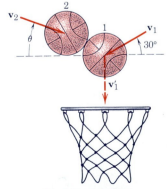

Problem 3/265

3/265 During a pregame warmup period, two basketballs collide above the hoop when in the positions shown. Just before impact, ball 1 has a velocity $\mathbf{v}_1$ that makes a 30° angle with the horizontal. If the velocity $\mathbf{v}_2$ of ball 2 just before impact has the same magnitude as $\mathbf{v}_1$, determine the two possible values of the angle θ, measured from the horizontal, that will cause ball 1 to go directly through the center of the basket. The coefficient of restitution is $e = 0.8$.

$$\text{Ans. } \theta = 82.3° \text{ or } -22.3°$$

3/266 In a game of pool, the eight ball is to be struck by the cue ball A so that the eight ball enters the right corner pocket B. Specify the distance x from the center of the left corner pocket C to the point where the cue ball strikes the cushion after hitting the eight ball. The equal-mass balls are 2 in. in diameter, and the coefficient of restitution is $e = 0.9$.

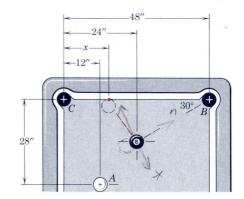

Problem 3/266

3/267 A child throws a ball from point A with a speed of 50 ft/sec. It strikes the wall at point B and then returns exactly to point A. Determine the necessary angle α if the coefficient of restitution in the wall impact is $e = 0.5$.   *Ans.* $\alpha = 11.37°$ or $78.6°$

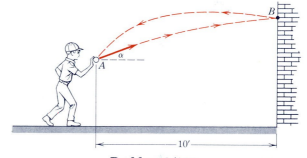

Problem 3/267

3/268 The 2-kg sphere is projected horizontally with a velocity of 10 m/s against the 10-kg carriage that is backed up by the spring with a stiffness of 1600 N/m. The carriage is initially at rest with the spring uncompressed. If the coefficient of restitution is 0.6, calculate the rebound velocity v', the rebound angle θ, and the maximum travel δ of the carriage after impact.

Ans. $v' = 6.04$ m/s, $\theta = 85.9°$, $\delta = 165.0$ mm

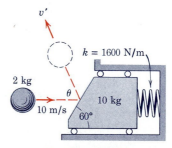

Problem 3/268

3/13 CENTRAL-FORCE MOTION

When a particle moves under the influence of a force directed toward a fixed center of attraction, the motion is called *central-force motion*. The most common example of central-force motion is found with the orbits of planets and satellites. The laws that govern this motion were deduced from observation of the motions of the planets by J. Kepler (1571–1630). The dynamics of central-force motion is basic to the design of high-altitude rockets, earth satellites, and space vehicles of all types.

Consider a particle of mass m, Fig. 3/18, moving under the action of the central gravitational attraction

$$F = G\,\frac{mm_0}{r^2}$$

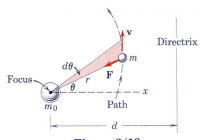

Figure 3/18

where m_0 is the mass of the attracting body assumed to be fixed, G is the universal gravitational constant, and r is the distance between the centers of the masses. The particle of mass m could represent the earth moving about the sun, the moon moving about the earth, or a satellite in its orbital motion about the earth above the atmosphere. The most convenient coordinate system to use is polar coordinates in the plane of motion since $\mathbf{F}$ will always be in the negative r-direction and there is no force in the θ-direction.

Equations 3/8 may be applied directly for the r- and θ-directions to give

$$-G\,\frac{mm_0}{r^2} = m(\ddot{r} - r\dot{\theta}^2)$$

$$0 = m(r\ddot{\theta} + 2\dot{r}\dot{\theta}) \tag{3/33}$$

The second of the two equations when multiplied by r/m is seen to be the same as $d(r^2\dot{\theta})/dt = 0$, which is integrated to give

$$r^2\dot{\theta} = h, \qquad \text{a constant} \tag{3/34}$$

The physical significance of Eq. 3/34 is made clear when we note that the angular momentum $\mathbf{r} \times m\mathbf{v}$ of m about m_0 has the magnitude $mr^2\dot{\theta}$. Thus, Eq. 3/34 merely states that the angular momentum of m about m_0 remains constant, or is conserved. This statement is easily deduced from Eq. 3/27 where it is observed that the angular momentum $\mathbf{H}_O$ remains constant (is conserved) if there is no moment acting on the particle about a fixed point O.

We observe that during time dt, the radius vector sweeps out an area, shaded in Fig. 3/18, equal to $dA = (\frac{1}{2}r)(r\,d\theta)$. Therefore, the rate at which area is swept by the radius vector is $\dot{A} = \frac{1}{2}r^2\dot{\theta}$, which is constant according to Eq. 3/34. This conclusion is expressed in Kepler's *second law* of planetary motion which states that the areas swept through in equal times are equal.

The shape of the path followed by m may be obtained by solving the first of Eqs. 3/33, with the time t eliminated through combination with Eq. 3/34. To this end the mathematical substitution $r = 1/u$ is useful. Thus, $\dot{r} = -(1/u^2)\dot{u}$, which from Eq. 3/34 becomes $\dot{r} = -h(\dot{u}/\dot{\theta})$ or $\dot{r} = -h(du/d\theta)$. The second time derivative is $\ddot{r} = -h(d^2u/d\theta^2)\dot{\theta}$, which by combining with Eq. 3/34, becomes $\ddot{r} = -h^2u^2(d^2u/d\theta^2)$. Substitution into the first of Eqs. 3/33 now gives

$$-Gm_0u^2 = -h^2u^2 \frac{d^2u}{d\theta^2} - \frac{1}{u} h^2u^4$$

or

$$\frac{d^2u}{d\theta^2} + u = \frac{Gm_0}{h^2} \qquad (3/35)$$

which is a nonhomogeneous linear differential equation. The solution of this familiar second-order equation may be verified by direct substitution and is

$$u = \frac{1}{r} = C \cos(\theta + \delta) + \frac{Gm_0}{h^2}$$

where C and δ are the two integration constants. The phase angle δ may be eliminated by choosing the x-axis so that r is a minimum when $\theta = 0$. Thus,

$$\frac{1}{r} = C \cos \theta + \frac{Gm_0}{h^2} \qquad (3/36)$$

The interpretation of Eq. 3/36 requires a knowledge of the equations for conic sections. We recall that a conic section is formed by the locus of a point which moves so that the ratio e of its distance from a point (focus) to a line (directrix) is constant. Thus, from Fig. 3/18, $e = r/(d - r \cos \theta)$, which may be rewritten as

$$\frac{1}{r} = \frac{1}{d} \cos \theta + \frac{1}{ed} \qquad (3/37)$$

which is the same form as Eq. 3/36. Hence, it is seen that the motion of m is along a conic section with $d = 1/C$ and $ed = h^2/(Gm_0)$ or

$$e = \frac{h^2C}{Gm_0} \qquad (3/38)$$

There are three cases to be investigated corresponding to $e < 1$ (ellipse), $e = 1$ (parabola), and $e > 1$ (hyperbola). The trajectory for each of these cases in shown in Fig. 3/19.

Case I: ellipse ($e < 1$). From Eq. 3/37 it is seen that r is a minimum when $\theta = 0$ and is a maximum when $\theta = \pi$. Thus,

$$2a = r_{\min} + r_{\max} = \frac{ed}{1 + e} + \frac{ed}{1 - e} \quad \text{or} \quad a = \frac{ed}{1 - e^2}$$

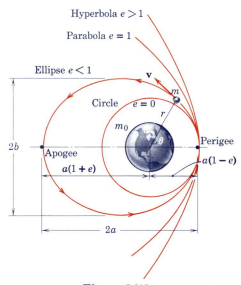

Figure 3/19

With the distance d expressed in terms of a, Eq. 3/37 and the maximum and minimum values of r may be written as

$$\frac{1}{r} = \frac{1 + e \cos \theta}{a(1 - e^2)}$$

(3/39)

$$r_{\min} = a(1 - e) \qquad r_{\max} = a(1 + e)$$

In addition, the relation $b = a\sqrt{1 - e^2}$, which comes from the geometry of the ellipse, gives the expression for the semiminor axis. It is seen that the ellipse becomes a circle with $r = a$ when $e = 0$. Equation 3/39 is an expression of Kepler's *first law* which says that the planets move in elliptical orbits around the sun as a focus.

The period τ for the elliptical orbit is the total area A of the ellipse divided by the constant rate $\dot{A}$ at which the area is swept through. Thus,

$$\tau = A/\dot{A} = \frac{\pi ab}{\frac{1}{2}r^2\dot{\theta}} \qquad \text{or} \qquad \tau = \frac{2\pi ab}{h}$$

by Eq. 3/34. Substitution of Eq. 3/38, the identity $d = 1/C$, the geometric relationships $a = ed/(1 - e^2)$ and $b = a\sqrt{1 - e^2}$ for the ellipse, and the equivalence $Gm_0 = gR^2$ yields upon simplification

$$\boxed{\tau = 2\pi \frac{a^{3/2}}{R\sqrt{g}}}$$

(3/40)

In this equation it is noted that R is the mean radius of the central attracting body and g is the absolute value of the acceleration due to gravity at the surface of the attracting body.

Equation 3/40 expresses Kepler's *third law* of planetary motion which states that the square of the period of motion is proportional to the cube of the semimajor axis of the orbit.

Case II: parabola ($e = 1$). Equations 3/37 and 3/38 become

$$\frac{1}{r} = \frac{1}{d}(1 + \cos \theta) \qquad \text{and} \qquad h^2C = Gm_0$$

The radius vector and the dimension a become infinite as θ approaches π.

Case III: hyperbola ($e > 1$). From Eq. 3/37 it is seen that the radial distance r becomes infinite for the two values of the polar angle θ_1 and $-\theta_1$ defined by $\cos \theta_1 = -1/e$. Only branch I corresponding to $-\theta_1 < \theta < \theta_1$, Fig. 3/20, represents a physically possible motion. Branch II corresponds to angles in the remaining sector (with r negative). For this branch positive r's may be used if θ is replaced by $\theta - \pi$ and $-r$ by r. Thus, Eq. 3/37 becomes

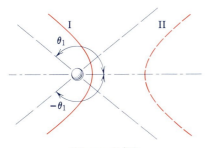

Figure 3/20

$$\frac{1}{-r} = \frac{1}{d}\cos(\theta - \pi) + \frac{1}{ed} \qquad \text{or} \qquad \frac{1}{r} = -\frac{1}{ed} + \frac{\cos\theta}{d}$$

But this expression contradicts the form of Eq. 3/36 where Gm_0/h^2 is necessarily positive. Hence, branch II does not exist (except for repulsive forces).

Now consider the energies of particle m. The system is conservative, and the constant energy E of m is the sum of its kinetic energy T and potential energy V. The kinetic energy is $T = \frac{1}{2}mv^2 = \frac{1}{2}m(\dot{r}^2 + r^2\dot{\theta}^2)$ and the potential energy from Eq. 3/15 is $V = -mgR^2/r$. Recall that g is the absolute acceleration due to gravity measured at the surface of the attracting body, R is the radius of the attracting body, and $Gm_0 = gR^2$. Thus,

$$E = \tfrac{1}{2}m(\dot{r}^2 + r^2\dot{\theta}^2) - \frac{mgR^2}{r}$$

This constant value of E can be determined from its value at $\theta = 0$, where $\dot{r} = 0$, $1/r = C + gR^2/h^2$ from Eq. 3/36, and $r\dot{\theta} = h/r$ from Eq. 3/34. Substitution into the expression for E and simplification yield

$$\frac{2E}{m} = h^2C^2 - \frac{g^2R^4}{h^2}$$

Now eliminate C by substitution of Eq. 3/38, which may be written as $h^2C = egR^2$, and obtain

$$e = +\sqrt{1 + \frac{2Eh^2}{mg^2R^4}} \qquad\qquad (3/41)$$

The plus value of the radical is mandatory since by definition e is positive. It is now seen that for the

$$\begin{array}{lll} \text{elliptical orbit} & e < 1, & E \text{ is negative} \\ \text{parabolic orbit} & e = 1, & E \text{ is zero} \\ \text{hyperbolic orbit} & e > 1, & E \text{ is positive} \end{array}$$

These conclusions, of course, depend on the arbitrary selection of the datum condition for zero potential energy ($V = 0$ when $r = \infty$).

The expression for the velocity v of m may be found from the energy equation, which is

$$\tfrac{1}{2}mv^2 - \frac{mgR^2}{r} = E$$

The total energy E is obtained from Eq. 3/41 by combining Eq. 3.38 and $1/C = d = a(1 - e^2)/e$ to give for the elliptical orbit

$$E = -\frac{gR^2m}{2a} \qquad\qquad (3/42)$$

Substitution into the energy equation yields

$$v^2 = 2gR^2 \left(\frac{1}{r} - \frac{1}{2a} \right)$$

(3/43)

from which the magnitude of the velocity may be computed for a particular orbit in terms of the radial distance r. Combination of the expressions for r_{min} and r_{max} corresponding to perigee and apogee, Eq. 3/39, with Eq. 3/43 gives for the respective velocities at these two positions for the elliptical orbit

$$v_P = R\sqrt{\frac{g}{a}}\sqrt{\frac{1+e}{1-e}} = R\sqrt{\frac{g}{a}}\sqrt{\frac{r_{max}}{r_{min}}}$$

$$v_A = R\sqrt{\frac{g}{a}}\sqrt{\frac{1-e}{1+e}} = R\sqrt{\frac{g}{a}}\sqrt{\frac{r_{min}}{r_{max}}}$$

(3/44)

Selected numerical data pertaining to the solar system are included in Appendix D and will be found useful in applying the foregoing relationships to problems in planetary motion.

The foregoing analysis is based on three assumptions:

1. The two bodies possess spherical mass symmetry so that they may be treated as if their masses were concentrated at their centers, that is, as if they were particles.

2. There are no forces present except the gravitational force which each mass exerts on the other.

3. Mass m_0 is fixed in space.

Assumption (1) is excellent for bodies that are distant from the central attracting body, which is the case for most heavenly bodies. A significant class of problems for which assumption (1) is poor is that of artificial satellites in the very near vicinity of oblate planets. As a comment on assumption (2), we note that aerodynamic drag on a low-altitude earth satellite is a force that usually cannot be ignored in the orbital analysis. For an artificial satellite in earth orbit, the error of assumption (3) is negligible because the ratio of the mass of the satellite to that of the earth is very small. On the other hand, for the earth–moon system, a small but significant error is introduced if assumption (3) is invoked—note that the lunar mass is about 1/81 times that of the earth.

Two-body problem. We now account for the motion of both masses and allow the presence of forces in addition to those of mutual attraction by considering the "perturbed two-body" problem.

Figure 3/21 depicts the major mass m_0, the minor mass m, their respective position vectors $\mathbf{r}_1$ and $\mathbf{r}_2$ measured relative to an inertial frame, the gravitational forces $\mathbf{F}$ and $-\mathbf{F}$, and a non-two-body force $\mathbf{P}$ that is exerted on mass m. The force $\mathbf{P}$ might be due to aerodynamic drag, solar pressure, the presence of a third body, on-board thrusting activities, a nonspherical gravitational field, or a combination of these and other sources. Application of Newton's second law to each mass results in

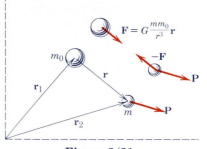

Figure 3/21

$$G\,\frac{mm_0}{r^3}\,\mathbf{r} = m_0\ddot{\mathbf{r}}_1 \qquad \text{and} \qquad -G\,\frac{mm_0}{r^3}\,\mathbf{r} + \mathbf{P} = m\ddot{\mathbf{r}}_2$$

Dividing the first equation by m_0, the second equation by m, and subtracting the first equation from the second give

$$-G\,\frac{(m_0 + m)}{r^3}\,\mathbf{r} + \frac{\mathbf{P}}{m} = \ddot{\mathbf{r}}_2 - \ddot{\mathbf{r}}_1 = \ddot{\mathbf{r}}$$

or

$$\ddot{\mathbf{r}} + G\,\frac{(m_0 + m)}{r^3}\,\mathbf{r} = \frac{\mathbf{P}}{m} \qquad (3/45)$$

Equation 3/45 is a second-order differential equation which, when solved, yields the relative position vector $\mathbf{r}$ as a function of time. Numerical techniques are usually required for the integration of the scalar differential equations that are equivalent to the vector equation 3/45, especially if $\mathbf{P}$ is nonzero. If $m_0 \gg m$ and $\mathbf{P} = \mathbf{0}$, we have the restricted two-body problem, the equation of motion of which is

$$\ddot{\mathbf{r}} + G\,\frac{m_0}{r^3}\,\mathbf{r} = \mathbf{0} \qquad (3/45a)$$

With $\mathbf{r}$ and $\ddot{\mathbf{r}}$ expressed in polar coordinates, Eq. 3/45a becomes

$$(\ddot{r} - r\dot{\theta}^2)\mathbf{e}_r + (r\ddot{\theta} + 2\dot{r}\dot{\theta})\mathbf{e}_\theta + G\,\frac{m_0}{r^3}\,(r\mathbf{e}_r) = \mathbf{0}$$

When we equate coefficients of like unit vectors, we recover Eqs. 3/33.

Comparison of Eq. 3/45 (with $\mathbf{P} = \mathbf{0}$) and Eq. 3/45a allows us to relax the assumption that mass m_0 is fixed in space. If we replace m_0 with $(m_0 + m)$ in the expressions derived with the assumption of m_0 fixed, then we have expressions that account for the motion of m_0. For example, the corrected expression for the period of elliptical motion of m about m_0 is, from Eq. 3/40,

$$\tau = 2\pi\,\frac{a^{3/2}}{\sqrt{G(m_0 + m)}} \qquad (3/45b)$$

where the equality $R^2g = Gm_0$ has been used.

Sample Problem 3/27

An artificial satellite is launched from point B on the equator by its carrier rocket and inserted into an elliptical orbit with a perigee altitude of 2000 km. If the apogee altitude is to be 4000 km, compute (a) the necessary perigee velocity v_P and the corresponding apogee velocity v_A, (b) the velocity at point C where the altitude of the satellite is 2500 km, and (c) the period τ for a complete orbit.

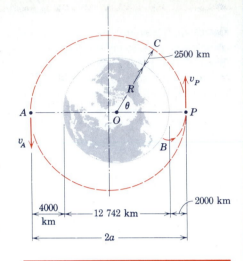

Solution. **(a)** The perigee and apogee velocities for specified altitudes are given by Eqs. 3/44, where

$$r_{\max} = 6371 + 4000 = 10\ 371 \text{ km}$$

①

$$r_{\min} = 6371 + 2000 = 8371 \text{ km}$$

$$a = (r_{\min} + r_{\max})/2 = 9371 \text{ km}$$

Thus,

$$v_P = R\sqrt{\frac{g}{a}}\ \sqrt{\frac{r_{\max}}{r_{\min}}} = 6371(10^3)\sqrt{\frac{9.825}{9371(10^3)}}\ \sqrt{\frac{10\ 371}{8371}}$$

$$= 7261 \text{ m/s} \quad \text{or} \quad 26\ 140 \text{ km/h} \quad Ans.$$

$$v_A = R\sqrt{\frac{g}{a}}\ \sqrt{\frac{r_{\min}}{r_{\max}}} = 6371(10^3)\sqrt{\frac{9.825}{9371(10^3)}}\ \sqrt{\frac{8371}{10\ 371}}$$

$$= 5861 \text{ m/s} \quad \text{or} \quad 21\ 099 \text{ km/h} \quad Ans.$$

(b) For an altitude of 2500 km the radial distance from the earth's center is $r = 6371 + 2500 = 8871$ km. From Eq. 3/43 the velocity at point C becomes

②

$$v_C{}^2 = 2gR^2\left(\frac{1}{r} - \frac{1}{2a}\right) = 2(9.825)[(6371)(10^3)]^2\left(\frac{1}{8871} - \frac{1}{18\ 742}\right)\frac{1}{10^3}$$

$$= 47.353(10^6)(\text{m/s})^2$$

$$v_C = 6881 \text{ m/s} \quad \text{or} \quad 24\ 773 \text{ km/h} \quad Ans.$$

(c) The period of the orbit is given by Eq. 3/40, which becomes

③

$$\tau = 2\pi\frac{a^{3/2}}{R\sqrt{g}} = 2\pi\frac{[(9371)(10^3)]^{3/2}}{(6371)(10^3)\sqrt{9.825}} = 9026 \text{ s}$$

$$\text{or} \quad \tau = 2.507 \text{ h} \quad Ans.$$

① The mean radius of $12\ 742/2 = 6371$ km from Table D/2 in Appendix D is used. Also the absolute acceleration due to gravity $g = 9.825$ m/s² from Art. 1/5 will be used.

② We must be careful with units. It is often safer to work in base units, meters in this case, and convert later.

③ We should observe here that the time interval between successive overhead transits of the satellite as recorded by an observer on the equator is longer than the period calculated here since the observer will have moved in space due to the counterclockwise rotation of the earth, as seen looking down on the north pole.

PROBLEMS

(Unless otherwise indicated, the velocities mentioned in the problems that follow are measured from a nonrotating reference frame moving with the center of the attracting body. Also, aerodynamic drag is to be neglected unless stated otherwise. Use $g = 9.825$ m/s² (32.23 ft/sec²) for the absolute gravitational acceleration at the surface of the earth and $R = 6371$ km (3959 mi) for the radius of the earth.)

Introductory problems

3/269 What velocity v must the space shuttle have in order to release the Hubble space telescope in a circular earth orbit 590 km above the earth?

Ans. $v = 7569$ m/s

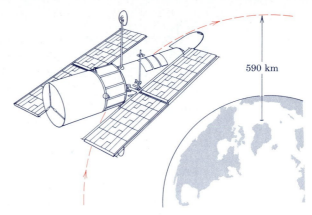

Problem 3/269

3/270 Calculate the velocity of a spacecraft that orbits the moon in a circular path of 80-km altitude.

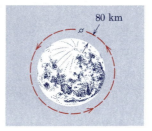

Problem 3/270

3/271 Determine the speed relative to the sun that a spacecraft must possess at a distance from the sun of $150(10^6)$ km (that is, in the vicinity of the orbit of the earth) in order to eventually escape from the solar system. Consult Appendix Table D/2 as needed.

Ans. $v = 151\ 400$ km/h

3/272 Show that the path of the moon is concave toward the sun at the position shown. Assume that the sun, earth, and moon are in the same line.

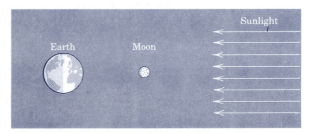

Problem 3/272

3/273 A satellite is in a circular earth orbit of radius $2R$, where R is the radius of the earth. What is the minimum velocity boost Δv necessary to reach point B, which is a distance of $6R$ from the earth's center? At what point in the original circular orbit should the velocity increment be added?

Ans. $\Delta v = 1257$ m/s

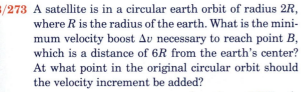

Problem 3/273

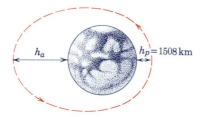

Problem 3/274

Problem 3/275

3/274 Determine the speed v required of an earth satellite at point A for (a) a circular orbit, (b) an elliptical orbit of eccentricity $e = 0.1$, (c) an elliptical orbit of eccentricity $e = 0.9$, and (d) a parabolic orbit. In cases (b) and (c), A is the orbit perigee.

3/275 The Mars orbiter for the Viking mission was designed to make one complete trip around the planet in exactly the same time that it takes Mars to revolve once about its own axis. This time is 24 h, 37 min, 23 s. In this way, it is possible for the orbiter to pass over the landing site of the lander capsule at the same time in each Martian day at the orbiter's minimum (periapsis) altitude. For the Viking I mission, the periapsis altitude of the orbiter was 1508 km. Make use of the data in Table D/2 in Appendix D and compute the maximum (apoapsis) altitude h_a for the orbiter in its elliptical path.

Ans. $h_a = 32\,600$ km

3/276 Of the four major satellites of Jupiter (which were first discovered by Galileo in 1610), Ganymede is the largest and is now known to have a mass of $1.490(10^{23})$ kg and an orbital radius of $1.070(10^6)$ km in its near-circular path around Jupiter. The mass of Jupiter is $1.900(10^{27})$ kg (which is 318 times the mass of the earth) and its equatorial diameter is 142 800 km. Calculate the gravitational force F exerted on Ganymede by Jupiter and determine the acceleration a_n of Ganymede with respect to the center of Jupiter. Use this result to calculate the period τ of its orbit and compare with the observed value of 7.16 sidereal days. (1 sidereal day = 23.93 h)

Representative problems

3/277 If the perigee altitude of an earth satellite is 240 km and the apogee altitude is 400 km, compute the eccentricity e of the orbit and the period τ of one complete orbit in space.

Ans. $e = 0.01196$, $\tau = 5446$ s

3/278 An earth satellite moves in a polar orbit with an eccentricity of 0.2 and a minimum altitude of 200 mi at the north pole. An observer A directly under the satellite as it passes over the equator at B notes a change $\Delta\beta$ in the longitude of the satellite when it returns to position B on its next trip around the earth. Calculate $\Delta\beta$.

Problem 3/278

3/279 A satellite passes over the north pole at the perigee altitude of 500 km in an elliptical orbit of eccentricity $e = 0.7$. Calculate the absolute velocity v of the satellite as it crosses the equator.

Ans. $v = 25\ 680$ km/h

3/280 A spacecraft is in a circular orbit at an altitude of 200 km. What increase Δv in velocity must be given to the spacecraft by its rocket engine in order to escape from the earth's gravity field?

3/281 In one of the orbits of the Apollo spacecraft about the moon, its distance from the lunar surface varied from 60 mi to 180 mi. Compute the maximum velocity of the spacecraft in this orbit.

Ans. $v_p = 3745$ mi/hr

3/282 Determine the energy difference ΔE between an 80 000-kg space-shuttle orbiter on the launch pad in Cape Canaveral (latitude 28.5°) and the same orbiter in a circular orbit of altitude $h = 300$ km.

3/283 A "drag-free" satellite is one that carries a small mass inside a chamber as shown. If the satellite speed decreases because of drag, the mass speed will not, and so the mass moves relative to the chamber as indicated. Sensors detect this change in the position of the mass within the chamber, and the satellite thruster is periodically fired to recenter the mass. In this manner, compensation is made for drag. If the satellite is in a circular earth orbit of 200-km altitude and a total thruster burn time of 300 seconds occurs during 10 orbits, determine the drag force D acting on the 100-kg satellite. The thruster force T is 2 N.   *Ans. $D = 0.01132$ N*

Problem 3/283

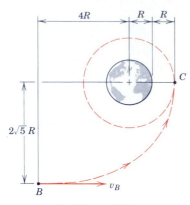

Problem 3/284

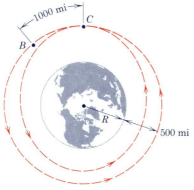

Problem 3/285

Problem 3/287

3/284 Determine the necessary velocity v_B in the direction indicated so that the spacecraft path will be tangent to the circular orbit of radius $2R$ at point C.

3/285 Two satellites B and C are in the same circular orbit of altitude 500 miles. Satellite B is 1000 mi ahead of satellite C as indicated. Show that C can catch up to B by "putting on the brakes." Specifically, by what amount Δv should the circular orbit velocity of C be reduced so that it will rendezvous with B after one period in its new elliptical orbit? Check to see that C does not strike the earth in the elliptical orbit.
Ans. $\Delta v = 302$ ft/sec

3/286 A satellite is moving in a circular orbit at an altitude of 200 mi above the surface of the earth. If a rocket motor on the satellite is activated to produce a velocity increase of 1000 ft/sec in the direction of its motion during a very short interval of time, calculate the altitude H of the satellite at its new apogee position.

3/287 An artificial satellite is injected into orbit at an altitude H with an absolute velocity v_0 at a flight angle β as shown. Determine the expression for the velocity v of the satellite during its orbit in terms of its radial distance r from the center of the earth.
$$\text{Ans. } v^2 = v_0{}^2 + 2gR^2 \left(\frac{1}{r} - \frac{1}{R + H} \right)$$

3/288 Prove that the velocity of a satellite traveling in an elliptical orbit equals $R\sqrt{g/a}$ when it reaches point C on the end of the semiminor axis. Also show that this velocity is the same as that for a circular orbit of radius a.

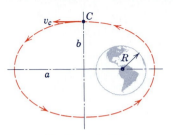

Problem 3/288

3/289 The 175,000-lb space shuttle orbiter is in a circular orbit of altitude 200 miles. The two orbital maneuvering system (OMS) engines, each of which has a thrust of 6000 lb, are fired in retro-thrust for 150 seconds. Determine the angle β that locates the intersection of the shuttle trajectory with the earth's surface. Assume that the shuttle position B corresponds to the completion of the OMS burn and that no loss of altitude occurs during the burn.

Ans. $\beta = 153.3°$

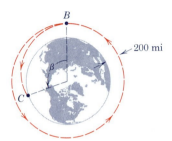

Problem 3/289

3/290 A satellite is placed in a circular polar orbit a distance H above the earth. As the satellite goes over the north pole at A, its retro-rocket is activated to produce a burst of negative thrust that reduces its velocity to a value that will ensure an equatorial landing. Derive the expression for the required reduction Δv_A of velocity at A. Note that A is the apogee of the elliptical path.

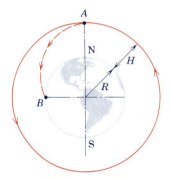

Problem 3/290

3/291 A satellite moving in a west-to-east equatorial orbit is observed by a tracking station located on the equator. If the satellite has a perigee altitude $H = 150$ km and velocity v directly over the station and an apogee altitude of 1500 km, determine an expression for the angular rate p (relative to the earth) at which the antenna dish must be rotated when the satellite is directly overhead. Compute p. The angular velocity of the earth is $\omega = 0.7292(10^{-4})$ rad/s.

Ans. $p = 0.0514$ rad/s

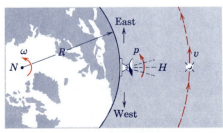

Problem 3/291

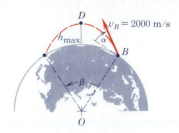

Problem 3/292

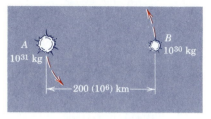

Problem 3/293

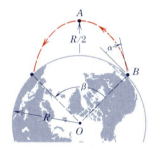

Problem 3/294

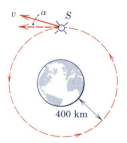

Problem 3/296

3/292 A projectile is launched from B with a speed of 2000 m/s at an angle α of 30° with the horizontal as shown. Determine the maximum altitude h_{max}.

3/293 The binary star system consists of stars A and B, both of which orbit about the system mass center. Compare the orbital period τ_f calculated with the assumption of a fixed star A with the period τ_{nf} calculated without this assumption.

Ans. $\tau_f = 21{,}760{,}000$ s, $\tau_{nf} = 20{,}740{,}000$ s

3/294 Compute the magnitude of the necessary launch velocity at B if the projectile trajectory is to intersect the earth's surface such that the angle β equals 90°. The altitude at the highest point of the trajectory is $0.5R$.

3/295 Compute the necessary launch angle α at point B for the trajectory prescribed in Prob. 3/294.

Ans. $\alpha = 38.8°$

3/296 The spacecraft S is to be injected into a circular orbit of altitude 400 km. Because of equipment malfunction, the injection speed v is correct for the circular orbit, but the initial velocity $\mathbf{v}$ makes an angle α with the intended direction. What is the maximum permissible error α in order that the spacecraft not strike the earth? Neglect atmospheric resistance.

3/297 A spacecraft with a mass of 800 kg is traveling in a circular orbit 6000 km above the earth. It is desired to change the orbit to an elliptical one with a perigee altitude of 3000 km as shown. The transition is made by firing its retro-engine at A with a reverse thrust of 2000 N. Calculate the required time t for the engine to be activated.   *Ans.* $t = 162$ s

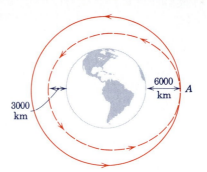

Problem 3/297

3/298 A spacecraft in an elliptical orbit has the position and velocity indicated in the figure at a certain instant. Determine the semimajor axis length a of the orbit and find the acute angle α between the semimajor axis and the line L. Does the spacecraft eventually strike the earth?

Ans. $a = 7462$ km, $\alpha = 72.8°$, no

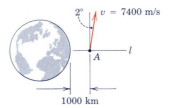

Problem 3/298

3/299 A space vehicle moving in a circular orbit of radius r_1 transfers to a larger circular orbit of radius r_2 by means of an elliptical path between A and B. (This transfer path is known as the Hohmann transfer ellipse.) The transfer is accomplished by a burst of speed Δv_A at A and a second burst of speed Δv_B at B. Write expressions for Δv_A and Δv_B in terms of the radii shown and the value g of the acceleration due to gravity at the earth's surface. If each Δv is positive, how can the velocity for path 2 be less than the velocity for path 1? Compute each Δv if $r_1 = (6371 + 500)$ km and $r_2 = (6371 + 35\,800)$ km. Note that r_2 has been chosen as the radius of a geosynchronous orbit.

$$Ans. \ \Delta v_A = R\sqrt{\frac{g}{r_1}}\left(\sqrt{\frac{2r_2}{r_1 + r_2}} - 1\right)$$
$$= 2370 \text{ m/s}$$
$$\Delta v_B = R\sqrt{\frac{g}{r_2}}\left(1 - \sqrt{\frac{2r_1}{r_1 + r_2}}\right)$$
$$= 1447 \text{ m/s}$$

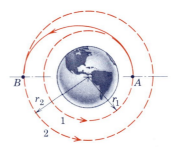

Problem 3/299

3/300 The satellite has a velocity at B of 3200 m/s in the direction indicated. Determine the angle β that locates the point C of impact with the earth.

Ans. $\beta = 109.1°$

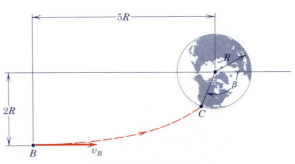

Problem 3/300

3/14 RELATIVE MOTION

Up to this point in our development of the kinetics of particle motion, we have applied Newton's second law and the equations of work-energy and impulse-momentum to problems where all measurements of motion were made with respect to a reference system that was considered fixed. The nearest we can come to a "fixed" reference system is the primary inertial system or astronomical frame of reference, which is an imaginary set of axes attached to the fixed stars. All other reference systems then are considered to have motion in space, including any reference system attached to the moving earth.

The accelerations of points attached to the earth as measured in the primary system are quite small, however, and we normally neglect them for most earth-surface measurements. For example, the acceleration of the center of the earth in its near-circular orbit around the sun considered fixed is 0.00593 m/s² (or 0.01946 ft/sec²), and the acceleration of a point on the equator at sea level with respect to the center of the earth considered fixed is 0.0339 m/s² (or 0.1113 ft/sec²). Clearly, these accelerations are small compared with g and with most other significant accelerations in engineering work. Thus, we make only a small error when we assume that our earth-attached reference axes are equivalent to a fixed reference system.

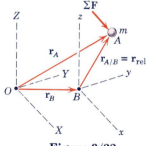

Figure 3/22

(a) Relative-motion equation.

We now consider a particle A of mass m, Fig. 3/22, whose motion is observed from a set of axes x-y-z that have translatory motion with respect to a fixed reference frame X-Y-Z. Thus, the x-y-z directions always remain parallel to the X-Y-Z directions. A discussion of motion relative to a rotating reference system is reserved for a later treatment in Arts. 5/7 and 7/7. The acceleration of the origin B of x-y-z is $\mathbf{a}_B$. The acceleration of A as observed from or relative to x-y-z is $\mathbf{a}_{\text{rel}} = \mathbf{a}_{A/B} = \ddot{\mathbf{r}}_{A/B}$, and by the relative-motion principle of Art. 2/8, the absolute acceleration of A is

$$\mathbf{a}_A = \mathbf{a}_B + \mathbf{a}_{\text{rel}}$$

Thus, Newton's second law $\Sigma \mathbf{F} = m\mathbf{a}_A$ becomes

$$\Sigma \mathbf{F} = m(\mathbf{a}_B + \mathbf{a}_{\text{rel}}) \tag{3/46}$$

The force sum $\Sigma \mathbf{F}$ is disclosed, as always, by a complete freebody diagram that will appear the same to an observer in x-y-z or to one in X-Y-Z as long as only the real forces acting on the particle are represented. We can conclude immediately that Newton's second law does not hold with respect to an accelerating system since $\Sigma \mathbf{F} \neq m\mathbf{a}_{\text{rel}}$.

D'Alembert's principle. When a particle is observed from a fixed set of axes X-Y-Z, Fig. 3/23a, its absolute acceleration $\mathbf{a}$ is measured and the familiar relation $\Sigma\mathbf{F} = m\mathbf{a}$ is applied. When the particle is observed from a moving system x-y-z to which it is attached at the origin, Fig. 3/23b, the particle necessarily appears to be at rest or in equilibrium in x-y-z. Thus, the observer who is accelerating with x-y-z concludes that a force $-m\mathbf{a}$ acts on the particle to balance $\Sigma\mathbf{F}$. This point of view, which permits the treatment of a dynamics problem by the methods of statics, was an outgrowth of the work of D'Alembert contained in his *Traité de Dynamique* published in 1743. This approach merely amounts to rewriting the equation of motion as $\Sigma\mathbf{F} - m\mathbf{a} = \mathbf{0}$, which assumes the form of a zero force summation if $-m\mathbf{a}$ is treated as a force. This fictitious force is known as the *inertia force*, and the artificial state of equilibrium created is known as *dynamic equilibrium*. The apparent transformation of a problem in dynamics to one in statics has become known as *D'Alembert's principle*.

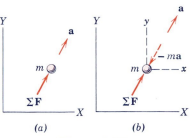

(a)   (b)

Figure 3/23

Opinion differs concerning the original interpretation of D'Alembert's principle, but the principle in the form in which it is generally known is regarded in this book as being mainly of historical interest. It was evolved during a time when understanding and experience with dynamics were extremely limited and was a means of explaining dynamics in terms of the principles of statics that during earlier times were more fully understood. This excuse for using an artificial situation to describe a real one is open to question, as there is today a wealth of knowledge and experience with the phenomena of dynamics to support strongly the direct approach of thinking in terms of dynamics rather than in terms of statics. It is somewhat difficult to justify the long persistence in the acceptance of statics as a way of understanding dynamics, particularly in view of the continued search for the understanding and description of physical phenomena in their most direct form.

Only one simple example of the method known as D'Alembert's principle will be cited. The conical pendulum of mass m, Fig. 3/24a, is swinging in a horizontal circle, with its radial line r having an angular velocity ω. In the straightforward application of the equation of motion $\Sigma\mathbf{F} = m\mathbf{a}_n$ in the direction n of the acceleration, the free-body diagram in the b-part of the figure shows that $T \sin \theta = mr\omega^2$. When the equilibrium requirement in the y-direction, $T \cos \theta - mg = 0$, is introduced, the unknowns T and θ can be found. But if the reference axes are attached to the particle, the particle will appear to be in equilibrium relative to these axes. Accordingly, the inertia force $-m\mathbf{a}$ must be added, which amounts to visualizing the application of $mr\omega^2$ in the direction opposite to the acceleration, as shown in the c-part of the figure. With this pseudo free-body diagram, a zero force summation in the n-direction gives $T \sin \theta - mr\omega^2 = 0$ which, of course, gives us the same result as

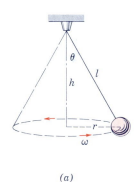

(a)

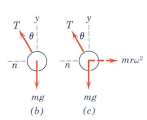

(b)   (c)

Figure 3/24

before. We may conclude that no advantage results from this alternative formulation. The authors recommend against using it since it introduces no simplification and adds a nonexistent force to the diagram. In the case of a particle moving in a circular path, this hypothetical inertia force is known as the *centrifugal force* since it is directed away from the center and is opposite to the direction of the acceleration. The student is urged to recognize that there is no actual centrifugal force acting on the particle. The only actual force that may properly be called centrifugal is the horizontal component of the tension T exerted *on* the cord *by* the particle.

(b) Constant-velocity translating systems.

In discussing particle motion relative to moving reference systems, it is of importance for us to note the special case where the reference system has a constant translational velocity. If the x-y-z axes of Fig. 3/22 have a constant velocity, then $\mathbf{a}_B = \mathbf{0}$ and the acceleration of the particle is $\mathbf{a}_A = \mathbf{a}_{\text{rel}}$. Therefore, we may write Eq. 3/46 as

$$\boxed{\Sigma \mathbf{F} = m\mathbf{a}_{\text{rel}}} \qquad (3/47)$$

which tells us that Newton's second law holds for measurements made in a system moving with a constant velocity. Such a system is known as an inertial system or as a Newtonian frame of reference. Observers in the moving system and in the fixed system will also agree on the designation of the resultant force acting on the particle from their identical free-body diagrams, provided they avoid the use of any so-called "inertia forces."

We will now examine the parallel question concerning the validity of the work-energy equation and the impulse-momentum equation relative to a constant-velocity, nonrotating system. Again, we take the x-y-z axes of Fig. 3/22 to be moving with a constant velocity $\mathbf{v}_B = \dot{\mathbf{r}}_B$ relative to the fixed axes X-Y-Z. The path of the particle A relative to x-y-z is governed by $\mathbf{r}_{\text{rel}}$ and is represented schematically in Fig. 3/25. The work done by $\Sigma\mathbf{F}$ relative to x-y-z is $dU_{\text{rel}} = \Sigma\mathbf{F} \cdot d\mathbf{r}_{\text{rel}}$. But $\Sigma\mathbf{F} = m\mathbf{a}_A = m\mathbf{a}_{\text{rel}}$ since $\mathbf{a}_B = \mathbf{0}$. Also $\mathbf{a}_{\text{rel}} \cdot d\mathbf{r}_{\text{rel}} = \mathbf{v}_{\text{rel}} \cdot d\mathbf{v}_{\text{rel}}$ for the same reason that $a_t\, ds = v\, dv$ in Art. 2/5 on curvilinear motion. Thus, we have

$$dU_{\text{rel}} = m\mathbf{a}_{\text{rel}} \cdot d\mathbf{r}_{\text{rel}} = mv_{\text{rel}}\, dv_{\text{rel}} = d(\tfrac{1}{2}mv_{\text{rel}}^2)$$

We define the kinetic energy relative to x-y-z as $T_{\text{rel}} = \tfrac{1}{2}mv_{\text{rel}}^2$ so that we now have

$$\boxed{dU_{\text{rel}} = dT_{\text{rel}}} \quad \text{or} \quad \boxed{U_{\text{rel}} = \Delta T_{\text{rel}}} \qquad (3/48)$$

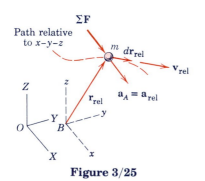

Figure 3/25

Path relative to x-y-z

which shows that the work-energy equation holds for measurements made relative to a constant-velocity system.

Relative to *x-y-z*, the impulse on the particle during time dt is $\Sigma F \, dt = m\mathbf{a}_A \, dt = m\mathbf{a}_{\text{rel}} \, dt$. But $m\mathbf{a}_{\text{rel}} \, dt = m \, d\mathbf{v}_{\text{rel}} = d(m\mathbf{v}_{\text{rel}})$ so

$$\Sigma \mathbf{F} \, dt = d(m\mathbf{v}_{\text{rel}})$$

We define the linear momentum of the particle relative to *x-y-z* as $\mathbf{G}_{\text{rel}} = m\mathbf{v}_{\text{rel}}$, which gives us $\Sigma \mathbf{F} \, dt = d\mathbf{G}_{\text{rel}}$. Dividing by dt and integrating give

$$\boxed{\Sigma \mathbf{F} = \dot{\mathbf{G}}_{\text{rel}}} \quad \text{and} \quad \boxed{\int \Sigma \mathbf{F} \, dt = \Delta \mathbf{G}_{\text{rel}}} \qquad \textbf{(3/49)}$$

Thus, the impulse-momentum equations for a fixed reference system also hold for measurements made relative to a constant-velocity system.

Finally, we define the relative angular momentum of the particle about a point in *x-y-z*, such as the origin *B*, as the moment of the relative linear momentum. Thus, $\mathbf{H}_{B_{\text{rel}}} = \mathbf{r}_{\text{rel}} \times \mathbf{G}_{\text{rel}}$. The time derivative gives $\dot{\mathbf{H}}_{B_{\text{rel}}} = \dot{\mathbf{r}}_{\text{rel}} \times \mathbf{G}_{\text{rel}} + \mathbf{r}_{\text{rel}} \times \dot{\mathbf{G}}_{\text{rel}}$. The first term is nothing more than $\mathbf{v}_{\text{rel}} \times m\mathbf{v}_{\text{rel}} = \mathbf{0}$ and the second term becomes $\mathbf{r}_{\text{rel}} \times \Sigma \mathbf{F} = \Sigma \mathbf{M}_B$, the sum of the moments about *B* of all forces on *m*. Thus, we have

$$\boxed{\Sigma \mathbf{M}_B = \dot{\mathbf{H}}_{B_{\text{rel}}}} \qquad \textbf{(3/50)}$$

which shows that the moment-angular momentum relation holds with respect to a constant-velocity system.

Although the work-energy and impulse-momentum equations hold relative to a system translating with a constant velocity, the individual expressions for work, kinetic energy, and momentum differ between the fixed and the moving systems. Thus,

$$(dU = \Sigma \mathbf{F} \cdot d\mathbf{r}_A) \neq (dU_{\text{rel}} = \Sigma \mathbf{F} \cdot d\mathbf{r}_{\text{rel}})$$

$$(T = \tfrac{1}{2}mv_A{}^2) \neq (T_{\text{rel}} = \tfrac{1}{2}mv_{\text{rel}}{}^2)$$

$$(\mathbf{G} = m\mathbf{v}_A) \neq (\mathbf{G}_{\text{rel}} = m\mathbf{v}_{\text{rel}})$$

Equations 3/47 through 3/50 are formal proof of the validity of the Newtonian equations of kinetics in any constant-velocity, nonrotating system. We might have surmised these conclusions from the fact that $\Sigma \mathbf{F} = m\mathbf{a}$ depends on acceleration and not velocity. We are also ready to conclude that there is no experiment that can be conducted in and relative to a constant-velocity system (Newtonian frame of reference) which discloses its absolute velocity. Any mechanical experiment will achieve the same results in any Newtonian system.

Sample Problem 3/28

A simple pendulum of mass m and length r is mounted on the flatcar that has a constant horizontal acceleration a_0 as shown. If the pendulum is released from rest relative to the flatcar at the position $\theta = 0$, determine the expression of the tension T in the supporting light rod for any value of θ. Also find T for $\theta = \pi/2$ and $\theta = \pi$.

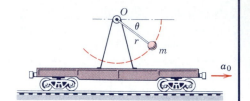

Solution. We attach our moving x-y coordinate system to the translating car with origin at O for convenience. Relative to this system, n- and t-coordinates are the natural ones to use since the motion is circular within x-y. The acceleration of m is given by the relative-acceleration equation

$$\mathbf{a} = \mathbf{a}_0 + \mathbf{a}_{\text{rel}}$$

where $\mathbf{a}_{\text{rel}}$ is the acceleration that would be measured by an observer riding with the car. He would measure an n-component equal to $r\dot{\theta}^2$ and a t-component equal to $r\ddot{\theta}$. The three components of the absolute acceleration of m are shown in the separate view.

First, we apply Newton's second law to the t-direction and get

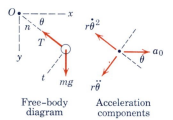

Free-body diagram   Acceleration components

① $[\Sigma F_t = ma_t]$ $mg \cos \theta = m(r\ddot{\theta} - a_0 \sin \theta)$

$$r\ddot{\theta} = g \cos \theta + a_0 \sin \theta$$

Integrating to obtain $\dot{\theta}$ as a function of θ yields

$[\dot{\theta}\, d\dot{\theta} = \ddot{\theta}\, d\theta]$ $\displaystyle \int_0^{\dot{\theta}} \dot{\theta}\, d\dot{\theta} = \int_0^{\theta} \frac{1}{r}(g \cos \theta + a_0 \sin \theta)\, d\theta$

$$\frac{\dot{\theta}^2}{2} = \frac{1}{r}[g \sin \theta + a_0(1 - \cos \theta)]$$

We now apply Newton's second law to the n-direction, noting that the n-component of the absolute acceleration is $r\dot{\theta}^2 - a_0 \cos \theta$.

② $[\Sigma F_n = ma_n]$ $T - mg \sin \theta = m(r\dot{\theta}^2 - a_0 \cos \theta)$

$$= m[2g \sin \theta + 2a_0(1 - \cos \theta) - a_0 \cos \theta]$$

$$T = m[3g \sin \theta + a_0(2 - 3 \cos \theta)] \qquad Ans.$$

For $\theta = \pi/2$ and $\theta = \pi$, we have

$$T_{\pi/2} = m[3g(1) + a_0(2 - 0)] = m(3g + 2a_0) \qquad Ans.$$

$$T_{\pi} = m[3g(0) + a_0(2 - 3[-1])] = 5ma_0 \qquad Ans.$$

① We choose the t-direction first since the n-direction equation, which contains the unknown T, will involve $\dot{\theta}^2$, which, in turn, is obtained from an integration of $\ddot{\theta}$.

② Be sure to recognize that $\dot{\theta}\, d\dot{\theta} = \ddot{\theta}\, d\theta$ may be obtained from $v\, dv = a_t\, ds$ by dividing by r^2.

Sample Problem   3/29

The flatcar moves with a constant speed v_0 and carries a winch that produces a constant tension P in the cable attached to the small carriage. The carriage has a mass m and rolls freely on the horizontal surface starting from rest relative to the flatcar at $x = 0$, at which instant $X = x_0 = b$. Apply the work-energy equation to the carriage, first, as an observer moving with the frame of reference of the car and, second, as an observer on the ground. Show the compatibility of the two expressions.

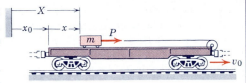

Solution. To the observer on the flatcar, the work done by P is

① $$U_{\text{rel}} = \int_0^x P \, dx = Px \qquad \text{for constant } P$$

The change in kinetic energy relative to the car is

$$\Delta T_{\text{rel}} = \tfrac{1}{2}m(\dot{x}^2 - 0)$$

The work-energy equation for the moving observer becomes

$$[U_{\text{rel}} = \Delta T_{\text{rel}}] \qquad Px = \tfrac{1}{2}m\dot{x}^2$$

To the observer on the ground, the work done by P is

$$U = \int_b^X P \, dX = P(X - b)$$

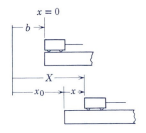

The change in kinetic energy for the ground measurement is

② $$\Delta T = \tfrac{1}{2}m(\dot{X}^2 - v_0^2)$$

The work-energy equation for the fixed observer gives

$$[U = \Delta T] \qquad P(X - b) = \tfrac{1}{2}m(\dot{X}^2 - v_0^2)$$

To reconcile this equation with that for the moving observer, we can make the following substitutions:

$$X = x_0 + x, \qquad \dot{X} = v_0 + \dot{x}, \qquad \ddot{X} = \ddot{x}$$

Thus,

③ $$P(X - b) = Px + P(x_0 - b) = Px + m\ddot{x}(x_0 - b)$$
$$= Px + m\ddot{x}v_0 t = Px + mv_0\dot{x}$$

and

$$\dot{X}^2 - v_0^2 = (v_0^2 + \dot{x}^2 + 2v_0\dot{x} - v_0^2) = \dot{x}^2 + 2v_0\dot{x}$$

The work-energy equation for the fixed observer now becomes

$$Px + mv_0\dot{x} = \tfrac{1}{2}m\dot{x}^2 + mv_0\dot{x}$$

which is merely $Px = \tfrac{1}{2}m\dot{x}^2$, as concluded by the moving observer. We see, therefore, that the difference between the two work-energy expressions is

$$U - U_{\text{rel}} = T - T_{\text{rel}} = mv_0\dot{x}$$

① The only coordinate that the moving observer can measure is x.

② To the ground observer, the initial velocity of the carriage is v_0 so its initial kinetic energy is $\tfrac{1}{2}mv_0^2$.

③ The symbol t stands for the time of motion from $x = 0$ to $x = x$. The displacement $x_0 - b$ of the carriage is its velocity v_0 times the time t or $x_0 - b = v_0 t$. Also, since the constant acceleration times the time equals the velocity change, $\ddot{x}t = \dot{x}$.

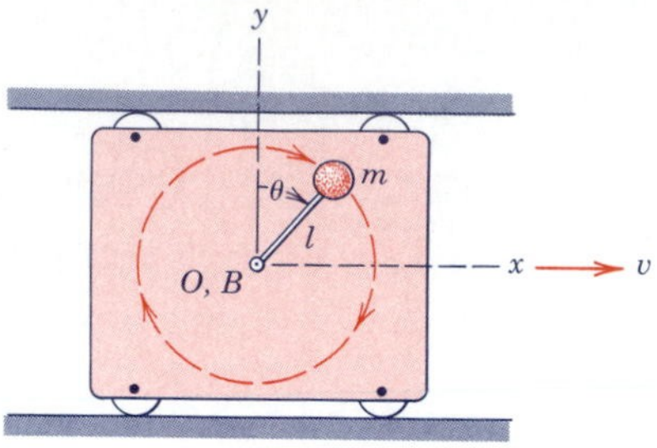

Problem 3/301

PROBLEMS

Introductory problems

3/301 The cart with attached x-y axes moves with an absolute speed $v = 2$ m/s to the right. Simultaneously, the light arm of length $l = 0.5$ m rotates about point B of the cart with angular velocity $\dot{\theta} = 2$ rad/s. If the mass of the sphere is $m = 3$ kg, determine the following quantities for the sphere when $\theta = 0$: $\mathbf{G}$, $\mathbf{G}_{\text{rel}}$, T, T_{rel}, $\mathbf{H}_O$, $\mathbf{H}_{B_{\text{rel}}}$, where the subscript "rel" indicates measurement relative to the x-y axes. Point O is an inertially fixed point coincident with point B at the instant under consideration.

Ans. $\mathbf{G} = 9\mathbf{i}$ kg·m/s, $\mathbf{G}_{\text{rel}} = 3\mathbf{i}$ kg·m/s

$T = 13.5$ J, $T_{\text{rel}} = 1.5$ J

$$\mathbf{H}_O = -4.5\mathbf{k}\,\frac{\text{kg}\cdot\text{m}^2}{\text{s}},\ \mathbf{H}_{B_{\text{rel}}} = -1.5\mathbf{k}\,\frac{\text{kg}\cdot\text{m}^2}{\text{s}}$$

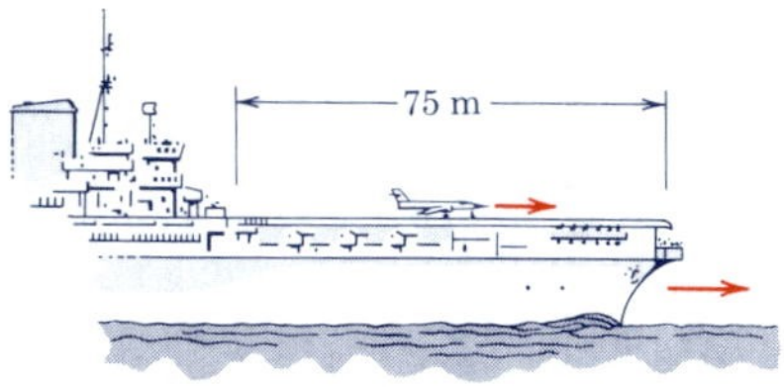

Problem 3/302

3/302 The aircraft carrier is moving at a constant speed and launches a jet plane with a mass of 3 Mg in a distance of 75 m along the deck by means of a steam-driven catapult. If the plane leaves the deck with a velocity of 240 km/h relative to the carrier and if the jet thrust is constant at 22 kN during takeoff, compute the constant force P exerted by the catapult on the airplane during the 75-m travel of the launch carriage.

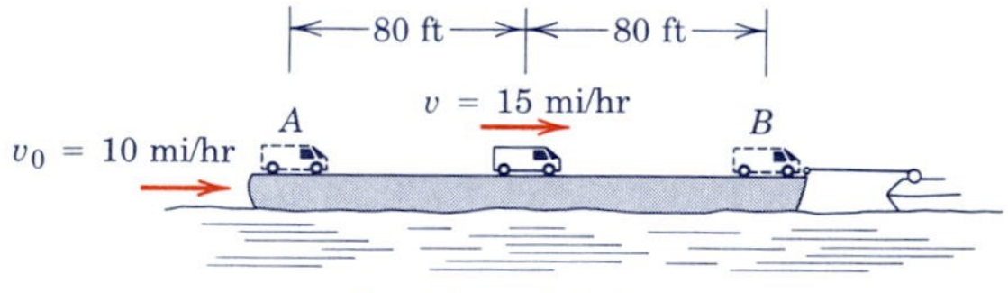

Problem 3/303

3/303 The 4000-lb van is driven from position A to position B on the barge that is towed at a constant speed $v_0 = 10$ mi/hr. The van starts from rest relative to the barge at A, accelerates to $v = 15$ mi/hr relative to the barge over a distance of 80 ft, and then stops with a deceleration of the same magnitude. Determine the magnitude of the net force F between the tires of the van and the barge during this maneuver.

Ans. $F = 376$ lb

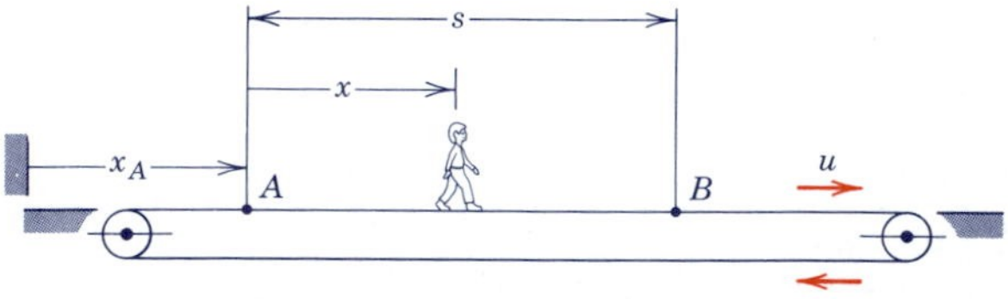

Problem 3/304

3/304 A boy of mass m is standing initially at rest relative to the moving walkway, which has a constant horizontal speed u. He decides to accelerate his progress and starts to walk from point A with a steadily increasing speed and reaches point B with a speed $\dot{x} = v$ relative to the walkway. During his acceleration he generates an average horizontal force F between his shoes and the walkway. Write the work-energy equations for his absolute and relative motions and explain the meaning of the term muv.

3/305 The escalator moves with a constant speed $u = 0.8$ m/s. A 50-kg woman is initially at rest relative to the moving stairs but at A begins to accelerate uniformly. By the time she reaches point B, which is a distance $s = 7$ m from point A, her speed is $v = 2$ m/s relative to the escalator. Determine the horizontal and vertical components of the average force F exerted by the stairs on the woman's feet.

Ans. $F_{\text{hor}} = 12.95$ N, $F_{\text{ver}} = 497$ N

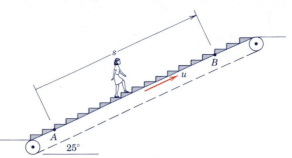

Problem 3/305

3/306 Two spacecraft A and B are traveling on the same straight-line path in deep space. Spacecraft B is 20 km directly ahead of A and has a forward velocity relative to A of 2 km/s. If it is desired to stabilize the separation distance at 1000 km, determine the magnitude F of the constant retro-thrust that must be applied to the 200-kg spacecraft B. Calculate the necessary time duration t of this thrusting activity.

Representative problems

3/307 The slider A has a mass of 2 kg and moves with negligible friction in the 30° slot in the vertical sliding plate. What horizontal acceleration a_0 should be given to the plate so that the absolute acceleration of the slider will be vertically down? What is the value of the corresponding force R exerted on the slider by the slot?   *Ans.* $a_0 = 16.99$ m/s², $R = 0$

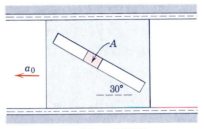

Problem 3/307

3/308 If the aircraft carrier of Prob. 3/302 is moving with a constant speed u and its catapult launches a plane of mass m with a velocity v relative to the carrier in the direction of u, the kinetic energy of the plane with respect to the land is $\frac{1}{2}m(v + u)^2 = \frac{1}{2}mv^2 + \frac{1}{2}mu^2 + muv$ when launching is complete. If the plane is launched by the same catapult with the same accelerating force when the carrier is not moving, the kinetic energy of the plane is $\frac{1}{2}mv^2$. Explain the difference between the two kinetic-energy expressions.

3/309 The launch catapult of the aircraft carrier gives the 7-Mg jet airplane a constant acceleration and launches the airplane in a distance of 100 m measured along the angled takeoff ramp. The carrier is moving at a steady speed $v_C = 16$ m/s. If an absolute aircraft speed of 90 m/s is desired for takeoff, determine the net force F supplied by the catapult and the aircraft engines.   *Ans.* $F = 194.0$ kN

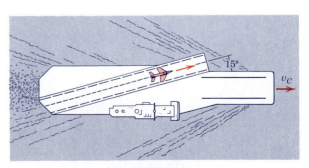

Problem 3/309

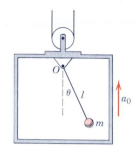

Problem 3/310

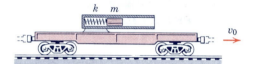

Problem 3/312

Problem 3/313

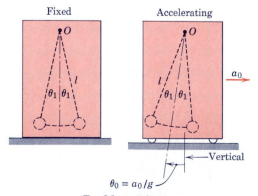

Problem 3/314

3/310 If the flatcar is given an acceleration $a_0 = 6$ ft/sec^2 starting from rest, compute the corresponding tension T in the cable attached to the 100-lb crate A. Neglect the mass of the pulley at B and its friction. The coefficient of friction between the crate and the horizontal surface of the flatcar is $\mu_k = 0.30$.

3/311 A simple pendulum is placed on an elevator that accelerates upward as shown. If the pendulum is displaced an amount θ_0 and released from rest relative to the elevator, find the tension T_0 in the supporting light rod when $\theta = 0$. Evaluate your result for $\theta_0 = \pi/2$.

$$Ans. \ T_0 = m(g + a_0)(3 - 2\cos\theta_0)$$

3/312 The capsule of mass m is released from rest relative to the horizontal tube, with the spring of stiffness k initially compressed a distance δ. The flatcar has a constant velocity v_0 to the right. Apply the work-energy equation to the motion of the capsule, both as an observer riding with the car and as an observer fixed to the ground. Reconcile the two equations (*Hint:* For the second case include the spring with the capsule as your system.)

3/313 The coefficients of friction between the flat bed of the truck and crate are $\mu_s = 0.8$ and $\mu_k = 0.7$. The coefficient of kinetic friction between the truck tires and the road surface is 0.9. If the truck stops from an initial speed of 15 m/s with maximum braking (wheels skidding), determine where on the bed the crate finally comes to rest or the velocity v_{rel} relative to the truck with which the craft strikes the wall at the forward edge of the bed.

$$Ans. \ v_{rel} = 2.46 \text{ m/s}$$

3/314 For small amplitudes θ_1 the simple pendulum with fixed support O will oscillate with simple harmonic motion about the vertical with a period $\tau = 2\pi\sqrt{l/g}$. If the support O is given a small but steady horizontal acceleration a_0, show that the pendulum will oscillate about the inclined axis $\theta_0 = a_0/g$ with the same period as with the fixed support, provided that the amplitude of motion θ_1 is again small.

3/315 A ball is released from rest relative to the elevator at a distance h_1 above the floor. The speed of the elevator at the time of ball release is v_0. Determine the bounce height h_2 of the ball (*a*) if v_0 is constant and (*b*) if an upward elevator acceleration $a = g/4$ begins at the instant the ball is released. The coefficient of restitution for the impact is *e*.

Ans. (*a*) and (*b*) $h_2 = e^2 h_1$

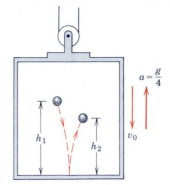

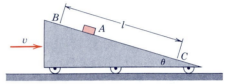

Problem 3/315

3/316 The small slider *A* moves with negligible friction down the inclined block that moves to the right with constant speed $v = v_0$. Use the principle of work-energy to determine the magnitude v_A of the absolute velocity of the slider as it passes point *C* if it is released at point *B* with no velocity relative to the block. Apply the equation, both as an observer fixed to the block and as an observer fixed to the ground, and reconcile the two relations.

Ans. $v_A = (v_0{}^2 + 2gl \sin \theta$
$+ 2v_0 \cos \theta \sqrt{2gl \sin \theta})^{1/2}$

Problem 3/316

3/317 Repeat Prob. 3/316, except let the inclined block have a small constant acceleration a_0 to the right. The velocity of the block is v_0 when the small slider is released with no velocity relative to the block.

Ans. $v_A{}^2 = v_0{}^2 + 2gl \sin \theta$

$+ 2 \sin \theta(g \cos \theta + a_0 \sin \theta)d$

where $d = v_0\sqrt{\dfrac{2l}{g \sin \theta - a_0 \cos \theta}}$

$+ \dfrac{a_0 l}{g \sin \theta - a_0 \cos \theta}$

3/318 When a particle is dropped from rest relative to the earth's surface at a latitude γ, the initial apparent acceleration is the relative acceleration due to gravity g_{rel}. The absolute acceleration due to gravity g is directed toward the center of the earth. Derive an expression for g_{rel} in terms of g, R, ω, and γ, where R is the radius of the earth treated as a sphere and ω is the constant angular velocity of the earth about the polar axis considered as fixed. (Although axes *x-y-z* are attached to the earth and hence rotate, we may use Eq. 3/46 as long as the particle has no velocity relative to *x-y-z*.) (*Hint:* Use the first two terms of the binomial expansion for the approximation.)

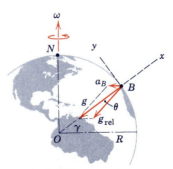

Problem 3/318

Ans. $g_{\text{rel}} = g - R\omega^2 \cos^2 \gamma \left(1 - \dfrac{R\omega^2}{2g}\right) + \cdots$
$= 9.825 - 0.03382 \cos^2 \gamma \ \text{m/s}^2$

3/15 *PROBLEM FORMULATION AND REVIEW*

In Chapter 3 we have developed the three basic methods of solution to problems in particle kinetics. This experience is central to the study of dynamics and lays the foundation for the subsequent study of rigid-body and nonrigid-body dynamics. First, we applied Newton's second law $\Sigma \mathbf{F} = m\mathbf{a}$ to determine the instantaneous relation between forces and the acceleration they produce. With the background of Chapter 2 for identifying the kind of motion and with the aid of our familiar free-body diagram to be certain that all forces are accounted for, we were able to solve a large variety of problems using x-y, n-t, and r-θ coordinates for plane-motion problems and x-y-z, r-θ-z, and R-θ-ϕ coordinates for space problems.

Second, we integrated the basic equation of motion $\Sigma \mathbf{F} = m\mathbf{a}$ with respect to displacement and derived the scalar equations for work and energy. These equations enabled us to relate the initial and final velocities to the work done during the interval by forces external to our defined system. We expanded this approach to include potential energy, both elastic and gravitational. With these tools we discovered that the energy approach is especially valuable for conservative systems, that is, systems wherein the loss of energy due to friction or other forms of dissipation is negligible.

Third, we rewrote Newton's second law in the form of force equals time rate of change of linear momentum and moment equals time rate of change of angular momentum. Then we integrated these relations with respect to time and derived the impulse and momentum equations. These equations were then applied to motion intervals where the forces were functions of the time. We also investigated the interactions between particles under conditions where the linear momentum is conserved and where the angular momentum is conserved.

In the final section of Chapter 3, we employed the three basic methods in specific application areas. In particular, we noted that the impulse-momentum method is convenient in developing the relations governing particle impacts. We observed that the direct application of Newton's second law allows the determination of the trajectory properties associated with a particle under central-force attraction. And lastly, we have seen that any of the three basic methods may be applied to particle motion relative to a translating frame of reference.

It is clear that the successful solution of problems in particle kinetics depends on knowledge of the prerequisite particle kinematics. Analogously, the analytical developments associated with particle systems and rigid bodies, which are covered in the remainder of *Dynamics*, are heavily dependent on the fundamental principles of particle kinetics as treated in Chapter 3.

REVIEW PROBLEMS

3/319 A 30-g tire-balance weight is attached to a vertical surface of the wheel rim by means of an adhesive backing. The tire-wheel unit is then given a final test on the tire-balance machine. If the adhesive can support a maximum shear force of 80 N, determine the maximum rotational speed N for which the weight remains fixed to the wheel. Assume very gradual speed changes.   *Ans.* $N = 1177$ rev/min

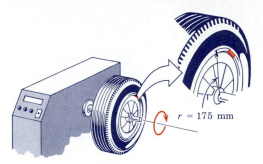

Problem 3/319

3/320 The small 2-kg carriage is moving freely along the horizontal with a speed of 4 m/s at time $t = 0$. A force applied to the carriage in the direction opposite to motion produces two impulse "peaks," one after the other, as shown by the graphical plot of the readings of the instrument that measured the force. Approximate the loading by the dotted lines and determine the velocity v of the carriage for $t = 1.5$ s.

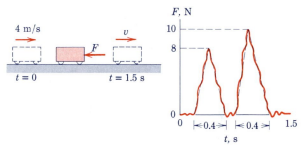

Problem 3/320

3/321 The crate is at rest at point A when it is nudged down the incline. If the coefficient of kinetic friction between the crate and the incline is 0.30 from A to B and 0.22 from B to C, determine its speeds at points B and C.
    Ans. $v_B = 2.87$ m/s, $v_C = 1.533$ m/s

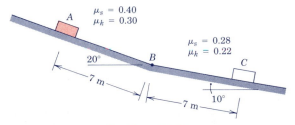

Problem 3/321

3/322 Six identical spheres are arranged as shown in the figure. The two spheres at the left end are released from the dotted positions and strike sphere 3 with speed v_1. Assuming that the common coefficient of restitution is $e = 1$, explain why two spheres leave the right end of the row with speed v_1 instead of one sphere leaving the right end with speed $2v_1$.

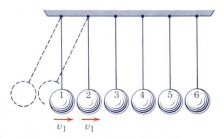

Problem 3/322

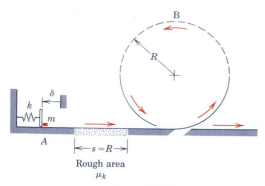

Problem 3/323

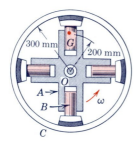

Problem 3/324

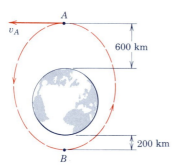

Problem 3/325

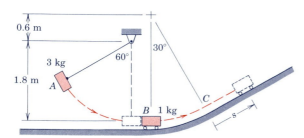

Problem 3/327

3/323 The spring of stiffness k is compressed and suddenly released, sending the particle of mass m sliding along the track. Determine the minimum spring compression δ for which the particle will not lose contact with the loop-the-loop track. The sliding surface is smooth except for the rough portion of length s equal to R, where the coefficient of kinetic friction is μ_k.

$$Ans. \quad \delta = \sqrt{\frac{mgR(5 + 2\mu_k)}{k}}$$

3/324 The figure shows a centrifugal clutch consisting in part of a rotating spider A that carries four plungers B. As the spider is made to rotate about its center with a speed ω, the plungers move outward and bear against the interior surface of the rim of wheel C, causing it to rotate. The wheel and spider are independent except for frictional contact. If each plunger has a mass of 2 kg with a center of mass at G, and if the coefficient of kinetic friction between the plungers and the wheel is 0.40, calculate the maximum moment M that can be transmitted to wheel C for a spider speed of 3000 rev/min.

3/325 For the elliptical orbit of a spacecraft around the earth, determine the speed v_A at point A that results in a perigee altitude at B of 200 km. What is the eccentricity e of the orbit?

$$Ans. \quad v_A = 7451 \text{ m/s}, \, e = 0.0295$$

3/326 The last two appearances of Comet Halley were in 1910 and 1986. The distance of its closest approach to the sun averages about one-half of the distance between the earth and the sun. Determine its maximum distance from the sun. Neglect the gravitational effects of the planets.

3/327 The 3-kg block A is released from rest in the 60° position shown and subsequently strikes the 1-kg cart B. If the coefficient of restitution for the collision is $e = 0.7$, determine the maximum displacement s of cart B beyond point C. Neglect friction.

$$Ans. \quad s = 2.28 \text{ m}$$

3/328 The 60-g bullet is fired at the two blocks resting on a surface where the coefficient of kinetic friction is 0.50. The bullet passes through the 8-kg block and lodges in the 6-kg block. The blocks slide the distances shown. Compute the initial velocity v of the bullet.

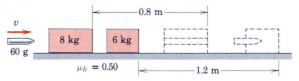

Problem 3/328

3/329 A person rolls a small ball with speed u along the floor from point A. If $x = 3R$, determine the required speed u so that the ball returns to A after rolling on the circular surface in the vertical plane from B to C and becoming a projectile at C. What is the minimum value of x for which the game could be played if contact must be maintained to point C? Neglect friction.   *Ans.* $u = \frac{5}{2}\sqrt{gR}$, $x_{min} = 2R$

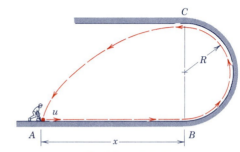

Problem 3/329

3/330 An automobile accident occurs as follows: The driver of a full-size car (vehicle A, 4000 lb) is traveling on a dry, level road and approaches a stationary compact car (vehicle B, 2000 lb). Just 50 feet before collision, he applies the brakes, skidding all wheels. After impact, vehicle A skids an additional 50 ft and vehicle B, whose driver had all brakes fully applied, skids 100 ft. The final positions of the vehicles are shown in the figure. If the coefficient of kinetic friction is 0.9, was the driver of vehicle A exceeding the speed limit of 55 mi/hr before the initial application of his brakes?

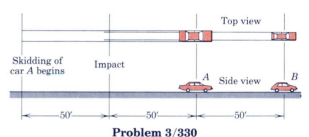

Problem 3/330

3/331 A small sphere of mass m is connected by a string to a swivel at O and moves in a circle of radius r on the smooth plane inclined an angle θ with the horizontal. If the sphere has a velocity u at the top position A, determine the tension in the string as the sphere passes the 90° position B and the bottom position C.

$$\textit{Ans. } T_B = m\left(\frac{u^2}{r} + 2g\sin\theta\right)$$

$$T_C = m\left(\frac{u^2}{r} + 5g\sin\theta\right)$$

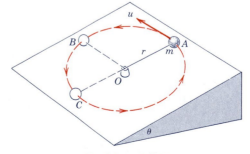

Problem 3/331

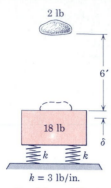

2 lb

6′

18 lb

δ

k k

$k = 3$ lb/in.

Problem 3/332

3/332 The 2-lb piece of putty is dropped 6 ft onto the 18-lb block initially at rest on the two springs, each with a stiffness $k = 3$ lb/in. Calculate the additional deflection δ of the springs due to the impact of the putty that adheres to the block upon contact.

3/333 A car of mass m is traveling at a road speed v_r along an equatorial east-west highway at sea level. If the road follows the curvature of the earth, derive an expression for the difference ΔP between the total force exerted by the road on the car for eastward travel and the total force for westward travel. Calculate ΔP for $m = 1500$ kg and $v_r = 200$ km/h. The angular velocity ω of the earth is $0.7292(10^{-4})$ rad/s. Neglect the motion of the center of the earth.

Ans. $\Delta P = -4m\omega v_r$, $\Delta P = -24.3$ N

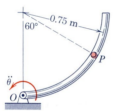

0.75 m

$60°$

P

$\ddot{\theta}$

O

Problem 3/334

3/334 The quarter-circular hollow tube of circular cross section starts from rest at time $t = 0$ and rotates about point O in a horizontal plane, with a constant counterclockwise angular acceleration $\ddot{\theta} = 2$ rad/s². At what time t will the 0.5-kg particle P slip relative to the tube? The coefficient of static friction between the particle and the tube is $\mu_s = 0.8$.

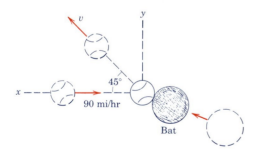

v

y

$45°$

x

90 mi/hr

Bat

Problem 3/335

3/335 A baseball pitcher delivers a fast ball with a near-horizontal velocity of 90 mi/hr. The batter hits a homerun over the centerfield fence. The 5-oz ball travels a horizontal distance of 350 ft, with an initial velocity in the 45° direction shown. Determine the magnitude F of the average force exerted by the bat on the ball during the 0.01 sec of contact between the bat and the ball. Neglect air resistance during the flight of the ball.   *Ans.* $F_{av} = 214$ lb

3/336 The system is released from rest with $\theta = 0$. The cord to the 1.5-kg cylinder is securely wound around the light 50-mm-diameter pulley at O, to which are attached the light arms and their 2-kg spheres. The centers of the spheres are 250 mm and 375 mm from the axis at O. Determine the downward velocity of the 1.5-kg cylinder when $\theta = 30°$.

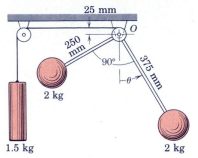

25 mm

250 mm

90°

375 mm

2 kg

1.5 kg   2 kg

Problem 3/337

3/337 A slider C has a speed of 3 m/s as it passes point A of the guide, which lies in a horizontal plane. The coefficient of kinetic friction between the slider and the guide is $\mu_k = 0.6$. Compute the tangential deceleration a_t of the slider just after it passes point A if (a) the slider hole and guide cross section are both circular and (b) the slider hole and guide cross section are both square. In case (b), the sides of the square are vertical and horizontal. Assume a slight clearance between the slider and the guide.

$$\text{Ans. } (a) \ a_t = -10.75 \text{ m/s}^2$$
$$(b) \ a_t = -14.89 \text{ m/s}^2$$

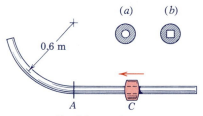

(a)   (b)

0.6 m

A C

Problem 3/336

3/338 "Bungee jumping" has become a popular sport in some circles. For the case illustrated, a 170-lb man jumps from the bridge at A with the bungee cord secured to his bound ankles. He falls 64 ft before the 54-ft length of elastic bungee cord begins to stretch. The 10 ft of rope above the elastic cord has no appreciable stretch. The man is observed to drop a total of 144 ft before being projected upward. Neglect any energy loss and calculate (a) the stiffness k of the bungee cord per foot of its length, (b) the maximum velocity v_{max} of the man during his fall, and (c) his maximum acceleration a_{max}. Treat the man as a particle located at the end of the bungee cord.

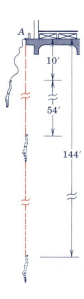

A

10'

54'

144'

Problem 3/338

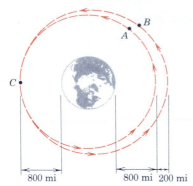

Problem 3/339

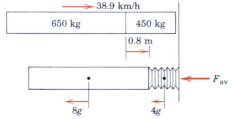

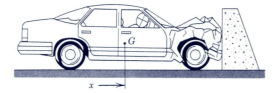

Problem 3/340

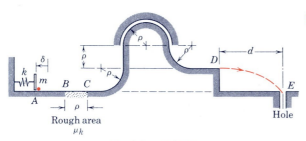

Rough area
μ_k

Problem 3/341

3/339 Satellite A moving in the circular orbit and satellite B moving in the elliptical orbit collide and become entangled at point C. If the masses of the satellites are equal, determine the maximum altitude h_{max} of the resulting orbit.   *Ans.* $h_{max} = 899$ mi

3/340 The crash-tested car of Prob. 2/30 is shown again here, where the impact velocity was found to be 38.9 km/h. If the car has a mass of 1100 kg and the front end of the car is crushed an amount $x = 0.8$ m, calculate the average force F_{av} exerted by the car on the barrier during the impact. Because the barrier does not move, F_{av} does no work. As a first approximation to the analysis, we consider the car. to be composed of two parts as shown. The undamaged portion of mass 650 kg travels 0.8 m during the crash, with an average deceleration of $8g$. The accordioned portion of the car has a mass of 450 kg whose mass center has a deceleration of $4g$. Determine F_{av} by analyzing each part separately and find the loss of energy ΔE during the crash.

3/341 The object of the pinball-type game is to project the particle so that it enters the hole at E. When the spring is compressed and suddenly released, the particle is projected along the track, which is smooth except for the rough portion between points B and C, where the coefficient of kinetic friction is μ_k. The particle becomes a projectile at point D. Determine the correct spring compression δ so that the particle enters the hole at E. State any necessary conditions relating the lengths d and ρ.

$$Ans. \ \ \delta = \sqrt{\frac{mg}{k}\left[\frac{d^2}{2\rho} + 2\rho(1 + \mu_k)\right]}$$

$$d \geq 2\sqrt{2}\rho$$

3/342 A long fly ball strikes the wall at point A (where $e_1 = 0.5$) and then hits the ground at B (where $e_2 = 0.3$). The outfielder likes to catch the ball when it is 4 ft above the ground and 2 ft in front of him as shown. Determine the distance x from the wall where he can catch the ball as described. Note the two possible solutions.

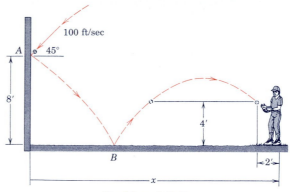

Problem 3/342

3/343 One of the functions of the space shuttle is to release communications satellites at low altitude. A booster rocket is fired at B, placing the satellite in an elliptical transfer orbit, the apogee of which is the altitude necessary for a geosynchronous orbit. (A geosynchronous orbit is an equatorial-plane circular orbit whose period is equal to the absolute rotational period of the earth. A satellite in such an orbit appears to remain stationary to an earth-fixed observer.) A second booster rocket is then fired at C, and the final circular orbit is achieved. On one of the early space-shuttle missions, a 1500-lb satellite was released from the shuttle at B, where $h_1 = 170$ miles. The booster rocket was to fire for $t = 90$ seconds, forming a transfer orbit with $h_2 = 22,300$ miles. The rocket failed during its burn. Radar observations determined the apogee altitude of the transfer orbit to be only 700 miles. Determine the actual time t' that the rocket motor operated before failure. Assume negligible mass change during the booster rocket firing.   *Ans.* $t' = 8.50$ sec

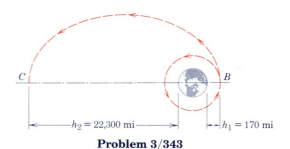

Problem 3/343

3/344 The retarding forces that act on the racecar are the drag force F_D and a nonaerodynamic force F_R. The drag force is $F_D = C_D(\frac{1}{2}\rho v^2)S$, where C_D is the drag coefficient, ρ is the air density, v is the car speed, and $S = 30$ ft^2 is the projected frontal area of the car. The nonaerodynamic force F_R is constant at 200 lb. With its sheet metal in good condition, the racecar has a drag coefficient $C_D = 0.3$ and it has a corresponding top speed $v = 200$ mi/hr. After a minor collision, the damaged front-end sheet metal causes the drag coefficient to be $C_D' = 0.4$. What is the corresponding top speed v' of the racecar?
Ans. $v' = 182.9$ mi/hr

Problem 3/344

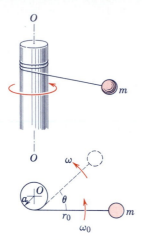

Problem 3/345

▶ **3/345** The small particle of mass m is given an initial high velocity in the horizontal plane and winds its cord around the fixed vertical shaft of radius a. All motion occurs essentially in the horizontal plane. If the angular velocity of the cord is ω_0 when the distance from the particle to the tangency point is r_0, determine the angular velocity ω of the cord and its tension T after it has turned through an angle θ. Does either of the principles of momentum conservation apply?

$$Ans. \quad \omega = \frac{\omega_0}{1 - \dfrac{a}{r_0}\theta}, \quad T = mr_0\omega_0\omega$$

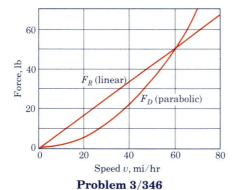

Problem 3/346

▶ **3/346** Extensive wind-tunnel and coast-down studies of a 2000-lb automobile reveal the aerodynamic drag force F_D and the total nonaerodynamic rolling resistance force F_R to vary with speed as shown in the plot. Determine (a) the power P required for steady speeds of 30 mi/hr and (b) the time t and the distance s required for the car to coast down to a speed of 5 mi/hr from an initial speed of 60 mi/hr. Assume a straight, level road and no wind.

$$Ans. \quad P_{30} = 3 \text{ hp}, \quad P_{60} = 16 \text{ hp}$$
$$t = 205 \text{ sec}, \quad s = 5898 \text{ ft}$$

Computer-oriented problems

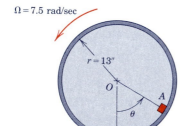

Problem 3/347

✻ **3/347** The rotating drum of Prob. 3/81 is repeated here. The 26-in. drum rotates about a horizontal axis with a constant angular velocity $\Omega = 7.5$ rad/sec. The small block A has no motion relative to the drum surface as it passes the bottom position $\theta = 0$. Determine the coefficient of static friction μ_s that would result in block slippage at an angular position θ; plot your expression for $0 \leq \theta \leq 180°$. Determine the minimum required coefficient value μ_{min} that would allow the block to remain fixed relative to the drum throughout a full revolution. For a friction coefficient slightly less than μ_{min}, at what angular position θ would slippage occur?

$$Ans. \quad \mu_{min} = 0.622, \quad \theta = 121.9°$$

3/348 The tennis player practices by hitting the ball against the wall at A. The ball bounces off the court surface at B and then up to its maximum height at C. For the conditions shown in the figure, plot the location of point C for values of the coefficient of restitution in the range $0.5 \leq e \leq 0.9$. (The value of e is common to both A and B.) For what value of e is $x = 0$ at point C, and what is the corresponding value of y?

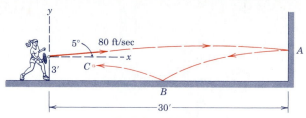

Problem 3/348

3/349 The simple pendulum of length $l = 0.5$ m has an angular velocity $\dot{\theta}_0 = 0.2$ rad/s at time $t = 0$ when $\theta = 0$. Derive an integral expression for the time t required to reach an arbitrary angle θ. Plot t vs. θ for $0 \leq \theta \leq \dfrac{\pi}{2}$ and state the value of t for $\theta = \dfrac{\pi}{2}$.

$$Ans. \quad t = 0.5 \int_0^\theta \frac{d\theta}{\sqrt{9.81 \sin \theta + 0.01}}, \quad t = 0.409 \text{ s}$$

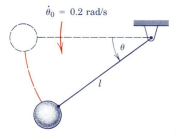

Problem 3/349

3/350 The amusement park ride of Prob. 3/68 is shown again here. Determine and plot the angle θ as a function of the rotational speed N for $0 \leq N \leq 10$ rev/min. Neglect the mass of the arms to which the gondolas are attached and treat each gondola as a particle. State the value of θ for $N = 8$ rev/min.

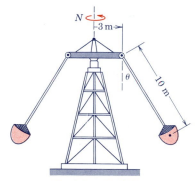

Problem 3/350

3/351 A system similar to that of Prob. 3/67 is shown here. The square plate is at rest in position A at time $t = 0$ and subsequently translates in a vertical circle according to $\theta = kt^2$, where $k = 1$ rad/s^2, the displacement θ is in radians, and time t is in seconds. A small 0.4-kg instrument P is temporarily fixed to the plate with adhesive. Plot the required shear force F vs. time t for $0 \leq t \leq 5$ s. If the adhesive fails when the shear force F reaches 30 N, determine the time t and angular position θ when failure occurs.

$$Ans. \quad t = 3.40 \text{ s}, \quad \theta = 663°$$

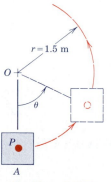

Problem 3/351

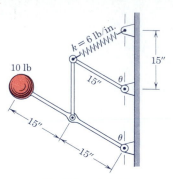

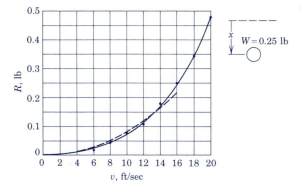

Problem 3/352

*3/352 The mechanism shown lies in a vertical plane and is released from rest in the position for which $\theta = 60°$, where the spring is unstretched. Plot the speed of the 10-lb sphere as a function of θ and determine (*a*) the maximum speed of the sphere with the corresponding value of θ and (*b*) the maximum angle θ reached by the link. Friction and the mass of the links are small and may be neglected.

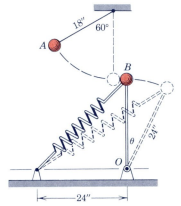

Problem 3/353

*3/353 A 20-lb sphere A is held at the 60° angle shown and released. It strikes the 10-lb sphere B. The coefficient of restitution for this collision is $e = 0.75$. Sphere B is attached to the end of a light rod that pivots freely about point O. If the spring of constant $k = 100$ lb/ft is initially unstretched, determine the maximum rotation angle θ of the light rod after impact. *Ans.* $\theta = 21.7°$

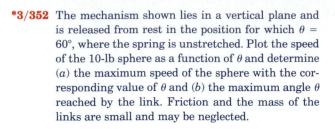

Problem 3/354

*3/354 Wind-tunnel tests of the resistance of a certain 4-oz sphere in a moving airstream for low velocities give the plotted curve shown in the full line. If the sphere is released from rest in still air, use these data to predict the velocity v that it will acquire after dropping 10 ft from rest. Multiply the equation of motion by dx and rewrite it as a finite-difference equation using 1-ft intervals in your solution. Next, solve for v by approximating the data with the analytic expression $R = kv^2$, where agreement at $v = 13$ ft/sec represents a fair average with the experimental data over the region considered. (Read the curve and get $k = 0.83(10^{-3})$ lb-sec^2/ft.^{2})

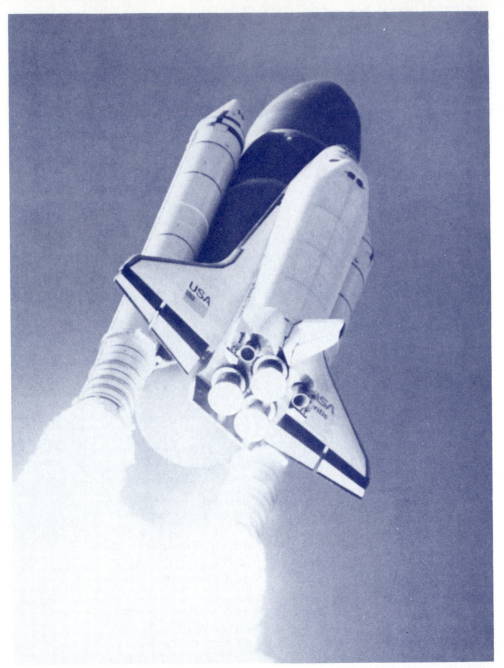

The dynamics of a group or system of masses is a generalization of the dynamics of a single mass particle and leads to the equations of translation and rotation which describe the motion of nonrigid bodies as well as rigid bodies. The space shuttle, together with its detachable rocket boosters and its expendable fuel, is a good example of a nonrigid system that is losing mass.

KINETICS OF SYSTEMS OF PARTICLES 4

4/1 INTRODUCTION

In the previous two chapters, we have dealt with the principles of dynamics for the motion of a particle. Although attention was focused primarily on the kinetics of a single particle in Chapter 3, reference to the motion of two particles, considered together as a system, was made in the discussions of work-energy and impulse-momentum. Our next major step in the development of the subject of dynamics is to extend these principles, which we applied to a single particle, to describe the motion of a general system of particles. This extension provides a unity to the remaining sections of dynamics and enables us to treat the motion of rigid bodies and the motion of nonrigid systems as well. We recall that a rigid body is a solid system of particles, wherein the distances between particles remain essentially unchanged. The overall motions found with machines, land and air vehicles, rockets and spacecraft, and many moving structures provide examples of rigid-body problems. A nonrigid body, on the other hand, may be a solid body where the object of investigation is the time dependence of the changes in shape due to elastic or nonelastic deformations. Or a nonrigid body may be a defined mass of liquid or gaseous particles that are flowing at a specified rate. Examples would be the air and fuel flowing through the turbine of an aircraft engine, the burned gases issuing from the nozzle of a rocket motor, or the water passing through a rotary pump.

Although the extension of the equations for single-particle motion to a general system of particles is accomplished without undue difficulty, it cannot be expected that the generality and significance of these extended principles will be fully understood without considerable problem experience. For this reason, we strongly recommend that the general results obtained in the following articles be reviewed frequently during the remainder of the study of dynamics. In this way, the unity contained in these broader principles of dynamics will be brought into sharper focus and a more fundamental view of the subject will be acquired.

4/2 *GENERALIZED NEWTON'S SECOND LAW*

Newton's second law of motion will now be extended to cover a general mass system that we model by considering n mass particles bounded by a closed surface in space, Fig. 4/1. This bounding envelope, for example, might be the exterior surface of a given rigid body, the bounding surface of an arbitrary portion of the body, the exterior surface of a rocket containing both rigid and flowing particles, or a particular volume of fluid particles. In each case, the system considered is the mass within the envelope, and that mass must be clearly defined and isolated.

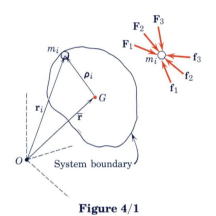

Figure 4/1

Figure 4/1 shows a representative particle of mass m_i of the system isolated with forces $\mathbf{F}_1, \mathbf{F}_2, \mathbf{F}_3, \dots$ acting on m_i from sources *external* to the envelope, and forces $\mathbf{f}_1, \mathbf{f}_2, \mathbf{f}_3, \dots$ acting on m_i from sources *internal* to the envelope. The external forces are due to contact with external bodies or to external gravitational, electric, or magnetic forces. The internal forces are forces of reaction with other mass particles within the envelope. A representative particle of mass m_i is located by its position vector $\mathbf{r}_i$ measured from the nonaccelerating origin O of a Newtonian set of reference axes.* The center of mass G of the system of particles isolated is located by the position vector $\bar{\mathbf{r}}$ that, from the definition of the mass center as covered in statics, is given by

$$m\bar{\mathbf{r}} = \Sigma m_i \mathbf{r}_i$$

where the total system mass is $m = \Sigma m_i$. The summation sign Σ represents the summation $\Sigma_{i=1}^{n}$ over all n particles. Newton's second law, Eq. 3/3, when applied to m_i gives

$$\mathbf{F}_1 + \mathbf{F}_2 + \mathbf{F}_3 + \cdots + \mathbf{f}_1 + \mathbf{f}_2 + \mathbf{f}_3 + \cdots = m_i \ddot{\mathbf{r}}_i$$

where $\ddot{\mathbf{r}}_i$ is the acceleration of m_i. A similar equation may be written for each of the particles of the system. If these equations written for *all* particles of the system are added together, there results

$$\Sigma \mathbf{F} + \Sigma \mathbf{f} = \Sigma m_i \ddot{\mathbf{r}}_i$$

The term $\Sigma \mathbf{F}$ then becomes the vector sum of *all* forces acting on all particles of the isolated system from sources external to the system, and $\Sigma \mathbf{f}$ becomes the vector sum of all forces on all particles produced by the internal actions and reactions between particles. This last sum is identically zero since all internal forces occur in pairs of equal and opposite actions and reactions. By differentiating the equation defining $\bar{\mathbf{r}}$ twice with time, we have $m\ddot{\bar{\mathbf{r}}} = \Sigma m_i \ddot{\mathbf{r}}_i$ where m has no time derivative as long as mass is not entering or leaving

---

*It was shown in Art. 3/14 that any nonrotating and nonaccelerating set of axes constitutes a Newtonian reference system in which the principles of Newtonian mechanics are valid.

the system.* Substitution into the summation of the equations of motion gives

$$\Sigma\mathbf{F} = m\ddot{\overline{\mathbf{r}}} \quad \text{or} \quad \Sigma\mathbf{F} = m\overline{\mathbf{a}} \qquad \textbf{(4/1)}$$

where $\overline{\mathbf{a}}$ is the acceleration $\ddot{\overline{\mathbf{r}}}$ of the center of mass of the system.

Equation 4/1 is the generalized Newton's second law of motion for a mass system and is referred to as the *equation of motion of m.* The equation states that the resultant of the external forces on *any* system of mass equals the total mass of the system times the acceleration of the center of mass. This law expresses the so-called *principle of motion of the mass center.* It should be observed that $\overline{\mathbf{a}}$ is the acceleration of the mathematical point that represents instantaneously the position of the mass center for the given n particles. For a nonrigid body, this acceleration need not represent the acceleration of any particular particle. It should be noted further that Eq. 4/1 holds for each instant of time and is therefore an instantaneous relationship. Equation 4/1 for the mass system had to be proved as it may not be inferred directly from Eq. 3/3 for the single particle. Equation 4/1 may be expressed in component form using *x-y-z* coordinates or whatever coordinate system is most convenient for the problem at hand. Thus,

$$\Sigma F_x = m\overline{a}_x \quad \Sigma F_y = m\overline{a}_y \quad \Sigma F_z = m\overline{a}_z \qquad (4/1a)$$

Although Eq. 4/1, as a vector equation, requires that the acceleration vector $\overline{\mathbf{a}}$ must have the same direction as the resultant external force $\Sigma\mathbf{F}$, it does not follow that $\Sigma\mathbf{F}$ necessarily passes through G. In general, in fact, $\Sigma\mathbf{F}$ does not pass through G, as will be shown later.

4/3 WORK-ENERGY

In Art. 3/6 the work-energy relation for a single particle was developed. Applicability to a system of two joined particles was also noted. We now turn our attention to the general system of Fig. 4/1, where the work-energy relation for the representative particle of mass m_i is $(U_{1\text{-}2})_i = \Delta T_i$. Here $(U_{1\text{-}2})_i$ is the work done on m_i during an interval of motion by all forces $\mathbf{F}_i = \mathbf{F}_1 + \mathbf{F}_2 + \mathbf{F}_3 + \cdots$ applied from sources external to the system and by all forces $\mathbf{f}_i = \mathbf{f}_1 + \mathbf{f}_2 + \mathbf{f}_3 + \cdots$ applied from sources internal to the system. The kinetic energy of m_i is $T_i = \frac{1}{2}m_i v_i^2$, where v_i is the magnitude of the particle velocity $\mathbf{v}_i = \dot{\mathbf{r}}_i$.

For the entire system, the sum of the work-energy equations written for all particles is $\Sigma(U_{1\text{-}2})_i = \Sigma\Delta T_i$, which may be repre-

---

*If m is a function of time, a more complex situation develops; this situation is discussed in Art. 4/7 on variable mass.

sented by the same expression as Eq. 3/11 of Art. 3/6, namely,

$$U_{1\text{-}2} = \Delta T \qquad \text{or} \qquad T_1 + U_{1\text{-}2} = T_2 \qquad \textbf{(4/2)}$$

where $U_{1\text{-}2} = \Sigma(U_{1\text{-}2})_i$, the work done by all forces on all particles, and ΔT is the change in the total kinetic energy $T = \Sigma T_i$ of the system.

For a rigid body or a system of rigid bodies joined by ideal frictionless connections, no net work is done by the internal interacting forces or moments in the connections. We see that the work done by all pairs of internal forces $\mathbf{f}_i$ and $-\mathbf{f}_i$ at a typical connection, Fig. 4/2, in the system is zero since their points of application have identical displacement components while the forces are equal but opposite. For this situation $U_{1\text{-}2}$ becomes the work done on the system by the external forces only.

For a nonrigid mechanical system that includes elastic members capable of storing energy, a part of the work done by the external forces goes into changing the internal elastic potential energy V_e. Also, if the work done by the gravity forces is *excluded* from the work term and is accounted for instead by the changes in gravitational potential energy V_g, then we may equate the work $U'_{1\text{-}2}$ done on the system during an interval of motion to the change ΔE in the mechanical energy of the system. Thus, $U'_{1\text{-}2} = \Delta E$ or

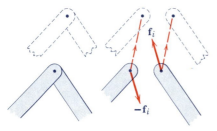

Figure 4/2

$$U'_{1\text{-}2} = \Delta T + \Delta V_g + \Delta V_e \qquad \textbf{(4/3)}$$

or

$$T_1 + V_{g_1} + V_{e_1} + U'_{1\text{-}2} = T_2 + V_{g_2} + V_{e_2} \qquad \textbf{(4/3}\textit{a}\textbf{)}$$

which are the same as Eqs. 3/17 and 3/17a.

We now examine the expression $T = \Sigma\frac{1}{2}m_i v_i^2$ for the kinetic energy of the mass system in more detail. By our principle of relative motion discussed in Art. 2/8, we may write the velocity of the representative particle as

$$\mathbf{v}_i = \bar{\mathbf{v}} + \dot{\boldsymbol{\rho}}_i$$

where $\bar{\mathbf{v}}$ is the velocity of the mass center G and $\dot{\boldsymbol{\rho}}_i$ is the velocity of m_i with respect to a translating reference frame moving with the mass center G. We recall the identity $v_i^2 = \mathbf{v}_i \cdot \mathbf{v}_i$ and write the kinetic energy of the system as

$$T = \Sigma\tfrac{1}{2}m_i\mathbf{v}_i\cdot\mathbf{v}_i = \Sigma\tfrac{1}{2}m_i(\bar{\mathbf{v}} + \dot{\boldsymbol{\rho}}_i)\cdot(\bar{\mathbf{v}} + \dot{\boldsymbol{\rho}}_i)$$
$$= \Sigma\tfrac{1}{2}m_i\bar{v}^2 + \Sigma\tfrac{1}{2}m_i|\dot{\boldsymbol{\rho}}_i|^2 + \Sigma m_i\bar{\mathbf{v}}\cdot\dot{\boldsymbol{\rho}}_i$$

Since $\boldsymbol{\rho}_i$ is measured from the mass center, $\Sigma m_i \boldsymbol{\rho}_i = \mathbf{0}$ and the third term is $\overline{\mathbf{v}} \cdot \Sigma m_i \dot{\boldsymbol{\rho}}_i = \overline{\mathbf{v}} \cdot \dfrac{d}{dt} \Sigma(m_i \boldsymbol{\rho}_i) = 0$. Also $\Sigma \frac{1}{2} m_i \overline{v}^2 = \frac{1}{2} \overline{v}^2 \Sigma m_i = \frac{1}{2} m \overline{v}^2$. Therefore, the total kinetic energy becomes

$$T = \tfrac{1}{2} m \overline{v}^2 + \Sigma \tfrac{1}{2} m_i |\dot{\boldsymbol{\rho}}_i|^2 \qquad (4/4)$$

This equation expresses the fact that the total kinetic energy of a mass system equals the energy of mass-center translation of the system as a whole plus the energy due to motion of all particles relative to the mass center.

4/4 IMPULSE-MOMENTUM

(a) *Linear momentum.* From our definition in Art. 3/8, the linear momentum of the representative particle of the system depicted in Fig. 4/1 is $\mathbf{G}_i = m_i \mathbf{v}_i$ where the velocity of m_i is $\mathbf{v}_i = \dot{\mathbf{r}}_i$. The linear momentum of the system is defined as the vector sum of the linear momenta of all of its particles or $\mathbf{G} = \Sigma m_i \mathbf{v}_i$. By substituting the relative velocity relation $\mathbf{v}_i = \overline{\mathbf{v}} + \dot{\boldsymbol{\rho}}_i$ and noting again that $\Sigma m_i \boldsymbol{\rho}_i = m \overline{\boldsymbol{\rho}} = \mathbf{0}$, we get

$$\mathbf{G} = \Sigma m_i (\overline{\mathbf{v}} + \dot{\boldsymbol{\rho}}_i) = \Sigma m_i \overline{\mathbf{v}} + \frac{d}{dt} \Sigma m_i \boldsymbol{\rho}_i$$

$$= \overline{\mathbf{v}} \Sigma m_i + \frac{d}{dt} (\mathbf{0})$$

or

$$\mathbf{G} = m \overline{\mathbf{v}} \qquad (4/5)$$

Thus, the linear momentum of any mass system of constant mass is the product of the mass and the velocity of its center of mass.

The time derivative of $\mathbf{G}$ is $m \dot{\overline{\mathbf{v}}} = m \overline{\mathbf{a}}$, which by Eq. 4/1 is the resultant external force acting on the system. Thus, we have

$$\Sigma \mathbf{F} = \dot{\mathbf{G}} \qquad (4/6)$$

which has the same form as Eq. 3/21 for a single particle. Equation 4/6 states that the resultant of the external forces on any mass system equals the time rate of change of the linear momentum of the system and is an alternative form of the generalized second law of motion, Eq. 4/1. As was noted at the end of the last article, $\Sigma \mathbf{F}$, in general, does not pass through the mass center G. In Eq. 4/6 the total mass was assumed to be constant during differentiation with time, so that the equation does not apply to systems whose mass changes with time.

(b) Angular momentum. The angular momentum of our general mass system will now be determined about the fixed point O, about the mass center G, and about an arbitrary point P, shown in Fig. 4/3, which may have an acceleration $\mathbf{a}_P = \ddot{\mathbf{r}}_P$.

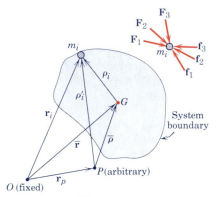

Figure 4/3

(About O). The angular momentum of the mass system about the point O, fixed in the Newtonian reference system, is defined as the vector sum of the moments of the linear momenta about O of all particles of the system and is

$$\mathbf{H}_O = \Sigma(\mathbf{r}_i \times m_i \mathbf{v}_i)$$

The time derivative of the vector product is $\dot{\mathbf{H}}_O = \Sigma(\dot{\mathbf{r}}_i \times m_i \mathbf{v}_i) + \Sigma(\mathbf{r}_i \times m_i \dot{\mathbf{v}}_i)$. The first summation vanishes since the cross product of two equal vectors $\dot{\mathbf{r}}_i$ and $\mathbf{v}_i$ is zero. The second summation is $\Sigma(\mathbf{r}_i \times m_i \mathbf{a}_i) = \Sigma(\mathbf{r}_i \times \mathbf{F}_i)$ which is the vector sum of the moments about O of all forces acting on all particles of the system. This moment sum $\Sigma\mathbf{M}_O$ represents only the moments of forces external to the system, since the internal forces cancel one another and their moments add up to zero. Thus, the moment sum is

$$\boxed{\Sigma\mathbf{M}_O = \dot{\mathbf{H}}_O} \qquad\qquad (4/7)$$

which has the same form as Eq. 3/27 for a single particle. Equation 4/7 states that the resultant vector moment about any fixed point of all external forces on any system of mass equals the time rate of change of angular momentum of the system about the fixed point. As in the linear-momentum case, Eq. 4/7 does not apply if the total mass of the system is changing with time.

(About G). The angular momentum of the mass system about the mass center G is the sum of the moments of the linear momenta about G of all particles and is

$$\mathbf{H}_G = \Sigma \boldsymbol{\rho}_i \times m_i \dot{\mathbf{r}}_i \qquad (4/8)$$

We may write the absolute velocity $\dot{\mathbf{r}}_i$ as $(\dot{\bar{\mathbf{r}}} + \dot{\boldsymbol{\rho}}_i)$ so that $\mathbf{H}_G$ becomes

$$\mathbf{H}_G = \Sigma \boldsymbol{\rho}_i \times m_i(\dot{\bar{\mathbf{r}}} + \dot{\boldsymbol{\rho}}_i) = \Sigma \boldsymbol{\rho}_i \times m_i \dot{\bar{\mathbf{r}}} + \Sigma \boldsymbol{\rho}_i \times m_i \dot{\boldsymbol{\rho}}_i$$

The first term on the right side of this equation may be rewritten as $-\dot{\bar{\mathbf{r}}} \times \Sigma m_i \boldsymbol{\rho}_i$, which is zero because $\Sigma m_i \boldsymbol{\rho}_i = \mathbf{0}$ by definition of the mass center. Hence, we have

$$\mathbf{H}_G = \Sigma \boldsymbol{\rho}_i \times m_i \dot{\boldsymbol{\rho}}_i \qquad (4/8a)$$

The expression of Eq. 4/8 is called the *absolute* angular momentum because the absolute velocity $\dot{\mathbf{r}}_i$ is used. The expression of Eq. 4/8a is called the *relative* angular momentum because the relative velocity $\dot{\boldsymbol{\rho}}_i$ is used. With the mass center G as a reference, the absolute and relative angular momenta are seen to be identical. We will see that this identity does not hold for an arbitrary reference point P; there is no distinction for a fixed reference point O.

Differentiating Eq. 4/8 with respect to time gives

$$\dot{\mathbf{H}}_G = \Sigma \dot{\boldsymbol{\rho}}_i \times m_i(\dot{\bar{\mathbf{r}}} + \dot{\boldsymbol{\rho}}_i) + \Sigma \boldsymbol{\rho}_i \times m_i \ddot{\mathbf{r}}_i$$

The first summation is expanded as $\Sigma \dot{\boldsymbol{\rho}}_i \times m_i \dot{\bar{\mathbf{r}}} + \Sigma \dot{\boldsymbol{\rho}}_i \times m_i \dot{\boldsymbol{\rho}}_i$. The first term may be rewritten as $-\dot{\bar{\mathbf{r}}} \times \Sigma m_i \dot{\boldsymbol{\rho}}_i = -\dot{\bar{\mathbf{r}}} \times \dfrac{d}{dt} \Sigma m_i \boldsymbol{\rho}_i$ which is zero from the definition of the mass center. The second term is zero because the cross product of parallel vectors is zero. With $\mathbf{F}_i$ representing the sum of all external forces acting on m_i and $\mathbf{f}_i$ the sum of all internal forces acting on m_i, the second summation by Newton's second law becomes $\Sigma \boldsymbol{\rho}_i \times (\mathbf{F}_i + \mathbf{f}_i) = \Sigma \boldsymbol{\rho}_i \times \mathbf{F}_i = \Sigma \mathbf{M}_G$, the sum of all external moments about point G. Recall that the sum of all internal moments $\Sigma \boldsymbol{\rho}_i \times \mathbf{f}_i$ is zero. Thus, we are left with

$$\boxed{\Sigma \mathbf{M}_G = \dot{\mathbf{H}}_G} \qquad \mathbf{(4/9)}$$

where we may use either the absolute or the relative angular momentum.

Equations 4/7 and 4/9 are among the most powerful of the governing equations in dynamics and apply to any defined system of mass—rigid or nonrigid.

(About P). The angular momentum about an arbitrary point P (which may have an acceleration $\ddot{\mathbf{r}}_P$) will now be expressed with the notation of Fig. 4/3. Thus,

$$\mathbf{H}_P = \Sigma \boldsymbol{\rho}_i' \times m_i \dot{\mathbf{r}}_i = \Sigma(\bar{\boldsymbol{\rho}} + \boldsymbol{\rho}_i) \times m_i \dot{\mathbf{r}}_i$$

The first term may be written as $\bar{\boldsymbol{\rho}} \times \Sigma m_i \dot{\mathbf{r}}_i = \bar{\boldsymbol{\rho}} \times \Sigma m_i \mathbf{v}_i = \bar{\boldsymbol{\rho}} \times m\bar{\mathbf{v}}$. The second term is $\Sigma \boldsymbol{\rho}_i \times m_i \dot{\mathbf{r}}_i = \mathbf{H}_G$. Thus, rearranging gives

$$\mathbf{H}_P = \mathbf{H}_G + \bar{\boldsymbol{\rho}} \times m\bar{\mathbf{v}} \qquad (4/10)$$

Equation 4/10 states that the absolute angular momentum about any point P equals the angular momentum about G plus the moment about P of the linear momentum $m\bar{\mathbf{v}}$ of the system considered concentrated at G.

We now make use of the principle of moments developed in our study of statics where we represented a force system by a resultant force through any point, such as G, and a corresponding couple. Figure 4/4 represents the resultants of the external forces acting on the system expressed in terms of the resultant force $\Sigma \mathbf{F}$ through

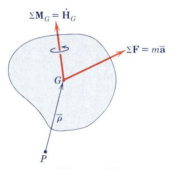

Figure 4/4

G and the corresponding couple $\Sigma \mathbf{M}_G$. We see that the sum of the moments about P of all forces external to the system must equal the moment of their resultants. Therefore, we may write

$$\Sigma \mathbf{M}_P = \Sigma \mathbf{M}_G + \bar{\boldsymbol{\rho}} \times \Sigma \mathbf{F}$$

which, by Eqs. 4/9 and 4/6, becomes

$$\Sigma \mathbf{M}_P = \dot{\mathbf{H}}_G + \bar{\boldsymbol{\rho}} \times m\bar{\mathbf{a}} \qquad (4/11)$$

Equation 4/11 enables us to write the moment equation about any convenient moment center P and is easily visualized with the aid of Fig. 4/4. This equation forms a rigorous basis for much of our treatment of planar rigid-body kinetics in Chapter 6.

We may also develop similar momentum relationships by using the momentum relative to P. Thus, from Fig. 4/3

$$(\mathbf{H}_P)_{\text{rel}} = \Sigma \boldsymbol{\rho}_i' \times m_i \dot{\boldsymbol{\rho}}_i'$$

where $\dot{\boldsymbol{\rho}}_i'$ is the velocity of m_i relative to P. With the substitution $\boldsymbol{\rho}_i' = \bar{\boldsymbol{\rho}} + \boldsymbol{\rho}_i$ and $\dot{\boldsymbol{\rho}}_i' = \dot{\bar{\boldsymbol{\rho}}} + \dot{\boldsymbol{\rho}}_i$ we may write

$$(\mathbf{H}_P)_{\text{rel}} = \Sigma \overline{\boldsymbol{\rho}} \times m_i \dot{\overline{\boldsymbol{\rho}}} + \Sigma \overline{\boldsymbol{\rho}} \times m_i \dot{\boldsymbol{\rho}}_i + \Sigma \boldsymbol{\rho}_i \times m_i \dot{\overline{\boldsymbol{\rho}}} + \Sigma \boldsymbol{\rho}_i \times m_i \dot{\boldsymbol{\rho}}_i$$

The first summation is $\overline{\boldsymbol{\rho}} \times m \overline{\mathbf{v}}_{\text{rel}}$. The second summation is $\overline{\boldsymbol{\rho}} \times \dfrac{d}{dt} \Sigma m_i \boldsymbol{\rho}_i$ and the third summation is $-\dot{\overline{\boldsymbol{\rho}}} \times \Sigma m_i \boldsymbol{\rho}_i$ where both are zero by definition of the mass center. The fourth summation is $(\mathbf{H}_G)_{\text{rel}}$. Rearranging gives us

$$(\mathbf{H}_P)_{\text{rel}} = (\mathbf{H}_G)_{\text{rel}} + \overline{\boldsymbol{\rho}} \times m \overline{\mathbf{v}}_{\text{rel}} \qquad (4/12)$$

where $(\mathbf{H}_G)_{\text{rel}}$ is the same as $\mathbf{H}_G$ (see Eqs. 4/8 and 4/8a). The similarity of Eqs. 4/12 and 4/10 should be noted.

The moment equation about P may now be expressed in terms of the angular momentum relative to P. We differentiate the definition $(\mathbf{H}_P)_{\text{rel}} = \Sigma \boldsymbol{\rho}_i' \times m_i \dot{\boldsymbol{\rho}}_i'$ with time and make the substitution $\ddot{\mathbf{r}}_i = \ddot{\mathbf{r}}_P + \ddot{\boldsymbol{\rho}}_i'$ to get

$$(\dot{\mathbf{H}}_P)_{\text{rel}} = \Sigma \dot{\boldsymbol{\rho}}_i' \times m_i \dot{\boldsymbol{\rho}}_i' + \Sigma \boldsymbol{\rho}_i' \times m_i \ddot{\mathbf{r}}_i - \Sigma \boldsymbol{\rho}_i' \times m_i \ddot{\mathbf{r}}_P$$

The first summation is identically zero, and the second summation is the sum $\Sigma \mathbf{M}_P$ of the moments of all external forces about P. The third summation becomes $\Sigma \boldsymbol{\rho}_i' \times m_i \mathbf{a}_P = -\mathbf{a}_P \times \Sigma m_i \boldsymbol{\rho}_i' = -\mathbf{a}_P \times m \overline{\boldsymbol{\rho}} = \overline{\boldsymbol{\rho}} \times m \mathbf{a}_P$. Substituting and rearranging terms give

$$\boxed{\Sigma \mathbf{M}_P = (\dot{\mathbf{H}}_P)_{\text{rel}} + \overline{\boldsymbol{\rho}} \times m \mathbf{a}_P} \qquad \textbf{(4/13)}$$

The form of Eq. 4/13 is convenient when a point P whose acceleration is known is used as a moment center. The equation reduces to the simpler form

$$\Sigma \mathbf{M}_P = (\dot{\mathbf{H}}_P)_{\text{rel}} \quad \text{if} \quad \begin{cases} 1. \ \mathbf{a}_P = \mathbf{0} \ \text{(equivalent to Eq. 4/7)} \\ 2. \ \overline{\boldsymbol{\rho}} = \mathbf{0} \ \text{(equivalent to Eq. 4/9)} \\ 3. \ \overline{\boldsymbol{\rho}} \ \text{and} \ \mathbf{a}_P \ \text{are parallel} \ (\mathbf{a}_P \ \text{directed} \\ \quad \text{toward or away from } G) \end{cases}$$

4/5 CONSERVATION OF ENERGY AND MOMENTUM

Conditions frequently occur when there is no net change in the total mechanical energy of a system during an interval of motion. Other conditions exist when there is no net change in the momentum of a system. These conditions are treated separately as follows.

(a) Conservation of energy.

A mass system is said to be *conservative* if it does not lose energy by virtue of internal friction forces that do negative work or by virtue of nonelastic members which dissipate energy upon cycling. If no work is done on a conservative system during an interval of motion by external forces (other than gravity or other potential forces), then no part of the energy of the system is lost. In this case, $\Delta E = 0$ or $E_{\text{initial}} = E_{\text{final}}$. Therefore, we may write Eq. 4/3 as

$$\Delta T + \Delta V_g + \Delta V_e = 0 \qquad\qquad (4/14)$$

or
$$T_1 + V_{g_1} + V_{e_1} = T_2 + V_{g_2} + V_{e_2} \qquad\qquad (4/14a)$$

which expresses the *law of conservation of dynamical energy*. This law holds only in the ideal case where internal kinetic friction is sufficiently small to be neglected.

(b) Conservation of momentum.   If, for a certain interval of time, the resultant external force $\Sigma\mathbf{F}$ acting on a conservative or nonconservative mass system is zero, Eq. 4/6 requires that $\dot{\mathbf{G}} = \mathbf{0}$, so that during this interval

$$\mathbf{G}_1 = \mathbf{G}_2 \qquad\qquad (4/15)$$

which expresses the *principle of conservation of linear momentum*. Thus, in the absence of an external impulse the linear momentum of a system remains unchanged.

Similarly, if the resultant moment about a fixed point O or about the mass center G of all external forces on any mass system is zero, Eq. 4/7 or 4/9 requires, respectively, that

$$(\mathbf{H}_O)_1 = (\mathbf{H}_O)_2 \qquad \text{or} \qquad (\mathbf{H}_G)_1 = (\mathbf{H}_G)_2 \qquad\qquad (4/16)$$

These relations express the *principle of conservation of angular momentum* for a general mass system in the absence of an angular impulse. Thus, if there is no angular impulse about a fixed point (or about the mass center), the angular momentum of the system about the fixed point (or about the mass center) remains unchanged. Either equation may hold without the other.

It was proved in Art. 3/14 that the basic laws of Newtonian mechanics hold for measurements made relative to a set of axes that translate with a constant velocity. Thus, Eqs. 4/1 through 4/16 are valid provided all quantities are expressed relative to the translating axes.

Equations 4/1 through 4/16 are among the most important of the basic derived laws of mechanics. In this chapter these laws have been derived for the most general system of constant mass in order that the generality of the laws will be established. These laws find repeated use when applied to specific mass systems such as rigid and nonrigid solids and certain fluid systems which are discussed in the articles that follow. The reader is urged to study these laws carefully and to compare them with their more restricted forms encountered earlier in Chapter 3.

Sample Problem 4/1

Each of the three balls has a mass m and is welded to the rigid equi-angular frame of negligible mass. The assembly rests on a smooth horizontal surface. If a force $\mathbf{F}$ is suddenly applied to the one bar as shown, determine (a) the acceleration of point O and (b) the angular acceleration $\ddot{\theta}$ of the frame.

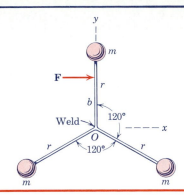

Solution. (a) Point O is the mass center of the system of the three balls, so that its acceleration is given by Eq. 4/1.

① $[\Sigma F = m\bar{a}]$ $\qquad F\mathbf{i} = 3m\bar{\mathbf{a}} \qquad \bar{\mathbf{a}} = \mathbf{a}_O = \dfrac{F}{3m}\mathbf{i}$ $\qquad$ *Ans.*

① We note that the result depends only on the magnitude and direction of $\mathbf{F}$ and not on b, which locates the line of action of $\mathbf{F}$.

(b) We determine $\ddot{\theta}$ from the moment principle, Eq. 4/9. To find $\mathbf{H}_G$ we note that the velocity of each ball relative to the mass center O as measured in the nonrotating axes x-y is $r\dot{\theta}$, where $\dot{\theta}$ is the common angular velocity of the spokes. The angular momentum of the system about O is the sum of the moments of the relative linear momenta as shown by Eq. 4/8, so it is expressed by

$$H_O = H_G = 3(mr\dot{\theta})r = 3mr^2\dot{\theta}$$

② Equation 4/9 now gives

$[\Sigma M_G = \dot{H}_G]$

$$Fb = \frac{d}{dt}(3mr^2\dot{\theta}) = 3mr^2\ddot{\theta} \quad \text{so} \quad \ddot{\theta} = \frac{Fb}{3mr^2} \qquad Ans.$$

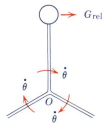

② Although $\dot{\theta}$ is initially zero, we need the expression for $H_O = H_G$ in order to get $\dot{H}_G$. We observe also that $\ddot{\theta}$ is independent of the motion of O.

Sample Problem 4/2

Consider the same conditions as for Sample Problem 4/1, except that the spokes are freely hinged at O and so do not constitute a rigid system. Explain the difference between the two problems.

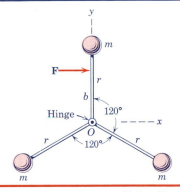

Solution. The generalized Newton's second law holds for any mass system, so that the acceleration $\bar{\mathbf{a}}$ of the mass center G is the same as with Sample Problem 4/1, namely,

$$\bar{\mathbf{a}} = \frac{F}{3m}\mathbf{i} \qquad Ans.$$

Although G coincides with O at the instant represented, the motion of the hinge O is not the same as the motion of G since O will not remain the center of mass as the angles between the spokes change.

① Both ΣM_G and $\dot{H}_G$ have the same values for the two problems at the instant represented. However, the angular motions of the spokes in this problem are all different and are not easily determined.

① This present system could be dismembered and the motion equations written for each of the parts, with the unknowns eliminated one by one. Or a more sophisticated method using the equations of Lagrange could be employed. (See the senior author's *Dynamics, 2nd Edition SI Version*, 1975, for a discussion of this approach.)

Sample Problem 4/3

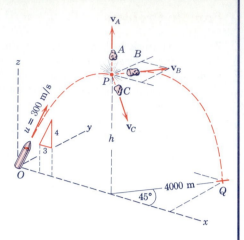

A shell with a mass of 20 kg is fired from point O, with a velocity $u = 300$ m/s in the vertical x-z plane at the inclination shown. When it reaches the top of its trajectory at P, it explodes into three fragments A, B, and C. Immediately after the explosion, fragment A is observed to rise vertically a distance of 500 m above P, and fragment B is seen to have a horizontal velocity $\mathbf{v}_B$ and eventually land at point Q. When recovered, the masses of the fragments A, B, and C are found to be 5, 9, and 6 kg, respectively. Calculate the velocity that fragment C has immediately after the explosion. Neglect atmospheric resistance.

Solution. From our knowledge of projectile motion (Sample Problem 2/6), the time required for the shell to reach P and its vertical rise are

$$t = u_z/g = 300(4/5)/9.81 = 24.5 \text{ s}$$

$$h = \frac{u_z^2}{2g} = \frac{[(300)(4/5)]^2}{2(9.81)} = 2936 \text{ m}$$

The velocity of A has the magnitude

$$v_A = \sqrt{2gh_A} = \sqrt{2(9.81)(500)} = 99.0 \text{ m/s}$$

With no z-component of velocity initially, fragment B requires 24.5 s to return to the ground. Thus, its horizontal velocity, which remains constant, is

$$v_B = s/t = 4000/24.5 = 163.5 \text{ m/s}$$

Since the force of the explosion is internal to the system of the shell and its three fragments, the linear momentum of the system remains unchanged during the explosion. Thus,

① $[\mathbf{G}_1 = \mathbf{G}_2] \quad m\mathbf{v} = m_A\mathbf{v}_A + m_B\mathbf{v}_B + m_C\mathbf{v}_C$

$$20(300)(\tfrac{3}{5})\mathbf{i} = 5(99.0\mathbf{k}) + 9(163.5)(\mathbf{i}\cos 45° + \mathbf{j}\sin 45°) + 6\mathbf{v}_C$$

$$6\mathbf{v}_C = 2560\mathbf{i} - 1040\mathbf{j} - 495\mathbf{k}$$

$$\mathbf{v}_C = 427\mathbf{i} - 173\mathbf{j} - 82.5\mathbf{k} \text{ m/s}$$

② $$v_C = \sqrt{(427)^2 + (173)^2 + (82.5)^2} = 468 \text{ m/s} \qquad Ans.$$

① The velocity $\mathbf{v}$ of the shell at the top of its trajectory is, of course, the constant horizontal component of its initial velocity $\mathbf{u}$, which becomes $u(3/5)$.

② We note that the mass center of the three fragments while still in flight continues to follow the same trajectory that the shell would have followed if it had not exploded.

Sample Problem 4/4

The 32.2-lb carriage A moves horizontally in its guide with a speed of 4 ft/sec and carries two assemblies of balls and light rods that rotate about a shaft at O in the carriage. Each of the four balls weighs 3.22 lb. The assembly on the front face rotates counterclockwise at a speed of 80 rev/min, and the assembly on the back side rotates clockwise at a speed of 100 rev/min. For the entire system, calculate (a) the kinetic energy T, (b) the magnitude G of the linear momentum, and (c) the magnitude H_O of the angular momentum about point O.

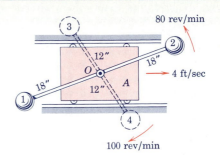

Solution. **(a) Kinetic energy.** The velocity of the balls with respect to O is

$$[|\dot{\boldsymbol{\rho}}_i| = v_{\text{rel}} = r\dot{\theta}] \quad (v_{\text{rel}})_{1,2} = \frac{18}{12}\frac{80(2\pi)}{60} = 12.57 \text{ ft/sec}$$

$$(v_{\text{rel}})_{3,4} = \frac{12}{12}\frac{100(2\pi)}{60} = 10.47 \text{ ft/sec}$$

The kinetic energy of the system is given by Eq. 4/4. The translational part is

① $$\tfrac{1}{2}m\bar{v}^2 = \frac{1}{2}\left(\frac{32.2}{32.2} + 4\frac{3.22}{32.2}\right)(4^2) = 11.20 \text{ ft-lb}$$

The rotational part of the kinetic energy depends on the squares of the relative velocities and is

② $$\Sigma\tfrac{1}{2}m_i|\dot{\boldsymbol{\rho}}_i|^2 = 2\left[\frac{1}{2}\frac{3.22}{32.2}(12.57)^2\right]_{(1,2)} + 2\left[\frac{1}{2}\frac{3.22}{32.2}(10.47)^2\right]_{(3,4)}$$

$$= 15.80 + 10.96 = 26.76 \text{ ft-lb}$$

The total kinetic energy is

$$T = \tfrac{1}{2}m\bar{v}^2 + \Sigma\tfrac{1}{2}m_i|\dot{\boldsymbol{\rho}}_i|^2 = 11.20 + 26.76 = 37.96 \text{ ft-lb} \quad \textit{Ans.}$$

(b) Linear momentum. The linear momentum of the system by Eq. 4/5 is the total mass times v_O, the velocity of the center of mass. Thus,

③ $$[\mathbf{G} = m\bar{\mathbf{v}}] \quad G = \left(\frac{32.2}{32.2} + 4\frac{3.22}{32.2}\right)(4) = 5.6 \text{ lb-sec}$$

(c) Angular momentum about O. The angular momentum about O is due to the moments of the linear momenta of the balls. Taking counterclockwise as positive, we have

$$H_O = \Sigma|\mathbf{r}_i \times m_i\mathbf{v}_i|$$

④ $$H_O = \left[2\left(\frac{3.22}{32.2}\right)\left(\frac{18}{12}\right)(12.57)\right]_{(1,2)} - \left[2\left(\frac{3.22}{32.2}\right)\left(\frac{12}{12}\right)(10.47)\right]_{(3,4)}$$

$$= 3.77 - 2.09 = 1.68 \text{ ft-lb-sec} \quad \textit{Ans.}$$

① Note that the mass m is the total mass, carriage plus the four balls, and that $\bar{v}$ is the velocity of the mass center O which is the carriage velocity.

② Note that the direction of rotation, clockwise or counterclockwise, makes no difference in the calculation of kinetic energy, which depends on the square of the velocity.

③ There is a temptation to overlook the contribution of the balls since their linear momenta relative to O in each pair are in opposite directions and cancel. However, each ball also has a velocity component $\bar{\mathbf{v}}$ and hence a momentum component $m_i\bar{\mathbf{v}}$.

④ Contrary to the case of kinetic energy where the direction of rotation was immaterial, angular momentum is a vector quantity and the direction of rotation must be accounted for.

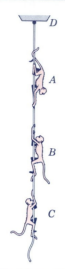

Problem 4/1

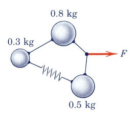

Problem 4/2

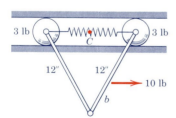

Problem 4/3

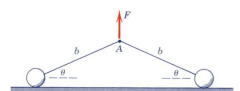

Problem 4/4

PROBLEMS

Introductory problems

4/1 Three monkeys A, B, and C weighing 20, 25, and 15 lb, respectively, are climbing up and down the rope suspended from D. At the instant represented, A is descending the rope with an acceleration of 5 ft/sec², and C is pulling himself up with an acceleration of 3 ft/sec². Monkey B is climbing up with a constant speed of 2 ft/sec. Treat the rope and monkeys as a complete system and calculate the tension T in the rope at D.   *Ans.* $T = 58.3$ lb

4/2 The three small spheres are connected by the cords and spring and are supported by a smooth horizontal surface. If a force $F = 6.4$ N is applied to one of the cords, find the acceleration $\bar{a}$ of the mass center of the spheres for the instant depicted.

4/3 The system consists of the two smooth spheres, each weighing 3 lb and connected by a light spring, and the two bars of negligible weight hinged freely at their ends and hanging in the vertical plane. The spheres are confined to slide in the smooth horizontal guide. If a horizontal force $F = 10$ lb is applied to the one bar at the position shown, what is the acceleration of the center C of the spring? Why does the result not depend on the dimension b?
Ans. $a_C = 53.7$ ft/sec²

4/4 The two small spheres, each of mass m, are connected by a cord of length $2b$ (measured from the centers of the spheres) and are initially at rest on a smooth horizontal surface in the position shown. If a vertical force of constant magnitude F is applied to the center A of the cord, determine the velocity v of each sphere when they collide as θ approaches 90°. What is the maximum value of F for which the spheres do not lose contact with the surface? (Analyze the system without dismembering it.)

4/5 The total linear momentum of a system of five particles at time $t = 2.2$ s is given by $\mathbf{G}_{2.2} = 3.4\mathbf{i} - 2.6\mathbf{j} + 4.6\mathbf{k}$ kg·m/s. At time $t = 2.4$ s, the linear momentum has changed to $\mathbf{G}_{2.4} = 3.7\mathbf{i} - 2.2\mathbf{j} + 4.9\mathbf{k}$ kg·m/s. Calculate the magnitude F of the time average of the resultant of the external forces acting on the system during the interval.

Ans. $F = 2.92$ N

4/6 The two small spheres, each of mass m, are rigidly connected by a rod of negligible mass. The center C of the rod has a velocity v in the x-direction, and the rod is rotating counterclockwise at the constant rate $\dot{\theta}$. For a given value of θ write the expressions for (a) the linear momentum of each sphere and (b) the linear momentum $\mathbf{G}$ of the system of the two spheres.

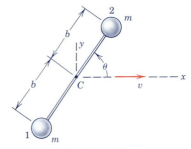

Problem 4/6

4/7 A centrifuge consists of four cylindrical containers, each of mass m, at a radial distance r from the rotation axis. Determine the time t required to bring the centrifuge to an angular velocity ω from rest under a constant torque M applied to the shaft. The diameter of each container is small compared with r, and the mass of the shaft and supporting arms is small compared with m.

Ans. $t = \dfrac{4mr^2\omega}{M}$

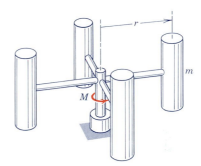

4/8 For the conditions of Prob. 4/6, write the expressions for (a) the absolute angular momentum of each sphere about C and (b) the absolute and relative angular momenta, $(\mathbf{H}_C)_{\text{abs}}$ and $(\mathbf{H}_C)_{\text{rel}}$, about C for the system.

Problem 4/7

Representative problems

4/9 Two steel balls, each of mass m, are welded to a light rod of length L and negligible mass and are initially at rest on a smooth horizontal surface. A horizontal force of magnitude F is suddenly applied to the rod as shown. Determine (a) the instantaneous acceleration $\bar{a}$ of the mass center G and (b) the corresponding rate $\ddot{\theta}$ at which the angular velocity of the assembly about G is changing with time.

Ans. (a) $\bar{a} = \dfrac{F}{2m}$, (b) $\ddot{\theta} = \dfrac{2Fb}{mL^2}$

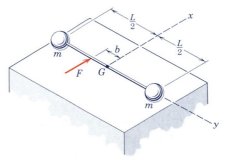

Problem 4/9

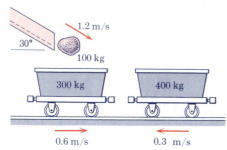

1.2 m/s

30°

100 kg

300 kg

400 kg

0.6 m/s   0.3 m/s

Problem 4/10

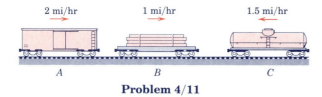

2 mi/hr   1 mi/hr   1.5 mi/hr

A   B   C

Problem 4/11

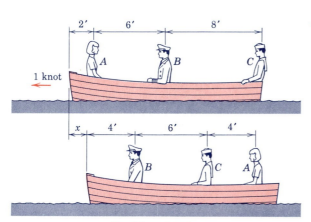

2′   6′   8′

A   B   C

1 knot

x   4′   6′   4′

B   C   A

Problem 4/12

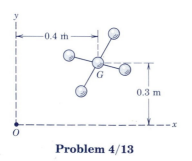

y

|← 0.4 m →|

G

0.3 m

O ----------------- x

Problem 4/13

4/10   The 300-kg and 400-kg mine cars are rolling in opposite directions along the horizontal track with the respective speeds of 0.6 m/s and 0.3 m/s. Upon impact the cars become coupled together. Just prior to impact, a 100-kg boulder leaves the delivery chute with a velocity of 1.2 m/s in the direction shown and lands in the 300-kg car. Calculate the velocity v of the system after the boulder has come to rest relative to the car. Would the final velocity be the same if the cars were coupled before the boulder dropped?

4/11   The three freight cars are rolling along the horizontal track with the velocities shown. After the impacts occur, the three cars become coupled together and move with a common velocity v. The weights of the loaded cars A, B, and C are 130,000, 100,000, and 150,000 lb, respectively. Determine v and calculate the percentage loss n of energy of the system due to coupling.

Ans. $v = 0.355$ mi/hr, $n = 95.0\%$

4/12   The young woman A, the captain B, and the sailor C weigh 120, 180, and 160 lb, respectively, and are sitting in the 300-lb skiff that is gliding through the water with a speed of 1 knot. If the three people change their positions as shown in the second figure, find the distance x from the skiff to the position where it would have been if the people had not moved. Neglect any resistance to motion afforded by the water. Does the sequence or timing of the change in positions affect the final result?

4/13   Each of the five connected particles has a mass of 0.6 kg, with G as the center of mass of the system. At a certain instant the angular momentum of the system about G is $1.20\mathbf{k}$ kg·m²/s, and the x- and y-components of the velocity of G are 3 m/s and 4 m/s, respectively. Calculate the angular momentum $\mathbf{H}_O$ of the system about O for this instant.

Ans. $\mathbf{H}_O = 3.3\mathbf{k}$ kg·m²/s

4/14 The two balls are attached to the light rigid rod, which is suspended by a cord from the support above it. If the balls and rod, initially at rest, are struck by the force $F = 12$ lb, calculate the corresponding acceleration $\bar{a}$ of the mass center and the rate $\ddot{\theta}$ at which the angular velocity of the bar is changing.

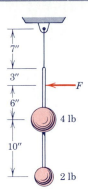

Problem 4/14

4/15 At a certain instant, particles A, B, and C with masses of 2, 3, and 4 kg, respectively, have position vectors $\mathbf{r}_A = 2\mathbf{i} + 3\mathbf{j}$ m, $\mathbf{r}_B = 2\mathbf{j} - \mathbf{k}$ m, and $\mathbf{r}_C = \mathbf{i} - 2\mathbf{k}$ m. The velocities at this same instant are $\mathbf{v}_A = \dot{\mathbf{r}}_A = 2\mathbf{j} - 2\mathbf{k}$ m/s, $\mathbf{v}_B = \dot{\mathbf{r}}_B = 2\mathbf{i} + 3\mathbf{j}$ m/s, and $\mathbf{v}_C = \dot{\mathbf{r}}_C = -2\mathbf{j} + \mathbf{k}$ m/s. Calculate the angular momentum $\mathbf{H}_O$ of the system of three particles about the origin O.

> Ans. $\mathbf{H}_O = -19\mathbf{i} - 2\mathbf{j} - 12\mathbf{k}$ kg·m²/s

4/16 The angular momentum of a system of six particles about a fixed point O at time $t = 4$ s is $\mathbf{H}_4 = 3.65\mathbf{i} + 4.27\mathbf{j} - 5.36\mathbf{k}$ kg·m²/s. At time $t = 4.1$ s, the angular momentum is $\mathbf{H}_{4.1} = 3.67\mathbf{i} + 4.30\mathbf{j} - 5.20\mathbf{k}$ kg·m²/s. Determine the average value of the resultant moment about point O of all forces acting on all particles during the 0.1-s interval.

4/17 Two projectiles, each weighing 20 lb, are fired simultaneously from the vehicle shown that weighs 2000 lb and is moving with an initial velocity $v_1 = 4$ ft/sec in the direction opposite to the firing. Each projectile has a muzzle velocity $v_r = 800$ ft/sec relative to the barrel. Calculate the velocity v_2 of the vehicle after the projectiles have been fired.

> Ans. $v_2 = 19.69$ ft/sec

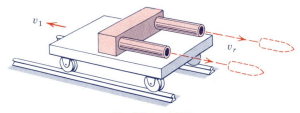

Problem 4/17

4/18 The small car that has a mass of 20 kg rolls freely on the horizontal track and carries the 5-kg sphere mounted on the light rotating rod with $r = 0.4$ m. A geared motor drive maintains a constant angular speed $\dot{\theta} = 4$ rad/s of the rod. If the car has a velocity $v = 0.6$ m/s when $\theta = 0$, calculate v when $\theta = 60°$. Neglect the mass of the wheels and any friction.

4/19 Determine the maximum value of the velocity v of the car in Prob. 4/18 and the corresponding angle θ.

> Ans. $v = 0.920$ m/s, $\theta = 90°$

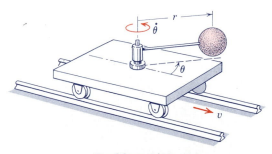

Problem 4/18

Problem 4/20

4/20 The cars of a roller-coaster ride have a speed of 30 km/h as they pass over the top of the circular track. Neglect any friction and calculate their speed v when they reach the horizontal bottom position. At the top position, the radius of the circular path of their mass centers is 18 m, and all six cars have the same mass.

4/21 A department-store escalator makes an angle of 30° with the horizontal and takes 40 seconds to transport a person from the first to the second floor with a vertical rise of 20 ft. At a certain instant, there are 10 people on the escalator averaging 150 lb per person and standing at rest relative to the moving steps. Additionally, 3 boys averaging 120 lb each are running down the escalator at a speed of 2 ft/sec relative to the moving steps. Calculate the power output P of the driving motor to maintain the constant speed of the escalator. The no-load power without passengers is 2.2 hp to overcome friction in the mechanism. *Ans. $P = 3.24$ hp*

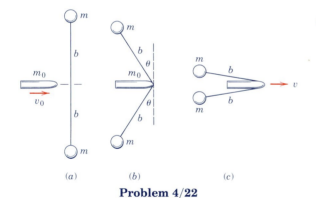

(a)   (b)   (c)

Problem 4/22

4/22 The two small spheres, each of mass m, are connected by a cord of length $2b$ (measured to the centers of the spheres) and are initially at rest on a smooth horizontal surface. A projectile of mass m_0 with a velocity v_0 perpendicular to the cord hits it in the middle, causing the deflection shown in the b-part of the figure. Determine the velocity v of m_0 as the two spheres near contact, with θ approaching 90° as indicated in the c-part of the figure. Also find $\dot{\theta}$ for this condition.

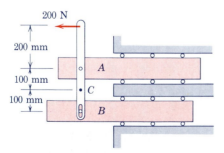

Problem 4/23

4/23 Each of the bars A and B has a mass of 10 kg and slides in its horizontal guideway with negligible friction. Motion is controlled by the lever of negligible mass connected to the bars as shown. Calculate the acceleration of point C on the lever when the 200-N force is applied as indicated. To verify your result, analyze the kinetics of each member separately and determine a_C by kinematic considerations from the calculated accelerations of the two bars.
Ans. $a_C = 10$ m/s^2

4/24 The three small spheres, each of mass m, are secured to the light rods to form a rigid unit supported
in the vertical plane by the smooth circular surface.
The force of constant magnitude P is applied perpendicular to the rod at its midpoint. If the unit
starts from rest at $\theta = 0$, determine (*a*) the minimum force P_{min} that will bring the unit to rest at
$\theta = 60°$ and (*b*) the common velocity v of spheres
1 and 2 when $\theta = 60°$ if $P = 2P_{min}$.

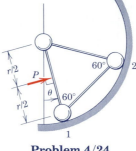

Problem 4/24

4/25 The three identical 2-kg spheres are welded to the
connecting rods of negligible mass and are hanging
by a cord from point A. The spheres are initially at
rest when a horizontal force $F = 16$ N is applied to
the upper sphere. Calculate the initial acceleration
$\bar{a}$ of the mass center of the spheres, the rate $\ddot{\theta}$ at
which the angular velocity is increasing, and the ini-
tial acceleration a of the top sphere.

$$Ans.\ \bar{a} = 2.67\ \text{m/s}^2$$
$$\ddot{\theta} = 15.40\ \text{rad/s}^2,\ a = 5.33\ \text{m/s}^2$$

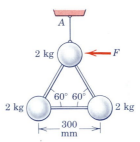

Problem 4/25

4/26 The three identical spheres, each of mass m, are supported in the vertical plane on the 30° incline. The
spheres are welded to the two connecting rods of
negligible mass. The upper rod, also of negligible
mass, is pivoted freely to the upper sphere and to
the bracket at A. If the stop at B is suddenly removed, determine the velocity v with which the upper sphere hits the incline. (Note that the corresponding velocity of the center sphere is $v/2$.)
Explain the loss of energy that occurs after all motion has ceased.

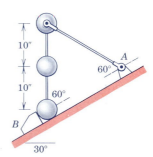

Problem 4/26

4/27 A flexible nonextensible rope of mass ρ per unit
length and length equal to 1/4 of the circumference
of the fixed drum of radius r is released from rest in
the horizontal dotted position, with end B secured
to the top of the drum. When the rope finally comes
to rest with end A at C, determine the loss of energy
ΔQ of the system. What becomes of the lost energy?

$$Ans.\ \Delta Q = 0.571\ \rho g r^2$$

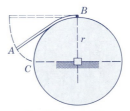

Problem 4/27

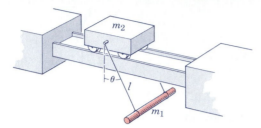

Problem 4/28

▶ **4/28** A horizontal bar of mass m_1 and small diameter is suspended by two wires of length l from a carriage of mass m_2, which is free to roll along the horizontal rails. If the bar and carriage are released from rest with the wires making an angle θ with the vertical, determine the velocity $v_{b/c}$ of the bar relative to the carriage and the velocity v_c of the carriage at the instant when $\theta = 0$. Neglect all friction and treat the carriage and the bar as particles in the vertical plane of motion.

Ans. $v_{b/c} = \sqrt{(1 + m_1/m_2)2gl(1 - \cos \theta)}$

$$v_c = \sqrt{\frac{2gl(1 - \cos \theta)}{(m_2/m_1)(1 + m_2/m_1)}}$$

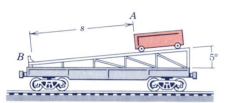

Problem 4/29

▶ **4/29** The 50,000-lb flatcar supports a 15,000-lb vehicle on a 5° ramp built on the flatcar. If the vehicle is released from rest with the flatcar also at rest, determine the velocity v of the flatcar when the vehicle has rolled $s = 40$ ft down the ramp just before hitting the stop at B. Neglect all friction and treat the vehicle and the flatcar as particles.

Ans. $v = 3.92$ ft/sec

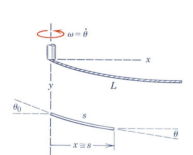

Problem 4/30

▶ **4/30** A flexible rope of length L and mass ρ per unit length is whirled around a fixed vertical axis with a constant angular velocity $\omega = \dot{\theta}$. Determine the shape $y = f(x)$ assumed by the rope for large ω so that the distance along the rope to any element may be approximated by the horizontal radius to the element. The rope may be considered a system of connected particles. (*Hint:* Draw the free-body diagram for the section of the rope shown and write its x- and y-equations of motion. Recognize that $dy/dx = \tan \theta$ and that the tension T in the rope is zero at the outer end where $x \cong L$.)

Ans. $y = \dfrac{2g}{\omega^2} \ln \left(1 + \dfrac{x}{L}\right)$

4/6 STEADY MASS FLOW

The momentum relation developed in Art. 4/4 for a general system of mass provides us with a direct means of analyzing the action of mass flow where a change of momentum occurs. The dynamics of mass flow is of great importance in the description of fluid machinery of all types including turbines, pumps, nozzles, air-breathing jet engines, and rockets. The treatment of mass flow in this article is not intended to take the place of a study of fluid mechanics, but merely to present the basic principles and equations of momentum that find important usage in fluid mechanics and in the general flow of mass whether the form be liquid, gaseous, or granular.

One of the most important cases of mass flow occurs during steady-flow conditions where the rate at which mass enters a given volume equals the rate at which mass leaves the same volume. The volume in question may be enclosed by a rigid container, fixed or moving, such as the nozzle of a jet aircraft or rocket, the space between blades in a gas turbine, the volume within the casing of a centrifugal pump, or the volume within the bend of a pipe through which a fluid is flowing at a steady rate. The design of such fluid machines is dependent on the analysis of the forces and moments that are developed through the corresponding momentum changes of the flowing mass.

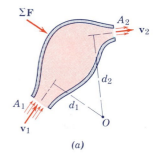

Consider a rigid container, shown in section in Fig. 4/5a, into which mass flows in a steady stream at the rate m' through the entrance section of area A_1. Mass leaves the container through the exit section of area A_2 at the same rate, so that there is no accumulation or depletion of the total mass within the container during the period of observation. The velocity of the entering stream is $\mathbf{v}_1$ normal to A_1 and that of the leaving stream is $\mathbf{v}_2$ normal to A_2. If ρ_1 and ρ_2 are the respective densities of the two streams, the continuity of flow requires that

$$\rho_1 A_1 v_1 = \rho_2 A_2 v_2 = m' \qquad (4/17)$$

The forces that act are described by isolating either the mass of fluid within the container or the entire container and the fluid within it. We would use the first approach if the forces between the container and the fluid were to be described, and we would adopt the second approach when the forces external to the container are desired. The latter situation is our primary interest, in which case, the *system isolated* consists of the fixed structure of the container and the fluid within it at a particular instant of time. This isolation is described by a free-body diagram of the mass within a closed space volume defined by the exterior surface of the container and the entrance and exit surfaces. We must account for all forces applied *externally* to this system, and in Fig. 4/5a the vector sum of this

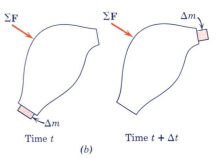

Figure 4/5

external force system is denoted by $\Sigma\mathbf{F}$. Included in $\Sigma\mathbf{F}$ are the forces exerted on the container at points of its attachment to other structures, including attachments at A_1 and A_2, if present; the forces acting on the fluid within the container at A_1 and A_2 due to any static pressure that may exist in the fluid at these positions; and the weight of the fluid and structure if appreciable. The resultant $\Sigma\mathbf{F}$ of all of these external forces must equal $\dot{\mathbf{G}}$, the time rate of change of the linear momentum of the isolated system, by Eq. 4/6 which was developed in Art. 4/4 for any system of constant mass, rigid or nonrigid.

The expression for $\dot{\mathbf{G}}$ may be obtained by an incremental analysis. Figure 4/5b illustrates the system at time t when the system mass is that of the container, the mass within it, and an increment Δm about to enter during time Δt. At time $t + \Delta t$ the same total mass is that of the container, the mass within it, and an equal increment Δm that leaves the container in time Δt. The linear momentum of the container and mass within it between the two sections A_1 and A_2 remains unchanged during Δt so that the change in momentum of the system in time Δt is

$$\Delta\mathbf{G} = (\Delta m)\mathbf{v}_2 - (\Delta m)\mathbf{v}_1 = \Delta m(\mathbf{v}_2 - \mathbf{v}_1)$$

Division by Δt and passage to the limit yield $\dot{\mathbf{G}} = m' \, \Delta\mathbf{v}$, where

$$m' = \lim_{\Delta t \to 0}\left(\frac{\Delta m}{\Delta t}\right) = \frac{dm}{dt}$$

Thus, by Eq. 4/6

$$\boxed{\Sigma\mathbf{F} = m'\Delta\mathbf{v}} \qquad\qquad (4/18)$$

Equation 4/18 establishes the relation between the resultant force on a steady-flow system and the corresponding mass flow rate and vector velocity increment.*

Alternatively, we may note that the time rate of change of linear momentum is the vector difference between the rate at which linear momentum leaves the system and the rate at which linear momentum enters the system. Thus, we may write $\dot{\mathbf{G}} = m'\mathbf{v}_2 - m'\mathbf{v}_1 = m'\Delta\mathbf{v}$, which agrees with the foregoing result.

We can now see one of the powerful applications of our general force-momentum equation that we derived for any mass system. Our system here includes particles that are rigid (the structural container for the mass stream) and particles that are in motion (the

---

*We must be careful not to interpret dm/dt as the time derivative of the mass of the isolated system. That derivative is zero since the system mass is constant for a steady-flow process. To help avoid confusion, the symbol m' rather than dm/dt is used to represent the steady mass flow rate.

flow of mass). By defining the boundary of the system, the mass within which is constant for steady flow conditions, we are able to utilize the generality of Eq. 4/6. However, we must be very careful to account for *all* external forces acting on the system, and they become clear if our free-body diagram is correct.

A similar formulation is obtained for the case of angular momentum in steady-flow systems. The resultant moment of all external forces about some fixed point O on or off the system, Fig. 4/5a, equals the time rate of change of angular momentum of the system about O. This fact was established in Eq. 4/7 which, for the case of steady flow in a single plane, becomes

$$\Sigma M_O = m'(v_2 d_2 - v_1 d_1) \tag{4/19}$$

When the velocities of the incoming and outgoing flows are not in the same plane, the equation may be written in vector form as

$$\boxed{\Sigma \mathbf{M}_O = m'(\mathbf{d}_2 \times \mathbf{v}_2 - \mathbf{d}_1 \times \mathbf{v}_1)} \qquad \textbf{(4/19}\boldsymbol{a}\textbf{)}$$

where $\mathbf{d}_1$ and $\mathbf{d}_2$ are the position vectors to the centers of A_1 and A_2 from the fixed reference O. In both relations, the mass center G may be used alternatively as a moment center by virtue of Eq. 4/9.

Equations 4/18 and 4/19 are very simple relations which find important use in describing relatively complex fluid actions. It should be noted that these equations relate *external* forces to the resultant changes in momentum and are independent of the flow path and momentum changes *internal* to the system.

The foregoing analysis may also be applied to systems moving with constant velocity by noting that the basic relations $\Sigma \mathbf{F} = \dot{\mathbf{G}}$ and $\Sigma \mathbf{M}_O = \dot{\mathbf{H}}_O$ or $\Sigma \mathbf{M}_G = \dot{\mathbf{H}}_G$ apply to systems moving with constant velocity as discussed in Arts. 3/12 and 4/4. The only restriction is that the mass within the system remain constant with respect to time.

Three examples of the analysis of steady mass flow are given in the following sample problems that illustrate the application of the principles embodied in Eqs. 4/18 and 4/19.

Sample Problem 4/5

The smooth vane shown diverts the open stream of fluid of cross-sectional area A, mass density ρ, and velocity v. (a) Determine the force components R and F required to hold the vane in a fixed position. (b) Find the forces when the vane is given a constant velocity u less than v and in the direction of v.

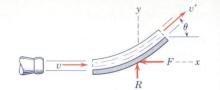

Fixed vane

Solution. **Part (a).** The free-body diagram of the vane together with the fluid portion undergoing the momentum change is shown. The momentum equation may be applied to the isolated system for the change in motion in both the x- and y-directions. With the vane stationary, the magnitude of the exit velocity v' equals that of the entering velocity v with fluid friction neglected. The changes in the velocity components are then

① $$\Delta v_x = v' \cos \theta - v = -v(1 - \cos \theta)$$

and

$$\Delta v_y = v' \sin \theta - 0 = v \sin \theta$$

The mass rate of flow is $m' = \rho A v$, and substitution into Eq. 4/18 gives

$$[\Sigma F_x = m' \Delta v_x] \qquad -F = \rho A v[-v(1 - \cos \theta)]$$

$$F = \rho A v^2 (1 - \cos \theta) \qquad Ans.$$

$$[\Sigma F_y = m' \Delta v_y] \qquad R = \rho A v[v \sin \theta]$$

$$R = \rho A v^2 \sin \theta \qquad Ans.$$

Moving vane

Part (b). In the case of the moving vane, the final velocity v' of the fluid upon exit is the vector sum of the velocity u of the vane plus the velocity of the fluid relative to the vane $v - u$. This combination is shown in the velocity diagram to the right of the figure for the exit conditions. The x-component of v' is the sum of the components of its two parts, so $v'_x = (v - u) \cos \theta + u$. The change in x-velocity of the stream is

$$\Delta v_x = (v - u) \cos \theta + (u - v) = -(v - u)(1 - \cos \theta)$$

The y-component of v' is $(v - u) \sin \theta$, so that the change in the y-velocity of the stream is $\Delta v_y = (v - u) \sin \theta$.

The mass rate of flow m' is the mass undergoing momentum change per unit of time. This rate is the mass flowing over the vane per unit time and *not* the rate of issuance from the nozzle. Thus,

$$m' = \rho A (v - u)$$

The impulse-momentum principle of Eq. 4/18 applied in the positive coordinate direction gives

② $$[\Sigma F_x = m' \Delta v_x] \qquad -F = \rho A(v - u)[-(v - u)(1 - \cos \theta)]$$

$$F = \rho A(v - u)^2(1 - \cos \theta) \qquad Ans.$$

$$[\Sigma F_y = m' \Delta v_y] \qquad R = \rho A(v - u)^2 \sin \theta \qquad Ans.$$

① Be careful with algebraic signs when using Eq. 4/18. The change in v_x is the final value minus the initial value measured in the positive x-direction. Also we must be careful to write $-F$ for ΣF_x.

② Observe that for given values of u and v, the angle for maximum force F is $\theta = 180°$.

Sample Problem 4/6

For the moving vane of Sample Problem 4/5, determine the optimum speed u of the vane for the generation of maximum power by the action of the fluid on the vane.

Solution. The force R shown with the figure for Sample Problem 4/5 is normal to the velocity of the vane so does no work. The work done by the force F shown is negative, but the power developed by the force (reaction to F) exerted by the fluid on the moving vane is

$$[P = Fu] \qquad P = \rho A(v - u)^2 u(1 - \cos \theta)$$

The velocity of the vane for maximum power for the one blade in the stream is specified by

$$\left[\frac{dP}{du} = 0\right] \qquad \rho A(1 - \cos \theta)(v^2 - 4uv + 3u^2) = 0$$

$$(v - 3u)(v - u) = 0 \qquad u = \frac{v}{3} \qquad \textit{Ans.}$$

① The second solution $u = v$ gives a minimum condition of zero power. An angle $\theta = 180°$ completely reverses the flow and clearly produces both maximum force and maximum power for any value of u.

① The result here applies to a single vane only. In the case of multiple vanes, such as the blades on a turbine disk, the rate at which fluid issues from the nozzles is the same rate at which fluid is undergoing momentum change. Thus, $m' = \rho Av$ rather than $\rho A(v - u)$. With this change, the optimum value of u turns out be $u = v/2$.

Sample Problem 4/7

The offset nozzle has a discharge area A at B and an inlet area A_0 at C. A liquid enters the nozzle at a static gage pressure p through the fixed pipe and issues from the nozzle with a velocity v in the direction shown. If the constant density of the liquid is ρ, write expressions for the tension T, shear Q, and bending moment M in the pipe at C.

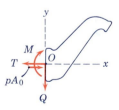

Solution. The free-body diagram of the nozzle and the fluid within it shows the tension T, shear Q, and bending moment M acting on the flange of the nozzle where it attaches to the fixed pipe. The force pA_0 on the fluid within the nozzle due to the static pressure is an additional external force.

Continuity of flow with constant density requires that

$$Av = A_0 v_0$$

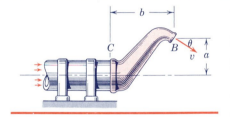

where v_0 is the velocity of the fluid at the entrance to the nozzle. The momentum principle of Eq. 4/18 applied to the system in the two coordinate directions gives

$$[\Sigma F_x = m' \Delta v_x] \qquad pA_0 - T = \rho Av(v \cos \theta - v_0)$$

$$T = pA_0 + \rho Av^2 \left(\frac{A}{A_0} - \cos \theta\right) \qquad \textit{Ans.}$$

$$[\Sigma F_y = m' \Delta v_y] \qquad -Q = \rho Av(-v \sin \theta - 0)$$

$$Q = \rho Av^2 \sin \theta \qquad \textit{Ans.}$$

The moment principle of Eq. 4/19 applied in the clockwise sense gives

$$[\Sigma M_O = m'(v_2 d_2 - v_1 d_1)] \qquad M = \rho Av(va \cos \theta + vb \sin \theta - 0)$$

$$M = \rho Av^2(a \cos \theta + b \sin \theta) \qquad \textit{Ans.}$$

① Again, be careful to observe the correct algebraic signs of the terms on both sides of Eqs. 4/18 and 4/19.

② The forces and moment acting on the pipe are equal and opposite to those shown acting on the nozzle.

Sample Problem 4/8

An air-breathing jet aircraft of total mass m flying with a constant speed v consumes air at the mass rate m'_a and exhausts burned gas at the mass rate m'_g with a velocity u relative to the aircraft. Fuel is consumed at the constant rate m'_f. The total aerodynamic forces acting on the aircraft are the lift L, normal to the direction of flight, and the drag D, opposite to the direction of flight. Any force due to the static pressure across the inlet and exhaust surfaces is assumed to be included in D. Write the equation for the motion of the aircraft and identify the thrust T.

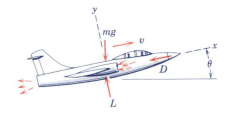

Solution. The free-body diagram of the aircraft together with the air, fuel, and exhaust gas within it is given and shows only the weight, lift, and drag forces as defined. We attach axes x-y to the aircraft and apply our momentum equation relative to the moving system.

The fuel will be treated as a steady stream entering the aircraft with no velocity relative to the system and leaving with a relative velocity u in the exhaust stream. We now apply Eq. 4/18 relative to the reference axes and treat the air and fuel flows separately. For the air flow, the change in velocity in the x-direction relative to the moving system is

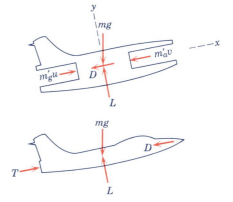

$$\Delta v_a = -u - (-v) = -(u - v)$$

and for the fuel flow the x-change in velocity relative to x-y is

$$\Delta v_f = -u - (0) = -u$$

Thus, we have

$$[\Sigma F_x = m' \Delta v_x] \qquad -mg \sin \theta - D = -m'_a(u - v) - m'_f u$$
$$= -m'_g u + m'_a v$$

where the substitution $m'_g = m'_a + m'_f$ has been made. Changing signs gives

$$m'_g u - m'_a v = mg \sin \theta + D$$

which is simply an equation of equilibrium of the aircraft.

If we modify the boundaries of our system to expose the interior surfaces on which the air and gas act, we will have the simulated model shown, where the air exerts a force $m'_a v$ on the interior of the turbine and the exhaust gas reacts against the interior surfaces with the force $m'_g u$.

The commonly used model is shown in the final diagram, where the net effect of air and exhaust momentum changes is replaced by a simulated thrust

$$T = m'_g u - m'_a v \qquad \qquad Ans.$$

applied to the aircraft from a presumed external source.

Inasmuch as m'_f is generally only 2 percent or less of m'_a, we can use the approximation $m'_g \cong m'_a$ and express the thrust as

$$T \cong m'_g(u - v) \qquad \qquad Ans.$$

We have analyzed the case of constant velocity. Although our Newtonian principles do not generally hold relative to accelerating axes, it can be shown that we may use the $F = ma$ equation for the simulated model and write $T - mg \sin \theta - D = m\dot{v}$ with virtually no error.

① Note that the boundary of the system cuts across the air stream at the entrance to the air scoop and across the exhaust stream at the nozzle.

② We are permitted to use moving axes that translate with constant velocity. See Arts. 3/14 and 4/2.

③ Riding with the aircraft, we observe the air entering our system with a velocity $-v$ measured in the plus x-direction and leaving the system with an x-velocity of $-u$. The final minus the initial values give the expression cited, namely, $-u - (-v) = -(u - v)$.

④ We now see that the "thrust" is, in reality, not a force external to the entire airplane shown in the first figure but can be modeled as an external force.

PROBLEMS

Introductory problems

4/31 A jet of air issues from the nozzle with a velocity of 300 ft/sec at the rate of 6.50 ft^3/sec and is deflected by the right-angled vane. Calculate the force F required to hold the vane in a fixed position. The specific weight of the air is 0.0753 lb/ft^3.

Ans. $F = 4.56$ lb

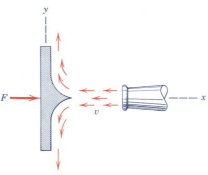

Problem 4/31

4/32 Fresh water issues from the nozzle with a velocity of 30 m/s at the rate of 0.05 m^3/s and is split into two equal streams by the fixed vane and deflected through 60° as shown. Calculate the force F required to hold the vane in place. The density of water is 1000 kg/m^3.

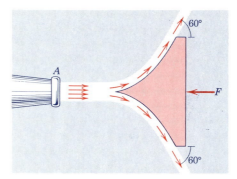

Problem 4/32

4/33 A jet-engine noise suppressor consists of a movable duct that is secured directly behind the jet exhaust by cable A and deflects the blast directly upward. During a ground test, the engine sucks in air at the rate of 43 kg/s and burns fuel at the rate of 0.8 kg/s. The exhaust velocity is 720 m/s. Determine the tension T in the cable.

Ans. $T = 32.6$ kN

Problem 4/33

4/34 The fire tug discharges a stream of salt water (specific weight 64.0 lb/ft^3) with a nozzle velocity of 120 ft/sec at the rate of 1200 gal/min. Calculate the propeller thrust T that must be developed by the tug to maintain a fixed position while pumping. (There are 231 in.^{3} in 1 gal.)

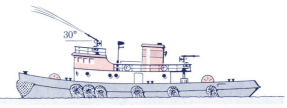

Problem 4/34

4/35 The jet aircraft has a mass of 4.6 Mg and a drag (air resistance) of 32 kN at a speed of 1000 km/h at a particular altitude. The aircraft consumes air at the rate of 106 kg/s through its intake scoop and uses fuel at the rate of 4 kg/s. If the exhaust has a rearward velocity of 680 m/s relative to the exhaust nozzle, determine the maximum angle of elevation α at which the jet can fly with a constant speed of 1000 km/h at the particular altitude in question.

Ans. $\alpha = 17.22°$

Problem 4/35

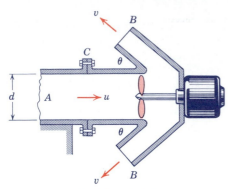

Problem 4/36

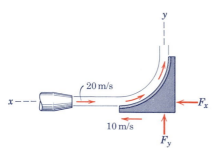

Problem 4/37

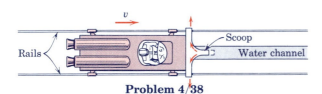

Problem 4/38

Problem 4/39

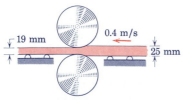

Problem 4/40

4/36 The pump shown draws air with a density ρ through the fixed duct A of diameter d with a velocity u and discharges it at high velocity v through the two outlets B. The pressure in the airstreams at A and B is atmospheric. Determine the expression for the tension T exerted on the pump unit through the flange at C.

4/37 The jet water ski has reached its maximum velocity of 70 km/h when operating in salt water. The water intake is in the horizontal tunnel in the bottom of the hull, so the water enters the intake at the velocity of 70 km/h relative to the ski. The motorized pump discharges water from the horizontal exhaust nozzle of 50-mm diameter at the rate of 0.082 m³/s. Calculate the resistance R of the water to the hull at the operating speed.   *Ans. R* = 1885 N

4/38 The figure shows the top view of an experimental rocket sled that is traveling at a speed of 1000 ft/sec when its forward scoop enters a water channel to act as a brake. The water is diverted at right angles relative to the motion of the sled. If the frontal flow area of the scoop is 15 in.², calculate the initial braking force. The specific weight of water is 62.4 lb/ft³.

Representative problems

4/39 The 90° vane moves to the left with a constant velocity of 10 m/s against a stream of fresh water issuing with a velocity of 20 m/s from the 25-mm-diameter nozzle. Calculate the forces F_x and F_y on the vane required to support the motion.
Ans. F_x = 442 N, F_y = 442 N

4/40 The 25-mm steel slab 1.2 m wide enters the rolls at the speed of 0.4 m/s and is reduced in thickness to 19 mm. Calculate the small horizontal thrust T on the bearings of each of the two rolls.

4/41 The axial-flow fan C pumps air through the duct of circular cross section and exhausts it with a velocity v at B. The air densities at A and B are ρ_A and ρ_B, respectively, and the corresponding pressures are p_A and p_B. The fixed deflecting blades at D restore axial flow to the air after it passes through the propeller blades C. Write an expression for the resultant horizontal force R exerted on the fan unit by the flange and bolts at A.

$$\text{Ans. } R = \frac{\pi d^2}{4}\left[\rho_B\left(1 - \frac{\rho_B}{\rho_A}\right)v^2 + (p_B - p_A)\right]$$

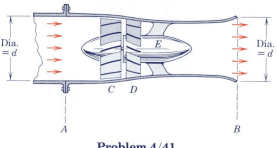

Problem 4/41

4/42 One of the most advanced methods for cutting metal plates uses a high-velocity water jet that carries an abrasive garnet powder. The jet issues from the 0.01-in.-diameter nozzle at A and follows the path shown through the thickness t of the plate. As the plate is slowly moved to the right, the jet makes a narrow precision slot in the plate. The water-abrasive mixture is used at the low rate of 1/2 gal/min and has a specific weight of 68 lb/ft^3. Water issues from the bottom of the plate with a velocity that is 60 percent of the impinging nozzle velocity. Calculate the horizontal force F required to hold the plate against the jet. (There are 231 in.^{3} in 1 gal.)

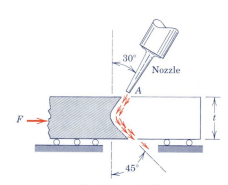

Problem 4/42

4/43 In the static test of a jet engine and exhaust nozzle assembly, air is sucked into the engine at the rate of 30 kg/s and fuel is burned at the rate of 1.6 kg/s. The flow area, static pressure, and axial-flow velocity for the three sections shown are as follows:

	Sec. A	Sec. B	Sec. C
Flow area, m^2	0.15	0.16	0.06
Static pressure, kPa	-14	140	14
Axial-flow velocity, m/s	120	315	600

Determine the tension T in the diagonal member of the supporting test stand and calculate the force F exerted on the nozzle flange at B by the bolts and gasket to hold the nozzle to the engine housing.

$$\text{Ans. } T = 21.1 \text{ kN}, F = 12.55 \text{ kN}$$

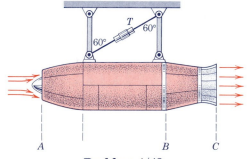

Problem 4/43

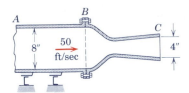

Problem 4/44

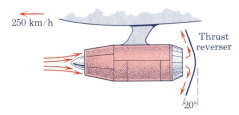

Problem 4/45

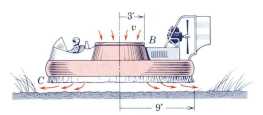

Problem 4/46

4/44 Air is pumped through the stationary duct A with a velocity of 50 ft/sec and exhausted through an experimental nozzle section BC. The average static pressure across section B is 150 lb/in.^{2} gage, and the specific weight of air at this pressure and at the temperature prevailing is 0.840 lb/ft^3. The average static pressure across the exit section C is measured to be 2 lb/in.^{2} gage, and the corresponding specific weight of air is 0.0760 lb/ft^3. Calculate the force T exerted on the nozzle flange at B by the bolts and the gasket to hold the nozzle in place.

4/45 Reverse thrust to decelerate a jet aircraft during landing is obtained by swinging the exhaust deflectors into place as shown, thus causing a partial reversal of the exhaust stream. For a jet engine that consumes 40 kg of air per second at a ground speed of 250 km/h and that uses 2 kg of fuel per second, determine the reverse thrust as a fraction n of the forward thrust with deflectors removed. The exhaust velocity relative to the nozzle is 600 m/s. Assume air enters the engine with a relative velocity equal to the ground speed of the aircraft.
Ans. $n = 0.508$

4/46 The experimental ground-effect machine has a total weight of 4200 lb. It hovers 1 or 2 ft off the ground by pumping air at atmospheric pressure through the circular intake duct at B and discharging it horizontally under the periphery of the skirt C. For an intake velocity v of 150 ft/sec, calculate the average air pressure p under the 18-ft-diameter machine at ground level. The specific weight of the air is 0.076 lb/ft^3.

4/47 For a jet acting on multiple vanes where each vane that enters the jet is followed immediately by another, as in a turbine or water wheel (see figures for Probs. 4/65 and 4/66), determine the maximum power P that can be developed for a given blade angle and the corresponding optimum peripheral speed u of the vanes in terms of the jet velocity v for maximum power. Modify Sample Problem 4/6 by assuming an infinite number of vanes so that the rate at which fluid leaves the nozzle equals the rate at which fluid passes over the vanes.
Ans. $P = \frac{1}{4}\rho Av^3(1 - \cos\theta)$, $u = v/2$

4/48 The pipe bend shown has a cross-sectional area A and is supported in its plane by the tension T applied to its flanges by the adjacent connecting pipes (not shown). If the velocity of the liquid is v, its density ρ, and its static pressure p, determine T and show that it is independent of the angle θ.

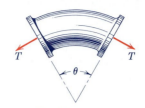

Problem 4/48

4/49 A commercial aircraft flying horizontally at 500 mi/hr encounters a heavy downpour of rain falling vertically at the rate of 20 ft/sec with an intensity equivalent to an accumulation of 1 in./hr on the ground. The upper surface area of the aircraft projected onto the horizontal plane is 2960 ft². Calculate the negligible downward force F of the rain on the aircraft. *Ans.* $F = 2.66$ lb

Problem 4/49

4/50 In a test of the operation of a "cherry picker" fire truck, the equipment is free to roll with its brakes released. For the position shown, the truck is observed to deflect the spring of stiffness $k = 15$ kN/m a distance of 150 mm because of the action of the horizontal stream of water issuing from the nozzle when the pump is activated. If the exit diameter of the nozzle is 30 mm, calculate the velocity v of the stream as it leaves the nozzle. Also determine the added moment M that the joint at A must resist when the pump is in operation with the nozzle in the position shown.

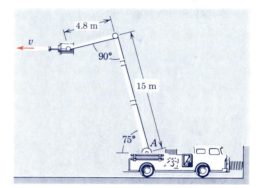

Problem 4/50

4/51 The sump pump has a net mass of 310 kg and pumps fresh water against a 6-m head at the rate of 0.125 m³/s. Determine the vertical force R between the supporting base and the pump flange at A during operation. The mass of water in the pump may be taken as the equivalent of a 200-mm-diameter column 6 m in height. *Ans.* $R = 5980$ N

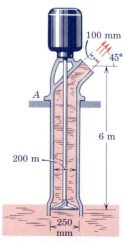

Problem 4/51

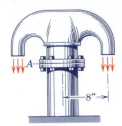

Problem 4/52

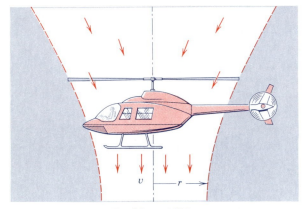

Problem 4/53

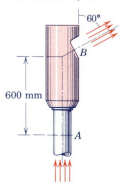

Problem 4/54

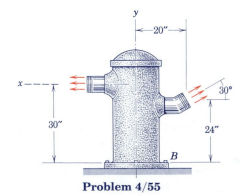

Problem 4/55

4/52 The 180° return pipe discharges salt water (specific weight 64.4 lb/ft³) into the atmosphere at a constant rate of 1.6 ft³/sec. The static pressure in the water at section A is 10 lb/in.² above atmospheric pressure. The flow area of the pipe at A is 20 in.² and that at each of the two outlets is 3.2 in.² If each of the six flange bolts is tightened with a torque wrench so that it is under a tension of 150 lb, determine the average pressure p on the gasket between the two flanges. The flange area in contact with the gasket in 16 in.² Also determine the bending moment M in the pipe at section A if the left-hand discharge is blocked off and the flow rate is cut in half. Neglect the weight of the pipe and the water within it.

4/53 The helicopter shown has a mass m and hovers in position by imparting downward momentum to a column of air defined by the slip-stream boundary shown. Find the downward velocity v given to the air by the rotor at a section in the stream below the rotor, where the pressure is atmospheric and the stream radius is r. Also find the power P required of the engine. Neglect the rotational energy of the air, any temperature rise due to air friction, and any change in air density ρ.

$$Ans. \quad v = \frac{1}{r} \sqrt{\frac{mg}{\pi\rho}}, \; P = \frac{mg}{2r} \sqrt{\frac{mg}{\pi\rho}}$$

4/54 Air enters the pipe at A at the rate of 6 kg/s under a pressure of 1400 kPa gage and leaves the whistle at atmospheric pressure through the opening at B. The entering velocity of the air at A is 45 m/s, and the exhaust velocity at B is 360 m/s. Calculate the tension T, shear V, and bending moment M in the pipe at A. The net flow area at A is 7500 mm².

4/55 The fire hydrant is tested under a high standpipe pressure of 120 lb/in.² The total flow of 10 ft³/sec is divided equally between the two outlets, each of which has a cross-sectional area of 0.040 ft². The inlet cross-sectional area at the base is 0.75 ft². Neglect the weight of the hydrant and water within it and compute the tension T, the shear V, and the bending moment M in the base of the standpipe at B. The specific weight of water is 62.4 lb/ft³. The static pressure of the water as it enters the base at B is 120 lb/in.²
Ans. $T = 12,610$ lb, $V = 162$ lb, $M = 1939$ lb-ft

4/56 A marine terminal for unloading bulk wheat from a ship is equipped with a vertical pipe with a nozzle at A that sucks wheat up the pipe and transfers it to the storage building. Calculate the x- and y-components of the force $\mathbf{R}$ required to change the momentum of the flowing mass in rounding the bend. Identify all forces applied externally to the bend and mass within it. Air flows through the 14-in.-diameter pipe at the rate of 18 tons per hour under a vacuum of 9 in. of mercury ($p = -4.42.$ lb/in.^{2}) and carries with it 150 tons of wheat per hour at a speed of 124 ft/sec.

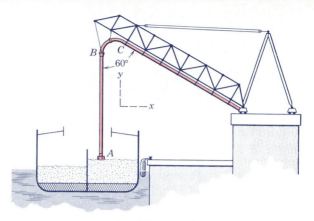

Problem 4/56

4/57 The industrial blower sucks in air through the axial opening A with a velocity v_1 and discharges it at atmospheric pressure and temperature through the 150-mm-diameter duct B with a velocity v_2. The blower handles 16 m^3 of air per minute with the motor and fan running at 3450 rev/min. If the motor requires 0.32 kW of power under no load (both ducts closed), calculate the power P consumed while air is being pumped.   *Ans. $P = 0.671$ kW*

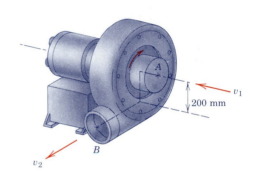

Problem 4/57

4/58 The feasibility of a one-passenger VTOL (vertical takeoff and landing) craft is under review. The preliminary design calls for a small engine with a high power-to-weight ratio driving an air pump that draws in air through the 70° ducts with an inlet velocity $v = 40$ m/s at a static gage pressure of -1.8 kPa across the inlet areas totaling 0.1320 m^2. The air is exhausted vertically down with a velocity $u = 420$ m/s. For a 90-kg passenger, calculate the maximum net mass m of the machine for which it can take off and hover. (See Table D/1 for air density.)

Problem 4/58

4/59 The military jet aircraft has a gross weight of 24,000 lb and is poised for takeoff with brakes set while the engine is revved up to maximum power. At this condition, air with a specific weight of 0.0753 lb/ft^3 is sucked into the intake ducts at the rate of 106 lb/sec with a static pressure of -0.30 lb/in.^{2} (gage) across the duct entrance. The total cross-sectional area of both intake ducts (one on each side) is 1800 in.^{2} The air–fuel ratio is 18, and the exhaust velocity u is 3100 ft/sec with zero back pressure (gage) across the exhaust nozzle. Compute the initial acceleration a of the aircraft upon release of the brakes.   *Ans. $a = 14.68$ ft/sec^2*

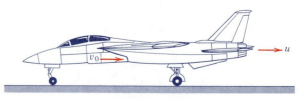

Problem 4/59

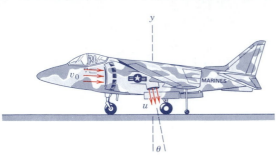

Problem 4/60

4/60 The VTOL (vertical takeoff and landing) military aircraft is capable of rising vertically under the action of its jet exhaust, which can be "vectored" from $\theta \cong 0$ for takeoff and hovering to $\theta = 90°$ for forward flight. The loaded aircraft has a mass of 8600 kg. At full takeoff power, its turbo-fan engine consumes air at the rate of 90 kg/s and has an air–fuel ratio of 18. Exhaust gas velocity is 1020 m/s with essentially atmospheric pressure across the exhaust nozzles. Air with a density of 1.206 kg/m³ is sucked into the intake scoops at a pressure of −2 kPa (gage) over the total inlet area of 1.10 m². Determine the angle θ for vertical takeoff and the corresponding vertical acceleration a_y of the aircraft.

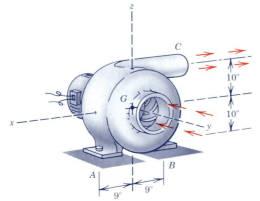

Problem 4/61

4/61 The blower is run by an attached electric motor that turns the impeller at 3450 rev/min and blows air from the 6-in.-diameter outlet at C at the rate of 800 ft³/min. Air enters the unit in the negative y-direction with a density of 0.0753 lbm/ft³ (pounds mass). The entire unit weighs 60 lb, with center of gravity at G directly above the midpoint between the hold-down bolts A and B. Calculate the net vertical force exerted on the blower supports at A and B.

Ans. $A = 32.4$ lb, $B = 27.6$ lb

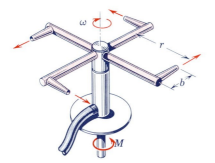

Problem 4/62

▶**4/62** The sprinkler is made to rotate at the constant angular velocity ω and distributes water at the volume rate Q. Each of the four nozzles has an exit area A. Write an expression for the torque M on the shaft of the sprinkler necessary to maintain the given motion. For a given pressure and, hence, flow rate Q, at what speed ω_0 will the sprinkler operate with no applied torque? Let ρ be the density of the water.

$$Ans. \ \ M = \rho Q \left[\frac{Qr}{4A} - (r^2 + b^2)\, \omega \right]$$

$$\omega_0 = \frac{Qr}{4A(r^2 + b^2)}$$

▶**4/63** An axial section of the suction nozzle *A* for the bulk wheat unloader described in Prob. 4/56 is shown here. The outer pipe is secured to the inner pipe by several longitudinal webs that do not restrict the flow of air. A vacuum of 9 in. of mercury ($p = -4.42$ lb/in.^{2} gage) is maintained in the inner pipe, and the pressure across the bottom of the outer pipe is atmospheric ($p = 0$). Air at 0.075 lb/ft^3 is drawn in through the space between the pipes at a rate of 18 tons/hr at atmospheric pressure and draws with it 150 tons of wheat per hour up the pipe at a velocity of 124 ft/sec. If the nozzle unit below section *A-A* weighs 60 lb, calculate the compression *C* in the connection at *A-A*.   *Ans.* $C = 100.3$ lb

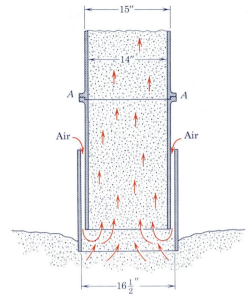

Problem 4/63

▶**4/64** The centrifugal pump handles 20 m^3 of fresh water per minute with inlet and outlet velocities of 18 m/s. The impeller is turned clockwise through the shaft at *O* by a motor that delivers 40 kw at a pump speed of 900 rev/min. With the pump filled but not turning, the vertical reactions at *C* and *D* are each 250 N. Calculate the forces exerted by the foundation on the pump at *C* and *D* while the pump is running. The tensions in the connecting pipes at *A* and *B* are exactly balanced by the respective forces due to the static pressure in the water. (*Suggestion:* Isolate the entire pump and water within it between sections *A* and *B* and apply the momentum principle to the entire system.)

          Ans. $C = 4340$ N up, $D = 3840$ N down

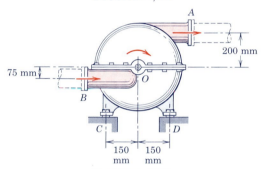

Problem 4/64

4/65 In the figure is shown a detail of the stationary nozzle diaphragm *A* and the rotating blades *B* of a gas turbine. The products of combustion pass through the fixed diaphragm blades at the 27° angle and impinge on the moving rotor blades. The angles shown are selected so that the velocity of the gas relative to the moving blade at entrance is at the 20° angle for minimum turbulence, corresponding to a mean blade velocity of 315 m/s at a radius of 375 mm. If gas flows past the blades at the rate of 15 kg/s, determine the theoretical power output *P* of the turbine. Neglect fluid and mechanical friction with the resulting heat-energy loss and assume that all the gases are deflected along the surfaces of the blades with a velocity of constant magnitude relative to the blade.   *Ans.* $P = 1.197$ MW

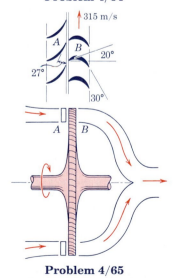

Problem 4/65

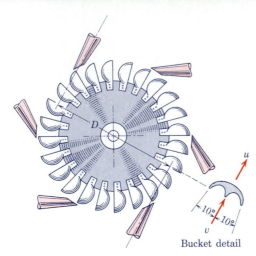

Problem 4/66

▶ **4/66** In the figure is shown an impulse-turbine wheel for a hydroelectric power plant that is to operate with a static head of water of 300 m at each of its six nozzles and is to rotate at the speed of 270 rev/min. Each wheel and generator unit is to develop an output power of 22 000 kW. The efficiency of the generator may be taken to be 0.90, and an efficiency of 0.85 for the conversion of the kinetic energy of the water jets to energy delivered by the turbine may be expected. The mean peripheral speed of such a wheel for greatest efficiency will be about 0.47 times the jet velocity. If each of the buckets is to have the shape shown, determine the necessary jet diameter d and wheel diameter D. Assume that the water acts on the bucket which is at the tangent point of each jet stream.   *Ans.* $d = 165.3$ mm, $D = 2.55$ m

4/7 VARIABLE MASS

In Art. 4/4 we extended the equations for the motion of a particle to include a system of particles. This extension led to the very general expressions $\Sigma \mathbf{F} = \dot{\mathbf{G}}$, $\Sigma \mathbf{M}_O = \dot{\mathbf{H}}_O$, and $\Sigma \mathbf{M}_G = \dot{\mathbf{H}}_G$, which are Eqs. 4/6, 4/7, and 4/9, respectively. In the derivation of these equations, the summations were taken over a fixed collection of particles, so that the mass of the system to be analyzed was constant. In Art. 4/6 these momentum principles were extended in Eqs. 4/18 and 4/19 to describe the action of forces on a system defined by a geometric volume through which passes a steady flow of mass. The amount of mass within the control volume was, therefore, constant with respect to time and we were, thus, permitted to use Eqs. 4/6, 4/7, and 4/9. When the mass within the boundary of a system under consideration is not constant, the foregoing relationships are no longer valid.*

We will now develop the equation for the linear motion of a system whose mass varies with time. Consider first a body that gains mass by overtaking and swallowing a stream of matter, Fig. 4/6a. The mass of the body and its velocity at any instant are m and v, respectively. The stream of matter is assumed to be moving in the same direction as m with a constant velocity v_0 less than v. By virtue of Eq. 4/18, the force exerted by m on the particles of the stream to accelerate them from a velocity v_0 to a greater velocity v is $R = m'(v - v_0) = \dot{m}u$, where the time rate of increase of m is $m' = \dot{m}$ and where u is the magnitude of the relative velocity with which the particles approach m. In addition to R, all other forces acting on m in the direction of its motion are denoted by ΣF. The equation

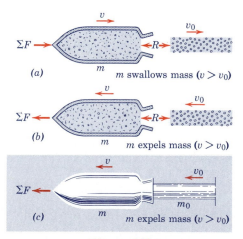

(a) m swallows mass ($v > v_0$)

(b) m expels mass ($v > v_0$)

(c) m expels mass ($v > v_0$)

Figure 4/6

---

*In relativistic mechanics the mass is found to be a function of velocity, and its time derivative has a meaning different from that in Newtonian mechanics.

of motion of m from Newton's second law is, therefore, $\Sigma F - R = m\dot{v}$ or

$$\boxed{\Sigma F = m\dot{v} + \dot{m}u} \qquad (4/20)$$

Similarly, if the body loses mass by expelling it rearward so that its velocity v_0 is less than v, Fig. 4/6b, the force R required to decelerate the particles from a velocity v to a lesser velocity v_0 is $R = m'(-v_0 - [-v]) = m'(v - v_0)$. But $m' = -\dot{m}$ since m is decreasing. Also the relative velocity with which the particles leave m is $u = v - v_0$. Thus, the force R becomes $R = -\dot{m}u$. If ΣF denotes the resultant of all other forces acting on m in the direction of its motion, Newton's second law requires $\Sigma F + R = m\dot{v}$ or

$$\Sigma F = m\dot{v} + \dot{m}u$$

which is the same relationship as in the case where m is gaining mass. We may use Eq. 4/20, therefore, as the equation of motion of m, whether it is gaining or losing mass.

A frequent error in the use of the force-momentum equation is to express the partial force sum ΣF as

$$\Sigma F = \frac{d}{dt}(mv) = m\dot{v} + \dot{m}v$$

From this expansion we see that the direct differentiation of the linear momentum gives the correct force ΣF *only* when the body picks up mass initially at rest or when it expels mass which is left with zero absolute velocity. In both instances, $v_0 = 0$ and $u = v$.

We may also obtain Eq. 4/20 by a direct differentiation of the momentum from the basic relation $\Sigma F = \dot{G}$, provided a proper system of constant total mass is chosen. To illustrate this approach, we take the case where m is losing mass and use Fig. 4/6c which shows the system of m and an arbitrary portion m_0 of the stream of ejected mass. The mass of this system is $m + m_0$ and is constant. The ejected stream of mass is assumed to move undisturbed once separated from m, and the only force external to the entire system is ΣF which is applied directly to m as before. The reaction $R = -\dot{m}u$ is internal to the system and is not disclosed as an external force on the system. With constant total mass, the momentum principle $\Sigma F = \dot{G}$ is applicable and we have

$$\Sigma F = \frac{d}{dt}(mv + m_0v_0) = m\dot{v} + \dot{m}v + \dot{m}_0v_0 + m_0\dot{v}_0$$

Clearly, $\dot{m}_0 = -\dot{m}$, and the velocity of the ejected mass with respect to m is $u = v - v_0$. Also $\dot{v}_0 = 0$ since m_0 moves undisturbed with no acceleration once free of m. Thus, the relation becomes

$$\Sigma F = m\dot{v} + \dot{m}u$$

which is identical to the result of the previous formulation, Eq. 4/20.

The case of m losing mass is clearly descriptive of rocket propulsion. Figure 4/7a shows a vertically ascending rocket, the system for which is the mass within the volume defined by the exterior surface of the rocket and the exit plane across the nozzle. External to this system, the free-body diagram discloses the instantaneous values of gravitational attraction mg, aerodynamic resistance R, and the force pA due to the average static pressure p across the nozzle exit plane of area A. Also the rate of mass flow is $m' = -\dot{m}$. Thus, we may write the equation of motion of the rocket $\Sigma F = m\dot{v} + \dot{m}u$ as $pA - mg - R = m\dot{v} + \dot{m}u$, or

$$m'u + pA - mg - R = m\dot{v} \qquad (4/21)$$

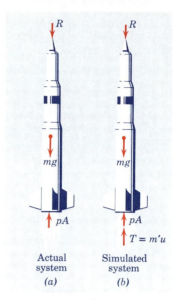

Actual system   Simulated system

(a)   (b)

Figure 4/7

Equation 4/21 is of the form "$\Sigma F = ma$" where the first term in "ΣF" is the thrust $T = m'u$. Thus, the rocket may be simulated as a body to which an external thrust T is applied, Fig. 4/7b, and the problem may then be analyzed like any other $F = ma$ problem, except that m is a function of time.

It may be observed that, during the initial stages of motion when the magnitude of the velocity v of the rocket is less than the relative exhaust velocity u, the absolute velocity v_0 of the exhaust gases will be directed rearward. On the other hand, when the rocket reaches a velocity v whose magnitude is greater than u, the absolute velocity v_0 of the exhaust gases will be directed forward. For a given mass rate of flow, the rocket thrust T depends only on the relative exhaust velocity u and not on the magnitude or on the direction of the absolute velocity v_0 of the exhaust gases.

In the foregoing treatment of bodies whose mass changes with time, we have assumed that all elements of the mass m of the body were moving with the same velocity v at any instant of time and that the particles of mass that were added to or expelled from the body underwent an abrupt transition of velocity upon entering or leaving the body. Thus, this velocity change has been modeled as a mathematical discontinuity. In reality, this change in velocity cannot be discontinuous even though the transition may be rapid. In the case of a rocket, for example, the velocity change occurs continuously in the space between the combustion zone and the exit plane of the exhaust nozzle. A more general analysis* of variable-mass dynamics removes this restriction of discontinuous velocity change and introduces a slight correction to Eq. 4/20.

*For a development of the equations that describe the general motion of a time-dependent system of mass, see Art. 53 of the senior author's *Dynamics, 2nd Edition SI Version*, 1975, John Wiley & Sons.

Sample Problem 4/9

The end of a chain of length L and mass ρ per unit length that is piled on a platform is lifted vertically with a constant velocity v by a variable force P. Find P as a function of the height x of the end above the platform. Also find the energy lost during the lifting of the chain.

Solution I (variable-mass approach). Equation 4/20 will be used and applied to the moving part of the chain of length x that is gaining mass. The force summation ΣF includes all forces acting on the moving part except the force exerted by the particles that are being attached. From the diagram we have

$$\Sigma F_x = P - \rho g x$$

The velocity is constant so that $\dot{v} = 0$. The rate of increase of mass is $\dot{m} = \rho v$, and the relative velocity with which the attaching particles approach the moving part is $u = v - 0 = v$. Thus, Eq. 4/20 becomes

① $[\Sigma F = m\dot{v} + \dot{m}u]$ $P - \rho g x = 0 + \rho v(v)$ $P = \rho(gx + v^2)$ *Ans.*

We now see that the force P consists of the two parts, $\rho g x$, which is the weight of the moving part of the chain, and ρv^2, which is the added force required to change the momentum of the links on the platform from a condition at rest to a velocity v.

Solution II (constant-mass approach). The principle of impulse and momentum for a system of particles expressed by Eq. 4/6 will be applied to the entire chain considered as the system of constant mass. The free-body diagram of the system shows the unknown force P, the total weight of all links $\rho g L$, and the force $\rho g(L - x)$ exerted by the platform on those links that are at rest on it. The momentum of the system at any position is $G_x = \rho x v$ and the momentum equation gives

② $\left[\Sigma F_x = \dfrac{dG_x}{dt}\right]$ $P + \rho g(L - x) - \rho g L = \dfrac{d}{dt}(\rho x v)$ $P = \rho(gx + v^2)$
Ans.

Again the force P is seen to be equal to the weight of the portion of the chain that is off the platform plus the added term that accounts for the time rate of increase of momentum of the chain.

Energy loss. Each link on the platform acquires its velocity abruptly through an impact with the link above it that lifts it off the platform. The succession of impacts gives rise to an energy loss ΔE. The work-energy equation may be written $U'_{1\text{-}2} = \Delta T + \Delta V_g + \Delta E$.

$$U'_{1\text{-}2} = \int P \, dx = \int_0^L (\rho g x + \rho v^2) \, dx = \tfrac{1}{2}\rho g L^2 + \rho v^2 L$$

$$\Delta T = \tfrac{1}{2}\rho L v^2 \qquad \Delta V_g = \rho g L \frac{L}{2} = \tfrac{1}{2}\rho g L^2$$

Substituting into the work-energy equation gives

$$\tfrac{1}{2}\rho g L^2 + \rho v^2 L = \tfrac{1}{2}\rho L v^2 + \tfrac{1}{2}\rho g L^2 + \Delta E \qquad \Delta E = \tfrac{1}{2}\rho L v^2 \quad \textit{Ans.}$$

Helpful Hints

① The model of Fig. 4/6a shows the mass being added to the leading end of the moving part. With the chain the mass is added to the trailing end, but the effect is the same.

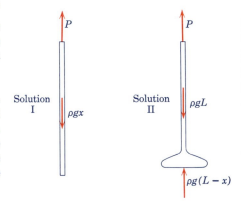

② We must be very careful not to use $\Sigma F = \dot{G}$ for a system whose mass is changing. Thus, we have taken the total chain as the system since its mass is constant.

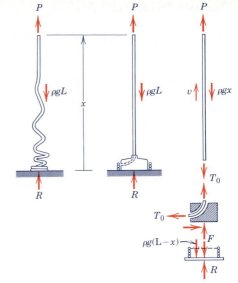

Sample Problem 4/10

Replace the open-link chain of Sample Problem 4/9 by a flexible but inextensible rope or bicycle-type chain of length L and mass ρ per unit length. Determine the force P required to elevate the end of the rope with a constant velocity v and determine the corresponding reaction R between the coil and the platform.

___

Solution. The free-body diagram of the coil and moving portion of the rope is shown in the left-hand figure. Because of some resistance

① to bending and some lateral motion, the transition from rest to vertical velocity v will occur over an appreciable segment of the rope. Nevertheless, assume first that all moving elements have the same velocity so that Eq. 4/6 for the system gives

② $$\left[\Sigma F_x = \frac{dG_x}{dt}\right] \quad P + R - \rho g L = \frac{d}{dt}(\rho x v) \quad P + R = \rho v^2 + \rho g L$$

We assume further that all elements of the coil of rope are at rest on the platform and transmit no force to the platform other than their weight, so that $R = \rho g(L - x)$. Substitution into the foregoing relation gives

$$P + \rho g(L - x) = \rho v^2 + \rho g L \quad \text{or} \quad P = \rho v^2 + \rho g x$$

which is the same result as that for the chain in Sample Problem 4/9.

The total work done on the rope by P becomes

$$U'_{1\text{-}2} = \int P \, dx = \int_0^x (\rho v^2 + \rho g x) \, dx = \rho v^2 x + \tfrac{1}{2}\rho g x^2$$

Substitution into the work-energy equation gives

$$[U'_{1\text{-}2} = \Delta T + \Delta V_g] \quad \rho v^2 x + \tfrac{1}{2}\rho g x^2 = \Delta T + \rho g x \frac{x}{2} \quad \Delta T = \rho x v^2$$

③ which is twice the kinetic energy $\tfrac{1}{2}\rho x v^2$ of vertical motion. Thus, an equal amount of kinetic energy is unaccounted for. This conclusion largely negates our assumption of one-dimensional x-motion.

In order to produce a one-dimensional model that retains the inextensible property assigned to the rope, it is necessary to impose a physical constraint at the base to guide the rope into vertical motion and at the

④ same time preserve a smooth transition from rest to upward velocity v without energy loss. Such a guide is included in the free-body diagram of the entire rope in the middle figure and is represented schematically in the middle free-body diagram of the right-hand figure.

For a conservative system, the work-energy equation gives

⑤ $$[dU = dT + dV_g] \quad P \, dx = d(\tfrac{1}{2}\rho x v^2) + d\left(\rho g x \frac{x}{2}\right)$$

$$P = \tfrac{1}{2}\rho v^2 + \rho g x$$

Substitution into the impulse-momentum equation gives

$$\tfrac{1}{2}\rho v^2 + \rho g x + R = \rho v^2 + \rho g L \quad R = \tfrac{1}{2}\rho v^2 + \rho g(L - x)$$

Although this force, which exceeds the weight by $\tfrac{1}{2}\rho v^2$, is unrealistic experimentally, it would be present in the idealized model.

Equilibrium of the vertical section requires

$$T_0 = P - \rho g x = \tfrac{1}{2}\rho v^2 + \rho g x - \rho g x = \tfrac{1}{2}\rho v^2$$

Because it requires a force of ρv^2 to change the momentum of the rope elements, the restraining guide must supply the balance $F = \tfrac{1}{2}\rho v^2$ that, in turn, is transmitted to the platform.

① Perfect flexibility would not permit any resistance to bending.

② Remember that v is constant and equals $\dot{x}$. Also note that this same relation applies to the chain of Sample Problem 4/9.

③ This added term of unaccounted for kinetic energy exactly equals the energy lost by the chain during the impact of its links

④ This restraining guide may be visualized as a canister of negligible mass rotating within the coil with an angular velocity v/r and connected to the platform through its shaft. As it turns, it feeds the rope from a rest position to an upward velocity v, as indicated in the accompanying figure.

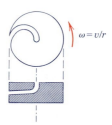

⑤ Note that the mass center of the section of length x is a distance $x/2$ above the base.

Sample Problem   4/11

A rocket of initial total mass m_0 is fired vertically up from the north pole and accelerates until the fuel, which burns at a constant rate, is exhausted. The relative nozzle velocity of the exhaust gas has a constant value u, and the nozzle exhausts at atmospheric pressure throughout the flight. If the residual mass of the rocket structure and machinery is m_b when burnout occurs, determine the expression for the maximum velocity reached by the rocket. Neglect atmospheric resistance and the variation of gravity with altitude.

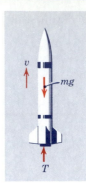

Solution I (F = ma solution).   We adopt the approach illustrated with Fig. 4/7b and treat the thrust as an external force on the rocket. With the neglect of the back pressure p across the nozzle and the atmospheric resistance R, Eq. 4/21 or Newton's second law gives

$$T = mg = m\dot{v}$$

But the thrust is $T = m'u = -\dot{m}u$ so that the equation of motion becomes

$$-\dot{m}u - mg = m\dot{v}$$

Multiplication by dt, division by m, and rearrangement give

$$dv = -u\,\frac{dm}{m} - g\,dt$$

which is now in a form that can be integrated. The velocity v corresponding to the time t is given by the integration

$$\int_0^v dv = -u \int_{m_0}^m \frac{dm}{m} - g \int_0^t dt$$

or

$$v = u \ln \frac{m_0}{m} - gt$$

Since the fuel is burned at the constant rate $m' = -\dot{m}$, the mass at any time t is $m = m_0 + \dot{m}t$. If we let m_b stand for the mass of the rocket when burnout occurs, then the time at burnout becomes $t_b = (m_b - m_0)/\dot{m} = (m_0 - m_b)/(-\dot{m})$. This time gives the condition for maximum velocity, which is

$$v_{\max} = u \ln \frac{m_0}{m_b} + \frac{g}{\dot{m}}(m_0 - m_b) \qquad Ans.$$

The quantity $\dot{m}$ is a negative number since the mass decreases with time.

Solution II (variable-mass solution).   If we use Eq. 4/20, then $\Sigma F = -mg$ and the equation becomes

$$[\Sigma F = m\dot{v} + \dot{m}u] \qquad -mg = m\dot{v} + \dot{m}u$$

But $\dot{m}u = -m'u = -T$ so that the equation of motion becomes

$$T - mg = m\dot{v}$$

which is the same as formulated with *Solution I*.

① The neglect of atmospheric resistance is not a bad assumption for a first approximation inasmuch as the velocity of the ascending rocket is smallest in the dense part of the atmosphere and greatest in the rarefied region. Also for an altitude of 320 km, the acceleration due to gravity is 91 percent of the value at the surface of the earth.

② Vertical launch from the north pole is taken only to eliminate any complication due to the earth's rotation in figuring the absolute trajectory of the rocket.

PROBLEMS

Introductory problems

4/67 The Saturn V rocket has a total launch weight of approximately $6.0(10^6)$ lb. At launch fuel is burned at the rate of $28(10^3)$ lb/sec with an exhaust velocity of 7800 ft/sec. Calculate the initial acceleration of the rocket. Assume atmospheric pressure across the exit plane of the exhaust nozzle.

$$Ans. \ a = 4.20 \ ft/sec^2$$

4/68 When the rocket reaches the position in its trajectory shown, it has a mass of 3 Mg and is beyond the effect of the earth's atmosphere. Gravitational acceleration is $9.60 \ m/s^2$. Fuel is being consumed at the rate of 130 kg/s, and the exhaust velocity relative to the nozzle is 600 m/s. Compute the n- and t-components of acceleration of the rocket.

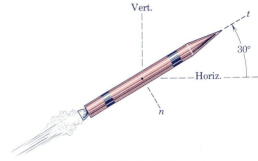

Problem 4/68

4/69 A tank truck for washing down streets has a total weight of 20,000 lb when its tank is full. With the spray turned on, 80 lb of water per second issue from the nozzle with a velocity of 60 ft/sec relative to the truck at the 30° angle shown. If the truck is to accelerate at the rate of 2 ft/sec² when starting on a level road, determine the required tractive force P between the tires and the road when (*a*) the spray is turned on and (*b*) when the spray is turned off.

$$Ans. \ (a) \ P = 1113 \ lb, \ (b) \ P = 1242 \ lb$$

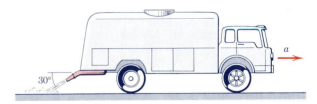

Problem 4/69

4/70 A tank, which has a mass of 50 kg when empty, is propelled to the left by a force P and scoops up fresh water from a stream flowing in the opposite direction with a velocity of 1.5 m/s. The entrance area of the scoop is 2000 mm², and water enters the scoop at a rate equal to the velocity of the scoop relative to the stream. Determine the force P at a certain instant for which 80 kg of water have been ingested and the velocity and acceleration of the tank are 2 m/s and 0.4 m/s², respectively. Neglect the small impact pressure at the scoop necessary to elevate the water in the tank.

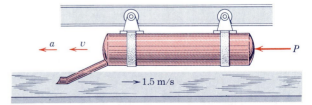

Problem 4/70

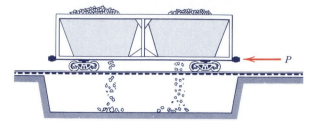

Problem 4/71

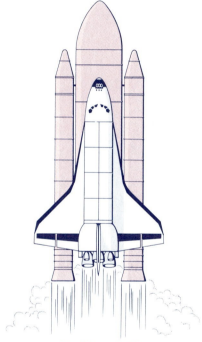

Problem 4/73

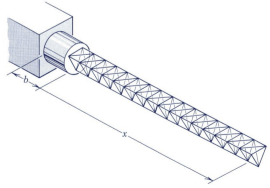

Problem 4/74

4/71 A railroad coal car weighs 54,600 lb empty and carries a total load of 180,000 lb of coal. The bins are equipped with bottom doors that permit discharging coal through an opening between the rails. If the car dumps coal at the rate of 20,000 lb/sec in a downward direction relative to the car, and if frictional resistance to motion is 4 lb per ton of total remaining weight, determine the coupler force P required to give the car an acceleration of 0.15 ft/sec² in the direction of P at the instant when half the coal has been dumped.   *Ans.* $P = 963$ lb

4/72 The mass m of a raindrop increases as it picks up moisture during its vertical descent through still air. If the air resistance to motion of the drop is R and its downward velocity is v, write the equation of motion for the drop and show that the relation $\Sigma F = d(mv)/dt$ is obeyed as a special case of the variable-mass equation.

4/73 The space shuttle, together with its central fuel tank and two booster rockets, has a total mass of $2.04(10^6)$ kg at lift-off. Each of the two booster rockets produces a thrust of $11.80(10^6)$ N, and each of the three main engines of the shuttle produces a thrust of $2.00(10^6)$ N. The specific impulse (ratio of exhaust velocity to gravitational acceleration) for each of the three main engines of the shuttle is 455 s. Calculate the initial vertical acceleration a of the assembly with all five engines operating and find the rate at which fuel is being consumed by each of the shuttle's three engines.
Ans. $a = 4.70$ m/s², $m' = 448$ kg/s

4/74 The magnetometer boom for a spacecraft consists of a large number of triangular-shaped units that spring into their deployed configuration upon release from the canister in which they were folded and packed prior to release. Write an expression for the force F that the base of the canister must exert on the boom during its deployment in terms of the increasing length x and its time derivatives. The mass of the boom per unit of deployed length is ρ. Consider the supporting base on the spacecraft as a fixed platform and assume that the deployment takes place outside of any gravitational field. Neglect the dimension b compared with x.

4/75 A small rocket of initial mass m_0 is fired vertically upward near the surface of the earth (g constant). If air resistance is neglected, determine the manner in which the mass m of the rocket must vary as a function of the time t after launching in order that the rocket may have a constant vertical acceleration a, with a constant relative velocity u of the escaping gases with respect to the nozzle.

$$Ans. \ m = m_0 e^{-\frac{a+g}{u}t}$$

Representative problems

4/76 The rocket has a mass of 8.5 Mg when launched. Fuel is burned at the rate of 200 kg/s with a nozzle exhaust velocity of 760 m/s. Calculate the initial acceleration a of the rocket along the ramp and find the angle θ with the horizontal made by the acceleration of the mass center of the rocket just after it clears the ramp.

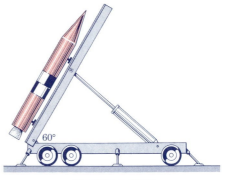

Problem 4/76

4/77 The upper end of the open-link chain of length L and mass ρ per unit length is lowered at a constant speed v by the force P. Determine the reading R of the platform scale in terms of x.

$$Ans. \ R = \rho g x + \rho v^2$$

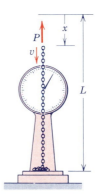

Problem 4/77

4/78 The upper end of the open-link chain of length L and mass ρ per unit length is released from rest with the lower end just touching the platform of the scale. Determine the expression for the force F read on the scale as a function of the distance x through which the upper end has fallen. (*Comment:* The chain acquires a free-fall velocity of $\sqrt{2gx}$ since the links on the scale exert no force on those above, which are still falling freely. Work the problem in two ways: first, by evaluating the time rate of change of momentum for the entire chain and, second, by considering the force F to be composed of the weight of the links at rest on the scale plus the force necessary to divert an equivalent stream of fluid.)

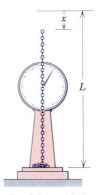

Problem 4/78

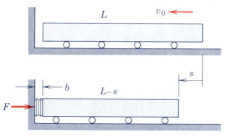

Problem 4/80

4/79 The rate m' in kilograms per second at which fuel is burned in a certain rocket mounted vertically in a test stand is programmed so that the momentum thrust balances the decreasing weight of the rocket for the duration of the test. If the initial mass of the rocket and fuel is 6000 kg and the relative exhaust velocity is 800 m/s regardless of fuel rate, determine the manner in which m' must vary with the time t in seconds after firing. Also find the mass m_f of fuel used in the first 10 s.

Ans. $m' = 73.6e^{-0.0123t}$ kg/s, $m_f = 692$ kg

4/80 The figure represents an idealized one-dimensional structure of uniform mass ρ per unit length moving horizontally with a velocity v_0 when its front end collides with an immovable barrier and crushes. The force F required to initiate and maintain an accordionlike deformation is constant. Neglect the length b of the collapsed portion of the structure compared with the movement s of the undeformed portion following the impact. The undeformed part may be viewed as a body of decreasing mass. Derive the differential equation that relates F to s, $\dot{s}$, and $\ddot{s}$ by using Eq. 4/20 carefully. Check your expression by applying Eq. 4/6 to both parts together as a system of constant mass.

Problem 4/81

4/81 The end of a pile of loose-link chain of mass ρ per unit length is being pulled horizontally along the surface by a constant force P. If the coefficient of kinetic friction between the chain and the surface is μ_k, determine the acceleration a of the chain in terms of x and $\dot{x}$.

Ans. $a = \dfrac{P}{\rho x} - \mu_k g - \dfrac{\dot{x}^2}{x}$

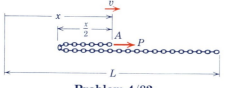

Problem 4/82

4/82 An open-link chain of length $L = 8$ m with a mass of 48 kg is resting on a smooth horizontal surface when end A is doubled back on itself by a force P applied to end A. (*a*) Calculate the required value of P to give A a constant velocity of 1.5 m/s. (*b*) Calculate the acceleration a of end A if $P = 20$ N and if $v = 1.5$ m/s when $x = 4$ m.

4/83 An ideal rope or bicycle-type chain of length L and mass ρ per unit length is resting on a smooth horizontal surface when end A is doubled back on itself by a force P applied to end A. End B of the rope is secured to a fixed support. Determine the force P required to give A a constant velocity v. (*Hint:* The action of the loop can be modeled by inserting a circular disk of negligible mass as shown in the separate sketch and then taking the disk radius as zero. It is easily shown that the tensions in the rope at C, D, and B are all equal to P under the ideal conditions imposed and with constant velocity.)

$$\text{Ans.} \quad P = \rho v^2/4$$

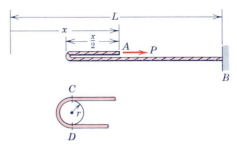

Problem 4/83

4/84 The open-link chain of total length L and of mass ρ per unit length is released from rest at $x = 0$ at the same instant that the platform starts from rest at $y = 0$ and moves vertically up with a constant acceleration a. Determine the expression for the total force R exerted on the platform by the chain t seconds after the motion starts.

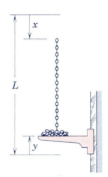

Problem 4/84

4/85 An open-link chain of mass ρ per unit length is released from rest on the incline with $x = 0$. The coefficient of kinetic friction between the chain and the incline is μ_k. The chain piles up in a bucket at the base of the incline. Determine as a function of x the added force R on the bucket required to hold it in place against the impact of the chain. Note that the last link in motion at the bottom of the incline has lost contact with the link that immediately preceded it.

$$\text{Ans.} \quad R = \rho g x(3 \sin \theta - 2\mu_k \cos \theta)$$
$$\text{with } \mu_k < \tan \theta$$

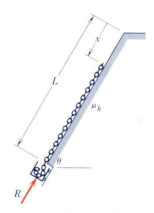

Problem 4/85

4/86 An open-link chain of length $\pi r/2$ and mass ρ per unit length is released from rest at $\theta = 0$ and slides down the smooth circular channel. The chain piles up at the bottom where the last moving link at $\beta = \pi/2$ has lost contact with the one before it, which has come to rest. Determine the expression for the tangential acceleration a_t common to all links as a function of θ. (*Hint:* Write the force equation of motion for a general element and integrate the tension from zero at $\beta = \theta$ to zero at $\beta = \pi/2$.) Also find the loss Q of energy during the motion from $\theta = 0$ to $\theta = \pi/2$.

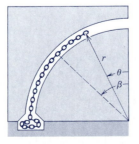

Last link
in motion
at $\beta = \pi/2$

Problem 4/86

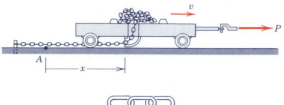

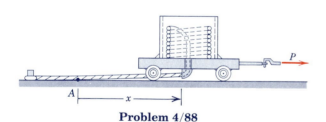

1   2   3

Transition link 2

Problem 4/87

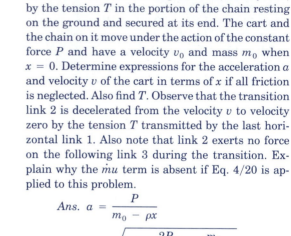

Problem 4/88

4/87 The cart carries a pile of open-link chain of mass ρ per unit length. The chain passes freely through the hole in the cart and is brought to rest, link by link, by the tension T in the portion of the chain resting on the ground and secured at its end. The cart and the chain on it move under the action of the constant force P and have a velocity v_0 and mass m_0 when $x = 0$. Determine expressions for the acceleration a and velocity v of the cart in terms of x if all friction is neglected. Also find T. Observe that the transition link 2 is decelerated from the velocity v to velocity zero by the tension T transmitted by the last horizontal link 1. Also note that link 2 exerts no force on the following link 3 during the transition. Explain why the $\dot{m}u$ term is absent if Eq. 4/20 is applied to this problem.

$$Ans. \quad a = \frac{P}{m_0 - \rho x}$$

$$v = \sqrt{v_0{}^2 + \frac{2P}{\rho} \ln \frac{m_0}{m_0 - \rho x}}$$

$$T = \rho v^2$$

4/88 The cart carries a coil of flexible cable of mass ρ per unit length. The cable unwinds and passes through the hole in the cart and is brought to rest by the tension T in the portion of the cable already on the ground and secured at its end. The cart and coil move under the action of the constant force P and have a mass m_0 when $x = 0$. Determine expressions for the acceleration a and the tension T in terms of x and the velocity v if all friction is neglected. Because the cart and entire rope are a conservative system, the work-energy equation may be applied for a differential displacement dx. The impulse-momentum equation for the entire system may then be written to obtain T. Compare this problem with Prob. 4/87. Assume that all elements of rope in the coil have the same horizontal velocity v.

4/89 In the figure is shown a system used to arrest the motion of an airplane landing on a field of restricted length. The plane of mass m rolling freely with a velocity v_0 engages a hook that pulls the ends of two heavy chains, each of length L and mass ρ per unit length, in the manner shown. A conservative calculation of the effectiveness of the device neglects the retardation of chain friction on the ground and any other resistance to the motion of the airplane. With these assumptions, compute the velocity v of the airplane at the instant that the last link of each chain is put in motion. Also determine the relation between displacement x and the time t after contact with the chain. Assume each link of the chain acquires its velocity v suddenly upon contact with the moving links.

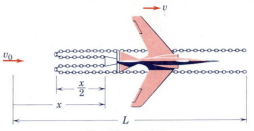

Problem 4/89

$$Ans. \quad v = \frac{v_0}{1 + 2\rho L/m}$$

$$x = \frac{m}{\rho}\left[\sqrt{1 + \frac{2v_0 t \rho}{m}} - 1\right]$$

▶ **4/90** A metal canister of mass m_0 contains a coil of flexible cord of mass ρ per unit length. The end of the cord is secured to a support at the top of an elevator shaft, and the canister is released from rest. A thumb screw on the neck of the canister adjusts a spring clamp that applies a constant friction force $F/2$ on each side of the cord as it passes through the neck. Determine an expression for the acceleration a of the canister and the force P at the support in terms of x and the velocity v. The total length of the cord is L.

Problem 4/90

$$Ans. \quad a = g + \frac{\rho v^2/2 - F}{m_0 + \rho(L - x)}$$

$$P = \rho g x + \rho v^2/2 + F$$

▶ **4/91** A rope or hinged-link bicycle-type chain of length L and mass ρ per unit length is released from rest with $x = 0$. Determine the expression for the total force R exerted on the fixed platform by the chain as a function of x. Note that the hinged-link chain is a conservative system during all but the last increment of motion. Compare the result with that of Prob. 4/84 if the upward motion of the platform in that problem is taken to be zero.

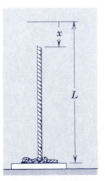

Problem 4/91

$$Ans. \quad R = \rho g x \frac{4L - 3x}{2(L - x)}$$

Problem 4/92

▶**4/92** The free end of the open-link chain of total length L and mass ρ per unit length is released from rest at $x = 0$. Determine the force R on the fixed end and the tension T_1 in the chain at the lower end of the nonmoving part in terms of x. Also find the total loss Q of energy when $x = L$.

> *Ans.* $R = \frac{1}{2}\rho g(L + 3x)$, $T_1 = \rho gx$, $Q = \frac{1}{4}\rho gL^2$

▶**4/93** Replace the chain of Prob. 4/92 by a flexible rope or bicycle chain of mass ρ per unit length and total length L. The free end is released from rest at $x = 0$ and falls under the influence of gravity. Determine the acceleration a of the free end, the force R at the fixed end, and the tension T_1 in the rope at the loop, all in terms of x. (Note that a is greater than g. What happens to the energy of the system when $x = L$?)

> *Ans.* $a = g\left[1 + \dfrac{x(L - x/2)}{(L - x)^2}\right]$
>
> $R = \frac{1}{2}\rho g\left[(L + x) + \dfrac{x(L - x/2)}{L - x}\right]$
>
> $T_1 = \frac{1}{2}\rho g\,\dfrac{x(L - x/2)}{L - x}$

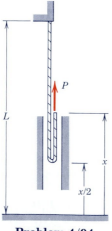

Problem 4/94

▶**4/94** The free end of the flexible and inextensible rope of mass ρ per unit length and total length L is given a constant upward velocity v. Write expressions for P, the force R supporting the fixed end, and the tension T_1 in the rope at the loop in terms of x. (For the loop of negligible size, the tension is the same on both sides.)

> *Ans.* $T_1 = \frac{1}{4}\rho v^2$
>
> $P = \frac{1}{2}\rho(\frac{1}{2}v^2 + gx)$
>
> $R = \frac{1}{4}\rho v^2 + \rho g\left(L - \dfrac{x}{2}\right)$

▶ **4/95** Replace the rope of Prob. 4/94 by an open-link chain with the same mass ρ per unit length. The free end is given a constant upward velocity v. Write expressions for P, the tension T_1 at the bottom of the moving part, and the force R supporting the fixed end in terms if x. Also find the energy loss Q in terms of x.

$$Ans. \quad T_1 = \tfrac{1}{2}\rho v^2$$
$$P = \tfrac{1}{2}\rho(v^2 + gx)$$
$$R = \rho g\left(L - \frac{x}{2}\right)$$
$$Q = \tfrac{1}{4}\rho x(v^2 + gx)$$

▶ **4/96** One end of the pile of chain falls through a hole in its support and pulls the remaining links after it in a steady flow. If the links that are initially at rest acquire the velocity of the chain suddenly and without frictional resistance or interference from the support or from adjacent links, find the velocity v of the chain as a function of x if $v = 0$ when $x = 0$. Also find the acceleration a of the falling chain and the energy Q lost from the system as the last link leaves the platform. (*Hint:* Apply Eq. 4/20 and treat the product xv as the variable when solving the differential equation. Also note at the appropriate step that $dx = v\, dt$.) The total length of the chain is L, and its mass per unit length is ρ.

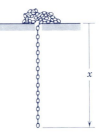

Problem 4/96

$$Ans. \quad v = \sqrt{\frac{2gx}{3}}, \; a = \frac{g}{3}, \; Q = \frac{\rho g L^2}{6}$$

▶ **4/97** Replace the pile of chain in Prob. 4/96 by a coil of rope of mass ρ per unit length and total length L as shown and determine the velocity of the falling section in terms of x if it starts from rest at $x = 0$. Show that the acceleration is constant at $g/2$. The rope is considered to be perfectly flexible in bending but inextensible and constitutes a conservative system (no energy loss). Rope elements acquire their velocity in a continuous manner from zero to v in a small transition section of the rope at the top of the coil. For comparison with the chain of Prob. 4/96, this transition section may be considered to have negligible length without violating the requirement that there be no energy loss in the present problem. Also determine the force R exerted by the platform on the coil in terms of x and explain why R becomes zero when $x = 2L/3$. Neglect the dimensions of the coil compared with x.

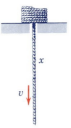

Problem 4/97

$$Ans. \quad v = \sqrt{gx}, \; R = \rho g\left(L - \frac{3x}{2}\right)$$

PART II

DYNAMICS OF
RIGID BODIES

Rigid-body kinematics describes the relation between the linear and angular motions of bodies that retain their shape. The design of gears, cams, connecting rods, and many other moving machine parts are kinematic problems. The design of the oil-well pumping rig is an example where the velocity and acceleration of the pump shaft is determined from the rotation of the driving crank.

PLANE KINEMATICS OF RIGID BODIES

5

5/1 INTRODUCTION

In Chapter 2 on particle kinematics, we developed the relationships governing the displacement, velocity, and acceleration of points as they moved along straight or curved paths. In rigid-body kinematics we use these same relationships but must also account for the rotational motion of the body as well. Thus, rigid-body kinematics involves both linear and angular quantities.

The description of the motion of rigid bodies is useful in two important ways. First, it is frequently necessary to generate, transmit, or control certain desired motions by the use of cams, gears, and linkages of various types. Here an analysis of the displacement, velocity, and acceleration of the motion is necessary to determine the design geometry of the mechanical parts. Furthermore, as a result of the motion generated, forces are frequently developed that must be accounted for in the design of the parts. Second, it is often necessary to determine the motion of a rigid body resulting from the forces applied to it. Calculation of the motion of a rocket under the influence of its jet thrust and gravitational attraction is an example of such a problem. In both types of problems, we need to have command of the principles of rigid-body kinematics. In this chapter we will cover the kinematics of motion that may be analyzed as taking place in a single plane. In Chapter 7 we will present a brief introduction to the kinematics of motion that requires three spatial coordinates.

A rigid body was defined in the previous chapter as a system of particles for which the distances between the particles remain unchanged. Thus, if each particle of such a body is located by a position vector from reference axes attached to and rotating with the body, there will be no change in any position vector as measured from these axes. This formulation is, of course, an ideal one since all solid materials change shape to some extent when forces are applied to them. Nevertheless, if the movements associated with the changes in shape are very small compared with the overall movements of the body as a whole, then the ideal concept of rigidity is quite acceptable. As mentioned in Chapter 1, the displacements due to the flutter of an aircraft wing, for instance, are of no consequence in the description of the flight path of aircraft as a whole, for which the rigid-body assumption is clearly in order. On the other hand, if the problem is one of describing, as a function of time, the internal wing stress due to wing flutter, then the relative motions of portions of the wing become of prime importance and cannot be neglected. In this case, the wing may not be considered a rigid body. In the present chapter and in the two that follow, essentially all of the material is based on the assumption of rigidity.

A rigid body executes plane motion when all parts of the body move in parallel planes. For convenience, we generally consider the *plane of motion* to be the plane that contains the mass center, and we treat the body as a thin slab with motion confined to the plane of the slab. This idealization fits a very large category of rigid-body motions encountered in engineering.

The plane motion of a rigid body may be divided into several categories, as represented in Fig. 5/1.

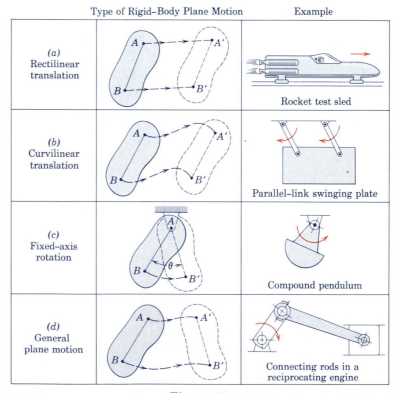

Figure 5/1

Translation is defined as any motion in which every line in the body remains parallel to its original position at all times. In translation there is *no rotation of any line in the body*. In *rectilinear translation*, part (*a*) of Fig. 5/1, all points in the body move in parallel straight lines. In *curvilinear translation*, part (*b*), all points move on congruent curves. We note that in each of the two cases of translation, the motion of the body is completely specified by the motion of any point in the body, since all points have the same motion. Thus, our earlier study of the motion of a point (particle) in Chapter 2 enables us to describe completely the translation of a rigid body.

Rotation about a fixed axis, part (*c*) of Fig. 5/1, is the angular motion about the axis. It follows that all particles move in circular paths about the axis of rotation, and all lines in the body which are perpendicular to the axis of rotation (including those that do not pass through the axis) rotate through the same angle in the same time. Again, our discussion in Chapter 2 on the circular motion of a point enables us to describe the motion of a rotating rigid body, which is treated in the next article.

General plane motion of a rigid body, part (*d*) of Fig. 5/1, is a combination of translation and rotation. We will utilize the principles of relative motion covered in Art. 2/8 in the description of general plane motion.

Note that in each of the examples cited, the actual paths of all particles in the body are projected onto the single plane of motion as represented in each figure.

Determination of the plane motion of rigid bodies is accomplished either by directly calculating the absolute displacements and their time derivatives from the geometry involved or by utilizing the principles of relative motion. Each method is important and useful and will be taken up in turn in the articles that follow.

5/2 ROTATION

The rotation of a rigid body is described by its angular motion. Figure 5/2 shows a rigid body that is rotating as it undergoes plane motion in the plane of the figure. The angular positions of any two

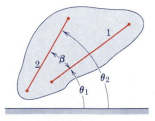

Figure 5/2

lines 1 and 2 attached to the body are specified by θ_1 and θ_2 measured from any fixed reference direction that is convenient. Because the angle β is invariant, the relation $\theta_2 = \theta_1 + \beta$ upon differentiation with respect to time gives $\dot{\theta}_2 = \dot{\theta}_1$ and $\ddot{\theta}_2 = \ddot{\theta}_1$ or, during a finite interval, $\Delta\theta_2 = \Delta\theta_1$. Thus, *all lines on a rigid body in its plane of motion have the same angular displacement, the same angular velocity, and the same angular acceleration.* It should be noted that the angular motion of a line depends only on its angular displacement with respect to any arbitrary fixed reference and on the time derivatives of the displacement. Angular motion does not require the presence of a fixed axis, normal to the plane of motion, about which the line and the body rotate.

(a) Angular motion relations. The angular velocity ω and angular acceleration α of a rigid body in plane rotation are, respectively, the first and second time derivatives of the angular position coordinate θ of any line in the plane of motion of the body. These definitions give

$$\omega = \frac{d\theta}{dt} = \dot{\theta}$$

$$\alpha = \frac{d\omega}{dt} = \dot{\omega} \qquad \text{or} \qquad \alpha = \frac{d^2\theta}{dt^2} = \ddot{\theta} \qquad \textbf{(5/1)}$$

$$\omega \, d\omega = \alpha \, d\theta \qquad \text{or} \qquad \dot{\theta} \, d\dot{\theta} = \ddot{\theta} \, d\theta$$

The third relation is obtained by eliminating dt from the first two. In each of these relations, the positive direction for ω and α, clockwise or counterclockwise, is the same as that chosen for θ. Equations 5/1 should be recognized as analogous to the defining equations for the rectilinear motion of a particle, expressed by Eqs. 2/1, 2/2, and 2/3. In fact, all relations that were described for rectilinear motion in Art. 2/2 apply to the case of rotation in a plane if the linear quantities s, v, and a are replaced by their respective equivalent angular quantities θ, ω, and α. As we proceed further with rigid-body dynamics, we will find that the analogies between the relationships for linear and angular motion are almost complete throughout kinematics and kinetics. These relations are important to recognize, as they help to demonstrate the symmetry and unity found throughout mechanics.

For rotation with *constant* angular acceleration, the integrals of Eqs. 5/1 become

$$\omega = \omega_0 + \alpha t$$

$$\omega^2 = \omega_0{}^2 + 2\alpha(\theta - \theta_0)$$

$$\theta = \theta_0 + \omega_0 t + \tfrac{1}{2}\alpha t^2$$

Here θ_0 and ω_0 are the values of the angular position coordinate and angular velocity, respectively, at $t = 0$ and t is the duration of the interval of motion considered. The student should be able to carry out these integrations easily, as they are completely analogous to the corresponding equations for rectilinear motion with constant acceleration covered in Art. 2/2.

The graphical relationships described for s, v, a, and t in Figs. 2/3 and 2/4 may be used for θ, ω, and α merely by the substitution of corresponding symbols. The student should sketch these graphical relations for plane rotation. The mathematical procedures for obtaining velocity and displacement from acceleration described for rectilinear motion may be applied to rotation by merely replacing the linear quantities by their corresponding angular quantities.

(b) Rotation about a fixed axis. When a body rotates about a fixed axis, all points other than those on the axis move in concentric circles about the fixed axis. Thus, for the rigid body in Fig. 5/3 rotating about a fixed axis normal to the plane of the figure through O, any point such as A moves in a circle of radius r. From our previous discussion in Art. 2/5, we are already familiar with the relationships between the linear motion of A and the angular motion of

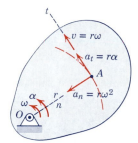

Figure 5/3

the line normal to its path, which is also the angular motion of the rigid body. With the notation $\omega = \dot{\theta}$ and $\alpha = \dot{\omega} = \ddot{\theta}$ for the angular velocity and angular acceleration, respectively, of the body we have Eqs. 2/11, rewritten as

$$v = r\omega$$
$$a_n = r\omega^2 = v^2/r = v\omega \qquad\qquad \text{(5/2)}$$
$$a_t = r\alpha$$

These quantities may be expressed alternatively using the cross-product relationship of vector notation. The vector formulation is especially important in three-dimensional analysis of motion.

The angular velocity of the rotating body may be expressed by the vector $\boldsymbol{\omega}$ normal to the plane of rotation and having a sense governed by the right-hand rule, as shown in Fig. 5/4a. From the definition of the vector cross product, we see that the vector $\mathbf{v}$ is ob-

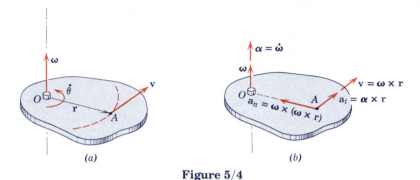

Figure 5/4

tained by crossing $\boldsymbol{\omega}$ into $\mathbf{r}$. This cross product gives the correct magnitude and direction for $\mathbf{v}$ and we write

$$\mathbf{v} = \dot{\mathbf{r}} = \boldsymbol{\omega} \times \mathbf{r}$$

The order of the vectors to be crossed must be retained. The reverse order gives $\mathbf{r} \times \boldsymbol{\omega} = -\mathbf{v}$.

The acceleration of A is obtained by differentiating the cross-product expression for $\mathbf{v}$, which gives

$$\mathbf{a} = \dot{\mathbf{v}} = \boldsymbol{\omega} \times \dot{\mathbf{r}} + \dot{\boldsymbol{\omega}} \times \mathbf{r}$$
$$= \boldsymbol{\omega} \times (\boldsymbol{\omega} \times \mathbf{r}) + \dot{\boldsymbol{\omega}} \times \mathbf{r}$$
$$= \boldsymbol{\omega} \times \mathbf{v} + \boldsymbol{\alpha} \times \mathbf{r}$$

Here $\boldsymbol{\alpha} = \dot{\boldsymbol{\omega}}$ stands for the angular acceleration of the body. Thus, the vector equivalents to Eqs. 5/2 are

$$\begin{aligned} \mathbf{v} &= \boldsymbol{\omega} \times \mathbf{r} \\ \mathbf{a}_n &= \boldsymbol{\omega} \times (\boldsymbol{\omega} \times \mathbf{r}) \\ \mathbf{a}_t &= \boldsymbol{\alpha} \times \mathbf{r} \end{aligned} \qquad (5/3)$$

and are shown in Fig. 5/4b.

For three-dimensional motion of a rigid body, the angular velocity vector $\boldsymbol{\omega}$ may change direction as well as magnitude, and in this case, the angular acceleration, which is the time derivative of angular velocity, $\boldsymbol{\alpha} = \dot{\boldsymbol{\omega}}$, will no longer be in the same direction as $\boldsymbol{\omega}$.

Sample Problem 5/1

A flywheel rotating freely at 1800 rev/min clockwise is subjected to a variable counterclockwise torque that is first applied at time $t = 0$. The torque produces a counterclockwise angular acceleration $\alpha = 4t$ rad/s^2, where t is the time in seconds during which the torque is applied. Determine (a) the time required for the flywheel to reduce its clockwise angular speed to 900 rev/min, (b) the time required for the flywheel to reverse its direction of rotation, and (c) the total number of revolutions, clockwise plus counterclockwise, turned by the flywheel during the first 14 seconds of torque application.

Solution. The counterclockwise direction will be taken arbitrarily as positive.

(1) **(a)** Since α is a known function of the time, we may integrate it to obtain angular velocity. With the initial angular velocity of $-1800(2\pi)/60 = -60\pi$ rad/s, we have

$$[d\omega = \alpha \, dt] \qquad \int_{-60\pi}^{\omega} d\omega = \int_{0}^{t} 4t \, dt \qquad \omega = -60\pi + 2t^2$$

Substituting the clockwise angular speed of 900 rev/min or $\omega = -900(2\pi)/60 = -30\pi$ rad/s gives

$$-30\pi = -60\pi + 2t^2 \qquad t^2 = 15\pi \qquad t = 6.86 \text{ s} \qquad \textit{Ans.}$$

(b) The flywheel changes direction when its angular velocity is momentarily zero. Thus,

$$0 = -60\pi + 2t^2 \qquad t^2 = 30\pi \qquad t = 9.71 \text{ s} \qquad \textit{Ans.}$$

(c) The total number of revolutions through which the flywheel turns during 14 seconds is the number of clockwise turns N_1 during the first 9.71 seconds, plus the number of counterclockwise turns N_2 during the remainder of the interval. Integrating the expression for ω in terms of t gives us the angular displacement in radians. Thus, for the first interval

$$[d\theta = \omega \, dt] \qquad \int_{0}^{\theta_1} d\theta = \int_{0}^{9.71} (-60\pi + 2t^2) \, dt$$

(2) $$\theta_1 = [-60\pi t + \tfrac{2}{3}t^3]_{0}^{9.71} = -1220 \text{ rad}$$

or $N_1 = 1220/2\pi = 194.2$ revolutions clockwise.

(3) For the second interval

$$\int_{0}^{\theta_2} d\theta = \int_{9.71}^{14} (-60\pi + 2t^2) \, dt$$

$$\theta_2 = [-60\pi t + \tfrac{2}{3}t^3]_{9.71}^{14} = 410 \text{ rad}$$

or $N_2 = 410/2\pi = 65.3$ revolutions counterclockwise. Thus, the total number of revolutions turned during the 14 seconds is

$$N = N_1 + N_2 = 194.2 + 65.3 = 259.5 \text{ rev} \qquad \textit{Ans.}$$

We have plotted ω versus t and we see that θ_1 is represented by the negative area and θ_2 by the positive area. If we had integrated over the entire interval in one step, we would have obtained $|\theta_2| - |\theta_1|$.

(1) We must be very careful to be consistent with our algebraic signs. The lower limit is the negative (clockwise) value of the initial angular velocity. Also we must convert revolutions to radians since α is in radian units.

(2) Again note that the minus sign signifies clockwise in this problem.

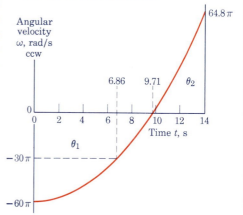

(3) We could have converted the original expression for α into the units of rev/s^2, in which case our integrals would have come out directly in revolutions.

Sample Problem 5/2

The pinion A of the hoist motor drives gear B that is attached to the hoisting drum. The load L is lifted from its rest position and acquires an upward velocity of 3 ft/sec in a vertical rise of 4 ft with constant acceleration. As the load passes this position, compute (a) the acceleration of point C on the cable in contact with the drum and (b) the angular velocity and angular acceleration of the pinion A.

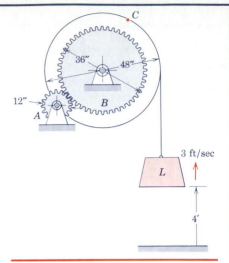

Solution. *(a)* If the cable does not slip on the drum, the vertical velocity and acceleration of the load L are, of necessity, the same as the tangential velocity v and tangential acceleration a_t of point C. For the rectilinear motion of L with constant acceleration, the n- and t-components of the acceleration of C become

$[v^2 = 2as]$ $\qquad a = a_t = v^2/2s = 3^2/[2(4)] = 1.125$ ft/sec²

① $[a_n = v^2/r]$ $\qquad a_n = 3^2/(24/12) = 4.5$ ft/sec²

$[a = \sqrt{a_n^2 + a_t^2}]$ $\qquad a_C = \sqrt{(4.5)^2 + (1.125)^2} = 4.64$ ft/sec² *Ans.*

① Recognize that a point on the cable changes the direction of its velocity after it contacts the drum and acquires a normal component of acceleration.

(b) The angular motion of gear A is determined from the angular motion by gear B by the velocity v_1 and tangential acceleration a_1 of their common point of contact. First, the angular motion of gear B is determined from the motion of point C on the attached drum. Thus,

$[v = r\omega]$ $\qquad \omega_B = v/r = 3/(24/12) = 1.5$ rad/sec

$[a_t = r\alpha]$ $\qquad \alpha_B = a_t/r = 1.125/(24/12) = 0.563$ rad/sec²

Then from $v_1 = r_A\omega_A = r_B\omega_B$ and $a_1 = r_A\alpha_A = r_B\alpha_B$, we have

$$\omega_A = \frac{r_B}{r_A}\omega_B = \frac{18/12}{6/12}\,1.5 = 4.5 \text{ rad/sec CW} \qquad Ans.$$

$$\alpha_A = \frac{r_B}{r_A}\alpha_B = \frac{18/12}{6/12}\,0.563 = 1.69 \text{ rad/sec}^2 \text{ CW} \qquad Ans.$$

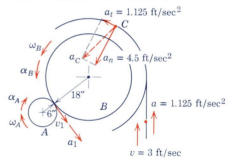

Sample Problem 5/3

The right-angled bar rotates clockwise with an angular velocity that is decreasing at the rate of 4 rad/s². Write the vector expressions for the velocity and acceleration of point A when $\omega = 2$ rad/s.

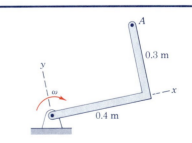

Solution. Using the right-hand rule gives

$$\boldsymbol{\omega} = -2\mathbf{k} \text{ rad/s} \quad \text{and} \quad \boldsymbol{\alpha} = +4\mathbf{k} \text{ rad/s}^2$$

The velocity and acceleration of A become

$[\mathbf{v} = \boldsymbol{\omega} \times \mathbf{r}]$ $\quad \mathbf{v} = -2\mathbf{k} \times (0.4\mathbf{i} + 0.3\mathbf{j}) = 0.6\mathbf{i} - 0.8\mathbf{j}$ m/s *Ans.*

$[\mathbf{a}_n = \boldsymbol{\omega} \times (\boldsymbol{\omega} \times \mathbf{r})]$ $\quad \mathbf{a}_n = -2\mathbf{k} \times (0.6\mathbf{i} - 0.8\mathbf{j}) = -1.6\mathbf{i} - 1.2\mathbf{j}$ m/s²

$[\mathbf{a}_t = \boldsymbol{\alpha} \times \mathbf{r}]$ $\quad \mathbf{a}_t = 4\mathbf{k} \times (0.4\mathbf{i} + 0.3\mathbf{j}) = -1.2\mathbf{i} + 1.6\mathbf{j}$ m/s²

$[\mathbf{a} = \mathbf{a}_n + \mathbf{a}_t]$ $\quad \mathbf{a} = -2.8\mathbf{i} + 0.4\mathbf{j}$ m/s² *Ans.*

The magnitudes of $\mathbf{v}$ and $\mathbf{a}$ are

$$v = \sqrt{0.6^2 + 0.8^2} = 1.0 \text{ m/s} \quad \text{and} \quad a = \sqrt{2.8^2 + 0.4^2} = 2.83 \text{ m/s}^2$$

PROBLEMS

Introductory problems

5/1 The rigid link moves from position *ABC* to position *A'B'C'* while end *A* moves 40 in. to the left with a constant velocity of 10 in./sec. Determine the average angular velocity ω_{av} of the arm *BC* during this interval. Assume counterclockwise motion.

Ans. $\omega_{av} = 0.393$ rad/sec

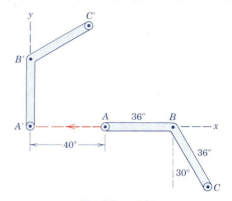

Problem 5/1

5/2 A torque applied to a flywheel causes it to accelerate uniformly from a speed of 200 rev/min to a speed of 800 rev/min in 4 seconds. Determine the number of revolutions *N* through which the wheel turns during this interval. (*Suggestion:* Use revolutions and minutes for units in your calculations.)

5/3 The small cart is released from rest in position 1 and requires 0.638 seconds to reach position 2 at the bottom of the path where its center *G* has a velocity of 14.20 ft/sec. Determine the angular velocity ω of line *AB* in position 2 and the average angular velocity ω_{av} of *AB* during the interval.

Ans. $\omega = 3.55$ rad/sec, $\omega_{av} = 1.231$ rad/sec

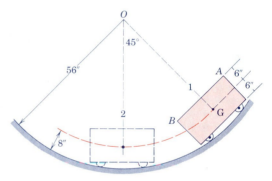

Problem 5/3

5/4 The rectangular plate is rotating about its corner axis through *O* with a constant angular velocity $\omega = 10$ rad/s. Determine the magnitudes of the velocity **v** and acceleration **a** of corner *A* by (*a*) using the scalar relations and (*b*) using the vector relations.

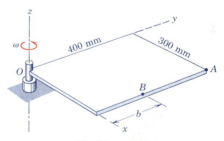

Problem 5/4

5/5 The circular disk rotates about its center *O* in the direction shown. At a certain instant point *P* on the rim has an acceleration given by $\mathbf{a} = -3\mathbf{i} - 4\mathbf{j}$ m/s². For this instant determine the angular velocity $\boldsymbol{\omega}$ and angular acceleration $\boldsymbol{\alpha}$ of the disk.

Ans. $\boldsymbol{\omega} = -\sqrt{8}\mathbf{k}$ rad/s, $\boldsymbol{\alpha} = 6\mathbf{k}$ rad/s²

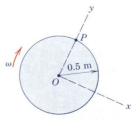

Problem 5/5

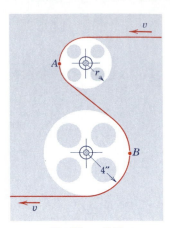

Problem 5/6

5/6 Magnetic tape is being fed over and around the light pulleys mounted in a computer. If the speed v of the tape is constant and if the magnitude of the acceleration of point A on the tape is 4/3 times that of point B, calculate the radius r of the smaller pulley.

5/7 If the rectangular plate of Prob. 5/4 starts from rest and point B has an initial acceleration of 5.5 m/s^2, determine the distance b if the plate reaches an angular speed of 300 rev/min in 2 seconds with a constant angular acceleration.   *Ans. b = 180.6 mm*

5/8 The angular position of a radial line in a rotating disk is given by the clockwise angle $\theta = 2t^3 - 3t^2 + 4$, where θ is in radians and t is in seconds. Calculate the angular displacement $\Delta\theta$ of the disk during the interval in which its angular acceleration increases from 42 rad/s^2 to 66 rad/s^2.

Representative problems

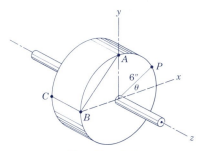

Problem 5/9

5/9 The solid cylinder rotates about its z-axis. At the instant represented, point P on the rim has a velocity whose x-component is -4.2 ft/sec, and $\theta = 20°$. Determine the angular velocity $\boldsymbol{\omega}$ of line AB on the face of the cylinder. Does the element line BC have an angular velocity?

Ans. $\boldsymbol{\omega} = 24.6\mathbf{k}$ rad/sec, No

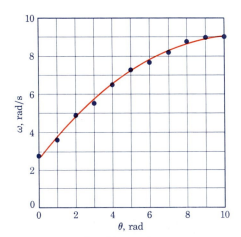

Problem 5/10

5/10 Experimental data for a rotating control element reveal the plotted relation between angular velocity and the angular coordinate θ as shown. Approximate the angular acceleration α of the element when $\theta = 6$ rad.

5/11 The rotating arm starts from rest and acquires a rotational speed $N = 600$ rev/min in 2 seconds with constant angular acceleration. Find the time t after starting before the acceleration vector of end P makes an angle of 45° with the arm OP.

Ans. $t = 0.1784$ sec

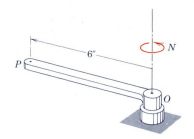

Problem 5/11

5/12 The rectangular plate rotates clockwise about its fixed bearing at O. If edge BC has a constant angular velocity of 6 rad/s, determine the vector expressions for the velocity and acceleration of point A using the coordinates given.

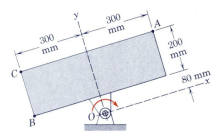

Problem 5/12

5/13 A shaft is accelerated from rest at a constant rate to a speed of 3600 rev/min and then is immediately decelerated to rest at a constant rate within a total time of 10 seconds. How many revolutions N has the shaft turned during this interval?

Ans. $N = 300$ rev

5/14 A shaft is turning clockwise with a rotational speed of 200 rev/min when a torque is applied to it causing a constant counterclockwise angular acceleration α. If the speed of the shaft is 400 rev/min in the counterclockwise direction after 3 seconds of torque application, compute the angular acceleration α of the shaft and the total number N of revolutions (CW plus CCW) through which the shaft turns during the 3 seconds.

5/15 The two V-belt pulleys form an integral unit and rotate about the fixed axis at O. At a certain instant, point A on the belt of the smaller pulley has a velocity $v_A = 1.5$ m/s, and point B on the belt of the larger pulley has an acceleration $a_B = 45$ m/s² as shown. For this instant determine the magnitude of the acceleration $\mathbf{a}_C$ of point C and sketch the vector in your solution. *Ans.* $a_C = 149.6$ m/s²

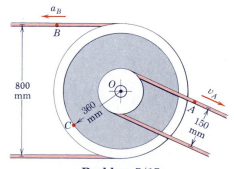

Problem 5/15

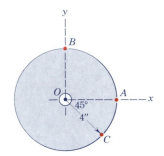

Problem 5/16

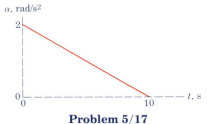

Problem 5/17

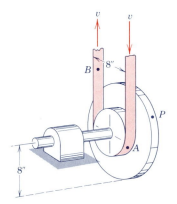

Problem 5/18

Problem 5/19

5/16 The circular disk rotates about its center O. For the instant represented, the velocity of A is $\mathbf{v}_A = 8\mathbf{j}$ in./sec and the tangential acceleration of B is $(\mathbf{a}_B)_t = 6\mathbf{i}$ in./sec^2. Write the vector expressions for the angular velocity $\boldsymbol{\omega}$ and angular acceleration $\boldsymbol{\alpha}$ of the disk. Use these results to write the vector expression for the acceleration of point C.

5/17 A gear rotating at a clockwise speed of 200 rev/min is subjected to a torque that gives it a clockwise angular acceleration α which varies with the time as shown. Find the speed N of the gear when $t = 5$ s.
Ans. $N = 272$ rev/min

5/18 The rotational speed of a control device decreases from 2000 rev/min to 500 rev/min in the manner shown during the 6-second interval. Determine the total number of revolutions N through which the device turns during the 6 seconds. Also find the angular acceleration α in rad/sec^2 during the last 2 seconds of the interval.

5/19 The two attached pulleys are driven by the belt with increasing speed. When the belt reaches a speed $v = 2$ ft/sec, the total acceleration of point P is 26 ft/sec^2. For this instant determine the angular acceleration α of the pulleys and the acceleration of point B on the belt.
Ans. $\alpha = 15$ rad/sec^2, $a_B = 5$ ft/sec^2

5/20 The circular disk rotates about its *z*-axis with an angular velocity in the direction shown. At a certain instant the magnitude of the velocity of point *A* is 10 ft/sec and is decreasing at the rate of 24 ft/sec². Write the vector expressions for the angular acceleration $\boldsymbol{\alpha}$ of the disk and the total acceleration of point *B* at this instant.

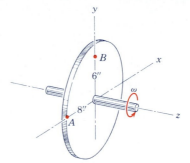

Problem 5/20

5/21 A clockwise variable torque is applied to a flywheel at time $t = 0$ causing its clockwise angular acceleration to decrease linearly with angular displacement θ during 20 revolutions of the wheel as shown. If the clockwise speed of the flywheel was 300 rev/min at $t = 0$, determine its speed *N* after turning the 20 revolutions. (*Suggestion:* Use units of revolutions instead of radians.)   *Ans. N* = 513 rev/min

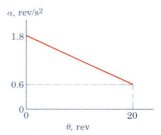

Problem 5/21

5/22 A V-belt speed-reduction drive is shown where pulley *A* drives the two integral pulleys *B* which in turn drive pulley *C*. If *A* starts from rest at time $t = 0$ and is given a constant angular acceleration α_1, derive expressions for the angular velocity of *C* and the magnitude of the acceleration of a point *P* on the belt, both at time *t*.

$$\textit{Ans. } \omega_C = \left(\frac{r_1}{r_2}\right)^2 \alpha_1 t$$

$$a_P = \frac{r_1^2}{r_2}\alpha_1 \sqrt{1 + \left(\frac{r_1}{r_2}\right)^4 \alpha_1^2 t^4}$$

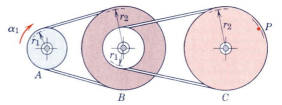

Problem 5/22

5/3 ABSOLUTE MOTION

We now develop the approach of absolute-motion analysis to describe the plane kinematics of rigid bodies. In this approach, we make use of the geometric relations that define the configuration of the body involved and then proceed to take the time derivatives of the defining geometric relations to obtain velocities and accelerations. In Art. 2/9 of Chapter 2 on particle kinematics, we introduced the application of absolute-motion analysis for the constrained motion of connected particles. For the pulley configurations treated, the relevant velocities and accelerations were determined by successive differentiation of the lengths of the connecting cables. In this earlier treatment, the geometric relations were quite simple, and no angular quantities had to be considered. Now that we will be dealing with rigid-body motion, however, we find that our defining geometric relations include both linear and angular measurements and, therefore, the time derivatives of these quantities will involve both linear and angular velocities and linear and angular accelerations.

In absolute-motion analysis, it is essential that we be totally consistent with the mathematics of the description. For example, if the angular position of a moving line in the plane of motion is specified by its counterclockwise angle θ measured from some convenient fixed reference axis, then the positive sense for both angular velocity $\dot{\theta}$ and angular acceleration $\ddot{\theta}$ will also be counterclockwise. A negative sign for either quantity will, of course, indicate a clockwise angular motion. The defining relations for linear motion, Eqs. 2/1, 2/2, and 2/3, and the relations involving angular motion, Eqs. 5/1 and 5/2 or 5/3, will find repeated use in the motion analysis and should be mastered thoroughly.

The absolute-motion approach to rigid-body kinematics is quite straightforward, provided the configuration lends itself to a geometric description that is not overly complex. If the geometric configuration is awkward or complex, analysis by the principles of relative motion may be preferable. Relative-motion analysis is treated in this chapter beginning with Art. 5/4. The choice between absolute- and relative-motion analyses is best made after experience has been gained with both approaches.

The three sample problems that follow illustrate the application of absolute-motion analysis to three commonly encountered situations, and they should be studied thoroughly. The kinematics of a rolling wheel, treated in Sample Problem 5/4, is especially basic and will find repeated use in the problem work because the rolling wheel in various forms is such a common element in mechanical systems.

Sample Problem 5/4

A wheel of radius r rolls on a flat surface without slipping. Determine the angular motion of the wheel in terms of the linear motion of its center O. Also determine the acceleration of a point on the rim of the wheel as the point comes into contact with the surface on which the wheel rolls.

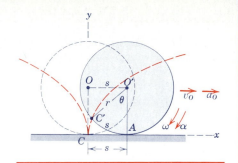

Solution. The figure shows the wheel rolling to the right from the dotted to the full position without slipping. The linear displacement of the center O is s, which is also the arc length $C'A$ along the rim on which the wheel rolls. The radial line CO rotates to the new position $C'O'$ through the angle θ, where θ is measured from the vertical direction. If the wheel does not slip, the arc $C'A$ must equal the distance s. Thus, the displacement relationship and its two time derivatives give

$$s = r\theta$$

$$v_O = r\omega \qquad \textit{Ans.}$$

① $$a_O = r\alpha$$

where $v_O = \dot{s}$, $a_O = \dot{v}_O = \ddot{s}$, $\omega = \dot{\theta}$, and $\alpha = \dot{\omega} = \ddot{\theta}$. The angle θ, of course, must be in radians. The acceleration a_O will be directed in the sense opposite to that of v_O if the wheel is slowing down. In this event, the angular acceleration α will have the opposite sense to ω.

The origin of fixed coordinates is taken arbitrarily but conveniently at the point of contact between C on the rim of the wheel and the ground. When point C has moved along its cycloidal path to C', its new coordinates and their time derivatives become

$$x = s - r \sin \theta = r(\theta - \sin \theta) \qquad y = r - r \cos \theta = r(1 - \cos \theta)$$

$$\dot{x} = r\dot{\theta}(1 - \cos \theta) = v_O(1 - \cos \theta) \quad \dot{y} = r\dot{\theta} \sin \theta = v_O \sin \theta$$

$$\ddot{x} = \dot{v}_O(1 - \cos \theta) + v_O\dot{\theta} \sin \theta \qquad \ddot{y} = \dot{v}_O \sin \theta + v_O\dot{\theta} \cos \theta$$

$$= a_O(1 - \cos \theta) + r\omega^2 \sin \theta \qquad = a_O \sin \theta + r\omega^2 \cos \theta$$

For the desired instant of contact, $\theta = 0$ and

② $$\ddot{x} = 0 \qquad \text{and} \qquad \ddot{y} = r\omega^2 \qquad \textit{Ans.}$$

Thus, the acceleration of the point C on the rim at the instant of contact with the ground depends only on r and ω and is directed toward the center of the wheel. If desired, the velocity and acceleration of C at any position θ may be obtained by writing the expressions $\mathbf{v} = \dot{x}\mathbf{i} + \dot{y}\mathbf{j}$ and $\mathbf{a} = \ddot{x}\mathbf{i} + \ddot{y}\mathbf{j}$.

Application of the kinematic relationships for a wheel that rolls without slipping should be recognized for various configurations of rolling wheels such as those illustrated on the right. If a wheel slips as it rolls, the foregoing relations are no longer valid.

① These three relations are not entirely unfamiliar at this point, and their application to the rolling wheel should be mastered thoroughly.

② Clearly, when $\theta = 0$, the point of contact has zero velocity so that $\dot{x} = \dot{y} = 0$. The acceleration of the contact point on the wheel will also be obtained by the principles of relative motion in Art. 5/6.

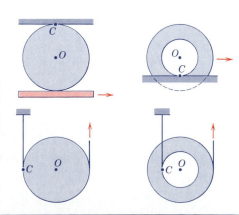

Sample Problem 5/5

The load L is being hoisted by the pulley and cable arrangement shown. Each cable is wrapped securely around its respective pulley so it does not slip. The two pulleys to which L is attached are fastened together to form a single rigid body. Calculate the velocity and acceleration of the load L and the corresponding angular velocity ω and angular acceleration α of the double pulley under the following conditions:

Case (a)   Pulley 1: $\omega_1 = \dot{\omega}_1 = 0$ (pulley at rest)

Pulley 2: $\omega_2 = 2$ rad/sec, $\alpha_2 = \dot{\omega}_2 = -3$ rad/sec^2

Case (b)   Pulley 1: $\omega_1 = 1$ rad/sec, $\alpha_1 = \dot{\omega}_1 = 4$ rad/sec^2

Pulley 2: $\omega_2 = 2$ rad/sec, $\alpha_2 = \dot{\omega}_2 = -2$ rad/sec^2

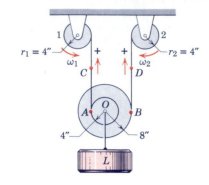

Solution. The tangential displacement, velocity, and acceleration of a point on the rim of pulley 1 or 2 equal the corresponding vertical motions of point A or B since the cables are assumed to be inextensible.

Case (a). With A momentarily at rest, line AB rotates to AB' through the angle $d\theta$ during time dt. From the diagram we see that the displacements and their time derivatives give

①
$$ds_B = \overline{AB}\, d\theta \qquad v_B = \overline{AB}\omega \qquad (a_B)_t = \overline{AB}\alpha$$
$$ds_O = \overline{AO}\, d\theta \qquad v_O = \overline{AO}\omega \qquad a_O = \overline{AO}\alpha$$

With $v_D = r_2\omega_2 = 4(2) = 8$ in./sec, $a_D = r_2\alpha_2 = 4(-3) = -12$ in./sec^2, we have for the angular motion of the double pulley

②
$$\omega = v_B/\overline{AB} = v_D/\overline{AB} = 8/12 = 2/3 \text{ rad/sec (CCW)} \qquad Ans.$$
$$\alpha = (a_B)_t/\overline{AB} = a_D/\overline{AB} = -12/12 = -1 \text{ rad/sec}^2 \text{ (CW)} \qquad Ans.$$

The corresponding motion of O and the load L is

③
$$v_O = \overline{AO}\omega = 4(2/3) = 8/3 \text{ in./sec} \qquad Ans.$$
$$a_O = \overline{AO}\alpha = 4(-1) = -4 \text{ in./sec}^2 \qquad Ans.$$

Case (b). With point C, and hence point A, in motion, line AB moves to $A'B'$ during time dt. From the diagram for this case, we see that the displacements and their time derivatives give

$$ds_B - ds_A = \overline{AB}\, d\theta \qquad v_B - v_A = \overline{AB}\omega \qquad (a_B)_t - (a_A)_t = \overline{AB}\alpha$$
$$ds_O - ds_A = \overline{AO}\, d\theta \qquad v_O - v_A = \overline{AO}\omega \qquad a_O - (a_A)_t = \overline{AO}\alpha$$

With $v_C = r_1\omega_1 = 4(1) = 4$ in./sec, $v_D = r_2\omega_2 = 4(2) = 8$ in./sec,

$a_C = r_1\alpha_1 = 4(4) = 16$ in./sec^2 $a_D = r_2\alpha_2 = 4(-2) = -8$ in./sec^2

we have for the angular motion of the double pulley

④
$$\omega = \frac{v_B - v_A}{\overline{AB}} = \frac{v_D - v_C}{\overline{AB}} = \frac{8-4}{12} = 1/3 \text{ rad/sec (CCW)} \qquad Ans.$$

$$\alpha = \frac{(a_B)_t - (a_A)_t}{\overline{AB}} = \frac{a_D - a_C}{\overline{AB}} = \frac{-8-16}{12} = -2 \text{ rad/sec}^2 \text{ (CW)} \qquad Ans.$$

The corresponding motion of O and the load L is

$$v_O = v_A + \overline{AO}\omega = v_C + \overline{AO}\omega = 4 + 4(1/3) = 16/3 \text{ in./sec} \qquad Ans.$$
$$a_O = (a_A)_t + \overline{AO}\alpha = a_C + \overline{AO}\alpha = 16 + 4(-2) = 8 \text{ in./sec}^2 \qquad Ans.$$

① Recognize that the inner pulley is a wheel rolling along the fixed line of the left-hand cable. Thus, the expressions of Sample Problem 5/4 hold.

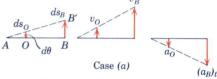

Case (a)

② Since B moves along a curved path, in addition to its tangential component of acceleration $(a_B)_t$, it will also have a normal component of acceleration toward O that does not affect the angular acceleration of the pulley.

③ The diagrams show these quantities and the simplicity of their linear relationships. The visual picture of the motion of O and B as AB rotates through the angle $d\theta$ should clarify the analysis.

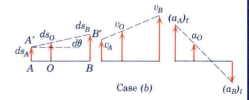

Case (b)

④ Again, as in case (a), the differential rotation of line AB as seen from the figure establishes the relation between the angular velocity of the pulley and the linear velocities of points A, O, and B. The negative sign for $(a_B)_t = a_D$ produces the acceleration diagram shown but does not destroy the linearity of the relationships.

Sample Problem 5/6

Motion of the equilateral triangular plate ABC in its plane is controlled by the hydraulic cylinder D. If the piston rod in the cylinder is moving upward at the constant rate of 0.3 m/s during an interval of its motion, calculate for the instant when $\theta = 30°$ the velocity and acceleration of the center of the roller B in the horizontal guide and the angular velocity and angular acceleration of edge CB.

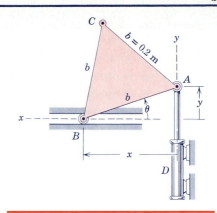

Solution. With the x-y coordinates chosen as shown, the given motion of A is $v_A = \dot{y} = 0.3$ m/s and $a_A = \ddot{y} = 0$. The accompanying motion of B is given by x and its time derivatives which may be obtained from $x^2 + y^2 = b^2$. Differentiating gives

$$x\dot{x} + y\dot{y} = 0 \qquad \dot{x} = -\frac{y}{x}\dot{y}$$

①

$$x\ddot{x} + \dot{x}^2 + y\ddot{y} + \dot{y}^2 = 0 \qquad \ddot{x} = -\frac{\dot{x}^2 + \dot{y}^2}{x} - \frac{y}{x}\ddot{y}$$

With $y = b \sin\theta$, $x = b \cos\theta$, and $\ddot{y} = 0$, the expressions become

$$v_B = \dot{x} = -v_A \tan\theta$$

$$a_B = \ddot{x} = -\frac{v_A{}^2}{b}\sec^3\theta$$

Substituting the numerical values $v_A = 0.3$ m/s and $\theta = 30°$ gives

$$v_B = -0.3\left(\frac{1}{\sqrt{3}}\right) = -0.1732 \text{ m/s} \qquad\qquad Ans.$$

$$a_B = -\frac{(0.3)^2(2/\sqrt{3})^3}{0.2} = -0.693 \text{ m/s}^2 \qquad\qquad Ans.$$

The negative signs indicate that the velocity and acceleration of B are both to the right since x and its derivatives are positive to the left.

The angular motion of CB is the same as that of every line on the plate, including AB. Differentiating $y = b \sin\theta$ gives

$$\dot{y} = b\dot{\theta}\cos\theta \qquad \omega = \dot{\theta} = \frac{v_A}{b}\sec\theta$$

The angular acceleration is

$$\alpha = \dot{\omega} = \frac{v_A}{b}\dot{\theta}\sec\theta\tan\theta = \frac{v_A{}^2}{b^2}\sec^2\theta\tan\theta$$

Substitution of the numerical values gives

$$\omega = \frac{0.3}{0.2}\frac{2}{\sqrt{3}} = 1.732 \text{ rad/s} \qquad\qquad Ans.$$

$$\alpha = \frac{(0.3)^2}{(0.2)^2}\left(\frac{2}{\sqrt{3}}\right)^2\frac{1}{\sqrt{3}} = 1.732 \text{ rad/s}^2 \qquad\qquad Ans.$$

Both ω and α are counterclockwise since their signs are positive in the sense of the positive measurement of θ.

① Observe that it is simpler to differentiate a product than a quotient. Thus, differentiate $x\dot{x} + y\dot{y} = 0$ rather than $\dot{x} = -y\dot{y}/x$.

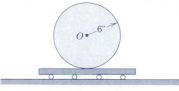

Problem 5/23

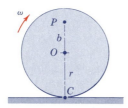

Problem 5/24

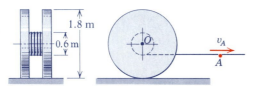

Problem 5/25

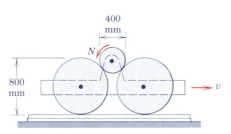

Problem 5/26

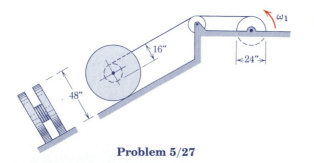

Problem 5/27

PROBLEMS

Introductory problems

5/23 The wheel rolls on the flat slab without slipping. If the center O moves 4 in. to the right while the slab moves 3 in. to the left, calculate the angular displacement $\Delta\theta$ of the wheel.

Ans. $\Delta\theta = 1.167$ rad

5/24 The wheel rolls without slipping with an angular velocity ω. By allowing line POC to rotate through a differential angle $d\theta$ during time dt, show that the velocity of point P equals its distance from the contact point C times the angular velocity of the wheel. Also express the velocity of P in terms of the velocity of the center O.

5/25 The telephone-cable reel rolls without slipping on the horizontal surface. If point A on the cable has a velocity $v_A = 0.8$ m/s to the right, compute the velocity of the center O and the angular velocity ω of the reel. (Be careful not to make the mistake of assuming that the reel rolls to the left.)

Ans. $v_O = 1.2$ m/s, $\omega = 1.333$ rad/s CW

5/26 The small vehicle rides on rails and is driven by the 400-mm-diameter friction wheel turned by an electric motor. Determine the speed v of the vehicle if the friction-drive wheel is rotating at a speed of 300 rev/min and if no slipping occurs.

5/27 The telephone-cable reel is rolled down the incline by the cable leading from the upper drum and wrapped around the inner hub of the reel. If the upper drum is turned at the constant rate $\omega_1 = 2$ rad/sec, calculate the time required for the center of the reel to move 100 ft along the incline. No slipping occurs.

Ans. $t = 66.7$ sec

5/28 The elements of a wheel-and-disk mechanical integrator are shown in the figure. The integrator wheel A turns about its fixed shaft and is driven by friction from disk B with no slipping occurring tangent to its rim. The distance y is a variable and can be controlled at will. Show that the angular displacement of the integrator wheel is given by $z = (1/b) \int y \, dx$, where x is the angular displacement of the disk B.

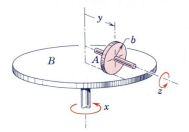

Problem 5/28

5/29 The cables at A and B are wrapped securely around the rims and the hub of the integral pulley as shown. If the cables at A and B are given upward velocities of 3 ft/sec and 4 ft/sec, respectively, calculate the velocity of the center O and the angular velocity of the pulley.

$\qquad$ *Ans.* $v_O = 3.4$ ft/sec, $\omega = 1.2$ rad/sec

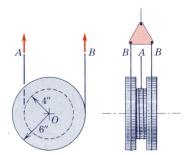

Problem 5/29

5/30 The central wheel is fixed and does not rotate. Arm AOB carries the two smaller wheels, which roll on the large wheel without slipping. If AOB is given a counterclockwise angular velocity ω_0, determine the angular velocity ω of each smaller wheel.

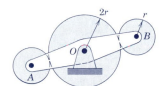

Problem 5/30

Representative problems

5/31 As end A of the slender bar is pulled to the right with a velocity v, the bar slides on the surface of the fixed half-cylinder. Determine the angular velocity $\omega = \dot{\theta}$ of the bar in terms of x.

$\qquad$ *Ans.* $\omega = -\dfrac{v}{x} \dfrac{r}{\sqrt{x^2 - r^2}}$

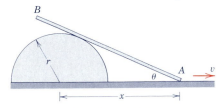

Problem 5/31

5/32 Calculate the angular velocity ω of the slender bar AB as a function of the distance x and the constant angular velocity ω_0 of the drum.

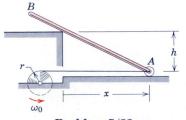

Problem 5/32

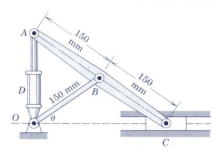

Problem 5/33

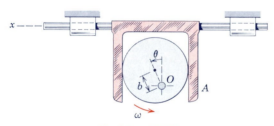

Problem 5/34

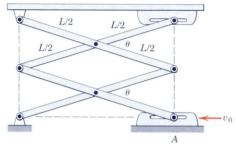

Problem 5/35

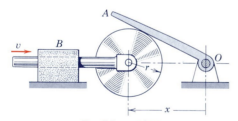

Problem 5/36

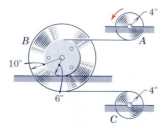

Problem 5/37

5/33 The hydraulic cylinder D is causing the distance OA to increase at the rate of 50 mm/s. Calculate the velocity of the pin at C in its horizontal guide for the instant when $\theta = 50°$.   *Ans.* $v_C = 59.6$ mm/s

5/34 Vertical motion of the work platform is controlled by the horizontal motion of pin A. If A has a velocity v_0 to the left, determine the vertical velocity v of the platform for any value of θ.

5/35 The circular cam is mounted eccentrically about its fixed bearing at O and turns counterclockwise at the constant angular velocity ω. The cam causes the fork A and attached control rod to oscillate in the horizontal x-direction. Write the expressions for the velocity v_x and acceleration a_x of the control rod in terms of the angle θ measured from the vertical. The contact surfaces of the fork are vertical.
    Ans. $v_x = e\omega \cos \theta$, $a_x = -e\omega^2 \sin \theta$

5/36 Rotation of the lever OA is controlled by the motion of the contacting circular disk whose center is given a horizontal velocity v. Determine the expression for the angular velocity ω of the lever OA in terms of x.

5/37 The cable from drum A turns the double wheel B, which rolls on its hubs without slipping. Determine the angular velocity ω and angular acceleration α of drum C for the instant when the angular velocity and angular acceleration of A are 4 rad/sec and 3 rad/sec^2, respectively, both in the counterclockwise direction.
    Ans. $\omega = \frac{4}{3}$rad/sec CCW, $\alpha = 1$ rad/sec^2 CCW

5/38 The wheel rolls without slipping, and its center O has a constant velocity v_O to the right. Choose an appropriate fixed reference system for expressing the coordinates of point A and derive equations for the velocity v_A and acceleration a_A of point A in terms of r, θ, and v_O.

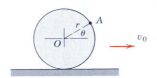

Problem 5/38

5/39 The telescoping link is hinged at O, and its end A is given a constant upward velocity of 200 mm/s by the piston rod of the fixed hydraulic cylinder B. Calculate the angular velocity $\dot{\theta}$ and the angular acceleration $\ddot{\theta}$ of link OA for the instant when $y = 600$ mm.

$Ans.$ $\dot{\theta} = 0.1639$ rad/s, $\ddot{\theta} = -0.0645$ rad/s^2

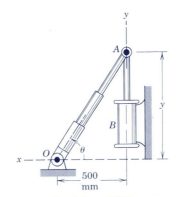

Problem 5/39

5/40 Derive an expression for the upward velocity v of the car hoist in terms of θ. The piston rod of the hydraulic cylinder is extending at the rate $\dot{s}$.

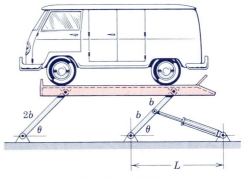

Problem 5/40

5/41 Film passes through the guide rollers shown and is being wound onto the reel, which is turned at a constant angular velocity ω. Determine the acceleration $a = \dot{v}$ of the film as it enters the rollers. The thickness of the film is t, and s is sufficiently large so that the change in the angle made by the film with the horizontal is negligible.

$Ans.$ $a = \dfrac{tv^2}{2\pi r^2}$

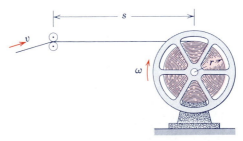

Problem 5/41

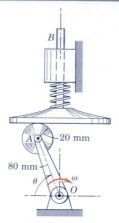

Problem 5/42

5/42 Determine the acceleration of the shaft B for $\theta = 60°$ if the crank OA has an angular acceleration $\ddot{\theta} = 8$ rad/s² and an angular velocity $\dot{\theta} = 4$ rad/s at this position. The spring maintains contact between the roller and the surface of the plunger.

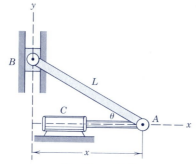

Problem 5/43

5/43 The hydraulic cylinder C gives end A of link AB a constant velocity v_0 in the negative x-direction. Determine expressions for the angular velocity $\omega = \dot{\theta}$ and angular acceleration $\alpha = \ddot{\theta}$ of the link in terms of x.

$$Ans. \quad \omega = \frac{v_0}{\sqrt{L^2 - x^2}}, \quad \alpha = \frac{-xv_0{}^2}{(L^2 - x^2)^{3/2}}$$

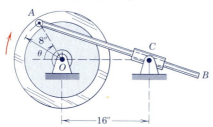

Problem 5/44

5/44 The flywheel turns clockwise with a constant speed of 600 rev/min. The connecting link AB slides through the pivoted collar at C. Calculate the angular velocity ω of AB for the instant when $\theta = 60°$.

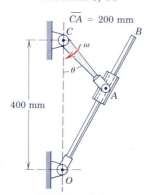

Problem 5/45

5/45 The rod OB slides through the collar pivoted to the rotating link at A. If OA has an angular velocity $\omega = 3$ rad/s for an interval of motion, calculate the angular velocity of OB when $\theta = 45°$.

$$Ans. \quad \omega_{OB} = 0.572 \text{ rad/s CCW}$$

5/46 Activation of the hydraulic cylinder causes *OB* to elongate at the constant rate of 0.260 m/s. Calculate the normal acceleration of point *A* in its circular path around *C* for the instant when $\theta = 60°$.

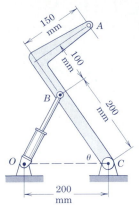

Problem 5/46

5/47 The rotation of link *AO* is controlled by the piston rod of hydraulic cylinder *BC*, which is elongating at the constant rate $\dot{s} = k$ for an interval of motion. Write the vector expression for the acceleration of end *A* for a given value of θ using unit vectors $\mathbf{e}_n$ and $\mathbf{e}_t$ with n–t coordinates.

$$Ans. \quad \mathbf{a}_A = \frac{k^2 l}{b^2 \cos^2 \frac{\theta}{2}} \left(\mathbf{e}_n + \frac{1}{2} \tan \frac{\theta}{2} \mathbf{e}_t \right)$$

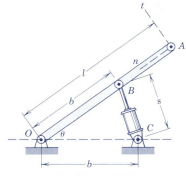

Problem 5/47

5/48 A variable-speed belt drive consists of the two pulleys each of which is constructed of two cones which turn as a unit but are capable of being drawn together or separated so as to change the effective radius of the pulley. If the angular velocity ω_1 of pulley 1 is constant, determine the expression for the angular acceleration $\alpha_2 = \dot{\omega}_2$ of pulley 2 in terms of the rates of change $\dot{r}_1$ and $\dot{r}_2$ of the effective radii.

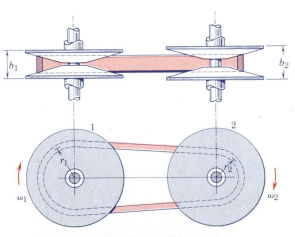

Problem 5/48

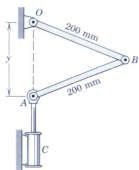

Problem 5/49

5/49 For the instant when $y = 200$ mm, the piston rod of the hydraulic cylinder C imparts a vertical motion to the pin A of $\dot{y} = 400$ mm/s and $\ddot{y} = -100$ mm/s^2. For this instant determine the angular velocity ω and the angular acceleration α of link AB.

Ans. $\omega = 1.155$ rad/s CCW
$\alpha = 0.481$ rad/s^2 CCW

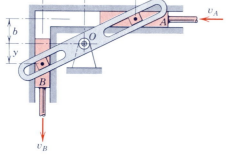

Problem 5/50

5/50 The slotted arm pivots about O and maintains the relation between the motions of sliders A and B and their control rods. Each small pivoted block is pinned to its respective slider and is constrained to slide in its rotating slot. Show that the displacement x is proportional to the reciprocal of y. Then establish the relation between the velocities v_A and v_B. Also, if v_A is constant for a short interval of motion, determine the acceleration of B.

5/51 For the conditions of Prob. 5/46, determine the tangential component of the acceleration of point A in its circular path around C for the instant when $\theta = 60°$. *Ans.* $(a_A)_t = 0.218$ m/s^2

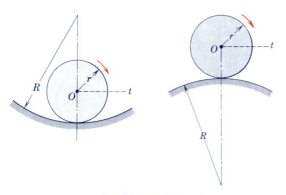

Problem 5/52

5/52 Show that the expressions $v = r\omega$ and $a_t = r\alpha$ hold for the motion of the center O of the wheel which rolls on the concave or convex circular arc, where ω and α are the absolute angular velocity and acceleration, respectively, of the wheel. (*Hint:* Follow the example of Sample Problem 5/4 and allow the wheel to roll a small distance. Be very careful to identify the correct *absolute* angle through which the wheel turns in each case in determining its angular velocity and angular acceleration.)

5/53 The Geneva wheel is a mechanism for producing intermittent rotation. Pin P in the integral unit of wheel A and locking plate B engages the radial slots in wheel C thus turning wheel C one-fourth of a revolution for each revolution of the pin. At the engagement position shown, $\theta = 45°$. For a constant clockwise angular velocity $\omega_1 = 2$ rad/s of wheel A, determine the corresponding counterclockwise angular velocity ω_2 of wheel C for $\theta = 20°$. (Note that the motion during engagement is governed by the geometry of triangle $O_1 O_2 P$ with changing θ.)

Ans. $\omega_2 = 1.923$ rad/s

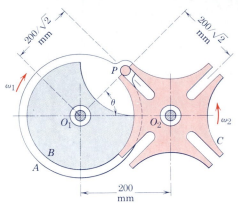

Problem 5/53

5/54 The rod AB slides through the pivoted collar as end A moves along the slot. If A starts from rest at $x = 0$ and moves to the right with a constant acceleration of 4 in./sec², calculate the angular acceleration α of AB at the instant when $x = 6$ in.

Ans. $\alpha = 0.1408$ rad/sec² CCW

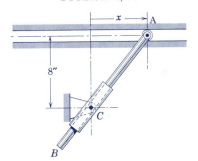

Problem 5/54

5/55 The punch is operated by a simple harmonic oscillation of the pivoted sector given by $\theta = \theta_0 \sin 2\pi t$ where the amplitude is $\theta_0 = \pi/12$ rad (15°) and the time for one complete oscillation is 1 second. Determine the acceleration of the punch when (*a*) $\theta = 0$ and (*b*) $\theta = \pi/12$.

Ans. (*a*) $a = 0.909$ m/s² up
(*b*) $a = 0.918$ m/s² down

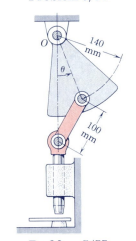

Problem 5/55

5/56 One of the most common mechanisms is the slider-crank. Express the angular velocity ω_{AB} and angular acceleration α_{AB} of the connecting rod AB in terms of the crank angle θ for a given constant crank speed ω_0. Take ω_{AB} and α_{AB} to be positive counterclockwise.

Ans. $\omega_{AB} = \dfrac{r\omega_0}{l} \dfrac{\cos \theta}{\sqrt{1 - \dfrac{r^2}{l^2} \sin^2 \theta}}$

$\alpha_{AB} = \dfrac{r\omega_0{}^2}{l} \sin \theta \dfrac{\dfrac{r^2}{l^2} - 1}{\left(1 - \dfrac{r^2}{l^2} \sin^2 \theta\right)^{3/2}}$

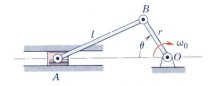

Problem 5/56

5/4 *RELATIVE VELOCITY*

The second approach to rigid-body kinematics is to use the principles of relative motion. In Art. 2/8 we developed these principles for measurements relative to translating axes and applied the relative velocity equation

$$\mathbf{v}_A = \mathbf{v}_B + \mathbf{v}_{A/B} \qquad (2/20)$$

to the motions of two particles A and B.

We now choose two points on the *same* rigid body for our two particles. The consequence of this choice is that the motion of one point as seen by an observer translating with the other point must be circular since the radial distance to the observed point from the reference point does not change. This observation is the *key* to the successful understanding of a large majority of problems in the plane motion of rigid bodies. This concept is illustrated in Fig. 5/5*a*, which shows a rigid body moving in the plane of the figure from position AB to $A'B'$ during time Δt. This movement may be visualized as occurring in two parts. First, the body translates to the parallel position $A''B'$ with the displacement $\Delta\mathbf{r}_B$. Second, the body rotates about B' through the angle $\Delta\theta$. From the nonrotating reference axes x'-y' attached to the reference point B', it is seen that this remaining motion of the body is one of simple rotation about B', giving rise to the displacement $\Delta\mathbf{r}_{A/B}$ of A with respect to B. To the nonrotating observer attached to B, the body appears to undergo fixed-axis rotation about B with A executing circular motion as emphasized in Fig. 5/5*b*. Therefore, the relationships developed for circular motion in Arts. 2/5 and 5/2 and cited as Eqs. 2/11 and 5/2 (or 5/3) describe the relative portion of A's motion.

Point B was arbitrarily chosen as the reference point for attachment of our nonrotating reference axes x-y. Point A could have been used just as well, in which case we observe B to have circular motion about A considered fixed as shown in Fig. 5/5*c*. We see that the sense of the rotation, counterclockwise in this example, is the

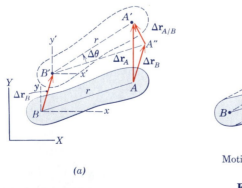

(a)

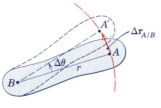

Motion relative to B
(b)

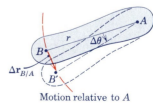

Motion relative to A
(c)

Figure 5/5

same whether we choose A or B as the reference, and we see that $\Delta \mathbf{r}_{B/A} = -\Delta \mathbf{r}_{A/B}$.

With B as the reference point, the total displacement of A is seen from Fig. 5/5a to be

$$\Delta \mathbf{r}_A = \Delta \mathbf{r}_B + \Delta \mathbf{r}_{A/B}$$

where $\Delta \mathbf{r}_{A/B}$ has the magnitude $r \Delta \theta$ as $\Delta \theta$ approaches zero. We note that the *relative linear motion* $\Delta \mathbf{r}_{A/B}$ is accompanied by the *absolute angular motion* $\Delta \theta$, as seen from the translating axes x'-y'. Dividing the expression for $\Delta \mathbf{r}_A$ by the corresponding time interval Δt and passing to the limit give the relative-velocity equation

$$\boxed{\mathbf{v}_A = \mathbf{v}_B + \mathbf{v}_{A/B}} \tag{5/4}$$

This expression is the same as Eq. 2/20, with the one restriction that the distance r between A and B remains constant. The magnitude of the relative velocity is thus seen to be $v_{A/B} = \lim_{\Delta t \to 0} (|\Delta \mathbf{r}_{A/B}|/\Delta t) = \lim_{\Delta t \to 0} (r \Delta \theta / \Delta t)$ that, with $\omega = \dot{\theta}$, becomes

$$\boxed{v_{A/B} = r\omega} \tag{5/5}$$

With $\mathbf{r}$ representing the vector $\mathbf{r}_{A/B}$, from the first of Eqs. 5/3 we may write the relative velocity as the vector

$$\boxed{\mathbf{v}_{A/B} = \boldsymbol{\omega} \times \mathbf{r}} \tag{5/6}$$

where $\boldsymbol{\omega}$ is the angular-velocity vector normal to the plane of the motion in the sense determined by the right-hand rule. A critical observation seen from Figs. 5/5b and c is that the relative linear velocity is always perpendicular to the line joining the two points in question.

The application of Eq. 5/4 is clarified by visualizing the separate translation and rotation components of the equation. These components are emphasized in Fig. 5/6, which shows a rigid body

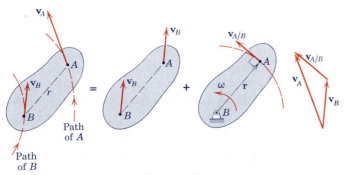

Figure 5/6

in plane motion. With B chosen as the reference point, the velocity of A is the vector sum of the translational portion $\mathbf{v}_B$, plus the rotational portion $\mathbf{v}_{A/B} = \boldsymbol{\omega} \times \mathbf{r}$ that has the magnitude $v_{A/B} = r\omega$, where $|\boldsymbol{\omega}| = \dot{\theta}$, the *absolute* angular velocity of AB. The fact that the *relative linear velocity* is *always perpendicular* to the line joining the two points in question is an important key to the solution of many problems. The student should draw the equivalent diagram where point A is used as the reference point rather than B.

A second use of Eq. 5/4 for relative-velocity problems in plane motion may be made for constrained sliding contact between two links in a mechanism. In this case, we choose points A and B as coincident points, one on each link, for the instant under consideration. In contrast to the previous example, in this case, the two points are on different bodies so they are not a fixed distance apart. This second use of the relative-velocity equation is illustrated in Sample Problem 5/10.

Solution of the relative-velocity equation may be carried out by scalar or vector algebra, or a graphical analysis may be employed. In any event, a sketch of the vector polygon that represents the vector equation should be made to reveal the physical relationships involved. From this sketch, scalar component equations may be written by projecting the vectors along convenient directions. Usually, a simultaneous solution may be avoided by a careful choice of the projections. Alternatively, each term in the relative-motion equation may be written in terms of its $\mathbf{i}$- and $\mathbf{j}$-components from which two scalar equations result when the equality is applied, separately, to the coefficients of the $\mathbf{i}$- and $\mathbf{j}$-terms. Many problems lend themselves to a graphical solution, particularly when the given geometry results in an awkward mathematical expression. In this case, the known vectors are first constructed in their correct positions using a convenient scale. Next, the unknown vectors, which complete the polygon and satisfy the vector equation, are measured directly from the drawing. The choice of method to be used depends on the particular problem at hand, the accuracy required, and individual preference and experience. All three of the approaches are illustrated in the sample problems that follow.

Regardless of which method of solution we employ, we note that the single vector equation in two dimensions is equivalent to two scalar equations, so that two scalar unknowns at the most can be determined. The unknowns, for instance, might be the magnitude of one vector and the direction of another. A systematic identification of the knowns and unknowns should be made before a solution is attempted.

Sample Problem 5/7

The wheel of radius $r = 300$ mm rolls to the right without slipping and has a velocity $v_O = 3$ m/s of its center O. Calculate the velocity of point A on the wheel for the instant represented.

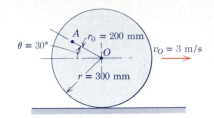

Solution I (scalar-geometric). The center O is chosen as the reference point for the relative-velocity equation since its motion is given. We therefore write

$$\mathbf{v}_A = \mathbf{v}_O + \mathbf{v}_{A/O}$$

where the relative-velocity term is observed from the translating axes x-y attached to O. The angular velocity of AO is the same as that of the wheel which, from Sample Problem 5/4, is $\omega = v_O/r = 3/0.3 = 10$ rad/s. Thus, from Eq. 5/5 we have

$$[v_{A/O} = r_0\dot{\theta}] \qquad v_{A/O} = 0.2(10) = 2 \text{ m/s}$$

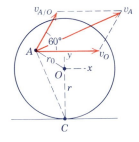

① which is normal to AO as shown. The vector sum $\mathbf{v}_A$ is shown on the diagram and may be calculated from the law of cosines. Thus,

② $$v_A{}^2 = 3^2 + 2^2 + 2(3)(2)\cos 60° = 19 \text{ (m/s)}^2 \quad v_A = 4.36 \text{ m/s} \quad Ans.$$

The contact point C momentarily has zero velocity and can be used alternatively as the reference point, in which case the relative-velocity equation becomes $\mathbf{v}_A = \mathbf{v}_C + \mathbf{v}_{A/C} = \mathbf{v}_{A/C}$ where

$$v_{A/C} = \overline{AC}\omega = \frac{\overline{AC}}{\overline{OC}}v_O = \frac{0.436}{0.300}(3) = 4.36 \text{ m/s} \quad v_A = v_{A/C} = 4.36 \text{ m/s}$$

The distance $\overline{AC} = 436$ mm is calculated separately. We see that $\mathbf{v}_A$ is
③ normal to AC since A is momentarily rotating about point C.

Solution II (vector). We will now use Eq. 5/6 and write

$$\mathbf{v}_A = \mathbf{v}_O + \mathbf{v}_{A/O} = \mathbf{v}_O + \boldsymbol{\omega} \times \mathbf{r}_0$$

where

④ $$\boldsymbol{\omega} = -10\mathbf{k} \text{ rad/s}$$

$$\mathbf{r}_0 = 0.2(-\mathbf{i}\cos 30° + \mathbf{j}\sin 30°) = -0.173\mathbf{i} + 0.1\mathbf{j} \text{ m}$$

$$\mathbf{v}_O = 3\mathbf{i} \text{ m/s}$$

We now solve the vector equation

$$\mathbf{v}_A = 3\mathbf{i} + \begin{vmatrix} \mathbf{i} & \mathbf{j} & \mathbf{k} \\ 0 & 0 & -10 \\ -0.173 & 0.1 & 0 \end{vmatrix} = 3\mathbf{i} + 1.73\mathbf{j} + 1.0\mathbf{i}$$

$$= 4\mathbf{i} + 1.73\mathbf{j} \text{ m/s} \qquad Ans.$$

The magnitude $v_A = \sqrt{4^2 + (1.73)^2} = \sqrt{19} = 4.36$ m/s and direction agree with the previous solution.

Helpful Hints

① Be sure to visualize $v_{A/O}$ as the velocity that A appears to have in its circular motion relative to O.

② The vectors may also be laid off to scale graphically and the magnitude and direction of v_A measured directly from the diagram.

③ The velocity of any point on the wheel is easily determined by using the contact point C as the reference point. The student should construct the velocity vectors for a number of points on the wheel for practice.

④ The vector $\boldsymbol{\omega}$ is directed into the paper by the right-hand rule, whereas the positive z-direction is out from the paper; hence, the minus sign.

350

Sample Problem 5/8

Crank CB oscillates about C through a limited arc, causing crank OA to oscillate about O. When the linkage passes the position shown with CB horizontal and OA vertical, the angular velocity of CB is 2 rad/s counterclockwise. For this instant, determine the angular velocities of OA and AB.

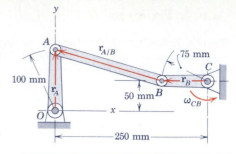

Solution I (vector). The relative-velocity equation $\mathbf{v}_A = \mathbf{v}_B + \mathbf{v}_{A/B}$ is rewritten as

$$\boldsymbol{\omega}_{OA} \times \mathbf{r}_A = \boldsymbol{\omega}_{CB} \times \mathbf{r}_B + \boldsymbol{\omega}_{AB} \times \mathbf{r}_{A/B}$$

where $\boldsymbol{\omega}_{OA} = \omega_{OA}\mathbf{k}$, $\boldsymbol{\omega}_{CB} = 2\mathbf{k}$ rad/s, $\boldsymbol{\omega}_{AB} = \omega_{AB}\mathbf{k}$,

$$\mathbf{r}_A = 100\mathbf{j} \text{ mm} \qquad \mathbf{r}_B = -75\mathbf{i} \text{ mm} \qquad \mathbf{r}_{A/B} = -175\mathbf{i} + 50\mathbf{j} \text{ mm}$$

Substitution gives

$$\omega_{OA}\mathbf{k} \times 100\mathbf{j} = 2\mathbf{k} \times (-75\mathbf{i}) + \omega_{AB}\mathbf{k} \times (-175\mathbf{i} + 50\mathbf{j})$$

$$-100\omega_{OA}\mathbf{i} = -150\mathbf{j} - 175\omega_{AB}\mathbf{j} - 50\omega_{AB}\mathbf{i}$$

Matching coefficients of the respective $\mathbf{i}$- and $\mathbf{j}$-terms gives

$$-100\omega_{OA} + 50\omega_{AB} = 0 \qquad 25(6 + 7\omega_{AB}) = 0$$

the solutions of which are

$$\omega_{AB} = -6/7 \text{ rad/s} \qquad \text{and} \qquad \omega_{OA} = -3/7 \text{ rad/s} \qquad \textit{Ans.}$$

Solution II (scalar-geometric). Solution by the scalar geometry of the vector triangle is particularly simple here since $\mathbf{v}_A$ and $\mathbf{v}_B$ are at right angles for this special position of the linkages. First, we compute v_B, which is

$$[v = r\omega] \qquad v_B = 0.075(2) = 0.150 \text{ m/s}$$

and represent it in its correct direction as shown. The vector $\mathbf{v}_{A/B}$ must be perpendicular to AB, and the angle θ between $\mathbf{v}_{A/B}$ and $\mathbf{v}_B$ is also the angle made by AB with the horizontal direction. This angle is given by

$$\tan \theta = \frac{100 - 50}{250 - 75} = \frac{2}{7}$$

The horizontal vector $\mathbf{v}_A$ completes the triangle for which we have

$$v_{A/B} = v_B/\cos \theta = 0.150/\cos \theta$$

$$v_A = v_B \tan \theta = 0.150(2/7) = 0.30/7 \text{ m/s}$$

The angular velocities become

$$[\omega = v/r] \qquad \omega_{AB} = v_{A/B}/\overline{AB} = \frac{0.150}{\cos \theta} \frac{\cos \theta}{0.250 - 0.075}$$

$$= 6/7 \text{ rad/s CW} \qquad \textit{Ans.}$$

$$[\omega = v/r] \qquad \omega_{OA} = v_A/\overline{OA} = \frac{0.30}{7} \frac{1}{0.100} = 3/7 \text{ rad/s CW} \qquad \textit{Ans.}$$

① We are using here the first of Eqs. 5/3 and Eq 5/6.

② The minus signs in the answers indicate that the vectors $\boldsymbol{\omega}_{AB}$ and $\boldsymbol{\omega}_{OA}$ are in the negative $\mathbf{k}$-direction. Hence, the angular velocities are clockwise.

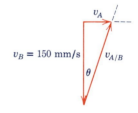

③ Always make certain that the sequence of vectors in the vector polygon agrees with the equality of vectors specified by the vector equation.

Sample Problem 5/9

The common configuration of a reciprocating engine is that of the slider-crank mechanism shown. If the crank OB has a clockwise rotational speed of 1500 rev/min, determine for the position where $\theta = 60°$ the velocity of the piston A, the velocity of point G on the connecting rod, and the angular velocity of the connecting rod.

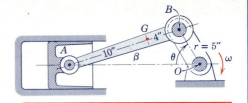

Solution. The velocity of the crank pin B as a point on AB is easily found, so that B will be used as the reference point for determining the velocity of A. The relative-velocity equation may now be written

$$\mathbf{v}_A = \mathbf{v}_B + \mathbf{v}_{A/B}$$

The crank-pin velocity is

① $[v = r\omega]$ $\qquad v_B = \dfrac{5}{12} \dfrac{1500(2\pi)}{60} = 65.4 \text{ ft/sec}$

① Remember always to convert ω to radians per unit time when using $v = r\omega$.

and is normal to OB. The direction of $\mathbf{v}_A$ is, of course, along the horizontal cylinder axis. The direction of $\mathbf{v}_{A/B}$ must be perpendicular to the line AB as explained in the present article and as indicated on the diagram, where the reference point B is shown as fixed. We obtain this direction by computing angle β from the law of sines, which gives

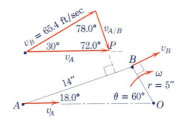

$$\frac{5}{\sin \beta} = \frac{14}{\sin 60°} \qquad \beta = \sin^{-1} 0.3093 = 18.0°$$

We now complete the sketch of the velocity triangle where the angle between $\mathbf{v}_{A/B}$ and $\mathbf{v}_A$ is $90° - 18.0° = 72.0°$ and the third angle is $180° - 30° - 72.0° = 78.0°$. Vectors $\mathbf{v}_A$ and $\mathbf{v}_{A/B}$ are shown with their proper sense such that the head-to-tail sum of $\mathbf{v}_B$ and $\mathbf{v}_{A/B}$ equals $\mathbf{v}_A$. The magnitudes of the unknowns are now calculated from the trigonometry of the vector triangle or are scaled from the diagram if a graphical solution is used. Solving for $\mathbf{v}_A$ and $\mathbf{v}_{A/B}$ by the law of sines gives

② $\qquad \dfrac{v_A}{\sin 78.0°} = \dfrac{65.4}{\sin 72.0°} \qquad v_A = 67.3 \text{ ft/sec}$ $\qquad$ *Ans.*

② A graphical solution to this problem is the quickest to achieve, although its accuracy is limited. Solution by vector algebra can, of course, be used but would involve somewhat more labor in this problem.

$$\frac{v_{A/B}}{\sin 30°} = \frac{65.4}{\sin 72.0°} \qquad v_{A/B} = 34.4 \text{ ft/sec}$$

The angular velocity of AB is counterclockwise, as revealed by the sense of $\mathbf{v}_{A/B}$, and is

$[\omega = v/r]$ $\qquad \omega_{AB} = \dfrac{v_{A/B}}{AB} = \dfrac{34.4}{14/12} = 29.5 \text{ rad/sec}$ $\qquad$ *Ans.*

We now determine the velocity of G by writing

$$\mathbf{v}_G = \mathbf{v}_B + \mathbf{v}_{G/B}$$

where $\quad v_{G/B} = \overline{GB}\,\omega_{AB} = \dfrac{\overline{GB}}{\overline{AB}}\,v_{A/B} = \dfrac{4}{14}(34.4) = 9.83 \text{ ft/sec.}$

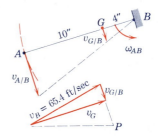

As seen from the diagram, $\mathbf{v}_{G/B}$ has the same direction as $\mathbf{v}_{A/B}$. The vector sum is shown on the last diagram. We can calculate v_G with some geometric labor or simply measure its magnitude and direction from the velocity diagram drawn to scale. For simplicity we adopt the latter procedure here and obtain

$$v_G = 64.1 \text{ ft/sec} \qquad \textit{Ans.}$$

As seen, the diagram may be superposed directly on the first velocity diagram.

Sample Problem   5/10

The power screw turns at a speed that gives the threaded collar C a velocity of 0.8 ft/sec vertically down. Determine the angular velocity of the slotted arm when $\theta = 30°$.

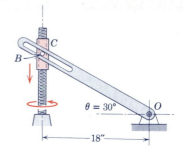

Solution.   The angular velocity of the arm can be found if the velocity of a point on the arm is known. We choose a point A on the arm coincident with the pin B of the collar for this purpose. If we use B as our reference point and write $\mathbf{v}_A = \mathbf{v}_B + \mathbf{v}_{A/B}$, we see from the diagram, which shows the arm and points A and B an instant before and an instant after coincidence, that $\mathbf{v}_{A/B}$ has a direction along the slot away from O.

① Physically, of course, this point does not exist, but we can imagine such a point in the middle of the slot and attached to the arm.

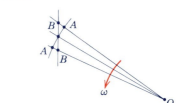

The magnitudes of $\mathbf{v}_A$ and $\mathbf{v}_{A/B}$ are the only unknowns in the vector equation, so that it may now be solved. We draw the known vector $\mathbf{v}_B$ and then obtain the intersection P of the known directions of $\mathbf{v}_{A/B}$ and $\mathbf{v}_A$. The solution gives

$$v_A = v_B \cos\theta = 0.8 \cos 30° = 0.693 \text{ ft/sec}$$

$$[\omega = v/r] \qquad \omega = v_A/\overline{OA} = \frac{0.693}{(\frac{18}{12})/\cos 30°}$$

$$= 0.400 \text{ rad/sec CCW} \qquad \textit{Ans.}$$

② Always identify the knowns and unknowns before attempting the solution of a vector equation.

We note the difference between this problem of constrained sliding contact between two links and the three preceding sample problems of relative velocity, where no sliding contact occurred and where the points A and B were located on the same rigid body in each case.

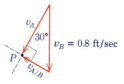

PROBLEMS

Introductory problems

5/57 The cart has a velocity of 4 ft/sec to the right. Determine the angular speed N of the wheel so that point A on the top of the rim has a velocity (a) equal to 4 ft/sec to the left, (b) equal to zero, and (c) equal to 8 ft/sec to the right.

$$\text{Ans. } (a) \ N \ = \ 91.7 \text{ rev/min CCW}$$
$$(b) \ N \ = \ 45.8 \text{ rev/min CCW}$$
$$(c) \ N \ = \ 45.8 \text{ rev/min CW}$$

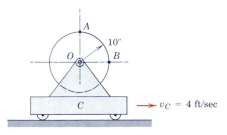

Problem 5/57

5/58 The magnitude of the absolute velocity of point A on the automobile tire is 12 m/s when A is in the position shown. What are the corresponding velocity v_O of the car and the angular velocity ω of the wheel? (The wheel rolls without slipping.)

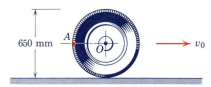

Problem 5/58

5/59 A control element in a special-purpose mechanism undergoes motion in the plane of the figure. If the velocity of B with respect to A has a magnitude of 0.926 m/s at a certain instant, what is the corresponding magnitude of the velocity of C with respect to D? *Ans.* $v_{C/D} = 0.579$ m/s

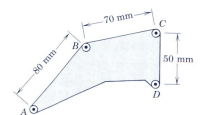

Problem 5/59

5/60 If v_A and v_B are the instantaneous velocities of the respective ends of the rigid link, show that the angular velocity of the link is given by $(v_A \sin \theta + v_B \sin \beta)/L$ and that $v_A \cos \theta = v_B \cos \beta$.

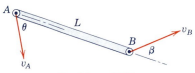

Problem 5/60

5/61 For the instant represented, end B of the 0.5-m link has a velocity of 3 m/s in the direction shown. Determine the minimum possible velocity of end A and the corresponding angular velocity of the link.
Ans. $(v_A)_{min} = 1.5$ m/s, $\omega_{AB} = 5.20$ rad/s CCW

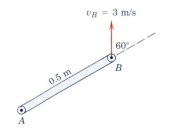

Problem 5/61

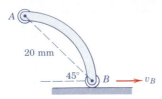

Problem 5/62

5/62 For the instant represented the curved link has a counterclockwise angular velocity of 4 rad/s, and the roller at B has a velocity of 40 mm/s along the constraining surface as shown. Determine the magnitude v_A of the velocity of A.

5/63 If point B on the wheel of Prob. 5/57 has a velocity whose magnitude is 5 ft/sec, determine the corresponding angular speed N of the wheel. What can be said about the direction of rotation?

        Ans. $N = 34.4$ rev/min CW or CCW

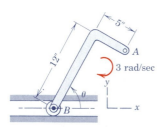

Problem 5/64

5/64 The right-angled link AB has a clockwise angular velocity of 3 rad/sec at the instant when $\theta = 60°$. Express the velocity of A with respect to B in vector notation for this instant.

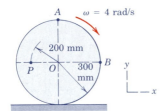

Problem 5/65

5/65 The circular disk rolls without slipping with a clockwise angular velocity $\omega = 4$ rad/s. For the instant represented, write the vector expressions for the velocity of A with respect to B and for the velocity of P.   *Ans.* $\mathbf{v}_{A/B} = 1.2(\mathbf{i} + \mathbf{j})$ m/s

                                                $\mathbf{v}_P = 4(0.3\mathbf{i} + 0.2\mathbf{j})$ m/s

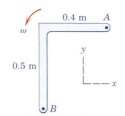

Problem 5/66

5/66 The right-angled link has a counterclockwise angular velocity of 3 rad/s at the instant represented, and point B has a velocity $\mathbf{v}_B = 2\mathbf{i} - 0.3\mathbf{j}$ m/s. Determine the velocity of A using vector notation. Sketch the vector polygon that corresponds to the terms in the relative-velocity equation and estimate or measure the magnitude of $\mathbf{v}_A$.

Representative problems

5/67 At the instant represented the triangular plate ABD has a clockwise angular velocity of 3 rad/sec. For this instant determine the angular velocity ω_{BC} of link BC.   *Ans.* $\omega_{BC} = 3$ rad/sec CW

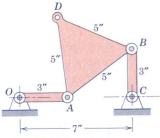

Problem 5/67

5/68 Determine the angular velocity of the telescoping link *AB* for the position shown where the driving links have the angular velocities indicated.

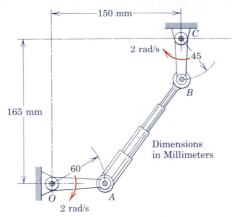

Problem 5/68

5/69 The rider of the bicycle shown pumps steadily to maintain a constant speed of 16 km/h against a slight head wind. Calculate the maximum and minimum magnitudes of the absolute velocity of the pedal *A*.   *Ans.* $(v_A)_{max} = 5.33$ m/s
$(v_A)_{min} = 3.56$ m/s

Problem 5/69

5/70 For an interval of its motion the piston rod of the hydraulic cylinder has a velocity $v_A = 4$ ft/sec as shown. At a certain instant $\theta = \beta = 60°$. For this instant determine the angular velocity ω_{BC} of link *BC*.

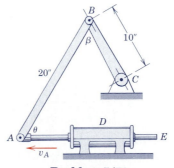

Problem 5/70

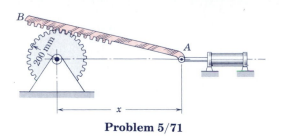

Problem 5/71

5/71 The rotation of the gear is controlled by the horizontal motion of end A of the rack AB. If the piston rod has a constant velocity $\dot{x} = 300$ mm/s during a short interval of motion, determine the angular velocity ω_0 of the gear and the angular velocity ω_{AB} of AB at the instant when $x = 800$ mm.

Ans. $\omega_{AB} = 0.0968$ rad/s CCW
$\omega_0 = 1.452$ rad/s CW

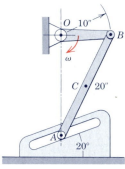

Problem 5/72

5/72 For the instant represented, crank OB has a clockwise angular velocity $\omega = 0.8$ rad/sec and is passing the horizontal position. Determine the corresponding velocity of the guide roller A in the 20° slot and the velocity of point C midway between A and B.

5/73 Motion of the triangular plate ABD is controlled by links O_1A and O_2B. Determine the angular velocity ω of the plate and the magnitude of the velocity of point D. The drawing is represented to scale so that lengths and angles may be measured from the drawing to within available accuracy.

Ans. $\omega \cong 1.04$ rad/sec CCW, $v_D \cong 12.7$ in./sec

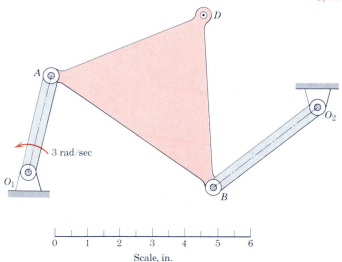

Problem 5/73

5/74 If link *OA* has a clockwise angular velocity of 2 rad/s in the position for which $x = 75$ mm, determine the velocity of the slider at *B*.

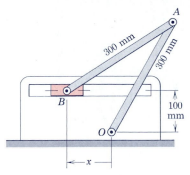

Problem 5/74

5/75 Rod *CB* slides through the pivoted collar attached to link *OA*. If *CB* has a clockwise angular velocity of 2 rad/s, determine the angular velocity ω_{OA} of link *OA* when $\theta = 60°$.   *Ans.* $\omega_{OA} = 4$ rad/s CW

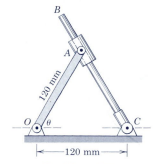

Problem 5/75

5/76 The mechanism of Prob. 5/44 is repeated here. The flywheel turns clockwise with a constant speed of 600 rev/min, and the connecting rod *AB* slides through the pivoted collar at *C*. For the position of $\theta = 45°$, determine the angular velocity ω of *AB* by using the relative velocity relations. (*Suggestion:* Choose a point *D* on *AB* coincident with *C* as a reference point whose direction of velocity is known.)

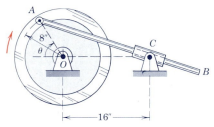

Problem 5/76

5/77 Determine the velocity of point *D* that will produce a counterclockwise angular velocity of 40 rad/s for link *AB* in the position shown for the four-bar linkage.   *Ans.* $v_D = 9$ m/s

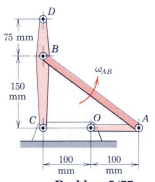

Problem 5/77

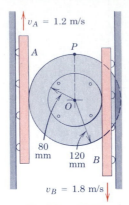

Problem 5/78

5/78 The sliding rails A and B engage the rims of the double wheel without slipping. For the specified velocities of A and B, determine the angular velocity ω of the wheel and the magnitude of the velocity of point P.

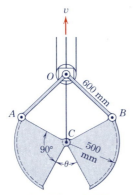

Problem 5/79

5/79 The elements of a simplified clam-shell bucket for a dredge are shown. The cable which opens and closes the bucket passes through the block at O. With O as a fixed point, determine the angular velocity ω of the bucket jaws when $\theta = 45°$ as they are closing. The upward velocity of the control cable is 0.5 m/s as it passes through the block.

Ans. $\omega = 0.722$ rad/s

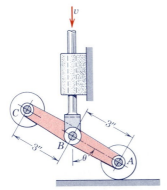

Problem 5/80

5/80 The elements of a switching device are shown. If the vertical control rod has a downward velocity v of 3 ft/sec when $\theta = 60°$ and if roller A is in continuous contact with the horizontal surface, determine the magnitude of the velocity of C for this instant.

5/81 Motion of the rectangular plate P is controlled by the two links which cross without touching. For the instant represented where the links are perpendicular to each other, the plate has a counterclockwise angular velocity $\omega_P = 2$ rad/s. Determine the corresponding angular velocities of the two links.

$Ans.\ \boldsymbol{\omega}_{AO} = 1.333\mathbf{k}$ rad/s, $\boldsymbol{\omega}_{BD} = 1.20\mathbf{k}$ rad/s

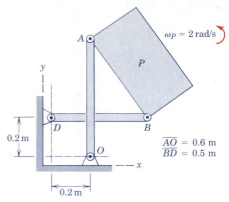

Problem 5/81

5/82 The mechanism is part of a latching device where rotation of link AOB is controlled by the rotation of slotted link D about C. If member D has a clockwise angular velocity of 1.5 rad/s when the slot is parallel to OC, determine the corresponding angular velocity of AOB. Solve graphically or geometrically.

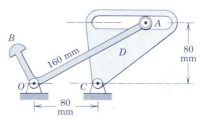

Problem 5/82

5/83 The elements of the mechanism for deployment of a spacecraft magnetometer boom are shown. Determine the angular velocity of the boom when the driving link OB crosses the y-axis with an angular velocity $\omega_{OB} = 0.5$ rad/s if at this instant $\tan\theta = 4/3$. $Ans.\ \boldsymbol{\omega}_{CA} = 0.429\mathbf{k}$ rad/s

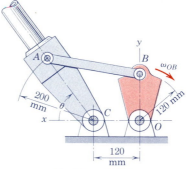

Problem 5/83

5/84 In the four-bar linkage shown control link OA has a counterclockwise angular velocity $\omega_0 = 10$ rad/s during a short interval of motion. When link CB passes the vertical position shown, point A has coordinates $x = -60$ mm and $y = 80$ mm. By means of vector algebra determine the angular velocity of AB and BC.

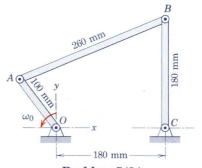

Problem 5/84

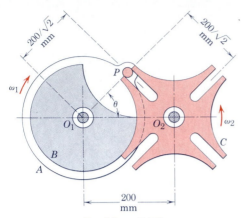

Problem 5/85

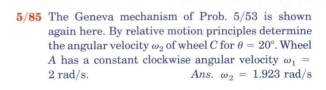

5/85 The Geneva mechanism of Prob. 5/53 is shown again here. By relative motion principles determine the angular velocity ω_2 of wheel C for $\theta = 20°$. Wheel A has a constant clockwise angular velocity $\omega_1 = 2$ rad/s.   *Ans.* $\omega_2 = 1.923$ rad/s

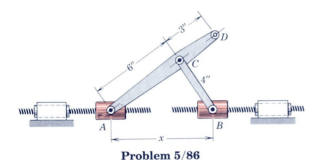

Problem 5/86

5/86 Motion of the threaded collars A and B is controlled by the rotation of their respective lead screws. If A has a velocity to the right of 3 in./sec and B has a velocity to the left of 2 in./sec when $x = 6$ in., determine the angular velocity ω of ACD at this instant.

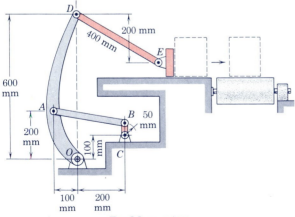

Problem 5/87

5/87 A mechanism for pushing small boxes from an assembly line onto a conveyor belt is shown with arm OD and crank CB in their vertical positions. The crank revolves clockwise at a constant rate of 1 revolution every 2 seconds. For the position shown, determine the speed at which the box is being shoved horizontally onto the conveyor belt.

Ans. $v_E = 0.514$ m/s

▶ **5/88** Ends A and C of the connected links are controlled by the vertical motion of the piston rods of the hydraulic cylinders. For a short interval of motion, A has an upward velocity of 3 m/s, and C has a downward velocity of 2 m/s. Determine the velocity of B for the instant when $y = 150$ mm.

Ans. $v_B = 3.97$ m/s

Problem 5/88

▶ **5/89** At the instant represented, $a = 150$ mm and $b = 125$ mm, and the distance $a + b$ between A and C is decreasing at the rate of 0.2 m/s. Determine the common velocity v of points B and D for this instant.

Ans. $v = 0.0536$ m/s

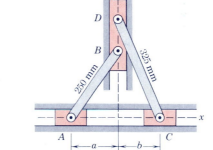

Problem 5/89

5/90 The wheel rolls without slipping. For the instant portrayed, when O is directly under point C, link OA has a velocity $v = 1.5$ m/s to the right and $\theta = 30°$. Determine the angular velocity ω of the slotted link.

Ans. $\omega = 18.22$ rad/s CCW

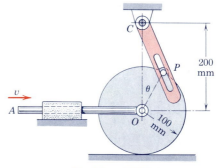

Problem 5/90

5/5 *INSTANTANEOUS CENTER OF ZERO VELOCITY*

In the previous article, we determined the velocity of a point on a rigid body in plane motion by adding the relative velocity due to rotation about a convenient reference point to the velocity of the reference point. In the present article, we will solve the problem by choosing a unique reference point that momentarily has zero velocity. As far as velocities are concerned, the body may be considered to be in pure rotation about an axis, normal to the plane of motion, passing through this point. This axis is called the *instantaneous axis of zero velocity*, and the intersection of this axis with the plane of motion is known as the *instantaneous center* of zero velocity. This approach provides us with a valuable means for visualizing and analyzing velocities in plane motion.

The existence of the instantaneous center is easily shown. For the body in Fig. 5/7*a*, let us assume that the directions of the absolute velocities of any two points *A* and *B* on the body are known and are not parallel. If there is a point about which *A* has absolute circular motion at the instant considered, this point must lie on the normal to $\mathbf{v}_A$ through *A*. Similar reasoning applies to *B*, and the intersection *C* of these two perpendiculars fulfills the requirement for an absolute center of rotation *at the instant considered*. Point *C* is the instantaneous center of zero velocity and may lie on or off the body. If it lies off the body, it may be visualized as lying on the body extended. The instantaneous center is not a fixed point in the body nor a fixed point in the plane.

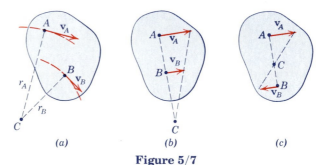

Figure 5/7

If the magnitude of the velocity of one of the points, say, v_A, is also known, the angular velocity ω of the body and the linear velocity of every point in the body are easily obtained. Thus, the angular velocity of the body, Fig. 5/7*a*, is

$$\omega = \frac{v_A}{r_A}$$

which, of course, is also the angular velocity of *every* line in the body.

Therefore, the velocity of B is $v_B = r_B\omega = (r_B/r_A)v_A$. Once the instantaneous center is located, the direction of the instantaneous velocity of every point in the body is readily found since it must be perpendicular to the radial line joining the point in question with C.

If the velocities of two points in a body having plane motion are parallel, Fig. 5/7*b* or 5/7*c*, the line joining the points is perpendicular to the direction of the velocities, and the instantaneous center C is located by direct proportion as shown. We can readily see from Fig. 5/7*b* that as the parallel velocities approach equality in magnitude, the instantaneous center C moves farther away from the body and approaches infinity in the limit as the body gives up its angular velocity and translates only.

As the body changes its position, the instantaneous center C also changes its position both in space and on the body. The locus of the instantaneous centers in space is known as the *space centrode*, and the locus of the positions of the instantaneous centers on the body is known as the *body centrode*. At the instant considered, the two curves are tangent at the position of point C. It may be shown that the body centrode curve rolls on the space centrode curve during the motion of the body, as indicated schematically in Fig. 5/8.

Whereas the instantaneous center of zero velocity as a point on the body is momentarily at rest, its acceleration generally is *not* zero. Thus, this point may *not* be used as an instantaneous center of zero acceleration in a manner analogous to its use for finding velocity. An instantaneous center of zero acceleration does exist for bodies in general plane motion, but its location and use represent a specialized topic in mechanism kinematics and will not be treated here.

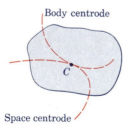

Figure 5/8

Sample Problem 5/11

The wheel of Sample Problem 5/7, shown again here, rolls to the right without slipping, with its center O having a velocity $v_O = 3$ m/s. Locate the instantaneous center of zero velocity and use it to find the velocity of point A for the position indicated.

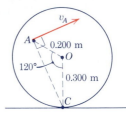

Solution. The point on the rim of the wheel in contact with the ground has no velocity if the wheel is not slipping; it is, therefore, the instantaneous center C of zero velocity. The angular velocity of the wheel becomes

$$[\omega = v/r] \qquad \omega = v_O/\overline{OC} = 3/0.300 = 10 \text{ rad/s}$$

The distance from A to C is

① $\overline{AC} = \sqrt{(0.300)^2 + (0.200)^2 - 2(0.300)(0.200)\cos 120°} = 0.436$ m

The velocity of A becomes

② $[v = r\omega] \qquad v_A = \overline{AC}\omega = 0.436(10) = 4.36 \text{ m/s}$ *Ans.*

The direction of $\mathbf{v}_A$ is perpendicular to AC as shown.

① Be sure to recognize that the cosine of 120° is itself negative.

② From the results of this problem, you should be able to visualize and sketch the velocities of all points on the wheel.

Sample Problem 5/12

Arm OB of the linkage has a clockwise angular velocity of 10 rad/sec in the position shown where $\theta = 45°$. Determine the velocity of A, the velocity of D, and the angular velocity of link AB for the instant shown.

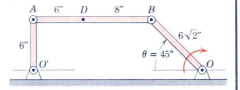

Solution. The directions of the velocities of A and B are tangent to their circular paths about the fixed centers O' and O as shown. The intersection of the two perpendiculars to the velocities from A and B ① locates the instantaneous center C for the link AB. The distances $\overline{AC}$, $\overline{BC}$, and $\overline{DC}$ shown on the diagram are computed or scaled from the drawing. The angular velocity of BC, considered as a line on the body extended, is equal to the angular velocity of AC, DC, and AB and is

$$[\omega = v/r] \qquad \omega_{BC} = \frac{v_B}{\overline{BC}} = \frac{\overline{OB}\omega_{OB}}{\overline{BC}} = \frac{6\sqrt{2}(10)}{14\sqrt{2}}$$

$$= 4.29 \text{ rad/sec CCW} \qquad \textit{Ans.}$$

Thus, the velocities of A and D are

$$[v = r\omega] \qquad v_A = \frac{14}{12}(4.29) = 5.00 \text{ ft/sec} \qquad \textit{Ans.}$$

$$v_D = \frac{15.23}{12}(4.29) = 5.44 \text{ ft/sec} \qquad \textit{Ans.}$$

in the directions shown.

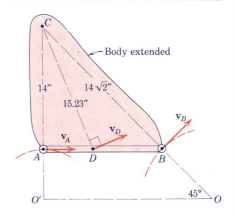

① For the instant depicted, we should visualize link AB and its body extended to be rotating as a single unit about point C.

PROBLEMS

Introductory problems

5/91 The bar *AB* has a clockwise angular velocity of 5 rad/sec. Construct and determine the vector velocity of each end if the instantaneous center of zero velocity is (*a*) at C_1 and (*b*) at C_2.

Ans. (*a*) $\mathbf{v}_A = -20\mathbf{j}$ in./sec, $\mathbf{v}_B = 40\mathbf{j}$ in./sec

(*b*) $\mathbf{v}_A = 15\mathbf{j}$ in./sec, $\mathbf{v}_B = 75\mathbf{j}$ in./sec

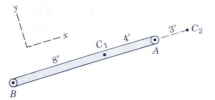

Problem 5/91

5/92 For the instant represented, the instantaneous center of zero velocity for the rectangular plate in plane motion is located at *C*. If the plate has a counterclockwise angular velocity of 4 rad/s at this instant, determine the magnitude of the velocity $\mathbf{v}_O$ of the center *O* of the plate.

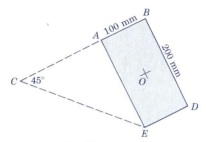

Problem 5/92

5/93 For the instant represented, when crank *OA* passes the horizontal position, determine the velocity of the center *G* of link *AB* by the method of Art. 5/5.

Ans. $v_G = 277$ mm/s

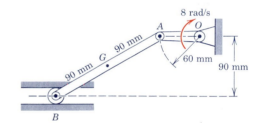

Problem 5/93

5/94 At a certain instant vertex *B* of the right-triangular plate has a velocity of 200 mm/s in the direction shown. If the instantaneous center of zero velocity for the plate is 40 mm from point *B* and if the angular velocity of the plate is clockwise, determine the velocity of point *D*.

5/95 Solve Prob. 5/81 by the method of Art. 5/5.

Ans. $\omega_{BD} = 1.2$ rad/s CCW

$\omega_{AO} = 1.333$ rad/s CCW

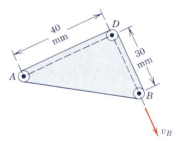

Problem 5/94

5/96 The rear wheels of a $\frac{1}{2}$-ton pickup truck have a diameter of 26 in. and are slipping on an icy road as torque is applied to them through the differential. If the body centrode for each wheel is a circle of diameter 2 in. and if each wheel is turning at the rate of 400 rev/min, determine the velocity of the truck. Also determine the velocity v_r with which the rubber slips on the icy surface.

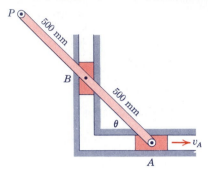

Problem 5/97

5/97 Motion of the bar is controlled by the constrained paths of A and B. If the angular velocity of the bar is 2 rad/s counterclockwise as the position $\theta = 45°$ is passed, determine the velocity v_A and the magnitude of the velocity of P.

  Ans. $v_A = 0.707$ m/s, $v_P = 1.581$ m/s

Representative problems

5/98 Solve Prob. 5/72 by the method of Art. 5/5.

5/99 Determine the velocity of the piston in Sample Problem 5/9 for a crank angle $\theta = 135°$.

  Ans. $v_A = 34.2$ ft/sec

5/100 Solve Prob. 5/79 by the method of Art. 5/5.

5/101 Solve Prob. 5/80 by the method of Art. 5/5.

  Ans. $v_C = 6.24$ ft/sec

5/102 Solve Prob. 5/83 by the method of Art. 5/5.

5/103 Solve Prob. 5/70 by the method of Art. 5/5.

  Ans. $\omega_{BC} = 2.77$ rad/sec CCW

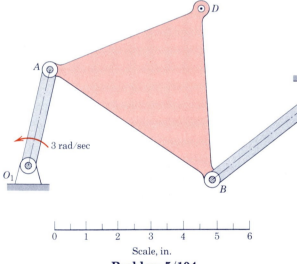

Problem 5/104

5/104 The scale drawing of the linkage of Prob. 5/73 is shown again here. Locate the instantaneous center of zero velocity directly on the drawing and use the scale shown to determine the velocity of point D to a close approximation for the instant depicted. Also determine the angular velocity ω of the plate. After working the problem, compare your results to those cited with Prob. 5/73.

5/105 Solve Prob. 5/78 by the method of Art. 5/5.

  Ans. $\omega = 15$ rad/s, $v_P = 1.897$ m/s

5/106 The gear *D* (teeth not shown) rotates clockwise about *O* with a constant angular velocity of 4 rad/s. The 90° sector *AOB* is mounted on an independent shaft at *O*, and each of the small gears at *A* and *B* meshes with gear *D*. If the sector has a counterclockwise angular velocity of 3 rad/s at the instant represented, determine the corresponding angular velocity ω of each of the small gears.

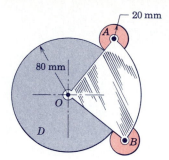

Problem 5/106

5/107 Solve Prob. 5/76 by the method of Art. 5/5.
Ans. ω_{AB} = 19.38 rad/sec CW

5/108 The hydraulic cylinder produces a limited horizontal motion of point *A*. If v_A = 4 m/s when θ = 45°, determine the magnitude of the velocity of *D* and the angular velocity ω of *ABD* for this position.

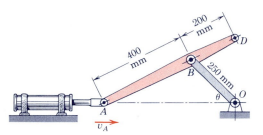
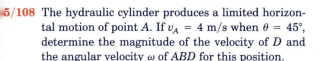
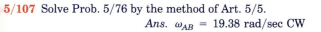

Problem 5/108

5/109 A device which tests the resistance to wear of two materials *A* and *B* is shown. If the link *EO* has a velocity of 4 ft/sec to the right when θ = 45°, determine the rubbing velocity v_A.
Ans. v_A = 9.19 ft/sec

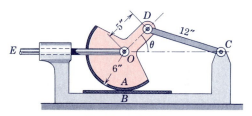
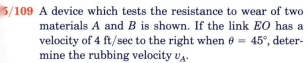

Problem 5/109

5/110 The rectangular body *B* is pivoted to the crank *OA* at *A* and is supported by the wheel at *D*. If *OA* has a counterclockwise angular velocity of 2 rad/s, determine the velocity of point *E* and the angular velocity of body *B* when the crank *OA* passes the vertical position shown.

5/111 Solve Prob. 5/74 by the method of Art. 5/5.
Ans. v_B = 269 mm/s

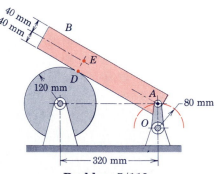
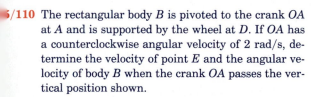
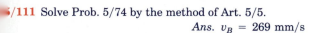

Problem 5/110

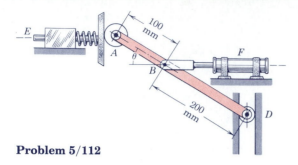

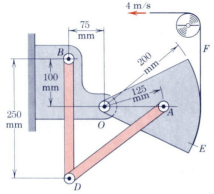

Problem 5/112

Problem 5/113

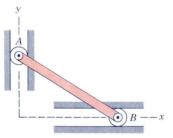

Problem 5/114

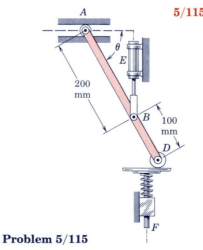

Problem 5/115

5/112 Horizontal oscillation of the spring-loaded plunger *E* is controlled by varying the air pressure in the horizontal pneumatic cylinder *F*. If the plunger has a velocity of 2 m/s to the right when $\theta = 30°$, determine the downward velocity v_D of roller *D* in the vertical guide and find the angular velocity ω of *ABD* for this position.

5/113 The flexible band *F* is attached at *E* to the rotating sector and leads over the guide pulley. Determine the angular velocities of *AD* and *BD* for the position shown if the band has a velocity of 4 m/s.

  Ans. $\omega_{AD} = 12.5$ rad/s, $\omega_{BD} = 7.5$ rad/s

5/114 Construct the space centrode and the body centrode for the link *AB* within the limits of its constrained motion. Show that the motion may be described by rolling the body centrode on the space centrode.

5/115 Vertical oscillation of the spring-loaded plunger *F* is controlled by a periodic change in pressure in the vertical hydraulic cylinder *E*. For the position $\theta = 60°$, determine the angular velocity of *AD* and the velocity of the roller *A* in its horizontal guide if the plunger *F* has a downward velocity of 2 m/s.

  Ans. $\omega_{AD} = 13.33$ rad/s CW, $v_A = 2.31$ m/s

5/116 Determine the angular velocity ω of the ram head *AE* of the rock crusher in the position for which $\theta = 60°$. The crank *OB* has an angular speed of 60 rev/min. When *B* is at the bottom of its circle, *D* and *E* are on a horizontal line through *F*, and lines *BD* and *AE* are vertical. The dimensions are $\overline{OB} = 4$ in., $\overline{BD} = 30$ in., and $\overline{AE} = \overline{ED} = \overline{DF} = 15$ in. Carefully construct the configuration graphically, and use the method of Art. 5/5.

Ans. $\omega = 1.10$ rad/sec CW

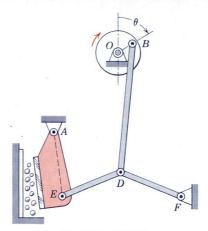

Problem 5/116

5/117 The shaft at *O* drives the arm *OA* at a clockwise speed of 90 rev/min about the fixed bearing at *O*. Use the method of the instantaneous center of zero velocity to determine the rotational speed of gear *B* (gear teeth not shown) if (*a*) ring gear *D* is fixed and (*b*) ring gear *D* rotates counterclockwise about *O* with a speed of 80 rev/min.

Ans. (*a*) $\omega_B = 360$ rev/min

(*b*) $\omega_B = 600$ rev/min

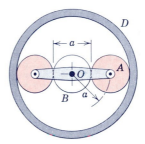

Problem 5/117

5/118 The large roller bearing rolls to the left on its outer race with a velocity of its center *O* of 0.9 m/s. At the same time the central shaft and inner race rotate counterclockwise with an angular speed of 240 rev/min. Determine the angular velocity ω of each of the rollers.

Ans. $\omega = 10.73$ rad/s CW

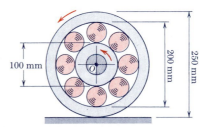

Problem 5/118

5/6 RELATIVE ACCELERATION

The relative-acceleration relationship may be obtained from the equation $\mathbf{v}_A = \mathbf{v}_B + \mathbf{v}_{A/B}$ for relative velocities using nonrotating reference axes for two points A and B in plane motion by differentiation with respect to time to get $\dot{\mathbf{v}}_A = \dot{\mathbf{v}}_B + \dot{\mathbf{v}}_{A/B}$ or

$$\mathbf{a}_A = \mathbf{a}_B + \mathbf{a}_{A/B} \tag{5/7}$$

In words, Eq. 5/7 states that the acceleration of point A equals the acceleration of point B plus (vectorially) the acceleration that A appears to have to a nonrotating observer moving with B. If points A and B are located on the same rigid body in the plane of motion, the distance r between them remains constant so that the observer moving with B perceives A to have circular motion about B, as we saw in Art. 5/4 with the relative-velocity relationship. With the relative motion being circular, it follows that the relative-acceleration term will have both a normal component directed from A toward B due to the change of direction of $\mathbf{v}_{A/B}$ and a tangential component perpendicular to AB due to the change in magnitude of $\mathbf{v}_{A/B}$. These acceleration components for circular motion, cited in Eqs. 5/2, were covered earlier in Art. 2/5 and should be thoroughly familiar by now. Thus, we may write

$$\mathbf{a}_A = \mathbf{a}_B + (\mathbf{a}_{A/B})_n + (\mathbf{a}_{A/B})_t \tag{5/8}$$

where the magnitudes of the relative-acceleration components are

$$(a_{A/B})_n = v_{A/B}{}^2/r = r\omega^2$$
$$(a_{A/B})_t = \dot{v}_{A/B} = r\alpha \tag{5/9}$$

or in vector notation the acceleration components become

$$(\mathbf{a}_{A/B})_n = \boldsymbol{\omega} \times (\boldsymbol{\omega} \times \mathbf{r})$$
$$(\mathbf{a}_{A/B})_t = \boldsymbol{\alpha} \times \mathbf{r} \tag{5/9a}$$

In these relationships, $\boldsymbol{\omega}$ is the angular velocity and $\boldsymbol{\alpha}$ is the angular acceleration of the body. The vector locating A from B is $\mathbf{r}$. It is important to observe that the *relative* acceleration terms depend on the respective *absolute* angular velocity and *absolute* angular acceleration.

The meaning of Eqs. 5/8 and 5/9 is illustrated in Fig. 5/9, which shows a rigid body in plane motion with points A and B moving along separate curved paths with absolute accelerations $\mathbf{a}_A$ and $\mathbf{a}_B$. Contrary to the case with velocities, the accelerations $\mathbf{a}_A$ and $\mathbf{a}_B$

are, in general, not tangent to the paths described by A and B when these paths are curvilinear. The figure shows the acceleration of A to be composed of two parts: the acceleration of B and the acceleration of A with respect to B. A sketch showing the reference point as fixed is useful in disclosing the correct sense of each of the two components of the relative-acceleration term.

Alternatively, we may write the acceleration equation in the reverse order, which puts the nonrotating reference axes on A rather than B. This order gives

$$\mathbf{a}_B = \mathbf{a}_A + \mathbf{a}_{B/A}$$

Here $\mathbf{a}_{B/A}$ and its n- and t-components are the negatives of $\mathbf{a}_{A/B}$ and its n- and t-components. The student should make a sketch corresponding to Fig. 5/9 for this reversed sequence of terms.

As in the case of the relative-velocity equation, we can handle the solution to Eq. 5/8 in three different ways, namely, by scalar algebra and geometry, by vector algebra, or by graphical construction. The student will be well advised to become familiar with all three techniques. In any event, a sketch of the vector polygon representing the vector equation should be made with close attention paid to the head-to-tail combination of vectors that agrees with the equation. Known vectors should be added first, and the unknown vectors will become the closing legs of the vector polygon. It is vital that we visualize the vectors in their geometrical sense, as only then can the full significance of the acceleration equation be perceived. Before attempting a solution, we should identify the knowns and unknowns, keeping in mind that a solution to a vector equation in two dimensions can be carried out when the unknowns have been reduced to two scalar quantities. These quantities may be the magnitude or direction of any of the terms of the equation. We observe that when both points move on curved paths, there will, in general, be six scalar quantities to account for in Eq. 5/8.

Because the normal acceleration components depend on velocities, it is generally necessary to solve for the velocities before the acceleration calculations can be made. The reference point in the relative-acceleration equation is chosen as some point on the body in question whose acceleration is either known or can be easily found. Care must be exercised *not* to use the instantaneous center of zero velocity as the reference point unless its acceleration is known and accounted for. An instantaneous center of zero acceleration exists for a rigid body in general plane motion, but will not be discussed here since its use is somewhat specialized.

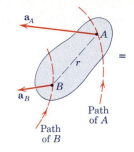

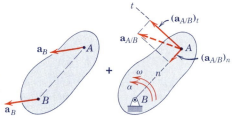

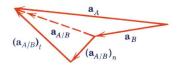

Figure 5/9

Sample Problem 5/13

The wheel of radius r rolls to the left without slipping and, at the instant considered, the center O has a velocity $\mathbf{v}_O$ and an acceleration $\mathbf{a}_O$ to the left. Determine the acceleration of points A and C on the wheel for the instant considered.

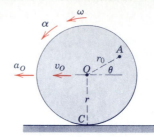

Solution. From our previous analysis of Sample Problem 5/4, we know that the angular velocity and angular acceleration of the wheel are

$$\omega = v_O/r \quad \text{and} \quad \alpha = a_O/r$$

The acceleration of A is written in terms of the given acceleration of O. Thus,

$$\mathbf{a}_A = \mathbf{a}_O + \mathbf{a}_{A/O} = \mathbf{a}_O + (\mathbf{a}_{A/O})_n + (\mathbf{a}_{A/O})_t$$

The relative-acceleration terms are viewed as though O were fixed, and for this relative circular motion they have the magnitudes

$$(a_{A/O})_n = r_0\omega^2 = r_0\left(\frac{v_O}{r}\right)^2$$

$$(a_{A/O})_t = r_0\alpha = r_0\left(\frac{a_O}{r}\right)$$

① and the directions shown.

Adding the vectors head-to-tail gives $\mathbf{a}_A$ as shown. In a numerical problem, we may obtain the combination algebraically or graphically. The algebraic expression for the magnitude of $\mathbf{a}_A$ is found from the square root of the sum of the squares of its components. If we use n- and t-directions, we have

②
$$\begin{aligned}
a_A &= \sqrt{(a_A)_n{}^2 + (a_A)_t{}^2} \\
&= \sqrt{[a_O\cos\theta + (a_{A/O})_n]^2 + [a_O\sin\theta + (a_{A/O})_t]^2} \\
&= \sqrt{(r\alpha\cos\theta + r_0\omega^2)^2 + (r\alpha\sin\theta + r_0\alpha)^2} \qquad \textit{Ans.}
\end{aligned}$$

The direction of $\mathbf{a}_A$ can be computed if desired.

The acceleration of the instantaneous center C of zero velocity, considered as a point on the wheel, is obtained from the expression

$$\mathbf{a}_C = \mathbf{a}_O + \mathbf{a}_{C/O}$$

where the components of the relative-acceleration term are $(a_{C/O})_n = r\omega^2$ directed from C to O and $(a_{C/O})_t = r\alpha$ directed to the right on account of the counterclockwise angular acceleration of line CO about O. The terms are added together in the lower diagram, and it is seen that

③
$$a_C = r\omega^2 \qquad \textit{Ans.}$$

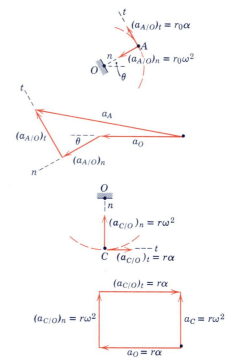

① The counterclockwise angular acceleration α of OA determines the positive direction of $(a_{A/O})_t$. The normal component $(a_{A/O})_n$ is, of course, directed toward the reference center O.

② If the wheel were rolling to the right with the same velocity v_O but still had an acceleration a_O to the left, note that the solution for a_A would be unchanged.

③ We note that the acceleration of the instantaneous center of zero velocity is independent of α and is directed toward the center of the wheel. This conclusion is a useful result to remember.

Sample Problem 5/14

The linkage of Sample Problem 5/8 is repeated here. Crank CB has a constant counterclockwise angular velocity of 2 rad/s in the position shown during a short interval of its motion. Determine the angular acceleration of links AB and OA for this position. Solve by using vector algebra.

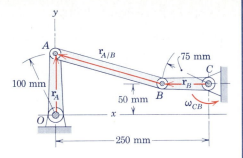

Solution. We first solve for the velocities that were obtained in Sample Problem 5/8. They are

$$\omega_{AB} = -6/7 \text{ rad/s} \quad \text{and} \quad \omega_{OA} = -3/7 \text{ rad/s}$$

where the counterclockwise direction ($+\mathbf{k}$-direction) is taken as positive. The acceleration equation is

$$\mathbf{a}_A = \mathbf{a}_B + (\mathbf{a}_{A/B})_n + (\mathbf{a}_{A/B})_t$$

where, from Eqs. 5/3 and 5/9a, we may write

①
$$\mathbf{a}_A = \boldsymbol{\alpha}_{OA} \times \mathbf{r}_A + \boldsymbol{\omega}_{OA} \times (\boldsymbol{\omega}_{OA} \times \mathbf{r}_A)$$
$$= \alpha_{OA}\mathbf{k} \times 100\mathbf{j} + (-\tfrac{3}{7}\mathbf{k}) \times (-\tfrac{3}{7}\mathbf{k} \times 100\mathbf{j})$$
$$= -100\alpha_{OA}\mathbf{i} - 100(\tfrac{3}{7})^2\mathbf{j} \text{ mm/s}^2$$

$$\mathbf{a}_B = \boldsymbol{\alpha}_{CB} \times \mathbf{r}_B + \boldsymbol{\omega}_{CB} \times (\boldsymbol{\omega}_{CB} \times \mathbf{r}_B)$$
$$= \mathbf{0} + 2\mathbf{k} \times (2\mathbf{k} \times [-75\mathbf{i}])$$
$$= 300\mathbf{i} \text{ mm/s}^2$$

②
$$(\mathbf{a}_{A/B})_n = \boldsymbol{\omega}_{AB} \times (\boldsymbol{\omega}_{AB} \times \mathbf{r}_{A/B})$$
$$= -\tfrac{6}{7}\mathbf{k} \times [(-\tfrac{6}{7}\mathbf{k}) \times (-175\mathbf{i} + 50\mathbf{j})]$$
$$= (\tfrac{6}{7})^2(175\mathbf{i} - 50\mathbf{j}) \text{ mm/s}^2$$

$$(\mathbf{a}_{A/B})_t = \boldsymbol{\alpha}_{AB} \times \mathbf{r}_{A/B}$$
$$= \alpha_{AB}\mathbf{k} \times (-175\mathbf{i} + 50\mathbf{j})$$
$$= -50\alpha_{AB}\mathbf{i} - 175\alpha_{AB}\mathbf{j} \text{ mm/s}^2$$

We now substitute these results into the relative-acceleration equation and equate separately the coefficients of the $\mathbf{i}$-terms and the coefficients of the $\mathbf{j}$-terms to give

$$-100\alpha_{OA} = 428.6 - 50\alpha_{AB}$$
$$-18.37 = -36.73 - 175\alpha_{AB}$$

The solutions are

$$\alpha_{AB} = -0.105 \text{ rad/s}^2 \quad \text{and} \quad \alpha_{OA} = -4.34 \text{ rad/s}^2 \quad \textit{Ans.}$$

Since the unit vector $\mathbf{k}$ points out from the paper in the positive z-direction, we see that the angular accelerations of AB and OA are both clockwise (negative).

It is recommended that the student sketch each of the acceleration vectors in its proper geometric relationship according to the relative-acceleration equation to help clarify the meaning of the solution.

① Remember to preserve the order of the factors in the cross products.

② In expressing the term $\mathbf{a}_{A/B}$, be certain that $\mathbf{r}_{A/B}$ is written as the vector from B to A and not the reverse.

Sample Problem 5/15

The slider-crank mechanism of Sample Problem 5/9 is repeated here. The crank OB has a constant clockwise angular speed of 1500 rev/min. For the instant when the crank angle θ is 60°, determine the acceleration of the piston A and the angular acceleration of the connecting rod AB.

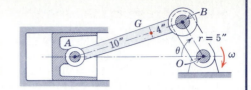

Solution. The acceleration of A may be expressed in terms of the acceleration of the crank pin B. Thus,

$$\mathbf{a}_A = \mathbf{a}_B + (\mathbf{a}_{A/B})_n + (\mathbf{a}_{A/B})_t$$

① Point B moves in a circle of 5-in. radius with a constant speed so that it has only a normal component of acceleration directed from B to O.

$$[a_n = r\omega^2] \qquad a_B = \frac{5}{12}\left(\frac{1500[2\pi]}{60}\right)^2 = 10{,}280 \text{ ft/sec}^2$$

The relative-acceleration terms are visualized with A rotating in a circle relative to B, which is considered fixed, as shown. From Sample Problem 5/9, the angular velocity of AB for these same conditions is $\omega_{AB} = 29.5$ rad/sec so that

② $[a_n = r\omega^2] \qquad (a_{A/B})_n = \frac{14}{12}(29.5)^2 = 1015 \text{ ft/sec}^2$

directed from A to B. The tangential component $(\mathbf{a}_{A/B})_t$ is known in direction only since its magnitude depends on the unknown angular acceleration of AB. We also know the direction of $\mathbf{a}_A$ since the piston is confined to move along the horizontal axis of the cylinder. There are now only two remaining unknowns left in the equation, namely, the magnitudes of $\mathbf{a}_A$ and $(\mathbf{a}_{A/B})_t$, so the solution may be carried out.

If we adopt an algebraic solution using the geometry of the acceleration polygon, we first compute the angle between AB and the horizontal. With the law of sines, this angle becomes 18.0°. Equating separately the horizontal components and the vertical components of the terms in the acceleration equation, as seen from the acceleration polygon, gives

$$a_A = 10{,}280 \cos 60° + 1015 \cos 18.0° - (a_{A/B})_t \sin 18.0°$$

$$0 = 10{,}280 \sin 60° - 1015 \sin 18.0° - (a_{A/B})_t \cos 18.0°$$

The solution to these equations gives the magnitudes

$$(a_{A/B})_t = 9030 \text{ ft/sec}^2 \qquad \text{and} \qquad a_A = 3310 \text{ ft/sec}^2 \qquad Ans.$$

With the sense of $(\mathbf{a}_{A/B})_t$ also determined from the diagram, the angular acceleration of AB is seen from the figure representing rotation relative to B to be

$$[\alpha = a_t/r] \qquad \alpha_{AB} = 9030/(14/12) = 7740 \text{ rad/sec}^2 \text{ clockwise} \qquad Ans.$$

③ If we adopt a graphical solution, we begin with the known vectors $\mathbf{a}_B$ and $(\mathbf{a}_{A/B})_n$ and add them head-to-tail using a convenient scale. Next we construct the direction of $(\mathbf{a}_{A/B})_t$ through the head of the last vector. The solution of the equation is obtained by the intersection P of this last line with a horizontal line through the starting point representing the known direction of the vector sum $\mathbf{a}_A$. Scaling the magnitudes from the diagram gives values that agree with the calculated results

$$a_A = 3310 \text{ ft/sec}^2 \qquad \text{and} \qquad (a_{A/B})_t = 9030 \text{ ft/sec}^2 \qquad Ans.$$

① If the crank OB had an angular acceleration, $\mathbf{a}_B$ would also have a tangential component of acceleration.

② Alternatively, the relation $a_n = v^2/r$ may be used for calculating $(a_{A/B})_n$, provided the relative velocity $v_{A/B}$ is used for v. The equivalence is easily seen when it is recalled that $v_{A/B} = r\omega$.

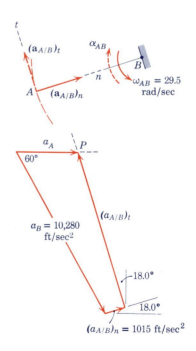

③ Except where extreme accuracy is required, do not hesitate to use a graphical solution, as it is quick and reveals the physical relationships among the vectors. The known vectors, of course, may be added in any order as long as the governing equation is satisfied.

PROBLEMS

Introductory problems

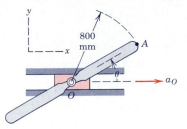

Problem 5/119

5/119 The two rotor blades of 800-mm radius rotate counterclockwise with a constant angular velocity $\omega = \dot{\theta} = 2$ rad/s about the shaft at O mounted in the sliding block. The acceleration of the block is $a_O = 3$ m/s². Determine the magnitude of the acceleration of the tip A of the blade when (a) $\theta = 0$, (b) $\theta = 90°$, and (c) $\theta = 180°$. Does the velocity of O or the sense of ω enter into the calculation?

$$Ans. \ (a) \ a_A = 0.2 \text{ m/s}^2$$
$$(b) \ a_A = 4.39 \text{ m/s}^2$$
$$(c) \ a_A = 6.2 \text{ m/s}^2$$

5/120 Refer to the rotor blades and sliding bearing block of Prob. 5/119 where $a_O = 3$ m/s². If $\ddot{\theta} = 5$ rad/s² and $\dot{\theta} = 0$ when $\theta = 0$, find the acceleration of point A for this instant.

5/121 The circular disk is rolling to the right without slipping. If $\mathbf{a}_{B/A} = 32\mathbf{i}$ in./sec², determine the velocity and acceleration of the center O of the disk.

$$Ans. \ \mathbf{v}_O = 12\mathbf{i} \text{ in./sec}, \ \mathbf{a}_O = -24\mathbf{i} \text{ in./sec}^2$$

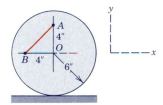

Problem 5/121

5/122 A container for waste materials is dumped by the hydraulic-actuated linkage shown. If the piston rod starts from rest in the position indicated and has an acceleration of 0.5 m/s² in the direction shown, compute the initial angular acceleration of the container.

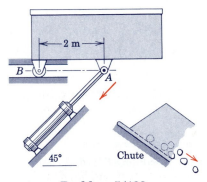

Problem 5/122

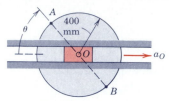

Problem 5/123

5/123 The center O of the wheel is mounted on the sliding block which has an acceleration $a_O = 8$ m/s^2 to the right. At the instant when $\theta = 45°$, $\dot{\theta} = 3$ rad/s and $\ddot{\theta} = -8$ rad/s^2. For this instant determine the magnitudes of the acceleration of points A and B.
    Ans. $a_A = 9.58$ m/s^2, $a_B = 9.09$ m/s^2

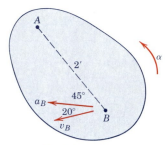

Problem 5/124

5/124 At the instant depicted point B on the plate has an acceleration of 3 ft/sec^2 and a velocity of 4 ft/sec in the directions shown. The plate also has a counter-clockwise angular acceleration of 1.5 rad/sec^2. What additional information is needed to determine the acceleration of point A? Choose a reasonable value for this unknown and proceed to find your value of a_A.

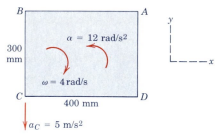

Problem 5/125

5/125 For the instant represented corner C of the rectangular plate has an acceleration of 5 m/s^2 in the negative y-direction, and the plate has a clockwise angular velocity of 4 rad/s which is decreasing by 12 rad/s each second. Determine the acceleration of A at this instant. Solve by scalar-geometric and by vector-algebraic methods.   *Ans.* $a_A = 11.18$ m/s^2

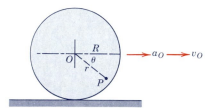

Problem 5/126

5/126 The wheel of radius R rolls without slipping, and its center O has an acceleration a_O. A point P on the wheel is a distance r from O. For given values of a_O, R, and r, determine the angle θ and the velocity v_O of the wheel for which P has no acceleration in this position.

Representative problems

5/127 Determine the acceleration of the piston of Sample Problem 5/15 for (*a*) $\theta = 0°$, (*b*) $\theta = 90°$, and (*c*) $\theta = 180°$. Take the positive x-direction to the right.   *Ans.* (*a*) $\mathbf{a}_A = 13{,}950\mathbf{i}$ ft/sec^2
    (*b*) $\mathbf{a}_A = -3930\mathbf{i}$ ft/sec^2
    (*c*) $\mathbf{a}_A = -6610\mathbf{i}$ ft/sec^2

5/128 The punch of Prob. 5/55 is shown again here where the sector is given an oscillation specified by $\theta = \theta_0 \sin 2\pi t$. Calculate the acceleration of the punch when $\theta = 0$ if $\theta_0 = \pi/12$ rad.

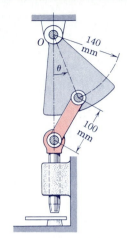

Problem 5/128

5/129 Determine the angular acceleration of link AB and the linear acceleration of A for $\theta = 90°$ if $\dot{\theta} = 0$ and $\ddot{\theta} = 3$ rad/s^2 at that position. Carry out your solution using vector notation.

Ans. $\boldsymbol{\alpha}_{AB} = -4\mathbf{k}$ rad/s^2, $\mathbf{a}_A = 1.6\mathbf{i}$ m/s^2

5/130 Determine the angular acceleration of AB and the linear acceleration of A for the linkage of Prob. 5/129 for the position $\theta = 90°$ if $\dot{\theta} = 4$ rad/s and $\ddot{\theta} = 0$ at that position. Carry out your solution using vector notation.

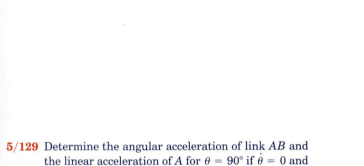

Problem 5/129

5/131 Work Prob. 5/130 by a scalar-geometric solution of the polygon of acceleration vectors.

Ans. $\alpha_{AB} = 37.9$ rad/s^2 CW, $a_A = 17.30$ m/s^2

5/132 The load L is lowered by the two pulleys which are fastened together and rotate as a single unit. For the instant represented, drum A has a counterclockwise angular velocity of 4 rad/sec, which is decreasing by 4 rad/sec each second. Simultaneously drum B has a clockwise angular velocity of 6 rad/sec, which is increasing by 2 rad/sec each second. Calculate the acceleration of points C and D and the load L.

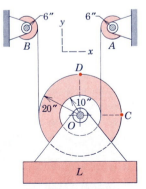

Problem 5/132

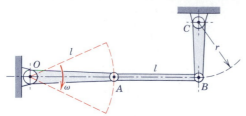

Problem 5/133

5/133 Crank *OA* oscillates between the dotted positions shown and causes small angular motion of crank *BC* through the connecting link *AB*. When *OA* crosses the horizontal position with *AB* horizontal and *BC* vertical, it has an angular velocity ω and zero angular acceleration. Determine the angular acceleration of *BC* for this position.

Ans. $\alpha = 2l\omega^2/r$ CW

Problem 5/134

5/134 The bicycle of Prob. 5/69 is repeated here. If the cyclist starts from rest and acquires a velocity of 2 m/s in a distance of 3 m with constant acceleration, compute the magnitude of the acceleration of pedal *A* for this condition if it is in the top position.

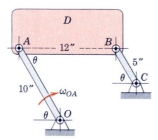

Problem 5/135

5/135 Calculate the angular acceleration of the plate in the position shown where control link *AO* has a constant angular velocity $\omega_{OA} = 4$ rad/sec and $\theta = 60°$ for both links.   *Ans.* $\alpha_{AB} = 15.40$ rad/sec^2 CW

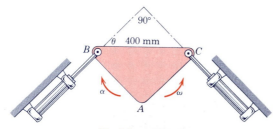

Problem 5/136

5/136 At the instant represented $\theta = 45°$, and the triangular plate *ABC* has a counterclockwise angular velocity of 20 rad/s and a clockwise angular acceleration of 100 rad/s^2. Determine the magnitudes of the corresponding velocity **v** and acceleration **a** of the piston rod of the hydraulic cylinder attached to *C*.

5/137 The linkage of Prob. 5/70 is shown again here. For the instant when $\theta = \beta = 60°$, the hydraulic cylinder gives A a velocity $v_A = 4$ ft/sec which is increasing by 3 ft/sec each second. For this instant determine the angular acceleration of link BC.

$$Ans. \ \alpha_{BC} = 2.08 \ rad/sec^2 \ CCW$$

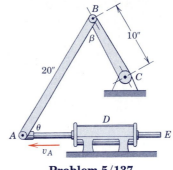

Problem 5/137

5/138 If the wheel in each case rolls on the circular surface without slipping, determine the acceleration of the point C on the wheel momentarily in contact with the circular surface. The wheel has an angular velocity ω and an angular acceleration α.

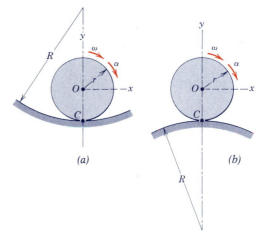

Problem 5/138

5/139 The figure for Prob. 5/77 is repeated here with reference axes added. If link AB has a constant counterclockwise angular velocity of 40 rad/s during an interval that includes the instant represented, determine the angular acceleration of AO and the acceleration of point D. Express your results in vector notation.

$$Ans. \ \mathbf{a}_D = -120(4\mathbf{i} + 3\mathbf{j}) \ m/s^2, \ \boldsymbol{\alpha}_{AO} = \mathbf{0}$$

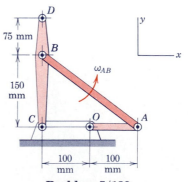

Problem 5/139

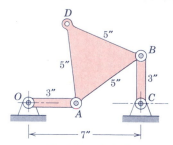

Problem 5/140

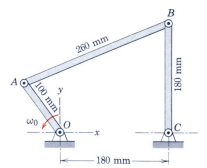

Problem 5/141

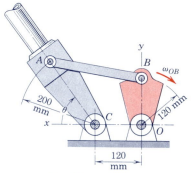

Problem 5/142

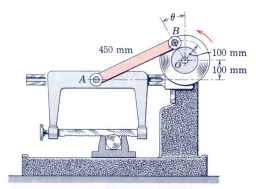

Problem 5/143

5/140 The mechanism of Prob. 5/67 is repeated here where the triangular plate has a clockwise angular velocity of 3 rad/sec and *OA* has zero angular acceleration for the instant represented. Determine the angular acceleration of *AD* and *BC* for this instant.

5/141 The deployment mechanism for the spacecraft magnetometer boom of Prob. 5/83 is shown again here. The driving link *OB* has a constant clockwise angular velocity ω_{OB} of 0.5 rad/s as it crosses the vertical position. Determine the angular acceleration α_{CA} of the boom for the position shown where $\tan \theta = 4/3$.   *Ans.* $\alpha_{CA} = -0.0758\mathbf{k}$ rad/s²

5/142 The linkage of Prob. 5/84 is shown again here. If *OA* has a constant counterclockwise angular velocity $\omega_0 = 10$ rad/s, calculate the angular acceleration of link *AB* for the position where the coordinates of *A* are $x = -60$ mm and $y = 80$ mm. Link *BC* is vertical for this position. Solve by vector algebra. (Use the results of Prob. 5/84 for the angular velocities of *AB* and *BC* which are $\omega_{BC} = 5.83\mathbf{k}$ rad/s and $\omega_{AB} = 2.5\mathbf{k}$ rad/s.)

5/143 The elements of a power hacksaw are shown in the figure. The saw blade is mounted in a frame which slides along the horizontal guide. If the motor turns the flywheel at a constant counterclockwise speed of 60 rev/min, determine the acceleration of the blade for the position where $\theta = 90°$, and find the corresponding angular acceleration of the link *AB*.   *Ans.* $a_A = 4.89$ m/s², $\alpha_{AB} = 0.467$ rad/s² CCW

5/144 The simplified clam-shell bucket of Prob. 5/79 is shown again here. With the block at O considered fixed and with the constant velocity of the control cable at C equal to 0.5 m/s, determine the angular acceleration α of the right-hand bucket jaw when $\theta = 45°$ as the bucket jaws are closing.

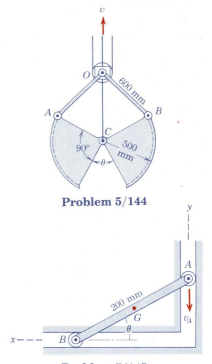

Problem 5/144

5/145 If end A of the constrained link has a constant downward velocity v_A of 2 m/s as the bar passes the position for which $\theta = 30°$, determine the acceleration of the mass center G in the middle of the link. Compare solution by vector algebra with solution by vector geometry.　　　*Ans.* $a_G = 15.40$ m/s^2

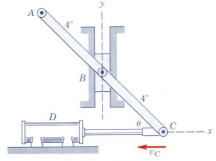

Problem 5/145

5/146 Motion of link ABC is controlled by the horizontal movement of the piston rod of the hydraulic cylinder D and by the vertical guide for the pinned slider at B. For the instant when $\theta = 45°$, the piston rod is retracting at the constant rate $v_C = 0.6$ ft/sec. Determine the acceleration of point A for this instant.

Problem 5/146

5/147 The mechanism of Prob. 5/113 is repeated here where the flexible band F attached to the sector at E is given a constant velocity of 4 m/s as shown. For the instant when BD is perpendicular to OA, determine the angular acceleration of BD.　　　*Ans.* $\alpha_{BD} = 46.9$ rad/s^2 CW

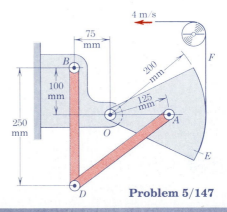

Problem 5/147

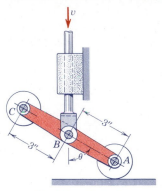

Problem 5/148

5/148 Elements of the switching device of Prob. 5/80 are shown again here. If the velocity v of the control rod is 3 ft/sec and is slowing down at the rate of 20 ft/sec² when $\theta = 60°$, determine the magnitude of the acceleration of C.

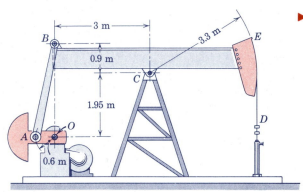

Problem 5/149

▶**5/149** An oil pumping rig is shown in the figure. The flexible pump rod D is fastened to the sector at E and is always vertical as it enters the fitting below D. The link AB causes the beam BCE to oscillate as the weighted crank OA revolves. If OA has a constant clockwise speed of 1 rev every 3 s, determine the acceleration of the pump rod D when the beam and the crank OA are both in the horizontal position shown.   *Ans.* $a_D = 0.568$ m/s² down

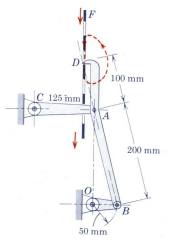

Problem 5/150

▶**5/150** An intermittent-drive mechanism for perforated tape F consists of the link DAB driven by the crank OB. The trace of the motion of the finger at D is shown by the dotted line. Determine the acceleration of D at the instant shown when both OB and CA are horizontal if OB has a constant clockwise rotational velocity of 120 rev/min.
                                Ans. $a_D = 1997$ mm/s²

5/7 MOTION RELATIVE TO ROTATING AXES

   In our discussion of the relative motion of particles in Art. 2/8 and in our use of the relative-motion equations for the plane motion of rigid bodies in this present chapter, we have made all of our relative-velocity and relative-acceleration measurements from *nonrotating* reference axes. The solution of many problems in kinematics where motion is generated within or observed from a system that itself is rotating is greatly facilitated by the use of rotating reference axes. An example of such a motion would be the movement of a fluid particle along the curved vane of a centrifugal pump, where the path relative to the vanes of the impeller becomes an important design consideration.

   We begin the description of motion using rotating axes by considering the plane motion of two particles A and B in the fixed X-Y plane, Fig. 5/10a. For the time being, we will consider A and B to be moving independently of one another for the sake of generality. The motion of A will now be observed from a moving reference frame x-y that has its origin attached to B and that rotates with an angular velocity $\omega = \dot{\theta}$. We may write this angular velocity as the vector $\boldsymbol{\omega} = \omega\mathbf{k} = \dot{\theta}\mathbf{k}$, where the vector is normal to the plane of motion and where its positive sense is in the positive z-direction (out from the paper), as established by the right-hand rule. The absolute position vector of A is given by

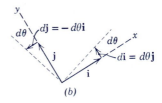

$$\mathbf{r}_A = \mathbf{r}_B + \mathbf{r} = \mathbf{r}_B + (x\mathbf{i} + y\mathbf{j}) \qquad (5/10)$$

where $\mathbf{i}$ and $\mathbf{j}$ are unit vectors attached to the x-y frame and $\mathbf{r} = x\mathbf{i} + y\mathbf{j}$ stands for $\mathbf{r}_{A/B}$, the position vector of A with respect to B. The velocity and acceleration equations require successive differentiation of the position-vector equation with respect to time. In contrast to the case of translating axes treated in Art. 2/8, the unit vectors $\mathbf{i}$ and $\mathbf{j}$ are now rotating with the x-y axes and, therefore, have time derivatives that must be evaluated. These derivatives may be seen from Fig. 5/10b, which shows the infinitesimal change in each unit vector during time dt as the reference axes rotate through an angle $d\theta = \omega\, dt$. The differential change in $\mathbf{i}$ is $d\mathbf{i}$, and it has the direction of $\mathbf{j}$ and a magnitude equal to the angle $d\theta$ times the length of the vector $\mathbf{i}$, which is unity. Thus, $d\mathbf{i} = d\theta\,\mathbf{j}$. Similarly, the unit vector $\mathbf{j}$ has an infinitesimal change $d\mathbf{j}$ that points in the negative x-direction, so that $d\mathbf{j} = -d\theta\,\mathbf{i}$. Dividing by dt and replacing $d\mathbf{i}/dt$ by $\dot{\mathbf{i}}$, $d\mathbf{j}/dt$ by $\dot{\mathbf{j}}$, and $d\theta/dt$ by $\dot{\theta} = \omega$ result in

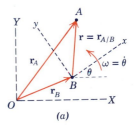

$$\dot{\mathbf{i}} = \omega\mathbf{j} \qquad \text{and} \qquad \dot{\mathbf{j}} = -\omega\mathbf{i}$$

When the cross product is introduced, we see from Fig. 5/10c that $\boldsymbol{\omega} \times \mathbf{i} = \omega\mathbf{j}$ and $\boldsymbol{\omega} \times \mathbf{j} = -\omega\mathbf{i}$. Thus, the time derivatives of the unit vectors may be written as

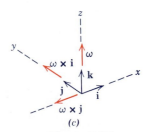

Figure 5/10

$$\dot{\mathbf{i}} = \boldsymbol{\omega} \times \mathbf{i} \quad \text{and} \quad \dot{\mathbf{j}} = \boldsymbol{\omega} \times \mathbf{j} \qquad (5/11)$$

(a) Relative velocity. We shall now use the expressions of Eqs. 5/11 when we take the time derivative of the position-vector equation for A and B to obtain the relative-velocity relation. Differentiation of Eq. 5/10 gives

$$\dot{\mathbf{r}}_A = \dot{\mathbf{r}}_B + \frac{d}{dt}(x\mathbf{i} + y\mathbf{j})$$

$$= \dot{\mathbf{r}}_B + (x\dot{\mathbf{i}} + y\dot{\mathbf{j}}) + (\dot{x}\mathbf{i} + \dot{y}\mathbf{j})$$

But $x\dot{\mathbf{i}} + y\dot{\mathbf{j}} = \boldsymbol{\omega} \times x\mathbf{i} + \boldsymbol{\omega} \times y\mathbf{j} = \boldsymbol{\omega} \times (x\mathbf{i} + y\mathbf{j}) = \boldsymbol{\omega} \times \mathbf{r}$. Also, since the observer in x-y measures velocity components $\dot{x}$ and $\dot{y}$, we see that $\dot{x}\mathbf{i} + \dot{y}\mathbf{j} = \mathbf{v}_{\text{rel}}$, which is the velocity relative to the x-y frame of reference. Thus, the relative-velocity equation becomes

$$\mathbf{v}_A = \mathbf{v}_B + \boldsymbol{\omega} \times \mathbf{r} + \mathbf{v}_{\text{rel}} \qquad (5/12)$$

Comparison of Eq. 5/12 with Eq. 2/20 for nonrotating reference axes shows that $\mathbf{v}_{A/B} = \boldsymbol{\omega} \times \mathbf{r} + \mathbf{v}_{\text{rel}}$, from which we conclude that the term $\boldsymbol{\omega} \times \mathbf{r}$ is the difference between the relative velocity as measured from nonrotating and rotating axes.

To illustrate further the meaning of the last two terms in Eq. 5/12, the motion of particle A relative to the rotating x-y plane is shown in Fig. 5/11 as taking place in a curved slot in a plate that represents the rotating x-y reference system. The velocity of A as measured relative to the plate, $\mathbf{v}_{\text{rel}}$, would be tangent to the path fixed in the x-y plate and would have a magnitude $\dot{s}$, where s is measured along the path. This relative velocity may also be viewed as the velocity $\mathbf{v}_{A/P}$ relative to a point P attached to the plate and coincident with A at the instant under consideration. The term $\boldsymbol{\omega} \times \mathbf{r}$ has a magnitude $r\dot{\theta}$ and a direction normal to $\mathbf{r}$ and is the velocity relative to B of point P as seen from nonrotating axes attached to B.

The following comparison will help to establish the equivalence of, and clarify the differences between, the relative-velocity equations written for rotating and nonrotating reference axes:

$$\mathbf{v}_A = \mathbf{v}_B + \boldsymbol{\omega} \times \mathbf{r} + \mathbf{v}_{\text{rel}}$$
$$\mathbf{v}_A = \underbrace{\mathbf{v}_B + \mathbf{v}_{P/B}} + \mathbf{v}_{A/P} \qquad (5/12a)$$
$$\mathbf{v}_A = \underbrace{\mathbf{v}_P} + \mathbf{v}_{A/P}$$
$$\mathbf{v}_A = \mathbf{v}_B + \mathbf{v}_{A/B}$$

In the second equation, the term $\mathbf{v}_{P/B}$ is measured from a nonrotating position—otherwise, it would be zero. The term $\mathbf{v}_{A/P}$ is the

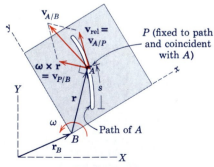

Figure 5/11

same as $\mathbf{v}_{\text{rel}}$ and is the velocity of A as measured in the x-y frame. In the third equation, $\mathbf{v}_P$ is the absolute velocity of P and represents the effect of the moving coordinate system, both translational and rotational. The fourth equation is the same as that developed for nonrotating axes, Eq. 2/20, and it is seen that $\mathbf{v}_{A/B} = \mathbf{v}_{P/B} + \mathbf{v}_{A/P} = \boldsymbol{\omega} \times \mathbf{r} + \mathbf{v}_{\text{rel}}$.

(b) Transformation of a time derivative.

Equation 5/12 represents a transformation of the time derivative of the position vector between rotating and nonrotating axes. We may easily generalize this result to apply to the time derivative of any vector quantity $\mathbf{V} = V_x\mathbf{i} + V_y\mathbf{j}$. Accordingly, the total time derivative as taken in X-Y is

$$\left(\frac{d\mathbf{V}}{dt}\right)_{XY} = (\dot{V}_x\mathbf{i} + \dot{V}_y\mathbf{j}) + (V_x\dot{\mathbf{i}} + V_y\dot{\mathbf{j}})$$

The first two terms in the expression represent that part of the total derivative of $\mathbf{V}$ that is measured relative to the x-y reference system, and the second two terms represent that part of the derivative due to the rotation of the reference system. With the expressions for $\dot{\mathbf{i}}$ and $\dot{\mathbf{j}}$ from Eqs. 5/11, we may now write

$$\boxed{\left(\frac{d\mathbf{V}}{dt}\right)_{XY} = \left(\frac{d\mathbf{V}}{dt}\right)_{xy} + \boldsymbol{\omega} \times \mathbf{V}} \qquad \textbf{(5/13)}$$

Here $\boldsymbol{\omega} \times \mathbf{V}$ represents the difference between the time derivative of the vector as measured in a fixed reference and its time derivative as measured in the rotating reference. As we shall see in Art. 7/2 where three-dimensional motion is introduced, Eq. 5/13 is valid in three dimensions, as well as in two dimensions.

The physical significance of Eq. 5/13 is illustrated in Fig. 5/12, which shows the vector $\mathbf{V}$ at time t as observed both in the fixed axes X-Y and in the rotating axes x-y. Since we are dealing with the effects of rotation only, the vector may be shown through the origin of coordinates without loss of generality. During time dt, the vector swings to position $\mathbf{V}'$, and the observer in x-y measures the components dV due to its change in magnitude and $V\,d\beta$ due to its rotation $d\beta$ relative to x-y. To the rotating observer, then, the derivative $(d\mathbf{V}/dt)_{xy}$ that he measures has the components dV/dt and $V\,d\beta/dt = V\dot{\beta}$. The remaining part of the total time derivative not measured by the rotating observer has the magnitude $V\,d\theta/dt$ and, expressed as a vector, is $\boldsymbol{\omega} \times \mathbf{V}$. Thus, we see physically from the diagram that

$$(\dot{\mathbf{V}})_{XY} = (\dot{\mathbf{V}})_{xy} + \boldsymbol{\omega} \times \mathbf{V}$$

which is Eq. 5/13.

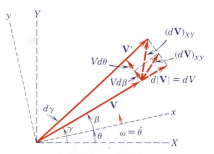

Figure 5/12

(c) *Relative acceleration.* The relative-acceleration equation may be obtained by differentiating the relative-velocity relation, Eq. 5/12. Thus,

$$\mathbf{a}_A = \mathbf{a}_B + \dot{\boldsymbol{\omega}} \times \mathbf{r} + \boldsymbol{\omega} \times \dot{\mathbf{r}} + \dot{\mathbf{v}}_{\text{rel}}$$

From the derivation of Eq. 5/12 we saw that

$$\dot{\mathbf{r}} = \frac{d}{dt}(x\mathbf{i} + y\mathbf{j}) = (\dot{x}\mathbf{i} + \dot{y}\mathbf{j}) + (x\dot{\mathbf{i}} + y\dot{\mathbf{j}})$$

$$= \boldsymbol{\omega} \times \mathbf{r} + \mathbf{v}_{\text{rel}}$$

Therefore, the third term on the right side of the acceleration equation becomes

$$\boldsymbol{\omega} \times \dot{\mathbf{r}} = \boldsymbol{\omega} \times (\boldsymbol{\omega} \times \mathbf{r} + \mathbf{v}_{\text{rel}}) = \boldsymbol{\omega} \times (\boldsymbol{\omega} \times \mathbf{r}) + \boldsymbol{\omega} \times \mathbf{v}_{\text{rel}}$$

With the aid of Eqs. 5/11, the last term on the right side of the equation for $\mathbf{a}_A$ becomes

$$\dot{\mathbf{v}}_{\text{rel}} = \frac{d}{dt}(\dot{x}\mathbf{i} + \dot{y}\mathbf{j}) = (\ddot{x}\mathbf{i} + \ddot{y}\mathbf{j}) + (\dot{x}\dot{\mathbf{i}} + \dot{y}\dot{\mathbf{j}})$$

$$= \boldsymbol{\omega} \times (\dot{x}\mathbf{i} + \dot{y}\mathbf{j}) + (\ddot{x}\mathbf{i} + \ddot{y}\mathbf{j})$$

$$= \boldsymbol{\omega} \times \mathbf{v}_{\text{rel}} + \mathbf{a}_{\text{rel}}$$

Substitution into the expression for $\mathbf{a}_A$ and collection of terms give

$$\boxed{\mathbf{a}_A = \mathbf{a}_B + \dot{\boldsymbol{\omega}} \times \mathbf{r} + \boldsymbol{\omega} \times (\boldsymbol{\omega} \times \mathbf{r}) + 2\boldsymbol{\omega} \times \mathbf{v}_{\text{rel}} + \mathbf{a}_{\text{rel}}} \quad \textbf{(5/14)}$$

Equation 5/14 is the general vector expression for the absolute acceleration of a particle A in terms of its acceleration $\mathbf{a}_{\text{rel}}$ measured relative to a moving coordinate system that rotates with an angular velocity $\boldsymbol{\omega}$ and an angular acceleration $\dot{\boldsymbol{\omega}}$. The terms $\dot{\boldsymbol{\omega}} \times \mathbf{r}$ and $\boldsymbol{\omega} \times (\boldsymbol{\omega} \times \mathbf{r})$ are shown in Fig. 5/13. They represent, respectively, the t- and n-components of the acceleration $\mathbf{a}_{P/B}$ of the coincident point P in its circular motion with respect to B. This motion would be observed from a set of nonrotating axes moving with B. The magnitude of $\dot{\boldsymbol{\omega}} \times \mathbf{r}$ is $r\ddot{\theta}$ and its direction is tangent to the circle. The magnitude of $\boldsymbol{\omega} \times (\boldsymbol{\omega} \times \mathbf{r})$ is $r\omega^2$ and its direction is from P to B along the normal to the circle. The acceleration of A relative to the plate along the path, $\mathbf{a}_{\text{rel}}$, may be expressed in rectangular, normal and tangential, or polar coordinates in the rotating system. Generally, n- and t-components are most frequently used, and these components are depicted in Fig. 5/13. The tangential component would have the magnitude $(a_{\text{rel}})_t = \ddot{s}$, where s is the distance measured along the path to A. The normal component would have the magnitude v_{rel}^2/ρ, where ρ is the radius of curvature of the path as measured in x-y. The sense of this vector would always be toward the center of curvature.

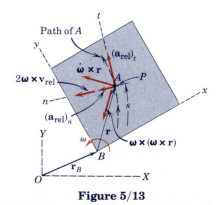

Figure 5/13

The term $2\boldsymbol{\omega} \times \mathbf{v}_{\mathrm{rel}}$, shown in Fig. 5/13, is known as the *Coriolis* acceleration* and represents the difference between the acceleration of A relative to P as measured from nonrotating axes and from rotating axes. The direction is always normal to the term $\mathbf{v}_{\mathrm{rel}}$, and the sense is established by the right-hand rule for the cross product.

The Coriolis acceleration is difficult to visualize because it is composed of two separate physical effects. To help with this visualization, we will consider the simplest possible motion in which this term appears. In Fig. 5/14*a* we have a rotating disk with a radial slot in which a small particle A is confined to slide. Let the disk turn with a constant angular velocity $\omega = \dot{\theta}$ and let the particle move along the slot with a constant speed $v_{\mathrm{rel}} = \dot{x}$ relative to the slot. The velocity of A has the components $\dot{x}$ due to motion along the slot and $x\omega$ due to the rotation of the slot. The changes in these two velocity components due to the rotation of the disk are shown in the *b*-part of the figure for the interval dt, during which the x-y axes rotate with the disk through the angle $d\theta$ to x'-y'. The velocity increment due to the change in direction of $\mathbf{v}_{\mathrm{rel}}$ is $\dot{x}\, d\theta$ and that due to the change in magnitude of $x\omega$ is $\omega\, dx$, both being in the y-direction normal to the slot. Dividing each increment by dt and adding give the sum $\omega\dot{x} + \dot{x}\omega = 2\dot{x}\omega$, which is the magnitude of the Coriolis acceleration $2\boldsymbol{\omega} \times \mathbf{v}_{\mathrm{rel}}$. Dividing the remaining velocity increment $x\omega\, d\theta$ due to the change in direction of $x\omega$ by dt gives $x\omega\dot{\theta}$ or $x\omega^2$, which is the acceleration of a point P fixed to the slot and momentarily coincident with the particle A. We now see how Eq. 5/14 fits these results. With the origin B in that equation taken at the fixed center O, $\mathbf{a}_B = \mathbf{0}$. With constant angular velocity, $\dot{\boldsymbol{\omega}} \times \mathbf{r} = \mathbf{0}$. With $\mathbf{v}_{\mathrm{rel}}$ constant in magnitude and no curvature to the slot, $\mathbf{a}_{\mathrm{rel}} = \mathbf{0}$. We are left with

$$\mathbf{a}_A = \boldsymbol{\omega} \times (\boldsymbol{\omega} \times \mathbf{r}) + 2\boldsymbol{\omega} \times \mathbf{v}_{\mathrm{rel}}$$

Replacing $\mathbf{r}$ by $x\mathbf{i}$, $\boldsymbol{\omega}$ by $\omega\mathbf{k}$, and $\mathbf{v}_{\mathrm{rel}}$ by $\dot{x}\mathbf{i}$ gives

$$\mathbf{a}_A = -x\omega^2\mathbf{i} + 2\dot{x}\omega\mathbf{j}$$

which checks our analysis from Fig. 5/14. We also note that this same result is contained in our polar-coordinate analysis of plane curvilinear motion in Eq. 2/14 when we let $\ddot{r} = 0$ and $\ddot{\theta} = 0$ and replace r by x and $\dot{\theta}$ by ω. If the slot in the disk of Fig. 5/14 had been curved, we would have had a normal component of acceleration relative to the slot so that $\mathbf{a}_{\mathrm{rel}}$ would not be zero.

The following comparison will help to establish the equivalence of, and clarify the differences between, the relative-acceleration equations written for rotating and nonrotating reference axes:

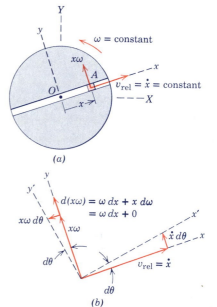

(a)

(b)

Figure 5/14

---

*Named after the French miliary engineer G. Coriolis (1792–1843), who was the first to call attention to this term.

$$\mathbf{a}_A = \mathbf{a}_B + \underbrace{\dot{\boldsymbol{\omega}} \times \mathbf{r} + \boldsymbol{\omega} \times (\boldsymbol{\omega} \times \mathbf{r})} + \underbrace{2\boldsymbol{\omega} \times \mathbf{v}_{\text{rel}} + \mathbf{a}_{\text{rel}}}$$

$$\mathbf{a}_A = \mathbf{a}_B + \underbrace{\qquad\qquad \mathbf{a}_{P/B} \qquad\qquad} + \qquad \mathbf{a}_{A/P}$$

$$\mathbf{a}_A = \qquad\qquad \mathbf{a}_P \qquad\qquad\qquad + \qquad \mathbf{a}_{A/P}$$ (5/14a)

$$\mathbf{a}_A = \mathbf{a}_B + \underbrace{\qquad\qquad\qquad \mathbf{a}_{A/B} \qquad\qquad\qquad}$$

The equivalence of $\mathbf{a}_{P/B}$ and $\dot{\boldsymbol{\omega}} \times \mathbf{r} + \boldsymbol{\omega} \times (\boldsymbol{\omega} \times \mathbf{r})$, as shown in the second equation, has already been described. From the third equation where $\mathbf{a}_B + \mathbf{a}_{P/B}$ has been combined to give $\mathbf{a}_P$, it is seen that the relative-acceleration term $\mathbf{a}_{A/P}$, unlike the corresponding relative-velocity term, is not equal to the relative acceleration $\mathbf{a}_{\text{rel}}$ measured from the rotating x-y frame of reference. The Coriolis term is, therefore, the difference between the acceleration $\mathbf{a}_{A/P}$ of A relative to P as measured in a nonrotating system and the acceleration $\mathbf{a}_{\text{rel}}$ of A relative to P as measured in a rotating system. From the fourth equation, it is seen that the acceleration $\mathbf{a}_{A/B}$ of A with respect to B as measured in a nonrotating system, Eq. 2/21, is a combination of the last four terms in the first equation for the rotating system.

The results expressed by Eq. 5/14 may be visualized somewhat more simply by writing the acceleration of A in terms of the acceleration of the coincident point P. Since the acceleration of P is $\mathbf{a}_P = \mathbf{a}_B + \dot{\boldsymbol{\omega}} \times \mathbf{r} + \boldsymbol{\omega} \times (\boldsymbol{\omega} \times \mathbf{r})$, we may rewrite Eq. 5/14 as

$$\boxed{\mathbf{a}_A = \mathbf{a}_P + 2\boldsymbol{\omega} \times \mathbf{v}_{\text{rel}} + \mathbf{a}_{\text{rel}}}$$ **(5/14b)**

When written in this form, point P may not be picked at random because it is the one point attached to the rotating reference frame coincident with A at the instant of analysis. Again, reference to Fig. 5/13 should be made to clarify the meaning of each of the terms in Eq. 5/14 and its equivalent, Eq. 5/14b.

In summary, once we have chosen our rotating reference system, we must recognize the following quantities:

$\mathbf{v}_B$ = absolute velocity of the origin B of the rotating axes

$\mathbf{a}_B$ = absolute acceleration of the origin B of the rotating axes

$\mathbf{r}$ = position vector of the coincident point P measured from B

$\boldsymbol{\omega}$ = angular velocity of the rotating axes

$\dot{\boldsymbol{\omega}}$ = angular acceleration of the rotating axes

$\mathbf{v}_{\text{rel}}$ = velocity of A measured relative to the rotating axes

$\mathbf{a}_{\text{rel}}$ = acceleration of A measured relative to the rotating axes

Also, it is well to be reminded that our vector analysis depends on the consistent use of a right-handed set of coordinate axes. Finally, note is made of the fact that Eqs. 5/12 and 5/14, developed here for plane motion, hold equally well for space motion, and this extension of generality will be covered in Art. 7/6.

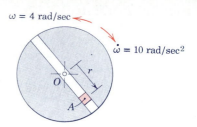

Sample Problem 5/16

At the instant represented, the disk with the radial slot is rotating about O with a counterclockwise angular velocity of 4 rad/sec which is decreasing at the rate of 10 rad/sec^2. The motion of slider A is separately controlled, and at this instant, $r = 6$ in., $\dot{r} = 5$ in./sec, and $\ddot{r} = 81$ in/sec^2. Determine the absolute velocity and acceleration of A for this position.

Solution. We have motion relative to a rotating path, so that a rotating coordinate system with origin at O is indicated. We attach x-y axes to the disk and use the unit vectors $\mathbf{i}$ and $\mathbf{j}$.

Velocity. With the origin at O, the term $\mathbf{v}_B$ of Eq. 5/12 disappears and we have

① $$\mathbf{v}_A = \boldsymbol{\omega} \times \mathbf{r} + \mathbf{v}_{\text{rel}}$$

The angular velocity as a vector is $\boldsymbol{\omega} = 4\mathbf{k}$ rad/sec, where $\mathbf{k}$ is the unit ② vector normal to the x-y plane in the $+z$-direction. Our relative-velocity equation becomes

$$\mathbf{v}_A = 4\mathbf{k} \times 6\mathbf{i} + 5\mathbf{i} = 24\mathbf{j} + 5\mathbf{i} \text{ in./sec} \qquad Ans.$$

in the direction indicated and has the magnitude

$$v_A = \sqrt{(24)^2 + (5)^2} = 24.5 \text{ in./sec} \qquad Ans.$$

Equation 5/14 written for zero acceleration of the origin of the rotating coordinate system is

$$\mathbf{a}_A = \boldsymbol{\omega} \times (\boldsymbol{\omega} \times \mathbf{r}) + \dot{\boldsymbol{\omega}} \times \mathbf{r} + 2\boldsymbol{\omega} \times \mathbf{v}_{\text{rel}} + \mathbf{a}_{\text{rel}}$$

The terms become

③ $$\boldsymbol{\omega} \times (\boldsymbol{\omega} \times \mathbf{r}) = 4\mathbf{k} \times (4\mathbf{k} \times 6\mathbf{i}) = 4\mathbf{k} \times 24\mathbf{j} = -96\mathbf{i} \text{ in./sec}^2$$

$$\dot{\boldsymbol{\omega}} \times \mathbf{r} = -10\mathbf{k} \times 6\mathbf{i} = -60\mathbf{j} \text{ in./sec}^2$$

$$2\boldsymbol{\omega} \times \mathbf{v}_{\text{rel}} = 2(4\mathbf{k}) \times 5\mathbf{i} = 40\mathbf{j} \text{ in./sec}^2$$

$$\mathbf{a}_{\text{rel}} = 81\mathbf{i} \text{ in./sec}^2$$

The total acceleration is, therefore,

$$\mathbf{a}_A = (81 - 96)\mathbf{i} + (40 - 60)\mathbf{j} = -15\mathbf{i} - 20\mathbf{j} \text{ in./sec}^2 \quad Ans.$$

in the direction indicated and has the magnitude

$$a_A = \sqrt{(15)^2 + (20)^2} = 25 \text{ in./sec}^2 \qquad Ans.$$

Vector notation is certainly not essential to the solution of this problem. The student should be able to work out the steps with scalar notation just as easily. The correct direction of the Coriolis-acceleration term may always be found by the direction in which the head of the $\mathbf{v}_{\text{rel}}$ vector would move if rotated about its tail in the sense of $\boldsymbol{\omega}$ as shown.

① This equation is the same as $\mathbf{v}_A = \mathbf{v}_P + \mathbf{v}_{A/P}$, where P is a point attached to the disk coincident with A at this instant.

② Note that the x-y-z axes chosen constitute a right-handed system.

③ Be sure to recognize that $\boldsymbol{\omega} \times (\boldsymbol{\omega} \times \mathbf{r})$ and $\dot{\boldsymbol{\omega}} \times \mathbf{r}$ represent the normal and tangential components of acceleration of a point P on the disk coincident with A. This description becomes that of Eq. 5/14b.

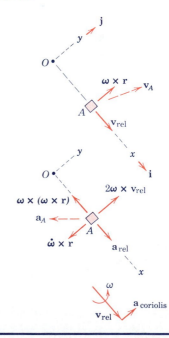

Sample Problem 5/17

The pin A of the hinged link AC is confined to move in the rotating slot of link OD. The angular velocity of OD is $\omega = 2$ rad/s clockwise and is constant for the interval of motion concerned. For the position where $\theta = 45°$ with AC horizontal, determine the velocity of pin A and the velocity of A relative to the rotating slot in OD.

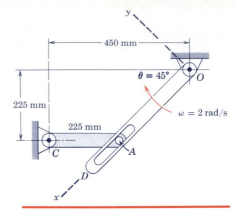

Solution. Motion of a point (pin A) along a rotating path (the slot) suggests the use of rotating coordinate axes x-y attached to arm OD. With the origin at the fixed point O, the term $\mathbf{v}_B$ of Eq. 5/12 vanishes, and we have $\mathbf{v}_A = \boldsymbol{\omega} \times \mathbf{r} + \mathbf{v}_{rel}$.

The velocity of A in its circular motion about C is

$$\mathbf{v}_A = \boldsymbol{\omega}_{CA} \times \mathbf{r}_{CA} = \omega_{CA}\mathbf{k} \times (225/\sqrt{2})(-\mathbf{i} - \mathbf{j}) = (225/\sqrt{2})\omega_{CA}(\mathbf{i} - \mathbf{j})$$

① where the angular velocity ω_{CA} is arbitrarily assigned in a clockwise sense in the positive z-direction $(+\mathbf{k})$.

The angular velocity $\boldsymbol{\omega}$ of the rotating axes is that of the arm OD and, by the right-hand rule, is $\boldsymbol{\omega} = \omega\mathbf{k} = 2\mathbf{k}$ rad/s. The vector from the origin to the point P on OD coincident with A is $\mathbf{r} = \overline{OP}\mathbf{i} = \sqrt{(450 - 225)^2 + (225)^2}\,\mathbf{i} = 225\sqrt{2}\mathbf{i}$ mm. Thus,

$$\boldsymbol{\omega} \times \mathbf{r} = 2\mathbf{k} \times 225\sqrt{2}\mathbf{i} = 450\sqrt{2}\mathbf{j} \text{ mm/s}$$

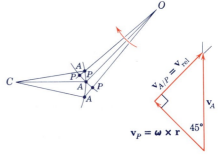

Finally, the relative-velocity term $\mathbf{v}_{rel}$ is the velocity measured by an observer attached to the rotating reference frame and is $\mathbf{v}_{rel} = \dot{x}\mathbf{i}$. Substitution into the relative-velocity equation gives

$$(225/\sqrt{2})\omega_{CA}(\mathbf{i} - \mathbf{j}) = 450\sqrt{2}\mathbf{j} + \dot{x}\mathbf{i}$$

Equating separately the coefficients of the $\mathbf{i}$ and $\mathbf{j}$ terms yields

$$(225/\sqrt{2})\omega_{CA} = \dot{x} \quad \text{and} \quad -(225/\sqrt{2})\omega_{CA} = 450\sqrt{2}$$

giving

$$\omega_{CA} = -4 \text{ rad/s} \quad \text{and} \quad \dot{x} = v_{rel} = -450\sqrt{2} \text{ mm/s} \quad \textit{Ans.}$$

With a negative value for ω_{CA}, the actual angular velocity of CA is counterclockwise, so the velocity of A is up with a magnitude of

$$v_A = 225(4) = 900 \text{ mm/s} \quad \textit{Ans.}$$

Geometric clarification of the terms is helpful and is easily shown. Using the equivalence between the third and the first of Eqs. 5/12a with $\mathbf{v}_B = \mathbf{0}$ enables us to write $\mathbf{v}_A = \mathbf{v}_P + \mathbf{v}_{A/P}$, where P is the point on the rotating arm OD coincident with A. Clearly, $v_P = \overline{OP}\omega = 225\sqrt{2}(2) = 450\sqrt{2}$ mm/s and its direction is normal to OD. The relative velocity $\mathbf{v}_{A/P}$, which is the same as $\mathbf{v}_{rel}$, is seen from the figure to be along the slot toward O. This conclusion becomes clear when it is observed that A is approaching P along the slot from below before coincidence and is receding from P upward along the slot following coincidence. The velocity of A is tangent to its circular arc about C. The vector equation may now be satisfied since there are only two remaining scalar unknowns, namely, the magnitude of $\mathbf{v}_{A/P}$ and the magnitude of $\mathbf{v}_A$. For the 45° position, the figure requires $v_{A/P} = 450\sqrt{2}$ mm/s and $v_A = 900$ mm/s, each in its direction shown. The angular velocity of AC is

$$[\omega = v/r] \quad \omega_{AC} = v_A/\overline{AC} = 900/225 = 4 \text{ rad/s counterclockwise}$$

① It is clear enough physically that CA will have a counterclockwise angular velocity for the conditions specified, so we anticipate a negative value for ω_{CA}.

② Solution of the problem is not restricted to the reference axes used. Alternatively, the origin of the x-y axes, still attached to CD, could be chosen at the coincident point P on CD. This choice would merely replace the $\boldsymbol{\omega} \times \mathbf{r}$ term by its equal, $\mathbf{v}_P$. As a further selection, all vector quantities could be expressed in terms of X-Y components using unit vectors $\mathbf{I}$ and $\mathbf{J}$.

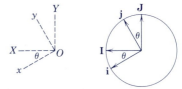

A direct conversion between the two reference systems is obtained from the geometry of the unit circle and gives

$$\mathbf{i} = \mathbf{I} \cos \theta - \mathbf{J} \sin \theta$$

and

$$\mathbf{j} = \mathbf{I} \sin \theta + \mathbf{J} \cos \theta$$

Sample Problem 5/18

For the conditions of Sample Problem 5/17, determine the angular acceleration of AC and the acceleration of A relative to the rotating slot in arm OD.

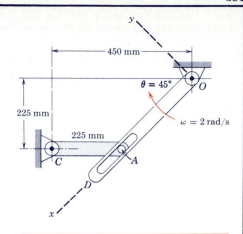

Solution. We attach the rotating coordinate system x-y to arm OD and use Eq. 5/14. With the origin at the fixed point O, the term $\mathbf{a}_B$ becomes zero so that

$$\mathbf{a}_A = \dot{\boldsymbol{\omega}} \times \mathbf{r} + \boldsymbol{\omega} \times (\boldsymbol{\omega} \times \mathbf{r}) + 2\boldsymbol{\omega} \times \mathbf{v}_{\text{rel}} + \mathbf{a}_{\text{rel}}$$

From the solution to Sample Problem 5/17, we make use of the values $\boldsymbol{\omega} = 2\mathbf{k}$ rad/s, $\boldsymbol{\omega}_{CA} = -4\mathbf{k}$ rad/s, $\mathbf{v}_{\text{rel}} = -450\sqrt{2}\mathbf{i}$ mm/s and write

$$\mathbf{a}_A = \dot{\boldsymbol{\omega}}_{CA} \times \mathbf{r}_{CA} + \boldsymbol{\omega}_{CA} \times (\boldsymbol{\omega}_{CA} \times \mathbf{r}_{CA})$$

$$= \dot{\omega}_{CA}\mathbf{k} \times \frac{225}{\sqrt{2}}(-\mathbf{i} - \mathbf{j}) - 4\mathbf{k} \times \left(-4\mathbf{k} \times \frac{225}{\sqrt{2}}[-\mathbf{i} - \mathbf{j}]\right)$$

$$\dot{\boldsymbol{\omega}} \times \mathbf{r} = \mathbf{0} \text{ since } \boldsymbol{\omega} = \text{constant}$$

$$\boldsymbol{\omega} \times (\boldsymbol{\omega} \times \mathbf{r}) = 2\mathbf{k} \times (2\mathbf{k} \times 225\sqrt{2}\mathbf{i}) = -900\sqrt{2}\mathbf{i} \text{ mm/s}^2$$

$$2\boldsymbol{\omega} \times \mathbf{v}_{\text{rel}} = 2(2\mathbf{k}) \times (-450\sqrt{2}\mathbf{i}) = -1800\sqrt{2}\mathbf{j} \text{ mm/s}^2$$

$$\mathbf{a}_{\text{rel}} = \ddot{x}\mathbf{i}$$

① Substitution into the relative-acceleration equation yields

$$\frac{1}{\sqrt{2}}(225\dot{\omega}_{CA} + 3600)\mathbf{i} + \frac{1}{\sqrt{2}}(-225\dot{\omega}_{CA} + 3600)\mathbf{j}$$

$$= -900\sqrt{2}\mathbf{i} - 1800\sqrt{2}\mathbf{j} + \ddot{x}\mathbf{i}$$

Equating separately the $\mathbf{i}$ and $\mathbf{j}$ terms gives

$$(225\dot{\omega}_{CA} + 3600)/\sqrt{2} = -900\sqrt{2} + \ddot{x}$$

and $$(-225\dot{\omega}_{CA} + 3600)/\sqrt{2} = -1800\sqrt{2}$$

Solving for the two unknowns gives

$$\dot{\omega}_{CA} = 32 \text{ rad/s}^2 \quad \text{and} \quad \ddot{x} = a_{\text{rel}} = 8910 \text{ mm/s}^2 \quad \textit{Ans.}$$

If desired, the acceleration of A may also be written as

$$\mathbf{a}_A = (225/\sqrt{2})(32)(\mathbf{i} - \mathbf{j}) + (3600/\sqrt{2})(\mathbf{i} + \mathbf{j}) = 7637\mathbf{i} - 2546\mathbf{j} \text{ mm/s}^2$$

We make use here of the geometric representation of the relative-acceleration equation to further clarify the problem. The geometric approach may be used as an alternative solution. Again, we introduce point P on OD coincident with A. The equivalent scalar terms are

$$(a_A)_t = |\dot{\boldsymbol{\omega}}_{CA} \times \mathbf{r}_{CA}| = r\dot{\omega}_{CA} = r\alpha_{CA} \text{ normal to } CA, \text{ sense unknown}$$

$$(a_A)_n = |\boldsymbol{\omega}_{CA} \times (\boldsymbol{\omega}_{CA} \times \mathbf{r}_{CA})| = r\omega_{CA}^2 \text{ from } A \text{ to } C$$

$$(a_P)_n = |\boldsymbol{\omega} \times (\boldsymbol{\omega} \times \mathbf{r})| = \overline{OP}\omega^2 \text{ from } P \text{ to } O$$

$$(a_P)_t = |\dot{\boldsymbol{\omega}} \times \mathbf{r}| = r\dot{\omega} = 0 \text{ since } \boldsymbol{\omega} = \text{constant}$$

$$|2\boldsymbol{\omega} \times \mathbf{v}_{\text{rel}}| = 2\omega v_{\text{rel}} \text{ directed as shown}$$

$$a_{\text{rel}} = \ddot{x} \text{ along } OD, \text{ sense unknown}$$

We start with the known vectors and add them head-to-tail for each side of the equation beginning at R and ending at S, where the intersection of the known directions of $(\mathbf{a}_A)_t$ and $\mathbf{a}_{\text{rel}}$ establishes the solution. Closure of the polygon determines the sense of each of the two unknown vectors, ② and their magnitudes are easily calculated from the figure geometry.

① If the slot had been curved with a radius of curvature ρ, the term $\mathbf{a}_{\text{rel}}$ would have had a component v_{rel}^2/ρ normal to the slot and directed toward the center of curvature in addition to its component along the slot.

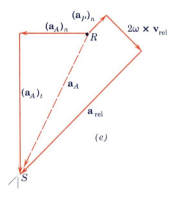

(e)

② It is always possible to avoid a simultaneous solution by projecting the vectors onto the perpendicular to one of the unknowns.

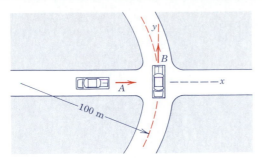

Problem 5/151

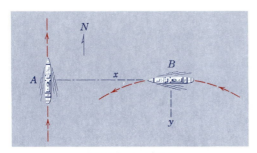

Problem 5/152

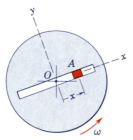

Problem 5/153

PROBLEMS

Introductory problems

5/151 Car *B* is rounding the curve with a constant speed of 54 km/h, and car *A* is approaching car *B* in the intersection with a constant speed of 72 km/h. Determine the velocity that car *A* appears to have to an observer riding in and turning with car *B*. The *x*-*y* axes are attached to car *B*. Is this apparent velocity the negative of the velocity that *B* appears to have to a nonrotating observer in car *A*? The distance separating the two cars at the instant depicted is 40 m. *Ans.* $\mathbf{v}_{\text{rel}} = 20\mathbf{i} - 9\mathbf{j}$ m/s, No

5/152 Ship *A* is proceeding north at a constant speed of 12 knots, while ship *B* is moving at 10 knots while turning to port (left) at the constant rate of 10 deg/min. When the ships are separated by a distance of 2 nautical miles in the relative positions shown, with *B* momentarily heading west and *A* crossing its bow, the navigator of *B* measures the apparent velocity of *A*. Find the magnitude of this velocity and specify the clockwise angle β that it makes with the north direction.

5/153 The disk with the radial slot rotates in a horizontal plane about *O* with an angular acceleration $\dot{\omega} = 20$ rad/s² while the 2-kg slider *A* moves with a constant speed $\dot{x} = 120$ mm/s relative to the slot during a certain interval of its motion. If the angular velocity of the disk is $\omega = 15$ rad/s at the instant when the slider crosses the center *O* of rotation of the disk, determine the acceleration of the slider at this instant by using both Eq. 5/14 and Eq. 2/14. *Ans.* $\mathbf{a}_A = 3.6\mathbf{j}$ m/s²

5/154 For the cars of Prob. 5/151 traveling with constant speed, determine the acceleration that car *A* appears to have to an observer riding in and turning with car *B*.

5/155 Consider a straight and level railroad track with a 500-ton railroad car moving along it at a constant speed of 50 mi/hr. Determine the horizontal force *R* exerted by the rails on the car if the track were located hypothetically at (*a*) the north pole and (*b*) the equator, oriented in a north-south direction. *Ans.* (*a*) $R = 33.2$ lb, (*b*) $R = 0$

5/156 A vehicle A travels west at high speed on a perfectly straight road B which is tangent to the earth's surface at the equator. The road has no curvature whatsoever in the vertical plane. Determine the necessary speed v_{rel} of the vehicle relative to the road that will give rise to zero acceleration of the vehicle in the vertical direction. Assume that the center of the earth has no acceleration.

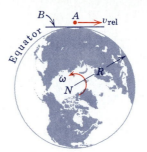

Problem 5/156

5/157 If the road of Prob. 5/156, instead of being straight with no curvature, followed the curvature of the earth's surface, determine the necessary speed v_{rel} of the vehicle to the west relative to the road that will give rise to zero vertical acceleration of the vehicle. Assume that the center of the earth has no acceleration.   *Ans.* $v_{rel} = 1674$ km/h

5/158 In negotiating an unbanked curve at a steady speed of 25 km/h, the center C of the railroad car follows a circular path of radius $\rho = 60$ m. The longitudinal axis of the car remains tangent to the circle. Determine the absolute velocity $\mathbf{v}$ of a person P who walks at the constant speed of 1.5 m/s relative to the car when he is at points A, B, and C. Use the x-y axes attached to the car as shown.

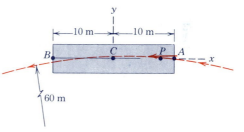

Problem 5/158

Representative problems

5/159 The slider A moves in the slot at the same time that the disk rotates about its center O with an angular speed ω positive in the counterclockwise sense. Determine the x- and y-components of the absolute acceleration of A if, at the instant represented, $\omega = 5$ rad/sec, $\dot{\omega} = -10$ rad/sec^2, $x = 4$ in., $\dot{x} = 6$ in./sec, and $\ddot{x} = 20$ in./sec^2.
        Ans. $(a_A)_x = -40$ in./sec^2, $(a_A)_y = 80$ in./sec^2

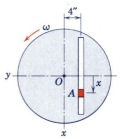

Problem 5/159

5/160 The disk with the circular slot of 200-mm radius rotates about O with a constant angular velocity $\omega = 15$ rad/s. Determine the acceleration of the slider A at the instant when it passes the center of the disk if, at that moment, $\dot{\theta} = 12$ rad/s and $\ddot{\theta} = 0$.

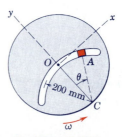

Problem 5/160

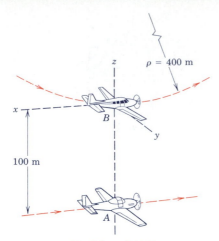

Problem 5/161

5/161 Aircraft B has a constant speed of 540 km/h at the bottom of a circular loop of 400-m radius. Aircraft A flying horizontally in the plane of the loop passes 100 m directly under B at a constant speed of 360 km/h. With coordinate axes attached to B as shown, determine the acceleration that A appears to have to the pilot of B for this instant.

Ans. $\mathbf{a}_{\text{rel}} = -4.69\mathbf{k}$ m/s^2

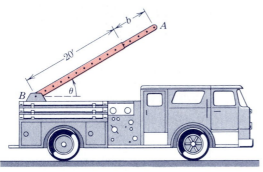

Problem 5/162

5/162 The fire truck is moving forward at a speed of 35 mi/hr and is decelerating at the rate of 10 ft/sec^2. Simultaneously, the ladder is being raised and extended. At the instant considered the angle θ is 30° and is increasing at the constant rate of 10 deg/sec. Also at this instant the extension b of the ladder is 5 ft, with $\dot{b} = 2$ ft/sec and $\ddot{b} = -1$ ft/sec^2. For this instant determine the acceleration of the end A of the ladder (a) with respect to the truck and (b) with respect to the ground.

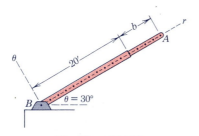

Problem 5/163

5/163 For an alternative solution to Prob. 5/162 assign r-θ coordinates with origin at B as shown. Then make use of the polar coordinate relations for the acceleration of A relative to B. The r- and θ- components of the absolute acceleration should coincide with the components along and normal to the ladder as found in Prob. 5/162.

Ans. $\mathbf{a}_A = -10.42\mathbf{e}_r + 5.70\mathbf{e}_\theta$ ft/sec^2

5/164 Two boys A and B are sitting on opposite sides of a horizontal turntable which rotates at a constant counterclockwise angular velocity ω as seen from above. Boy A throws a ball toward B by giving it a horizontal velocity $\mathbf{u}$ relative to the turntable toward B. Assume that the ball has no horizontal acceleration once released and write an expression for the acceleration $\mathbf{a}_{\mathrm{rel}}$ which B would observe the ball to have in the plane of the turntable just after it is thrown. Sketch the path of the ball on the turntable as observed by B.

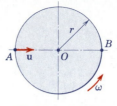

Problem 5/164

5/165 Determine the acceleration of the person described in Prob. 5/158 for each of the positions A, B, and C.

$$Ans. \quad \mathbf{a}_A = -0.1340\mathbf{i} - 1.151\mathbf{j} \text{ m/s}^2$$
$$\mathbf{a}_B = 0.1340\mathbf{i} - 1.151\mathbf{j} \text{ m/s}^2$$
$$\mathbf{a}_C = -1.151\mathbf{j} \text{ m/s}^2$$

5/166 Car B turns onto the circular off-ramp with a speed v. Car A, traveling with the same speed v, continues in a straight line. Prove that the velocity which A appears to have to an observer riding in and turning with car B is zero when car A passes the position shown regardless of the angle θ.

5/167 For the conditions and conclusion of Prob. 5/166 show that the acceleration that car A appears to have to an observer in and turning with car B is equal to v^2/R in the direction normal to the true velocity of A.

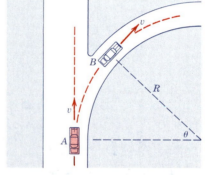

Problem 5/166

5/168 The slotted disk sector rotates with a constant counterclockwise angular velocity $\omega = 3$ rad/s. Simultaneously the slotted arm OC oscillates about the line OB (fixed to the disk) so that θ changes at the constant rate of 2 rad/s except at the extremities of the oscillation during reversal of direction. Determine the total acceleration of the pin A when $\theta = 30°$ and $\dot{\theta}$ is positive (clockwise).

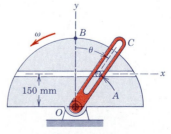

Problem 5/168

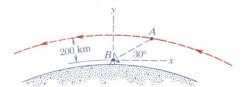

Problem 5/169

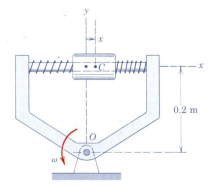

Problem 5/170

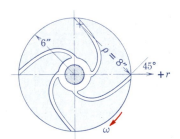

Problem 5/171

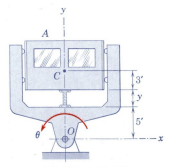

Problem 5/172

5/169 A radar station B situated at the equator observes a satellite A in a circular equatorial orbit of 200-km altitude and moving from west to east. When the satellite is 30° above the horizon, determine the difference between the velocity of the satellite relative to the radar station, as measured from a nonrotating frame of reference, and the velocity as measured relative to the reference frame of the radar station.

Ans. $\Delta\mathbf{v}_{\text{rel}} = -50.3\mathbf{i} + 87.1\mathbf{j}$ km/h

5/170 The spring-mounted collar oscillates on the shaft according to $x = 0.04 \sin \pi t$, where x is in meters and t is in seconds. Simultaneously the frame rotates about the bearing at O with an angular velocity $\omega = 2 \sin (\pi t/2)$ rad/s. Determine the acceleration of the center C of the collar (a) when $t = 3$ s and (b) when $t = 1/2$ s.

5/171 The figure shows the vanes of a centrifugal-pump impeller which turns with a constant clockwise speed of 200 rev/min. The fluid particles are observed to have an absolute velocity whose component in the r-direction is 10 ft/sec at discharge from the vane. Furthermore, the magnitude of the velocity of the particles measured relative to the vane is increasing at the rate of 80 ft/sec^2 just before they leave the vane. Determine the magnitude of the total acceleration of a fluid particle an instant before it leaves the impeller. The radius of curvature ρ of the vane at its end is 8 in.   *Ans.* $a = 156$ ft/sec^2

5/172 A test chamber A is used to study motion sickness. It is capable of oscillation about a horizontal axis through O according to $\theta = \theta_0 \sin 2\pi f_1 t$, and at the same time the chamber has a linear motion $y = y_0 \sin 2\pi f_2 t$ relative to the frame. For a certain series of tests, the amplitudes are set at $\theta_0 = \pi/4$ radians and $y_0 = 6$ in., while the frequencies are $f_1 = \frac{1}{4}$ cycle/sec and $f_2 = \frac{1}{2}$ cycle/sec. Determine the vector expression for the acceleration of point C in the chamber at the instant when $t = 2$ sec.

5/173 The crank *OA* revolves clockwise with a constant angular velocity of 10 rad/s within a limited arc of its motion. For the position $\theta = 30°$ determine the angular velocity of the slotted link *CB* and the acceleration of *A* as measured relative to the slot in *CB*.

Ans. $\omega = 5$ rad/s CW, $\mathbf{a}_{rel} = -8660\mathbf{i}$ mm/s^2

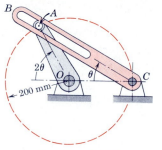

Problem 5/173

5/174 A tall building is situated on the equator. The north face of the building houses a standard 12-hr clock whose center is a distance h above the ground (which is essentially at sea level). Develop expressions for the velocity and acceleration of the tip A of the hour hand at 12 o'clock as measured from an earth-centered nonrotating coordinate system. Take the positive direction of the x-axis to be from the center of the earth radially outward toward the building and that of the z-axis toward the north. The hour hand has a length l and makes two complete rotations relative to the clock during one complete rotation of the earth. The radius and angular velocity of the earth are R and ω.

5/175 The Geneva wheel of Prob. 5/53 is shown again here. Determine the angular acceleration α_2 of wheel *C* for the instant when $\theta = 20°$. Wheel *A* has a constant clockwise angular velocity of 2 rad/s.

Ans. $\alpha_2 = 16.53$ rad/s^2 CCW

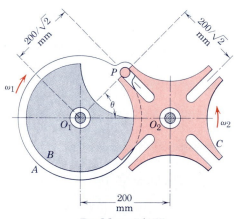

Problem 5/175

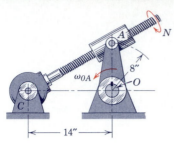

Problem 5/176

▶**5/176** The mechanism shown is a device to produce high torque in the shaft at O. The gear unit, pivoted at C, turns the right-handed screw at a constant speed $N = 100$ rev/min in the direction shown which advances the threaded collar at A along the screw toward C. Determine the time rate of change $\dot{\omega}_{AO}$ of the angular velocity of AO as it passes the vertical position shown. The screw has 8 single threads per inch of length.

Ans. $\dot{\omega}_{AO} = -3.88(10^{-4})$ rad/sec^2

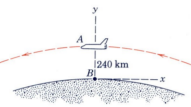

Problem 5/177

▶**5/177** The space shuttle A is in an equatorial circular orbit of 240-km altitude and is moving from west to east. Determine the velocity and acceleration which it appears to have to an observer B fixed to and rotating with the earth at the equator as the shuttle passes overhead. Use $R = 6378$ km for the radius of the earth. Also use Fig. 1/1 for the appropriate value of g and carry out your calculations to 4-figure accuracy.   *Ans.* $\mathbf{v}_{\text{rel}} = -26\ 220\mathbf{i}$ km/h

$\mathbf{a}_{\text{rel}} = -8.018\mathbf{j}$ m/s^2

(using $g = 9.814$ m/s^2)

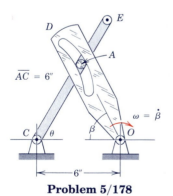

Problem 5/178

▶**5/178** Determine the angular acceleration of link EC in the position shown, where $\omega = \dot{\beta} = 2$ rad/sec and $\ddot{\beta} = 6$ rad/sec^2 when $\theta = \beta = 60°$. Pin A is fixed to link EC. The circular slot in link DO has a radius of curvature of 6 in. In the position shown, the tangent to the slot at the point of contact is parallel to AO.

Ans. $\alpha_{EC} = 12$ rad/sec^2 CCW

5/8 *PROBLEM FORMULATION AND REVIEW*

In Chapter 5 we have applied our knowledge of basic kinematics, which we developed in Chapter 2, to the plane motion of rigid bodies. We approached the problem essentially in two ways. First, we wrote an equation that described the general geometric configuration of a given problem in terms of knowns and variables. Then we differentiated the equation with respect to time to obtain velocities and accelerations, both linear and angular. We called this approach an absolute-motion analysis, and it is satisfactory when the geometric expression and its time derivatives can be easily written.

In our second approach, we applied the principles of relative motion to rigid bodies and found that this approach enabled us to solve many problems that would be too awkward to handle by mathematical differentiation. The relative-velocity equation, the instantaneous center of zero velocity, and the relative-acceleration equation all require that we visualize clearly and analyze correctly the case of circular motion of one point around another point, as viewed from nonrotating axes. Recognition of the relationships for circular motion is certainly an essential key without which little progress can be made.

The relative-velocity and relative-acceleration relationships are vector equations that we may solve in any one of three ways: by a scalar-geometric analysis of the vector polygon, by vector algebra, or by a graphical construction of the vector polygon. Each method has its advantages, and the student is urged to gain experience with all three. If one uses a vector-algebra approach, he should sketch the geometry of the vectors in any event, because the ability to visualize physical-geometrical relationships of force and motion is probably the most important experience of all in the study of mechanics.

In Chapter 6 we shall study the kinetics of rigid bodies in plane motion. Here we find that the ability to solve for the linear and angular accelerations of rigid bodies is absolutely necessary in order to apply the force and moment equations that relate the applied forces to the resulting motions. Thus, the material of Chapter 5 is essential to that in Chapter 6.

Finally, in Chapter 5 we introduced rotating coordinate systems that enable us to solve problems where the motion is observed relative to a rotating frame of reference. Whenever a point moves along a path that itself is turning, analysis by rotating axes is indicated if a relative-motion approach is used. In deriving Eq. 5/12 for relative velocity and Eq. 5/14 for relative acceleration, where the relative terms are measured from a rotating reference system, it was necessary for us to account for the time derivatives of the unit vectors **i** and **j**. Equations 5/12 and 5/14 also apply to spatial motion, as will be shown in Chapter 7.

Problem 5/179

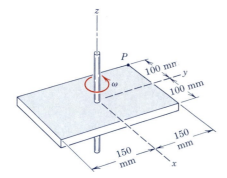

Problem 5/180

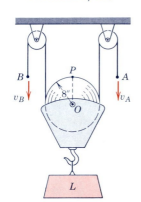

Problem 5/183

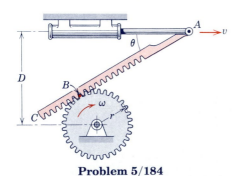

Problem 5/184

REVIEW PROBLEMS

5/179 The circular disk rotates about its z-axis with an angular velocity $\omega = 2$ rad/s. A point P located on the rim has a velocity given by $\mathbf{v} = -0.8\mathbf{i} - 0.6\mathbf{j}$ m/s. Determine the coordinates of P and the radius r of the disk.

   Ans. $x = -0.3$ m, $y = 0.4$ m, $r = 0.5$ m

5/180 The rectangular plate rotates about its fixed z-axis. At the instant considered its angular velocity is $\omega = 3$ rad/s and is decreasing at the rate of 6 rad/s per second. For this instant write the vector expressions for the velocity of P and its normal and tangential components of acceleration.

5/181 If the acceleration of point P in Prob. 5/179 is in the x-direction at the instant considered, determine the angular acceleration $\boldsymbol{\alpha}$ of the disk. Use the results cited with Prob. 5/179.   *Ans.* $\boldsymbol{\alpha} = -\frac{16}{3}\mathbf{k}$ rad/s^2

5/182 Determine the acceleration of the mass center G of the connecting rod AB for the slider-crank mechanism of Sample Problem 5/15 for (*a*) $\theta = 0$ and (*b*) $\theta = 180°$.

5/183 The load L is being elevated by the downward velocities of ends A and B of the cable. Determine the acceleration of point P on the top of the sheave for the instant when $v_A = 2$ ft/sec, $\dot{v}_A = 0.5$ ft/sec^2, $v_B = 3$ ft/sec, and $\dot{v}_B = -0.5$ ft/sec^2.

   Ans. $a_P = 5/8$ ft/sec^2

5/184 The hydraulic cylinder moves pin A to the right with a constant velocity v. Use the fact that the distance from A to B is invariant, where B is the point on AC momentarily in contact with the gear, and write an expression for the angular velocity ω of the gear and the angular velocity of the rack AC.

5/185 The right-angled link *ABC* has plane motion in the *x-y* plane. At the instant represented the acceleration of *B* with respect to *A* is 20**i** m/s², and the acceleration of *C* is zero. Determine the acceleration of point *A*.   *Ans.* $\mathbf{a}_A = -12.8\mathbf{i} - 9.6\mathbf{j}$ m/s²

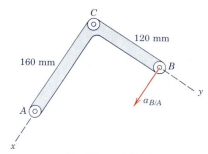

Problem 5/185

5/186 The wheel slips as it rolls. If $v_O = 4$ ft/sec and if the velocity of *A* with respect to *B* is $3\sqrt{2}$ ft/sec, locate the instantaneous center *C* of zero velocity and find the velocity of point *P*.

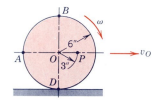

Problem 5/186

5/187 In the linkage shown *OC* has a constant clockwise angular velocity $\omega = 2$ rad/s during an interval of motion while the hydraulic cylinder gives pin *A* a constant velocity of 1.2 m/s to the right. For the position shown where *OC* is vertical and *BC* is horizontal calculate the angular velocity of *BC*. Solve by drawing the necessary velocity polygon.
     Ans. $\omega_{BC} = 2$ rad/s CW

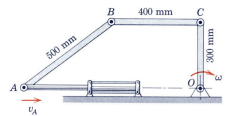

Problem 5/187

5/188 Solve Prob. 5/187 by vector algebra starting with the position-vector equation $\mathbf{r}_C + \mathbf{r}_{B/C} = \mathbf{r}_A + \mathbf{r}_{B/A}$ as shown in the separate diagram here.

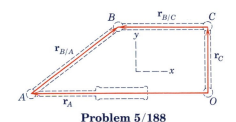

Problem 5/188

5/189 The large power-cable reel is rolled up the incline by the service vehicle as shown. The vehicle starts from rest with $x = 0$ for the reel and accelerates at the constant rate of 2 ft/sec². When $x = 6$ ft, calculate the acceleration of point *P* on the reel in the position shown.   *Ans.* $a_P = 11.14$ ft/sec²

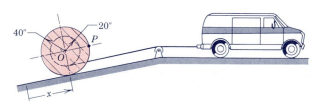

Problem 5/189

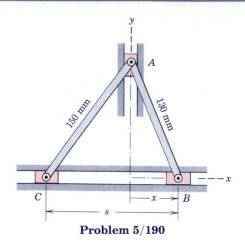

Problem 5/190

5/190 At the instant represented $x = 50$ mm and $\dot{s} = 1.6$ m/s. Determine the corresponding velocity of point B.

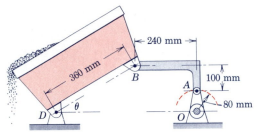

Problem 5/191

5/191 The tilting device maintains a sloshing water bath for washing vegetable produce. If the crank OA oscillates about the vertical and has a clockwise angular velocity of 4π rad/s when OA is vertical, determine the angular velocity of the basket in the position shown where $\theta = 30°$.

Ans. $\omega_{DB} = 3.24$ rad/s CW

5/192 The piston rod of the hydraulic cylinder gives A a constant vertical velocity v_A for an interval of motion. Roller B is confined to move in the horizontal guide. Determine the angular acceleration of link AB in terms of θ.

5/193 Determine the angular acceleration of the basket of the vegetable washer of Prob. 5/191 for the position where OA is vertical. In this position OA has an angular velocity of 4π rad/s and no angular acceleration.   *Ans.* $\alpha_{BD} = 41.2$ rad/s² CW

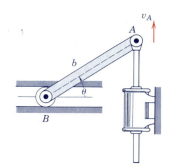

Problem 5/192

5/194 In the position shown the bar DC is rotating counterclockwise at the constant rate $N = 2$ rad/sec. Determine the angular velocity ω and the angular acceleration α of EBO at this instant.

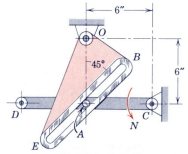

Problem 5/194

▶ **5/195** Near the end of its takeoff roll, the airplane is "rotating" (nose pitching up) just prior to lift-off. The velocity and acceleration of the aircraft, expressed in terms of the motion of the wheel assembly C, are v_C and a_C, both directed horizontally forward. The pitch angle is θ, and the pitch rate $\omega = \dot{\theta}$ is increasing at the rate $\alpha = \dot{\omega}$. If a person A is walking forward in the center aisle with a velocity and an acceleration v_{rel} and a_{rel}, both measured forward relative to the cabin, derive expressions for the velocity and acceleration of A as observed by a ground-fixed observer.

Ans. $\mathbf{v}_A = (v_C \cos\theta - \omega h + \dot{L})\mathbf{i} + (\omega L - v_C \sin\theta)\mathbf{j}$
$\mathbf{a}_A = (a_C \cos\theta - \alpha h - L\omega^2 + \ddot{L})\mathbf{i} +$
$(-a_C \sin\theta - h\omega^2 + L\alpha + 2\omega\dot{L})\mathbf{j}$

Problem 5/195

▶ **5/196** The crank OB revolves clockwise at the constant rate ω_0 of 5 rad/s. For the instant when $\theta = 90°$ determine the angular acceleration α of the rod BD, which slides through the pivoted collar at C.

Ans. $\alpha = 6.25$ rad/s^2 CW

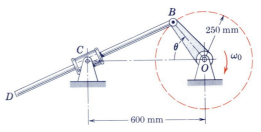

Problem 5/196

Computer-oriented problems

5/197 Link OA is given a constant counterclockwise angular velocity ω. Determine the angular velocity ω_{AB} of link AB as a function of θ. Compute and plot the ratio ω_{AB}/ω for the range $0 \le \theta \le 90°$. Indicate the value of θ for which the angular velocity of AB is half that of OA.

Ans. $\theta = 46.1°$

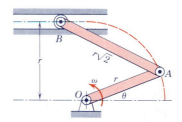

Problem 5/197

5/198 For the Geneva wheel of Prob. 5/53, shown again here, write the expression for the angular velocity ω_2 of the slotted wheel C during engagement of pin P and plot ω_2 for the range $-45° \le \theta \le 45°$. The driving wheel A has a constant angular velocity $\omega_1 = 2$ rad/s.

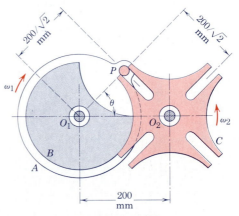

Problem 5/198

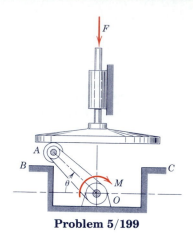

Problem 5/199

A constant torque M exceeds the moment about O due to the force F on the plunger, and an angular acceleration $\ddot{\theta} = 100(1 - \cos\theta)$ rad/s² results. If the crank OA is released from rest at B, where $\theta = 30°$, and strikes the stop at C, where $\theta = 150°$, plot the angular velocity $\dot{\theta}$ as a function of θ and find the time t for the crank to rotate from $\theta = 90°$ to $\theta = 150°$.

Ans. $\dot{\theta} = 10\sqrt{2}\sqrt{\theta - \sin\theta - 0.0236}$ rad/s

$t = 0.0701$ s

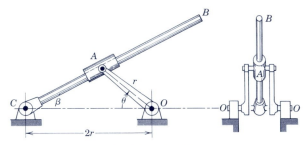

Problem 5/200

*5/200 The double crank is pivoted at O and permits complete rotation without interference with the pivoted rod CB as it slides through the collar A. If the crank has a constant angular velocity $\dot{\theta}$, determine and plot the ratio $\dot{\beta}/\dot{\theta}$ as a function of θ between $\theta = 0$ and $\theta = 180°$. By inspection determine the angle β for which $\dot{\beta} = 0$.

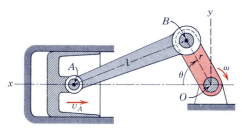

Problem 5/201

*5/201 For the slider-crank configuration shown, derive the expression for the velocity v_A of the piston (taken positive to the right) as a function of θ. Substitute the numerical data of Sample Problem 5/15 and calculate v_A as a function of θ for $0 \le \theta \le 180°$. Plot v_A versus θ and find its maximum magnitude and the corresponding value of θ. (By symmetry anticipate the results for $180° \le \theta \le 360°$.)

Ans. $(v_A)_{max} = 69.6$ ft/sec at $\theta = 72.3°$

*5/202 For the slider-crank of Prob. 5/201 derive the expression for the acceleration a_A of the piston (taken positive to the right) as a function of θ for $\omega = \dot{\theta} = $ constant. Substitute the numerical data of Sample Problem 5/15 and calculate a_A as a function of θ for $0 \le \theta \le 180°$. Plot a_A versus θ and find the value of θ for which $a_A = 0$. (By symmetry anticipate the results for $180° \le \theta \le 360°$.)

The forces that cause or accompany the motion of a rigid body, which may rotate and translate simultaneously, must be calculated. On takeoff, an airplane, which is accelerating and rotating upward, is acted upon by the thrust of its engines, its weight, and the aerodynamic forces on its exterior surfaces.

PLANE KINETICS OF RIGID BODIES

6

6/1 INTRODUCTION

The kinetics of rigid bodies treats the relationships between the forces that act on the bodies from sources external to their boundaries and the corresponding translational and rotational motions of the bodies. In Chapter 5 we developed the kinematic relationships for the plane motion of rigid bodies, and we will use these relationships extensively in this present chapter, where the effects of forces on the two-dimensional motion of rigid bodies are examined.

For our purpose in Chapter 6, a body that can be approximated as a thin slab with its motion confined to the plane of the slab will be treated under the category of plane motion. The plane of motion will contain the mass center, and all forces that act on the body will be projected onto the plane of motion. A body that has appreciable dimensions normal to the plane of motion but is symmetrical about that plane of motion through the mass center may be treated as having plane motion. These idealizations clearly fit a very large category of rigid-body motions.

In Chapter 3 we found that two force equations of motion were required to define the plane motion of a particle whose motion has two linear components. For the plane motion of a rigid body, an additional equation is needed to specify the state of rotation of the body. Thus, two force equations and one moment equation or their equivalent are required to determine the state of rigid-body plane motion.

The relationships that form the basis for most of the analysis of rigid-body motion were developed in Chapter 4 for a general system of particles. Frequent reference will be made to these equations as they are further developed in Chapter 6 and applied specifically to the plane motion of rigid bodies. It is advisable to refer to the developments of Chapter 4 frequently as Chapter 6 is studied. Also, it is inadvisable to proceed further until a firm grasp of the calculation of velocities and accelerations as developed in Chapter 5 for rigid-body plane motion is well in hand. Without the ability to determine accelerations correctly from the principles of kinematics, it is frequently useless to attempt application of the force and moment principles of motion. Consequently, it is essential to master the necessary kinematics, including the calculation of relative accelerations, before proceeding.

Basic to the approach to kinetics is the isolation of the body or system to be analyzed. This isolation was illustrated and used in Chapter 3 for particle kinetics and will be employed consistently in the present chapter. For problems involving the instantaneous relationships among force, mass, and acceleration or momentum, the body or system should be explicitly defined by isolating it with its *free-body diagram*. When the principles of work and energy are employed, an *active-force diagram* that shows only those external forces that do work on the system may be used in lieu of the free-body diagram. *No solution of a problem should be attempted without first defining the complete external boundary of the body or system and identifying all external forces that act on it.*

Chapter 6 is organized into the same three sections in which we treated the kinetics of particles in Chapter 3. Section A relates the forces and moments to the instantaneous linear and angular accelerations. Section B treats the solution of problems by the method of work and energy. Section C covers the methods of impulse and momentum. Virtually all of the basic concepts and approaches covered in these three sections were treated in Chapter 3 on particle kinetics. This repetition will result in accelerated progress in Chapter 6, provided the kinematics of rigid-body plane motion is well in hand. In each of the three sections, the three types of motion, namely, translation, fixed-axis rotation, and general plane motion, will be treated.

In the kinetics of rigid bodies that have angular motion, it is necessary to introduce a property of the body that accounts for the radial distribution of its mass with respect to a particular axis of rotation normal to the plane of motion. This property is known as the *mass moment of inertia* of the body, and it is essential that we be able to calculate this property in order to solve rotational problems. Familiarity with the calculation of mass moments of inertia is assumed of the reader at this point. Appendix B treats this topic for those who need instruction or review.

SECTION A. FORCE, MASS, AND ACCELERATION

6/2 *GENERAL EQUATIONS OF MOTION*

In Arts. 4/2 and 4/4 we derived the force and moment vector equations of motion for a perfectly general system of mass. We now apply these results by starting, first, with a general rigid body in three dimensions. The force equation, Eq. 4/1,

$$\Sigma \mathbf{F} = m\bar{\mathbf{a}} \qquad\qquad [4/1]$$

tells us that the resultant $\Sigma\mathbf{F}$ of the external forces acting on the body equals the mass m of the body times the acceleration $\bar{\mathbf{a}}$ of its mass center G. The moment equation taken about the mass center, Eq. 4/9,

$$\Sigma\mathbf{M}_G = \dot{\mathbf{H}}_G \qquad [4/9]$$

tells us that the resultant moment about the mass center of the external forces on the body equals the time rate of change of the angular momentum of the body about the mass center.

We recall from our study of statics that a general system of forces acting on a rigid body may be replaced by a resultant force applied at a chosen point and a corresponding couple. By replacing the external forces by their equivalent force–couple system with the resultant force acting through the mass center, we may then visualize the action of the forces and the corresponding dynamic response of the body with the aid of Fig. 6/1. The a-part of the figure represents the free-body diagram of the body. The b-part of the figure shows the equivalent force-couple system with the resultant force applied through G. The c-part of the figure represents the

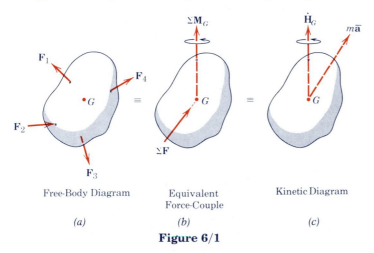

Free-Body Diagram   Equivalent   Kinetic Diagram
                          Force-Couple

(a)   (b)   (c)

Figure 6/1

resulting dynamic effects as specified by Eqs. 4/1 and 4/9 and is termed here the *kinetic diagram*. The equivalence between the free-body diagram and the kinetic diagram enables us to clearly visualize and easily remember the separate translational and rotational effects of the forces applied to a rigid body. We shall express this equivalence mathematically as we apply these results to the treatment of rigid-body plane motion that follows.

Plane-motion equations. We now apply the foregoing relationships to the case of plane motion. Figure 6/2 represents a rigid body moving with plane motion in the x-y plane. The mass center G has an acceleration $\bar{\mathbf{a}}$, and the body has an angular velocity

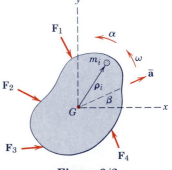

Figure 6/2

$\boldsymbol{\omega} = \omega\mathbf{k}$ and an angular acceleration $\boldsymbol{\alpha} = \alpha\mathbf{k}$, both taken in the positive z-direction. Because the z-direction of both $\boldsymbol{\omega}$ and $\boldsymbol{\alpha}$ remains perpendicular to the plane of motion, we may use scalar notation ω and $\alpha = \dot{\omega}$ to represent the angular velocity and angular acceleration.

The angular momentum about the mass center for the general system was expressed in Eq. 4/8a as $\mathbf{H}_G = \Sigma\boldsymbol{\rho}_i \times m_i\dot{\boldsymbol{\rho}}_i$ where $\boldsymbol{\rho}_i$ is the position vector relative to G of the representative particle of mass m_i. For our rigid body, the velocity of m_i relative to G is $\dot{\boldsymbol{\rho}}_i = \boldsymbol{\omega} \times \boldsymbol{\rho}_i$, which has a magnitude $\rho_i\omega$ and lies in the plane of motion normal to $\boldsymbol{\rho}_i$. The product $\boldsymbol{\rho}_i \times \dot{\boldsymbol{\rho}}_i$ is then a vector normal to the x-y plane in the sense of $\boldsymbol{\omega}$, and its magnitude is $\rho_i^2\omega$. Thus, the magnitude of $\mathbf{H}_G$ becomes $H_G = \Sigma\rho_i^2 m_i\omega = \omega\Sigma\rho_i^2 m_i$. The summation, which may also be written as $\int \rho^2\, dm$, is defined as the *mass moment of inertia* $\bar{I}$ of the body about the z-axis through G. (See Appendix B for a discussion of the calculation of mass moments of inertia.) We may now write

$$H_G = \bar{I}\omega$$

where $\bar{I}$ is a constant property of the body and is a measure of the rotational inertia or resistance to change in rotational velocity due to the radial distribution of mass around the z-axis through G. With this substitution, our moment equation, Eq. 4/9, becomes

$$\Sigma M_G = \dot{H}_G = \bar{I}\dot{\omega} = \bar{I}\alpha$$

where $\alpha = \dot{\omega}$ is the angular acceleration of the body.

The moment equation and the vector form of the generalized Newton's second law of motion, Eq. 4/1, are now written

$$\boxed{\begin{aligned} \Sigma\mathbf{F} &= m\bar{\mathbf{a}} \\ \Sigma M_G &= \bar{I}\alpha \end{aligned}} \qquad \textbf{(6/1)}$$

Equations 6/1 are the general equations of motion for a rigid body in plane motion. In applying Eqs. 6/1, the vector force equation will be expressed in terms of its two scalar components using x-y, n-t, or r-θ coordinates according to the description that is most convenient for the problem at hand.

As an alternative approach, it is instructive to derive the moment equation by referring directly to the forces that act on the representative particle of mass m_i, as shown in Fig. 6/3. The acceleration of m_i equals the vector sum of $\bar{a}$ and the relative terms $\rho_i\omega^2$ and $\rho_i\alpha$, where the mass center G is used as the reference point. It follows that the resultant of all forces on m_i has the components $m_i\bar{a}$, $m_i\rho_i\omega^2$, and $m_i\rho_i\alpha$ in the directions shown. The sum of the moments of these force components about G in the sense of α becomes

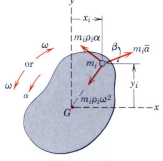

Figure 6/3

$$M_{G_i} = m_i\rho_i^2\alpha + (m_i\bar{a}\sin\beta)x_i - (m_i\bar{a}\cos\beta)y_i$$

Similar moment expressions exist for all particles in the body, and the sum of these moments about G for the resultant forces acting on all particles may be written as

$$\Sigma M_G = \Sigma m_i\rho_i^2\alpha + \bar{a}\sin\beta\,\Sigma m_i x_i - \bar{a}\cos\beta\,\Sigma m_i y_i$$

But the origin of coordinates is taken at the mass center, so that $\Sigma m_i x_i = m\bar{x} = 0$ and $\Sigma m_i y_i = m\bar{y} = 0$. Thus, the moment sum becomes

$$\Sigma M_G = \Sigma m_i\rho_i^2\alpha = \bar{I}\alpha$$

as before. The contribution to ΣM_G of the forces internal to the body is, of course, zero since they occur in pairs of equal and opposite forces of action and reaction between interacting particles. Thus, ΣM_G, as before, represents the sum of moments about the mass center G of only the external forces acting on the body, as disclosed by the free-body diagram. We note that the force component $m_i\rho_i\omega^2$ has no moment about G and conclude, therefore, that the angular velocity ω has no influence on the moment equation about the mass center G.

The results embodied in our basic equations of motion for a rigid body in plane motion, Eqs. 6/1, are represented diagrammatically in Fig. 6/4, which is the two-dimensional counterpart of the *a*- and *c*-parts of Fig. 6/1 for a general three-dimensional body. The free-body diagram discloses the forces and moments appearing on

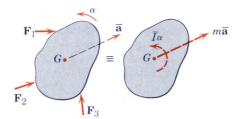

Free-Body Diagram   Kinetic Diagram

Figure 6/4

the left-hand side of our equations of motion. The kinetic diagram discloses the resulting dynamic response in terms of the translational term $m\bar{a}$ and the rotational term $\bar{I}\alpha$ that appear on the right-hand side of Eqs. 6/1. As previously mentioned, the translational term $m\bar{a}$ will be expressed by its x-y, n-t, or r-θ components once the appropriate inertial reference system is designated. The equivalence depicted in Fig. 6/4 is basic to our understanding of the kinetics of plane motion and will be employed frequently in the solution of problems. Representation of the resultants $m\bar{a}$ and $\bar{I}\alpha$ will help ensure that the force and moment sums determined from the free-body diagram are equated to their proper resultants.

Alternative moment equation. In Art. 4/4 of Chapter 4 on systems of particles, we developed a general moment equation about an arbitrary point P, Eq. 4/11, which is

$$\Sigma\mathbf{M}_P = \dot{\mathbf{H}}_G + \bar{\boldsymbol{\rho}} \times m\bar{\mathbf{a}} \qquad [4/11]$$

where $\bar{\boldsymbol{\rho}}$ is the vector from P to the mass center G and $\bar{\mathbf{a}}$ is the mass-center acceleration. As we have shown earlier in this article, for a rigid body in plane motion $\dot{\mathbf{H}}_G$ becomes $\bar{I}\alpha$. Also, the cross product $\bar{\boldsymbol{\rho}} \times m\bar{\mathbf{a}}$ is simply the moment of magnitude $m\bar{a}d$ of $m\bar{\mathbf{a}}$ about P. Therefore, for the two-dimensional body illustrated in Fig. 6/5 with its free-body diagram and kinetic diagram, we may rewrite Eq. 4/11 simply as

$$\boxed{\Sigma M_P = \bar{I}\alpha + m\bar{a}d} \qquad (6/2)$$

Clearly, all three terms are positive in the counterclockwise sense for the example shown, and the choice of P eliminates reference to $\mathbf{F}_1$ and $\mathbf{F}_3$. If we had wished to eliminate reference to $\mathbf{F}_2$ and $\mathbf{F}_3$, for example, by choosing their intersection as the reference point, then P would lie on the opposite side of the $m\bar{\mathbf{a}}$ vector, and the clockwise moment of $m\bar{\mathbf{a}}$ about P would be a negative term in the equation. Equation 6/2 is easily remembered as it is merely an expression of the familiar principle of moments, where the sum of the moments about P equals the combined moment about P of their sum, expressed by the resultant couple $\Sigma M_G = \bar{I}\alpha$ and the resultant force $\Sigma\mathbf{F} = m\bar{\mathbf{a}}$.

In Art. 4/4 we also developed an alternative moment equation about P, Eq. 4/13, which is

$$\Sigma\mathbf{M}_P = (\dot{\mathbf{H}}_P)_{\text{rel}} + \bar{\boldsymbol{\rho}} \times m\mathbf{a}_P \qquad [4/13]$$

For rigid-body plane motion, if P is chosen as a point *fixed* to the body, then in scalar form $(\dot{\mathbf{H}}_P)_{\text{rel}}$ becomes $I_P\alpha$, where I_P is the mass moment of inertia about an axis through P and α is the angular acceleration of the body. So we may write the equation as

$$\boxed{\Sigma\mathbf{M}_P = I_P\alpha + \bar{\boldsymbol{\rho}} \times m\mathbf{a}_P} \qquad (6/3)$$

where the acceleration of P is $\mathbf{a}_P$ and the position vector from P to G is $\bar{\boldsymbol{\rho}}$. When $\bar{\boldsymbol{\rho}} = \mathbf{0}$, point P becomes the mass center G, and Eq. 6/3 reduces to the scalar form $\Sigma M_G = \bar{I}\alpha$, previously derived. When $\mathbf{a}_P = \mathbf{0}$, point P becomes point O, fixed in an inertial reference system and attached to the body (or body extended), and Eq. 6/3 in scalar form reduces to

$$\boxed{\Sigma M_O = I_O\alpha} \qquad (6/4)$$

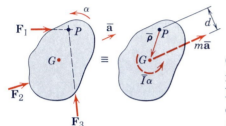

Free-Body Diagram   Kinetic Diagram

Figure 6/5

Equation 6/4 then applies to the rotation of a rigid body about a nonaccelerating point O fixed to the body and is the two-dimensional simplification of Eq. 4/7.

Unconstrained and constrained motion. The motion of a rigid body may be unconstrained or constrained. The rocket moving in a vertical plane, Fig. 6/6a, is an example of unconstrained motion as there are no physical confinements to its motion. The two components $\bar{a}_x$ and $\bar{a}_y$ of the mass-center acceleration and the angular acceleration α may be determined independently of one another by direct application of Eqs. 6/1. The bar in Fig. 6/6b, on the other hand, represents a constrained motion, where the vertical and horizontal guides for the ends of the bar impose a kinematic relationship between the acceleration components of the mass center and the angular acceleration of the bar. Thus, it is necessary to determine this kinematic relationship from the principles established in Chapter 5 and to combine it with the force and moment equations of motion before a solution can be carried out. In general, dynamics problems that involve physical constraints to motion require a kinematic analysis relating linear to angular acceleration before the force and moment equations of motion can be solved. It is for this reason that an understanding of the principles and methods of Chapter 5 is so vital to the work of Chapter 6.

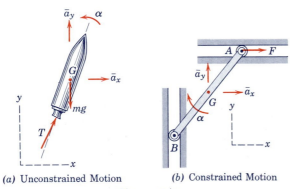

(a) Unconstrained Motion   (b) Constrained Motion

Figure 6/6

System of interconnected bodies. Upon occasion, in problems dealing with two or more connected rigid bodies whose motions are related kinematically, it is convenient to analyze the bodies as an entire system. Figure 6/7 illustrates two rigid bodies hinged at A and subjected to the external forces shown. The forces in the connection at A are internal to the system and are not disclosed. The resultant of all external forces must equal the vector sum of the two resultants $m_1\bar{\mathbf{a}}_1$ and $m_2\bar{\mathbf{a}}_2$, and the sum of the moments about some arbitrary point such as P of all external forces must equal the moment of the resultants, $\bar{I}_1\alpha_1 + \bar{I}_2\alpha_2 + m_1\bar{a}_1 d_1 + m_2\bar{a}_2 d_2$. Thus, we may state

$$\Sigma\mathbf{F} = \Sigma m\bar{\mathbf{a}}$$
$$\Sigma M_P = \Sigma\bar{I}\alpha + \Sigma m\bar{a}d$$

(6/5)

where the summations on the right-hand side of the equations represent as many terms as there are separate bodies. If there are more than three remaining unknowns in a system, however, the three independent scalar equations of motion, when applied to the system, are not sufficient to solve the problem. In this case, more advanced methods such as virtual work (Art. 6/7) or Lagrange's equations (not discussed in this book*) could be employed or else the system dismembered and each part analyzed separately with the resulting equations solved simultaneously.

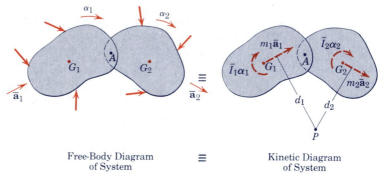

Free-Body Diagram
of System
$\equiv$
Kinetic Diagram
of System

Figure 6/7

Analysis procedure. In solving force-mass-acceleration problems for the plane motion of rigid bodies, the following steps should be taken after the conditions and requirements of the problem are clearly in mind.

(1) Kinematics. First, identify the class of motion and then solve for any needed linear and angular accelerations that can be determined solely from given kinematic information. In the case of constrained plane motion, it is usually necessary to establish the relation between the linear acceleration of the mass center and the angular acceleration of the body by first solving the appropriate relative-velocity and relative-acceleration equations. Again, it is em-

---

*When an interconnected system has more than one degree of freedom, that is, requires more than one coordinate to specify completely the configuration of the system, the more advanced equations of Lagrange are generally used. See the senior author's *Dynamics, 2nd Edition SI Version*, 1975, John Wiley & Sons, for a treatment of Lagrange's equations.

phasized that success in working force-mass-acceleration problems in this chapter is contingent on the ability to describe the necessary kinematics, so that frequent review of Chapter 5 is recommended.

(2) Diagrams. Always draw the complete free-body diagram of the body to be analyzed. Assign a convenient inertial coordinate system and label all known and unknown quantities. The kinetic diagram should also be constructed so as to clarify the equivalence between the applied forces and the resulting dynamic response.

(3) Equations of motion. Apply the three equations of motion from Eqs. 6/1, being consistent with the algebraic signs in relation to the choice of reference axes. Equation 6/2 or 6/3 may be employed as an alternative to one of Eqs. 6/1. Combine with the results from any needed kinematic analysis. Count the number of unknowns and be certain that there are an equal number of independent equations available. For a solvable rigid-body problem in plane motion, there can be no more than the five scalar unknowns that can be determined from the three scalar equations of motion, obtained from Eqs. 6/1, and the two scalar component relations that come from the relative-acceleration equation.

The sequence of steps to be followed in any given problem will depend on the conditions of the problem. To organize a solution in a logical sequence of steps, the student should recognize the dependency of a required quantity on intermediate quantities and then trace the dependencies back to the given information. This establishes the reverse order of the steps to be followed in carrying out the actual solution.

The foregoing developments will now be applied to the three cases of motion in a plane, namely, translation, fixed-axis rotation, and general plane motion in the three articles that follow.

6/3   TRANSLATION

Rigid-body translation in plane motion was described in Art. 5/1 and illustrated in Figs. 5/1*a* and 5/1*b*, where we saw that every line in a translating body remains parallel to its original position at all times. In rectilinear translation all points move in straight lines, whereas in curvilinear translation all points move on congruent curved paths. In either case, there is no angular motion of the translating body, so that both ω and α are zero. Therefore, from the moment relation of Eqs. 6/1, we see that all reference to the moment of inertia is eliminated for a translating body.

For a translating body then, our general equations for plane motion, Eqs. 6/1, may be written

$$\Sigma \mathbf{F} = m\bar{\mathbf{a}}$$
$$\Sigma M_G = \bar{I}\alpha = 0$$

(6/6)

For rectilinear translation, illustrated in Fig. 6/8*a*, if the *x*-axis is chosen in the direction of the acceleration, then the two scalar force equations become $\Sigma F_x = m\bar{a}_x$ and $\Sigma F_y = m\bar{a}_y = 0$. For curvilinear translation, Fig. 6/8*b*, if we use *n-t* coordinates, the two scalar force equations become $\Sigma F_n = m\bar{a}_n$ and $\Sigma F_t = m\bar{a}_t$. In both cases, $\Sigma M_G = 0$.

We may also employ the alternative moment equation, Eq. 6/2, with the aid of the kinetic diagram. For rectilinear translation we see that $\Sigma M_P = m\bar{a}d$ and $\Sigma M_A = 0$. For curvilinear translation the kinetic diagram permits us to write $\Sigma M_A = m\bar{a}_n d_A$ in the clockwise sense and $\Sigma M_B = m\bar{a}_t d_B$ in the counterclockwise sense. Thus, we have complete freedom for choosing a convenient moment center.

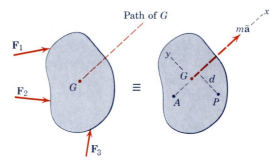

Free-Body Diagram   Kinetic Diagram
(a) Rectilinear Translation
$(\alpha = 0, \omega = 0)$

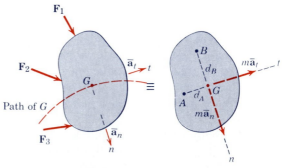

Free-Body Diagram   Kinetic Diagram
(b) Curvilinear Translation
$(\alpha = 0, \omega = 0)$

Figure 6/8

Sample Problem 6/1

The pickup truck weighs 3220 lb and reaches a speed of 30 mi/hr from rest in a distance of 200 ft up the 10-percent incline with constant acceleration. Calculate the normal force under each pair of wheels and the friction force under the rear driving wheels. The effective coefficient of friction between the tires and the road is known to be at least 0.8.

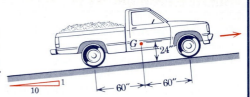

① Solution. We will assume that the mass of the wheels is negligible compared with the total mass of the truck. The truck may now be simulated by a single rigid body in rectilinear translation with an acceleration of

② $[v^2 = 2as]$ $\qquad \bar{a} = \dfrac{(44)^2}{2(200)} = 4.84 \text{ ft/sec}^2$

The free-body diagram of the complete truck shows the normal forces N_1 and N_2, the friction force F in the direction to oppose the slipping of the driving wheels, and the weight W represented by its two components. With $\theta = \tan^{-1} 1/10 = 5.71°$, these components are $W \cos \theta = 3220 \cos 5.71° = 3204$ lb and $W \sin \theta = 3220 \sin 5.71° = 320$ lb. The kinetic diagram shows the resultant, which passes through the mass center and is in the direction of its acceleration. Its magnitude is

$$m\bar{a} = \frac{3220}{32.2}(4.84) = 484 \text{ lb}$$

Applying the three equations of motion, Eqs. 6/1, for the three unknowns gives

③ $[\Sigma F_x = m\bar{a}_x]$ $\qquad F - 320 = 484 \qquad F = 804 \text{ lb}$ $\qquad$ *Ans.*

$[\Sigma F_y = m\bar{a}_y = 0]$ $\qquad N_1 + N_2 - 3204 = 0$ $\qquad\qquad$ (a)

$[\Sigma M_G = \bar{I}\alpha = 0]$ $\qquad 60N_1 + 804(24) - N_2(60) = 0$ $\qquad$ (b)

Solving (a) and (b) simultaneously gives

$$N_1 = 1441 \text{ lb} \qquad N_2 = 1763 \text{ lb} \qquad \textit{Ans.}$$

In order to support a friction force of 804 lb, a coefficient of friction of at least $F/N_2 = 804/1763 = 0.46$ is required. Since our coefficient of friction is at least 0.8, the surfaces are rough enough to support the calculated value of F so that our result is correct.

Alternative solution. From the kinetic diagram we see that N_1 and N_2 can be obtained independently of one another by writing separate moment equations about A and B.

④ $[\Sigma M_A = m\bar{a}d]$ $\qquad 120N_2 - 60(3204) - 24(320) = 484(24)$

$$N_2 = 1763 \text{ lb} \qquad \textit{Ans.}$$

$[\Sigma M_B = m\bar{a}d]$ $\qquad 3204(60) - 320(24) - 120N_1 = 484(24)$

$$N_1 = 1441 \text{ lb} \qquad \textit{Ans.}$$

---

① Without this assumption, we would be obliged to account for the relatively small additional forces that produce moments which give the wheels their angular acceleration.

② Recall that 30 mi/hr is 44 ft/sec.

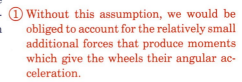

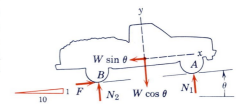

③ We must be careful not to use the friction equation $F = \mu N$ here since we do not have a case of slipping or impending slipping. If the given coefficient of friction were less than 0.46, the friction force would be μN_2, and the car would be unable to attain the acceleration of 4.84 ft/sec². In this case, the unknowns would be N_1, N_2, and a.

④ The left-hand side of the equation is evaluated from the free-body diagram, and the right-hand side from the kinetic diagram. The positive sense for the moment sum is arbitrary but must be the same for both sides of the equation. In this problem, we have taken the clockwise sense as positive for the moment of the resultant force about B.

Sample Problem   6/2

The vertical bar AB has a mass of 150 kg with center of mass G midway between the ends. The bar is elevated from rest at $\theta = 0$ by means of the parallel links of negligible mass, with a constant couple $M = 5$ kN·m applied to the lower link at C. Determine the angular acceleration α of the links as a function of θ and find the force B in the link DB at the instant when $\theta = 30°$.

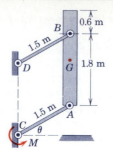

Solution. The motion of the bar is seen to be curvilinear translation since the bar itself does not rotate during the motion. With the circular motion of the mass center G, we choose n- and t-coordinates as the most convenient description. With negligible mass of the links, the tangential component A_t of the force at A is obtained from the free-body diagram of AC, where $\Sigma M_C \cong 0$ and $A_t = M/\overline{AC} = 5/1.5 = 3.33$ kN. The force at B is along the link. All applied forces are shown on the free-body diagram of the bar, and the kinetic diagram is also indicated, where the $m\overline{\mathbf{a}}$ resultant is shown in terms of its two components.

The sequence of solution is established by noting that A_n and B depend on the n-summation of forces and, hence, on $m\overline{r}\omega^2$ at $\theta = 30°$. The value of ω depends on the variation of $\alpha = \ddot{\theta}$ with θ. This dependency is established from a force summation in the t-direction for a general value of θ, where $\overline{a}_t = (\overline{a}_t)_A = \overline{AC}\alpha$. Thus, we begin with

$$[\Sigma F_t = m\overline{a}_t] \quad 3.33 - 0.15(9.81)\cos\theta = 0.15(1.5\alpha)$$

$$\alpha = 14.81 - 6.54\cos\theta \text{ rad/s}^2 \qquad Ans.$$

With α a known function of θ, the angular velocity ω of the links is obtained from

$$[\omega\, d\omega = \alpha\, d\theta] \quad \int_0^\omega \omega\, d\omega = \int_0^\theta (14.81 - 6.54\cos\theta)\, d\theta$$

$$\omega^2 = 29.6\theta - 13.08\sin\theta$$

Substitution for $\theta = 30°$ gives

$$(\omega^2)_{30°} = 8.97 \text{ (rad/s)}^2 \qquad \alpha_{30°} = 9.15 \text{ rad/s}^2$$

and

$$m\overline{r}\omega^2 = 0.15(1.5)(8.97) = 2.02 \text{ kN}$$

$$m\overline{r}\alpha = 0.15(1.5)(9.15) = 2.06 \text{ kN}$$

The force B may be obtained by a moment summation about A, which eliminates A_n and A_t and the weight. Or a moment summation may be taken about the intersection of A_n and the line of action of $m\overline{r}\alpha$, which eliminates A_n and $m\overline{r}\alpha$. Using A as a moment center gives

$$[\Sigma M_A = m\overline{a}d] \quad 1.8\cos 30° \, B = 2.02(1.2)\cos 30° + 2.06(0.6)$$

$$B = 2.14 \text{ kN} \qquad Ans.$$

The component A_n could be obtained from a force summation in the n-direction or from a moment summation about G or about the intersection of B and the line of action of $m\overline{r}\alpha$.

① Generally speaking, the best choice of reference axes is to make them coincide with the directions in which the components of the mass-center acceleration are expressed. Examine the consequences of choosing horizontal and vertical axes in this problem.

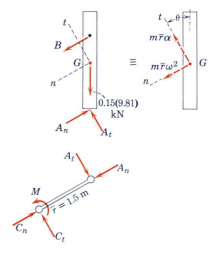

② The force and moment equations for a body of negligible mass become the same as the equations of equilibrium. Link BD, therefore, acts as a two-force member in equilibrium.

PROBLEMS

Introductory problems

6/1 For what acceleration a of the frame will the uniform slender rod maintain the orientation shown in the figure? Neglect the friction and mass of the small rollers at A and B. *Ans.* $a = g\sqrt{3}$

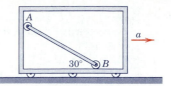

Problem 6/1

6/2 The homogeneous crate of mass m is mounted on small wheels as shown. Determine the maximum force P that can be applied without overturning the crate about (a) its lower front edge with $h = b$ and (b) its lower back edge with $h = 0$.

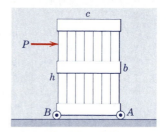

Problem 6/2

6/3 A uniform slender rod rests on a car seat as shown. Determine the deceleration a for which the rod will begin to tip forward. Assume that friction at B is sufficient to prevent slipping.
Ans. $a = 5.66$ m/s^2

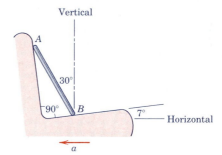

Problem 6/3

6/4 The uniform pole AB weighs 100 lb and is suspended in the horizontal position by the three wires shown. If wire CB breaks, calculate the tension in wire BD immediately after the break. (*Suggestion:* By a thoughtful choice of moment center solve by using only one equation of motion.)

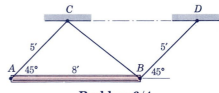

Problem 6/4

6/5 The rear-wheel-drive lawn mower, when placed into gear while at rest, is observed to momentarily spin its rear tires as it accelerates. If the coefficients of friction between the rear tires and the ground are $\mu_s = 0.7$ and $\mu_k = 0.5$, determine the forward acceleration a of the mower. The mass of the mower and attached bag is 50 kg with center of mass at G. Assume that the operator does not push on the handle so that $P = 0$. *Ans.* $a = 4.14$ m/s^2

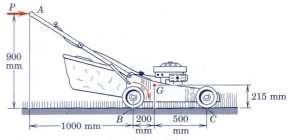

Problem 6/5

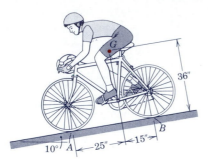

Problem 6/6

6/6 The bicyclist applies the brakes as she descends the 10° incline. What deceleration a would cause the dangerous condition of tipping about the front wheel A? The combined center of mass of the rider and bicycle is at G.

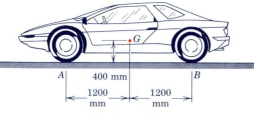

Problem 6/7

6/7 The 1650-kg car has its mass center at G. Calculate the normal forces N_A and N_B between the road and the front and rear pairs of wheels under conditions of maximum acceleration. The mass of the wheels is small compared with the total mass of the car. The coefficient of static friction between the road and the rear driving wheels is 0.8.

Ans. $N_A = 6.85$ kN, $N_B = 9.34$ kN

Representative problems

6/8 Arm AB of a classifying accelerometer has a weight of 0.25 lb with mass center at G and is pivoted freely to the frame F at A. The torsion spring at A is set to preload the arm with an applied clockwise moment of 2 lb-in. Determine the downward acceleration a of the frame at which the contacts at B will separate and break the electrical circuit.

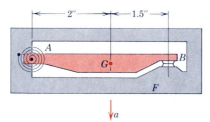

Problem 6/8

6/9 Determine the value of the force P that would cause the cabinet to begin to tip. What coefficient μ_s of static friction is necessary to ensure that tipping occurs without slipping?

Ans. $P = 392$ N, $\mu_s > \dfrac{a}{g} = \dfrac{2}{3}$

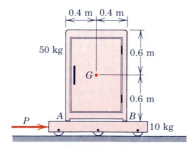

Problem 6/9

6/10 A passenger car of an overhead monorail system is driven by one of its two small wheels A or B. Select the one for which the car can be given the greater acceleration without slipping the driving wheel and compute the maximum acceleration if the effective coefficient of friction is limited to 0.25 between the wheels and the rail. Neglect the small mass of the wheels.

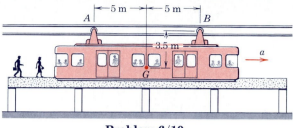

Problem 6/10

6/11 Solid homogeneous cylinders 400 mm high and 250 mm in diameter are supported by a flat conveyor belt that moves horizontally. If the speed of the belt increases according to $v = 1.2 + 0.9t^2$ m/s, where t is the time in seconds measured from the instant the increase begins, calculate the value of t for which the cylinders begin to tip over. Cleats on the belt prevent the cylinders from slipping.

Ans. $t = 3.41$ s

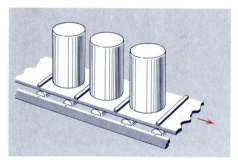

Problem 6/11

6/12 The uniform 30-kg bar OB is secured to the accelerating frame in the 30° position from the horizontal by the hinge at O and roller at A. If the horizontal acceleration of the frame is $a = 20$ m/s², compute the force F_A on the roller and the x- and y-components of the force supported by the pin at O.

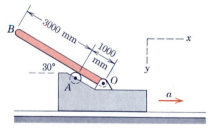

Problem 6/12

6/13 A drill press with mass center at G is transported on the flat bed of a truck. If the driver neglects to secure the drill press to the truck, determine the maximum acceleration a that the truck may have before the drill press slips or tips about its base at A. The coefficient of static friction between the truck bed and the drill press is 0.60.

Ans. $a = 4.91$ m/s²

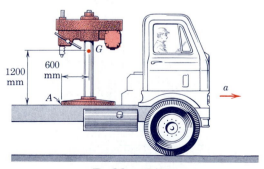

Problem 6/13

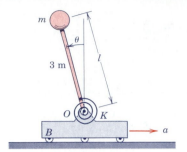

Problem 6/14

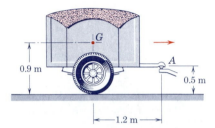

Problem 6/15

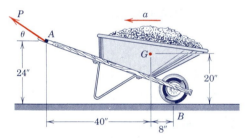

Problem 6/16

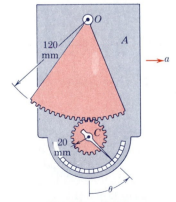

Problem 6/17

6/14 The cart B moves to the right with acceleration $a = 2g$. If the steady-state angular deflection of the uniform slender rod of mass $3m$ is observed to be 20°, determine the value of the torsional spring constant K. The spring, which exerts a moment of magnitude $M = K\theta$ on the rod, is undeformed when the rod is vertical. The values of m and l are 0.5 kg and 0.6 m, respectively. Treat the small end sphere of mass m as a particle.

6/15 The loaded trailer has a mass of 900 kg with center of mass at G and is attached at A to a rear-bumper hitch. If the car and trailer reach a velocity of 60 km/h on a level road in a distance of 30 m from rest with constant acceleration, compute the vertical component of the force supported by the hitch at A. Neglect the small friction force exerted on the relatively light wheels.   *Ans. $A_y = 1389$ N*

6/16 Determine the magnitude P and direction θ of the force required to impart a rearward acceleration $a = 5$ ft/sec² to the loaded wheelbarrow with no rotation from the position shown. The combined weight of the wheelbarrow and its load is 500 lb with center of gravity at G. Compare the normal force at B under acceleration with that for static equilibrium in the position shown. Neglect the friction and mass of the wheel.

6/17 The device shown consists of a vertical frame A to which are freely pivoted a geared sector at O and a balanced gear and attached pointer at C. Under a steady horizontal acceleration a to the right, the sector undergoes a clockwise angular displacement, thus causing the pointer to register a steady counterclockwise angle θ from the zero-acceleration position at $\theta = 0$. Determine the acceleration corresponding to an angle θ.

$$Ans. \ a = g \tan \frac{\theta}{6}$$

6/18 The 6-kg frame AC and 4-kg uniform slender bar AB of length l slide with negligible friction along the fixed horizontal rod under the action of the 80-N force. Calculate the tension T in wire BC and the x- and y-components of the force exerted on the bar by the pin at A. Motion lies in the vertical plane.

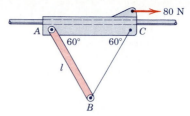

Problem 6/18

6/19 A laminate roller consists of the uniform 4-lb bar ACB with two light rollers that apply force to the laminates on the top and bottom of a countertop along its edge. Determine the force exerted by each roller on the laminate when a 10-lb force is applied normal to the bar in the position shown. Neglect all friction. *Ans.* $B = 15.94$ lb, $C = 27.1$ lb

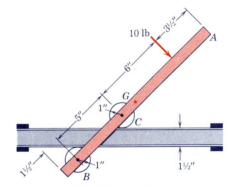

Problem 6/19

6/20 The uniform 20-kg beam ABC is carried by the pickup truck shown. End A is tied to the cab end of the truck bed by a short horizontal rope. The surfaces of contact between the beam and tailgate at B may be assumed to be smooth. If the truck starts with an initial acceleration $a = 3$ m/s^2, calculate the corresponding force at B.

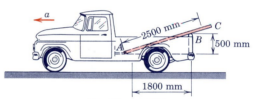

Problem 6/20

6/21 The uniform steel bar AD is rigidly mounted in the collar, which is given a vertical harmonic oscillation as indicated. Neglect the weight of the bar compared with other forces that act and determine the maximum bending moment M induced in the bar during its vertical oscillation for an amplitude $y_0 = 3$ mm and a frequency $n = 6$ cycles per second. *Ans.* $M = 0.417$ N·m

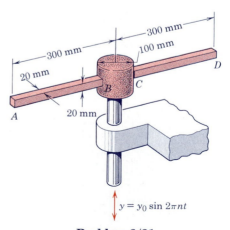

Problem 6/21

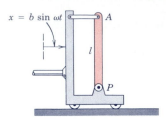

$x = b \sin \omega t$

Problem 6/22

6/22 The device shown oscillates horizontally according to $x = b \sin \omega t$, where b and ω are constants. Determine and plot the force T in the light link at A as a function of the time t. The mass of the uniform slender rod is m.

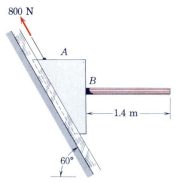

800 N

A

B

1.4 m

60°

Problem 6/23

6/23 The block A and attached rod have a combined mass of 60 kg and are confined to move along the 60° guide under the action of the 800-N applied force. The uniform horizontal rod has a mass of 20 kg and is welded to the block at B. Friction in the guide is negligible. Compute the bending moment M exerted by the weld on the rod at B.

Ans. M = 196.0 N·m

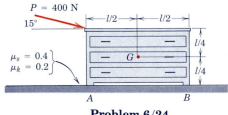

$P = 400$ N

15°

$l/2$ $l/2$

$l/4$

$\mu_s = 0.4$
$\mu_k = 0.2$

G

$l/4$

A   B

Problem 6/24

6/24 The force $P = 400$ N is applied to the 75-kg chest as shown. The mass center G of the chest is located at its geometric center. Determine the percent changes n_A and n_B in the normal forces at A and B compared with the static values when $P = 0$.

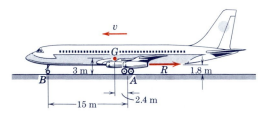

v

G

3 m

R

1.8 m

B

A

2.4 m

15 m

Problem 6/25

6/25 A jet transport with a landing speed of 200 km/h reduces its speed to 60 km/h with a negative thrust R from its jet thrust reversers in a distance of 425 m along the runway with constant deceleration. The total mass of the aircraft is 140 Mg with mass center at G. Compute the reaction N under the nose wheel B toward the end of the braking interval and prior to the application of mechanical braking. At the lower speed aerodynamic forces on the aircraft are small and may be neglected.   *Ans. N = 257 kN*

6/26 The mine skip has a loaded mass of 2000 kg and is attached to the towing vehicle by the light hinged link *CD*. If the towing vehicle has an acceleration of 3 m/s², calculate the corresponding reactions under the small wheels at *A* and *B*.

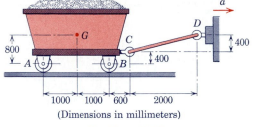

(Dimensions in millimeters)

Problem 6/26

6/27 The figure shows the Saturn V mobile launch platform *A* together with the umbilical tower *B*, unfueled rocket *C*, and crawler-transporter *D* that carries the system to the launch site. The approximate dimensions of the structure and locations of the mass centers *G* are given. The approximate masses are $m_A = 3$ Gg, $m_B = 3.3$ Gg, $m_C = 0.23$ Gg, and $m_D = 3$ Gg. The minimum stopping distance from the top speed of 1.5 km/h is 0.1 m. Compute the vertical component of the reaction under the front crawler unit *F* during the period of maximum deceleration. *Ans.* $F = 59.5$ MN

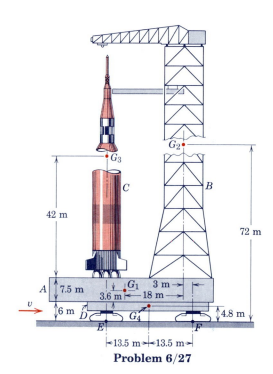

Problem 6/27

6/28 The uniform *L*-shaped bar pivots freely at point *P* of the slider, which moves along the horizontal rod. Determine the steady-state value of the angle θ if (*a*) $a = 0$ and (*b*) $a = g/2$. For what value of a would the steady-state value of θ be zero?

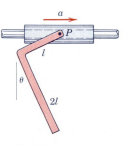

Problem 6/28

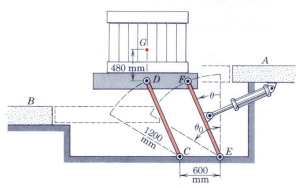

Problem 6/29

6/29 The parallelogram linkage is used to transfer crates from platform A to platform B and is hydraulically operated. The oil pressure in the cylinder is programmed to provide a smooth transition of motion from $\theta = 0$ to $\theta = \theta_0 = \pi/3$ rad given by $\theta = \dfrac{\pi}{6}\left(1 - \cos\dfrac{\pi t}{2}\right)$ where t is in seconds. Determine the force at D on the pin (a) just after the start of the motion with θ and t essentially zero and (b) when $t = 1$ s. The crate and platform have a combined mass of 200 kg with mass center at G. The mass of each link is small and may be neglected.

Ans. (a) $D = 1714$ N, (b) $D = 2178$ N

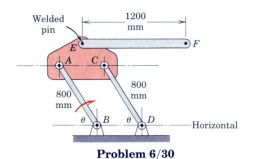

Problem 6/30

6/30 The parallelogram linkage shown moves in the vertical plane with the uniform 8-kg bar EF attached to the plate at E by a pin that is welded both to the plate and to the bar. A torque (not shown) is applied to link AB through its lower pin to drive the links in a clockwise direction. When θ reaches 60°, the links have an angular acceleration and an angular velocity of 6 rad/s² and 3 rad/s, respectively. For this instant calculate the magnitudes of the force F and torque M supported by the pin at E.

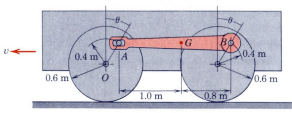

Problem 6/31

6/31 The two wheels of the vehicle are connected by a 20-kg link AB with center of mass at G. The link is pinned to the wheel at B, and the pin at A fits into a smooth horizontal slot in the link. If the vehicle has a constant speed of 4 m/s, determine the magnitude of the force supported by the pin at B for the position $\theta = 30°$.   *Ans.* $B = 188.3$ N

6/32 The semicircular plate of uniform thickness weighs 150 lb and is raised from rest by the parallel linkage of negligible weight under the action of a 500-lb-ft couple M applied to the end of the link. Calculate the components normal and tangent to AB of the shear force supported by the pin at A an instant after the couple M is applied.

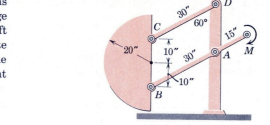

Problem 6/32

6/33 The 900-kg crate with center of mass G in its geometric center is suspended by the two cables at A and B and by a rope at C all in the vertical plane. If the rope is suddenly released with $\theta = 60°$, calculate the tension in the cable at A an instant after release. Solve, first, by writing three motion equations in the three unknowns, and, second, by writing only one motion equation. Compare the merits of the two approaches.   *Ans.* $T_A = 1616$ N

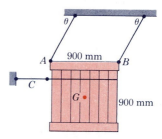

Problem 6/33

6/34 The uniform 200-kg bar AB is raised in the vertical plane by the application of a constant couple $M = 3$ kN·m applied to the link at C. The mass of the links is small and may be neglected. If the bar starts from rest at $\theta = 0$, determine the magnitude of the force supported by the pin at A as the position $\theta = 60°$ is passed.   *Ans.* $A = 2.02$ kN

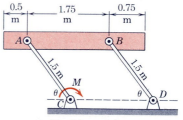

Problem 6/34

6/4 FIXED-AXIS ROTATION

Rotation of a rigid body about a fixed axis O was described in Art. 5/2 and illustrated in Fig. 5/1c. For this motion, we saw that all points in the body describe circles about the rotation axis, and all lines of the body in the plane of motion have the same angular velocity ω and angular acceleration α. The acceleration components of the mass center for circular motion are most easily expressed in n-t coordinates, so we have $\bar{a}_n = \bar{r}\omega^2$ and $\bar{a}_t = \bar{r}\alpha$, as shown in Fig. 6/9a for rotation of the rigid body about the fixed axis through O. The b-part of the figure represents the free-body diagram, and the equivalent kinetic diagram in the c-part of the figure shows the force resultant $m\bar{\mathbf{a}}$ in terms of its n- and t-components and the resultant couple $\bar{I}\alpha$. Our general equations for plane motion, Eqs. 6/1, are directly applicable and are repeated here.

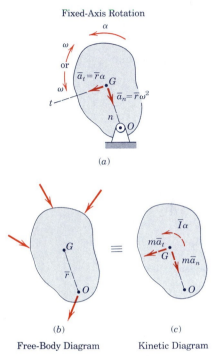

Fixed-Axis Rotation

(a)

$$\Sigma\mathbf{F} = m\bar{\mathbf{a}}$$
$$\Sigma M_G = \bar{I}\alpha$$

[6/1]

Thus, the two scalar components of the force equation become $\Sigma F_n = m\bar{r}\omega^2$ and $\Sigma F_t = m\bar{r}\alpha$. In applying the moment equation about G, it is essential to account for the moment of the force applied to the body at O, so this force must not be omitted from the free-body diagram.

For fixed-axis rotation, it is generally useful to apply a moment equation directly about the rotation axis O. We derived this equation previously as Eq. 6/4, which is repeated here.

(b)   *(c)*

Free-Body Diagram   Kinetic Diagram

Figure 6/9

$$\Sigma M_O = I_O\alpha$$

[6/4]

From the kinetic diagram in Fig. 6/9c, we may obtain Eq. 6/4 very easily by evaluating the moment of the resultants about O, which becomes $\Sigma M_O = \bar{I}\alpha + m\bar{a}_t\bar{r}$. Substitution of the transfer-of-axis relation for mass moments of inertia, $I_O = \bar{I} + m\bar{r}^2$, gives $\Sigma M_O = (I_O - m\bar{r}^2)\alpha + m\bar{r}^2\alpha = I_O\alpha$.

For the common case of rotation of a rigid body about a fixed axis through its mass center G, clearly, $\bar{\mathbf{a}} = \mathbf{0}$, and therefore $\Sigma\mathbf{F} = \mathbf{0}$. The resultant of the applied forces then is the couple $\bar{I}\alpha$.

We may combine the resultant force $m\bar{a}_t$ and resultant couple $\bar{I}\alpha$ by moving $m\bar{a}_t$ to a parallel position through point Q, Fig. 6/10, located by $m\bar{r}\alpha q = \bar{I}\alpha + m\bar{r}\alpha(\bar{r})$. Using the transfer-of-axis theorem and $I_O = k_O{}^2m$ gives $q = k_O{}^2/\bar{r}$. Point Q is called the *center of percussion* and has the unique property that the resultant of all forces applied to the body must pass through it. It follows that the sum of the moments of all forces about the center of percussion is always zero, $\Sigma M_Q = 0$.

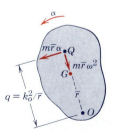

Figure 6/10

Sample Problem 6/3

The concrete block weighing 644 lb is elevated by the hoisting mechanism shown, where the cables are securely wrapped around the respective drums. The drums, which are fastened together and turn as a single unit about their mass center at O, have a combined weight of 322 lb and a radius of gyration about O of 18 in. If a constant tension P of 400 lb is maintained by the power unit at A, determine the vertical acceleration of the block and the resultant force on the bearing at O.

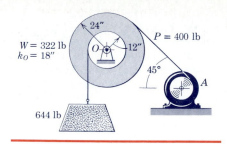

Solution I. The free-body and kinetic diagrams of the drums and concrete block are drawn showing all forces that act, including the components O_x and O_y of the bearing reaction. The resultant of the force system on the drums for centroidal rotation is the couple $\bar{I}\alpha = I_O\alpha$, where

$[I = k^2m]$ $\quad \bar{I} = I_O = \left(\dfrac{18}{12}\right)^2 \dfrac{322}{32.2} = 22.5 \text{ lb-ft-sec}^2$

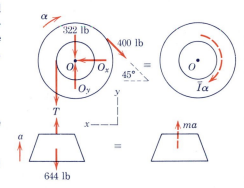

Taking moments about the mass center O for the pulley in the sense of the angular acceleration α gives

$[\Sigma M_G = \bar{I}\alpha]$ $\quad 400\left(\dfrac{24}{12}\right) - T\left(\dfrac{12}{12}\right) = 22.5\alpha$ $\quad$ (a)

The acceleration of the block is described by

$[\Sigma F_y = ma_y]$ $\quad T - 644 = \dfrac{644}{32.2}a$ $\quad$ (b)

From $a_t = r\alpha$, we have $a = (12/12)\alpha$. With this substitution, Eqs. (a) and (b) are combined to give

$\quad T = 717 \text{ lb} \quad \alpha = 3.67 \text{ rad/sec}^2 \quad a = 3.67 \text{ ft/sec}^2$ *Ans.*

The bearing reaction is computed from its components. Since $\bar{a} = 0$, we use the equilibrium equations

$[\Sigma F_x = 0] \quad O_x - 400\cos 45° = 0 \quad O_x = 283 \text{ lb}$
$[\Sigma F_y = 0] \quad O_y - 322 - 717 - 400\sin 45° = 0 \quad O_y = 1322 \text{ lb}$
$\quad O = \sqrt{(283)^2 + (1322)^2} = 1352 \text{ lb}$ *Ans.*

Solution II. We may use a more condensed approach by drawing the free-body diagram of the entire system, thus eliminating reference to T, which becomes internal to the new system. From the kinetic diagram for the system, we see that the moment sum about O must equal the resultant couple $\bar{I}\alpha$ for the drums, plus the moment of the resultant ma for the block. Thus, from the principle of Eq. 6/5 we have

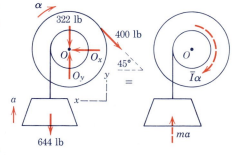

$[\Sigma M_O = \bar{I}\alpha + mad] \quad 400\left(\dfrac{24}{12}\right) - 644\left(\dfrac{12}{12}\right) = 22.5\alpha + \dfrac{644}{32.2}a\left(\dfrac{12}{12}\right)$

With $a = (12/12)\alpha$, the solution gives, as before, $a = 3.67 \text{ ft/sec}^2$.
We may equate the force sums on the entire system to the sums of the resultants. Thus,

$[\Sigma F_y = \Sigma m\bar{a}_y] \quad O_y - 322 - 644 - 400\sin 45° = \dfrac{322}{32.2}(0) + \dfrac{644}{32.2}(3.67)$

$\quad O_y = 1322 \text{ lb}$

$[\Sigma F_x = \Sigma m\bar{a}_x] \quad O_x - 400\cos 45° = 0 \quad O_x = 283 \text{ lb}$

① Be alert to the fact that the tension T is not 644 lb. If it were, the block would not accelerate.

② Do not overlook the need to express k_o in feet when using g in ft/sec^2.

Sample Problem 6/4

The pendulum has a mass of 7.5 kg with center of mass at G and has a radius of gyration about the pivot O of 295 mm. If the pendulum is released from rest at $\theta = 0$, determine the total force supported by the bearing at the instant when $\theta = 60°$. Friction in the bearing is negligible.

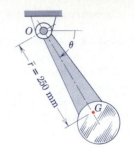

Solution. The free-body diagram of the pendulum in a general position is shown along with the corresponding kinetic diagram, where the components of the resultant force have been drawn through G.

The normal component O_n is found from a force equation in the n-direction that involves the normal acceleration $\bar{r}\omega^2$. Since the angular velocity ω of the pendulum is found from the integral of the angular acceleration and since O_t depends on the tangential acceleration $\bar{r}\alpha$, it follows that α should be obtained first. To this end with $I_O = k_O{}^2 m$, the moment equation about O gives

$[\Sigma M_O = I_O \alpha]$ $7.5(9.81)(0.25) \cos \theta = (0.295)^2 (7.5)\alpha$

$$\alpha = 28.2 \cos \theta \text{ rad/s}^2$$

and for $\theta = 60°$

$[\omega \, d\omega = \alpha \, d\theta]$ $\displaystyle\int_0^\omega \omega \, d\omega = \int_0^{\pi/3} 28.2 \cos \theta \, d\theta$

$$\omega^2 = 48.8 \text{ (rad/s)}^2$$

The remaining two equations of motion applied to the 60° position yield

$[\Sigma F_n = m\bar{r}\omega^2]$ $O_n - 7.5(9.81) \sin 60° = 7.5(0.25)(48.8)$

$$O_n = 155.2 \text{ N}$$

$[\Sigma F_t = m\bar{r}\alpha]$ $-O_t + 7.5(9.81) \cos 60° = 7.5(0.25)(28.2) \cos 60°$

$$O_t = 10.37 \text{ N}$$

$$O = \sqrt{(155.2)^2 + (10.37)^2} = 155.6 \text{ N} \qquad \textit{Ans.}$$

The proper sense for O_t may be observed at the outset by applying the moment equation $\Sigma M_G = \bar{I}\alpha$, where the moment about G due to O_t must be clockwise to agree with α. The force O_t may also be obtained initially by a moment equation about the center of percussion Q, shown in the lower figure, which avoids the necessity of computing α. First, we must obtain the distance q, which is

$[q = k_O{}^2 / \bar{r}]$ $q = \dfrac{(0.295)^2}{0.250} = 0.348 \text{ m}$

$[\Sigma M_Q = 0]$ $O_t(0.348) - 7.5(9.81)(\cos 60°)(0.348 - 0.250) = 0$

$$O_t = 10.37 \text{ N} \qquad \textit{Ans.}$$

① The acceleration components of G are, of course, $\bar{a}_n = \bar{r}\omega^2$ and $\bar{a}_t = \bar{r}\alpha$.

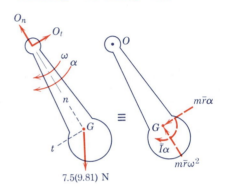

② Review the theory again and satisfy yourself that $\Sigma M_O = I_O \alpha = \bar{I}\alpha + m\bar{r}^2\alpha = m\bar{r}\alpha q$.

③ Note especially here that the force summations are taken in the positive direction of the acceleration components of the mass center G.

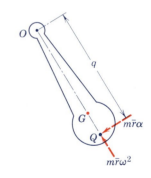

PROBLEMS

Introductory problems

6/35 The uniform 100-kg beam is freely hinged about its upper end A and is initially at rest in the vertical position with $\theta = 0$. Determine the initial angular acceleration α of the beam and the magnitude F_A of the force supported by the pin at A due to the application of a force $P = 300$ N on the attached cable.

Ans. $\alpha = 1.193$ rad/s^2, $F_A = 769$ N

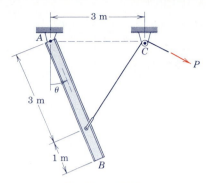

Problem 6/35

6/36 The uniform 20-kg slender bar is pivoted at O and swings freely in the vertical plane. If the bar is released from rest in the horizontal position, calculate the initial value of the force R exerted by the bearing on the bar an instant after release.

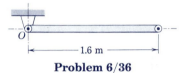

Problem 6/36

6/37 Each of the two drums and connected hubs of 8-in. radius weighs 200 lb and has a radius of gyration about its center of 15 in. Calculate the angular acceleration of each drum. Friction in each bearing is negligible.

Ans. $\alpha_a = 1.976$ rad/sec^2, $\alpha_b = 2.06$ rad/sec^2

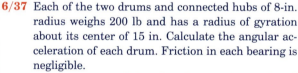

(a)   *(b)*

Problem 6/37

6/38 Calculate the downward acceleration a of the 10-kg cylinder. The drum is a uniform cylinder, and friction at the pivot is negligible. Compare your answer with that obtained by ignoring the inertia effects of the drum.

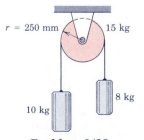

Problem 6/38

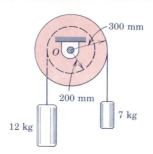

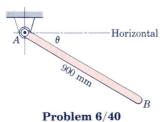

Problem 6/39

6/39 If the frictional moment at the pivot O is 2 N·m, determine the angular acceleration of the grooved drum, which has a mass of 8 kg and a radius of gyration $k_O = 225$ mm.

Ans. $\alpha = 0.622$ rad/s^2 CCW

Problem 6/40

6/40 The uniform slender bar AB has a mass of 8 kg and swings in a vertical plane about the pivot at A. If $\dot{\theta} = 2$ rad/s when $\theta = 30°$, compute the force supported by the pin at A at that instant.

Problem 6/41

6/41 The uniform slender bar is released from rest in the horizontal position shown. Determine the value of x for which the angular acceleration is a maximum, and determine the corresponding angular acceleration α.

Ans. $x = \dfrac{l}{2\sqrt{3}}, \ \alpha = \sqrt{3}\,\dfrac{g}{l}$

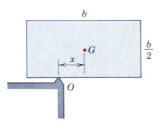

Problem 6/42

6/42 The uniform rectangular slab is released from rest in the position shown. Determine the value of x for which the angular acceleration is a maximum, and determine the corresponding angular acceleration. Compare your answers with those listed for Prob. 6/41.

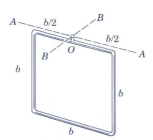

Problem 6/43

6/43 The square frame is composed of four equal lengths of uniform slender rod, and the ball attachment at O is suspended in a socket (not shown). Beginning from the position shown, the assembly is rotated $45°$ about axis A-A and released. Determine the initial angular acceleration of the frame. Repeat for a $45°$ rotation about axis B-B. Neglect the small mass, offset, and friction of the ball.

Ans. $\alpha = \dfrac{3\sqrt{2}}{5}\dfrac{g}{b}, \ \alpha = \dfrac{3\sqrt{2}}{7}\dfrac{g}{b}$

Representative problems

6/44 The bar is welded to the irregular slab at the mass center G of the slab as shown. Show that the angular acceleration that occurs when the string S is cut is independent of the angular orientation θ of the slab relative to the bar.

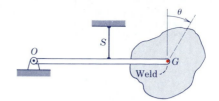

Problem 6/44

6/45 The automotive dynamometer is able to simulate road conditions for an acceleration of $0.5g$ for the loaded pickup truck with a gross weight of 5200 lb. Calculate the required moment of inertia of the dynamometer drum about its center O assuming that the drum turns freely during the acceleration phase of the test.   *Ans.* $I_O = 1453$ lb-ft-sec^2

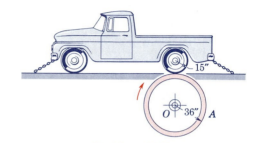

Problem 6/45

6/46 An *air table* is used to study the elastic motion of flexible spacecraft models. Pressurized air escaping from numerous small holes in the horizontal surface provide a supporting air cushion that largely eliminates friction. The model shown consists of a cylindrical hub of radius r and four appendages of length l and small thickness t. The hub and the four appendages all have the same depth d and are constructed of the same material of density ρ. Assume that the spacecraft is rigid and determine the moment M that must be applied to the hub to spin the model from rest to an angular velocity ω in a time period of τ seconds. (Note that for a spacecraft with highly flexible appendages, the moment must be judiciously applied to the rigid hub to avoid undesirable large elastic deflections of the appendages.)

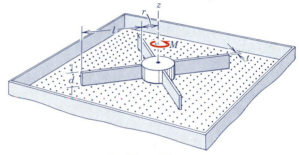

Problem 6/46

6/47 The ring of mean radius r and small cross-sectional dimensions is rotating in a horizontal plane with a constant angular velocity ω about its center O. Determine the tension T in the ring. Solve, first, by analyzing one half of the ring as a free body and, second, by analyzing a differential element of arc as a free body.   *Ans.* $T = \rho r^2 \omega^2$

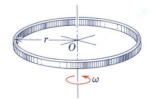

Problem 6/47

6/48 The spring is uncompressed when the uniform slender bar is in the vertical position shown. Determine the initial angular acceleration α of the bar when it is released from rest in a position where the bar has been rotated 30° clockwise from the position shown. Neglect any sag of the spring whose mass is negligible.

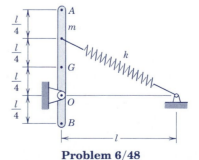

Problem 6/48

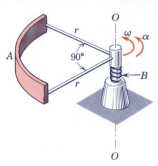

Problem 6/49

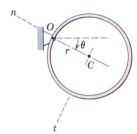

Problem 6/50

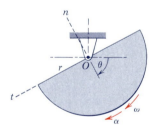

Problem 6/51

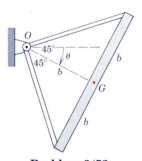

Problem 6/52

6/49 The bar A of mass m is formed into a 90° circular arc of radius r and is attached to the hub by the light rods. The curved bar oscillates about the vertical axis under the action of a torsional spring B. At the instant under consideration the angular velocity is ω and the angular acceleration is α. Write expressions for the moment M exerted by the spring on the hub and the horizontal force R exerted by the shaft on the hub.

$$\text{Ans. } M = mr^2\,\alpha,\, R = \frac{2\sqrt{2}mr}{\pi}\sqrt{\alpha^2 + \omega^4}$$

6/50 The narrow ring of mass m is free to rotate in the vertical plane about O. If the ring is released from rest at $\theta = 0$, determine expressions for the n- and t-components of the force at O in terms of θ.

6/51 The semicircular disk of mass m and radius r is released from rest at $\theta = 0$ and rotates freely in the vertical plane about its fixed bearing at O. Derive expressions for the n- and t-components of the force F on the bearing as functions of θ.

$$\text{Ans. } F_n = 1.721mg\,\sin\,\theta$$
$$F_t = 0.640mg\,\cos\,\theta$$

6/52 A uniform slender bar of mass m and length $2b$ is mounted in a right-angle frame of negligible mass. The bar and frame rotate in the vertical plane about a fixed axis at O. If the bar is released from rest in the vertical position ($\theta = 0$), derive an expression for the magnitude of the force exerted by the bearing at O on the frame as a function of θ.

6/53 The mass of gear A is 20 kg and its centroidal radius of gyration is 150 mm. The mass of gear B is 10 kg and its centroidal radius of gyration is 100 mm. Calculate the angular acceleration of gear B when a torque of 12 N·m is applied to the shaft of gear A. Neglect friction. *Ans.* $\alpha_B = 25.5$ rad/s² CCW

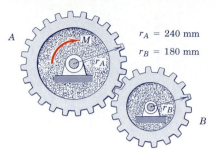

$r_A = 240$ mm
$r_B = 180$ mm

Problem 6/53

6/54 The solid cylindrical rotor B has a mass of 43 kg and is mounted on its central axis C-C. The frame A rotates about the fixed vertical axis O-O under the applied torque $M = 30$ N·m. The rotor may be unlocked from the frame by withdrawing the locking pin P. Calculate the angular acceleration α of the frame A if the locking pin is (*a*) in place and (*b*) withdrawn. Neglect all friction and the mass of the frame.

200 mm
250 mm

Problem 6/54

6/55 The figure shows a roll-off truck ramp for discharging loaded containers. The loaded 120-Mg container may be treated as a homogeneous solid rectangular block with mass center at G. If the supporting wheel A is restrained from movement, calculate the force F_B exerted by the ramp on the supporting wheel B when the truck starts from rest with a forward acceleration of 3 m/s². Neglect friction at B.
Ans. $F_B = 310$ kN

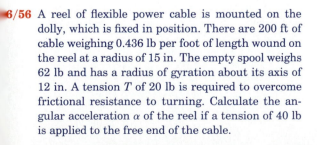

4 m
4 m
B
$\cdot G$
1.5 m
1.5 m
30° 45°
A
a

Problem 6/55

6/56 A reel of flexible power cable is mounted on the dolly, which is fixed in position. There are 200 ft of cable weighing 0.436 lb per foot of length wound on the reel at a radius of 15 in. The empty spool weighs 62 lb and has a radius of gyration about its axis of 12 in. A tension T of 20 lb is required to overcome frictional resistance to turning. Calculate the angular acceleration α of the reel if a tension of 40 lb is applied to the free end of the cable.

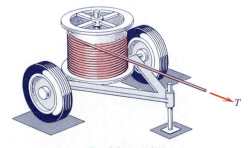

T

Problem 6/56

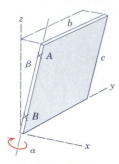

Problem 6/57

6/57 The uniform rectangular panel is hinged at A and B about a fixed axis that is inclined at an angle β from the vertical y-z plane. If the panel is released from rest with its upper and lower edges horizontal, determine its initial angular acceleration α.

$$Ans. \quad \alpha = \frac{3g}{2b} \sin \beta$$

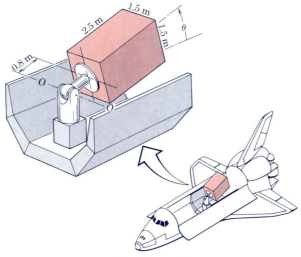

Problem 6/58

6/58 A gimbal pedestal supports a payload in the space shuttle and deploys it when the doors of the cargo bay are opened in orbit. The payload is modeled as a homogeneous rectangular block with a mass of 6000 kg. The torque on the gimbal axis O-O is 30 N·m supplied by a d-c brushless motor. With the shuttle orbiting in a "weightless" condition, determine the time t required to bring the payload from its stowed position at $\theta = 0$ to its deployed position at $\theta = 90°$ if the torque is applied for the first 45° of travel and then reversed for the remaining 45° to bring the payload to a stop ($\dot{\theta} = 0$).

Problem 6/59

6/59 A device for impact testing consists of a 34-kg pendulum with mass center at G and with radius of gyration about O of 620 mm. The distance b for the pendulum is selected so that the force on the bearing at O has the least possible value during impact with the specimen at the bottom of the swing. Determine b and calculate the magnitude of the total force R on the bearing O an instant after release from rest at $\theta = 60°$.   *Ans.* $b = 40.7$ mm, $R = 167.8$ N

6/60 The robotic device consists of the stationary pedestal *OA*, arm *AB* pivoted at *A*, and arm *BC* pivoted at *B*. The rotation axes are normal to the plane of the figure. Estimate (*a*) the moment M_A applied to arm *AB* required to rotate it about joint *A* at 4 rad/sec^2 counterclockwise from the position shown with joint *B* locked and (*b*) the moment M_B applied to arm *BC* required to rotate it about joint *B* at the same rate with joint *A* locked. The mass of arm *AB* is 25 kg and that of *BC* is 4 kg, with the stationary portion of joint *A* excluded entirely and the mass of joint *B* split between the two arms. Assume that the centers of mass G_1 and G_2 are in the geometric centers of the arms and model the arms as slender rods.

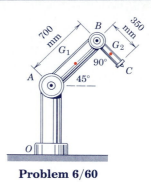

Problem 6/60

6/61 The uniform 72-ft mast weighs 600 lb and is hinged at its lower end to a fixed support at *O*. If the winch *C* develops a starting torque of 900 lb-ft, calculate the total force supported by the pin at *O* as the mast begins to lift off its support at *B*. Also find the corresponding angular acceleration α of the mast. The cable at *A* is horizontal, and the mass of the pulleys and winch is negligible.

Ans. $\alpha = 0.090$ rad/sec^2, $O = 1087$ lb

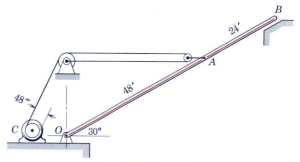

Problem 6/61

6/62 The 12-kg cylinder supported by the bearing brackets at *A* and *B* has a moment of inertia about the vertical z_0-axis through its mass center *G* equal to 0.080 kg·m^2. The disk and brackets have a moment of inertia about the vertical *z*-axis of rotation equal to 0.60 kg·m^2. If a torque $M = 16$ N·m is applied to the disk through its shaft with the disk initially at rest, calculate the horizontal *x*-components of force supported by the bearings at *A* and *B*.

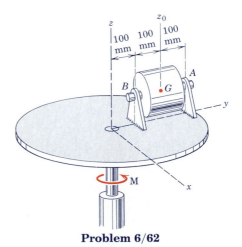

Problem 6/62

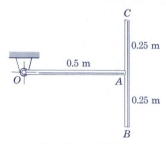

0.5 m

0.25 m

0.25 m

Problem 6/63

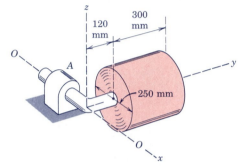

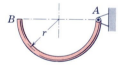

Problem 6/64

Problem 6/65

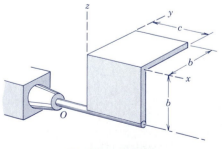

Problem 6/66

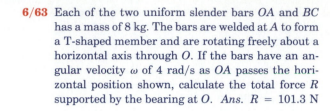

6/63 Each of the two uniform slender bars OA and BC has a mass of 8 kg. The bars are welded at A to form a T-shaped member and are rotating freely about a horizontal axis through O. If the bars have an angular velocity ω of 4 rad/s as OA passes the horizontal position shown, calculate the total force R supported by the bearing at O. *Ans.* $R = 101.3$ N

6/64 The solid homogeneous cylinder has a mass of 100 kg and is mounted on a right-angled shaft that turns freely about the horizontal O-O axis. If the cylinder is released from rest with its own axis in the horizontal plane, calculate the initial angular acceleration of the assembly and the resultant force F exerted by the bearing A on the shaft. The mass of the shaft may be neglected.

6/65 The uniform semicircular bar of mass m and radius r is hinged freely about a horizontal axis through A. If the bar is released from rest in the position shown where AB is horizontal, determine the initial angular acceleration α of the bar and the expression for the force exerted on the bar by the pin at A. (Note carefully that the initial tangential acceleration of the mass center is not vertical.)

$$Ans. \quad \alpha = \frac{g}{2r}, \quad A = 0.593mg$$

6/66 The right-angle plate is formed from a flat plate having a mass ρ per unit area and is welded to the horizontal shaft mounted in the bearing at O. If the shaft is free to rotate, determine the initial angular acceleration α of the plate when it is released from rest with the upper surface in the horizontal plane. Also determine the y- and z-components of the resultant force on the shaft at O.

6/67 A flexible cable 60 meters long with a mass of 0.160 kg per meter of length is wound around the reel. With $y = 0$, the weight of the 4-kg cylinder is required to start turning the reel to overcome friction in its bearings. Determine the downward acceleration a in meters per second squared of the cylinder as a function of y in meters. The empty reel has a mass of 16 kg with a radius of gyration about its bearing of 200 mm. *Ans. $a = 0.0758y$*

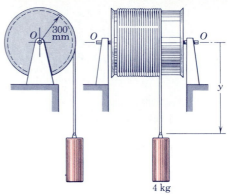

Problem 6/67

6/68 The link B weighs 0.80 lb with center of mass 2.20 in. from O-O and has a radius of gyration about O-O of 2.76 in. The link is welded to the steel tube and is free to rotate about the fixed horizontal shaft at O-O. The tube weighs 1.84 lb. If the tube is released from rest with the link in the horizontal position, calculate the initial angular acceleration α of the assembly and the corresponding reaction O exerted by the shaft on the link.

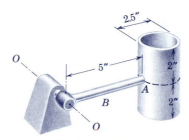

Problem 6/68

6/69 Prior to deployment of its two instrument arms AB, the spacecraft shown in the upper view is spinning at the constant rate of 1 revolution per second. Each instrument arm, shown in the lower view, has a mass of 10 kg with mass center at G. Calculate the tension T in the deployment cable prior to its release. Also find the magnitude of the force on the pin at A. Neglect any acceleration of the center O of the spacecraft. *Ans. $T = 987$ N, $A = 1.007$ kN*

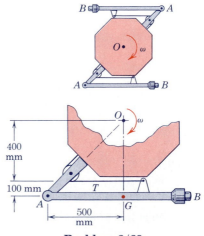

Problem 6/69

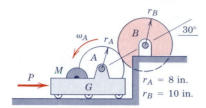

Problem 6/70

6/70 Disk B weighs 50 lb and has a centroidal radius of gyration of 8 in. The power unit C consists of a motor M and disk A, which is driven at a constant angular speed of 1600 rev/min. The coefficients of static and kinetic friction between the two disks are $\mu_s = 0.8$ and $\mu_k = 0.6$, respectively. Disk B is initially stationary when contact with disk A is established by application of the constant force $P = 3$ lb. Determine the angular acceleration α of B and the time t required for B to reach its steady-state speed.

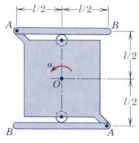

Problem 6/71

6/71 Two slender bars AB, each of mass m and length l, are pivoted at A to the plate. The plate rotates in the horizontal plane about a fixed vertical axis through its center O and is given a constant angular acceleration α. (*a*) Determine the force F exerted on each of the two rollers as the assembly starts to rotate. (*b*) Find the total force on the pin at A and show that it remains constant as long as $F > 0$. (*c*) Determine the angular velocity ω at which contact with the rollers ceases.   *Ans.* (*a*) $F = ml\alpha/6$

$$(b)\ A = \frac{\sqrt{10}}{6}\,ml\alpha$$

$$(c)\ \omega = \sqrt{\alpha/3}$$

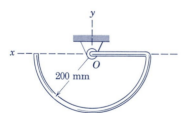

Problem 6/72

6/72 A uniform metal rod with a mass of 0.60 kg per meter of length is bent into the shape shown in the vertical plane and pivoted freely about a horizontal axis at O normal to the plane. If the rod is released from rest in the position shown with the straight section horizontal, compute the initial x- and y-components of the reaction $\mathbf{R}$ supported by the bearing at O.

Problem 6/73

6/73 The curved bar of mass m is hinged to the rotating disk at O and bears against one of the smooth pins A and B that are fastened to the disk. If the disk rotates about its vertical axis C, determine the force exerted on the bar by the hinge at O and the reaction A or B on the bar (*a*) if the disk has a constant angular velocity ω and (*b*) as the disk starts from rest with a counterclockwise angular acceleration α.

$$Ans.\ (a)\ O = \frac{2mr\omega^2}{\pi},\ A = \frac{2mr\omega^2}{\pi}$$

$$(b)\ O = mr\alpha\,\sqrt{1 + \frac{4}{\pi^2}},\ B = mr\alpha\left(1 - \frac{2}{\pi}\right)$$

6/74 The 16-ft I-beam weighs 2000 lb and is held in the horizontal position by the pin at O and by the vertical cable that passes around the pulley at A and around the drum of the 500-lb motorized winch at B. If the winch motor has an output starting torque of 600 lb-ft, calculate the initial vertical force supported by the pin at O. Treat the beam as a slender bar and the winch unit as a mass concentrated at the center of the pulley. (Is the horizontal component of the force on the pin zero?)

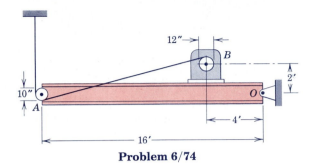

Problem 6/74

6/75 The uniform slender bar of mass m and length l is released from rest in the vertical position and pivots on its square end about the corner at O. (a) If the bar is observed to slip when $\theta = 30°$, find the coefficient of static friction μ_s between the bar and the corner. (b) If the end of the bar is notched so that it cannot slip, find the angle θ at which contact between the bar and the corner ceases.

Ans. (a) $\mu_s = 0.188$, (b) $\theta = 53.1°$

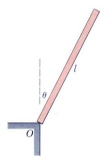

Problem 6/75

6/76 The 3-m slender beam has a mass of 50 kg and is released from rest in the horizontal position with $\theta = 0$. If the coefficient of static friction between the fixed support at O and the beam is 0.30, determine the angle θ at which slipping first occurs at O. Does the result depend on the mass of the beam?

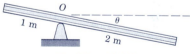

Problem 6/76

6/5 GENERAL PLANE MOTION

The dynamics of general plane motion of a rigid body combines translation and rotation. In Art. 6/2 we represented such a body in Fig. 6/4 with its free-body diagram and its kinetic diagram, which discloses the dynamic resultants of the applied forces. Figure 6/4 and Eq. 6/1, which apply to general plane motion, are repeated again here for convenient reference.

$$\Sigma \mathbf{F} = m\bar{\mathbf{a}}$$
$$\Sigma M_G = \bar{I}\alpha$$

[6/1]

As previously, the two scalar components of the force equation are expressed in whatever coordinate system most readily describes the acceleration of the mass center. Direct application of these equations expresses the equivalence between the externally applied forces, as disclosed by the free-body diagram, and their force and moment resultants, as represented by the kinetic diagram.

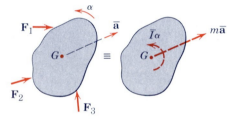

Free-Body Diagram   Kinetic Diagram

Figure 6/4

In Art. 6/2 we also showed, with the aid of Fig. 6/5, the application of the alternative relation for moments about any point P, Eq. 6/2. This figure and this equation are also repeated here for easy reference.

$$\Sigma M_P = \bar{I}\alpha + m\bar{a}d$$

[6/2]

In some instances, it may be more convenient to use the alternative moment relation of Eq. 6/3 when moments are taken about a point P whose acceleration is known. It is also noted that the equation for moments about a nonaccelerating point O on the body, Eq. 6/4, constitutes still another alternative moment relation and at times may be used to advantage.

In working a problem in general plane motion, we first observe whether the motion is unconstrained or constrained, as illustrated in the examples of Fig. 6/6. If the motion is constrained, we must account for the kinematic relationship between the linear and the angular accelerations and incorporate it into our force and moment equations of motion. If the motion is unconstrained, the accelerations can be determined independently of one another by direct application of the three motion equations, Eqs. 6/1.

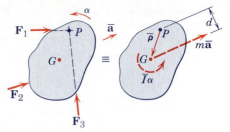

Free-Body Diagram   Kinetic Diagram

Figure 6/5

In order for a rigid-body problem to be solvable, the number of unknowns cannot exceed the number of independent equations available to describe them, and a check on the sufficiency of the relationships should always be made. At the most, for plane motion we have three scalar equations of motion and two scalar components of the vector relative-acceleration equation for constrained motion. Thus, we can handle as many as five unknowns.

Strong emphasis should be placed on the importance of a clear choice of the body to be isolated and the representation of this isolation by a correct free-body diagram. Only after this vital step has been completed can we properly evaluate the equivalence between the external forces and their resultants. Of equal importance in the analysis of plane motion is a clear understanding of the kinematics involved. Very often, the difficulties experienced at this point have to do with kinematics, and a thorough review of the relative acceleration relations for plane motion will be most helpful. In formulating the solution to a problem, we recognize that the directions of certain forces or accelerations may not be known at the outset, so that it may be necessary to make initial assumptions whose validity will be proved or disproved when the solution is carried out. It is essential, however, that all assumptions made be consistent with the principle of action and reaction and with any kinematical requirements, which are also called conditions of constraint. Thus, if a wheel is rolling on a horizontal surface, its center is constrained to move on a horizontal line. Furthermore, if the unknown linear acceleration a of the center of the wheel is assumed positive to the right, then the unknown angular acceleration α would be positive in a clockwise sense in order that $a = +r\alpha$, if we assume the wheel does not slip. Also, we note that for a wheel that rolls without slipping, the static friction force between the wheel and its supporting surface is generally *less* than its maximum value, so that $F \neq \mu_s N$. But if the wheel slips as it rolls, $a \neq r\alpha$, and a kinetic friction force is generated that is given by $F = \mu_k N$. It may be necessary to test the validity of either assumption, slipping or no slipping, in a given problem. The difference between the static and kinetic coefficients of friction, μ_s and μ_k, is sometimes ignored, in which case, μ is used for either or both coefficients.

Sample Problem   6/5

A metal hoop with a radius $r = 6$ in. is released from rest on the 20° incline. If the coefficients of static and kinetic friction are $\mu_s = 0.15$ and $\mu_k = 0.12$, determine the angular acceleration α of the hoop and the time t for the hoop to move a distance of 10 ft down the incline.

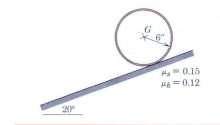

Solution.   The free-body diagram shows the unspecified weight mg, the normal force N, and the friction force F acting on the hoop at the contact point C with the incline. The kinetic diagram shows the resultant force $m\bar{a}$ through G in the direction of its acceleration and the couple $\bar{I}\alpha$. The counterclockwise angular acceleration requires a counterclockwise moment about G, so F must be up the incline.

Assume that the hoop rolls without slipping, so that $\bar{a} = r\alpha$. Application of the components of Eqs. 6/1 with x- and y-axes assigned gives

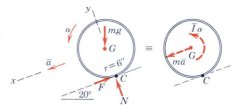

$[\Sigma F_x = m\bar{a}_x]$ $\qquad mg \sin 20° - F = m\bar{a}$

$[\Sigma F_y = m\bar{a}_y = 0]$ $\qquad N - mg \cos 20° = 0$

① $[\Sigma M_G = \bar{I}\alpha]$ $\qquad Fr = mr^2\alpha$

Elimination of F between the first and third equations and substitution of the kinematic assumption $\bar{a} = r\alpha$ give

② $\qquad \bar{a} = \dfrac{g}{2} \sin 20° = \dfrac{32.2}{2} (0.3420) = 5.51 \text{ ft/sec}^2$

① Because all of the mass of a hoop is a distance r from its center G, its moment of inertia about G must be mr^2.

Alternatively, with our assumption of $\bar{a} = r\alpha$ for pure rolling, a moment sum about C by Eq. 6/2 gives $\bar{a}$ directly. Thus,

② Note that $\bar{a}$ is independent of both m and r.

$[\Sigma M_C = \bar{I}\alpha + m\bar{a}d]$ $\quad mgr \sin 20° = mr^2\dfrac{\bar{a}}{r} + m\bar{a}r$ $\quad \bar{a} = \dfrac{g}{2} \sin 20°$

To check our assumption of no slipping, we calculate F and N and compare F with its limiting value. From the above equations,

$$F = mg \sin 20° - m\frac{g}{2} \sin 20° = 0.1710mg$$

$$N = mg \cos 20° = 0.9397mg$$

But the maximum possible friction force is

$[F_{max} = \mu_s N]$ $\qquad F_{max} = 0.15(0.9397mg) = 0.1410mg$

Because our calculated value of $0.1710mg$ exceeds the limiting value of $0.1410mg$, we conclude that our assumption of pure rolling was wrong. Therefore, the hoop slips as it rolls and $\bar{a} \neq r\alpha$. The friction force then becomes the kinetic value

$[F = \mu_k N]$ $\qquad F = 0.12(0.9397mg) = 0.1128mg$

The motion equations now give

$[\Sigma F_x = m\bar{a}_x]$ $\qquad mg \sin 20° - 0.1128mg = m\bar{a}$

$\qquad \bar{a} = 0.229(32.2) = 7.38 \text{ ft/sec}^2$

③ $[\Sigma M_G = \bar{I}\alpha]$ $\qquad 0.1128mg(r) = mr^2\alpha$

$\qquad \alpha = \dfrac{0.1128(32.2)}{6/12} = 7.26 \text{ rad/sec}^2$ $\qquad$ *Ans.*

③ Note that α is independent of m but dependent on r.

The time required for the center G of the hoop to move 10 ft from rest with constant acceleration is

$[x = \frac{1}{2}at^2]$ $\qquad t = \sqrt{\dfrac{2x}{\bar{a}}} = \sqrt{\dfrac{2(10)}{7.38}} = 1.646 \text{ sec}$ $\qquad$ *Ans.*

Sample Problem 6/6

The drum A is given a constant angular acceleration α_0 of 3 rad/s² and causes the 70-kg spool B to roll on the horizontal surface by means of the connecting cable, which wraps around the inner hub of the spool. The radius of gyration $\bar{k}$ of the spool about its mass center G is 250 mm, and the coefficient of static friction between the spool and the horizontal surface is 0.25. Determine the tension T in the cable and the friction force F exerted by the horizontal surface on the spool.

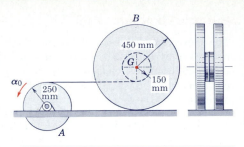

Solution.   The free-body diagram and the kinetic diagram of the spool are drawn as shown. The correct direction of the friction force may be assigned in this problem by observing from both diagrams that with counterclockwise angular acceleration, a moment sum about point G (and also about point D) must be counterclockwise. A point on the connecting cable has an acceleration $a_t = r\alpha = 0.25(3) = 0.75$ m/s², which is also the horizontal component of the acceleration of point D on the spool. It will be assumed initially that the spool rolls without slipping, in which case it has a counterclockwise angular acceleration $\alpha =$
① $(a_D)_x/\overline{DC} = 0.75/0.30 = 2.5$ rad/s². The acceleration of the mass center G is, therefore, $\bar{a} = r\alpha = 0.45(2.5) = 1.125$ m/s².

With the kinematics determined, we now apply the three equations of motion, Eqs. 6/1.

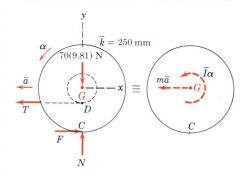

$$[\Sigma F_x = m\bar{a}_x] \qquad\qquad F - T = 70(-1.125) \qquad\qquad (a)$$

$$[\Sigma F_y = m\bar{a}_y] \qquad N - 70(9.81) = 0 \qquad N = 687 \text{ N}$$

② $$[\Sigma M_G = \bar{I}\alpha] \qquad F(0.450) - T(0.150) = 70(0.250)^2(2.5) \qquad (b)$$

Solving (a) and (b) simultaneously gives

$$F = 75.8 \text{ N} \qquad \text{and} \qquad T = 154.6 \text{ N} \qquad\qquad Ans.$$

To establish the validity of our assumption of no slipping, we see that the surfaces are capable of supporting a maximum friction force $F_{\max} = \mu_s N = 0.25(687) = 171.7$ N. Since only 75.8 N of friction force is required, we conclude that our assumption of rolling without slipping is valid.

If the coefficient of static friction had been 0.1, for example, then the friction force would have been limited to 0.1(687) = 68.7 N, which is less than 75.8 N, and the spool would slip. In this event, the kinematical relation $\bar{a} = r\alpha$ would no longer hold. With $(a_D)_x$ known, the angular
③ acceleration would be $\alpha = [\bar{a} - (a_D)_x]/\overline{GD}$. Using this relation along with $F = \mu_k N < 68.7$ N, we would then resolve the three equations of motion for the unknowns T, $\bar{a}$, and α.

Alternatively, with point C as a moment center, we may use Eq. 6/2 and obtain T directly. Thus,

$$[\Sigma M_C = \bar{I}\alpha + m\bar{a}r] \qquad 0.3T = 70(0.25)^2(2.5) + 70(1.125)(0.45)$$

$$T = 154.6 \text{ N} \qquad\qquad Ans.$$

where the kinematic condition of no slipping, $\bar{a} = \bar{r}\alpha$, has been incor-
④ porated. We could also write a moment equation about point D to obtain F directly.

① The relation between $\bar{a}$ and α is the kinematical requirement that accompanies the constraint to the motion imposed by the assumption that the spool rolls without slipping.

② Be careful not to make the mistake of using $\frac{1}{2}mr^2$ for $\bar{I}$ of the spool, which is not a uniform circular disk.

③ Our principles of relative acceleration are a necessity here. Hence, the relation $(a_{G/D})_t = \overline{GD}\alpha$ should be recognized.

④ The flexibility in the choice of moment centers provided by the kinetic diagram can usually be employed to simplify the analysis.

Sample Problem 6/7

The slender bar AB weighs 60 lb and moves in the vertical plane, with its ends constrained to follow the smooth horizontal and vertical guides. If the 30-lb force is applied at A with the bar initially at rest in the position for which $\theta = 30°$, calculate the resulting angular acceleration of the bar and the forces on the small end rollers at A and B.

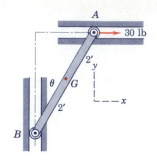

Solution. The bar undergoes constrained motion, so that we must establish the relationship between the mass-center acceleration and the angular acceleration. The relative-acceleration equation $\mathbf{a}_A = \mathbf{a}_B + \mathbf{a}_{A/B}$

① must be solved first, and then the equation $\bar{\mathbf{a}} = \mathbf{a}_G = \mathbf{a}_B + \mathbf{a}_{G/B}$ is next solved to obtain expressions relating $\bar{a}$ and α. With α assigned in its clockwise physical sense, the acceleration polygons that represent these equations are shown, and their solution gives

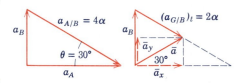

$$\bar{a}_x = \bar{a} \cos 30° = 2\alpha \cos 30° = 1.732\alpha \text{ ft/sec}^2$$

$$\bar{a}_y = \bar{a} \sin 30° = 2\alpha \sin 30° = 1.0\alpha \text{ ft/sec}^2$$

Next we construct the free-body diagram and the kinetic diagram as shown. With $\bar{a}_x$ and $\bar{a}_y$ now known in terms of α, the remaining unknowns are α and the forces A and B. We now apply Eqs. 6/1 which give

② $[\Sigma M_G = \bar{I}\alpha]$

$$30(2 \cos 30°) - A(2 \sin 30°) + B(2 \cos 30°) = \frac{1}{12}\frac{60}{32.2}(4^2)\alpha$$

$[\Sigma F_x = m\bar{a}_x]$ $\quad 30 - B = \dfrac{60}{32.2}(1.732\alpha)$

$[\Sigma F_y = m\bar{a}_y]$ $\quad A - 60 = \dfrac{60}{32.2}(1.0\alpha)$

Solving the three equations simultaneously gives us the results

$$A = 68.2 \text{ lb} \quad B = 15.74 \text{ lb} \quad \alpha = 4.42 \text{ rad/sec}^2 \quad Ans.$$

As an alternative solution, we can use Eq. 6/2 with point C as the moment center and avoid the necessity of solving three equations simultaneously. This choice eliminates reference to forces A and B and gives α directly. Thus,

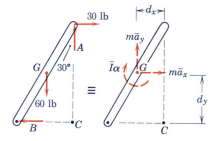

③ $[\Sigma M_C = \bar{I}\alpha + \Sigma m\bar{a}d]$

$$30(4 \cos 30°) - 60(2 \sin 30°) = \frac{1}{12}\frac{60}{32.2}(4^2)\alpha$$

$$+ \frac{60}{32.2}(1.732\alpha)(2 \cos 30°) + \frac{60}{32.2}(1.0\alpha)(2 \sin 30°)$$

$$43.9 = 9.94\alpha \quad \alpha = 4.42 \text{ rad/sec}^2 \quad Ans.$$

With α determined, we can now apply the force equations independently and get

$[\Sigma F_y = m\bar{a}_y]$ $\quad A - 60 = \dfrac{60}{32.2}(1.0)(4.42) \quad A = 68.2 \text{ lb} \quad Ans.$

$[\Sigma F_x = m\bar{a}_x]$ $\quad 30 - B = \dfrac{60}{32.2}(1.732)(4.42) \quad B = 15.74 \text{ lb} \quad Ans.$

① If the application of the relative-acceleration equations is not perfectly clear at this point, then review Art. 5/6. Note that the relative normal acceleration term is absent since there is no angular velocity of the bar.

② Recall that the moment of inertia of a slender rod about its center is $\frac{1}{12}ml^2$.

③ From the kinetic diagram, the terms $\Sigma m\bar{a}d$ are $m\bar{a}_x d_y + m\bar{a}_y d_x$. Since both are clockwise in the sense of $\bar{I}\alpha$, they are positive.

Sample Problem 6/8

A car door is inadvertently left slightly open when the brakes are applied to give the car a constant rearward acceleration a. Derive expressions for the angular velocity of the door as it swings pass the 90° position and the components of the hinge reactions for any value of θ. The mass of the door is m, its mass center is a distance $\bar{r}$ from the hinge axis O, and the radius of gyration about O is k_O.

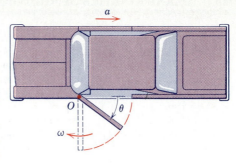

Solution. Because the angular velocity ω increases with θ, we need to find how the angular acceleration α varies with θ so that we may integrate it over the interval to obtain ω. We obtain α from a moment equation about O. First, we draw the free-body diagram of the door in the horizontal plane for a general position θ. The only forces in this plane are the components of the hinge reaction shown here in the x- and y-directions. On the kinetic diagram, in addition to the resultant couple $\bar{I}\alpha$ shown in the sense of α, we represent the resultant force $m\mathbf{\bar{a}}$ in terms of its components by using an equation of relative acceleration with respect to O. This equation becomes the kinematical equation of constraint ① and is

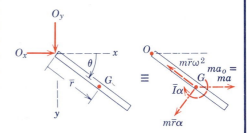

$$\mathbf{\bar{a}} = \mathbf{a}_G = \mathbf{a}_O + (\mathbf{a}_{G/O})_n + (\mathbf{a}_{G/O})_t$$

The magnitudes of the $m\mathbf{\bar{a}}$ components are then

② $$ma_O = ma \qquad m(a_{G/O})_n = m\bar{r}\omega^2 \qquad m(a_{G/O})_t = m\bar{r}\alpha$$

where $\omega = \dot{\theta}$ and $\alpha = \ddot{\theta}$.

For a given angle θ, the three unknowns are α, O_x, and O_y. We can eliminate O_x and O_y by a moment equation about O, which gives

③ $[\Sigma M_O = \bar{I}\alpha + \Sigma m\bar{a}d]$ $0 = m(k_O{}^2 - \bar{r}^2)\alpha + m\bar{r}\alpha(\bar{r}) - ma(\bar{r}\sin\theta)$

④ Solving for α gives $$\alpha = \frac{a\bar{r}}{k_O{}^2}\sin\theta$$

Now we integrate α first to a general position and get

$[\omega\,d\omega = \alpha\,d\theta]$ $$\int_0^\omega \omega\,d\omega = \int_0^\theta \frac{a\bar{r}}{k_O{}^2}\sin\theta\,d\theta$$

$$\omega^2 = \frac{2a\bar{r}}{k_O{}^2}(1 - \cos\theta)$$

For $\theta = \pi/2$, $$\omega = \frac{1}{k_O}\sqrt{2a\bar{r}}$$ *Ans.*

⑤   To find O_x and O_y for any given value of θ, the force equations give

$[\Sigma F_x = m\bar{a}_x]$ $O_x = ma - m\bar{r}\omega^2\cos\theta - m\bar{r}\alpha\sin\theta$

$$= m\left[a - \frac{2a\bar{r}^2}{k_O{}^2}(1 - \cos\theta)\cos\theta - \frac{a\bar{r}^2}{k_O{}^2}\sin^2\theta\right]$$

$$= ma\left[1 - \frac{\bar{r}^2}{k_O{}^2}(1 + 2\cos\theta - 3\cos^2\theta)\right]$$ *Ans.*

$[\Sigma F_y = m\bar{a}_y]$ $O_y = m\bar{r}\alpha\cos\theta - m\bar{r}\omega^2\sin\theta$

$$= m\bar{r}\frac{a\bar{r}}{k_O{}^2}\sin\theta\cos\theta - m\bar{r}\frac{2a\bar{r}}{k_O{}^2}(1 - \cos\theta)\sin\theta$$

$$= \frac{ma\bar{r}^2}{k_O{}^2}(3\cos\theta - 2)\sin\theta$$ *Ans.*

① Point O is chosen because it is the only point on the door whose acceleration is known.

② Be careful to place $m\bar{r}\alpha$ in the sense of positive α with respect to rotation about O.

③ The free-body diagram shows that there is zero moment about O. We use the transfer-of-axis theorem here and substitute $k_O{}^2 = \bar{k}^2 + \bar{r}^2$. If this relation is not totally familiar, review Art. B/1 in Appendix B.

④ We may also use Eq. 6/3 with O as a moment center

$$\Sigma \mathbf{M}_O = I_O\alpha + \bar{\boldsymbol{\rho}} \times m\mathbf{a}_O$$

where the scalar values of the terms are $I_O\alpha = mk_O{}^2\alpha$ and $\bar{\boldsymbol{\rho}} \times m\mathbf{a}_O$ becomes $-\bar{r}ma\sin\theta$.

⑤ The kinetic diagram shows clearly the terms that make up $m\bar{a}_x$ and $m\bar{a}_y$.

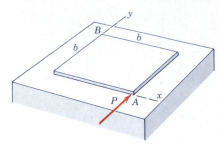

Problem 6/77

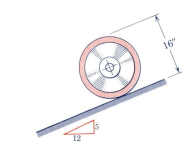

Problem 6/78

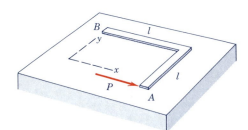

Problem 6/79

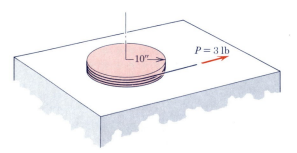

Problem 6/80

PROBLEMS

Introductory problems

6/77 The uniform square plate of mass m is lying motionless on the horizontal surface when the force P is applied at A as shown. Determine the resulting initial acceleration of point B. Friction is negligible.

$$Ans. \quad \mathbf{a}_B = \frac{P}{2m}(-3\mathbf{i} - \mathbf{j})$$

6/78 The L-shaped bar of mass m is lying motionless on the horizontal surface when the force P is applied at A as shown. Determine the initial acceleration of point A. Neglect friction and neglect the cross-sectional dimensions of the bar.

6/79 The 50-lb wheel has a centroidal radius of gyration of 5 in. If friction is sufficient to prevent slipping, compute the friction force F acting on the wheel during its downhill roll. What is the minimum coefficient of static friction μ_s to prevent slipping?

$$Ans. \quad F = 5.40 \text{ lb}, \ \mu_s = 0.117$$

6/80 The 64.4-lb solid circular disk is initially at rest on the horizontal surface when a 3-lb force P, constant in magnitude and direction, is applied to the cord wrapped securely around its periphery. Friction between the disk and the surface is negligible. Calculate the angular velocity ω of the disk after the 3-lb force has been applied for 2 seconds and find the linear velocity v of the center of the disk after it has moved 3 feet from rest.

6/81 The spacecraft is spinning with a constant angular velocity ω about the z-axis at the same time that its mass center O is traveling with a velocity v_O in the y-direction. If a tangential hydrogen-peroxide jet is fired when the craft is in the position shown, determine the expression for the absolute acceleration of point A on the spacecraft rim at the instant the jet force is F. The radius of gyration of the craft about the z-axis is k, and its mass is m.

$$Ans. \quad \mathbf{a}_A = -\frac{Fr^2}{mk^2}\mathbf{i} - \left(\frac{F}{m} - r\omega^2\right)\mathbf{j}$$

Problem 6/81

6/82 The solid homogeneous cylinder is released from rest on the ramp. If $\theta = 40°$, $\mu_s = 0.30$, and $\mu_k = 0.20$, determine the acceleration of the mass center G and the friction force exerted by the ramp on the cylinder.

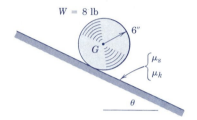

6/83 Repeat Prob. 6/82, except let $\theta = 30°$, $\mu_s = 0.15$, and $\mu_k = 0.10$.

$$Ans. \quad a = 13.31 \text{ ft/sec}^2, \ F = 0.693 \text{ lb}$$

Problem 6/82

6/84 The circular disk of mass m and radius r is released from rest with θ essentially zero and rolls without slipping on the circular guide of radius R. Determine the expression for the normal force N between the disk and the guide in terms of θ and calculate the angle θ' at which contact between the disk and the guide ceases.

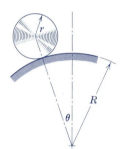

Problem 6/84

6/85 Above the earth's atmosphere at an altitude of 400 km where the acceleration due to gravity is 8.69 m/s^2 a certain rocket has a total remaining mass of 300 kg and is directed 30° from the vertical. If the thrust T from the rocket motor is 4 kN and if the rocket nozzle is tilted through an angle of 1° as shown, calculate the angular acceleration α and the x- and y-components of the acceleration of the mass center G. The rocket has a centroidal radius of gyration of 1.5 m.

$$Ans. \quad \alpha = 0.310 \text{ rad/s}^2$$
$$\bar{a}_x = 6.87 \text{ m/s}^2$$
$$\bar{a}_y = 2.74 \text{ m/s}^2$$

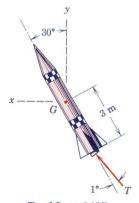

Problem 6/85

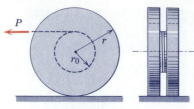

Problem 6/86

Representative problems

6/86 What should be the radius r_0 of the circular groove in order that there be no friction force acting between the wheel and the horizontal surface regardless of the magnitude of the force P applied to the cord? The centroidal radius of gyration of the wheel is $\bar{k}$.

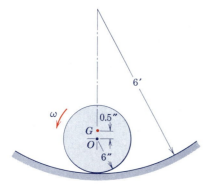

Problem 6/87

6/87 The mass center G of the 20-lb wheel is off center by 0.50 in. If G is in the position shown as the wheel rolls without slipping through the bottom of the circular path of 6-ft radius with an angular velocity ω of 10 rad/sec, compute the force P exerted by the path on the wheel. (Be careful to use the correct mass-center acceleration.) *Ans.* $P = 20.2$ lb

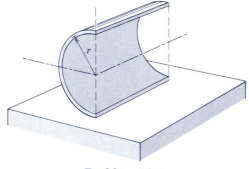

Problem 6/88

6/88 The uniform semicylindrical shell of mass m and radius r is released from rest in the position shown with its lower edge resting on the horizontal surface. Determine the minimum coefficient of static friction μ_s which is necessary to prevent any initial slipping of the shell.

6/89 Determine the angular acceleration of each of the two wheels as they roll without slipping down the inclines. For wheel A investigate the case where the mass of the rim and spokes is negligible and the mass of the bar is concentrated along its centerline. For wheel B assume that the thickness of the rim is negligible compared with its radius so that all of the mass is concentrated in the rim. Also specify the minimum coefficient of static friction μ_s required to prevent each wheel from slipping.

$$\textit{Ans.}\ \ \alpha_A = \frac{g}{r}\sin\theta,\ \mu_s = 0$$

$$\alpha_B = \frac{g}{2r}\sin\theta,\ \mu_s = \frac{1}{2}\tan\theta$$

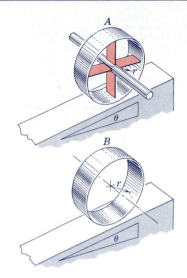

Problem 6/89

6/90 The circular disk of 200-mm radius has a mass of 25 kg with centroidal radius of gyration $\overline{k} = 175$ mm and has a concentric circular groove of 75-mm radius cut into it. A steady force T is applied at an angle θ to a cord wrapped around the groove as shown. If $T = 30$ N, $\theta = 0$, $\mu_s = 0.10$, and $\mu_k = 0.08$, determine the angular acceleration α of the disk, the acceleration a of its mass center G, and the friction force F that the surface exerts on the disk.

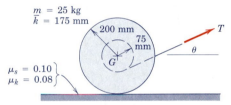

Problem 6/90

6/91 Repeat Prob. 6/90, except let $T = 50$ N and $\theta = 30°$.

$$\textit{Ans.}\ \ \alpha = 0.295\ \text{rad/s}^2\ \text{CCW}$$
$$a = 1.027\ \text{m/s}^2\ \text{right}$$
$$F = 17.62\ \text{N}$$

6/92 Repeat Prob. 6/90, except let $T = 30$ N and $\theta = 70°$.

6/93 Small ball-bearing rollers mounted on the ends of the slender bar of mass m and length l constrain the motion of the bar in the horizontal x-y slots. If a couple M is applied to the bar initially at rest at $\theta = 45°$, determine the forces exerted on the rollers at A and B as the bar starts to move.

$$\textit{Ans.}\ \ \mathbf{A} = \frac{3M}{2\sqrt{2}l}\,\mathbf{i},\ \ \mathbf{B} = -\frac{3M}{2\sqrt{2}l}\,\mathbf{j}$$

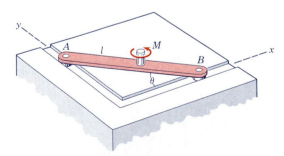

Problem 6/93

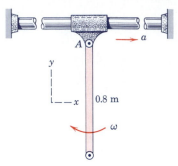

Problem 6/94

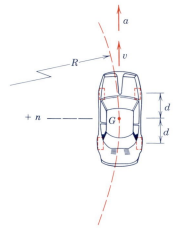

Problem 6/95

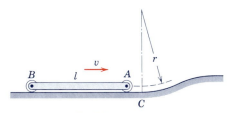

Problem 6/96

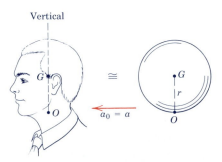

Problem 6/97

6/94 End A of the uniform 5-kg bar is pinned freely to the collar, which has an acceleration $a = 4$ m/s^2 along the fixed horizontal shaft. If the bar has a clockwise angular velocity $\omega = 2$ rad/s as it swings past the vertical, determine the components of the force on the bar at A for this instant.

6/95 During a test, the car travels in a horizontal circle of radius R and has a forward tangential acceleration a. Determine the lateral reactions at the front and rear wheel pairs if (a) the car speed $v = 0$ and (b) the speed $v \neq 0$. The car mass is m and its polar moment of inertia (about a vertical axis through G) is $\bar{I}$. Assume that $R \gg d$.

$$Ans. \ (a) \ R_f = \frac{\bar{I}a}{2dR}, \ R_r = -\frac{\bar{I}a}{2dR} \ \text{(a couple)}$$

$$(b) \ R_f = \frac{mv^2d + \bar{I}a}{2Rd}, \ R_r = \frac{mv^2d - \bar{I}a}{2Rd}$$

6/96 The uniform heavy bar AB of mass m is moving on its light end rollers along the horizontal with a velocity v when end A passes point C and begins to move on the curved portion of the path with radius r. Determine the force exerted by the path on the roller A immediately after it passes C.

6/97 In an investigation of whiplash resulting from rear-end collisions, sudden rotation of the head is modeled by a homogeneous solid sphere of mass m and radius r pivoted about a tangent axis (at the neck). If this axis at O is given a constant acceleration a with the head initially at rest, determine expressions for the initial angular acceleration α of the head and its angular velocity ω as a function of the angle θ of rotation. Assume that the neck is relaxed so that no moment is applied to the head at O.

$$Ans. \ \alpha = \frac{5a}{7r}, \ \omega = \sqrt{\frac{10}{7r}} \sqrt{g(1 - \cos \theta) + a \sin \theta}$$

6/98 The uniform steel beam of mass m and length l is suspended by the two cables at A and B. If the cable at B suddenly breaks, determine the tension T in the cable at A immediately after the break occurs. Treat the beam as a slender rod and show that the result is independent of the length of the beam.

Problem 6/98

6/99 Determine the maximum horizontal force P that may be applied to the cart of mass M for which the wheel will not slip as it begins to roll on the cart. The wheel has mass m, rolling radius r, and radius of gyration $\bar{k}$. The coefficients of static and kinetic friction between the wheel and the cart are μ_s and μ_k, respectively.

$$Ans. \ P = \mu_s g \left[m + M\left(1 + \frac{r^2}{\bar{k}^2} \right) \right]$$

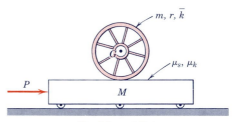

Problem 6/99

6/100 The uniform 12-kg square panel is suspended from point C by the two wires at A and B. If the wire at B suddenly breaks, calculate the tension T in the wire at A an instant after the break occurs.

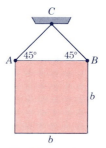

Problem 6/100

6/101 The compound pendulum of mass m and radius of gyration k_O about O is freely hinged to the trolley, which is given a constant horizontal acceleration a from rest with the pendulum initially at rest with $\theta = 0$. Determine an expression for the angular acceleration $\ddot{\theta}$ and the n- and t-components of the force at O as functions of θ. Calculate the maximum value reached by θ if $a = 0.5g$.
Ans.

$$\ddot{\theta} = \frac{\bar{r}}{k_O{}^2} (a \cos \theta - g \sin \theta)$$

$$O_t = m(g \sin \theta - a \cos \theta)\left(1 - \frac{\bar{r}^2}{k_O{}^2} \right)$$

$$O_n = m\left[(g \cos \theta + a \sin \theta)\left(1 + \frac{2\bar{r}^2}{k_O{}^2} \right) - 2g \frac{\bar{r}^2}{k_O{}^2} \right]$$

$$\theta_{max} = 53.1°$$

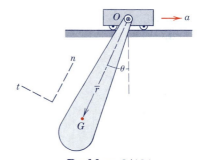

Problem 6/101

6/102 The truck, initially at rest with a solid cylindrical roll of paper in the position shown, moves forward with a constant acceleration a. Find the distance s that the truck goes before the paper rolls off the edge of its horizontal bed. Friction is sufficient to prevent slipping.

Problem 6/102

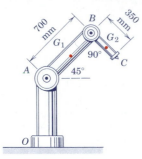

Problem 6/103

6/103 The robotic device of Prob. 6/60 is repeated here. Member AB is rotating about joint A with a counterclockwise angular velocity of 2 rad/s, and this rate is increasing at 4 rad/s². Determine the moment M_B exerted by arm AB on arm BC if joint B is held in a locked condition. The mass of arm BC is 4 kg, and the arm may be treated as a uniform slender rod.   *Ans.* $M_B = 3.55$ N·m CCW

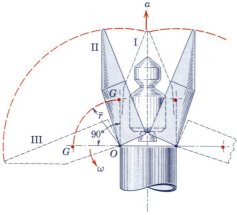

Problem 6/104

6/104 The fairing that covers the spacecraft package in the nose of the booster rocket is jettisoned when the rocket is in space where gravitational attraction is negligible. A mechanical actuator moves the two halves slowly from the closed position I to position II at which point the fairings are released to rotate freely about their hinges at O under the influence of a constant acceleration a of the rocket. When position III is reached, the hinge at O is released and the fairings drift away from the rocket. Determine the angular velocity ω of the fairings at the 90° position. The mass of each fairing is m with center of mass at G and radius of gyration k_O about O.

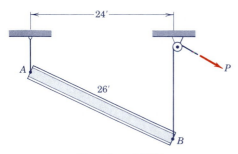

Problem 6/105

6/105 The steel I-beam weighs 650 lb and is supported in the fixed position shown by the force $P = 325$ lb. If the cable at B suddenly breaks, calculate the tension T in the vertical cable at A immediately after the break.   *Ans.* $T = 182.8$ lb

6/106 The uniform 12-ft pole is hinged to the truck bed and released from the vertical position as the truck starts from rest with an acceleration of 3 ft/sec². If the acceleration remains constant during the motion of the pole, calculate the angular velocity ω of the pole as it reaches the horizontal position.

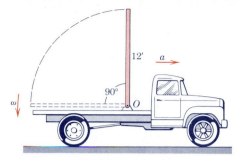

Problem 6/106

6/107 The truck carries a 1500-mm-diameter spool of cable with a mass of 0.75 kg per meter of length. There are 150 turns on the full spool. The empty spool has a mass of 140 kg with radius of gyration of 530 mm. The truck alone has a mass of 2030 kg with mass center at G. If the truck starts from rest with an initial acceleration of $0.2g$, determine (a) the tension T in the cable where it attaches to the wall and (b) the normal reaction under each pair of wheels. Neglect the rotational inertia of the truck wheels.

$$Ans. \ T = 1177 \ N$$
$$N_A = 17 \ 980 \ N$$
$$N_B = 8500 \ N$$

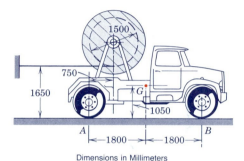

Dimensions in Millimeters

Problem 6/107

6/108 The uniform slender bar AB has a mass of 0.8 kg and is driven by crank OA and constrained by link CB of negligible mass. If OA has an angular acceleration $\alpha_0 = 4$ rad/s² and an angular velocity $\omega_0 = 2$ rad/s when both OA and CB are normal to AB, calculate the force in CB for this instant. (*Suggestion:* Consider the use of Eq. 6/3 with A as a moment center.)

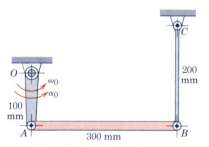

Problem 6/108

6/109 The crank OA rotates in the vertical plane with a constant clockwise angular velocity ω_0 of 4.5 rad/s. For the position where OA is horizontal, calculate the force under the light roller B of the 10-kg slender bar AB.   *Ans.* $B = 36.4$ N

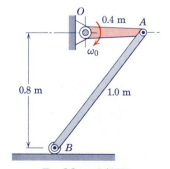

Problem 6/109

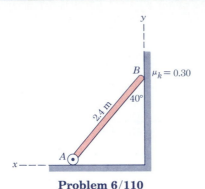

Problem 6/110

6/110 The uniform 15-kg bar is supported on the horizontal surface at A by a small roller of negligible mass. If the coefficient of kinetic friction between end B and the vertical surface is 0.30, calculate the initial acceleration of end A as the bar is released from rest in the position shown.

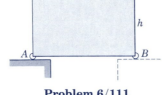

Problem 6/111

6/111 The rectangular block, which is solid and homogeneous, is supported at its corners by small rollers resting on horizontal surfaces. If the supporting surface at B is suddenly removed, determine the expression for the initial acceleration of corner A.

$$\text{Ans. } a_A = \frac{3g}{4\dfrac{b}{h} + \dfrac{h}{b}}$$

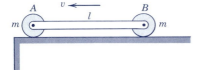

Problem 6/112

6/112 Two disk-shaped wheels, each of mass m, are connected by a light but rigid bar of length l. The unit is projected from a horizontal surface with speed v as shown. If v is sufficiently large so that the bar does not touch the corner after A has left the surface, determine an approximate value for the angular velocity ω of the unit once B has cleared the surface. Treat the wheels as concentrated masses, and state any other simplifying assumptions.

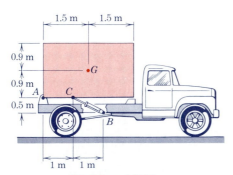

Problem 6/113

6/113 The hydraulic cylinder BC of the dump truck is broken and is disconnected. The driver (who has passed a course in dynamics) decides to calculate the minimum acceleration a of the truck required to tilt the dump about its pivot at A. He then proceeds to calculate the initial angular acceleration α of the dump if the truck is given an acceleration of $1.2a$. What are his correct answers for a and α and would he be able to carry out the experiment? The dump container may be modeled as a homogeneous and solid rectangular block with mass center at G.

$$\text{Ans. } a = 16.35 \text{ m/s}^2 \text{ (not feasible)}$$
$$\alpha = 0.72 \text{ rad/s}^2$$

6/114 The uniform slender bar of mass m and length L with small end rollers is released from rest in the position shown with the lower roller in contact with the horizontal plane. Determine the normal force N under the lower roller and the angular acceleration α of the bar immediately after release.

Problem 6/114

6/115 The connecting rod AB of a certain internal combustion engine weighs 1.2 lb with mass center at G and has a radius of gyration about G of 1.12 in. The piston and piston pin A together weigh 1.80 lb. The engine is running at a constant speed of 3000 rev/min, so that the angular velocity of the crank is $3000(2\pi)/60 = 100\pi$ rad/sec. Neglect the weights of the components and the force exerted by the gas in the cylinder compared with the dynamic forces generated and calculate the magnitude of the force on the piston pin A for the crank angle $\theta = 90°$. (*Suggestion:* Use the alternative moment relation, Eq. 6/3, with B as the moment center.)

Ans. $A = 347$ lb

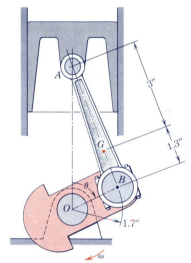

Problem 6/115

6/116 The overhead garage door is a homogeneous rectangular panel of mass m and is guided by its corner rollers that run in the tracks shown (dotted). If the door is released from rest in the position shown, determine the force exerted on the door by each of the rollers at A and B. Neglect any friction.

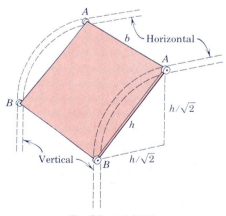

Problem 6/116

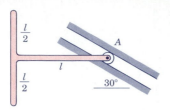

Problem 6/117

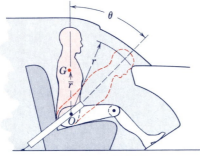

Problem 6/118

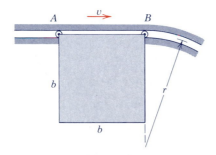

Problem 6/119

6/117 The T-shaped body of mass m is composed of two identical slender bars welded together. If the body is released from rest in the vertical plane in the position shown, determine the initial acceleration of point A. Neglect the small mass and friction of the roller.

$$Ans. \; a_A = \frac{14}{109} g \text{ down the slot}$$

6/118 In a study of head injury against the instrument panel of a car during sudden or crash stops where lap belts without shoulder straps are used, the segmented human model shown in the figure is analyzed. The hip joint O is assumed to remain fixed relative to the car, and the torso above the hip is treated as a rigid body of mass m freely pivoted at O. The center of mass of the torso is at G with the initial position of OG taken as vertical. The radius of gyration of the torso about O is k_O. If the car is brought to a sudden stop with a constant deceleration a, determine the velocity v relative to the car with which the model's head strikes the instrument panel. Substitute the values $m = 50$ kg, $\bar{r} = 450$ mm, $r = 800$ mm, $k_O = 550$ mm, $\theta = 45°$, and $a = 10g$ and compute v.

6/119 The uniform square plate of mass m is suspended in the vertical plane by its corner rollers A and B as it moves horizontally to the right with a velocity v. Determine the force that each roller exerts on the plate an instant after corner B has entered the circular portion of its guide.

$$Ans. \; A = \frac{m}{2}\left(g - \frac{v^2}{6r}\right), \; B = \frac{m}{2}\left(g - \frac{5v^2}{6r}\right)$$

6/120 The small end rollers of the 8-lb uniform slender bar are constrained to move in the slots that lie in a vertical plane. At the instant when $\theta = 30°$, the angular velocity of the bar is 2 rad/sec counterclockwise. Determine the angular acceleration of the bar, the reactions at A and B, and the accelerations of points A and B under the action of the 6-lb force P. Neglect the friction and mass of the small rollers.

$$Ans. \quad \alpha = 18.18 \text{ rad/sec}^2$$
$$R_A = 1.128 \text{ lb}$$
$$R_B = 0.359 \text{ lb}$$
$$a_A = 65.0 \text{ ft/sec}^2$$
$$a_B = 56.9 \text{ ft/sec}^2$$

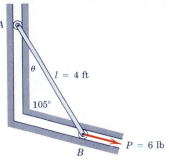

Problem 6/120

6/121 The small rollers at the ends of the uniform slender bar are confined to the circular slot in the vertical surface. If the bar is released from rest in the position shown, determine the initial angular acceleration α. Neglect the mass and friction of the rollers.

$$Ans. \quad \alpha = \frac{3}{2}\frac{g}{l} \text{ CW}$$

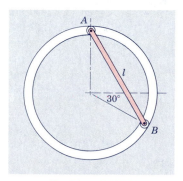

Problem 6/121

6/122 The roll-off truck ramp of Prob. 6/55 is shown again here. The 120-Mg loaded container may be treated as a homogeneous rectangular solid with mass center at G. In the position shown, the pair of wheels at A are blocked, and the brakes of the truck are applied to prevent movement of the truck. If the wheels at A are suddenly released to allow the container to roll, calculate the total friction force F applied to the truck wheels immediately after wheels A are unblocked. *Ans.* $F = 340$ kN

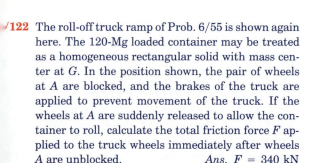

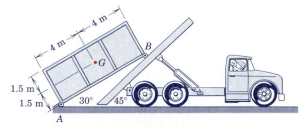

Problem 6/122

SECTION B. WORK AND ENERGY

6/6 *WORK-ENERGY RELATIONS*

In our study of the kinetics of particles in Arts. 3/6 and 3/7, we developed and applied the principles of work and energy to the motion of a particle and to selected cases of connected particles. We found that these principles were especially useful in describing motion resulting from the cumulative effect of forces acting through distances. Furthermore, when the forces were conservative, we were able to determine velocity changes by analyzing the energy conditions at the beginning and end of the motion interval. For finite displacements, the work-energy method eliminates the necessity for determining the acceleration and integrating it over the interval to obtain the velocity change. These same advantages are realized when we extend the work-energy principles to describe rigid-body motion. Before carrying out this extension we strongly recommend that the definitions and concepts of work, kinetic energy, gravitational and elastic potential energy, conservative forces, and power treated in Arts. 3/6 and 3/7 be carefully reviewed. We shall assume familiarity with these quantities as we apply them to rigid-body problems. In Arts. 4/3 and 4/4 on the kinetics of systems of particles, we extended the principles of Arts. 3/6 and 3/7 to encompass any general system of mass particles, which includes rigid bodies, and this discussion should also be reviewed.

(a) ***Work of forces and couples.***   The work done by a force **F** has been treated in detail in Art. 3/6 and is given by

$$U = \int \mathbf{F} \cdot d\mathbf{r} \qquad \text{or} \qquad U = \int (F \cos \alpha)\, ds$$

where $d\mathbf{r}$ is the infinitesimal vector displacement of the point of application of **F** during time dt. In the equivalent scalar form of the integral, α is the angle between **F** and the direction of the displacement and ds is the magnitude of the vector displacement $d\mathbf{r}$. We frequently have occasion to evaluate the work done by a couple M that acts on a rigid body during its motion. Figure 6/11 shows a couple $M = Fb$ acting on a rigid body that moves in the plane of the couple. During time dt the body rotates through an angle $d\theta$, and line AB moves to $A'B'$. We may consider this motion in two parts, first a translation to $A'B''$ and then a rotation $d\theta$ about A'. We see immediately that during the translation the work done by one of the forces cancels that done by the other force, so that the net work done is $dU = F(b\, d\theta) = M\, d\theta$ due to the rotational part of the motion. If the couple acts in the sense opposite to the rotation, the work done is negative. During a finite rotation, the work done by a couple M whose plane is parallel to the plane of motion is, therefore,

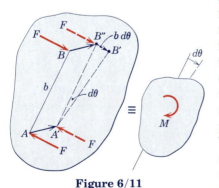

Figure 6/11

$$U = \int M \, d\theta$$

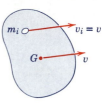

(b) Kinetic energy. We now use the familiar expression for the kinetic energy of a particle to develop expressions for the kinetic energy of a rigid body for each of the three classes of rigid-body plane motion illustrated in Fig. 6/12.

(a) Translation

(Translation) The translating rigid body of Fig. 6/12a has a mass m and all of its particles have a common velocity v. The kinetic energy of any particle of mass m_i of the body is $T_i = \frac{1}{2}m_i v^2$, so for the entire body $T = \Sigma \frac{1}{2}m_i v^2 = \frac{1}{2}v^2 \Sigma m_i$ or

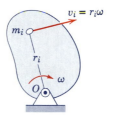

$$\boxed{T = \tfrac{1}{2}mv^2} \tag{6/7}$$

This expression holds for both rectilinear and curvilinear translation.

(b) Fixed–Axis Rotation

(Fixed-axis rotation) The rigid body in Fig. 6/12b rotates with an angular velocity ω about the fixed axis through O. The kinetic energy of a representative particle of mass m_i is $T_i = \frac{1}{2}m_i(r_i \omega)^2$. Thus, for the entire body $T = \frac{1}{2}\omega^2 \Sigma m_i r_i^2$. But the moment of inertia of the body about O is $I_O = \Sigma m_i r_i^2$, so

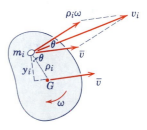

$$\boxed{T = \tfrac{1}{2}I_O \omega^2} \tag{6/8}$$

The similarity in form of the kinetic energy expressions for translation and rotation should be noted. The reader should verify that the dimensions of the two expressions are identical.

(c) General Plane Motion

Figure 6/12

(General plane motion) The rigid body in Fig. 6/12c executes plane motion where, at the instant considered, the velocity of its mass center G is $\bar{v}$ and the angular velocity is ω. The velocity v_i of a representative particle of mass m_i may be expressed in terms of the mass-center velocity $\bar{v}$ and the velocity $\rho_i \omega$ relative to the mass center as shown. With the aid of the law of cosines, we write the kinetic energy of the body as the sum ΣT_i of the kinetic energies of all its particles. Thus,

$$T = \Sigma \tfrac{1}{2}m_i v_i^2 = \Sigma \tfrac{1}{2}m_i(\bar{v}^2 + \rho_i^2 \omega^2 + 2\bar{v}\rho_i \omega \cos \theta)$$

Since ω and $\bar{v}$ are common to all terms in the third summation, we may factor them out. Thus, the third term in the expression for T becomes

$$\omega \bar{v} \Sigma m_i \rho_i \cos \theta = \omega \bar{v} \Sigma m_i y_i = 0$$

since $\Sigma m_i y_i = m\bar{y} = 0$. The kinetic energy of the body is then $T = \frac{1}{2}\bar{v}^2 \Sigma m_i + \frac{1}{2}\omega^2 \Sigma m_i \rho_i^2$ or

$$T = \tfrac{1}{2}m\bar{v}^2 + \tfrac{1}{2}\bar{I}\omega^2 \qquad\qquad (6/9)$$

where $\bar{I}$ is the moment of inertia of the body about its mass center. This expression for kinetic energy clearly shows the separate contributions to the total kinetic energy resulting from the translational velocity $\bar{v}$ of the mass center and the rotational velocity ω about the mass center.

The kinetic energy of plane motion may also be expressed in terms of the rotational velocity about the instantaneous center C of zero velocity. Since C momentarily has zero velocity, the proof leading to Eq. 6/8 for the fixed point O holds equally well for point C, so that, alternatively, we may write the kinetic energy of a rigid body in plane motion as

$$T = \tfrac{1}{2}I_C\omega^2 \qquad\qquad (6/10)$$

In Art. 4/3 we derived Eq. 4/4 for the kinetic energy of any system of mass. We now see that this expression becomes equivalent to Eq. 6/9 when the mass system is rigid. For a rigid body, the quantity $\dot{\boldsymbol{\rho}}_i$ in Eq. 4/4 is the velocity of the representative particle relative to the mass center and is the vector $\boldsymbol{\omega} \times \boldsymbol{\rho}_i$ that has the magnitude $\rho_i\omega$. The summation term in Eq. 4/4 becomes $\Sigma\tfrac{1}{2}m_i(\rho_i\omega)^2 = \tfrac{1}{2}\omega^2\Sigma m_i\rho_i^2 = \tfrac{1}{2}\bar{I}\omega^2$, which brings Eq. 4/4 into agreement with Eq. 6/9.

(c) Potential energy. Gravitational potential energy V_g and elastic potential energy V_e were covered in detail in Art. 3/7. Recall that the symbol U' (rather than U) is used to denote the work done by all forces except the weight and elastic forces, which are accounted for in the potential energy terms.

Work-energy equation. The work-energy relation, Eq. 3/17, was introduced in Art. 3/7 for particle motion and was generalized in Art. 4/3 to include the motion of a general system of particles. This equation

$$U'_{1\text{-}2} = \Delta T + \Delta V_g + \Delta V_e \qquad\qquad [4/3]$$

applies to any conservative mechanical system. For application to the motion of a single rigid body, the term ΔV_e disappears, and the total work $U'_{1\text{-}2}$ done on the body by the external forces (other than gravity forces) equals the corresponding change ΔT in the kinetic energy of the body plus the change ΔV_g in its potential energy of position in the gravitational field. Alternatively, the equation may be written $U_{1\text{-}2} = \Delta T$, provided that the work of gravitational forces is included in the expression for U.

When applied to an interconnected conservative system of rigid bodies, Eq. 4/3 will include the change ΔV_e in the stored elastic energy in the connections. The term $U'_{1\text{-}2}$ will include the work of all forces external to the system (other than gravitational forces), including the negative work of internal friction forces if any. The term ΔT is the sum of the changes in kinetic energy of all moving parts during the interval of motion in question, and ΔV_g is the sum of the changes in gravitational potential energy for the various members.

The work-energy relationship may be written alternatively in the form of Eq. 3/17*a* where the terms correspond to the natural sequence of events.

When the work-energy principle is applied to a single rigid body, either a *free-body diagram* or an *active-force diagram* should be used. In the case of an interconnected system of rigid bodies, an active-force diagram of the entire system should be drawn in order to isolate the system and disclose all forces that do work on the system. Diagrams should also be drawn to disclose the initial and final positions of the system for the given interval of motion.

The work-energy equation provides a direct relationship between the forces that do work and the corresponding changes in the motion of a mechanical system. However, if there is appreciable internal mechanical friction, then the system must be dismembered in order to disclose the kinetic-friction forces and account for the negative work that they do. When the system is dismembered, one of the primary advantages of the work-energy approach is automatically lost. The work-energy method is most useful for analyzing conservative systems of interconnected bodies, where energy loss due to the negative work of friction forces is negligible.

(d) Power. The concept of power was discussed in Art. 3/6 on work-energy for particle motion. It is recalled that power is the time rate at which work is performed. For a force **F** acting on a rigid body in plane motion, the power developed by that force at a given instant was given by Eq. 3/12 and is the rate at which it is doing work or

$$P = \frac{dU}{dt} = \frac{\mathbf{F} \cdot d\mathbf{r}}{dt} = \mathbf{F} \cdot \mathbf{v}$$

where $d\mathbf{r}$ and **v** are, respectively, the differential displacement and the velocity of the point of application of the force. Similarly, for a couple M acting on the body, the power developed by the couple at a given instant is the rate at which it is doing work or

$$P = \frac{dU}{dt} = \frac{M \, d\theta}{dt} = M\omega$$

where $d\theta$ and ω are, respectively, the differential angular displacement and the angular velocity of the body. If the senses of M and ω are the same, the power is positive and energy is supplied to the body. Conversely, if M and ω have opposite senses, the power is negative and energy is removed from the body. If the force $\mathbf{F}$ and the couple M act simultaneously, the total instantaneous power is

$$P = \mathbf{F} \cdot \mathbf{v} + M\omega$$

We may also express power by evaluating the rate at which the total mechanical energy of a rigid body or a system of rigid bodies is changing. The work-energy relation, Eq. 4/3, for an infinitesimal displacement is

$$dU' = dT + dV_g + dV_e$$

where dU' is the work of the active forces and couples applied to the body or to the system of bodies. Excluded from dU' are the works of gravitational forces and spring forces that are accounted for in the dV_g and dV_e terms. Dividing by dt gives the total power of the active forces and couples as

$$P = \frac{dU'}{dt} = \dot{T} + \dot{V}_g + \dot{V}_e = \frac{d}{dt}(T + V)$$

Thus, we see that the power developed by the active forces and couples equals the rate of change of the total mechanical energy of the body or system of bodies. We note from Eq. 6/9 that, for a given body, the first term may be written

$$\dot{T} = \frac{dT}{dt} = \frac{d}{dt}\left(\frac{1}{2}m\overline{\mathbf{v}} \cdot \overline{\mathbf{v}} + \frac{1}{2}\bar{I}\omega^2\right)$$

$$= \frac{1}{2}m(\overline{\mathbf{a}} \cdot \overline{\mathbf{v}} + \overline{\mathbf{v}} \cdot \overline{\mathbf{a}}) + \bar{I}\omega\dot{\omega}$$

$$= m\overline{\mathbf{a}} \cdot \overline{\mathbf{v}} + \bar{I}\alpha(\omega) = \mathbf{R} \cdot \overline{\mathbf{v}} + \overline{M}\omega$$

where $\mathbf{R}$ is the resultant of *all* forces acting on the body and $\overline{M}$ is the resultant moment about the mass center G of *all* forces. The dot product accounts for the case of curvilinear motion of the mass center, where $\overline{\mathbf{a}}$ and $\overline{\mathbf{v}}$ are not in the same direction.

Sample Problem 6/9

The wheel rolls up the incline on its hubs without slipping and is pulled by the 100-N force applied to the cord wrapped around its outer rim. If the wheel starts from rest, compute its angular velocity ω after its center has moved a distance of 3 m up the incline. The wheel has a mass of 40 kg with center of mass at O and has a centroidal radius of gyration of 150 mm. Determine the power input from the 100-N force at the end of the 3-m motion interval.

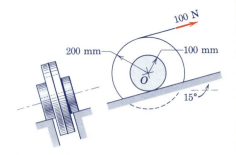

Solution. Of the four forces shown on the free-body diagram of
① the wheel, only the 100-N pull and the weight of $40(9.81) = 392$ N do work. The friction force does no work as long as the wheel does not slip. By use of the concept of the instantaneous center C of zero velocity, we see that a point A on the cord to which the 100-N force is applied has a velocity $v_A = [(200 + 100)/100]v$. Hence, point A on the cord moves a distance of $(200 + 100)/100 = 3$ times as far as the center O. Thus, with the effect of the weight included in the U-term, the work done on the wheel becomes

$$② \quad U_{1\text{-}2} = 100\,\frac{200 + 100}{100}\,(3) - (392\sin 15°)(3) = 595 \text{ J}$$

The wheel has general plane motion, so that the change in its kinetic energy is

$$③ \quad [T = \tfrac{1}{2}m\bar{v}^2 + \tfrac{1}{2}\bar{I}\omega^2] \quad \Delta T = \left[\frac{1}{2}\,40(0.10\omega)^2 + \frac{1}{2}\,40(0.15)^2\omega^2\right] - 0$$
$$= 0.650\omega^2$$

The work-energy equation gives

$$[U_{1\text{-}2} = \Delta T] \qquad 595 = 0.650\omega^2 \qquad \omega = 30.3 \text{ rad/s}$$

Alternatively, the kinetic energy of the wheel may be written

$$④ \quad [T = \tfrac{1}{2}I_C\omega^2] \qquad T = \frac{1}{2}\,40[(0.15)^2 + (0.10)^2]\omega^2 = 0.650\omega^2$$

The power input from the 100-N force when $\omega = 30.3$ rad/s is

$$⑤ \quad [P = \mathbf{F}\cdot\mathbf{v}] \qquad P_{100} = 100(0.3)(30.3) = 908 \text{ W} \qquad\qquad Ans.$$

① Since the velocity of the instantaneous center C on the wheel is zero, it follows that the rate at which the friction force does work is continuously zero. Hence, F does no work as long as the wheel does not slip. If the wheel were rolling on a moving platform, however, the friction force would do work, even though the wheel were not slipping.

② Note that the component of the weight down the plane does negative work.

③ Be careful to use the correct radius in the expression $v = r\omega$ for the velocity of the center of the wheel.

④ Recall that $I_C = \bar{I} + m\overline{OC}^2$, where $\bar{I} = I_O = mk_O^2$.

⑤ The velocity here is that of the application point of the 100-N force.

Sample Problem 6/10

The 4-ft slender bar weighs 40 lb with mass center at B and is released from rest in the position for which θ is essentially zero. Point B is confined to move in the smooth vertical guide, while end A moves in the smooth horizontal guide and compresses the spring as the bar falls. Determine (a) the angular velocity of the bar as the position $\theta = 30°$ is passed and (b) the velocity with which B strikes the horizontal surface if the stiffness of the spring is 30 lb/in.

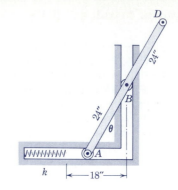

Solution. With the friction and mass of the small rollers at A and B neglected, the system may be treated as being conservative.

Part (a). For the first interval of motion from $\theta = 0$ to $\theta = 30°$, the spring is not engaged, so that there is no V_e term in the energy equation. If we adopt the alternative of treating the work of the weight in the ① V_g term, then there are no other forces that do work and $U'_{1\text{-}2} = 0$.

Since we have a constrained plane motion, there is a kinematic relation between the velocity v_B of the center of mass and the angular velocity ω of the bar. This relation is easily obtained by using the instantaneous center C of zero velocity and noting that $v_B = \overline{CB}\omega$. Thus, the kinetic energy of the bar in the 30° position becomes

$$[T = \tfrac{1}{2}m\bar{v}^2 + \tfrac{1}{2}\bar{I}\omega^2]$$

$$T = \frac{1}{2}\frac{40}{32.2}\left(\frac{12}{12}\omega\right)^2 + \frac{1}{2}\left(\frac{1}{12}\frac{40}{32.2}4^2\right)\omega^2 = 1.449\omega^2$$

The change in gravitational potential energy is the weight times the change in height of the mass center and equals

$$[\Delta V_g = W\Delta h] \qquad \Delta V_g = 40(2\cos 30° - 2) = -10.72 \text{ ft-lb}$$

We now substitute into the energy equation and get

$$[U'_{1\text{-}2} = \Delta T + \Delta V_g] \quad 0 = 1.449\omega^2 - 10.72 \quad \omega = 2.72 \text{ rad/sec} \qquad Ans.$$

Part (b). For the entire interval of motion, we include the spring as a part of the system, where

$$② \quad [V_e = \tfrac{1}{2}kx^2] \qquad \Delta V_e = \frac{1}{2}(30)(24-18)^2\frac{1}{12} - 0 = 45 \text{ ft-lb}$$

In the final horizontal position, point A has no velocity, so that the bar is, in effect, rotating about A. Hence, its kinetic energy is

$$[T = \tfrac{1}{2}I_A\omega^2] \qquad T = \frac{1}{2}\left(\frac{1}{3}\frac{40}{32.2}4^2\right)\left(\frac{v_B}{24/12}\right)^2 = 0.828v_B{}^2$$

The change in gravitational potential energy is

$$[\Delta V_g = W\Delta h] \qquad \Delta V_g = 40(-2) = -80 \text{ ft-lb}$$

Substituting into the energy equation gives

$$[U'_{1\text{-}2} = \Delta T + \Delta V_g + \Delta V_e] \quad 0 = (0.828v_B{}^2 - 0) - 80 + 45$$

$$v_B = 6.50 \text{ ft/sec} \qquad Ans.$$

Alternatively, if the bar alone constitutes the system, the active-force diagram shows the weight, which does positive work, and the spring force kx, which does negative work. We would then write

$$[U_{1\text{-}2} = \Delta T] \qquad 80 - 45 = 0.828v_B{}^2$$

which is identical with the previous result.

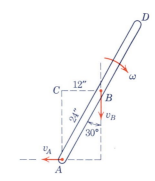

① We recognize that the forces acting on the bar at A and B are normal to the respective directions of motion and, hence, do no work.

② If we convert k to lb/ft, we have

$$\Delta V_e = \frac{1}{2}\left(30\,\frac{\text{lb}}{\text{in.}}\right)\left(12\,\frac{\text{in.}}{\text{ft}}\right)\left(\frac{24-18}{12}\text{ ft}\right)^2$$

$$= 45 \text{ ft-lb}$$

Always check the consistency of your units.

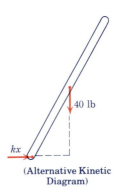

(Alternative Kinetic Diagram)

Sample Problem 6/11

In the mechanism shown, each of the two wheels has a mass of 30 kg and a centroidal radius of gyration of 100 mm. Each link OB has a mass of 10 kg and may be treated as a slender bar. The 7-kg collar at B slides on the fixed vertical shaft with negligible friction. The spring has a stiffness $k = 30$ kN/m and is contacted by the bottom of the collar when the links reach the horizontal position. If the collar is released from rest at the position $\theta = 45°$ and if friction is sufficient to prevent the wheels from slipping, determine (a) the velocity v_B of the collar as it first strikes the spring and (b) the maximum deformation x of the spring.

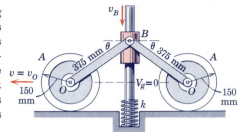

Solution. The mechanism executes plane motion and is conservative with the neglect of kinetic friction losses. The datum for zero gravitational potential energy V_g is conveniently taken through O as shown.

(a) For the interval from $\theta = 45°$ to $\theta = 0$, we note that ΔT_{wheels} is zero since each wheel starts from rest and momentarily comes to rest at $\theta = 0$. Also, at the lower position each link is merely rotating about its point O so that

$$\Delta T = [2(\tfrac{1}{2}I_O\omega^2) - 0]_{\text{links}} + [\tfrac{1}{2}mv^2 - 0]_{\text{collar}}$$

$$= \frac{1}{3}\,10(0.375)^2\left(\frac{v_B}{0.375}\right)^2 + \frac{1}{2}\,7v_B^2 = 6.83v_B^2$$

The collar at B drops a distance $0.375/\sqrt{2} = 0.265$ m so that

$$\Delta V = \Delta V_g = 0 - 2(10)(9.81)\frac{0.265}{2} - 7(9.81)(0.265) = -44.2 \text{ J}$$

① Also, $U'_{1\text{-}2} = 0$. Hence,

$$[U'_{1\text{-}2} = \Delta T + \Delta V] \qquad 0 = 6.83v_B^2 - 44.2 \qquad v_B = 2.54 \text{ m/s} \qquad Ans.$$

(b) At the condition of maximum deformation x of the spring, all parts are momentarily at rest, which makes $\Delta T = 0$. Thus,

$$[U'_{1\text{-}2} = \Delta T + \Delta V_g + \Delta V_e]$$

$$0 = 0 - 2(10)(9.81)\left(\frac{0.265}{2} + \frac{x}{2}\right) - 7(9.81)(0.265 + x) + \tfrac{1}{2}(30)(10^3)x^2$$

Solution for the positive value of x gives

$$x = 60.1 \text{ mm} \qquad Ans.$$

It should be noted that the results of parts (a) and (b) involve a very simple net energy change despite the fact that the mechanism has undergone a fairly complex sequence of motions. Solution of this and similar problems by other than a work-energy approach is not an inviting prospect.

① With the work of the weight of the collar B included in the ΔV_g term, there are no other forces external to the system that do work. The friction force acting under each wheel does no work since the wheel does not slip, and, of course, the normal force does no work here. Hence, $U'_{1\text{-}2} = 0$.

PROBLEMS

(In the following problems neglect any energy loss due to kinetic friction unless otherwise indicated.)

Introductory problems

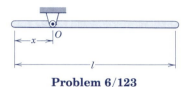

Problem 6/123

6/123 The uniform slender bar is released from rest in the position shown. If $x = l/4$, determine the angular velocity as the bar passes the vertical position. Friction at the pivot O is negligible.

$$Ans. \quad \omega = \sqrt{\frac{24g}{7l}}$$

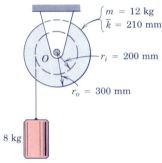

$m = 12$ kg
$\bar{k} = 210$ mm

$r_i = 200$ mm

$r_o = 300$ mm

8 kg

Problem 6/124

6/124 The velocity of the 8-kg cylinder is 0.3 m/s at a certain instant. What is its speed v after dropping an additional 1.5 m? The mass of the grooved drum is 12 kg, its centroidal radius of gyration is $\bar{k} = 210$ mm, and the radius of its groove is $r_i = 200$ mm. The frictional moment at O is a constant 3 N·m.

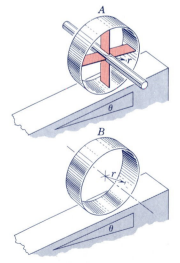

A

r

θ

B

r

θ

Problem 6/125

6/125 The two wheels of Prob. 6/89, shown again here, represent two extreme conditions of distribution of mass. For case A all of the mass m is assumed to be concentrated in the center of the hoop in the axial bar of negligible diameter. For case B all of the mass m is assumed to be concentrated in the rim. Determine the velocity of the center of each hoop after it has traveled a distance x down the incline from rest. The hoops roll without slipping.

$$Ans. \quad \text{Case } A: v_A = \sqrt{2gx \sin \theta}$$
$$\text{Case } B: v_B = \sqrt{gx \sin \theta}$$

6/126 The uniform rectangular plate is released from rest in the position shown. Determine the maximum angular velocity ω during the ensuing motion. Friction at the pivot is negligible.

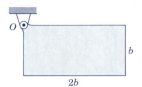

Problem 6/126

6/127 A 1200-kg flywheel with a radius of gyration of 400 mm has its speed reduced from 5000 to 3000 rev/min during a 2-min interval. Calculate the average power supplied by the flywheel. Express your answer both in kilowatts and in horsepower.
$\qquad$ *Ans.* $P = 140.4$ kW, $P = 188$ hp

6/128 The 15-kg slender bar OA is released from rest in the vertical position and compresses the spring of stiffness $k = 20$ kN/m as the horizontal position is passed. Determine the proper setting of the spring, by specifying the distance h, that will result in the bar having an angular velocity $\omega = 4$ rad/s as it crosses the horizontal position. What is the effect of x on the dynamics of the problem?

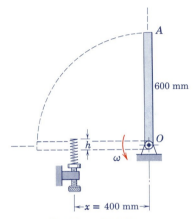

Problem 6/128

6/129 The 50-kg flywheel has a radius of gyration about its shaft axis of $\bar{k} = 0.4$ m and is subjected to the torque $M = 2(1 - e^{-0.1\theta})$ N·m, where θ is in radians. If the flywheel is at rest when $\theta = 0$, determine its angular velocity after 5 revolutions.
$\qquad$ *Ans.* $\omega = 3.31$ rad/s

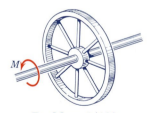

Problem 6/129

6/130 The uniform 12-lb disk pivots freely about a horizontal axis through O. A 4-lb slender bar is fastened to the disk as shown. If the system is nudged from rest while in the position shown, determine its angular velocity after it has rotated 180°.

6/131 For the pivoted slender rod of Prob. 6/123, determine the distance x for which the angular velocity will be a maximum as the bar passes the vertical position after being released in the horizontal position shown. State the corresponding angular velocity.
$\qquad$ *Ans.* $x = 0.211l$, $\omega_{max} = 1.861\sqrt{\dfrac{g}{l}}$

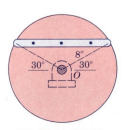

Problem 6/130

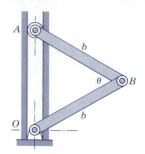

Problem 6/132

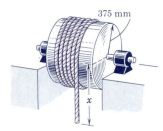

Problem 6/133

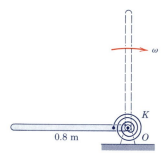

Problem 6/134

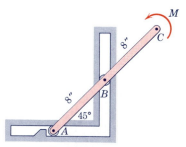

Problem 6/135

Representative problems

6/132 The two identical links, each of length b and mass m, may be treated as uniform slender bars. If they are released from rest in the position shown with end A constrained by the smooth vertical guide, determine the velocity v with which A reaches O with θ essentially zero.

6/133 The drum of 375-mm radius and its shaft have a mass of 41 kg and a radius of gyration of 300 mm about the axis of rotation. A total of 18 m of flexible steel cable with a mass of 3.08 kg per meter of length is wrapped around the drum with one end secured to the surface of the drum. The free end of the cable has an initial overhang $x = 0.6$ m as the drum is released from rest. Determine the angular velocity ω of the drum for the instant when $x = 6$ m. Assume that the center of mass of the portion of cable remaining on the drum at any time lies on the shaft axis. Neglect friction.   *Ans.* $\omega = 9.68$ rad/s

6/134 The torsional spring has a stiffness of 30 N·m/rad and is undeflected when the 6-kg uniform slender bar is in the upright position. If the bar is released from rest in the horizontal position shown, determine its angular velocity ω as it passes the vertical position. Friction is negligible.

6/135 The uniform bar ABC weighs 6 lb and is initially at rest with end A bearing against the stop in the horizontal guide. When a constant couple $M = 72$ lb-in. is applied to end C, the bar rotates causing end A to strike the side of the vertical guide with a velocity of 10 ft/sec. Calculate the loss of energy ΔE due to friction in the guides and rollers. The mass of the rollers may be neglected.
Ans. $\Delta E = 0.435$ ft-lb

6/136 The disk B rolls without slipping down the incline A, which moves with speed v. Describe the work done by the normal and friction forces which act on the disk for the cases (a) $v = 0$ and (b) $v \neq 0$.

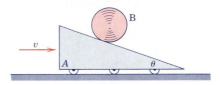

Problem 6/136

6/137 The homogeneous solid semicylinder is released from rest in the position shown. If friction is sufficient to prevent slipping, determine the maximum angular velocity ω reached by the cylinder as it rolls on the horizontal surface.

$$Ans. \quad \omega = 4\sqrt{\frac{g(1 - \cos \theta)}{(9\pi - 16)r}}$$

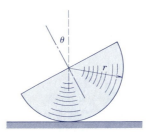

Problem 6/137

6/138 A facility for testing the performance of motorized golf carts consists of an endless belt where the angle θ can be adjusted. The cart of mass m is slowly brought up to its rated ground speed v with the braking torque M on the upper pulley constantly adjusted so that the cart remains in a fixed position A on the test stand. With no cart on the belt, a torque M_0 is required on the pulley to overcome friction and turn the pulleys regardless of speed. Friction is sufficient to prevent the wheels from slipping on the belt. Determine an expression for the power P absorbed by the braking torque M. Do the static friction forces between the wheels and the belt do work?

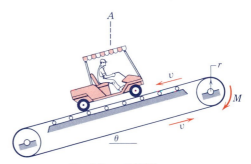

Problem 6/138

6/139 The 8-kg crank OA, with mass center at G and radius of gyration about O of 0.22 m, is connected to the 12-kg uniform slender bar AB. If the linkage is released from rest in the position shown, compute the velocity v of end B as OA swings through the vertical. *Ans.* $v = 2.29$ m/s

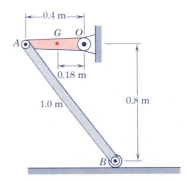

Problem 6/139

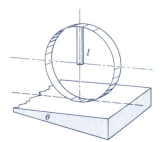

Problem 6/140

6/140 A slender rod of length l and mass m is welded to the rim of a hoop of radius l. If the hoop is released from rest in the position shown, determine the speed v of the center of the hoop after it has made one and one-half revolutions. Assume no slipping and continuous contact between the hoop and its supporting surface. Also neglect the mass of the hoop.

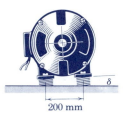

Problem 6/141

6/141 The electric motor shown is delivering 4 kW at 1725 rev/min to a pump which it drives. Calculate the angle δ through which the motor deflects under load if the stiffness of each of its four spring mounts is 15 kN/m. In what direction does the motor shaft turn?

$\qquad$ *Ans.* $\delta = 2.11°$, motor shaft turns clockwise

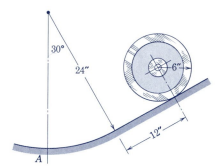

Problem 6/142

6/142 The center of the 200-lb wheel with centroidal radius of gyration of 4 in. has a velocity of 2 ft/sec down the incline in the position shown. Calculate the normal reaction N under the wheel as it rolls past position A. Assume that no slipping occurs.

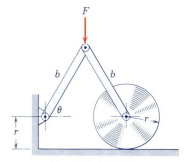

Problem 6/143

6/143 A constant force F is applied in the vertical direction to the symmetrical linkage starting from the rest position shown. Determine the angular velocity ω that the links acquire as they reach the position $\theta = 0$. Each link has a mass m_0. The wheel is a solid circular disk of mass m and rolls on the horizontal surface without slipping.

$$\textit{Ans.} \quad \omega = \sqrt{\frac{3(F + m_0 g)\sin\theta}{m_0 b}}$$

6/144 A lid-support mechanism is being designed for a storage chest to limit the angular velocity of the 10-lb uniform lid to 1.5 rad/sec for $\theta = 0$ when it is released from rest with θ essentially equal to 90°. Two identical mechanisms are included as indicated on the pictorial sketch. Specify the necessary stiffness k of each of the two springs, which are compressed 2 in. upon closure. Neglect the weight of the links and any friction in the sliding collars C. Also, the thickness of the lid is small compared with its other dimensions.

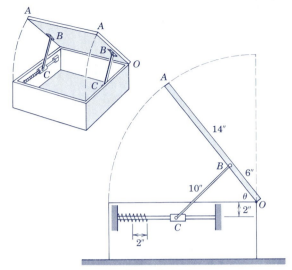

Problem 6/144

6/145 The sheave of 400-mm radius has a mass of 50 kg and a radius of gyration of 300 mm. The sheave and its 100-kg load are suspended by the cable and the spring, which has a stiffness of 1.5 kN/m. If the system is released from rest with the spring initially stretched 100 mm, determine the velocity of O after it has dropped 50 mm. *Ans.* $v = 0.757$ m/s

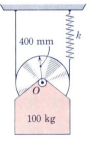

Problem 6/145

6/146 The uniform 50-kg bar OA of length $2b = 2.4$ m is released from rest in the vertical position with $\theta = 90°$. The spring of stiffness $k = 1.6$ kN/m is unstretched when $\theta = 90°$. Neglect the mass of each of the two connecting links and calculate the angular velocity $\omega = -\dot{\theta}$ of the bar as it reaches the collapsed position of $\theta = 0$.

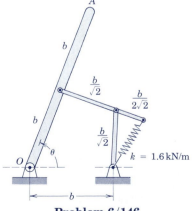

Problem 6/146

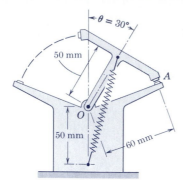

Problem 6/147

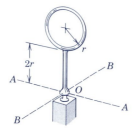

Problem 6/148

Problem 6/149

6/147 Specify the unstretched length l_0 of the spring of stiffness $k = 1400$ N/m that will result in a velocity of 0.25 m/s for the contact at A if the toggle is given a slight nudge from its null position at $\theta = 0$. The toggle has a mass of 1.5 kg and a radius of gyration about O of 55 mm. Motion occurs in the horizontal plane. *Ans.* $l_0 = 90.0$ mm

6/148 The body shown is constructed of uniform slender rod and consists of a ring of radius r attached to a straight rod section of length $2r$. The body pivots freely about a ball-and-socket joint at O. If the body is at rest in the vertical position shown and is given a slight nudge, compute its angular velocity ω after a 90° rotation about (*a*) axis *A-A* and (*b*) axis *B-B*.

6/149 Motive power for the experimental 10-Mg bus comes from the energy stored in a rotating flywheel which it carries. The flywheel has a mass of 1500 kg and a radius of gyration of 500 mm and is brought up to a maximum speed of 4000 rev/min. If the bus starts from rest and acquires a speed of 72 km/h at the top of a hill 20 m above the starting position, compute the reduced speed N of the flywheel. Assume that 10 percent of the energy taken from the flywheel is lost. Neglect the rotational energy of the wheels of the bus. *Ans.* $N = 3720$ rev/min

6/150 The figure shows the cross section of a garage door which is a uniform rectangular panel 8 by 8 ft and weighing 200 lb. The door carries two spring assemblies, one on each side of the door, like the one shown. Each spring has a stiffness of 50 lb/ft and is unstretched when the door is in the open position shown. If the door is released from rest in this position, calculate the velocity of the edge at *A* as it strikes the garage floor.

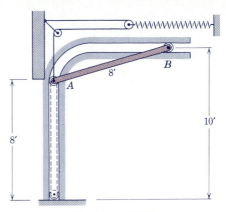

Problem 6/150

6/151 The uniform slender rod of mass *m* and length *L*, initially at rest in the centered horizontal position on the circular surface of radius *R*, is rocked to the dotted position and released from rest. Determine an expression for the angular velocity ω of the rod as it crosses the horizontal position. Friction is sufficient to prevent any slipping.

$$Ans. \quad \omega = \frac{2}{L}\sqrt{6gR(\theta \sin \theta + \cos \theta - 1)}$$

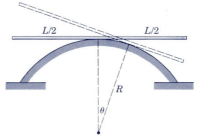

Problem 6/151

6/152 The small vehicle is designed for high-speed travel over the snow. The endless tread for each side of the vehicle has a mass ρ per unit length and is driven by the front wheels. Determine that portion *M* of the constant front-axle torque required to give both vehicle treads their motion corresponding to a vehicle velocity *v* achieved with constant acceleration in a distance *s* from rest on level terrain.

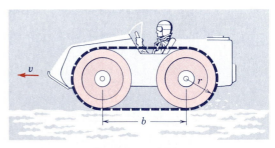

Problem 6/152

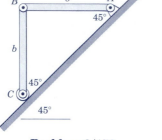

Problem 6/153

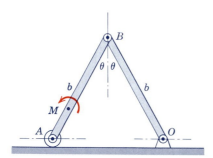

Problem 6/154

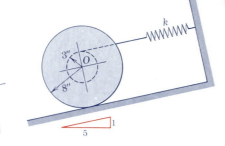

Problem 6/155

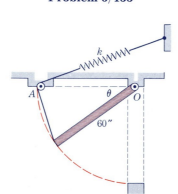

Problem 6/156

6/153 The two identical uniform bars are released from rest from the position shown in the vertical plane. Determine the angular velocity ω of AB when the bars become collinear.

$$Ans. \quad \omega = \sqrt{\frac{3g}{2b}\,(2\sqrt{2}-1)}$$

6/154 The double wheel weighs 20 lb with radius of gyration of 5 in. about O and is connected to the spring of stiffness $k = 40$ lb/ft by a cord that is wrapped securely around the inner hub. If the wheel is released from rest on the incline with the spring stretched 9 in., calculate the maximum velocity v of its center O during the ensuing motion. The wheel rolls without slipping.

6/155 The two slender bars each of mass m and length b are pinned together and move in the vertical plane. If the bars are released from rest in the position shown and move together under the action of a couple M of constant magnitude applied to AB, determine the velocity of A as it strikes O.

$$Ans. \quad v_A = \sqrt{3\left[\frac{M\theta}{m} - gb(1-\cos\theta)\right]}$$

6/156 The figure shows the cross section of a uniform 200-lb ventilator door hinged about its upper horizontal edge at O. The door is controlled by the spring-loaded cable that passes over the small pulley at A. The spring has a stiffness of 15 lb per foot of stretch and is undeformed when $\theta = 0$. If the door is released from rest in the horizontal position, determine the maximum angular velocity ω reached by the door and the corresponding angle θ.

6/157 Each of the two hinges at A and B of the uniform lid of mass m of a child's toy chest contains a torsion spring that exerts a resisting moment $M = K\theta$ on the lid as it is being closed. (*a*) Specify the torsional stiffness K of each spring which will result in zero angular velocity of the lid as it reaches the horizontal closed position ($\theta = \pi/2$) when released from rest at $\theta = 0$. (*b*) What would be the angular acceleration α of the lid when it is released from rest in the closed position? Would these hinges be a practical solution?

$$\text{Ans. } K = \frac{2mgl}{\pi^2}, \ \alpha = 0.410 \frac{g}{l}, \text{ No}$$

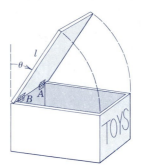

Problem 6/157

6/158 A small experimental vehicle has a total mass m of 500 kg including wheels and driver. Each of the four wheels has a mass of 40 kg and a centroidal radius of gyration of 400 mm. Total frictional resistance R to motion is 400 N and is measured by towing the vehicle at a constant speed on a level road with engine disengaged. Determine the power output of the engine for a speed of 72 km/h up the 10-percent grade (*a*) with zero acceleration and (*b*) with an acceleration of 3 m/s^2. (*Hint:* Power equals the time rate of increase of the total energy of the vehicle plus the rate at which frictional work is overcome.)

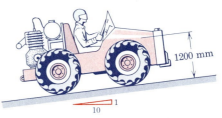

Problem 6/158

6/159 Each of the two links has a mass of 2 kg and a centroidal radius of gyration of 60 mm. The slider at B has a mass of 3 kg and moves freely in the vertical guide. The spring has a stiffness of 6 kN/m. If a constant torque $M = 20$ N·m is applied to link OA through its shaft at O starting from the rest position at $\theta = 45°$, determine the angular velocity ω of OA when $\theta = 0$.   *Ans.* $\omega = 9.51$ rad/s

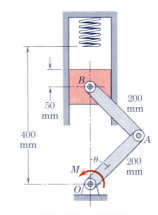

Problem 6/159

6/160 A solid roll of wrapping paper with an initial radius r_0 is released from rest on the incline and allowed to unroll with the free end clamped at the top. Determine the velocity of the roll in terms of the distance x that it has moved down the slope. The total length of the paper in the roll is L. Can you reconcile the difference between the initial energy of the roll and the final energy of the paper after all motion has ceased?

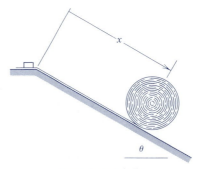

Problem 6/160

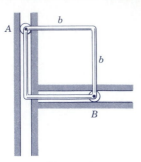

Problem 6/161

6/161 The open square frame is constructed of four identical slender rods each of length b. If the frame is released from rest in the position shown, determine the speed of corner A (a) after A has dropped a distance b and (b) after A has dropped a distance $2b$. The small wheels roll without friction in the slots of the vertical surface.

$$\text{Ans.} \quad (a) \text{ and } (b) \ v_A = \sqrt{\frac{12}{5}gb}$$

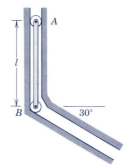

Problem 6/162

6/162 The uniform slender rod is released from rest in the position shown. With what speed v_A does end A strike the 30° incline? Neglect the small mass and friction of the end rollers.

Problem 6/163

▶**6/163** Calculate the constant force P required to give the center of the pulley a velocity of 4 ft/sec in an upward movement of the center of 3 ft from the rest position shown. The pulley weighs 30 lb with a radius of gyration of 10 in., and the cable has a total length of 15 ft with a weight of 2 lb/ft.

$$\text{Ans.} \quad P = 38.6 \text{ lb}$$

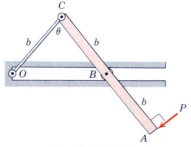

Problem 6/164

▶**6/164** The uniform bar ABC has a mass m and starts from rest with $\theta = 180°$ where A, B, C, and O are collinear. If the applied force P is constant in magnitude, determine the angular velocity ω of the bar as B reaches O with $\theta = 0$. The mass of the roller at B and the mass of the strut OC are negligible. (*Hint:* Replace P by a force P at B and a couple.)

$$\text{Ans.} \quad \omega = \sqrt{\frac{6P\pi}{13mb}}$$

6/7 ACCELERATION FROM WORK-ENERGY; VIRTUAL WORK

In addition to the determination of velocities resulting from the action of forces acting over finite displacements, we may use the work-energy equation to establish the instantaneous accelerations of the members of a system of interconnected bodies as a result of the active forces applied. We may also modify the equation to establish the configuration of such a system when it undergoes a constant acceleration.

For an infinitesimal interval of motion, Eq. 4/3 becomes

$$dU' = dT + dV$$

The term dU' represents the total work done by all active non-potential forces acting on the system under consideration during the infinitesimal displacement of the system. The work of potential forces is included in the dV-term. If we use the subscript i to denote a representative body of the interconnected system, the differential change in kinetic energy T for the entire system becomes

$$dT = d(\Sigma \tfrac{1}{2}m_i\bar{v}_i{}^2 + \Sigma \tfrac{1}{2}\bar{I}_i\omega_i{}^2) = \Sigma m_i\bar{v}_i\,d\bar{v}_i + \Sigma \bar{I}_i\omega_i\,d\omega_i$$

where $d\bar{v}_i$ and $d\omega_i$ are the respective changes in the magnitudes of the velocities and where the summation is taken over all bodies of the system. But for each body, $m_i\bar{v}_i\,d\bar{v}_i = m_i\bar{\mathbf{a}}_i\cdot d\bar{\mathbf{s}}_i$ and $\bar{I}_i\omega_i\,d\omega_i = \bar{I}_i\alpha_i\,d\theta_i$, where $d\bar{\mathbf{s}}_i$ represents the infinitesimal linear displacement of the center of mass and where $d\theta_i$ represents the infinitesimal angular displacement of the body in the plane of motion. We note that $\bar{\mathbf{a}}_i\cdot d\bar{\mathbf{s}}_i$ is identical to $(\bar{a}_i)_t\,d\bar{s}_i$, where $(\bar{a}_i)_t$ is the component of $\bar{\mathbf{a}}_i$ along the tangent to the curve described by the mass center of the body in question. Also α_i represents $\ddot{\theta}_i$, the angular acceleration of the representative body. Consequently, for the entire system

$$dT = \Sigma m_i\bar{\mathbf{a}}_i\cdot d\bar{\mathbf{s}}_i + \Sigma \bar{I}_i\alpha_i\,d\theta_i$$

This change may also be written as

$$dT = \Sigma \mathbf{R}_i\cdot d\bar{\mathbf{s}}_i + \Sigma \mathbf{M}_{G_i}\cdot d\boldsymbol{\theta}_i$$

where $\mathbf{R}_i$ and $\mathbf{M}_{G_i}$ are the resultant force and resultant couple acting on body i and where $d\boldsymbol{\theta}_i = d\theta_i\mathbf{k}$. These last two equations merely show us that the differential change in kinetic energy equals the differential work done on the system by the resultant forces and resultant couples acting on all the bodies of the system.

The term dV represents the differential change in the total gravitational potential energy V_g and the total elastic potential energy V_e and has the form

$$dV = d(\Sigma m_i g h_i + \Sigma \tfrac{1}{2}k_j x_j{}^2) = \Sigma m_i g\,dh_i + \Sigma k_j x_j\,dx_j$$

where h_i represents the vertical distance of the center of mass of the representative body of mass m_i above any convenient datum plane and where x_j stands for the deformation, tensile or compressive, of a representative elastic member of the system (spring) whose stiffness is k_j.

The complete expression for dU' may now be written as

$$dU' = \Sigma m_i \bar{\mathbf{a}}_i \cdot d\bar{\mathbf{s}}_i + \Sigma \bar{I}_i \alpha_i \, d\theta_i + \Sigma m_i g \, dh_i + \Sigma k_j x_j \, dx_j \quad \textbf{(6/11)}$$

In the application of Eq. 6/11 to a system of one degree of freedom, the terms $m_i \bar{\mathbf{a}}_i \cdot d\bar{\mathbf{s}}_i$ and $\bar{I}_i \alpha_i \, d\theta_i$ will be positive if the accelerations are in the same direction as the respective displacements and negative if they are in the opposite direction. Equation 6/11 has the advantage of relating the accelerations to the active forces directly, which eliminates the need for dismembering the system and then eliminating the internal forces and reactive forces by simultaneous solution of the force-mass-acceleration equations for each member.

Virtual work. In Eq. 6/11 the differential motions are differential changes in the real or actual displacements which occur. For a mechanical system that assumes a steady-state configuration during constant acceleration, we often find it convenient to introduce the concept of *virtual work*. The concepts of virtual work and virtual displacement have been introduced and used to establish equilibrium configurations for static systems of interconnected bodies (see Chapter 7 of *Vol. 1 Statics*). A virtual displacement is any assumed and arbitrary displacement, linear or angular, away from the natural or actual position. For a system of connected bodies, the virtual displacements must be consistent with the constraints of the system. For example, when one end of a link is hinged about a fixed pivot, the virtual displacement of the other end must be normal to the line joining the two ends. Such requirements for displacements consistent with the constraints are purely kinematical and provide what are known as the *equations of constraint*. If a set of virtual displacements satisfying the equations of constraint and therefore consistent with the constraints is given to a mechanical system, the proper relationship between the coordinates which specify the configuration of the system will be established by applying the work-energy relationship of Eq. 6/11, expressed in terms of virtual changes. Thus,

$$\delta U' = \Sigma m_i \bar{\mathbf{a}}_i \cdot \delta \bar{\mathbf{s}}_i + \Sigma \bar{I}_i \alpha_i \, \delta\theta_i + \Sigma m_i g \, \delta h_i + \Sigma k_j x_j \, \delta x_j \quad (6/11a)$$

It is customary to use the differential symbol d to refer to differential changes in the *real* displacements, whereas the symbol δ is used to signify differential changes that are assumed or *virtual* changes.

Sample Problem 6/12

The movable rack A has a mass of 3 kg, and rack B is fixed. The gear has a mass of 2 kg and a radius of gyration of 60 mm. In the position shown, the spring, which has a stiffness of 1.2 kN/m, is stretched a distance of 40 mm. For the instant represented, determine the acceleration a of rack A under the action of the 80-N force. The plane of the figure is vertical.

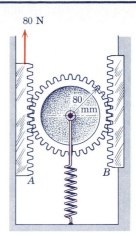

80 N

Solution. The given figure represents the active-force diagram for
① the entire system, which is conservative.

During an infinitesimal upward displacement dx of rack A, the work dU' done on the system is $80\ dx$, where x is in meters, and this work equals the sum of the corresponding changes in the total energy of the system. These changes, which appear in Eq. 6/11 are as follows:

$$[dT = \Sigma m_i \bar{\mathbf{a}}_i \cdot d\bar{\mathbf{s}}_i + \Sigma \bar{I}_i \alpha_i\, d\theta_i]$$

$$dT_{\text{rack}} = 3a\ dx$$

② $$dT_{\text{gear}} = 2\frac{a}{2}\frac{dx}{2} + 2(0.06)^2\frac{a/2}{0.08}\frac{dx/2}{0.08} = 0.781a\ dx$$

The change in potential energies of the system from Eq. 6/11 becomes

$$[dV = \Sigma m_i g\, dh_i + \Sigma k_j x_j\, dx_j]$$

$$dV_{\text{rack}} = 3g\ dx = 3(9.81)\ dx = 29.4\ dx$$

$$dV_{\text{gear}} = 2g(dx/2) = g\ dx = 9.81\ dx$$

③ $$dV_{\text{spring}} = k_j x_j\, dx_j = 1200(0.04)\ dx/2 = 24\ dx$$

Substitution into Eq. 6/11 gives us

$$80\ dx = 3a\ dx + 0.781a\ dx + 29.4\ dx + 9.81\ dx + 24\ dx$$

Canceling dx and solving for a give

$$a = 16.76/3.781 = 4.43\ \text{m/s}^2 \qquad \textit{Ans.}$$

We see that using the work-energy method for an infinitesimal displacement has given us the direct relation between the applied force and the resulting acceleration. It was unnecessary to dismember the system, draw two free-body diagrams, apply $\Sigma F = m\bar{a}$ twice, apply $\Sigma M_G = \bar{I}\alpha$ and $F = kx$, eliminate unwanted terms, and finally solve for a.

① Note that none of the remaining forces external to the system do any work. The work done by the weight and by the spring is accounted for in the potential energy terms.

② Note that $\bar{a}_i$ for the gear is its mass-center acceleration, which is half that for the rack A. Also its displacement is $dx/2$. For the rolling gear, the angular acceleration from $a = r\alpha$ becomes $\alpha_1 = (a/2)/0.08$, and the angular displacement from $ds = r\ d\theta$ becomes $d\theta_i = (dx/2)/0.08$.

③ Note here that the displacement of the spring is one-half that of the rack. Hence, $x_i = x/2$.

Sample Problem   6/13

A constant force P is applied to end A of the two identical and uniform links and causes them to move to the right in their vertical plane with a horizontal acceleration a. Determine the steady-state angle θ made by the bars with one another.

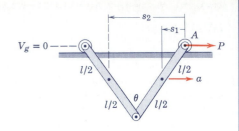

Solution.   The figure constitutes the active-force diagram for the system. To find the steady-state configuration, consider a virtual displacement of each bar from the natural position assumed during the acceleration. Measurement of the displacement with respect to end A eliminates any work done by force P during the virtual displacement. Thus,

$$\delta U' = 0$$

The virtual change in kinetic energy from Eq. 6/11a is

$$\delta T = \Sigma m\bar{\mathbf{a}} \cdot \delta\bar{\mathbf{s}} = ma(-\delta s_1) + ma(-\delta s_2)$$

$$= -ma\left[\delta\left(\frac{l}{2}\sin\frac{\theta}{2}\right) + \delta\left(\frac{3l}{2}\sin\frac{\theta}{2}\right)\right]$$

$$= -ma\left(l\cos\frac{\theta}{2}\,\delta\theta\right)$$

We choose the horizontal line through A as the datum for zero potential energy. Thus, the potential energy of the links is

$$V_g = 2mg\left(-\frac{l}{2}\cos\frac{\theta}{2}\right)$$

and the virtual change in potential energy becomes

$$\delta V_g = \delta\left(-2mg\frac{l}{2}\cos\frac{\theta}{2}\right) = \frac{mgl}{2}\sin\frac{\theta}{2}\,\delta\theta$$

Substitution into the work-energy equation for virtual changes, Eq. 6/11a, gives

$$[\delta U' = \delta T + \delta V_g]\qquad 0 = -mal\cos\frac{\theta}{2}\,\delta\theta + \frac{mgl}{2}\sin\frac{\theta}{2}\,\delta\theta$$

from which

$$\theta = 2\tan^{-1}\frac{2a}{g}\qquad\qquad Ans.$$

Again, in this problem we see that the work-energy approach obviated the necessity for dismembering the system, drawing separate free-body diagrams, applying motion equations, eliminating unwanted terms, and solving for θ.

① Note that we use the symbol δ to refer to an assumed or virtual differential change rather than the symbol d, which refers to an infinitesimal change in the real displacement.

② In evaluating δT, we are finding the variation of T resulting from a virtual change in displacement. Therefore, we must use the expression $m\mathbf{a}\cdot\delta\mathbf{s}$ and not the expression $\delta(\frac{1}{2}mv^2)$ which represents the variation of T resulting from a variation in v.

③ We have chosen to use the angle θ to describe the configuration of the links, although we could have used the distance between the two ends of the links just as well.

PROBLEMS

Introductory problems

/165 The load of mass m is supported by the light parallel links and the fixed stop A. Determine the initial angular acceleration α of the links due to the application of the couple M to one end as shown.

$$\text{Ans.} \quad \alpha = \frac{M}{mb^2} - \frac{g}{b} \sin \theta$$

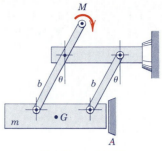

Problem 6/165

/166 The uniform slender bar of mass m is shown in its equilibrium configuration before the force P is applied. Compute the initial angular acceleration of the bar upon application of P.

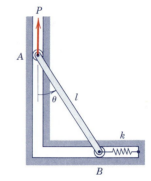

Problem 6/166

/167 Links A and B each weigh 8 lb, and bar C weighs 12 lb. Calculate the angle θ assumed by the links if the body to which they are pinned is given a steady horizontal acceleration a of 4 ft/sec^2.

$$\text{Ans.} \quad \theta = 7.1°$$

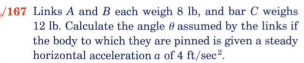

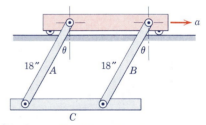

Problem 6/167

/168 The box and load of the dump truck have a mass m with mass center at G and a moment of inertia I_A about the pivot at A. Determine the angular acceleration α of the box when it is started from rest in the position shown under the application of the couple M to link CD. Neglect the mass of the links.

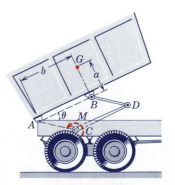

Problem 6/168

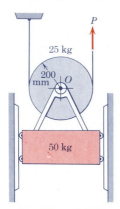

Problem 6/169

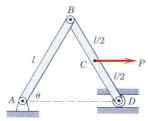

Problem 6/170

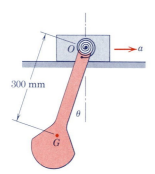

Problem 6/171

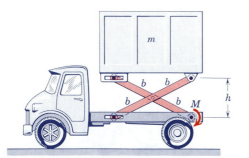

Problem 6/172

6/169 If the force P required to maintain equilibrium of the 25-kg sheave and attached 50-kg load is doubled, calculate the acceleration a of the center O of the sheave. The radius of gyration of the sheave about its center is 150 mm.   *Ans.* $a = 8.26$ m/s^2

Representative problems

6/170 The uniform slender links, each having mass m, move in a vertical plane. If the system starts from rest in the position $\theta = 60°$, determine the initial angular acceleration α of the links.

6/171 The sliding block is given a horizontal acceleration to the right that is slowly increased to a steady value a. The attached pendulum of mass m and mass center G assumes a steady angular deflection θ. The torsion spring at O exerts a moment $M = K\theta$ on the pendulum to oppose the angular deflection. Determine the torsional stiffness K that will allow a steady deflection θ.

$$Ans.\ K = \frac{m\bar{r}}{\theta}(a \cos\theta - g \sin\theta)$$

6/172 The cargo box of the food-delivery truck for aircraft servicing has a loaded mass m and is elevated by the application of a couple M on the lower end of the link that is hinged to the truck frame. The horizontal slots allow the linkage to unfold as the cargo box is elevated. Determine the upward acceleration of the box in terms of h for a given value of M. Neglect the mass of the links.

6/173 Each of the three identical uniform panels of a segmented industrial door has a mass m and is guided in the tracks (shown dotted). Determine the horizontal acceleration a of the upper panel under the action of the force P. Neglect any friction in the guide rollers.

Ans. $a = \dfrac{3}{8}\left(\dfrac{P}{m} - \dfrac{3g}{2}\right)$

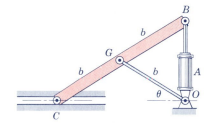

Problem 6/173

6/174 The hydraulic cylinder A exerts an upward force P on pin B of the uniform bar BC of mass m. Neglect the mass of link OG and determine the angular acceleration α of BC as the bar starts from rest in the position shown. Motion occurs in the vertical plane.

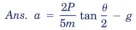

Problem 6/174

6/175 The load of mass m is given an upward acceleration a from its supported rest position by the application of the forces P. Neglect the mass of the links compared with m and determine the initial acceleration a.

Ans. $a = \dfrac{2P}{5m}\tan\dfrac{\theta}{2} - g$

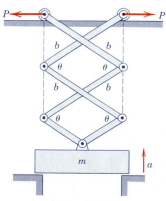

Problem 6/175

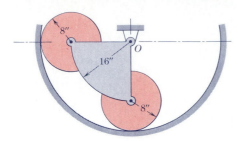

Problem 6/176

6/176 The sector and attached wheels are released from rest in the position shown in the vertical plane. Each wheel is a solid circular disk weighing 12 lb and rolls on the fixed circular path without slipping. The sector weighs 18 lb and is closely approximated by one-fourth of a solid circular disk of 16-in. radius. Determine the initial angular acceleration α of the sector.

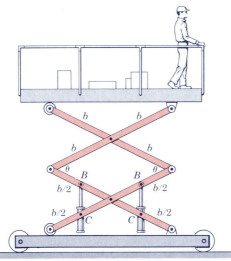

Problem 6/177

6/177 The portable work platform is elevated by means of the two hydraulic cylinders articulated at points C. The pressure in each cylinder produces a force F. The platform, man, and load have a combined mass m, and the mass of the linkage is small and may be neglected. Determine the upward acceleration a of the platform and show that it is independent of both b and θ.

$$Ans. \quad a = \frac{F}{2m} - g$$

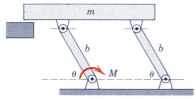

Problem 6/178

6/178 The horizontal platform of mass m is supported by four uniform links, two on each side of the platform as shown. In the rest position the left end is supported by the fixed shelf. Determine the moment M of the couple applied to one of the links required to produce an initial angular acceleration α of the links. Each of the four links may be treated as a slender bar of mass m_0.

6/179 Each of the uniform bars *OA* and *OB* weighs 4 lb and is freely hinged at *O* to the vertical shaft, which is given an upward acceleration $a = g/2$. The links that connect the light collar *C* to the bars have negligible weight, and the collar slides freely on the shaft. The spring has a stiffness $k = 0.75$ lb/in. and is uncompressed for the position equivalent to $\theta = 0$. Calculate the angle θ assumed by the bars under conditions of steady acceleration.

Ans. $\theta = 60°$

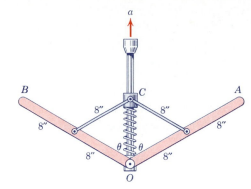

Problem 6/179

6/180 Pinion *A* of the electric motor turns gear *B* and its attached drum *C*. The motor rotor and pinion have a combined mass of 30 kg and a radius of gyration of 150 mm. The gear *B* and drum *C* have a combined mass of 95 kg with a radius of gyration of 300 mm. The motor receives 1.2 kW of electrical power, 94 percent of which is converted into mechanical power. For an instant when the upward velocity of the 80-kg cylinder is 0.9 m/s, calculate the corresponding upward acceleration *a* of the cylinder.

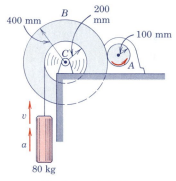

Problem 6/180

6/181 The mechanical tachometer measures the rotational speed *N* of the shaft by the horizontal motion of the collar *B* along the rotating shaft. This movement is caused by the centrifugal action of the two 12-oz weights *A*, which rotate with the shaft. Collar *C* is fixed to the shaft. Determine the rotational speed *N* of the shaft for a reading $\beta = 15°$. The stiffness of the spring is 5 lb/in., and it is uncompressed when $\theta = 0$ and $\beta = 0$. Neglect the weights of the links.

Ans. $N = 133.0$ rev/min

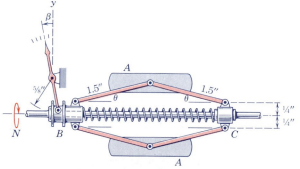

Problem 6/181

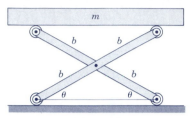

Problem 6/182

6/182 The bar of mass m is supported in the vertical plane by the two uniform links each of mass m_0. Without dismembering the system, determine the initial downward acceleration a of the bar if the wire connecting the lower ends of the links suddenly breaks. Treat the links as slender rods and neglect all friction.

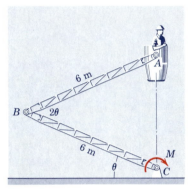

Problem 6/183

6/183 The aerial tower shown is designed to elevate a workman in a vertical direction. An internal mechanism at B maintains the angle between AB and BC at twice the angle θ between BC and the ground. If the combined mass of the man and the cab is 200 kg and if all other masses are neglected, determine the torque M applied to BC at C and the torque M_B in the joint at B required to give the cab an initial vertical acceleration of 1.2 m/s² when it is started from rest in the position $\theta = 30°$.

Ans. $M_B = 11.44$ kN·m
$M = 0$

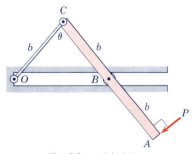

Problem 6/184

▶**6/184** The sliding bar of mass m in Prob. 6/164 is repeated here. Determine the angular acceleration of AC due to the action of the force P as the bar starts from rest with any value θ. Link OC has negligible mass, and the smooth guide is horizontal.

Ans. $\alpha = \left(\dfrac{3P}{mb}\right)\dfrac{2 + \cos\theta}{7 + 6\cos\theta}$

SECTION C. IMPULSE AND MOMENTUM

6/8 *IMPULSE-MOMENTUM EQUATIONS*

The principles of impulse and momentum were developed and used in Articles 3/9 and 3/10 for the description of particle motion. In this treatment, we observed that these principles were of particular importance when the applied forces were expressible as functions of the time and when interactions between particles occurred during short periods of time, such as with impact. Similar advantages result when the impulse-momentum principles are applied to the motion of rigid bodies.

In Art. 4/2 the impulse-momentum principles were extended to cover any defined system of mass particles without restriction as to the connections between the particles of the system. These extended relations all apply to the motion of a rigid body, which is merely a special case of a general system of mass. We will now apply these equations directly to rigid-body motion in two dimensions.

(a) Linear momentum. In Art. 4/4 we defined the linear momentum of a mass system as the vector sum of the linear momenta of all of its particles and wrote $\mathbf{G} = \Sigma m_i \mathbf{v}_i$. With $\mathbf{r}_i$ representing the position vector to m_i, we have $\mathbf{v}_i = \dot{\mathbf{r}}_i$ and $\mathbf{G} = \Sigma m_i \dot{\mathbf{r}}_i$ which, for a system whose total mass is constant, may be written as $\mathbf{G} = d(\Sigma m_i \mathbf{r}_i)/dt$. When we substitute the principle of moments $m\bar{\mathbf{r}} = \Sigma m_i \mathbf{r}_i$ to locate the mass center, the momentum becomes $\mathbf{G} = d(m\bar{\mathbf{r}})/dt = m\dot{\bar{\mathbf{r}}}$, where $\dot{\bar{\mathbf{r}}}$ is the velocity $\bar{\mathbf{v}}$ of the mass center. Therefore, as before, we find that the linear momentum of any mass system, rigid or nonrigid, is

$$\mathbf{G} = m\bar{\mathbf{v}} \qquad [4/5]$$

In the derivation of Eq. 4/5, we note that it was unnecessary to employ the kinematic condition for a rigid body, Fig. 6/13, which is $\mathbf{v}_i = \bar{\mathbf{v}} + \boldsymbol{\omega} \times \boldsymbol{\rho}_i$. In that case, we obtain the same result by writing $\mathbf{G} = \Sigma m_i(\bar{\mathbf{v}} + \boldsymbol{\omega} \times \boldsymbol{\rho}_i)$. The first sum is $\bar{\mathbf{v}}\Sigma m_i = m\bar{\mathbf{v}}$, and the second sum becomes $\boldsymbol{\omega} \times \Sigma m_i \boldsymbol{\rho}_i = \boldsymbol{\omega} \times m\bar{\boldsymbol{\rho}} = \mathbf{0}$ since $\boldsymbol{\rho}_i$ is measured from the mass center, making $\bar{\boldsymbol{\rho}}$ zero.

Next in Art. 4/4 we rewrote Newton's generalized second law as Eq. 4/6. This equation and its integrated form are

$$\Sigma\mathbf{F} = \dot{\mathbf{G}} \quad \text{and} \quad \int_{t_1}^{t_2} \Sigma\mathbf{F}\, dt = \mathbf{G}_2 - \mathbf{G}_1 \qquad \text{(6/12)}$$

Equation 6/12 may be written in its scalar component form which, for plane motion in the x-y plane, gives

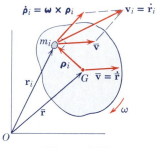

Figure 6/13

$$\boxed{\begin{aligned}\Sigma F_x &= \dot{G}_x \\ \Sigma F_y &= \dot{G}_y\end{aligned}} \quad \text{and} \quad \boxed{\begin{aligned}\int_{t_1}^{t_2} \Sigma F_x \, dt &= G_{x_2} - G_{x_1} \\ \int_{t_1}^{t_2} \Sigma F_y \, dt &= G_{y_2} - G_{y_1}\end{aligned}} \qquad \textbf{(6/12a)}$$

In words, the first of Eqs. 6/12 and 6/12a states that the resultant force equals the time rate of change of momentum. The integrated form of Eqs. 6/12 and 6/12a states that the linear impulse on the body during the interval $t_2 - t_1$ equals the corresponding change in linear momentum. The impulse-momentum relationships may be alternatively written in the form corresponding to Eq. 3/23a, where the order of the terms corresponds to the natural sequence of events. As in the force-mass-acceleration formulation, the force summations in Eqs. 6/12 and 6/12a must include *all* forces acting externally on the body considered. We emphasize, therefore, that in the use of the impulse-momentum equations, it is essential to construct the complete free-body diagram so as to disclose all forces that appear in the force summation. In contrast to the method of work and energy, all forces exert impulses, whether they do work or not.

(b) Angular momentum.

Angular momentum is defined as the moment of linear momentum. In Art. 4/4 we expressed the angular momentum about the mass center of any prescribed system of mass as $\mathbf{H}_G = \Sigma \boldsymbol{\rho}_i \times m_i \mathbf{v}_i$, which is merely the vector sum of the moments about G of the linear momenta of all particles. We showed in Art. 4/4 that this vector sum could also be written as $\mathbf{H}_G = \Sigma \boldsymbol{\rho}_i \times m_i \dot{\boldsymbol{\rho}}_i$, where $\dot{\boldsymbol{\rho}}_i$ is the velocity of m_i with respect to G. Although we have simplified this expression in Art. 6/2 in the course of deriving the moment equation of motion, we will pursue this same expression again for sake of emphasis by using the rigid body in plane motion represented in Fig. 6/13. The relative velocity becomes $\dot{\boldsymbol{\rho}}_i = \boldsymbol{\omega} \times \boldsymbol{\rho}_i$, where the angular velocity of the body is $\boldsymbol{\omega} = \omega \mathbf{k}$. The unit vector $\mathbf{k}$ is directed into the paper for the sense of $\boldsymbol{\omega}$ shown. Since $\boldsymbol{\rho}_i$, $\dot{\boldsymbol{\rho}}_i$, and $\boldsymbol{\omega}$ are at right angles to one another, the magnitude of $\dot{\boldsymbol{\rho}}_i$ is $\rho_i \omega$, and the magnitude of $\boldsymbol{\rho}_i \times m_i \dot{\boldsymbol{\rho}}_i$ is $\rho_i^2 \omega m_i$. Thus, we may write $\mathbf{H}_G = \Sigma \rho_i^2 m_i \omega \mathbf{k} = \bar{I} \omega \mathbf{k}$, where $\bar{I} = \Sigma m_i \rho_i^2$ is the mass moment of inertia of the body about its mass center. Because the angular-momentum vector is always normal to the plane of motion, vector notation is generally unnecessary, and we may write the angular momentum about the mass center as the scalar

$$\boxed{H_G = \bar{I} \omega} \qquad \textbf{(6/13)}$$

This angular momentum appears in the moment-angular-momentum relation, Eq. 4/9, which in scalar notation for plane motion, along with its integrated form, is

$$\boxed{\Sigma M_G = \dot{H}_G} \quad \text{and} \quad \boxed{\int_{t_1}^{t_2} \Sigma M_G \, dt = H_{G_2} - H_{G_1}} \quad (6/14)$$

In words, the first of Eqs. 6/14 states that the sum of the moments about the mass center of *all* forces acting on the body equals the time rate of change of angular momentum about the mass center. The integrated form of Eq. 6/14 states that the angular impulse about the mass center of all forces acting on the body during the interval $t_2 - t_1$ equals the corresponding change in the angular momentum about G. Again, as with the linear case, the angular-impulse and angular momentum relationship may be alternatively written in the form equivalent to Eq. 3/29a. The sense for positive rotation must be clearly established, and the algebraic signs of ΣM_G, H_{G_2}, and H_{G_1} must be consistent with this choice. Again, a free-body diagram is essential.

With the moments about G of the linear momenta of all particles accounted for by $H_G = \bar{I}\omega$, it follows that we may represent the linear momentum $\mathbf{G} = m\bar{\mathbf{v}}$ as a vector through the mass center G, as shown in Fig. 6/14a. Thus, $\mathbf{G}$ and $\mathbf{H}_G$ have vector properties analogous to those of the resultant force and couple.

With the establishment of the linear and angular momentum resultants in Fig. 6/14a, which represents the momentum diagram, the angular momentum H_O about any point O is easily written as

$$\boxed{H_O = \bar{I}\omega + m\bar{v}d} \quad (6/15)$$

This expression holds at any particular instant of time about O, which may be a fixed or moving point on or off the body.

When a body rotates about a fixed point O on the body or body extended, as shown in Fig. 6/14b, the relations $\bar{v} = \bar{r}\omega$ and $d = \bar{r}$ may be substituted into the expression for H_O, giving $H_O = (\bar{I}\omega + m\bar{r}^2\omega)$. But $\bar{I} + m\bar{r}^2 = I_O$ so that

$$\boxed{H_O = I_O\omega} \quad (6/16)$$

In Art. 4/2 we derived Eq. 4/7, which is the moment-angular-momentum equation about a fixed point O. This equation, written in scalar notation for plane motion along with its integrated form, is

$$\boxed{\Sigma M_O = I_O\dot{\omega}} \quad \text{and} \quad \boxed{\int_{t_1}^{t_2} \Sigma M_O \, dt = I_O(\omega_2 - \omega_1)} \quad (6/17)$$

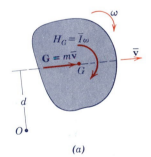

(a)

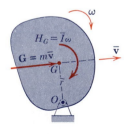

(b)

Figure 6/14

In addition to drawing a complete free-body diagram in order that the force and moment summations may be correctly evaluated when we apply the linear and angular impulse-momentum equations for rigid-body motion, we find it useful to draw the momentum diagram, which indicates the resultant linear-momentum vector and the angular-momentum couple. We caution the reader here not to add linear momentum and angular momentum for the same reason that force and moment cannot be added directly.

(c) Interconnected rigid bodies.   The equations of impulse and momentum may also be used for a system of interconnected rigid bodies since the momentum principles are applicable to any general system of constant mass. In Fig. 6/15 are shown the combined free-body diagram and momentum diagram for two interconnected bodies. Equations 4/6 and 4/7, which are $\Sigma\mathbf{F} = \dot{\mathbf{G}}$ and $\Sigma\mathbf{M}_O = \dot{\mathbf{H}}_O$ where O is a fixed reference point, may be written for each member of the system and added. The sums would be

$$\Sigma\mathbf{F} = \dot{\mathbf{G}}_1 + \dot{\mathbf{G}}_2 + \cdots \tag{6/18}$$
$$\Sigma\mathbf{M}_O = \dot{\mathbf{H}}_{O_1} + \dot{\mathbf{H}}_{O_2} + \cdots$$

In integrated form for a finite time interval, these expressions are

$$\int_{t_1}^{t_2}\Sigma\mathbf{F}\,dt = (\Delta\mathbf{G})_{\text{system}} \qquad \int_{t_1}^{t_2}\Sigma\mathbf{M}_O\,dt = (\Delta\mathbf{H}_O)_{\text{system}} \tag{6/19}$$

We note that the equal and opposite actions and reactions in the connections are internal to the system and cancel one another so they will not be involved in the force and moment summations. Also, point O is one fixed reference point for the entire system.

(d) Conservation of momentum.   In Part (*b*) of Art. 4/5, the principles of conservation of momentum for a general mass system were expressed by Eqs. 4/15 and 4/16. These principles are applicable to either a single rigid body or to a system of interconnected rigid bodies. Thus, if $\Sigma\mathbf{F} = \mathbf{0}$ for a given interval of time, then

$$\boxed{\Delta\mathbf{G} = \mathbf{0}} \tag{4/15}$$

which says that the linear momentum vector undergoes no change in the absence of a resultant linear impulse. For the system of interconnected rigid bodies, there may be linear momentum changes of individual parts of the system during the interval, but there will be no resultant momentum change for the system if there is no resultant linear impulse.

Similarly, if the resultant moment about a given fixed point O or about the mass center is zero during a particular interval of time

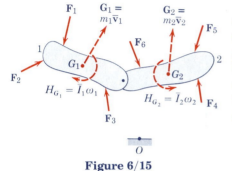

Figure 6/15

for a single rigid body or for a system of interconnected rigid bodies, then

$$\boxed{\Delta \mathbf{H}_O = 0} \quad \text{or} \quad \boxed{\Delta \mathbf{H}_G = 0} \qquad \textbf{[4/16]}$$

which says that the angular momentum either about the fixed point or about the mass center undergoes no change in the absence of a corresponding resultant angular impulse. Again, in the case of the interconnected system, there may be angular-momentum changes of individual components during the interval, but there will be no resultant angular-momentum change for the system if there is no resultant angular impulse about the fixed point or the mass center. Either of Eqs. 4/16 may hold without the other. In the case of an interconnected system, the use of the center of mass for the system is generally inconvenient. As was illustrated previously in Articles 3/9 and 3/10 in the chapter on particle motion, the use of momentum principles greatly facilitates the analysis of situations where forces and couples act for very short periods of time.

(e) *Impact of rigid bodies.*   Impact phenomena involve a fairly complex interrelationship of energy and momentum transfer, energy dissipation, elastic and plastic deformation, relative impact velocity, and body geometry. In Art. 3/12 we treated the impact of bodies modeled as particles and considered only the case of central impact, where the contact forces of impact passed through the mass centers of the bodies, as would always happen with colliding smooth spheres, for example. To relate the conditions after impact to those before impact required the introduction of the so-called coefficient of restitution e or impact coefficient, which compares the relative separation velocity to the relative approach velocity measured along the direction of the contact forces. Although in the classical theory of impact, e was considered a constant for given materials, more modern investigations show that e is highly dependent on geometry and impact velocity as well as on materials. At best, even for spheres and rods under direct central and longitudinal impact, the coefficient of restitution is a complex and variable factor of limited use.

Any attempt to extend this simplified theory of impact utilizing a coefficient of restitution for the noncentral impact of rigid bodies of varying shape is a gross oversimplification that has little practical value. For this reason, we shall not include such an exercise in this book, even though such a theory is easily developed and appears in certain references. We can and do, however, make full use of the principles of conservation of linear and angular momentum when applicable in discussing the impact and other interactions of rigid bodies.

Sample Problem 6/14

The force P, which is applied to the cable wrapped around the central hub of the symmetrical wheel, is increased slowly according to $P = 1.50t$, where P is in pounds and t is the time in seconds after P is first applied. Determine the angular velocity ω of the wheel 10 sec after P is applied if the wheel is rolling to the left, with a velocity of its center of 3 ft/sec at time $t = 0$. The wheel weighs 120 lb with a radius of gyration about its center of 10 in. and rolls without slipping.

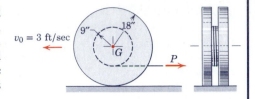

$v_0 = 3$ ft/sec

Solution. The free-body diagram of the wheel for any position within the interval is shown. Also indicated are the initial linear and angular momenta at time $t = 0$ and the final linear and angular momenta at time $t = 10$ sec. The correct direction of the friction force F is that to oppose the slipping which would occur without friction.

Application of the linear impulse-momentum equation and the angular impulse-momentum equation over the *entire* interval gives

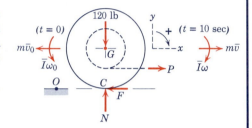

(120 lb) $(t = 0)$ $(t = 10 \text{ sec})$

$$\left[\int_{t_1}^{t_2} \Sigma F_x \, dt = G_{x_2} - G_{x_1}\right] \qquad \int_0^{10} (1.5t - F) \, dt = \frac{120}{32.2}\left[\frac{18}{12}\omega - (-3)\right]$$

$$\left[\int_{t_1}^{t_2} \Sigma M_G \, dt = H_{G_2} - H_{G_1}\right]$$

$$\int_0^{10}\left[\frac{18}{12} F - \frac{9}{12}(1.5t)\right] dt = \frac{120}{32.2}\left(\frac{10}{12}\right)^2\left[\omega - \left(-\frac{3}{18/12}\right)\right]$$

Since the force F is variable, it must remain under the integral sign. We eliminate F between the two equations by multiplying the second one by $\frac{12}{18}$ and adding to the first one. Integrating and solving for ω give

$$\omega = 3.13 \text{ rad/sec clockwise} \qquad\qquad \textit{Ans.}$$

Alternative Solution. We could avoid the necessity of a simultaneous solution by applying the second of Eqs. 6/17 about a fixed point O on the horizontal surface. The moments of the 120-lb weight and the equal and opposite force N cancel one another, and F is eliminated since its moment about O is zero. Thus, the angular momentum about O becomes $H_O = \bar{I}\omega + m\bar{v}r = m\bar{k}^2\omega + mr^2\omega = m(\bar{k}^2 + r^2)\omega$, where $\bar{k}$ is the centroidal radius of gyration and r is the 18-in. rolling radius. Thus, we see that $H_O = H_C$ since $\bar{k}^2 + r^2 = k_C^2$ and $H_C = I_C\omega = mk_C^2\omega$. Equation 6/17 now gives

$$\left[\int_{t_1}^{t_2} \Sigma M_O \, dt = H_{O_2} - H_{O_1}\right]$$

$$\int_0^{10} 1.5t\left(\frac{18 - 9}{12}\right) dt = \frac{120}{32.2}\left[\left(\frac{10}{12}\right)^2 + \left(\frac{18}{12}\right)^2\right]\left[\omega - \left(-\frac{3}{18/12}\right)\right]$$

Solution of this one equation is equivalent to the simultaneous solution of the two previous equations.

① Also, we note the clockwise imbalance of moments about C, which causes a clockwise angular acceleration as the wheel rolls without slipping. Since the moment sum about G must also be in the clockwise sense of α, the friction force must act to the left to provide it.

② Note carefully the signs of the momentum terms. The final linear velocity is assumed in the positive x-direction, so G_{x_2} is positive. The initial linear velocity is negative, so G_{x_1} is negative. When we subtract the negative term, we get the double minus sign.

③ Since the wheel rolls without slipping, a positive x-velocity requires a clockwise angular velocity, and vice versa. Again, we subtract a minus quantity.

Sample Problem 6/15

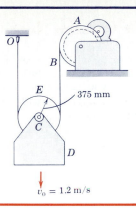

The sheave E of the hoisting rig shown has a mass of 30 kg and a centroidal radius of gyration of 250 mm. The 40-kg load D that is carried by the sheave has an initial downward velocity $v_0 = 1.2$ m/s at the instant when a clockwise torque is applied to the hoisting drum A to maintain essentially a constant force $F = 380$ N in the cable at B. Compute the angular velocity ω of the sheave 5 s after the torque is applied to the drum and find the tension T in the cable at O during the interval. Neglect all friction.

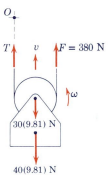

Solution. The load and the sheave taken together constitute the system, and its free-body diagram is shown. The tension T in the cable at O and the final angular velocity ω of the sheave are the two unknowns. We eliminate T initially by applying the moment-angular-momentum equation about the fixed point O, taking counterclockwise as positive.

$$\left[\int_{t_1}^{t_2} \Sigma M_O \, dt = H_{O_2} - H_{O_1} \right]$$

$$\int_{t_1}^{t_2} \Sigma M_O \, dt = \int_0^5 [380(0.750) - (30 + 40)(9.81)(0.375)] \, dt$$

$$= 137.4 \text{ N·m·s}$$

① $(H_{O_2} - H_{O_1})_D = m v_2 d - m v_1 d = md(v_2 - v_1)$

$$= 40(0.375)[v - (-1.2)] = 15(0.375\omega + 1.2)$$

$$= 5.63\omega + 18 \text{ N·m·s}$$

$(H_{O_2} - H_{O_1})_E = \bar{I}(\omega_2 - \omega_1) + md(\bar{v}_2 - \bar{v}_1)$

$$= 30(0.250)^2[\omega - (-1.2/0.375)]$$

$$+ 30(0.375)[0.375\omega - (-1.2)]$$

$$= 6.09\omega + 19.50 \text{ N·m·s}$$

Substituting into the momentum equation gives

$$137.4 = 5.63\omega + 18 + 6.09\omega + 19.50$$

$$\omega = 8.53 \text{ rad/s counterclockwise} \qquad Ans.$$

The force-linear-momentum equation is now applied to the system to determine T. With the positive direction up, we have

$$\left[\int_{t_1}^{t_2} \Sigma F \, dt = G_2 - G_1 \right]$$

$$\int_0^5 [T + 380 - 70(9.81)] \, dt = 70[0.375(8.53) - (-1.2)]$$

$$5T = 1841 \qquad T = 368 \text{ N} \qquad Ans.$$

If we had taken our moment equation around the center C of the sheave instead of point O, it would contain both unknowns T and ω, and we would be obliged to solve it simultaneously with the foregoing force equation that would also contain the same two unknowns.

① Watch for the double minus sign in subtracting a negative quantity. Also, the units of angular momentum, which are those of angular impulse, may be written as kg·m^2·s^{-1}.

Sample Problem 6/16

The uniform rectangular block of dimensions shown is sliding to the left on the horizontal surface with a velocity v when it strikes the small step in the surface. Assume negligible rebound at the step and compute the minimum value of v that will permit the block to pivot about the edge of the step and just reach the standing position A with no velocity. Compute the percentage energy loss $\Delta E/E$ for $b = c$.

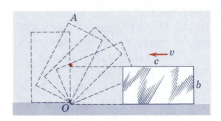

Solution. It will be assumed that the edge of the step O acts as a latch on the corner of the block, so that the block pivots about O. Furthermore, the height of the step is assumed to be negligible compared with the dimensions of the block. During impact the only force that exerts a moment about O is the weight mg, but the angular impulse due to the weight is extremely small since the time of impact is negligible. Thus, we may assume that the angular momentum about O is conserved.

The initial angular momentum of the block about O just before impact is the moment of its linear momentum and is $H_O = mv(b/2)$. The velocity of the center of mass G immediately after impact is $\bar{v}$, and the angular velocity is $\omega = \bar{v}/\bar{r}$. The angular momentum about O just after impact when the block is starting its rotation about O is

$$[H_O = I_O\omega] \qquad H_O = \left\{ \frac{1}{12} m(b^2 + c^2) + m\left[\left(\frac{c}{2}\right)^2 + \left(\frac{b}{2}\right)^2 \right] \right\} \omega$$

$$= \frac{m}{3}(b^2 + c^2)\omega$$

Conservation of angular momentum gives

$$[\Delta H_O = 0] \qquad \frac{m}{3}(b^2 + c^2)\omega = mv\frac{b}{2} \qquad \omega = \frac{3vb}{2(b^2 + c^2)}$$

This angular velocity will be sufficient to raise the block just past position A if the kinetic energy of rotation equals the increase in potential energy. Thus,

$$[\Delta T + \Delta V_g = 0] \qquad \frac{1}{2} I_O\omega^2 - mg\left[\sqrt{\left(\frac{b}{2}\right)^2 + \left(\frac{c}{2}\right)^2} - \frac{b}{2} \right] = 0$$

$$\frac{1}{2}\frac{m}{3}(b^2 + c^2)\left[\frac{3vb}{2(b^2 + c^2)} \right]^2 - \frac{mg}{2}(\sqrt{b^2 + c^2} - b) = 0$$

$$v = 2\sqrt{\frac{g}{3}\left(1 + \frac{c^2}{b^2}\right)}(\sqrt{b^2 + c^2} - b) \qquad\qquad \textit{Ans.}$$

The percentage loss of energy is

$$\frac{\Delta E}{E} = \frac{\frac{1}{2}mv^2 - \frac{1}{2}I_O\omega^2}{\frac{1}{2}mv^2} = 1 - \frac{k_O^2\omega^2}{v^2} = 1 - \left(\frac{b^2 + c^2}{3}\right)\left[\frac{3b}{2(b^2 + c^2)} \right]^2$$

$$= 1 - \frac{3}{4\left(1 + \frac{c^2}{b^2}\right)}, \qquad \Delta E/E = 62.5\% \qquad \text{for } b = c \qquad \textit{Ans.}$$

① If the corner of the block struck a spring instead of the rigid step, then the time of the interaction during compression of the spring could become appreciable, and the angular impulse about the fixed point at the end of the spring due to the moment of the weight would have to be accounted for.

② Notice the abrupt change in direction and magnitude of the velocity of G during the impact.

③ Be sure to use the transfer theorem $I_O = \bar{I} + m\bar{r}^2$ correctly here.

PROBLEMS

Introductory problems

6/185 The mass center G of the slender bar of mass 0.8 kg and length 0.4 m is falling vertically with a velocity $v = 2$ m/s at the instant depicted. Calculate the angular momentum H_O of the bar about point O if the angular velocity of the bar is (a) $\omega_a = 10$ rad/s clockwise and (b) $\omega_b = 10$ rad/s counterclockwise.

$$Ans. \ (a) \ H_O = 0.587 \ \text{kg·m}^2/\text{s}$$
$$(b) \ H_O = 0.373 \ \text{kg·m}^2/\text{s}$$

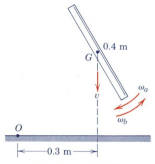

Problem 6/185

6/186 A constant horizontal force P is applied to the center O of the uniform circular disk of mass m through the light yoke as shown. The disk starts from rest and rolls for t seconds without slipping on the horizontal surface. Determine the velocity v of the center O in terms of t.

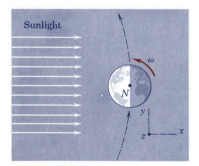

Problem 6/186

6/187 Determine the angular momentum of the earth about the center of the sun. Assume a homogeneous earth and a circular earth orbit of radius $149.6(10^6)$ km; consult Table D/2 for other needed information. Comment on the relative contributions of the terms $\bar{I}\omega$ and $m\bar{v}d$.

$$Ans. \ \bar{H} = 2.66(10^{40}) \ \text{kg·m}^2/\text{s}$$

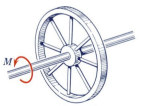

Problem 6/187

6/188 The 75-kg flywheel has a radius of gyration about its shaft axis of $\bar{k} = 0.50$ m and is subjected to the torque $M = 10(1 - e^{-t})$ N·m, where t is in seconds. If the flywheel is at rest at time $t = 0$, determine its angular velocity ω at $t = 3$ s.

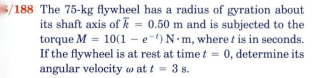

Problem 6/188

$m = 14$ kg, $\bar{k} = 225$ mm
$r_o = 325$ mm, $r_i = 215$ mm

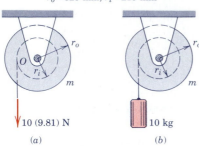

10 (9.81) N   10 kg
(a)   (b)

Problem 6/189

6/189 The grooved drums in the two systems shown are identical. In each case, (a) and (b), the system is at rest at time $t = 0$. Determine the angular velocity of the grooved drum at time $t = 4$ s. Neglect friction at the pivot O.

   Ans. (a) $\omega = 119.0$ rad/s, (b) $\omega = 72.0$ rad/s

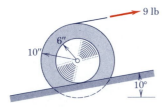

Problem 6/190

6/190 The constant 9-lb force is applied to the 80-lb stepped cylinder as shown. The centroidal radius of gyration of the cylinder is $\bar{k} = 8$ in. and it rolls on the incline without slipping. If the cylinder is at rest when the force is first applied, determine its angular velocity ω eight seconds later.

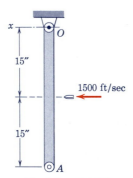

Problem 6/191

6/191 The 2-oz bullet has a horizontal velocity of 1500 ft/sec as it strikes the 20-lb slender bar OA, which is suspended from point O and is initially at rest. Calculate the angular velocity ω which the bar with its embedded bullet has acquired immediately after impact.   Ans. $\omega = 5.60$ rad/sec

6/192 If the bullet of Prob. 6/191 takes 0.001 sec to embed itself in the bar, calculate the time average of the horizontal force O_x exerted by the pin on the bar at O during the interaction between the bullet and the bar. Use the results cited for Prob. 6/191.

Representative problems

6/193 The homogeneous circular cylinder of mass m and radius R carries a slender rod of mass $m/2$ attached to it as shown. If the cylinder rolls on the surface without slipping with a velocity v_O of its center O, determine the angular momenta H_G and H_O of the system for the instant shown.

   Ans. $H_G = \frac{11}{16}mRv_O$, $H_O = \frac{37}{32}mRv_O$

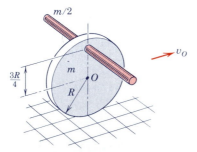

Problem 6/193

6/194 The small block of mass m slides along the radial slot of the disk while the disk rotates in the horizontal plane about its center O. The block is released from rest relative to the disk and moves outward with an increasing velocity $\dot{r}$ along the slot as the disk turns. Determine the expression in terms of r and $\dot{r}$ for the torque M that must be applied to the disk to maintain a constant angular velocity ω of the disk.

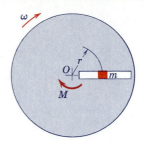

Problem 6/194

6/195 A uniform slender bar of mass M and length L is translating on the smooth horizontal x-y plane with a velocity v_M when a particle of mass m traveling with a velocity v_m as shown strikes and becomes embedded in the bar. Determine the final linear and angular velocities of the bar with its embedded particle.

$$Ans. \ v_x = \frac{Mv_M}{M + m}, \ v_y = \frac{mv_m}{M + m}$$
$$\omega = \frac{12v_m}{L}\left(\frac{m}{4M + 7m}\right) \ CCW$$

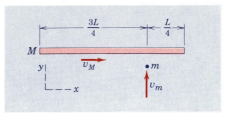

Problem 6/195

6/196 The unbalanced wheel is made to roll to the right without slipping with a constant velocity of 0.9 m/s of its center O. The wheel has a mass of 8 kg with center of mass at G and has a radius of gyration about O of 150 mm. Determine the angular momentum H_O of the wheel about O at the instant (a) when G passes directly over O with $\theta = 0$ and (b) when G passes the horizontal line through O where $\theta = 90°$.

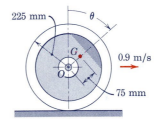

Problem 6/196

6/197 The uniform circular disk of 200-mm radius has a mass of 25 kg and is mounted on the rotating bar OA in three different ways. In each case the bar rotates about its vertical shaft at O with a clockwise angular velocity $\omega = 4$ rad/s. In case (a) the disk is welded to the bar. In case (b) the disk, which is pinned freely at A, moves with curvilinear translation and therefore has no rigid-body rotation. In case (c) the relative angle between the disk and the bar is increasing at the rate $\dot{\theta} = 8$ rad/s. Calculate the angular momentum of the disk about point O for each case.   *Ans.* (a) $H_O = 18$ kg·m²/s
(b) $H_O = 16$ kg·m²/s
(c) $H_O = 14$ kg·m²/s

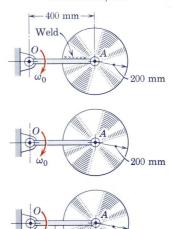

Problem 6/197

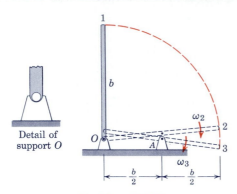

Problem 6/198

6/198 The slender bar of mass m and length b is pivoted at its lower end at O in the manner shown in the separate detail of the support O. The bar is released from rest in the vertical position 1. When the middle of the bar strikes the pivot at A in position 2, it becomes latched to the pivot, and simultaneously the connection at O becomes disengaged. Determine the angular velocity ω_3 of the bar just after it engages the pivot at A in position 3.

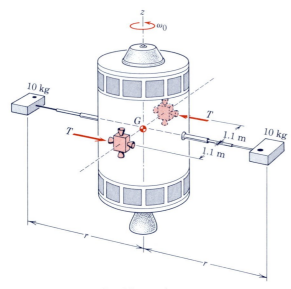

Problem 6/199

6/199 Two small variable-thrust jets are actuated to keep the spacecraft angular velocity about the z-axis constant at $\omega_0 = 1.25$ rad/s as the two telescopic booms are extended from $r_1 = 1.2$ m to $r_2 = 4.5$ m at a constant rate over a two-minute period. Determine the necessary thrust T for each jet as a function of time where $t = 0$ is the time when the telescoping action is begun. The small 10-kg experiment modules at the ends of the booms may be treated as particles, and the mass of the rigid booms is negligible.
Ans. $T = 0.750 + 0.01719t$ N

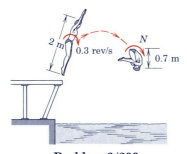

Problem 6/200

6/200 Just after leaving the platform, the diver's fully-extended 80-kg body has a rotational speed of 0.3 rev/s about an axis normal to the plane of the trajectory. Estimate the angular velocity N later in the dive when the diver has assumed the tuck position. Make reasonable assumptions concerning the mass moment of inertia of the body in each configuration.

6/201 With the gears initially at rest and the couple M equal to zero, the forces exerted by the frame on the shafts of the gears at A and B are 30 and 16 lb, respectively, both upward to support the weights of the two gears. A couple $M = 60$ lb-in. is now applied to the larger gear through its shaft at A. After 4 sec the larger gear has a clockwise angular momentum of 12 ft-lb-sec, and the smaller gear has a counterclockwise angular momentum of 4 ft-lb-sec. Calculate the new values of the forces R_A and R_B exerted by the frame on the shafts during the 4-sec interval. Isolate the two gears together as the system.

Ans. $R_A = 27.23$ lb, $R_B = 18.77$ lb

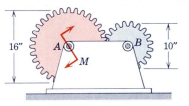

Problem 6/201

6/202 The elements of a spacecraft with axial mass symmetry and a reaction-wheel control system are shown in the figure. When the motor exerts a torque on the reaction wheel, an equal and opposite torque is exerted on the spacecraft, thereby changing its angular momentum in the z-direction. If all system elements start from rest and the motor exerts a constant torque M for a time period t, determine the final angular velocity of (a) the spacecraft and (b) the wheel relative to the spacecraft. The mass moment of inertia about the z-axis of the entire spacecraft, including the wheel, is I and that of the wheel alone is I_w. The spin axis of the wheel is coincident with the z-axis of symmetry of the spacecraft.

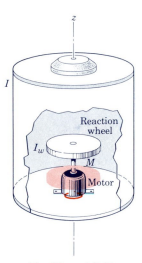

Problem 6/202

6/203 A 55-kg dynamics instructor is demonstrating the principles of angular momentum to her class. She stands on a freely rotating platform with her body aligned with the vertical platform axis. With the platform not rotating, she holds a modified bicycle wheel so that its axis is vertical. She then turns the wheel axis to a horizontal orientation without changing the 600-mm distance from the centerline of her body to the wheel center, and her students observe a platform rotation rate of 30 rev/min. If the rim-weighted wheel has a mass of 10 kg, a centroidal radius of gyration $\bar{k} = 300$ mm, and is spinning at a fairly constant rate of 250 rev/min, estimate the mass moment of inertia I of the instructor (in the posture shown) about the vertical platform axis.

Ans. $I = 3.45$ kg·m^2

Problem 6/203

6/204 If the dynamics instructor of Prob. 6/203 reorients the wheel axis by 180° with respect to its initial vertical position, what rotational speed N will her students observe? All the given information and the stated answer of Prob. 6/203 may be utilized.

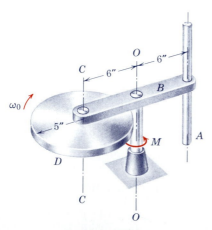

Problem 6/205

6/205 The 165-lb ice skater with arms extended horizontally spins about a vertical axis with a rotational speed of 1 rev/sec. Estimate his rotational speed N if he fully retracts his arms, bringing his hands very close to the centerline of his body. As a reasonable approximation, model the extended arms as uniform slender rods, each of which are 27 in. long and weigh 15 lb. Model the torso as a solid 135-lb cylinder 13 in. in diameter. With arms retracted, treat the man as a solid 165-lb cylinder of 13-in. diameter. Neglect friction at the skate-ice interface.

Ans. $N = 4.78$ rev/sec

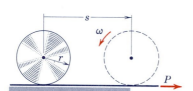

Problem 6/206

6/206 The solid circular cylinder of radius r is at rest on the flat belt when a force P is applied to the belt. If P is sufficient to cause slipping between the belt and the cylinder at all times, determine the time t required for the cylinder to reach the dotted position. Also determine the angular velocity ω of the cylinder in this same position. The coefficient of friction between the cylinder and the belt is μ_k.

Problem 6/207

6/207 The uniform circular disk D weighs 8 lb and is free to turn about the bearing axis C-C. The arm B weighs 5 lb and is fastened to the vertical shaft O-O. The arm may be approximated as a slender bar 12 in. long, and the moment of inertia of the vertical shaft O-O may be neglected. The rod A weighs 6 lb and is fastened securely to the arm. If the initial angular velocity of D is $\omega_0 = 7$ rad/sec in the direction shown and the arm B is at rest, determine the angular velocity ω_B of the arm after a torque $M = 15$ lb-in. has been applied to the shaft for 4 sec.

Ans. $\omega_B = 41.1$ rad/sec

6/208 Each of the two 300-mm uniform rods A has a mass of 1.5 kg and is hinged at its end to the rotating base B. The 4-kg base has a radius of gyration of 40 mm and is initially rotating freely about its vertical axis with a speed of 300 rev/min and with the rods latched in the vertical positions. If the latches are released and the rods assume the horizontal dotted positions, calculate the new rotational speed N of the assembly.

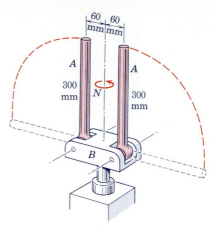

Problem 6/208

6/209 The homogeneous sphere of mass m and radius r is projected along the incline of angle θ with an initial speed v_0 and no angular velocity ($\omega_0 = 0$). If the coefficient of kinetic friction is μ_k, determine the time duration t of the period of slipping. In addition, state the velocity v of the mass center G and the angular velocity ω at the end of the period of slipping.

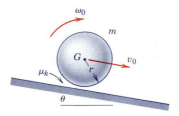

Problem 6/209

$$Ans. \quad t = \frac{v_0}{g}\,\frac{2}{7\mu_k\cos\theta - 2\sin\theta}$$

$$v = \frac{5v_0\mu_k}{7\mu_k - 2\tan\theta}$$

$$\omega = \frac{5v_0\mu_k/r}{7\mu_k - 2\tan\theta}$$

6/210 The homogeneous sphere of Prob. 6/209 is placed on the incline with a clockwise angular velocity ω_0 but no linear velocity of its center ($v_0 = 0$). Determine the time duration t of the period of slipping. In addition, state the velocity v and angular velocity ω at the end of the period of slipping.

6/211 The slender bar of mass m and length l is released from rest in the horizontal position shown. If point A of the bar becomes attached to the pivot at B upon impact, determine the angular velocity ω of the bar immediately after impact in terms of the distance x. Evaluate your expression for $x = 0$, $l/2$, and l.

$$Ans. \quad \omega = \left(\frac{l}{2} - x\right)\sqrt{2gh}/(\tfrac{1}{3}l^2 - lx + x^2)$$

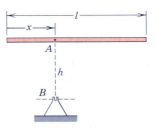

Problem 6/211

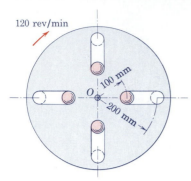

Problem 6/212

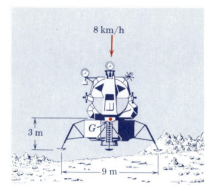

Problem 6/213

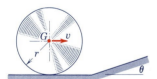

Problem 6/214

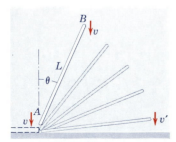

Problem 6/215

6/212 The slotted circular disk whose mass is 6 kg has a radius of gyration about O of 175 mm. The disk carries the four steel balls, each of mass 0.15 kg and located as shown, and rotates freely about a vertical axis through O with an angular speed of 120 rev/min. Each of the small balls is held in place by a latching device not shown. If the balls are released while the disk is rotating and come to rest in the dotted positions relative to the slots, compute the new angular velocity ω of the disk. Also find the magnitude $|\Delta E|$ of the energy loss due to the impact of the balls with the ends of the slots. Neglect the diameter of the balls and discuss this approximation.

6/213 The 17.5-Mg lunar landing module with center of mass at G has a radius of gyration of 1.8 m about G. The module is designed to contact the lunar surface with a vertical free-fall velocity of 8 km/h. If one of the four legs hits the lunar surface on a small incline and suffers no rebound, compute the angular velocity ω of the module immediately after impact as it pivots about the contact point. The 9-m dimension is the distance across the diagonal of the square formed by the four feet as corners.

Ans. $\omega = 0.308$ rad/s

6/214 A uniform circular disk that rolls with a velocity v without slipping encounters an abrupt change in the direction of its motion as it rolls onto the incline θ. Determine the new velocity v' of the center of the disk as it starts up the incline, and find the fraction n of the initial energy that is lost because of contact with the incline if $\theta = 10°$.

6/215 A uniform pole of length L is dropped at an angle θ with the vertical, and both ends have a velocity v as end A hits the ground. If end A pivots about its contact point during the remainder of the motion, determine the velocity v' with which end B hits the ground.

Ans. $v' = \sqrt{\dfrac{9v^2}{4} \sin^2 \theta + 3gL \cos \theta}$

6/216 The body of the spacecraft weighs 322 lb on earth and has a radius of gyration about its z-axis of 1.5 ft. Each of the two solar panels may be treated as a uniform flat plate weighing 16.1 lb. If the spacecraft is rotating about its z-axis at the angular rate of 1.0 rad/sec with $\theta = 0$, determine the angular rate ω after the panels are rotated to the position $\theta = \pi/2$ by an internal mechanism. Neglect the small momentum change of the body about the y-axis.

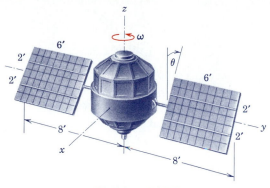

Problem 6/216

6/217 The small gear is made to rotate in a horizontal plane about the large stationary gear by means of the torque M applied to the arm OA. The small gear weighs 6 lb and may be treated as a circular disk. The arm OA weighs 4 lb and has a radius of gyration about the fixed bearing at O of 6 in. Determine the constant torque M required to give the arm OA an absolute angular velocity of 20 rad/sec in 3 sec, starting from rest. Neglect friction.

Ans. $M = 1.255$ lb-ft

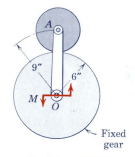

Problem 6/217

6/218 A froze-juice can rests on the horizontal rack of a freezer door as shown. With what maximum angular velocity Ω can the door be "slammed" shut against its seal and not dislodge the can? Assume that the can rolls without slipping on the corner of the rack, and neglect the dimension d compared with the 500-mm distance.

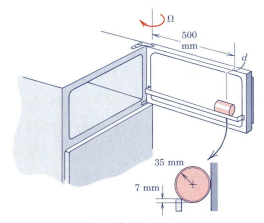

Problem 6/218

6/219 Determine the minimum velocity v that the wheel may have and just roll over the obstruction. The centroidal radius of gyration of the wheel is k, and it is assumed that the wheel does not slip.

Ans. $v = \dfrac{r}{k^2 + r^2 - rh} \sqrt{2gh(k^2 + r^2)}$

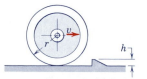

Problem 6/219

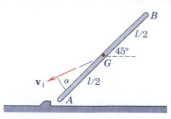

Problem 6/220

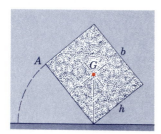

Problem 6/221

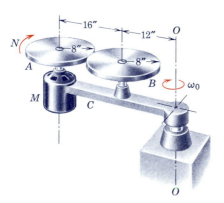

Problem 6/222

6/220 The uniform slender bar of mass m and length l has no angular velocity as end A strikes the ground against the stop with no rebound. If $\alpha = 15°$, what is the minimum magnitude of the initial velocity $\mathbf{v}_1$ for which the bar will rotate about A to the vertical position?

6/221 The uniform stone block with $b = 1.2$ m and $h = 0.9$ m is released from rest with its center of mass G almost directly above the supporting corner. Determine the angular velocity ω' of the block about corner A immediately after impact assuming that A remains in contact with the ground. Also assume that contact occurs at the corners only and that no slipping takes place. What fraction $|\Delta E|/E$ of the energy is lost in the impact?

Ans. $\omega' = 0.1121$ rad/s, $|\Delta E|/E = 0.998$

▶**6/222** The motor M drives disk A which turns disk B with no slipping. Disk A and its attached shaft and motor armature weigh 36 lb and have a combined radius of gyration of 3.4 in. Disk B weighs 10 lb and has a radius of gyration of 5.6 in. The motor housing and attached arm C together weigh 48 lb and have a radius of gyration about the axis O-O of 18 in. Before the motor is turned on, the entire assembly is rotating as a unit about O-O with a rotational speed $\omega_0 = 30$ rev/min in the direction shown. The motor M has an operating speed of 1720 rev/min in the direction shown as measured with C fixed. Determine the new rotational speed of arm C if the motor is turned on.

Ans. $N = 26.2$ rev/min clockwise

6/9 PROBLEM FORMULATION AND REVIEW

In Chapter 6 we have made use of essentially all the elements of dynamics studied so far. We have found that a knowledge of kinematics, using both absolute- and relative-motion analysis, is an essential part of the solution to problems in rigid-body kinetics. Our approach in Chapter 6 paralleled Chapter 3, where we developed the kinetics of particles using force-mass-acceleration, work-energy, and impulse-momentum methods.

The following outline will help to summarize important considerations in the solution of rigid-body problems in plane motion:

(a) Identification of body or system. It is essential to make an unambiguous decision as to which body or system of bodies is to be analyzed and then isolate the selected body or system by drawing its free-body diagram, kinetic diagram, or active-force diagram, whichever is appropriate.

(b) Type of motion. Next identify the category of motion as rectilinear translation, curvilinear translation, fixed-axis rotation, or general plane motion. Always see that the kinematics of the problem is properly described before attempting to solve the kinetic equations.

(c) Coordinate system. Choose an appropriate coordinate system. The geometry of the particular kinematics involved is usually the deciding factor. Designate the positive sense for moment and force summations and be consistent with the choice.

(d) Principle and method. If the instantaneous relationship between the applied forces and the acceleration is desired, then the equivalence between the forces and their $m\bar{\mathbf{a}}$ and $\bar{I}\alpha$ resultants, as disclosed by the free-body and kinetic diagrams, will indicate the most direct approach to a solution.

When motion occurs over an interval of displacement, the work-energy approach is indicated, and we relate initial to final velocities without calculating the acceleration. We have seen the advantage of this approach for interconnected mechanical systems with negligible internal friction.

If the interval of motion is specified in terms of time rather than displacement, the impulse-momentum approach is indicated. When the angular motion of a rigid body is suddenly changed, we use the principle of conservation of angular momentum.

(e) Assumptions and approximations. By now we should have acquired a feel for the practical significance of certain assumptions and approximations, such as treating a rod as an ideal slender bar and neglecting friction when minimal. These and other idealizations are important to the process of obtaining solutions to real problems.

REVIEW PROBLEMS

6/223 Make a list of the basic linear quantities of mechanics and their corresponding analogous angular quantities. List the SI units for each quantity.

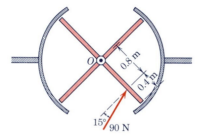

Problem 6/224

6/224 A person who walks through the revolving door exerts a 90-N horizontal force on one of the four door panels. If each panel is modeled by a 60-kg uniform rectangular plate which is 1.2 m in length as viewed from above, determine the angular acceleration of the door unit. Neglect friction.

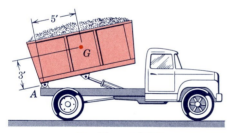

Problem 6/225

6/225 The dump truck carries 6 yd³ of dirt which weighs 110 lb/ft³, and the elevating mechanism rotates the dump about the pivot A at a constant angular rate of 4 deg/sec. The mass center of the dump and load is at G. Determine the maximum power P required during the tilting of the load.

Ans. $P = 11.31$ hp

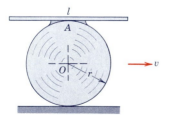

Problem 6/226

6/226 A slender rod of mass m_0 and length l is welded at its midpoint A to the rim of the solid circular disk of mass m and radius r. The center of the disk, which rolls without slipping, has a velocity v at the instant when A is at the top of the disk with the rod parallel to the ground. For this instant determine the angular momentum of the combined body about O.

6/227 A space telescope is shown in the figure. One of the reaction wheels of its attitude-control system is spinning as shown at 10 rad/s, and at this speed the friction in the wheel bearing causes an internal moment of 10^{-6} N·m. Both the wheel speed and the friction moment may be considered constant over a time span of several hours. If the mass moment of inertia of the entire spacecraft about the x-axis is $150(10^3)$kg·m², determine how much time passes before the line of sight of the initially stationary spacecraft drifts by 1 arc-second, which is 1/3600 degree. All other elements are fixed relative to the spacecraft, and no torquing of the reaction wheel shown is performed to correct the attitude drift. Neglect external torques.   *Ans. t = 1206 s*

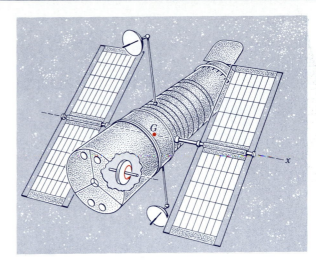

Problem 6/227

6/228 The uniform slender bar weighs 60 lb and is released from rest in the near-vertical position shown where the spring of stiffness 10 lb/ft is unstretched. Calculate the velocity with which end A strikes the horizontal surface.

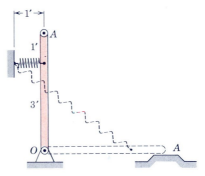

Problem 6/228

6/229 Four identical slender rods each of mass m are welded at their ends to form a square, and the corners are then welded to a light metal hoop of radius r. If the rigid assembly of rods and hoop is allowed to roll down the incline, determine the minimum value of the coefficient of static friction that will prevent slipping.   *Ans.* $\mu_s = \dfrac{2}{5} \tan \theta$

Problem 6/229

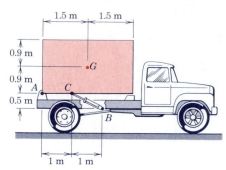

1.5 m   1.5 m

0.9 m

•G

0.9 m

A   C

0.5 m

B

1 m   1 m

Problem 6/230

6/230 The dump truck of Prob. 6/113 is shown again here. The loaded container has a mass of 8 Mg and may be modeled as a solid homogeneous block. The hydraulic cylinder applies a force of 300 kN to the container in the starting position shown. Calculate the initial angular acceleration α of the container (*a*) if the truck is not moving and (*b*) if the truck has a forward acceleration of 3 m/s².

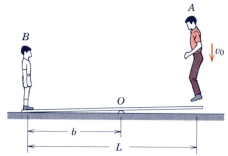

A

B

$\downarrow v_0$

O

b

L

Problem 6/231

6/231 In an acrobatic stunt, man *A* of mass m_A drops from a raised platform onto the end of the light but strong beam with a velocity v_0. The boy of mass m_B is propelled upward with a velocity v_B. For a given ratio $n = m_B/m_A$ determine *b* in terms of *L* to maximize the upward velocity of the boy. Assume that both man and boy act as rigid bodies.

$$Ans.\ b = \frac{L}{1 + \sqrt{n}}$$

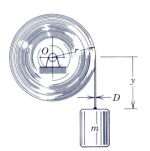

O   r

y

D

m

Problem 6/232

6/232 A solid roll of heavy paper mounted on a horizontal central axis *O* is unwound by the action of a falling mass *m* to which its end is attached. The paper has a mass ρ per unit circumferential length and a thickness *D*, and the roll has an initial radius r_0 when $y = 0$. If the system is released from rest with $y = 0$, determine the final velocity v_f of *m* as *r* approaches zero. Neglect the small horizontal motion of *m* and of the paper.

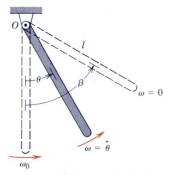

O

l

θ β

$\omega = 0$

$\omega = \dot\theta$ β

ω_0

Problem 6/233

6/233 The uniform slender rod of mass *m* and length *l* is freely hinged about a horizontal axis through its end *O* and is given an initial angular velocity ω_0 as it crosses the vertical position where $\theta = 0$. If the rod swings through a maximum angle $\beta < 90°$, derive an expression in integral form for the time *t* from release at $\theta = 0$ until $\theta = \beta$ is reached. (Express ω_0 in terms of β.)

$$Ans.\ t = \sqrt{\frac{l}{3g}} \int_0^\beta \frac{d\theta}{\sqrt{\cos\theta - \cos\beta}}$$

6/234 The car with standard rear-wheel drive weighs 3450 lb with center of gravity at G. The effective coefficient of friction between the tires and the road is 0.80. (*a*) Treat the car and wheels as a single rigid body by neglecting the rotational inertia of the wheels and calculate the maximum acceleration a that the car is capable of reaching. (*b*) Calculate the torque M applied to each wheel by its axle. Each rear wheel weighs 70 lb and has a diameter of 25 in. and a radius of gyration of 8.5 in.

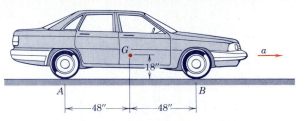

Problem 6/234

6/235 A gate in the form of a uniform rectangular panel of mass m is mounted on corner hinges A and B with the hinge axis inclined at an angle β with the vertical z-axis. Hinge A is capable of supporting force along the hinge axis as well as transverse to it. Hinge B is capable of supporting force only transverse to the hinge axis. With friction neglected, determine expressions for (*a*) the initial angular acceleration α of the gate when it is released from rest in the position shown, (*b*) the maximum angular velocity ω of the gate, and (*c*) the force R supported by hinge B when the angular velocity is maximum.

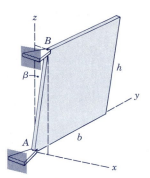

Ans. (*a*) $\alpha = \dfrac{3g}{2b} \sin \beta$

     (*b*) $\omega = \sqrt{(3g \sin \beta)/b}$

     (*c*) $R = \dfrac{mg}{2}\left[\dfrac{5}{2}\sin \beta + \dfrac{b}{h}\cos \beta\right]$

Problem 6/235

6/236 The link OA and pivoted circular disk are released from rest in the position shown and swing in the vertical plane about the fixed bearing at O. Link OA weighs 12 lb and has a radius of gyration about O of 15 in. The disk weighs 18 lb. The two bearings are assumed to be frictionless. Find the force F_O exerted at O on the link (*a*) just after release and (*b*) as OA swings through the vertical position OA'.

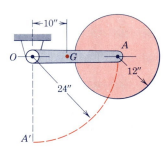

Problem 6/236

6/237 The light circular hoop of radius r carries a heavy uniform band of mass m around half of its circumference and is released from rest on the incline in the upper position shown. After the hoop has rolled one-half of a revolution, (*a*) determine its angular velocity ω and (*b*) find the normal force N under the hoop if $\theta = 10°$.

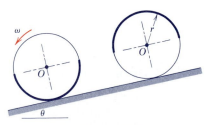

Ans. (*a*) $\omega = \sqrt{\dfrac{g}{r}\dfrac{4\cos \theta + \pi^2 \sin \theta}{\pi - 2}}$

     (*b*) $N = 4.14mg$

Problem 6/237

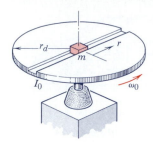

Problem 6/238

6/238 The small block of mass m slides in the smooth radial slot of the disk, which turns freely in its bearing. If the block is displaced slightly from the center position when the angular velocity of the disk is ω_0, determine its radial velocity v_r as a function of the radial distance r. The mass moment of inertia of the disk about its axis of rotation is I_O.

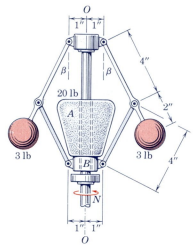

Problem 6/239

6/239 The mechanical flyball governor operates with a vertical shaft O-O. As the shaft speed N is increased, the rotational radius of the two 3-lb balls tends to increase, and the 20-lb weight A is lifted up by the collar B. Determine the steady-state value of β for a rotational speed of 150 rev/min. Neglect the mass of the arms and collar.   *Ans.* $\beta = 19.26°$

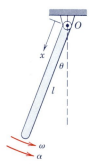

Problem 6/240

▶ **6/240** The uniform slender bar of mass m and length l is hinged about a horizontal axis through O and swings as a compound pendulum in the vertical plane. If the bar is released from rest in the horizontal position with $\theta = 90°$, write expressions for the tension T, shear force V, and bending moment M in the bar in terms of x for a given position θ. Neglect all friction.

$$Ans. \quad T = \frac{5l^2 - 2lx - 3x^2}{2l^2} mg \cos \theta$$

$$V = \frac{(l - x)(l - 3x)}{4l^2} mg \sin \theta$$

$$M = \frac{(l - x)^2 x}{4l^2} mg \sin \theta$$

▶ **6/241** The free end of a flexible rope of mass ρ per unit length is being pulled vertically up with a force P as shown. At the instant considered the free end has an upward acceleration a and an upward velocity v. Prove that, in the limit as the size of the loop becomes negligibly small, the tensions T_1 and T_2 in the rope on both sides of the loop are the same and equal $\frac{1}{4}\rho v^2$. (*Hint:* Model the loop as a rigid semicircular hoop attached to a massless wheel of radius r as shown. Next let r approach zero.)

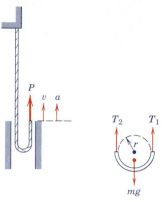

Problem 6/241

6/242 The split ring of radius r is rotating about a vertical axis through its center O with a constant angular velocity ω. Use a differential element of the ring and derive expressions for the shear force N and rim tension T in the ring in terms of the angle θ. Determine the bending moment M_C at point C by using one-half of the ring as a free body. The mass of the ring per unit length of rim is ρ.

$$Ans. \quad N = \rho r^2 \omega^2 \sin\theta$$
$$T = \rho r^2 \omega^2 (1 + \cos\theta)$$
$$M_C = 2\rho r^3 \omega^2$$

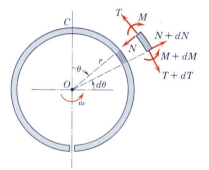

Problem 6/242

6/243 Before it hits the ground a falling chimney, such as the one shown, will usually crack at the point where the bending moment is greatest. Show that the position of maximum moment occurs at the center of percussion relative to the upper end for a slender chimney of constant cross section. Neglect any restraining moment at the base.

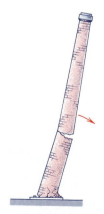

Problem 6/243

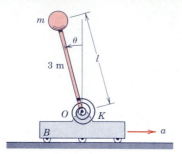

Problem 6/244

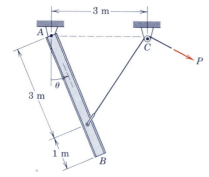

Problem 6/245

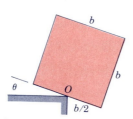

Problem 6/246

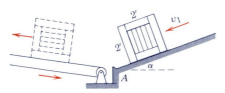

Problem 6/247

Computer-oriented problems

*6/244 The system of Prob. 6/14 is repeated here. The cart B moves to the right with acceleration $a = 2g$. If $m = 0.5$ kg, $l = 0.6$ m, and $K = 75$ N·m/rad, determine the steady-state angular deflection θ of the uniform slender rod of mass $3m$. Treat the small end sphere of mass m as a particle. The spring, which exerts a moment of magnitude $M = K\theta$ on the rod, is undeformed when the rod is vertical.

*6/245 The figure for Prob. 6/35 is shown again here. The uniform 100-kg beam AB is hanging initially at rest with $\theta = 0$ when the constant force $P = 300$ N is applied to the cable. Determine (a) the maximum angular velocity reached by the beam with the corresponding angle θ and (b) the maximum angle θ_{max} reached by the beam.

Ans. $\omega_{max} = 0.680$ rad/s at $\theta = 22.4°$
$\theta_{max} = 45.9°$

*6/246 The homogeneous square block of mass m is released from rest at θ essentially zero and pivots at the midpoint of its base about the fixed corner at O. Determine and plot the normal and tangential forces, expressed in dimensionless form N/mg and F/mg, exerted on the block by the corner as functions of θ. (a) If a small notch at O prevents the block from slipping, determine the angle θ at which contact with the corner ceases. (b) In the absence of a notch and with a coefficient of static friction of 0.8, determine the angle θ at which slipping first occurs.

*6/247 The crate slides down the incline with velocity v_1 and its corner strikes a small obstacle at A. Determine the minimum required velocity v_1 if the crate is to rotate about A so that it travels on the conveyor belt on its side as indicated in the figure. Plot the variation of v_1 with α for $0 \le \alpha \le 45°$.

Ans. $v_1 = 15.58\sqrt{1 - \cos(45° - \alpha)}$ ft/sec

*6/248 The 4-kg bar *OA* with mass center *G* is hinged in the vertical plane to the moving carriage at *O*. The attached spring is unstretched when $\theta = 0$. Determine the steady-state angle θ assumed by the bar during a steady horizontal acceleration $a = 4$ m/s^2 of the carriage. (*Hint:* There are two values of θ—one stable and the other unstable.)

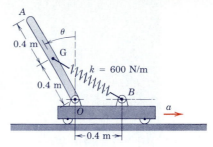

Problem 6/248

*6/249 The 4-kg bar of Prob. 6/248 is initially at rest in the vertical position with the attached spring of stiffness $k = 600$ N/m unstretched. If the carriage is given a constant acceleration $a = 0.4g$ from rest, determine and plot the angular velocity ω of the bar as a function of θ for $0 \le \theta \le 90°$.

> *Ans.* $\omega = \sqrt{36.8 f_1(\theta) - 225 f_2(\theta)}$
>
> where $f_1(\theta) = 0.4 \sin \theta + 1 - \cos \theta$
>
> $$f_2(\theta) = \sin \theta - 2\left(\sin \frac{\theta}{2} + \cos \frac{\theta}{2} - 1 \right)$$

*6/250 The 30-kg slender bar has an initial angular velocity $\omega_0 = 4$ rad/s in the vertical position, where the spring is unstretched. Determine the minimum angular velocity ω_{min} reached by the bar and the corresponding angle θ. Also find the angular velocity of the bar as it strikes the horizontal surface.

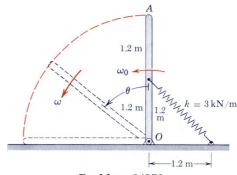

Problem 6/250

*6/251 The compound pendulum is composed of a uniform slender rod of length l and mass $2m$ to which is fastened a uniform disk of diameter $l/2$ and mass m. The body pivots freely about a horizontal axis through O. If the pendulum has a clockwise angular velocity of 3 rad/s when $\theta = 0$ at time $t = 0$, determine the time t at which the pendulum passes the position $\theta = 90°$. The pendulum length is $l = 0.8$ m.

> *Ans.* $t = 0.302$ s

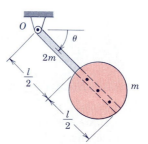

Problem 6/251

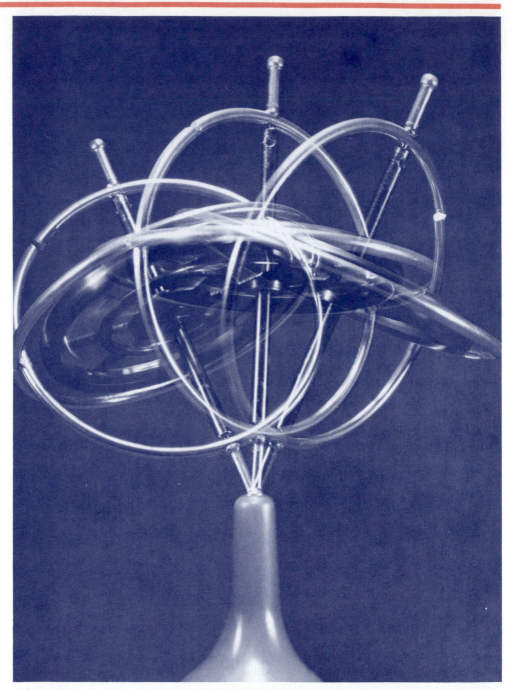

There is an increasing number of rigid bodies whose motions are three-dimensional. Adding a third dimension considerably complicates the mathematics. Gyroscopic motion with both spin and precession is one such example, as shown by the demonstration gyro which is gyrating about the vertical axis.

INTRODUCTION TO THREE-DIMENSIONAL DYNAMICS OF RIGID BODIES

7

7/1 INTRODUCTION

Although a large percentage of dynamics problems in engineering lend themselves to solution by means of the principles of plane motion, modern developments have focused increasing attention on problems that call for the analysis of motion in three dimensions. Inclusion of the third dimension adds considerable complexity to the kinematic and kinetic relationships. Not only does the added dimension introduce a third component to vectors that represent force, linear velocity, linear acceleration, and linear momentum, but the introduction of the third dimension adds the possibility of two additional components for vectors representing angular quantities including moments of forces, angular velocity, angular acceleration, and angular momentum. It is in three-dimensional motion that the full power of vector analysis is utilized.

A good background in the dynamics of plane motion is extremely useful in the study of three-dimensional dynamics, as the approach to problems and many of the terms are the same as or analogous to those in two dimensions. If the study of three-dimensional dynamics is undertaken without the benefit of prior study of plane-motion dynamics, more time will be required to master the principles and to become familiar with the approach to problems.

The treatment presented in Chapter 7 is not intended as a complete development of the three-dimensional motion of rigid bodies but merely as a basic introduction to the subject. This introduction should, however, be sufficient to solve many of the more common problems in three-dimensional motion and also to lay the foundation for more advanced study. We shall proceed as we did for particle motion and for rigid-body plane motion by first examining the necessary kinematics and then proceeding to the kinetics.

SECTION A. KINEMATICS

7/2 *TRANSLATION*

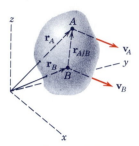

Figure 7/1

Figure 7/1 shows a rigid body translating in three-dimensional space. Any two points in the body, such as A and B, will move along parallel straight lines if the motion is one of *rectilinear translation* or will move along congruent curves if the motion is one of *curvilinear translation*. In either case, every line in the body, such as AB, remains parallel to its original position. The position vectors and their first and second time derivatives are

$$\mathbf{r}_A = \mathbf{r}_B + \mathbf{r}_{A/B} \qquad \mathbf{v}_A = \mathbf{v}_B \qquad \mathbf{a}_A = \mathbf{a}_B$$

where $\mathbf{r}_{A/B}$ remains constant and therefore has no time derivative. Thus, all points in the body have the same velocity and the same acceleration. The kinematics of translation presents no special difficulty, and further elaboration is unnecessary.

7/3 *FIXED-AXIS ROTATION*

Consider now the *rotation* of a rigid body about a fixed axis $n\text{-}n$ in space with an angular velocity $\boldsymbol{\omega}$, as shown in Fig. 7/2. The angular velocity is a vector in the direction of the rotation axis with a sense established by the familiar right-hand rule. For fixed-axis rotation, $\boldsymbol{\omega}$ does not change its direction since it lies along the axis. We choose the origin O of the fixed coordinate system on the rotation axis for convenience. Any point such as A that is not on the axis moves in a circular arc in a plane normal to the axis and has a velocity

$$\boxed{\mathbf{v} = \boldsymbol{\omega} \times \mathbf{r}} \qquad\qquad \textbf{(7/1)}$$

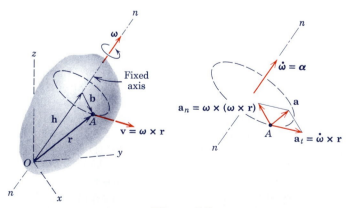

Figure 7/2

which may be seen by replacing $\mathbf{r}$ by $\mathbf{h} + \mathbf{b}$ and noting that $\boldsymbol{\omega} \times \mathbf{h} = \mathbf{0}$. The acceleration of A is given by the time derivative of Eq. 7/1. Thus,

$$\mathbf{a} = \dot{\boldsymbol{\omega}} \times \mathbf{r} + \boldsymbol{\omega} \times (\boldsymbol{\omega} \times \mathbf{r}) \qquad (7/2)$$

where $\dot{\mathbf{r}}$ has been replaced by its equal, $\mathbf{v} = \boldsymbol{\omega} \times \mathbf{r}$. The normal and tangential components of $\mathbf{a}$ for the circular motion have the familiar magnitudes $a_n = |\boldsymbol{\omega} \times (\boldsymbol{\omega} \times \mathbf{r})| = b\omega^2$ and $a_t = |\dot{\boldsymbol{\omega}} \times \mathbf{r}| = b\alpha$, where $\alpha = \dot{\omega}$. Inasmuch as both $\mathbf{v}$ and $\mathbf{a}$ are perpendicular to $\boldsymbol{\omega}$ and $\dot{\boldsymbol{\omega}}$, it follows that $\mathbf{v} \cdot \boldsymbol{\omega} = 0$, $\mathbf{v} \cdot \dot{\boldsymbol{\omega}} = 0$, $\mathbf{a} \cdot \boldsymbol{\omega} = 0$, and $\mathbf{a} \cdot \dot{\boldsymbol{\omega}} = 0$ for fixed-axis rotation.

7/4 PARALLEL-PLANE MOTION

When all points in a rigid body move in planes that are parallel to a fixed plane P, Fig. 7/3, we have a general form of plane motion. The reference plane is customarily taken through the mass center G and is referred to as the plane of motion. Since each point in the body, such as A', has a motion identical with the motion of point A in plane P, it follows that the kinematics of plane motion covered in Chapter 5 provides a complete description of the motion when applied to the reference plane.

7/5 ROTATION ABOUT A FIXED POINT

When a body rotates about a fixed point, the angular velocity vector no longer remains fixed in direction, and this change calls for a more general concept of rotation.

We must first examine the conditions under which rotation vectors obey the parallelogram law of addition and may, therefore, be treated as proper vectors. Consider a solid sphere, Fig. 7/4, that is cut from a rigid body confined to rotate about the fixed point O.

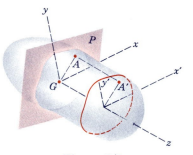

Figure 7/3

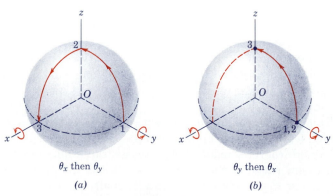

θ_x then θ_y	θ_y then θ_x
(a)	*(b)*

Figure 7/4

The x-y-z axes here are taken fixed in space and do not rotate with the body. In the a-part of the figure, two successive 90° rotations of the sphere about, first, the x-axis and, second, the y-axis result in the motion of a point that is initially on the y-axis in position 1, to positions 2 and 3, successively. On the other hand, if the order of the rotations is reversed, the point suffers no motion during the y-rotation but moves to point 3 during the 90° rotation about the x-axis. Thus, the two cases do not yield the same final position, and it is evident from this one special example that finite rotations do not generally obey the parallelogram law of vector addition and are not commutative. Thus, finite rotations may *not* be treated as proper vectors.

Infinitesimal rotations, however, do obey the parallelogram law of vector addition. This fact is shown in Fig. 7/5 that represents the combined effect of two infinitesimal rotations $d\boldsymbol{\theta}_1$ and $d\boldsymbol{\theta}_2$ of a rigid body about the respective axes through the fixed point O. As a result of $d\boldsymbol{\theta}_1$, point A has a displacement $d\boldsymbol{\theta}_1 \times \mathbf{r}$, and likewise $d\boldsymbol{\theta}_2$ causes a displacement $d\boldsymbol{\theta}_2 \times \mathbf{r}$ of point A. Either order of addition of these

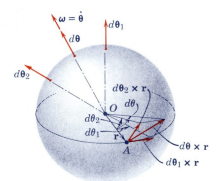

Figure 7/5

infinitesimal displacements clearly produces the same resultant displacement, which is $d\boldsymbol{\theta}_1 \times \mathbf{r} + d\boldsymbol{\theta}_2 \times \mathbf{r} = (d\boldsymbol{\theta}_1 + d\boldsymbol{\theta}_2) \times \mathbf{r}$. Hence, the two rotations are equivalent to the single rotation $d\boldsymbol{\theta} = d\boldsymbol{\theta}_1 + d\boldsymbol{\theta}_2$. It follows that the angular velocities $\boldsymbol{\omega}_1 = \dot{\boldsymbol{\theta}}_1$ and $\boldsymbol{\omega}_2 = \dot{\boldsymbol{\theta}}_2$ may be added vectorially to give $\boldsymbol{\omega} = \dot{\boldsymbol{\theta}} = \boldsymbol{\omega}_1 + \boldsymbol{\omega}_2$. We conclude, therefore, that at any instant of time a body with one fixed point is rotating instantaneously about a particular axis passing through the fixed point.

To aid in visualizing the concept of the instantaneous axis of rotation, we will cite a specific example. Figure 7/6 represents a solid cylindrical rotor made of clear plastic containing many black particles embedded in the plastic. The rotor is spinning about its shaft axis at the steady rate ω_1, and its shaft, in turn, is rotating about the fixed vertical axis at the steady rate ω_2, with rotations in

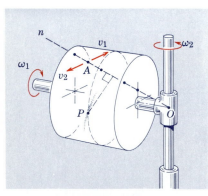

Figure 7/6

the directions indicated. If the rotor is photographed at a certain instant during its motion, the resulting picture would show one line of black dots in sharp focus indicating that, momentarily, their velocity was zero. This line of points with no velocity establishes the instantaneous position of the axis of rotation O-n. Any dot on this line, such as A, would have equal and opposite velocity components, v_1 due to ω_1 and v_2 due to ω_2. All other dots, such as the one at P, would appear blurred, and their movements would show as short streaks in the form of small circular arcs in planes normal to the axis O-n. Thus, all particles of the body, except those on line O-n, are momentarily rotating in circular arcs about the instantaneous axis of rotation. If a succession of photographs were taken, we would observe that the rotation axis would be defined by a new series of dots in focus and that the axis would change position both in space and relative to the body. For rotation of a rigid body about a fixed point, then, it is seen that the rotation axis is, in general, not a line fixed in the body.

(a) Body and space cones. Relative to the plastic cylinder of Fig. 7/6, the instantaneous axis of rotation O-A-n generates a right-circular cone about the cylinder axis called the *body cone*. As the two rotations continue and the cylinder swings around the vertical axis, the instantaneous axis of rotation also generates a right-circular cone about the vertical axis called the *space cone*. These cones are shown in Fig. 7/7 for this particular example. We see that the body cone rolls on the space cone and that the angular velocity

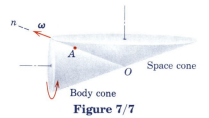

Figure 7/7

$\boldsymbol{\omega}$ of the body is a vector that lies along the common element of the two cones. For a more general case where the rotations are not steady, the space and body cones are not right-circular cones, Fig. 7/8, but the body cone still rolls on the space cone.

(b) Angular acceleration. The angular acceleration $\boldsymbol{\alpha}$ of a rigid body in three-dimensional motion is the time derivative of its angular velocity, $\boldsymbol{\alpha} = \dot{\boldsymbol{\omega}}$. In contrast to the case of rotation in a single plane where the scalar α measures only the change in magnitude of the angular velocity, in three-dimensional motion the vector $\boldsymbol{\alpha}$ reflects the change in direction of $\boldsymbol{\omega}$, as well as its change in magnitude. Thus in Fig. 7/8 where the tip of the angular velocity

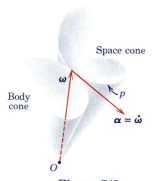

Figure 7/8

vector $\boldsymbol{\omega}$ follows the space curve p and changes both in magnitude and direction, the angular acceleration $\boldsymbol{\alpha}$ becomes a vector tangent to this curve in the direction of the change in $\boldsymbol{\omega}$.

When the magnitude of $\boldsymbol{\omega}$ remains constant, the angular acceleration $\boldsymbol{\alpha}$ is normal to $\boldsymbol{\omega}$. For this case, if we let $\boldsymbol{\Omega}$ stand for the angular velocity with which the vector $\boldsymbol{\omega}$ itself rotates (*precesses*) as it forms the space cone, the angular acceleration may be written

$$\boxed{\boldsymbol{\alpha} = \boldsymbol{\Omega} \times \boldsymbol{\omega}} \tag{7/3}$$

This relation is easily seen from Fig. 7/9 where the vectors $\boldsymbol{\alpha}$, $\boldsymbol{\omega}$, and $\boldsymbol{\Omega}$ in the lower figure bear exactly the same relationship to each other, as do the vectors $\mathbf{v}$, $\mathbf{r}$, and $\boldsymbol{\omega}$ in the upper figure for relating the velocity of a point A on a rigid body to its position vector from O and the angular velocity of the body.

If we use Fig. 7/2 to represent a rigid body rotating about a fixed point O with the instantaneous axis of rotation n-n, we see that the velocity $\mathbf{v}$ and acceleration $\mathbf{a} = \dot{\mathbf{v}}$ of any point A in the body are given by the same expressions as apply to the case in which the axis is fixed, namely,

$$\boxed{\begin{aligned} \mathbf{v} &= \boldsymbol{\omega} \times \mathbf{r} \\ \mathbf{a} &= \dot{\boldsymbol{\omega}} \times \mathbf{r} + \boldsymbol{\omega} \times (\boldsymbol{\omega} \times \mathbf{r}) \end{aligned}} \qquad \begin{aligned} &[7/1] \\ &[7/2] \end{aligned}$$

The one difference between the case of rotation about a fixed axis and rotation about a fixed point lies in the fact that for rotation about a fixed point, the angular acceleration $\boldsymbol{\alpha} = \dot{\boldsymbol{\omega}}$ will have a component normal to $\boldsymbol{\omega}$ due to the change in direction of $\boldsymbol{\omega}$, as well as a component in the direction of $\boldsymbol{\omega}$ to reflect any change in the magnitude of $\boldsymbol{\omega}$. Although any point on the rotation axis n-n momentarily will have zero velocity, it will *not* have zero acceleration as long as $\boldsymbol{\omega}$ is changing its direction. On the other hand, for rotation about a fixed axis, $\boldsymbol{\alpha} = \dot{\boldsymbol{\omega}}$ has only the one component along the fixed axis to reflect the change in the magnitude of $\boldsymbol{\omega}$. Furthermore, points that lie on the fixed rotation axis clearly have no velocity or acceleration.

Although the discussion in this article has been developed for the case of rotation about a fixed point, we observe that rotation is a function *solely* of angular change, so that the expressions for $\boldsymbol{\omega}$ and $\boldsymbol{\alpha}$ do not depend on the fixity of the point around which rotation occurs. Thus, rotation may take place independently of the linear motion of the rotation point. This conclusion is the three-dimensional counterpart of the concept of rotation of a rigid body in plane motion described in Art. 5/2 and used throughout Chapters 5 and 6.

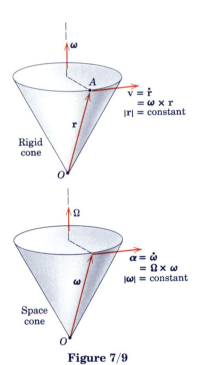

Figure 7/9

Sample Problem 7/1

The 0.8-m arm OA for a remote-control mechanism is pivoted about the horizontal x-axis of the clevis, and the entire assembly rotates about the z-axis with a constant speed $N = 60$ rev/min. Simultaneously, the arm is being raised at the constant rate $\dot{\beta} = 4$ rad/s. For the position where $\beta = 30°$, determine (a) the angular velocity of OA, (b) the angular acceleration of OA, (c) the velocity of point A, and (d) the acceleration of point A. If, in addition to the motion described, the vertical shaft and point O had a linear motion, say, in the z-direction, would that motion change the angular velocity or angular acceleration of OA?

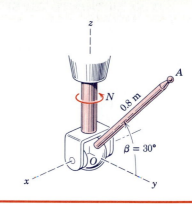

Solution. (a) Since the arm OA is rotating about both the x- and the z-axes, it has the components $\omega_x = \dot{\beta} = 4$ rad/s and $\omega_z = 2\pi N/60 = 2\pi(60)/60 = 6.283$ rad/s. The angular velocity is

$$\boldsymbol{\omega} = \boldsymbol{\omega}_x + \boldsymbol{\omega}_z = 4\mathbf{i} + 6.283\mathbf{k} \text{ rad/s} \qquad Ans.$$

(b) The angular acceleration of OA is

$$\boldsymbol{\alpha} = \dot{\boldsymbol{\omega}} = \dot{\boldsymbol{\omega}}_x + \dot{\boldsymbol{\omega}}_z$$

Since $\boldsymbol{\omega}_z$ is not changing in magnitude or direction, $\dot{\boldsymbol{\omega}}_z = \mathbf{0}$. But $\boldsymbol{\omega}_x$ is changing direction and thus has a derivative which, from Eq. 7/3, is

$$\dot{\boldsymbol{\omega}}_x = \boldsymbol{\omega}_z \times \boldsymbol{\omega}_x = 6.283\mathbf{k} \times 4\mathbf{i} = 25.13\mathbf{j} \text{ rad/s}^2$$

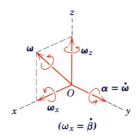

$(\omega_x = \dot{\beta})$

① Therefore,

$$\boldsymbol{\alpha} = 25.13\mathbf{j} + \mathbf{0} = 25.13\mathbf{j} \text{ rad/s}^2 \qquad Ans.$$

(c) With the position vector of A given by $\mathbf{r} = 0.693\mathbf{j} + 0.4\mathbf{k}$ m, the velocity of A from Eq. 7/1 becomes

$$\mathbf{v} = \boldsymbol{\omega} \times \mathbf{r} = \begin{vmatrix} \mathbf{i} & \mathbf{j} & \mathbf{k} \\ 4 & 0 & 6.283 \\ 0 & 0.693 & 0.4 \end{vmatrix} = -4.35\mathbf{i} - 1.60\mathbf{j} + 2.77\mathbf{k} \text{ m/s} \quad Ans.$$

(d) The acceleration of A from Eq. 7/2 is

$$\mathbf{a} = \dot{\boldsymbol{\omega}} \times \mathbf{r} + \boldsymbol{\omega} \times (\boldsymbol{\omega} \times \mathbf{r})$$

$$= \boldsymbol{\alpha} \times \mathbf{r} + \boldsymbol{\omega} \times \mathbf{v}$$

$$= \begin{vmatrix} \mathbf{i} & \mathbf{j} & \mathbf{k} \\ 0 & 25.13 & 0 \\ 0 & 0.693 & 0.4 \end{vmatrix} + \begin{vmatrix} \mathbf{i} & \mathbf{j} & \mathbf{k} \\ 4 & 0 & 6.283 \\ -4.35 & -1.60 & 2.77 \end{vmatrix}$$

$$= (10.05\mathbf{i}) + (10.05\mathbf{i} - 38.44\mathbf{j} - 6.40\mathbf{k})$$

② $$= 20.11\mathbf{i} - 38.44\mathbf{j} - 6.40\mathbf{k} \text{ m/s}^2 \qquad Ans.$$

The angular motion of OA depends only on the angular changes N and $\dot{\beta}$, so any linear motion of O does not affect $\boldsymbol{\omega}$ and $\boldsymbol{\alpha}$.

① Alternatively, consider axes x-y-z to be attached to the vertical shaft and clevis so that they rotate. The derivative of $\boldsymbol{\omega}_x$ becomes $\dot{\boldsymbol{\omega}}_x = 4\dot{\mathbf{i}}$. But from Eq. 5/11, we have $\dot{\mathbf{i}} = \boldsymbol{\omega}_z \times \mathbf{i} = 6.283\mathbf{k} \times \mathbf{i} = 6.283\mathbf{j}$. Thus, $\boldsymbol{\alpha} = \dot{\boldsymbol{\omega}}_x = 4(6.283)\mathbf{j} = 25.13\mathbf{j}$ rad/s^2 as before.

② To compare methods, it is suggested that these results for $\mathbf{v}$ and $\mathbf{a}$ be obtained by applying Eqs. 2/18 and 2/19 for particle motion in spherical coordinates, changing symbols as necessary.

Sample Problem 7/2

The electric motor with an attached disk is running at a constant low speed of 120 rev/min in the direction shown. Its housing and mounting base are initially at rest. The entire assembly is next set in rotation about the vertical Z-axis at the constant rate $N = 60$ rev/min with a fixed angle γ of 30°. Determine (a) the angular velocity and angular acceleration of the disk, (b) the space and body cones, and (c) the velocity and acceleration of point A at the top of the disk for the instant shown.

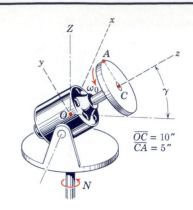

$\overline{OC} = 10''$
$\overline{CA} = 5''$

Solution. The axes x-y-z with unit vectors $\mathbf{i}, \mathbf{j}, \mathbf{k}$ are attached to the motor frame, with the z-axis coinciding with the rotor axis and the x-axis coinciding with the horizontal axis through O about which the motor tilts. The Z-axis is vertical and carries the unit vector $\mathbf{K} = \mathbf{j} \cos \gamma + \mathbf{k} \sin \gamma$.

(a) The rotor and disk have two components of angular velocity: $\omega_0 = 120(2\pi)/60 = 4\pi$ rad/sec about the z-axis and $\Omega = 60(2\pi)/60 = 2\pi$ rad/sec about the Z-axis. Thus, the angular velocity becomes

① $$\boldsymbol{\omega} = \boldsymbol{\omega}_0 + \boldsymbol{\Omega} = \omega_0 \mathbf{k} + \Omega \mathbf{K}$$
$$= \omega_0 \mathbf{k} + \Omega(\mathbf{j} \cos \gamma + \mathbf{k} \sin \gamma) = (\Omega \cos \gamma)\mathbf{j} + (\omega_0 + \Omega \sin \theta)\mathbf{k}$$
$$= (2\pi \cos 30°)\mathbf{j} + (4\pi + 2\pi \sin 30°)\mathbf{k} = \pi(\sqrt{3}\mathbf{j} + 5.0\mathbf{k}) \text{ rad/sec}$$
Ans.

The angular acceleration of the disk from Eq. 7/3 is

② $$\boldsymbol{\alpha} = \dot{\boldsymbol{\omega}} = \boldsymbol{\Omega} \times \boldsymbol{\omega}$$
$$= \Omega(\mathbf{j} \cos \gamma + \mathbf{k} \sin \gamma) \times [(\Omega \cos \gamma)\mathbf{j} + (\omega_0 + \Omega \sin \gamma)\mathbf{k}]$$
$$= \Omega(\omega_0 \cos \gamma + \Omega \sin \gamma \cos \gamma)\mathbf{i} - (\Omega^2 \sin \gamma \cos \gamma)\mathbf{i}$$
③ $$= (\Omega \omega_0 \cos \gamma)\mathbf{i} = \mathbf{i}(2\pi)(4\pi) \cos 30° = 68.4\mathbf{i} \text{ rad/sec}^2$$
Ans.

(b) The angular velocity vector $\boldsymbol{\omega}$ is the common element of the space and body cones that may now be constructed as shown.

(c) The position vector of point A for the instant considered is

$$\mathbf{r} = 5\mathbf{j} + 10\mathbf{k} \text{ in.}$$

From Eq. 7/1 the velocity of A is

$$\mathbf{v} = \boldsymbol{\omega} \times \mathbf{r} = \begin{vmatrix} \mathbf{i} & \mathbf{j} & \mathbf{k} \\ 0 & \sqrt{3}\pi & 5\pi \\ 0 & 5 & 10 \end{vmatrix} = -7.68\pi\mathbf{i} \text{ in./sec}$$
Ans.

From Eq. 7/2 the acceleration of point A is

$$\mathbf{a} = \dot{\boldsymbol{\omega}} \times \mathbf{r} + \boldsymbol{\omega} \times (\boldsymbol{\omega} \times \mathbf{r}) = \boldsymbol{\alpha} \times \mathbf{r} + \boldsymbol{\omega} \times \mathbf{v}$$
$$= 68.4\mathbf{i} \times (5\mathbf{j} + 10\mathbf{k}) + \pi(\sqrt{3}\mathbf{j} + 5\mathbf{k}) \times (-7.68\pi\mathbf{i})$$
$$= -1063\mathbf{j} + 473\mathbf{k} \text{ in./sec}^2$$
Ans.

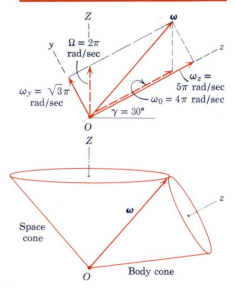

① Note that $\boldsymbol{\omega}_0 + \boldsymbol{\Omega} = \boldsymbol{\omega} = \boldsymbol{\omega}_y + \boldsymbol{\omega}_z$ as shown on the vector diagram.

② Remember that Eq. 7/3 gives the complete expression for $\boldsymbol{\alpha}$ only for steady precession where $|\boldsymbol{\omega}|$ is constant, which applies to this problem.

③ Since the magnitude of $\boldsymbol{\omega}$ is constant, $\boldsymbol{\alpha}$ must be tangent to the base circle of the space cone, which puts it in the plus x-direction in agreement with our calculated conclusion.

PROBLEMS

Introductory problems

7/1 Place your textbook on your desk, with fixed axes oriented as shown. Rotate the book about the *x*-axis through a 90° angle and then from this new position rotate it 90° about the *y*-axis. Sketch the final position of the book. Repeat the process but reverse the order of rotation. From your results, state your conclusion concerning the vector addition of finite rotations. Reconcile your observations with Fig. 7/4.

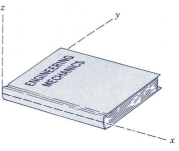

Problem 7/1

7/2 Repeat the experiment of Prob. 7/1 but use a small angle of rotation, say, 5°. Note the near-equal final positions for the two different rotation sequences. What does this observation lead you to conclude for the combination of infinitesimal rotations and for the time derivatives of angular quantities? Reconcile your observations with Fig. 7/5.

7/3 A timing mechanism consists of the rotating distributor arm *AB* and the fixed contact *C*. If the arm rotates about the fixed axis *OA* with a constant angular velocity $\boldsymbol{\omega} = 30(3\mathbf{i} + 2\mathbf{j} + 6\mathbf{k})$ rad/s, and if the coordinates of the contact *C* expressed in millimeters are (20, 30, 80), determine the magnitude of the acceleration of the tip *B* of the distributor arm as it passes point *C*.   *Ans.* $a = 1285$ m/s²

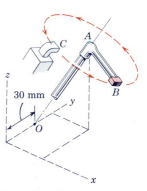

Problem 7/3

7/4 The disk rotates with a spin velocity of 15 rad/s about its horizontal *z*-axis first in the direction (*a*) and second in the direction (*b*). The assembly rotates with the velocity $N = 10$ rad/s about the vertical axis. Construct the space and body cones for each case.

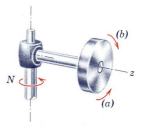

Problem 7/4

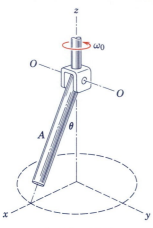

Problem 7/5

7/5 The rod is hinged about the axis *O-O* of the clevis, which is attached to the end of the vertical shaft. The shaft rotates with a constant angular velocity ω_0 as shown. If θ is decreasing at the constant rate $-\dot{\theta} = p$, write expressions for the angular velocity $\boldsymbol{\omega}$ and angular acceleration $\boldsymbol{\alpha}$ of the rod.

Ans. $\boldsymbol{\omega} = p\mathbf{j} + \omega_0\mathbf{k}, \ \boldsymbol{\alpha} = -p\omega_0\mathbf{i}$

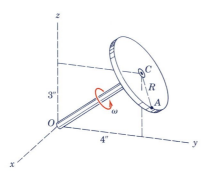

Problem 7/6

7/6 The circular disk is rotating about the fixed axis *OC* with a constant angular velocity $\omega = 20$ rad/sec. At a certain instant, point *A* on the rim passes a point whose *x-y-z* coordinates are 1.5, 4.75, and 2 inches, respectively. Calculate the magnitudes of the velocity $\mathbf{v}$ and acceleration $\mathbf{a}$ of point *A*. Also find the radius *R* of the disk.

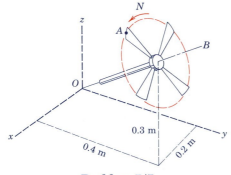

Problem 7/7

7/7 The four-bladed fan rotates about the fixed axis *OB* with a constant angular speed $N = 1200$ rev/min. Write the vector expressions for the velocity $\mathbf{v}$ and acceleration $\mathbf{a}$ of the tip *A* of the fan blade for the instant when its *x-y-z* coordinates are 0.260, 0.240, and 0.473 m, respectively.

Ans. $\mathbf{v} = 27.3\mathbf{i} - 3.87\mathbf{j} - 13.07\mathbf{k}$ m/s
$\mathbf{a} = -949\mathbf{i} + 2520\mathbf{j} - 2730\mathbf{k}$ m/s^2

7/8 A circular disk rotates about a fixed axis with a constant angular velocity $\omega = 10(\mathbf{i} + 2\mathbf{j} + 2\mathbf{k})$ rad/sec. At a certain instant, a point P on its rim has a velocity whose x- and y-components are 120 in./sec and -80 in./sec, respectively. Determine the magnitude v of the velocity of P and the radial distance R from P to the rotation axis. Also find the magnitude a of the acceleration of P.

7/9 A rigid body rotates about a fixed axis with a constant angular velocity $\omega = 2\mathbf{i} + 2\mathbf{j} - 4\mathbf{k}$ rad/s. If the x- and y-components of the acceleration of a point A on the body are 10 m/s^2 and 8 m/s^2, respectively, determine the magnitude a of the acceleration of point A and the radial distance R from A to the rotation axis.

> *Ans.* $a = 15.65$ m/s^2, $R = 0.652$ m

7/10 A rigid body rotates about a fixed axis with an angular acceleration $\alpha = 2\mathbf{i} + \mathbf{j} - 3\mathbf{k}$ rad/s^2. At a certain instant, the tangential acceleration of a point A on the body has y- and z-components equal to -0.5 m/s^2 and 0.4 m/s^2, respectively. Determine the magnitude of the tangential acceleration of A and the distance R from A to the rotation axis.

Representative problems

7/11 The motor of Sample Problem 7/2 is shown again here. If the motor pivots about the x-axis at the constant rate $\dot{\gamma} = 3\pi$ rad/sec with no rotation about the Z-axis ($N = 0$), determine the angular acceleration α of the rotor and disk as the position $\gamma = 30°$ is passed. The constant speed of the motor is 120 rev/min. Also find the velocity and acceleration of a point A, which is on the top of the disk for this position.   *Ans.* $\alpha = 12\pi^2\mathbf{j}$ rad/sec^2

$$\mathbf{v} = 5\pi(-4\mathbf{i} + 6\mathbf{j} - 3\mathbf{k}) \text{ in./sec}$$
$$\mathbf{a} = -5\pi^2(25\mathbf{j} + 18\mathbf{k}) \text{ in./sec}^2$$

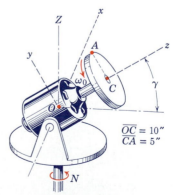

$\overline{OC} = 10''$
$\overline{CA} = 5''$

Problem 7/11

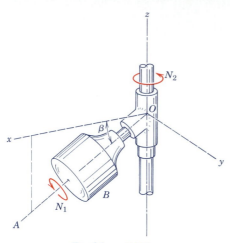

Problem 7/12

7/12 The rotor B spins about its inclined axis OA at the speed $N_1 = 200$ rev/min, where $\beta = 30°$. Simultaneously, the assembly rotates about the vertical z-axis at the rate N_2. If the total angular velocity of the rotor has a magnitude of 40 rad/s, determine N_2.

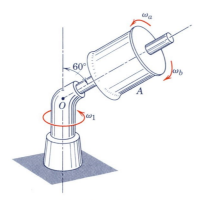

Problem 7/13

7/13 The spool A rotates about its axis with an angular velocity of 20 rad/s, first in the sense of ω_a and second in the sense of ω_b. Simultaneously, the assembly rotates about the vertical axis with an angular velocity $\omega_1 = 10$ rad/s. Determine the magnitude ω of the total angular velocity of the spool and construct the body and space cones for the spool for each case.
Ans. $\omega = 26.5$ rad/s, 17.32 rad/s

7/14 Solve Prob. 7/13 for the case of $\omega_1 = 5$ rad/s and $\omega_b = 20$ rad/s.

7/15 In manipulating the dumbbell, the jaws of the robotic device have an angular velocity $\omega_p = 2$ rad/s about the axis OG with γ fixed at 60°. The entire assembly rotates about the vertical Z-axis at the constant rate $\Omega = 0.8$ rad/s. Determine the angular velocity $\boldsymbol{\omega}$ and angular acceleration $\boldsymbol{\alpha}$ of the dumbbell. Express the results in terms of the given orientation of axes x-y-z, where the y-axis is parallel to the Y-axis.
Ans. $\boldsymbol{\omega} = -0.4\mathbf{i} + 2.69\mathbf{k}$ rad/s, $\boldsymbol{\alpha} = 0.8\mathbf{j}$ rad/s^2

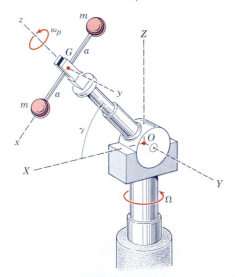

Problem 7/15

7/16 Determine the angular acceleration α of the dumb-bell of Prob. 7/15 for the conditions stated, except that Ω is increasing at the rate of 3 rad/s^2 for the instant under consideration.

7/17 The robot shown has five degrees of rotational freedom. The x-y-z axes are attached to the base ring, which rotates about the z-axis at the rate ω_1. The arm O_1O_2 rotates about the x-axis at the rate $\omega_2 = \dot{\theta}$. The control arm O_2A rotates about axis O_1-O_2 at the rate ω_3 and about a perpendicular axis through O_2 that is momentarily parallel to the x-axis at the rate $\omega_4 = \dot{\beta}$. Finally, the jaws rotate about axis O_2-A at the rate ω_5. The magnitudes of all angular rates are constant. For the configuration shown, determine the magnitude ω of the total angular velocity of the jaws for $\theta = 60°$ and $\beta = 45°$ if $\omega_1 = 2$ rad/s, $\dot{\theta} = 1.5$ rad/s, and $\omega_3 = \omega_4 = \omega_5 = 0$. Also express the angular acceleration α of arm O_1O_2 as a vector.
Ans. $\omega = 2.5$ rad/s, $\alpha = 3\mathbf{j}$ rad/s^2

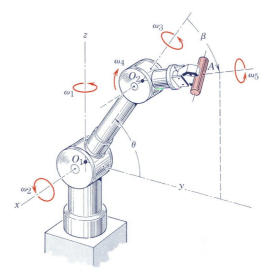

Problem 7/17

7/18 For the robot of Prob. 7/17 determine the angular velocity ω of the jaws for $\theta = 60°$ and $\beta = 45°$ if $\omega_3 = 3$ rad/s, $\omega_5 = 2$ rad/s, and $\omega_1 = \omega_2 = \omega_4 = 0$. Also find the angular acceleration α of the jaws.

7/19 For the robot of Prob. 7/17 determine the total angular velocity ω of the jaws for $\theta = 60°$ and $\beta = 45°$ if $\omega_1 = 2$ rad/s, $\omega_2 = 1.5$ rad/s, $\omega_4 = 3$ rad/s, and $\omega_3 = \omega_5 = 0$. Also find the angular acceleration α of the arm O_1O_2.
Ans. $\omega = -1.5\mathbf{i} + 2\mathbf{k}$ rad/s, $\alpha = 3\mathbf{j}$ rad/s^2

7/20 The electric motor and attached disk of Sample Problem 7/2 are shown again here. If the motor reaches a speed of 3000 rev/min in 2 seconds from rest with constant acceleration, determine the total angular acceleration of the rotor and disk $\frac{1}{3}$ second after it is turned on if the turntable is rotating at a constant rate $N = 30$ rev/min. The angle $\gamma = 30°$ is constant.

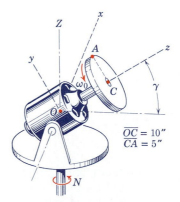

$\overline{OC} = 10''$
$\overline{CA} = 5''$

Problem 7/20

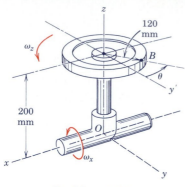

Problem 7/21

7/21 The circular disk of 120-mm radius rotates about the z-axis at the constant rate $\omega_z = 20$ rad/s, and the entire assembly rotates about the fixed x-axis at the constant rate $\omega_x = 10$ rad/s. Calculate the magnitudes of the velocity **v** and acceleration **a** of point B for the instant when $\theta = 30°$.

Ans. $v = 3.95$ m/s, $a = 72.2$ m/s^2

7/22 For the figure for Prob. 7/20, assume that the motor shaft and attached disk are rotating about the z-axis with a spin ω_0 of 120 rev/min which is increasing at the rate of 10 rad/sec^2 at the position $\gamma = 30°$. At the same time, $\dot{\gamma}$ is 12 rad/sec and is increasing at the rate of 15 rad/sec^2. Determine the vector expression for the angular acceleration α of the rotor at this instant. There is no rotation about the Z-axis.

7/23 The crane has a boom of length $OP = 24$ m and is revolving about the vertical axis at the constant rate of 2 rev/min in the direction shown. Simultaneously, the boom is being lowered at the constant rate $\dot{\beta} = 0.10$ rad/s. Calculate the magnitudes of the velocity and acceleration of the end P of the boom for the instant when it passes the position $\beta = 30°$.

Ans. $v = 3.48$ m/s, $a = 1.104$ m/s^2

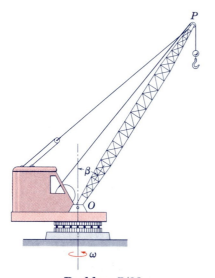

Problem 7/23

7/24 The vertical shaft and attached clevis rotate about the z-axis at the constant rate $\Omega = 4$ rad/s. Simultaneously, the shaft B revolves about its axis OA at the constant rate $\omega_0 = 3$ rad/s, and the angle γ is decreasing at the constant rate of $\pi/4$ rad/s. Determine the angular velocity $\boldsymbol{\omega}$ and the magnitude of the angular acceleration $\boldsymbol{\alpha}$ of shaft B when $\gamma = 30°$. The x-y-z axes are attached to the clevis and rotate with it.

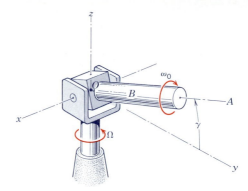

Problem 7/24

7/25 The wheel rolls without slipping in a circular arc of radius R and makes one complete turn about the vertical y-axis with constant speed in time τ. Determine the vector expression for the angular acceleration $\boldsymbol{\alpha}$ of the wheel and construct the space and body cones.

$$Ans. \ \boldsymbol{\alpha} = -\left(\frac{2\pi}{\tau}\right)^2 \frac{R}{r}\,\mathbf{i}$$

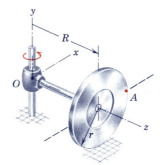

Problem 7/25

7/26 Determine expressions for the velocity **v** and acceleration **a** of point A on the wheel of Prob. 7/25 for the position shown, where A crosses the horizontal line through the center of the wheel.

▶**7/27** The right-circular cone A rolls on the fixed right-circular cone B at a constant rate and makes one complete trip around B every 4 s. Compute the magnitude of the angular acceleration $\boldsymbol{\alpha}$ of cone A during its motion. $Ans. \ \alpha = 6.32$ rad/s^2

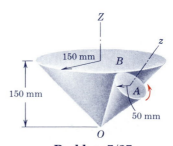

150 mm

150 mm

50 mm

Problem 7/27

▶ **7/28** For the robot of Prob. 7/17 determine the magnitude ω of the angular velocity of the jaws for $\theta = 60°$ and $\beta = 45°$ if $\omega_2 = 1.5$ rad/s, $\omega_3 = 2$ rad/s, $\omega_4 = 0.5$ rad/s, and $\omega_1 = \omega_5 = 0$. Also find the angular acceleration $\boldsymbol{\alpha}$ of the jaws.

Ans. $\omega = 2.24$ rad/s, $\boldsymbol{\alpha} = -3.46\mathbf{j} + 2\mathbf{k}$ rad/s^2

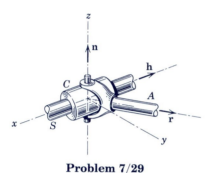

Problem 7/29

▶ **7/29** The end of link A is welded to the yoke that is pivoted about the z-axis to the collar C. The collar may rotate about the x-axis of the fixed shaft. Link A and its yoke can rotate about both the x- and z-axes but not about the y-axis. Regardless of the motion of the other end of link A, show that the angular velocity $\boldsymbol{\omega}$ of link A and its yoke must satisfy the relation $\boldsymbol{\omega} \cdot \mathbf{h} \times (\mathbf{r} \times \mathbf{h}) = 0$. Vectors $\mathbf{r}$ and $\mathbf{h}$ are any vectors directed, respectively, along the link and along the fixed shaft. The axis of the link A is normal to the yoke axis (z-direction) and, hence, lies in the x-y plane.

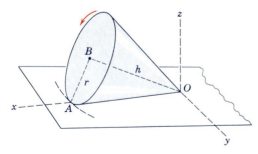

Problem 7/30

▶ **7/30** The solid right-circular cone of base radius r and height h rolls on a flat surface without slipping. The center B of the circular base moves in a circular path around the z-axis with a constant speed v. Determine the angular velocity $\boldsymbol{\omega}$ and the angular acceleration $\boldsymbol{\alpha}$ of the solid cone.

$$Ans. \ \boldsymbol{\omega} = v\sqrt{\frac{1}{r^2} + \frac{1}{h^2}}\ \mathbf{i}$$

$$\boldsymbol{\alpha} = -\frac{v^2}{h^2}\left(\frac{r}{h} + \frac{h}{r}\right)\mathbf{j}$$

7/6 GENERAL MOTION

The kinematic analysis of a rigid body that has general three-dimensional motion is best accomplished with the aid of our principles of relative motion. These principles have been applied to problems in plane motion and will now be extended to space motion. We will make use of both translating reference axes and rotating reference axes.

(a) Translating reference axes. In Fig. 7/10 is shown a rigid body which has an angular velocity $\boldsymbol{\omega}$. We choose any convenient point B as the origin of a translating reference system x-y-z. The velocity $\mathbf{v}$ and acceleration $\mathbf{a}$ of any other point A in the body are given by the relative-velocity and relative-acceleration expressions

$$\mathbf{v}_A = \mathbf{v}_B + \mathbf{v}_{A/B} \qquad [5/4]$$

$$\mathbf{a}_A = \mathbf{a}_B + \mathbf{a}_{A/B} \qquad [5/7]$$

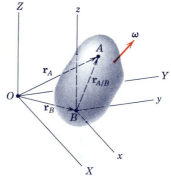

Figure 7/10

which were developed in Arts. 5/4 and 5/6 for the plane motion of rigid bodies and which hold equally well for any plane in space in which the three vectors for each of the equations must lie.

In applying these relations to rigid-body motion in space, we note from Fig. 7/10 that the distance $\overline{AB}$ remains constant. Thus, from an observer's position on x-y-z, the body appears to rotate about the point B and point A appears to lie on a spherical surface with B as the center. Consequently, we may view the general motion as a translation of the body with the motion of B plus a rotation of the body about B.

The relative motion terms represent the effect of the rotation about B and are identical to the velocity and acceleration expressions discussed in the previous article for rotation of a rigid body about a fixed point. Therefore, the relative-velocity and relative-acceleration equations may be written

$$\boxed{\begin{aligned} \mathbf{v}_A &= \mathbf{v}_B + \boldsymbol{\omega} \times \mathbf{r}_{A/B} \\ \mathbf{a}_A &= \mathbf{a}_B + \dot{\boldsymbol{\omega}} \times \mathbf{r}_{A/B} + \boldsymbol{\omega} \times (\boldsymbol{\omega} \times \mathbf{r}_{A/B}) \end{aligned}} \qquad (7/4)$$

where $\boldsymbol{\omega}$ is the instantaneous angular velocity of the body.

The selection of the reference point B is quite arbitrary in theory. In practice, point B is chosen for convenience as some point in the body whose motion is known in whole or in part. If point A is chosen as the reference point, the relative motion equations become

$$\mathbf{v}_B = \mathbf{v}_A + \boldsymbol{\omega} \times \mathbf{r}_{B/A}$$

$$\mathbf{a}_B = \mathbf{a}_A + \dot{\boldsymbol{\omega}} \times \mathbf{r}_{B/A} + \boldsymbol{\omega} \times (\boldsymbol{\omega} \times \mathbf{r}_{B/A})$$

where $\mathbf{r}_{B/A} = -\mathbf{r}_{A/B}$. It should be clear that $\boldsymbol{\omega}$ and, hence, $\dot{\boldsymbol{\omega}}$ are the same vectors for either formulation since the absolute angular motion of the body is independent of the choice of reference point. When we come to the kinetic equations for general motion, we will see that the mass center of a body is frequently the most convenient reference point to choose.

If points A and B in Fig. 7/10 represent the ends of a rigid control link in a spatial mechanism where the end connections act as ball-and-socket joints (as in Sample Problem 7/3), it is necessary to impose certain kinematical requirements. Clearly, any rotation of the link about its own axis AB does not affect the action of the link. Thus, the angular velocity $\boldsymbol{\omega}_n$ whose vector is normal to the link describes its action. It is necessary, therefore, that $\boldsymbol{\omega}_n$ and $\mathbf{r}_{A/B}$ be at right angles, and this condition is satisfied if $\boldsymbol{\omega}_n \cdot \mathbf{r}_{A/B} = 0$. Similarly, it is only the component $\boldsymbol{\alpha}_n$* of the angular acceleration of the link normal to AB that affects its action, so that $\boldsymbol{\alpha}_n \cdot \mathbf{r}_{A/B} = 0$ must also hold.

(b) Rotating reference axes. A more general formulation of the motion of a rigid body in space calls for the use of reference axes which rotate as well as translate. The description of Fig. 7/10 is modified in Fig. 7/11 to show reference axes whose origin is attached to the reference point B as before, but which rotate with an absolute angular velocity $\boldsymbol{\Omega}$ that may be different from the absolute angular velocity $\boldsymbol{\omega}$ of the body.

We now make use of Eqs. 5/11, 5/12, 5/13, and 5/14 developed in Art. 5/7 for describing the plane motion of a rigid body with the use of rotating axes. The extension of these relations from two to three dimensions is easily accomplished by merely including the z-component of the vectors, and this step is left to the student to carry out. Replacing $\boldsymbol{\omega}$ in these equations by the angular velocity $\boldsymbol{\Omega}$ of our rotating x-y-z axes gives us

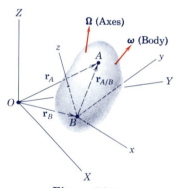

Figure 7/11

$$\dot{\mathbf{i}} = \boldsymbol{\Omega} \times \mathbf{i} \qquad \dot{\mathbf{j}} = \boldsymbol{\Omega} \times \mathbf{j} \qquad \dot{\mathbf{k}} = \boldsymbol{\Omega} \times \mathbf{k} \qquad (7/5)$$

for the time derivatives of the rotating unit vectors attached to x-y-z. The expressions for the velocity and acceleration of point A become

$$\mathbf{v}_A = \mathbf{v}_B + \boldsymbol{\Omega} \times \mathbf{r}_{A/B} + \mathbf{v}_{\text{rel}}$$
$$\mathbf{a}_A = \mathbf{a}_B + \dot{\boldsymbol{\Omega}} \times \mathbf{r}_{A/B} + \boldsymbol{\Omega} \times (\boldsymbol{\Omega} \times \mathbf{r}_{A/B}) + 2\boldsymbol{\Omega} \times \mathbf{v}_{\text{rel}} + \mathbf{a}_{\text{rel}} \qquad (7/6)$$

---

*It may be shown that $\boldsymbol{\alpha}_n = \dot{\boldsymbol{\omega}}_n$ if the angular velocity of the link about its own axis is not changing. See the senior author's *Dynamics, 2nd Edition SI Version*, 1975, John Wiley & Sons, Art. 37.

where $\mathbf{v}_{rel} = \dot{x}\mathbf{i} + \dot{y}\mathbf{j} + \dot{z}\mathbf{k}$ and $\mathbf{a}_{rel} = \ddot{x}\mathbf{i} + \ddot{y}\mathbf{j} + \ddot{z}\mathbf{k}$ are, respectively, the velocity and acceleration of point A measured relative to x-y-z by an observer attached to x-y-z. Again, we note that $\boldsymbol{\Omega}$ is the angular velocity of the axes and may be different from the angular velocity $\boldsymbol{\omega}$ of the body. Also we note that $\mathbf{r}_{A/B}$ remains constant in magnitude for a rigid body, but it will change direction with respect to x-y-z when the angular velocity $\boldsymbol{\Omega}$ of the axes is different from the angular velocity $\boldsymbol{\omega}$ of the body. We observe further that, if x-y-z are rigidly attached to the body, $\boldsymbol{\Omega} = \boldsymbol{\omega}$ and $\mathbf{v}_{rel}$ and $\mathbf{a}_{rel}$ are both zero, which makes the equations identical to Eqs. 7/4.

In Art. 5/7 we also developed the relationship (Eq. 5/13) between the time derivative of a vector $\mathbf{V}$ as measured in the fixed X-Y system and the time derivative of $\mathbf{V}$ as measured relative to the rotating x-y system. For our three-dimensional case, this relation becomes

$$\left(\frac{d\mathbf{V}}{dt}\right)_{XYZ} = \left(\frac{d\mathbf{V}}{dt}\right)_{xyz} + \boldsymbol{\Omega} \times \mathbf{V} \qquad (7/7)$$

When we apply this transformation to the relative position vector $\mathbf{r}_{A/B} = \mathbf{r}_A - \mathbf{r}_B$ for our rigid body of Fig. 7/11, we get

$$\left(\frac{d\mathbf{r}_A}{dt}\right)_{XYZ} = \left(\frac{d\mathbf{r}_B}{dt}\right)_{XYZ} + \left(\frac{d\mathbf{r}_{A/B}}{dt}\right)_{xyz} + \boldsymbol{\Omega} \times \mathbf{r}_{A/B}$$

or

$$\mathbf{v}_A = \mathbf{v}_B + \mathbf{v}_{rel} + \boldsymbol{\Omega} \times \mathbf{r}_{A/B}$$

which gives us the first of Eqs. 7/6.

Equations 7/6 are particularly useful when the reference axes are attached to a moving body within which relative motion occurs.

Equation 7/7 may be recast as the vector operator

$$\left(\frac{d[\ \ ]}{dt}\right)_{XYZ} = \left(\frac{d[\ \ ]}{dt}\right)_{xyz} + \boldsymbol{\Omega} \times [\ \ ] \qquad (7/7a)$$

where [ ] stands for any vector $\mathbf{V}$ expressible both in X-Y-Z and in x-y-z. If we apply the operator to itself, we obtain the second time derivative which becomes

$$\left(\frac{d^2[\ \ ]}{dt^2}\right)_{XYZ} = \left(\frac{d^2[\ \ ]}{dt^2}\right)_{xyz} + \dot{\boldsymbol{\Omega}} \times [\ \ ] + \boldsymbol{\Omega} \times (\boldsymbol{\Omega} \times [\ \ ])$$

$$+ \, 2\boldsymbol{\Omega} \times \left(\frac{d[\ \ ]}{dt}\right)_{xyz} \qquad (7/7b)$$

This exercise is left to the student, and the form of Eq. 7/7b is observed to be the same as that of the second of Eqs. 7/6 expressed for $\mathbf{a}_{A/B} = \mathbf{a}_A - \mathbf{a}_B$.

Sample Problem 7/3

Crank CB rotates about the horizontal axis with an angular velocity $\omega_1 = 6$ rad/s which is constant for a short interval of motion that includes the position shown. The link AB has a ball-and-socket fitting on each end and connects crank DA with CB. For the instant shown, determine the angular velocity ω_2 of crank DA and the angular velocity ω_n of link AB.

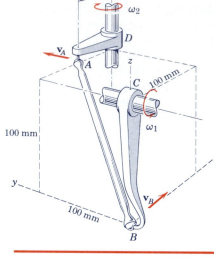

Solution. The relative-velocity relation, Eq. 7/4, will be solved ① first using translating reference axes attached to B. The equation is

$$\mathbf{v}_A = \mathbf{v}_B + \boldsymbol{\omega}_n \times \mathbf{r}_{A/B}$$

② where $\boldsymbol{\omega}_n$ is the angular velocity of link AB taken normal to AB. The velocities of A and B are

$$[v = r\omega] \qquad \mathbf{v}_A = 50\omega_2\mathbf{j} \qquad \mathbf{v}_B = 100(6)\mathbf{i} = 600\mathbf{i} \text{ mm/s}$$

Also $\mathbf{r}_{A/B} = 50\mathbf{i} + 100\mathbf{j} + 100\mathbf{k}$ mm. Substitution into the velocity relation gives

$$50\omega_2\mathbf{j} = 600\mathbf{i} + \begin{vmatrix} \mathbf{i} & \mathbf{j} & \mathbf{k} \\ \omega_{nx} & \omega_{ny} & \omega_{nz} \\ 50 & 100 & 100 \end{vmatrix}$$

Expanding the determinant and equating the coefficients of the $\mathbf{i}, \mathbf{j}, \mathbf{k}$ terms give

$$-6 = + \omega_{ny} - \omega_{nz}$$
$$\omega_2 = -2\omega_{nx} + \omega_{nz}$$
$$0 = 2\omega_{nx} - \omega_{ny}$$

These equations may be solved for ω_2, which becomes

$$\omega_2 = 6 \text{ rad/s} \qquad\qquad Ans.$$

As they stand, the three equations incorporate the fact that $\boldsymbol{\omega}_n$ is normal to $\mathbf{v}_{A/B}$, but they cannot be solved until the requirement that $\boldsymbol{\omega}_n$ be nor-
③ mal to $\mathbf{r}_{A/B}$ is included. Thus,

$$[\boldsymbol{\omega}_n \cdot \mathbf{r}_{A/B} = 0] \qquad 50\omega_{nx} + 100\omega_{ny} + 100\omega_{nz} = 0$$

Combination with two of the three previous equations yields the solutions

$$\omega_{nx} = -\tfrac{4}{3} \text{ rad/s} \qquad \omega_{ny} = -\tfrac{8}{3} \text{ rad/s} \qquad \omega_{nz} = \tfrac{10}{3} \text{ rad/s}$$

Thus,

$$\boldsymbol{\omega}_n = \tfrac{2}{3}(-2\mathbf{i} - 4\mathbf{j} + 5\mathbf{k}) \text{ rad/s}$$

with

$$\omega_n = \tfrac{2}{3}\sqrt{2^2 + 4^2 + 5^2} = 2\sqrt{5} \text{ rad/s} \qquad\qquad Ans.$$

① We select B as the reference point since its motion can easily be determined from the given angular velocity ω_1 of CB.

② The angular velocity $\boldsymbol{\omega}$ of AB is taken as a vector $\boldsymbol{\omega}_n$ normal to AB since any rotation of the link about its own axis AB has no influence on the behavior of the linkage.

③ The relative-velocity equation may be written as $\mathbf{v}_A - \mathbf{v}_B = \mathbf{v}_{A/B} = \boldsymbol{\omega}_n \times \mathbf{r}_{A/B}$, which requires that $\mathbf{v}_{A/B}$ be perpendicular to both $\boldsymbol{\omega}_n$ and $\mathbf{r}_{A/B}$. This equation alone does not incorporate the additional requirement that $\boldsymbol{\omega}_n$ be perpendicular to $\mathbf{r}_{A/B}$. Thus, we must also satisfy $\boldsymbol{\omega}_n \cdot \mathbf{r}_{A/B} = 0$.

Sample Problem 7/4

Determine the angular acceleration $\dot{\omega}_2$ of crank AD in Sample Problem 7/3 for the conditions cited. Also find the angular acceleration $\dot{\omega}_n$ of link AB.

Solution. The accelerations of the links may be found from the second of Eqs. 7/4, which may be written

$$\mathbf{a}_A = \mathbf{a}_B + \dot{\boldsymbol{\omega}}_n \times \mathbf{r}_{A/B} + \boldsymbol{\omega}_n \times (\boldsymbol{\omega}_n \times \mathbf{r}_{A/B})$$

where $\boldsymbol{\omega}_n$, as in Sample Problem 7/3, is the angular velocity of AB taken
① normal to AB. The angular acceleration of AB is written as $\dot{\boldsymbol{\omega}}_n$.

In terms of their normal and tangential components, the accelerations of A and B are

$$\mathbf{a}_A = 50\omega_2{}^2\mathbf{i} + 50\dot{\omega}_2\mathbf{j} = 1800\mathbf{i} + 50\dot{\omega}_2\mathbf{j} \text{ mm/s}^2$$

$$\mathbf{a}_B = 100\omega_1{}^2\mathbf{k} + (0)\mathbf{i} = 3600\mathbf{k} \text{ mm/s}^2$$

Also

$$\boldsymbol{\omega}_n \times (\boldsymbol{\omega}_n \times \mathbf{r}_{A/B}) = -\omega_n{}^2\mathbf{r}_{A/B} = -20(50\mathbf{i} + 100\mathbf{j} + 100\mathbf{k}) \text{ mm/s}^2$$

$$\dot{\boldsymbol{\omega}}_n \times \mathbf{r}_{A/B} = (100\dot{\omega}_{ny} - 100\dot{\omega}_{nz})\mathbf{i}$$
$$+ (50\dot{\omega}_{nz} - 100\dot{\omega}_{nx})\mathbf{j} + (100\dot{\omega}_{nx} - 50\dot{\omega}_{ny})\mathbf{k}$$

Substitution into the relative acceleration equation and equating respective coefficients of $\mathbf{i}, \mathbf{j}, \mathbf{k}$ give

$$28 = \dot{\omega}_{ny} - \dot{\omega}_{nz}$$

$$\dot{\omega}_2 + 40 = -2\dot{\omega}_{nx} + \dot{\omega}_{nz}$$

$$-32 = 2\dot{\omega}_{nx} - \dot{\omega}_{ny}$$

Solution of these equations for $\dot{\omega}_2$ gives

$$\dot{\omega}_2 = -36 \text{ rad/s}^2 \qquad \textit{Ans.}$$

② The vector $\dot{\boldsymbol{\omega}}_n$ is normal to $\mathbf{r}_{A/B}$ but is not normal to $\mathbf{v}_{A/B}$, as was the case with $\boldsymbol{\omega}_n$.

$$[\dot{\boldsymbol{\omega}}_n \cdot \mathbf{r}_{A/B} = 0] \qquad 2\dot{\omega}_{nx} + 4\dot{\omega}_{ny} + 4\dot{\omega}_{nz} = 0$$

which, when combined with the preceding relations for these same quantities, gives

$$\dot{\omega}_{nx} = -8 \text{ rad/s}^2 \qquad \dot{\omega}_{ny} = 16 \text{ rad/s}^2 \qquad \dot{\omega}_{nz} = -12 \text{ rad/s}^2$$

Thus,

$$\dot{\boldsymbol{\omega}}_n = 4(-2\mathbf{i} + 4\mathbf{j} - 3\mathbf{k}) \text{ rad/s}^2 \qquad \textit{Ans.}$$

and

$$|\dot{\boldsymbol{\omega}}_n| = 4\sqrt{2^2 + 4^2 + 3^2} = 4\sqrt{29} \text{ rad/s}^2 \qquad \textit{Ans.}$$

① If the link AB had an angular velocity component along AB, then a change in both magnitude and direction of this component could occur that would contribute to the actual angular acceleration of the link as a rigid body. However, since any rotation about its own axis AB has no influence on the motion of the cranks at C and D, we shall concern ourselves only with $\dot{\boldsymbol{\omega}}_n$.

② The component of $\dot{\boldsymbol{\omega}}_n$ which is not normal to $\mathbf{v}_{A/B}$ gives rise to the change in direction of $\mathbf{v}_{A/B}$.

Sample Problem 7/5

The motor housing and its bracket rotate about the Z-axis at the constant rate $\Omega = 3$ rad/s. The motor shaft and disk have a constant angular velocity of spin $p = 8$ rad/s with respect to the motor housing in the direction shown. If γ is constant at 30°, determine the velocity and acceleration of point A at the top of the disk and the angular acceleration α of the disk.

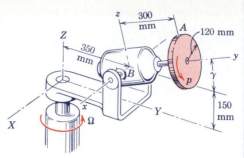

Solution. The rotating reference axes x-y-z are attached to the motor housing, and the rotating base for the motor has the momentary orientation shown with respect to the fixed axes X-Y-Z. We will use both X-Y-Z components with unit vectors $\mathbf{I}$, $\mathbf{J}$, $\mathbf{K}$ and x-y-z components with unit vectors $\mathbf{i}$, $\mathbf{j}$, $\mathbf{k}$. The angular velocity of the x-y-z axes becomes
$\Omega = \Omega\mathbf{K} = 3\mathbf{K}$ rad/s.

① This choice for the reference axes provides a simple description for the motion of the disk relative to these axes.

Velocity. The velocity of A is given by the first of Eqs. 7/6
$$\mathbf{v}_A = \mathbf{v}_B + \Omega \times \mathbf{r}_{A/B} + \mathbf{v}_{\text{rel}}$$
where
$$\mathbf{v}_B = \Omega \times \mathbf{r}_B = 3\mathbf{K} \times 0.350\mathbf{J} = -1.05\mathbf{I} = -1.05\mathbf{i} \text{ m/s}$$
$$\Omega \times \mathbf{r}_{A/B} = 3\mathbf{K} \times (0.300\mathbf{j} + 0.120\mathbf{k})$$
$$= (-0.9 \cos 30°)\mathbf{i} + (0.36 \sin 30°)\mathbf{i} = -0.599\mathbf{i} \text{ m/s}$$
$$\mathbf{v}_{\text{rel}} = \mathbf{p} \times \mathbf{r}_{A/B} = 8\mathbf{j} \times (0.300\mathbf{j} + 0.120\mathbf{k}) = 0.960\mathbf{i} \text{ m/s}$$
Thus,
$$\mathbf{v}_A = -1.05\mathbf{i} - 0.599\mathbf{i} + 0.960\mathbf{i} = -0.689\mathbf{i} \text{ m/s} \qquad Ans.$$

Acceleration. The acceleration of A is given by the second of Eqs. 7/6
$$\mathbf{a}_A = \mathbf{a}_B + \dot{\Omega} \times \mathbf{r}_{A/B} + \Omega \times (\Omega \times \mathbf{r}_{A/B}) + 2\Omega \times \mathbf{v}_{\text{rel}} + \mathbf{a}_{\text{rel}}$$
where
$$\mathbf{a}_B = \Omega \times (\Omega \times \mathbf{r}_B) = 3\mathbf{K} \times (3\mathbf{K} \times 0.350\mathbf{J}) = -3.15\mathbf{J}$$
$$= 3.15(-\mathbf{j} \cos 30° + \mathbf{k} \sin 30°) = -2.73\mathbf{j} + 1.58\mathbf{k} \text{ m/s}^2$$
$$\dot{\Omega} = 0$$
$$\Omega \times (\Omega \times \mathbf{r}_{A/B}) = 3\mathbf{K} \times [3\mathbf{K} \times (0.300\mathbf{j} + 0.120\mathbf{k})]$$
$$= 3\mathbf{K} \times (-0.599\mathbf{i}) = -1.557\mathbf{j} + 0.899\mathbf{k} \text{ m/s}^2$$
$$2\Omega \times \mathbf{v}_{\text{rel}} = 2(3\mathbf{K}) \times 0.960\mathbf{i} = 5.76\mathbf{J}$$
$$= 5.76(\mathbf{j} \cos 30° - \mathbf{k} \sin 30°) = 4.99\mathbf{j} - 2.88\mathbf{k} \text{ m/s}^2$$
$$\mathbf{a}_{\text{rel}} = \mathbf{p} \times (\mathbf{p} \times \mathbf{r}_{A/B}) = 8\mathbf{j} \times [8\mathbf{j} \times (0.300\mathbf{j} + 0.120\mathbf{k})]$$
$$= -7.68\mathbf{k} \text{ m/s}^2$$

② Note that $\mathbf{K} \times \mathbf{i} = \mathbf{J} = \mathbf{j} \cos \gamma - \mathbf{k} \sin \gamma$, $\mathbf{K} \times \mathbf{j} = -\mathbf{i} \cos \gamma$, and $\mathbf{K} \times \mathbf{k} = \mathbf{i} \sin \gamma$.

Substitution into the expression for $\mathbf{a}_A$ and collecting terms give us
$$\mathbf{a}_A = 0.703\mathbf{j} - 8.086\mathbf{k} \text{ m/s}^2$$
and
$$a_A = \sqrt{(0.703)^2 + (8.086)^2} = 8.12 \text{ m/s}^2 \qquad Ans.$$

Angular acceleration. Since the precession is steady, we may use Eq. 7/3 to give us
$$\alpha = \dot{\omega} = \Omega \times \omega = 3\mathbf{K} \times (3\mathbf{K} + 8\mathbf{j})$$
$$= 0 + (-24 \cos 30°)\mathbf{i} = -20.8\mathbf{i} \text{ rad/s}^2 \qquad Ans.$$

PROBLEMS

Introductory problems

7/31 The helicopter is nosing over at the constant rate q rad/s. If the rotor blades revolve at the constant speed p rad/s, write the expression for the angular acceleration $\boldsymbol{\alpha}$ of the rotor. Take the y-axis to be attached to the fuselage and pointing forward perpendicular to the rotor axis.   *Ans.* $\boldsymbol{\alpha} = pq\mathbf{j}$

Problem 7/31

7/32 An unmanned radar-radio controlled aircraft with tilt-rotor propulsion is being designed for reconnaissance purposes. Vertical rise begins with $\theta = 0$ and is followed by horizontal flight as θ approaches 90°. If the rotors turn at a constant speed N of 360 rev/min, determine the angular acceleration $\boldsymbol{\alpha}$ of rotor A for $\theta = 30°$ if $\dot{\theta}$ is constant at 0.2 rad/s.

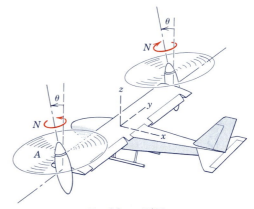

Problem 7/32

7/33 The collar at O and attached shaft OC rotate about the fixed x_0-axis at the constant rate $\Omega = 4$ rad/s. Simultaneously, the circular disk rotates about OC at the constant rate $p = 10$ rad/s. Determine the magnitude of the total angular velocity $\boldsymbol{\omega}$ of the disk and find its angular acceleration $\boldsymbol{\alpha}$.   *Ans.* $\omega = 10.77$ rad/s, $\boldsymbol{\alpha} = -40\mathbf{j}$ rad/s^2

7/34 If the angular rate p of the disk in Prob. 7/33 is increasing at the rate of 6 rad/s per second and if Ω remains constant at 4 rad/s, determine the angular acceleration $\boldsymbol{\alpha}$ of the disk at the instant when p reaches 10 rad/s.

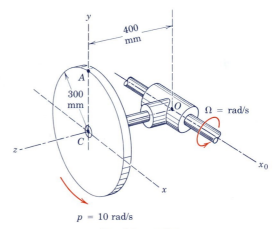

Problem 7/33

7/35 For the conditions of Prob. 7/33, determine the velocity $\mathbf{v}_A$ and acceleration $\mathbf{a}_A$ of point A on the disk as it passes the position shown. Reference axes x-y-z are attached to the collar at O and its shaft OC.
    Ans. $\mathbf{v}_A = -3\mathbf{i} - 1.6\mathbf{j} + 1.2\mathbf{k}$ m/s
        $\mathbf{a}_A = -34.8\mathbf{j} - 6.4\mathbf{k}$ m/s^2

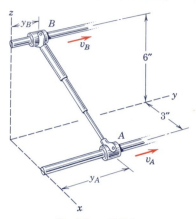

Problem 7/36

7/36 The collars at the ends of the telescoping link AB slide along the fixed shafts shown. During an interval of motion, $v_A = 5$ in./sec and $v_B = 2$ in./sec. Determine the vector expression for the angular velocity $\boldsymbol{\omega}_n$ of the centerline of the link for the position where $y_A = 4$ in. and $y_B = 2$ in.

Representative problems

7/37 The spacecraft is revolving about its z-axis, which has a fixed space orientation, at the constant rate $p = \frac{1}{10}$ rad/s. Simultaneously its solar panels are unfolding at the rate $\dot{\beta}$ which is programmed to produce the variation with β shown in the graph. Determine the angular acceleration $\boldsymbol{\alpha}$ of panel A an instant (a) before and an instant (b) after it reaches the position $\beta = 18°$.

$$Ans. \quad (a) \; \boldsymbol{\alpha} = -(3.88\mathbf{i} + 3.49\mathbf{j})10^{-3} \text{ rad/s}^2$$
$$(b) \; \boldsymbol{\alpha} = -3.49(10^{-3})\mathbf{j} \text{ rad/s}^2$$

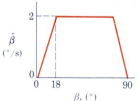

Problem 7/37

7/38 The collar and clevis A are given a constant upward velocity of 8 in./sec for an interval of motion and cause the ball end of the bar to slide in the radial slot in the rotating disk. Determine the angular acceleration of the bar when the bar passes the position for which $z = 3$ in. The disk turns at the constant rate of 2 rad/sec.

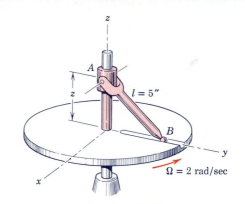

Problem 7/38

7/39 The circular disk of 100-mm radius rotates about its z-axis at the constant speed $p = 240$ rev/min, and arm OCB rotates about the Y-axis at the constant speed $N = 30$ rev/min. Determine the velocity $\mathbf{v}$ and acceleration $\mathbf{a}$ of point A on the disk as it passes the position shown. Use reference axes x-y-z attached to the arm OCB.

Ans. $\mathbf{v} = \pi(0.1\mathbf{i} + 0.8\mathbf{j} + 0.08\mathbf{k})$ m/s
$\mathbf{a} = -\pi^2(6.32\mathbf{i} + 0.1\mathbf{k})$ m/s^2

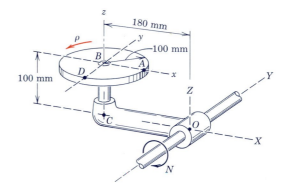

Problem 7/39

7/40 Solve Prob. 7/39 by attaching the reference axes x-y-z to the rotating disk.

7/41 The robot of Prob. 7/17 is shown again here, where the coordinate system x-y-z with origin at O_2 rotates about the X-axis at the rate $\dot{\theta}$. Nonrotating axes X-Y-Z oriented as shown have their origin at O_1. If $\omega_2 = \dot{\theta} = 3$ rad/s constant, $\omega_3 = 1.5$ rad/s constant, $\omega_1 = \omega_5 = 0$, $\overline{O_1O_2} = 1.2$ m, and $\overline{O_2A} = 0.6$ m, determine the velocity of the center A of the jaws for the instant when $\theta = 60°$. The angle β lies in the y-z plane and is constant at $45°$.

Ans. $\mathbf{v}_A = -0.636\mathbf{i} - 4.873\mathbf{j} + 1.273\mathbf{k}$ m/s

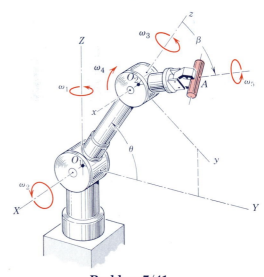

Problem 7/41

Problem 7/42

7/42 The center O of the spacecraft is moving through space with a constant velocity. During the period of motion prior to stabilization, the spacecraft has a constant rotational rate $\Omega = \frac{1}{2}$ rad/sec about its z-axis. The x-y-z axes are attached to the body of the craft, and the solar panels rotate about the y-axis at the constant rate $\dot{\theta} = \frac{1}{4}$ rad/sec with respect to the spacecraft. If $\boldsymbol{\omega}$ is the absolute angular velocity of the solar panels, determine $\dot{\boldsymbol{\omega}}$. Also find the acceleration of point A when $\theta = 30°$.

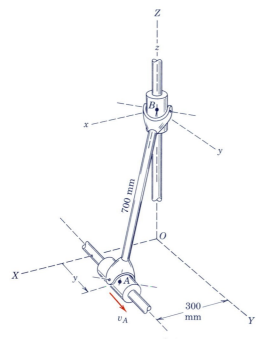

Problem 7/43

7/43 For a short interval of motion, collar A moves along its fixed shaft with a velocity $v_A = 2$ m/s in the Y-direction. Collar B, in turn, slides along its fixed vertical shaft. Link AB is 700 mm in length and can turn within the clevis at A to allow for the angular change between the clevises. For the instant when A passes the position where $y = 200$ mm, determine the velocity of collar B using nonrotating axes attached to B and find the component $\boldsymbol{\omega}_n$, normal to AB, of the angular velocity of the link. Also solve for $\mathbf{v}_B$ by differentiating the appropriate $x^2 + y^2 + z^2 = l^2$ relation.

$$Ans. \quad \mathbf{v}_B = -\tfrac{2}{3}\mathbf{k} \text{ m/s}$$
$$\boldsymbol{\omega}_n = \tfrac{10}{49}(\tfrac{40}{3}\mathbf{i} - 2\mathbf{j} + 6\mathbf{k}) \text{ rad/s}$$

7/44 Link AB is secured to the rotating arm OA and to the slider at B by a ball-and-socket joint at each end. Arm OA is confined to rotate about the fixed vertical shaft, and the slider is confined to move along the fixed rectangular bar. If the velocity of the slider is $v_B = 0.5$ m/s when $s = 50$ mm with OA in the position shown, calculate the corresponding angular velocity $\boldsymbol{\omega}_n$ of link AB. Use the nonrotating axes x-y-z attached to B.

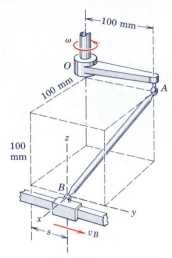

Problem 7/44

7/45 The thin circular disk of mass m and radius r is rotating about its z-axis with a constant angular velocity p, and the yoke in which it is mounted rotates about the X-axis through OB with a constant angular velocity ω_1. Simultaneously, the entire assembly rotates about the fixed $\overline{Y}$-axis through O with a constant angular velocity ω_2. Determine the velocity $\mathbf{v}$ and acceleration $\mathbf{a}$ of point A on the rim of the disk as it passes the position shown where the x-y plane of the disk coincides with the X-Y plane. The x-y-z axes are attached to the yoke.

Ans. $\mathbf{v} = -rp\mathbf{i} - (r\omega_1 + b\omega_2)\mathbf{k}$
$\mathbf{a} = -\omega_2(b\omega_2 + 2r\omega_1)\mathbf{i}$
$\quad - r(\omega_1{}^2 + p^2)\mathbf{j} + 2rp\omega_2\mathbf{k}$

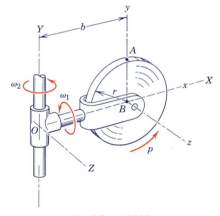

Problem 7/45

7/46 For the robot shown with Prob. 7/41, determine the velocity of the center A of the jaws for the following conditions: $\omega_1 = 2$ rad/s constant, $\omega_2 = \dot{\theta} = -3$ rad/s constant, $\omega_4 = \dot{\beta} = 1.5$ rad/s constant, $\omega_3 = 0$, $\overline{O_1O_2} = 1.2$ m, $\overline{O_2A} = 0.6$ m, $\theta = 60°$, and $\beta = 45°$.

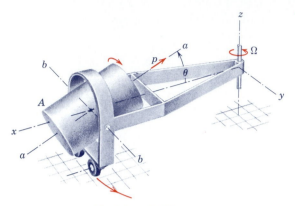

Problem 7/47

7/47 The test chamber of a flight simulator consists of a drum that spins about the *a-a* axis at a constant angular rate p relative to the cylindrical housing A. The housing, in turn, is mounted on transverse horizontal bearings and rotates about the axis *b-b* at the constant rate $\dot\theta$. The entire assembly is made to rotate about the fixed vertical *z*-axis at the constant rate Ω. Write the expression for the angular acceleration $\boldsymbol\alpha$ of the spinning drum during the compounded motion for a given value of θ.

Ans. $\boldsymbol\alpha = \dot\theta(p \sin\theta - \Omega)\mathbf{i}$
$- (p\Omega \cos\theta)\mathbf{j} + (p\dot\theta \cos\theta)\mathbf{k}$

Problem 7/48

7/48 A simulator for perfecting the docking procedure for spacecraft consists of the frame A that is mounted on four air-bearing pads so that it can translate and rotate freely on the horizontal surface. Mounted in the frame is a drum B that can rotate about the horizontal axis of the frame A. The coordinate axes *x-y-z* are attached to the drum B, and the *z*-axis of the drum makes an angle β with the horizontal. Inside the drum is the simulated command module C that can rotate within the drum about its *z*-axis at a rate p. For a certain test run, frame A is rotating on the horizontal surface in the direction shown with a constant angular velocity of 0.2 rad/s. Simultaneously, the drum B is rotating about the *x*-axis at the constant rate $\dot\beta = 0.15$ rad/s, and the module C is turning inside the drum at the constant rate $p = 0.9$ rad/s in the direction indicated. For these conditions, determine the angular velocity $\boldsymbol\omega$ and the angular acceleration $\boldsymbol\alpha$ of the simulator C as it passes the position $\beta = 0$.

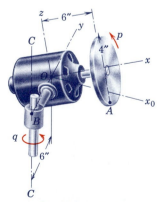

Problem 7/49

▶ **7/49** The motor turns the disk at the constant speed $p = 30$ rad/sec. The motor is also swiveling about the horizontal axis BO (*y*-axis) at the constant speed $\dot\theta = 2$ rad/sec. Simultaneously, the entire assembly is rotating about the vertical axis C-C at the constant rate $q = 8$ rad/sec. For the instant when $\theta = 30°$, determine the angular acceleration $\boldsymbol\alpha$ of the disk and the acceleration $\mathbf{a}$ of point A at the bottom of the disk. Axes *x-y-z* are attached to the motor housing, and plane O-x_0-y is horizontal.

Ans. $\mathbf{a} = -2090\mathbf{i} - 369\mathbf{j} + 4810\mathbf{k}$ in./sec^2
$\boldsymbol\alpha = 8\sqrt{3}\mathbf{i} + 120\sqrt{3}\mathbf{j} + 52\mathbf{k}$ rad/sec^2

▶ **7/50** For the conditions specified with Sample Problem 7/2, except that γ is increasing at the steady rate of 3π rad/sec, determine the angular velocity $\boldsymbol{\omega}$ and the angular acceleration $\boldsymbol{\alpha}$ of the rotor when the position $\gamma = 30°$ is passed. (*Suggestion:* Apply Eq. 7/7 to the vector $\boldsymbol{\omega}$ to find $\boldsymbol{\alpha}$. Note that $\boldsymbol{\Omega}$ in Sample Problem 7/2 is no longer the complete angular velocity of the axes.)

> *Ans.* $\boldsymbol{\omega} = \pi(-3\mathbf{i} + \sqrt{3}\mathbf{j} + 5\mathbf{k})$ rad/sec
> $\boldsymbol{\alpha} = \pi^2(4\sqrt{3}\mathbf{i} + 9\mathbf{j} + 3\sqrt{3}\mathbf{k})$ rad/sec^2

▶ **7/51** The gyro rotor shown is spinning at the constant rate of 100 rev/min relative to the *x-y-z* axes in the direction indicated. If the angle γ between the gimbal ring and the horizontal *X-Y* plane is made to increase at the constant rate of 4 rad/s and if the unit is forced to precess about the vertical at the constant rate $N = 20$ rev/min, calculate the magnitude of the angular acceleration $\boldsymbol{\alpha}$ of the rotor when $\gamma = 30°$. Solve by using Eq. 7/7 applied to the angular velocity of the rotor.

> *Ans.* $\alpha = 42.8$ rad/s^2

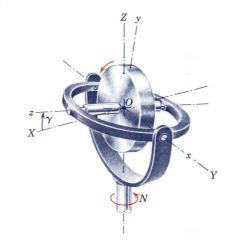

Problem 7/51

▶ **7/52** The wheel of radius r is free to rotate about the bent axle *CO* that turns about the vertical axis at the constant rate p rad/s. If the wheel rolls without slipping on the horizontal circle of radius R, determine the expressions for the angular velocity $\boldsymbol{\omega}$ and angular acceleration $\boldsymbol{\alpha}$ of the wheel. The *x*-axis is always horizontal.

> *Ans.* $\boldsymbol{\omega} = p\left[\mathbf{j} \cos \theta + \mathbf{k} \left(\sin \theta + \dfrac{R}{r} \right) \right]$
>
> $\boldsymbol{\alpha} = \left(\dfrac{Rp^2}{r} \cos \theta \right) \mathbf{i}$

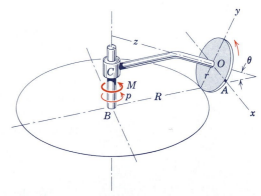

Problem 7/52

SECTION B. KINETICS

7/7 ANGULAR MOMENTUM

The force equation for a mass system, rigid or nonrigid, Eq. 4/1 or 4/6, is the generalization of Newton's second law for the motion of a particle and should require no further explanation. The moment equation for three-dimensional motion, however, is not nearly as simple as the third of Eqs. 6/1 for plane motion since the change of angular momentum has a number of additional components that are absent in plane motion.

We consider now a rigid body moving with any general motion in space, Fig. 7/12a. Axes x-y-z are *attached* to the body with origin at the mass center G. Thus, the angular velocity $\boldsymbol{\omega}$ of the body becomes the angular velocity of the x-y-z axes as observed from the fixed reference axes X-Y-Z. The absolute angular momentum $\mathbf{H}_G$ of the body about its mass center G is the sum of the moments about G of the linear momenta of all elements of the body and was expressed in part (b) of Art. 4/4 as $\mathbf{H}_G = \Sigma(\boldsymbol{\rho}_i \times m_i \mathbf{v}_i)$, where $\mathbf{v}_i$ is the absolute velocity of the mass element m_i. But for the rigid body $\mathbf{v}_i = \bar{\mathbf{v}} + \boldsymbol{\omega} \times \boldsymbol{\rho}_i$, where $\boldsymbol{\omega} \times \boldsymbol{\rho}_i$ is the relative velocity of m_i with respect to G as seen from nonrotating axes. Thus, we may write

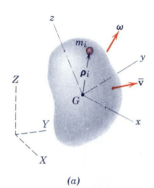

(a)

$$\mathbf{H}_G = -\bar{\mathbf{v}} \times \Sigma m_i \boldsymbol{\rho}_i + \Sigma[\boldsymbol{\rho}_i \times m_i(\boldsymbol{\omega} \times \boldsymbol{\rho}_i)]$$

where we have factored out $\bar{\mathbf{v}}$ from the first summation terms by reversing the order of the cross product and changing the sign. With origin at the mass center G, the first term in $\mathbf{H}_G$ is zero since $\Sigma m_i \boldsymbol{\rho}_i = m\bar{\boldsymbol{\rho}} = \mathbf{0}$. The second term with the substitution of dm for m_i and $\boldsymbol{\rho}$ for $\boldsymbol{\rho}_i$ gives

$$\mathbf{H}_G = \int [\boldsymbol{\rho} \times (\boldsymbol{\omega} \times \boldsymbol{\rho})]\, dm \qquad (7/8)$$

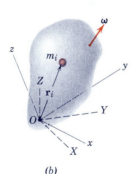

(b)

Figure 7/12

Before expanding the integrand of Eq. 7/8, we consider also the case of a rigid body rotating about a fixed point O, Fig. 7/12b. The x-y-z axes are attached to the body, and both body and axes have an angular velocity $\boldsymbol{\omega}$. The angular momentum about O was expressed in part (b) of Art. 4/4 and is $\mathbf{H}_O = \Sigma(\mathbf{r}_i \times m_i \mathbf{v}_i)$, where, for the rigid body, $\mathbf{v}_i = \boldsymbol{\omega} \times \mathbf{r}_i$. Thus, with the substitution of dm for m_i and $\mathbf{r}$ for $\mathbf{r}_i$, the angular momentum is

$$\mathbf{H}_O = \int [\mathbf{r} \times (\boldsymbol{\omega} \times \mathbf{r})]\, dm \qquad (7/9)$$

We observe now that for the two cases of Figs. 7/12a and 7/12b, the position vectors $\boldsymbol{\rho}_i$ and $\mathbf{r}_i$ are given by the same expression $x\mathbf{i} + y\mathbf{j} + z\mathbf{k}$. Thus, Eqs. 7/8 and 7/9 are identical in form, and the symbol $\mathbf{H}$ will be used here for either case. We now carry out the expansion of the integrand in the two expressions for angular momentum, with recognition of the fact that the components of $\boldsymbol{\omega}$ are invariant with respect to the integrals over the body and, hence, become constant multipliers of the integrals. The cross-product expansion applied to the triple vector product upon collection of terms gives

$$
\begin{aligned}
d\mathbf{H} = \; &\mathbf{i}[(y^2 + z^2)\omega_x && -xy\omega_y && -\,xz\omega_z]\,dm \\
+\,&\mathbf{j}[&& -yx\omega_x + (z^2 + x^2)\omega_y && -yz\omega_z]\,dm \\
+\,&\mathbf{k}[&& -zx\omega_x && -zy\omega_y + (x^2 + y^2)\omega_z]\,dm
\end{aligned}
$$

Now let

$$
\boxed{
\begin{aligned}
I_{xx} &= \int (y^2 + z^2)\,dm & I_{xy} &= \int xy\,dm \\[2mm]
I_{yy} &= \int (z^2 + x^2)\,dm & I_{xz} &= \int xz\,dm \\[2mm]
I_{zz} &= \int (x^2 + y^2)\,dm & I_{yz} &= \int yz\,dm
\end{aligned}
}
\qquad (7/10)
$$

The quantities I_{xx}, I_{yy}, I_{zz} are known as the *moments of inertia* of the body about the respective axes, and I_{xy}, I_{xz}, I_{yz} are known as the *products of inertia* with respect to the coordinate axes. These quantities describe the manner in which the mass of a rigid body is distributed with respect to the chosen axes. The calculation of moments and products of inertia is explained fully in Appendix B. The double subscripts for the moments and products of inertia preserve a symmetry of notation that has special meaning in their description by tensor notation.* It is observed that $I_{xy} = I_{yx}$, $I_{xz} = I_{zx}$, $I_{yz} = I_{zy}$. With the substitutions of Eqs. 7/10, the expression for $\mathbf{H}$ becomes

$$
\boxed{
\begin{aligned}
\mathbf{H} = \; &(\;\;I_{xx}\omega_x - I_{xy}\omega_y - I_{xz}\omega_z)\mathbf{i} \\
+&(-I_{yx}\omega_x + I_{yy}\omega_y - I_{yz}\omega_z)\mathbf{j} \\
+&(-I_{zx}\omega_x - I_{zy}\omega_y + I_{zz}\omega_z)\mathbf{k}
\end{aligned}
}
\qquad (7/11)
$$

and the components of $\mathbf{H}$ are clearly

$$
\boxed{
\begin{aligned}
H_x &= \;\;\;I_{xx}\omega_x - I_{xy}\omega_y - I_{xz}\omega_z \\
H_y &= -I_{yx}\omega_x + I_{yy}\omega_y - I_{yz}\omega_z \\
H_z &= -I_{zx}\omega_x - I_{zy}\omega_y + I_{zz}\omega_z
\end{aligned}
}
\qquad (7/12)
$$

*See, for example, the senior author's *Dynamics, 2nd Edition SI Version*, 1975, John Wiley & Sons, Art 41.

Equation 7/11 is the general expression for the angular momentum about either the mass center G or about a fixed point O for a rigid body rotating with an instantaneous angular velocity $\boldsymbol{\omega}$.

Attention is called to the fact that in each of the two cases represented, the reference axes x-y-z are *attached* to the rigid body. This attachment makes the moment-of-inertia integrals and the product-of-inertia integrals of Eqs. 7/10 invariant with time. If the x-y-z axes were to rotate with respect to an irregular body, then these inertia integrals would be functions of the time, which would introduce an undesirable complexity into the angular-momentum relations. An important exception occurs when a rigid body is spinning about an axis of symmetry, in which case, the inertia integrals are not affected by the angular position of the body about its spin axis. Thus, it is frequently convenient to permit a body with axial symmetry to rotate relative to the reference system about one of the coordinate axes. In addition to the momentum components due to the angular velocity $\boldsymbol{\Omega}$ of the reference axes, then an added angular-momentum component along the spin axis due to the relative spin about the axis would have to be accounted for.

In Eq. 7/12 the array of moments and products of inertia

$$
\begin{bmatrix}
I_{xx} & -I_{xy} & -I_{xz} \\
-I_{yz} & I_{yy} & -I_{yz} \\
-I_{zx} & -I_{zy} & I_{zz}
\end{bmatrix}
$$

is known as the *inertia matrix* or *inertia tensor*. As we change the orientation of the axes relative to the body, the moments and products of inertia will also change in value. It can be shown* that there is one unique orientation of axes x-y-z for a given origin for which the products of inertia vanish and the moments of inertia I_{xx}, I_{yy}, I_{zz} take on stationary values. For this orientation, the inertia matrix takes the form

$$
\begin{bmatrix}
I_{xx} & 0 & 0 \\
0 & I_{yy} & 0 \\
0 & 0 & I_{zz}
\end{bmatrix}
$$

and is said to be diagonalized. The axes x-y-z for which the products of inertia vanish are called the *principal axes of inertia*, and I_{xx}, I_{yy}, and I_{zz} are called the *principal moments of inertia*. The principal moments of inertia for a given origin represent the maximum, the minimum, and an intermediate value of the moments of inertia.

---

*See, for example, the senior author's *Dynamics, 2nd Edition SI Version*, 1975, John Wiley & Sons, Art 41.

If the coordinate axes coincide with the principal axes of inertia, Eq. 7/11 for the angular momentum about the mass center or about a fixed point becomes

$$\mathbf{H} = I_{xx}\omega_x\mathbf{i} + I_{yy}\omega_y\mathbf{j} + I_{zz}\omega_z\mathbf{k} \qquad (7/13)$$

It is always possible to locate the principal axes of inertia for a general three-dimensional rigid body. Thus, we can express its angular momentum by Eq. 7/13, although it may not always be convenient to do so for geometric reasons. Except when the body rotates about one of the principal axes of inertia or when $I_{xx} = I_{yy} = I_{zz}$, the vectors $\mathbf{H}$ and $\boldsymbol{\omega}$ have different directions.

The momentum properties of a rigid body may be represented by the resultant linear-momentum vector $\mathbf{G} = m\bar{\mathbf{v}}$ through the mass center and the resultant angular-momentum vector $\mathbf{H}_G$ about the mass center, as shown in Fig. 7/13. Although $\mathbf{H}_G$ has the properties of a free vector, we represent it through G for convenience. These vectors have properties analogous to those of a force and a couple. Thus, the angular momentum about any point A that may or may not be fixed to the body, equals the free vector $\mathbf{H}_G$, plus the moment of the linear-momentum vector about A. Therefore, we may write

$$\mathbf{H}_A = \mathbf{H}_G + \bar{\mathbf{r}} \times \mathbf{G} = \mathbf{H}_G + \bar{\mathbf{r}} \times m\bar{\mathbf{v}} \qquad (7/14)$$

Equation 7/14 constitutes a transfer theorem for angular momentum.

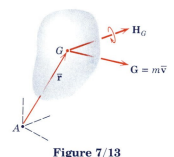

Figure 7/13

7/8 KINETIC ENERGY

In Art. 4/3 of the chapter on the dynamics of systems of particles, we developed the expression for the kinetic energy T of any general system of mass, rigid or nonrigid, and obtained the result

$$T = \tfrac{1}{2}m\bar{v}^2 + \Sigma\tfrac{1}{2}m_i|\dot{\boldsymbol{\rho}}_i|^2 \qquad [4/4]$$

where $\bar{v}$ is the velocity of the mass center and $\boldsymbol{\rho}_i$ is the position vector of a representative element of mass m_i with respect to the mass center. We identified the first term as the kinetic energy due to the translation of the system and the second term as the kinetic energy associated with the motion relative to the mass center. The translational term may be written alternatively as

$$\tfrac{1}{2}m\bar{v}^2 = \tfrac{1}{2}m\dot{\bar{\mathbf{r}}}\cdot\dot{\bar{\mathbf{r}}} = \tfrac{1}{2}\bar{\mathbf{v}}\cdot\mathbf{G}$$

where $\dot{\bar{\mathbf{r}}}$ is the velocity $\bar{\mathbf{v}}$ of the mass center and $\mathbf{G}$ is the linear momentum of the body.

For a rigid body, the relative term becomes the kinetic energy due to rotation about the mass center. Since $\dot{\boldsymbol{\rho}}_i$ is the velocity of the representative particle with respect to the mass center, then for the rigid body we may write it as $\dot{\boldsymbol{\rho}}_i = \boldsymbol{\omega} \times \boldsymbol{\rho}_i$, where $\boldsymbol{\omega}$ is the angular velocity of the body. With this substitution, the relative term in the kinetic energy expression becomes

$$\Sigma \tfrac{1}{2} m_i |\dot{\boldsymbol{\rho}}_i|^2 = \Sigma \tfrac{1}{2} m_i (\boldsymbol{\omega} \times \boldsymbol{\rho}_i) \cdot (\boldsymbol{\omega} \times \boldsymbol{\rho}_i)$$

If we use the fact that the dot and the cross may be interchanged in the triple scalar product, that is, $\mathbf{P} \times \mathbf{Q} \cdot \mathbf{R} = \mathbf{P} \cdot \mathbf{Q} \times \mathbf{R}$, we may write

$$(\boldsymbol{\omega} \times \boldsymbol{\rho}_i) \cdot (\boldsymbol{\omega} \times \boldsymbol{\rho}_i) = \boldsymbol{\omega} \cdot \boldsymbol{\rho}_i \times (\boldsymbol{\omega} \times \boldsymbol{\rho}_i)$$

Since $\boldsymbol{\omega}$ is the same factor in all terms of the summation, it may be factored out to give

$$\Sigma \tfrac{1}{2} m_i |\dot{\boldsymbol{\rho}}_i|^2 = \tfrac{1}{2} \boldsymbol{\omega} \cdot \Sigma \boldsymbol{\rho}_i \times m_i (\boldsymbol{\omega} \times \boldsymbol{\rho}_i) = \tfrac{1}{2} \boldsymbol{\omega} \cdot \mathbf{H}_G$$

where $\mathbf{H}_G$ is the same as the integral expressed by Eq. 7/8. Thus, the general expression for the kinetic energy of a rigid body moving with mass-center velocity $\bar{\mathbf{v}}$ and angular velocity $\boldsymbol{\omega}$ is

$$\boxed{T = \tfrac{1}{2} \bar{\mathbf{v}} \cdot \mathbf{G} + \tfrac{1}{2} \boldsymbol{\omega} \cdot \mathbf{H}_G} \qquad (7/15)$$

Expansion of this vector equation by substitution of the expression for $\mathbf{H}_G$ written from Eq. 7/11 yields

$$\begin{aligned} T = {} & \tfrac{1}{2} m \bar{v}^2 + \tfrac{1}{2} (\bar{I}_{xx} \omega_x^2 + \bar{I}_{yy} \omega_y^2 + \bar{I}_{zz} \omega_z^2) \\ & - (\bar{I}_{xy} \omega_x \omega_y + \bar{I}_{xz} \omega_x \omega_z + \bar{I}_{yz} \omega_y \omega_z) \end{aligned} \qquad (7/16)$$

If the axes coincide with the principal axes of inertia, the kinetic energy is merely

$$T = \tfrac{1}{2} m \bar{v}^2 + \tfrac{1}{2} (\bar{I}_{xx} \omega_x^2 + \bar{I}_{yy} \omega_y^2 + \bar{I}_{zz} \omega_z^2) \qquad (7/17)$$

When a rigid body is pivoted about a fixed point O or when there is a point O in the body that momentarily has zero velocity, the kinetic energy is $T = \Sigma \tfrac{1}{2} m_i \dot{\mathbf{r}}_i \cdot \dot{\mathbf{r}}_i$. This expression reduces to

$$\boxed{T = \tfrac{1}{2} \boldsymbol{\omega} \cdot \mathbf{H}_O} \qquad (7/18)$$

where $\mathbf{H}_O$ is the angular momentum about O, as may be seen by replacing $\boldsymbol{\rho}_i$ in the previous derivation by $\mathbf{r}_i$, the position vector from O. Equations 7/15 and 7/18 are the three-dimensional counterparts of Eqs. 6/9 and 6/8 for plane motion.

Sample Problem 7/6

The bent plate has a mass of 70 kg per square meter of surface area and revolves about the z-axis at the rate $\omega = 30$ rad/s. Determine (a) the angular momentum **H** of the plate about point O and (b) the kinetic energy T of the plate. Neglect the mass of the hub and the thickness of the plate compared with its surface dimensions.

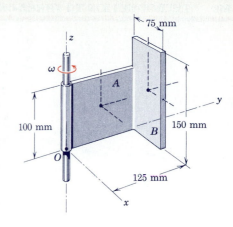

Solution. The moments and products of inertia are written with the aid of Eqs. B/3 and B/9 in Appendix B by transfer from the parallel centroidal axes for each part. First, the mass of each part is $m_A = (0.100)(0.125)(70) = 0.875$ kg, $m_B = (0.075)(0.150)(70) = 0.788$ kg.

Part A

$[I_{xx} = \bar{I}_{xx} + md^2]$ $\quad I_{xx} = \dfrac{0.875}{12}[(0.100)^2 + (0.125)^2]$

$$+ \; 0.875[(0.050)^2 + (0.0625)^2] = 0.007\ 47 \text{ kg·m}^2$$

$[I_{yy} = \frac{1}{3}ml^2]$ $\quad I_{yy} = \dfrac{0.875}{3}(0.100)^2 = 0.002\ 92 \text{ kg·m}^2$

$[I_{zz} = \frac{1}{3}ml^2]$ $\quad I_{zz} = \dfrac{0.875}{3}(0.125)^2 = 0.004\ 56 \text{ kg·m}^2$

$\left[I_{xy} = \displaystyle\int xy \, dm, \quad I_{xz} = \int xz \, dm \right] \qquad I_{xy} = 0 \qquad I_{xz} = 0$

$[I_{yz} = \bar{I}_{yz} + md_y d_z]$ $\quad I_{yz} = 0 + 0.875(0.0625)(0.050) = 0.002\ 73 \text{ kg·m}^2$

Part B

$[I_{xx} = \bar{I}_{xx} + md^2]$ $\quad I_{xx} = \dfrac{0.788}{12}(0.150)^2 + 0.788[(0.125)^2 + (0.075)^2]$

$$= 0.018\ 21 \text{ kg·m}^2$$

$[I_{yy} = \bar{I}_{yy} + md^2]$ $\quad I_{yy} = \dfrac{0.788}{12}[(0.075)^2 + (0.150)^2]$

$$+ \; 0.788[(0.0375)^2 + (0.075)^2] = 0.007\ 38 \text{ kg·m}^2$$

$[I_{zz} = \bar{I}_{zz} + md^2]$ $\quad I_{zz} = \dfrac{0.788}{12}(0.075)^2 + 0.788[(0.125)^2 + (0.0375)^2]$

$$= 0.013\ 78 \text{ kg·m}^2$$

$[I_{xy} = \bar{I}_{xy} + md_x d_y]$ $\quad I_{xy} = 0 + 0.788(0.0375)(0.125) = 0.003\ 69 \text{ kg·m}^2$

$[I_{xz} = \bar{I}_{xz} + md_x d_z]$ $\quad I_{xz} = 0 + 0.788(0.0375)(0.075) = 0.002\ 21 \text{ kg·m}^2$

$[I_{yz} = \bar{I}_{yz} + md_y d_z]$ $\quad I_{yz} = 0 + 0.788(0.125)(0.075) = 0.007\ 38 \text{ kg·m}^2$

The sum of the respective inertia terms gives for the two plates together

$$I_{xx} = 0.0257 \text{ kg·m}^2 \qquad I_{xy} = 0.003\ 69 \text{ kg·m}^2$$
$$I_{yy} = 0.010\ 30 \text{ kg·m}^2 \qquad I_{xz} = 0.002\ 21 \text{ kg·m}^2$$
$$I_{zz} = 0.018\ 34 \text{ kg·m}^2 \qquad I_{yz} = 0.010\ 12 \text{ kg·m}^2$$

(a) The angular momentum of the body is given by Eq. 7/11, where $\omega_z = 30$ rad/s and ω_x and ω_y are zero. Thus,

$$\mathbf{H}_O = 30(-0.002\ 21\mathbf{i} - 0.010\ 12\mathbf{j} + 0.018\ 34\mathbf{k}) \text{ N·m·s} \qquad Ans.$$

(b) The kinetic energy from Eq. 7/18 becomes

$$T = \tfrac{1}{2}\boldsymbol{\omega} \cdot \mathbf{H}_O = \tfrac{1}{2}(30\mathbf{k}) \cdot 30(-0.002\ 21\mathbf{i} - 0.010\ 12\mathbf{j} + 0.018\ 34\mathbf{k})$$
$$= 8.25 \text{ J}$$

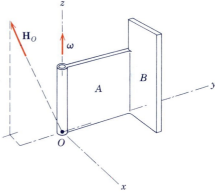

① The parallel-axis theorems for transferring moments and products of inertia from centroidal axes to parallel axes are explained in Appendix B and are most useful relations.

② Recall that the units of angular momentum may also be written in the base units as kg·m²/s.

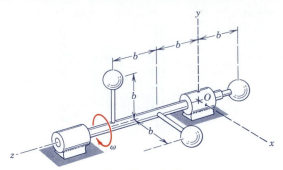

Problem 7/53

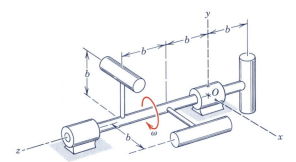

Problem 7/54

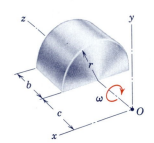

Problem 7/55

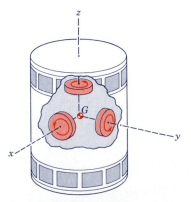

Problem 7/56

PROBLEMS

Introductory problems

7/53 The three small spheres, each of mass m, are rigidly mounted to the horizontal shaft that rotates with the angular velocity ω as shown. Neglect the radius of each sphere compared with the other dimensions and write expressions for the magnitudes of their linear momentum $\mathbf{G}$ and their angular momentum $\mathbf{H}_O$ about the origin O of the coordinates.

Ans. $G = mb\omega\sqrt{2}$
$H_O = 3mb\omega^2$

7/54 The spheres of Prob. 7/53 are replaced by three rods, each of mass m and length l, mounted at their centers to the shaft, which rotates with the angular velocity ω as shown. The axes of the rods are, respectively, in the x-, y-, and z-directions, and their diameters are negligible compared with the other dimensions. Determine the angular momentum $\mathbf{H}_O$ of the three rods with respect to the coordinate origin O.

7/55 The solid half-circular cylinder of mass m revolves about the z-axis with an angular velocity ω as shown. Determine its angular momentum $\mathbf{H}$ with respect to the x-y-z axes.

Ans. $\mathbf{H} = mr\omega \left[-\dfrac{2(2c + b)}{3\pi}\mathbf{j} + \dfrac{r}{2}\mathbf{k} \right]$

7/56 The elements of a reaction-wheel attitude-control system for a spacecraft are shown in the figure. Point G is the center of mass for the system of the spacecraft and wheels, and x, y, z are principal axes for the system. Each wheel has a mass m and a moment of inertia I about its own axis and spins with a relative angular velocity p in the direction indicated. The center of each wheel, which may be treated as a thin disk, is a distance b from G. If the spacecraft has angular velocity components Ω_x, Ω_y, and Ω_z, determine the angular momentum $\mathbf{H}_G$ of the three wheels as a unit.

7/57 The solid cube of mass m and side a revolves about an axis M-M through a diagonal with an angular velocity ω. Write the expression for the angular momentum $\mathbf{H}$ of the cube with respect to the axes indicated.

$$Ans.\ \mathbf{H} = \frac{ma^2\omega}{6\sqrt{3}}(\mathbf{i} + \mathbf{j} + \mathbf{k})$$

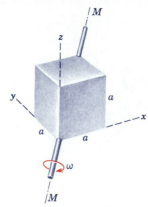

Problem 7/57

7/58 The slender rod of mass m and length l rotates about the y-axis as the element of a right-circular cone. If the angular velocity about the y-axis is ω, determine the expression for the angular momentum of the rod with respect to the x-y-z axes for the particular position shown.

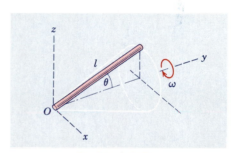

Problem 7/58

7/59 The bent rod has a mass m and revolves about the x-axis with an angular velocity ω. Determine the angular momentum of the rod about the origin O of the coordinates for the position shown. Also find the kinetic energy of the rod.

$$Ans.\ \mathbf{H}_O = mb^2\omega(\tfrac{8}{9}\mathbf{i} - \tfrac{1}{2}\mathbf{j} - \tfrac{1}{6}\mathbf{k}),\ T = \tfrac{4}{9}mb^2\omega^2$$

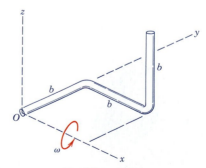

Problem 7/59

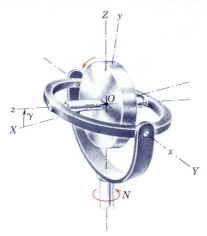

Problem 7/60

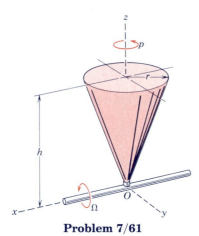

Problem 7/61

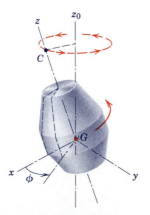

Problem 7/62

Representative problems

7/60 The gyro rotor of Prob. 7/51, repeated here, is spinning at the constant rate $p = 100$ rev/min relative to the *x-y-z* axes in the direction indicated. If the angle γ between the gimbal ring and horizontal *X-Y* plane is made to increase at the rate of 4 rad/sec and if the unit is forced to precess about the vertical at the constant rate $N = 20$ rev/min, calculate the angular momentum $\mathbf{H}_O$ of the rotor when $\gamma = 30°$. The axial and transverse moments of inertia are $I_{zz} = 5(10^{-3})$ lb-ft-sec^2 and $I_{xx} = I_{yy} = 2.5(10^{-3})$ lb-ft-sec^2.

7/61 The right-circular cone of height h and base radius r spins about its axis of symmetry with an angular rate p. Simultaneously, the entire cone revolves about the *x*-axis with angular rate Ω. Determine the angular momentum $\mathbf{H}_O$ of the cone about the origin O of the *x-y-z* axes and the kinetic energy T for the position shown. The mass of the cone is m.

$$Ans. \ \mathbf{H}_O = \frac{3}{10} \, mr^2 \left[\left(\frac{1}{2} + 6\frac{h^2}{r^2} \right) \Omega\mathbf{i} + p\mathbf{k} \right]$$

$$T = \frac{3}{10} \, mr^2 \left[\left(\frac{1}{4} + \frac{h^2}{r^2} \right) \Omega^2 + \frac{1}{2} \, p^2 \right]$$

7/62 The space capsule shown has a mass m with mass center G. Its radius of gyration about its *z*-axis of rotational symmetry is k and that about either the *x*- or *y*-axis is k'. In space, the capsule spins within its *x-y-z* reference frame at the rate $p = \dot{\phi}$. Simultaneously, a point C on the *z*-axis moves in a circle about the z_0-axis with a frequency f (rotations per unit time). The z_0-axis has a constant direction in space. Determine the angular momentum $\mathbf{H}_G$ of the capsule relative to the axes designated. Note that the *x*-axis always lies in the z-z_0 plane and that the *y*-axis is therefore normal to z_0.

7/63 The circular disk of mass m and radius r is mounted on the vertical shaft with an angle α between its plane and the plane of rotation of the shaft. Determine an expression for the angular momentum $\mathbf{H}$ of the disk about O. Find the angle β that the angular momentum $\mathbf{H}$ makes with the shaft if $\alpha = 10°$.

Ans. $\mathbf{H} = \frac{1}{4}mr^2\omega[(-\sin \alpha \cos \alpha)\mathbf{i}$
$+ (\sin^2 \alpha + 2 \cos^2 \alpha)\mathbf{k}]$
$\beta = 4.96°$

Problem 7/63

7/64 The assembly, consisting of the solid sphere of mass m and the uniform rod of length $2c$ and equal mass m, revolves about the vertical z-axis with an angular velocity ω. The rod of length $2c$ has a diameter that is small compared with its length and is perpendicular to the horizontal rod to which it is welded with the inclination β shown. Determine the combined angular momentum $\mathbf{H}_O$ of the sphere and inclined rod.

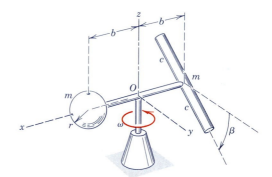

Problem 7/64

7/65 Each of the slender rods of length l and mass m is welded to the circular disk that rotates about the vertical z-axis with an angular velocity ω. Each rod makes an angle β with the vertical and lies in a plane parallel to the y-z plane. Determine an expression for the angular momentum $\mathbf{H}_O$ of the two rods about the origin O of the axes.

Ans. $\mathbf{H}_O = 2m(\frac{1}{3}l^2 \sin^2 \beta + b^2)\omega\mathbf{k}$

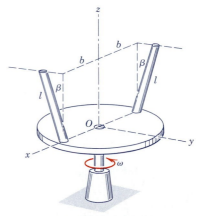

Problem 7/65

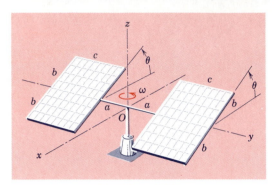

Problem 7/66

7/66 In a test of the solar panels for a spacecraft, the model shown is rotated about the vertical axis at the angular rate ω. If the mass per unit area of panel is ρ, write the expression for the angular momentum $\mathbf{H}_O$ of the assembly about the axes shown in terms of θ. Also determine the maximum, minimum, and intermediate values of the moment of inertia about the axes through O.

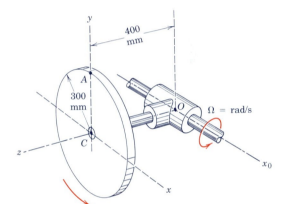

Problem 7/67

7/67 The circular disk of Prob. 7/33 with the two components of angular velocity is shown again here. If the disk has a mass of 60 kg, calculate its kinetic energy T and the magnitude H_O of its angular momentum with respect to point O.

$$Ans.\ H_O = 51.5\ \text{kg} \cdot \text{m}^2/\text{s},\ T = 223\ \text{J}$$

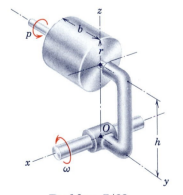

Problem 7/68

7/68 The solid circular cylinder of mass m, radius r, and length b revolves about its geometric axis at an angular rate p rad/s. Simultaneously, the bracket and attached shaft revolve about the x-axis at the rate ω rad/s. Write the expression for the angular momentum $\mathbf{H}_O$ of the cylinder about O with reference axes as shown.

7/69 The uniform circular disk of Prob. 7/39, repeated here, has a mass of 2.5 kg. With the rotational velocities $p = 240$ rev/min and $N = 30$ rev/min, determine the angular momentum $\mathbf{H}_O$ of the disk about the fixed point O and its kinetic energy T. The thickness of the disk is small compared with its radius.

$$Ans. \ \mathbf{H}_O = \pi(0.1123\mathbf{j} + 0.1\mathbf{k}) \ \text{kg}\cdot\text{m}^2/\text{s}$$
$$T = 0.456\pi^2 \ \text{J}$$

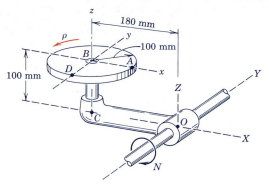

Problem 7/69

7/70 The solid circular disk of mass $m = 2$ kg and radius $r = 100$ mm rolls in a circle of radius $b = 200$ mm on the horizontal plane without slipping. If the centerline OC of the axle of the wheel rotates about the z-axis with an angular velocity $\omega = 4\pi$ rad/s, determine the expression for the angular momentum of the disk with respect to the fixed point O. Also compute the kinetic energy of the wheel.

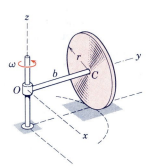

Problem 7/70

7/71 The uniform circular disk of Prob. 7/45 with the three components of angular velocity is shown again here. Determine the kinetic energy T and the angular momentum $\mathbf{H}_O$ with respect to O of the disk for the instant shown when its x-y plane coincides with the X-Y plane. The mass of the disk is m.

$$Ans. \ \mathbf{H}_O = \frac{1}{4} mr^2 \left[-\omega_1\mathbf{i} + \left(1 + \frac{4b^2}{r^2}\right) \omega_2\mathbf{j} + 2p\mathbf{k} \right]$$
$$T = \frac{1}{8} mr^2 \left[\omega_1^2 + \left(1 + \frac{4b^2}{r^2}\right) \omega_2^2 + 2p^2 \right]$$

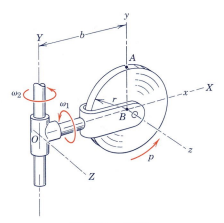

Problem 7/71

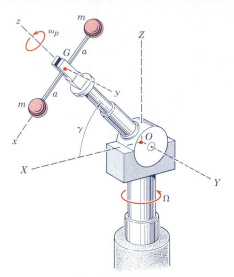

Problem 7/72

▶ **7/72** The robot of Prob. 7/15 is shown again here. In manipulating the dumbbell-shaped object, the robotic device rotates about its base at the angular rate Ω. Simultaneously, the cylindrical arm rotates upward at the rate $\dot{\gamma}$ and the grips rotate about the line OG at the rate ω_p. For given values of the constant length $\overline{OG} = l$ and the angle γ, determine the angular momentum $\mathbf{H}_G$ of the dumbbell about its mass center G and its angular momentum $\mathbf{H}_O$ about point O. Express $\mathbf{H}_G$ in terms of components along the x-y-z axes attached to the dumbbell and $\mathbf{H}_O$ in terms of components along the X-Y-Z axes. The moments of inertia of the dumbbell about the y- and z-axes are $2ma^2$ and that about the x-axis is negligible.
Ans.

$$H_{G_x} = 0$$
$$H_{G_y} = -2ma^2\dot{\gamma}$$
$$H_{G_z} = 2ma^2(\omega_p + \Omega \sin \gamma)$$
$$H_{O_X} = 2m \cos \gamma [a^2(\omega_p + \Omega \sin \gamma) - l^2\Omega \sin \gamma]$$
$$H_{O_Y} = -2m\dot{\gamma}(a^2 + l^2)$$
$$H_{O_Z} = 2m[a^2 \sin \gamma(\omega_p + \Omega \sin \gamma) + l^2\Omega \cos^2\gamma]$$

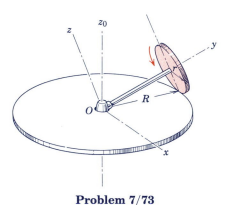

Problem 7/73

▶ **7/73** The uniform circular disk of mass m and radius r is mounted on its shaft, which is pivoted at O about the vertical z_0-axis. If the disk rolls without slipping and makes one complete trip around the large fixed disk in time τ, write the expression for the angular momentum of the disk with respect to the x-y-z axes through O. (*Hint:* The spin relative to x-y-z is $R\omega/r$, where $\omega = 2\pi/\tau$.)

$$\text{Ans. } \mathbf{H}_O = \frac{2\pi mr^2}{\tau} \left[\frac{1}{2} \left(\frac{r}{R} - \frac{R}{r} \right) \mathbf{j} \right.$$
$$\left. + \left(\frac{R^2}{r^2} - \frac{3}{4} \right) \sqrt{1 - \frac{r^2}{R^2}} \, \mathbf{k} \right]$$

Problem 7/74

▶ **7/74** The half-cylindrical shell of mass m, radius r, and length b revolves about one edge along the z-axis with a constant rate ω as shown. Determine the angular momentum $\mathbf{H}$ of the shell with respect to the x-y-z axes.

$$\text{Ans. } \mathbf{H} = mr\omega \left(\frac{b}{2} \mathbf{i} + \frac{b}{\pi} \mathbf{j} + 2r\mathbf{k} \right)$$

7/9 *MOMENTUM AND ENERGY EQUATIONS OF MOTION*

With the description of angular momentum, inertial properties, and kinetic energy of a rigid body established in the previous two articles, we are ready to apply the general momentum and energy equations of motion.

(a) Momentum equations. In Art. 4/4 of Chapter 4, we established the general linear- and angular-momentum equations for a system of constant mass. These equations are

$$\Sigma\mathbf{F} = \dot{\mathbf{G}} \qquad\qquad\qquad \textbf{[4/6]}$$

$$\Sigma\mathbf{M} = \dot{\mathbf{H}} \qquad\qquad \textbf{[4/7] or [4/9]}$$

The general moment relation, Eq. 4/7 or 4/9, is expressed here by the single equation $\Sigma\mathbf{M} = \dot{\mathbf{H}}$, where the terms are taken about either a fixed point O or about the mass center G. In the derivation of the moment principle, the derivative of $\mathbf{H}$ was taken with respect to an absolute coordinate system. When $\mathbf{H}$ is expressed in terms of components measured relative to a moving coordinate system x-y-z which has an angular velocity $\mathbf{\Omega}$, then by Eq. 7/7 the moment relation becomes

$$\Sigma\mathbf{M} = \left(\frac{d\mathbf{H}}{dt}\right)_{xyz} + \mathbf{\Omega} \times \mathbf{H}$$

$$= (\dot{H}_x\mathbf{i} + \dot{H}_y\mathbf{j} + \dot{H}_z\mathbf{k}) + \mathbf{\Omega} \times \mathbf{H}$$

The terms in parentheses represent that part of $\dot{\mathbf{H}}$ due to the change in magnitude of the components of $\mathbf{H}$, and the cross-product term represents that part due to the changes in direction of the components of $\mathbf{H}$. Expansion of the cross product and rearrangement of terms give

$$\Sigma\mathbf{M} = (\dot{H}_x - H_y\Omega_z + H_z\Omega_y)\mathbf{i}$$
$$+ (\dot{H}_y - H_z\Omega_x + H_x\Omega_z)\mathbf{j} \qquad\qquad \textbf{(7/19)}$$
$$+ (\dot{H}_z - H_x\Omega_y + H_y\Omega_x)\mathbf{k}$$

Equation 7/19 is the most general form of the moment equation about a fixed point O or about the mass center G. The Ω's are the angular velocity components of rotation of the reference axes, and the H-components in the case of a rigid body are as defined in Eq. 7/12, where the ω's are the components of the angular velocity of the body. We now apply Eq. 7/19 to a rigid body where the coordinate axes are *attached* to the body. Under these conditions when expressed in the x-y-z coordinates, the *moments and products of inertia are invariant with time,* and $\mathbf{\Omega} = \boldsymbol{\omega}$. Thus for axes attached to the body, the three scalar components of Eq. 7/19 become

$$\Sigma M_x = \dot{H}_x - H_y \omega_z + H_z \omega_y$$
$$\Sigma M_y = \dot{H}_y - H_z \omega_x + H_x \omega_z \qquad \textbf{(7/20)}$$
$$\Sigma M_z = \dot{H}_z - H_x \omega_y + H_y \omega_x$$

Equations 7/20 are the general moment equations for rigid-body motion with axes *attached to the body*. They hold with respect to axes through a fixed point O or through the mass center G.

In Art. 7/7 it was mentioned that, in general, for any origin fixed to a rigid body, there are three principal axes of inertia with respect to which the products of inertia vanish. If the reference axes coincide with the principal axes of inertia with origin at the mass center G or at a point O fixed to the body and fixed in space, the factors I_{xy}, I_{yz}, I_{xz} will be zero, and Eqs. 7/20 become

$$\Sigma M_x = I_{xx} \dot{\omega}_x - (I_{yy} - I_{zz}) \omega_y \omega_z$$
$$\Sigma M_y = I_{yy} \dot{\omega}_y - (I_{zz} - I_{xx}) \omega_z \omega_x \qquad \textbf{(7/21)}$$
$$\Sigma M_z = I_{zz} \dot{\omega}_z - (I_{xx} - I_{yy}) \omega_x \omega_y$$

These relations, known as *Euler's equations**, are extremely useful in the study of rigid-body motion.

(b) Energy equations.   The resultant of all external forces acting on a rigid body may be replaced by the resultant force $\Sigma \mathbf{F}$ acting through the mass center and a resultant couple $\Sigma \mathbf{M}_G$ acting about the mass center. Work is done by the resultant force and the resultant couple at the respective rates $\Sigma \mathbf{F} \cdot \overline{\mathbf{v}}$ and $\Sigma \mathbf{M}_G \cdot \boldsymbol{\omega}$, where $\overline{\mathbf{v}}$ is the linear velocity of the mass center and $\boldsymbol{\omega}$ is the angular velocity of the body. Integration over the time from condition 1 to condition 2 gives the total work done during the time interval. Equating the works done to the respective changes in kinetic energy as expressed in Eq. 7/15 gives

$$\int_{t_1}^{t_2} \Sigma \mathbf{F} \cdot \overline{\mathbf{v}} \, dt = \tfrac{1}{2} \overline{\mathbf{v}} \cdot \mathbf{G} \bigg]_1^2 \qquad \int_{t_1}^{t_2} \Sigma \mathbf{M}_G \cdot \boldsymbol{\omega} \, dt = \tfrac{1}{2} \boldsymbol{\omega} \cdot \mathbf{H}_G \bigg]_1^2 \qquad \textbf{(7/22)}$$

These equations express the change in translational kinetic energy and the change in rotational kinetic energy, respectively, for the interval during which $\Sigma \mathbf{F}$ or $\Sigma \mathbf{M}_G$ acts, and the sum of the two expressions equals ΔT.

The work-energy relationship, developed in Chapter 4 for a general system of particles and given by

---

*Named after Leonhard Euler (1707–1783), a Swiss mathematician.

$$U'_{1\text{-}2} = \Delta T + \Delta V_e + \Delta V_g \qquad [4/3]$$

was used in Chapter 6 for rigid bodies in plane motion. The equation is equally applicable to rigid-body motion in three dimensions. As we have seen previously, the work-energy approach is of great advantage when analyzing end-point conditions of motion (initial and final). Here the work $U'_{1\text{-}2}$ done during the interval by all active forces external to the body or system is equated to the sum of the corresponding changes in kinetic energy ΔT, elastic potential energy ΔV_e, and gravitational potential energy ΔV_g. The potential-energy changes are determined in the usual way, as described previously in Art. 3/7.

We shall limit our application of the equations developed in this article to two problems of special interest, parallel-plane motion and gyroscopic motion, discussed in the next two articles.

7/10 PARALLEL-PLANE MOTION

When all particles of a rigid body move in planes that are parallel to a fixed plane, the body has a general form of plane motion, as described in Art. 7/4 and pictured in Fig. 7/3. Every line in such a body that is normal to the fixed plane remains parallel to itself at all times. We take the mass center G as the origin of coordinates x-y-z that are attached to the body, with the x-y plane coinciding with the plane of motion P. The components of the angular velocity of both body and attached axes become $\omega_x = \omega_y = 0$, $\omega_z \neq 0$. For this case, the angular-momentum components from Eq. 7/12 become

$$H_x = -I_{xz}\omega_z \qquad H_y = -I_{yz}\omega_z \qquad H_z = I_{zz}\omega_z$$

and the moment relations of Eqs. 7/20 reduce to

$$\boxed{\begin{aligned} \Sigma M_x &= -I_{xz}\dot{\omega}_z + I_{yz}\omega_z{}^2 \\ \Sigma M_y &= -I_{yz}\dot{\omega}_z - I_{xz}\omega_z{}^2 \\ \Sigma M_z &= I_{zz}\dot{\omega}_z \end{aligned}} \qquad \textbf{(7/23)}$$

It is seen that the third moment equation is equivalent to the second of Eqs. 6/1, where the z-axis passes through the mass center or to Eq. 6/4 if the z-axis passes through a fixed point.

Equations 7/23 hold for an origin of coordinates at the mass center, as shown in Fig. 7/3, or for any origin on a fixed axis of rotation. The three independent force equations of motion that also apply to parallel-plane motion are clearly

$$\Sigma F_x = m\bar{a}_x \qquad \Sigma F_y = m\bar{a}_y \qquad \Sigma F_z = 0$$

Equations 7/23 find special use in describing the effect of dynamic unbalance in rotating shafts and in rolling bodies.

Sample Problem 7/7

The two circular disks, each of mass m_1, are connected by the curved bar bent into quarter-circular arcs and welded to the disks. The bar has a mass m_2. The total mass of the assembly is $m = 2m_1 + m_2$. If the disks roll without slipping on a horizontal plane with a constant velocity v of the disk centers, determine the value of the friction force under each disk at the instant represented when the plane of the curved bar is horizontal.

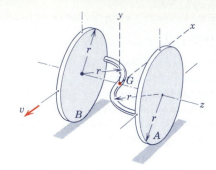

Solution. The motion is identified as parallel-plane motion since the planes of motion of all parts of the system are parallel. The free-body diagram shows the normal forces and friction forces at A and B and the total weight mg acting through the mass center G, which we take as the origin of coordinates that rotate with the body.

We now apply Eqs. 7/23, where $I_{yz} = 0$ and $\dot{\omega}_z = 0$. The moment equation about the y-axis requires determination of I_{xz}. From the diagram showing the geometry of the curved rod and with ρ standing for the mass of the rod per unit length, we have

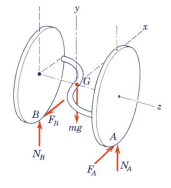

① $\left[I_{xz} = \int xz \, dm \right]$ $I_{xz} = \int_0^{\pi/2} (r \sin \theta)(-r + r \cos \theta)\rho r \, d\theta$

$$+ \int_0^{\pi/2} (-r \sin \theta)(r - r \cos \theta)\rho r \, d\theta$$

Evaluating the integrals gives

$$I_{xz} = -\rho r^3/2 - \rho r^3/2 = -\rho r^3 = -\frac{m_2 r^2}{\pi}$$

The second of Eqs. 7/23 with $\omega_z = v/r$ and $\dot{\omega}_z = 0$ gives

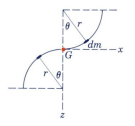

$[\Sigma M_y = -I_{xz}\omega_z^2]$ $F_A r + F_B r = -\left(-\frac{m_2 r^2}{\pi} \right) \frac{v^2}{r^2}$

$$F_A + F_B = \frac{m_2 v^2}{\pi r}$$

But with $\bar{v} = v$ constant, $\bar{a}_x = 0$ so that

$[\Sigma F_x = 0]$ $F_A - F_B = 0$ $F_A = F_B$

Thus,

$$F_A = F_B = \frac{m_2 v^2}{2\pi r} \qquad \textit{Ans.}$$

We also note for the given position that with $I_{yz} = 0$ and $\dot{\omega}_z = 0$, the moment equation about the x-axis gives

② $[\Sigma M_x = 0]$ $-N_A r + N_B r = 0$ $N_A = N_B = mg/2$

① We must be very careful to observe the correct signs for each of the coordinates of the mass element dm that make up the product xz.

② When the plane of the curved bar is not horizontal, the normal forces under the disks are no longer equal.

PROBLEMS

Introductory problems

7/75 Each of the two rods of mass m is welded to the face of the disk that rotates about the vertical axis with a constant angular velocity ω. Determine the bending moment M acting on each rod at its base.

Ans. $M = \frac{1}{2}mbl\omega^2$

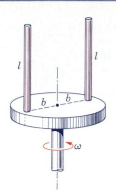

Problem 7/75

7/76 The slender shaft carries two offset particles, each of mass m, and rotates with the constant angular rate ω as indicated. Determine the x- and y-components of the bearing reactions at A and B due to the dynamic imbalance of the shaft for the position shown.

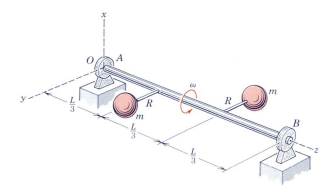

Problem 7/76

7/77 The paint stirrer shown in the figure is made from a rod of length $7b$ and mass ρ per unit length. Before immersion in the paint, the stirrer is rotating freely at a constant high angular velocity ω about its z-axis. Determine the bending moment M in the rod at the base O of the chuck. Ans. $M = M_x = -2\rho b^3 \omega^2$

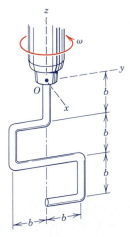

Problem 7/77

Problem 7/78

7/78 The 6-kg circular disk and attached shaft rotate at a constant speed $\omega = 10\ 000$ rev/min. If the center of mass of the disk is 0.05 mm off center, determine the magnitudes of the horizontal forces A and B supported by the bearings because of the rotational imbalance.

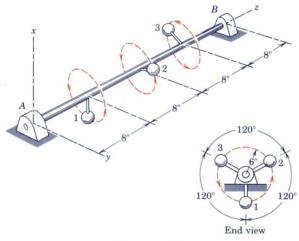

Problem 7/79

7/79 The dynamic imbalance of a certain crankshaft is approximated by the physical model shown, where the shaft carries three small 1.5-lb spheres attached by rods of negligible mass. If the shaft rotates at the constant speed of 1200 rev/min, calculate the forces R_A and R_B acting on the bearings. Neglect the gravitational forces.   *Ans.* $R_A = R_B = 159.3$ lb

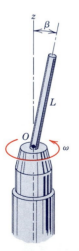

Problem 7/80

7/80 The slender rod of mass m and length L is mounted in a rotating chuck with the rod axis misaligned from the z-axis of rotation by the angle β. Determine the bending moment M_O in the rod at its base O if ω is constant. Neglect the moment due to the weight of the rod.

Representative problems

7/81 The irregular rod has a mass ρ per unit length and rotates about the shaft z-axis at the constant angular velocity ω. Determine the bending moment M in the rod at A. Neglect the small moment due to the weight of the rod. *Ans.* $M = \sqrt{13}\rho b^3 \omega^2$

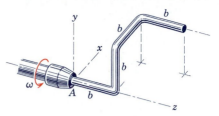

Problem 7/81

7/82 If the rod of Prob. 7/81 starts from rest under the action of a torque M_O applied to it by the collar at A about the z-axis, determine the initial bending moment M in the rod at A. Neglect the small moment due to the weight of the rod.

7/83 Each 200-mm leg of the right-angled rods that are welded to the vertical shaft has a mass of 0.12 kg. Calculate the bending moment M in the shaft at O due to rotation of the assembly about the vertical shaft at the constant speed of 1200 rev/min. Neglect the small moment due to the weight of the rods. *Ans.* $M = 268$ N·m

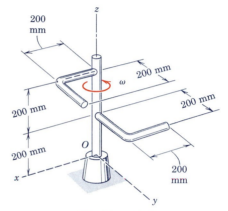

Problem 7/83

7/84 Calculate the bending moment M in the vertical shaft at O for the assembly of Prob. 7/83 due to its angular acceleration as it starts from rest under the action of a torque of 64 N·m applied to the shaft about the z-axis. Neglect the small moment due to the weight of the rods.

7/85 The thin circular disk of mass m and radius R is hinged about its horizontal tangent axis to the end of a shaft rotating about its vertical axis with an angular velocity ω. Determine the steady-state angle β assumed by the plane of the disk with the vertical axis. *Ans.* $\beta = \cos^{-1} \dfrac{4g}{5R\omega^2}$

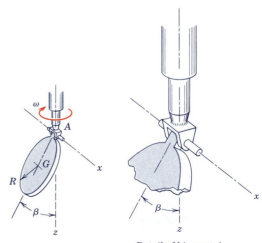

Detail of hinge at A

Problem 7/85

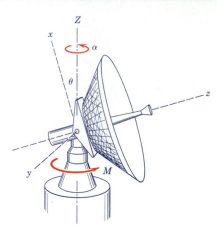

Problem 7/86

7/86 The large satellite-tracking antenna has a moment of inertia I about its z-axis of symmetry and a moment of inertia I_O about each of the x- and y-axes. Determine the angular acceleration α of the antenna about the vertical Z-axis caused by a torque M applied about Z by the drive mechanism for a given orientation θ.

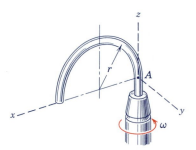

Problem 7/87

7/87 Determine the bending moment M at the tangency point A in the semicircular rod of radius r and mass m as it rotates about the tangent axis with a constant and large angular velocity ω. Neglect the moment mgr produced by the weight of the rod.

$$Ans. \quad M = \frac{2}{\pi} mr^2 \omega^2$$

7/88 If the semicircular rod of Prob. 7/87 starts from rest under the action of a torque M_O applied through the collar about its z-axis of rotation, determine the initial bending moment M in the rod at A.

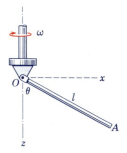

Problem 7/89

7/89 The slender rod OA of mass m and length l is pivoted freely about a horizontal axis through O. The pivot at O and attached shaft rotate with a constant angular speed ω about the vertical z-axis. Write the expression for the angle θ assumed by the rod. What minimum value must ω attain before the rod will assume other than a vertical position?

$$Ans. \quad \theta = \cos^{-1} \frac{3g}{2l\omega^2}, \quad \omega_{min} = \sqrt{\frac{3g}{2l}}$$

7/90 If the rod of Prob. 7/89 is welded to O at an angle θ with the vertical, determine the expression for the bending moment M in the rod at O due to the combined effect of the angular velocity ω of the shaft and the weight of the rod.

7/91 The plate has a mass of 3 kg and is welded to the fixed vertical shaft that rotates at the constant speed of 20π rad/s. Compute the moment $\mathbf{M}$ applied *to* the shaft *by* the plate due to dynamic imbalance.

Ans. $\mathbf{M} = -79.0\mathbf{i}$ N·m

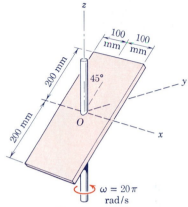

Problem 7/91

7/92 The circular disk of mass m and radius r, repeated from Prob. 7/63, is mounted on the vertical shaft with a small angle α between its plane and the plane of rotation of the shaft. Determine the expression for the bending moment $\mathbf{M}$ acting *on* the shaft due to the wobble of the disk at a shaft speed of ω rad/s.

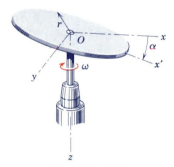

Problem 7/92

7/93 The uniform square flaps, each of mass m, are freely hinged at A and B to the square plate and attached shaft that rotate about the vertical z-axis with a constant angular velocity ω. Determine the angular velocity ω required to maintain a specified positive angle θ.

Ans. $\omega = \sqrt{\dfrac{1}{b}\dfrac{6g\tan\theta}{4\sin\theta + 3}}$

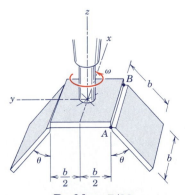

Problem 7/93

7/94 If the mechanism of Prob. 7/93 rotates with a constant angular velocity greater than that specified in the answer to that problem, determine the frictional moment M_f that the hinge pins must support to maintain the flaps at the specified angle θ.

7/95 The uniform slender rod of length l is welded to the bracket at A on the underside of the disk B. The disk rotates about a vertical axis with a constant angular velocity ω. Determine the value of ω that will result in a zero moment supported by the weld at A for the position $\theta = 60°$ with $b = l/4$.

$$Ans. \quad \omega = 2\sqrt{\frac{\sqrt{3}\,g}{l}}$$

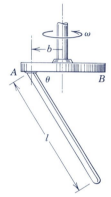

Problem 7/95

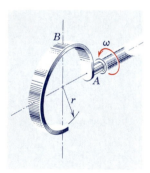

Problem 7/96

7/96 The partial ring has a mass ρ per unit length of rim and is welded to the shaft at A. Determine the bending moment M in the ring at A due to the rotation of the ring for an angular velocity ω.

7/97 For the ring section of Prob. 7/96, determine the torsional moment T in the ring at B (moment about an axis tangent to the rim) as the ring starts from rest under the action of a torque M transmitted by the shaft to the ring at A. The mass per unit length of rim is ρ.   *Ans.* $T = \frac{2}{3}M$

▶ **7/98** Each of the two circular disks has a mass m and is welded to the end of the rigid rod of mass m_0 so that the disks have a common z-axis and are separated by a distance b. A couple M, applied to one of the disks with the assembly initially at rest, gives the centers of the disks an acceleration $\mathbf{a} = +a\mathbf{i}$. Friction is sufficient to prevent slipping. Derive expressions for the normal forces N_A and N_B exerted by the horizontal surface on the disks as they begin to roll. Express the results in terms of the acceleration a rather than the moment M.

$$Ans.\ N_A = mg + \frac{m_0 g}{2}\left(1 + \frac{a}{3g}\right)$$

$$N_B = mg + \frac{m_0 g}{2}\left(1 - \frac{a}{3g}\right)$$

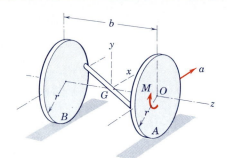

Problem 7/98

▶ **7/99** The homogeneous thin triangular plate of mass m is welded to the horizontal shaft that rotates freely in the bearings at A and B. If the plate is released from rest in the horizontal position shown, determine the magnitude of the bearing reaction at A for the instant just after release.   *Ans.* $A = mg/6$

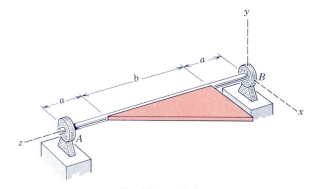

Problem 7/99

7/100 If the homogeneous triangular plate of Prob. 7/99 is released from rest in the position shown, determine the magnitude of the bearing reaction at A after the plate has rotated 90°.

$$Ans.\ A = \frac{mg}{3}\left[\frac{7a + 2b}{2a + b}\right]$$

7/11 GYROSCOPIC MOTION: STEADY PRECESSION

One of the most interesting of all problems in dynamics is that of gyroscopic motion. This motion occurs whenever the axis about which a body is spinning is itself rotating about another axis. Although the complete description of this motion involves considerable complexity, the most common and useful examples of gyroscopic motion occur when the axis of a rotor spinning at constant speed turns (precesses) about another axis at a steady rate. Our discussion in this article will focus on this special case.

The gyroscope has important engineering applications. With a mounting in gimbal rings (see Fig. 7/19*b*), the gyro is free from external moments, and its axis will retain a fixed direction in space regardless of the rotation of the structure to which it is attached. In this way, the gyro is used for inertial guidance systems and other directional control devices. With the addition of a pendulous mass to the inner gimbal ring, the earth's rotation causes the gyro to precess so that the spin axis will always point north, and this action forms the basis of the gyro compass. The gyroscope has also found important use as a stabilizing device. The controlled precession of a large gyro mounted in a ship is used to produce a gyroscopic moment to counteract the rolling of a ship at sea. The gyroscopic effect is also an extremely important consideration in the design of bearings for the shafts of rotors that are subjected to forced precessions.

We shall first describe gyroscopic action with a simple physical approach that relies on our previous experience with the vector changes encountered in plane dynamics. This approach will help us gain a direct physical insight into gyroscopic action. Next, we will make use of the general momentum relation, Eq. 7/19, for a more complete description.

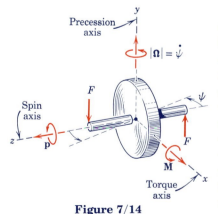

Figure 7/14

(a) Simplified approach. Figure 7/14 shows a symmetrical rotor spinning about the *z*-axis with a large angular velocity **p**, known as the *spin velocity*. If we should apply two forces F to the rotor axle to form a couple **M** whose vector is directed along the *x*-axis, we will find that the rotor shaft will rotate in the *x-z* plane about the *y*-axis in the sense indicated with a relatively slow angular velocity $\Omega = \dot{\psi}$ known as the *precession velocity*. Thus, we identify the spin axis (**p**), the torque axis (**M**), and the precession axis (**Ω**), where the usual right-hand rule identifies the sense of the rotation vectors. The rotor shaft does *not* turn about the *x*-axis in the sense of **M**, as it would if the rotor were not spinning. To aid understanding of this phenomenon, a direct analogy may be made between the rotation vectors and the familiar vectors that describe the curvilinear motion of a particle.

Figure 7/15*a* shows a particle of mass *m* moving in the *x-z* plane with constant speed $|\mathbf{v}| = v$. The application of a force **F** normal to its linear momentum $\mathbf{G} = m\mathbf{v}$ causes a change $d\mathbf{G} = d(m\mathbf{v})$ in its momentum. We see that $d\mathbf{G}$, and hence $d\mathbf{v}$, is a vector in the direction of the normal force $\mathbf{F}$ according to Newton's second law $\mathbf{F} = \dot{\mathbf{G}}$, which may be written as $\mathbf{F}\, dt = d\mathbf{G}$. From Fig. 7/15b we see that, in the limit, $\tan d\theta = d\theta = F dt/mv$ or $F = mv\dot{\theta}$. In vector notation with $\boldsymbol{\omega} = \dot{\theta}\mathbf{j}$, the force becomes

$$\mathbf{F} = m\boldsymbol{\omega} \times \mathbf{v}$$

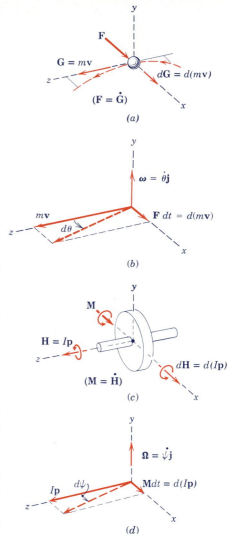

(a)

(b)

(c)

(d)

Figure 7/15

which is the vector equivalent of our familiar scalar relation $F_n = ma_n$ for the normal force on the particle as treated extensively in Chapter 3.

With these relations in mind, we now turn to our problem of rotation. Recall now the analogous equation $\mathbf{M} = \dot{\mathbf{H}}$ that we developed for any prescribed mass system, rigid or nonrigid, referred to its mass center (Eq. 4/9) or to a fixed point O (Eq. 4/7). We now apply this relation to our symmetrical rotor, as shown in Fig. 7/15c. For a high rate of spin $\mathbf{p}$ and a low precession rate $\boldsymbol{\Omega}$ about the y-axis, the angular momentum is represented by the vector $\mathbf{H} = I\mathbf{p}$, where $I = I_{zz}$ is the moment of inertia of the rotor about the spin axis. Initially, we are neglecting the small component of angular momentum about the y-axis that accompanies the slow precession. The application of the couple $\mathbf{M}$ normal to $\mathbf{H}$ causes a change $d\mathbf{H} = d(I\mathbf{p})$ in the angular momentum. We see that $d\mathbf{H}$, and hence $d\mathbf{p}$, is a vector in the direction of the couple $\mathbf{M}$ since $\mathbf{M} = \dot{\mathbf{H}}$, which may also be written $\mathbf{M}\, dt = d\mathbf{H}$. Just as the change in the linear-momentum vector of the particle is in the direction of the applied force, so is the change in the angular-momentum vector of the gyro in the direction of the couple. Thus, we see that the vectors $\mathbf{M}$, $\mathbf{H}$, and $d\mathbf{H}$ are analogous to the vectors $\mathbf{F}$, $\mathbf{G}$, and $d\mathbf{G}$. With this insight, it is no longer strange to see the rotation vector undergo a change that is in the direction of $\mathbf{M}$, thereby causing the axis of the rotor to precess about the y-axis.

In Fig. 7/15d we see that during time dt the angular-momentum vector $I\mathbf{p}$ has swung through the angle $d\psi$, so that in the limit with $\tan d\psi = d\psi$, we have

$$d\psi = \frac{M\, dt}{Ip} \qquad \text{or} \qquad M = I\frac{d\psi}{dt}p$$

Substituting $\Omega = d\psi/dt$ for the magnitude of the precession velocity gives us

$$\boxed{M = I\Omega p} \qquad\qquad (7/24)$$

We note that $\mathbf{M}$, $\boldsymbol{\Omega}$, and $\mathbf{p}$ as vectors are mutually perpendicular, and that their vector relationship may be represented by writing the equation in the cross-product form

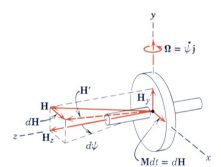

Figure 7/16

$$\boxed{\mathbf{M} = I\boldsymbol{\Omega} \times \mathbf{p}} \qquad (7/24a)$$

which is completely analogous to the foregoing relation $\mathbf{F} = m\boldsymbol{\omega} \times \mathbf{v}$ for the curvilinear motion of a particle as developed from Figs. 7/15*a* and *b*. Equations 7/24 and 7/24*a* apply to moments taken about the mass center or about a fixed point on the axis of rotation.

The correct spatial relationship among the three vectors may be remembered from the fact that $d\mathbf{H}$, and hence $d\mathbf{p}$, is in the direction of $\mathbf{M}$, which establishes the correct sense for the precession $\boldsymbol{\Omega}$. Thus, the spin vector $\mathbf{p}$ always tends to rotate toward the torque vector $\mathbf{M}$. Figure 7/16 represents three orientations of the three vectors that are consistent with their correct order. Unless we establish this order correctly in a given problem, we are likely to arrive at a conclusion directly opposite to the correct one. It is emphasized that Eq. 7/24, like $\mathbf{F} = m\mathbf{a}$ and $M = I\alpha$, is an equation of motion, so that the couple $\mathbf{M}$ represents the couple due to *all* forces acting *on* the rotor, as disclosed by a correct *free-body diagram of the rotor*. Also it is pointed out that, when a rotor is forced to precess, as occurs with the turbine in a ship that is executing a turn, the motion will generate a *gyroscopic couple* $\mathbf{M}$ that obeys Eq. 7/24*a* both in magnitude and sense.

In the foregoing discussion of gyroscopic motion, it was assumed that the spin was large and the precession was small. Although we can see from Eq. 7/24 that for given values of I and M the precession $\boldsymbol{\Omega}$ must be small if p is large, let us now examine the influence of $\boldsymbol{\Omega}$ on the momentum relations. Again, we restrict our attention to steady precession, where $\boldsymbol{\Omega}$ has a constant magnitude. Figure 7/17 shows our same rotor again. Because it has a moment of inertia about the y-axis and an angular velocity of precession about this axis, there will be an additional component of angular momentum about the y-axis. Thus, we have the two components $H_z = Ip$ and $H_y = I_0\Omega$, where I_0 stands for I_{yy} and, again, I stands for I_{zz}. The total angular momentum is $\mathbf{H}$ as shown. The change in $\mathbf{H}$ remains $d\mathbf{H} = \mathbf{M}\,dt$ as previously, and the precession during time dt is the angle $d\psi = M\,dt/H_z = M\,dt/(Ip)$ as before. Thus, Eq. 7/24 is still valid and for steady precession is an exact description of the motion as long as the spin axis is perpendicular to the axis around which precession occurs.

Consider now the steady precession of a symmetrical top, Fig. 7/18, spinning about its axis with a high angular velocity p and supported at its point O. Here the spin axis makes an angle θ with the vertical Z-axis around which precession occurs. Again, we will neglect the small angular-momentum component due to the precession and consider H equal to $I\mathbf{p}$, the angular momentum about the axis of the top associated with the spin only. The moment about O

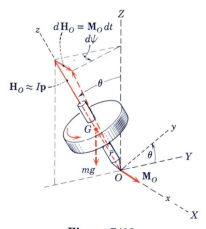

Figure 7/17

Figure 7/18

is due to the weight and is $mg\bar{r} \sin \theta$, where $\bar{r}$ is the distance from O to the mass center G. From the diagram, we see that the angular-momentum vector $\mathbf{H}_O$ has a change $d\mathbf{H}_O = \mathbf{M}_O \, dt$ in the direction of $\mathbf{M}_O$ during time dt and that θ is unchanged. The increment in precessional angle around the Z-axis is

$$d\psi = \frac{M_O \, dt}{Ip \sin \theta}$$

Substituting the values $M_O = mg\bar{r} \sin \theta$ and $\Omega = d\psi/dt$ gives

$$mg\bar{r} \sin \theta = I\Omega p \sin \theta \quad \text{or} \quad mg\bar{r} = I\Omega p$$

which is independent of θ. Introducing the radius of gyration so that $I = mk^2$ and solving for the precessional velocity give

$$\boxed{\Omega = \frac{g\bar{r}}{k^2 p}} \qquad\qquad \textbf{(7/25)}$$

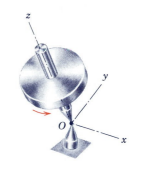

(a)

Unlike Eq. 7/24, which is an exact description for the rotor of Fig. 7/17 with precession confined to the x-z plane, Eq. 7/25 is an approximation based on the assumption that the angular momentum associated with Ω is negligible, compared with that associated with p. We shall see the amount of the error in part (b) of this article. On the basis of our analysis, the top will have a steady precession at the constant angle θ only if it is set in motion with a value of Ω that satisfies Eq. 7/25. When these conditions are not met, the precession becomes unsteady, and θ may oscillate with an amplitude that increases as the spin velocity decreases. The corresponding rise and fall of the rotation axis is known as *nutation*.

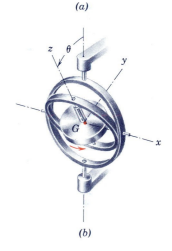

(b)

(b) More complete analysis. We will now make direct use of Eq. 7/19, which is the general angular-momentum equation for a rigid body, by applying it to a body spinning about its axis of rotational symmetry. This equation is valid for rotation about a fixed point or for rotation about the mass center. A spinning top, the rotor of a gyroscope, and a space capsule are examples of bodies whose motions can be described by the equations for rotation about a point. The general moment equations for this class of problems are fairly complex, and their complete solutions involve the use of elliptic integrals and somewhat lengthy computations. A large fraction of engineering problems where the motion is one of rotation about a pin involve the steady precession of bodies of revolution that are spinning about their axes of symmetry. These conditions introduce simplification that greatly facilitate solution of the equations. Consider a body with axial symmetry, Fig. 7/19a, rotating about a fixed point O on its axis, which is taken to be the z-direction. With O as origin, the x- and y-axes automatically become principal

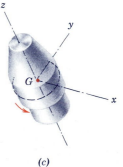

(c)

Figure 7/19

axes of inertia along with the z-axis. This same description may be used for the rotation of a similar symmetrical body about its center of mass G, which is taken as the origin of coordinates as shown with the gimbaled gyroscope rotor of Fig. 7/19b. Again, the x- and y-axes are principal axes of inertia for point G. The same description may also be used to represent the rotation about the mass center of an axially symmetric body in space, such as the space capsule in Fig. 7/19c. In each case, we note that, regardless of the rotation of the axes or of the body relative to the axes (spin about the z-axis), the moments of inertia about the x- and y-axes remain constant with time. The principal moments of inertia are again designated $I_{zz} = I$ and $I_{xx} = I_{yy} = I_0$. The products of inertia are, of course, zero.

Before applying Eq. 7/19, we introduce a set of coordinates that provide a natural description for our problem. These coordinates are shown in Fig. 7/20 for the example of rotation about a fixed point O. The aces X-Y-Z are fixed in space, and plane A contains the X-Y axes and the fixed point O on the rotor axis. Plane B contains

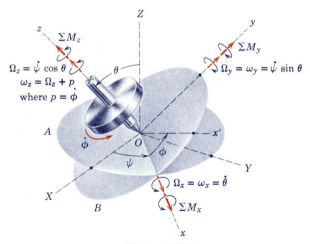

Figure 7/20

point O and is always normal to the rotor axis. Angle θ measures the inclination of the rotor axis from the vertical Z-axis and is also a measure of the angle between planes A and B. The intersection of the two planes is the x-axis, which is located by the angle ψ from the X-axis. The y-axis lies in plane B, and the z-axis coincides with the rotor axis. The angles θ and ψ completely specify the position of the rotor axis. The angular displacement of the rotor with respect to axes x-y-z is specified by the angle ϕ measured from the x-axis to the x'-axis, which is attached to the rotor. The spin velocity becomes $p = \dot{\phi}$.

The components of the angular velocity $\boldsymbol{\omega}$ of the rotor and the angular velocity $\boldsymbol{\Omega}$ of the axes *x-y-z* from Fig. 7/20 become

$$\Omega_x = \dot{\theta} \qquad\qquad \omega_x = \dot{\theta}$$
$$\Omega_y = \dot{\psi} \sin\theta \qquad \omega_y = \dot{\psi} \sin\theta$$
$$\Omega_z = \dot{\psi} \cos\theta \qquad \omega_z = \dot{\psi} \cos\theta + p$$

It is important to note that the axes and the body have identical *x*- and *y*-components of angular velocity, but that the *z*-components differ by the relative angular velocity p.

The angular-momentum components from Eq. 7/12 become

$$H_x = I_{xx}\omega_x = I_0\dot{\theta}$$
$$H_y = I_{yy}\omega_y = I_0\dot{\psi} \sin\theta$$
$$H_z = I_{zz}\omega_z = I(\dot{\psi} \cos\theta + p)$$

Substitution of the angular-velocity and angular-momentum components into Eq. 7/19 yields

$$\Sigma M_x = I_0(\ddot{\theta} - \dot{\psi}^2 \sin\theta \cos\theta) + I\dot{\psi}(\dot{\psi} \cos\theta + p) \sin\theta$$
$$\Sigma M_y = I_0(\ddot{\psi} \sin\theta + 2\dot{\psi}\dot{\theta} \cos\theta) - I\dot{\theta}(\dot{\psi} \cos\theta + p) \qquad \textbf{(7/26)}$$
$$\Sigma M_z = I \frac{d}{dt} (\dot{\psi} \cos\theta + p)$$

Equations 7/26 are the general equations of rotation of a symmetrical body about either a fixed point O or the mass center G. In a given problem, the solution to the equations will depend on the moment sums applied to the body about the three coordinate axes. We will confine our use of these equations to two particular cases of rotation about a point that are described in the following sections.

(c) Steady-state precession. We now examine the conditions under which the rotor precesses at a steady rate $\dot{\psi}$ at a constant angle θ and with constant spin velocity p. Thus,

$$\dot{\psi} = \text{constant}, \quad \ddot{\psi} = 0$$
$$\theta = \text{constant}, \quad \dot{\theta} = \ddot{\theta} = 0$$
$$p = \text{constant}, \quad \dot{p} = 0$$

and Eqs. 7/26 become

$$\Sigma M_x = \dot{\psi} \sin\theta[I(\dot{\psi} \cos\theta + p) - I_0\dot{\psi} \cos\theta]$$
$$\Sigma M_y = 0 \qquad\qquad\qquad\qquad\qquad\qquad\qquad \textbf{(7/27)}$$
$$\Sigma M_z = 0$$

From these results, we see that the required moment acting on the rotor about O (or about G) must be in the x-direction since the y- and z-components are zero. Furthermore, with the constant values of θ, $\dot{\psi}$, and p, the moment is constant in magnitude. It is also important to observe that the moment axis is perpendicular to the plane defined by the precession axis (Z-axis) and the spin axis (z-axis).

We may also obtain Eqs. 7/27 by observing that the components of $\mathbf{H}$ remain constant as observed in x-y-z so that $(\dot{\mathbf{H}})_{xyz} = \mathbf{0}$. Since in general $\Sigma\mathbf{M} = (\dot{\mathbf{H}})_{xyz} + \mathbf{\Omega} \times \mathbf{H}$, we have for the case of steady precession

$$\boxed{\Sigma\mathbf{M} = \mathbf{\Omega} \times \mathbf{H}} \qquad (7/28)$$

which reduces to Eqs. 7/27 upon substitution of the values of $\mathbf{\Omega}$ and $\mathbf{H}$.

By far the most common engineering examples of gyroscopic motion occur when precession takes place about an axis that is normal to the rotor axis, as in Fig. 7/14. Thus with the substitution $\theta = \pi/2$, $\omega_z = p$, $\dot{\psi} = \Omega$, and $\Sigma M_x = M$, we have from Eqs. 7/27

$$M = I\Omega p \qquad [7/24]$$

which we derived initially in this article from a direct analysis of this special case.

Now let us examine the steady precession of the rotor (symmetrical top) of Fig. 7/20 for any constant value of θ other than $\pi/2$. The moment ΣM_x about the x-axis is due to the weight of the rotor and is $mg\bar{r} \sin \theta$. Substitution into Eqs. 7/27 and rearrangement of terms give us

$$mg\bar{r} = I\dot{\psi}p - (I_0 - I)\dot{\psi}^2 \cos \theta$$

We see that $\dot{\psi}$ is small when p is large, so that the second term on the right-hand side of the equation becomes very small compared with $I\dot{\psi}p$. If we neglect this smaller term, we have $\dot{\psi} = mg\bar{r}/(Ip)$ which, upon use of the previous substitutions $\Omega = \dot{\psi}$ and $mk^2 = I$, becomes

$$\Omega = \frac{g\bar{r}}{k^2 p} \qquad [7/25]$$

We derived this same relation earlier by assuming that the angular momentum was entirely along the spin axis.

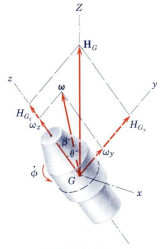

Figure 7/21

(d) Steady precession with zero moment. Consider now the motion of a symmetrical rotor with no external moment about its mass center. Such motion is encountered with spacecraft and projectiles that both spin and precess during flight. Figure 7/21 represents such a body. Here the Z-axis, which has a fixed direction in space, is chosen to coincide with the direction of the angular momentum $\mathbf{H}_G$ which is constant since $\Sigma\mathbf{M}_G = \mathbf{0}$. The x-y-z axes are attached in the manner described in Fig. 7/20. From Fig. 7/21 the three components of momentum are $H_{G_x} = 0$, $H_{G_y} = H_G \sin\theta$, $H_{G_z} = H_G \cos\theta$. From the defining relations, Eqs. 7/12, with the notation of this article, these components are also given by $H_{G_x} = I_0\omega_x$, $H_{G_y} = I_0\omega_y$, $H_{G_z} = I\omega_z$. Thus, $\omega_x = \Omega_x = 0$ so that θ is constant. This result means that the motion is one of steady precession about the constant $\mathbf{H}_G$ vector. With no x-component, the angular velocity $\boldsymbol{\omega}$ of the rotor lies in the y-z plane along with the Z-axis and makes an angle β with the z-axis. The relationship between β and θ is obtained from $\tan\theta = H_{G_y}/H_{G_z} = I_0\omega_y/(I\omega_z)$, which is

$$\tan\theta = \frac{I_0}{I}\tan\beta \qquad (7/29)$$

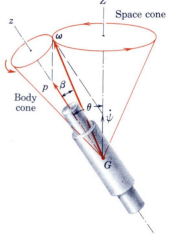

Direct precession $I_0 > I$

(a)

Thus, the angular velocity $\boldsymbol{\omega}$ makes a constant angle β with the spin axis.

The rate of precession is easily obtained from Eq. 7/27 with $M = 0$, which gives

$$\dot{\psi} = \frac{Ip}{(I_0 - I)\cos\theta} \qquad (7/30)$$

It is clear from this relation that the direction of the precession depends on the relative magnitudes of the two moments of inertia.

If $I_0 > I$, then $\beta < \theta$, as indicated in Fig. 7/22a, and the precession is said to be *direct*. Here the body cone rolls on the outside of the space cone.

If $I > I_0$, then $\theta < \beta$, as indicated in Fig. 7/22b, and the precession is said to be *retrograde*. In this instance, the space cone is internal to the body cone, and $\dot{\psi}$ and p have opposite signs.

If $I = I_0$, then $\theta = \beta$ from Eq. 7/29, and Fig. 7/22 shows that both angles must be zero to be equal. For this case, the body has no precession and merely rotates with an angular velocity $\mathbf{p}$. This condition occurs for a body with point symmetry, such as with a homogeneous sphere.

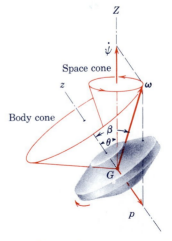

Retrograde precession $I_0 < I$

(b)

Figure 7/22

Sample Problem 7/8

The turbine rotor in a ship's power plant has a mass of 1000 kg, with center of mass at G and a radius of gyration of 200 mm. The rotor shaft is mounted in bearings A and B with its axis in the horizontal fore-and-aft direction and turns counterclockwise at a speed of 5000 rev/min when viewed from the stern. Determine the vertical components of the bearing reactions at A and B if the ship is making a turn to port (left) of 400-m radius at a speed of 25 knots (1 knot = 0.514 m/s). Does the bow of the ship tend to rise or fall because of the gyroscopic action?

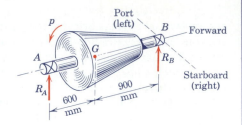

Solution. The vertical component of the bearing reactions will equal the static reactions R_1 and R_2 due to the weight of the rotor, plus or minus the increment ΔR due to the gyroscopic effect. The moment principle from statics easily gives $R_1 = 5886$ N and $R_2 = 3924$ N. The given directions of the spin velocity **p** and the precession velocity **Ω** are shown with the free-body diagram of the rotor. Since the spin axis always tends to rotate toward the torque axis, we see that the torque axis **M** points in the starboard direction as shown. The sense of the ΔR's is, therefore, up at B and down at A to produce the couple **M**. Thus, the bearing reactions at A and B are

$$R_A = R_1 - \Delta R \quad \text{and} \quad R_B = R_2 + \Delta R$$

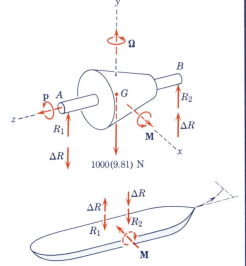

The precession velocity Ω is the speed of the ship divided by the radius of its turn.

$$[v = \rho\Omega] \qquad \Omega = \frac{25(0.514)}{400} = 0.0321 \text{ rad/s}$$

Equation 7/24 is now applied around the mass center G of the rotor to give

$$[M = I\Omega p] \qquad 1.500 \, \Delta R = 1000(0.200)^2(0.0321)\left[\frac{5000(2\pi)}{60}\right]$$

$$\Delta R = 449 \text{ N}$$

The required bearing reactions become

$$R_A = 5886 - 449 = 5437 \text{ N} \quad \text{and} \quad R_B = 3924 + 449 = 4373 \text{ N}$$

$$\textit{Ans.}$$

We now observe from the principle of action and reaction that the forces just computed are those exerted *on* the rotor shaft *by* the structure of the ship. Consequently, the equal and opposite reactions are applied to the ship *by* the rotor shaft, as shown in the bottom sketch. Therefore, the effect of the gyroscopic couple is to generate the increments ΔR shown, and the bow will tend to fall and the stern to rise (but only slightly).

① If the ship is making a left turn, the rotation is counterclockwise when viewed from above, and the precession vector **Ω** is up by the right-hand rule.

② After figuring the correct sense of **M** *on* the rotor, the common mistake is to apply it to the ship in the same sense, forgetting the action-and-reaction principle. Clearly, the results are then reversed. (Be certain not to make this mistake when operating a vertical gyro stabilizer in your yacht to counteract its roll!)

Sample Problem 7/9

A proposed space station is closely approximated by four uniform spherical shells, each of mass m and radius r. The mass of the connecting structure and internal equipment may be neglected as a first approximation. If the station is designed to rotate about its z-axis at the rate of one revolution every 4 seconds, determine (a) the number n of complete cycles of precession for each revolution about the z-axis if the plane of rotation deviates only slightly from a fixed orientation, and (b) find the period τ of precession if the spin axis z makes an angle of $20°$ with respect to the axis of fixed orientation about which precession occurs. Draw the space and body cones for this latter condition.

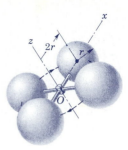

Solution. (a) The number of precession cycles or wobbles for each revolution of the station about the z-axis would be the ratio of the precessional velocity $\dot{\psi}$ to the spin velocity p which, from Eq. 7/30, is

$$\frac{\dot{\psi}}{p} = \frac{I}{(I_0 - I)\cos\theta}$$

The moments of inertia are

$$I_{zz} = I = 4[\tfrac{2}{3}mr^2 + m(2r)^2] = \tfrac{56}{3}mr^2$$

$$I_{xx} = I_0 = 2(\tfrac{2}{3})mr^2 + 2[\tfrac{2}{3}mr^2 + m(2r)^2] = \tfrac{32}{3}mr^2$$

With θ very small, $\cos\theta \cong 1$, and the ratio of angular rates becomes

$$n = \frac{\dot{\psi}}{p} = \frac{\tfrac{56}{3}}{\tfrac{32}{3} - \tfrac{56}{3}} = -\frac{7}{3} \qquad Ans.$$

The minus sign indicates retrograde precession where, in the present case, $\dot{\psi}$ and p are essentially of opposite sense. Thus, the station will make seven wobbles for every three revolutions.

(b) For $\theta = 20°$ and $p = 2\pi/4$ rad/s, the period of precession or wobble is $\tau = 2\pi/|\dot{\psi}|$, so that from Eq. 7/30

$$\tau = \frac{2\pi}{2\pi/4}\left|\frac{I_0 - I}{I}\cos\theta\right| = 4(\tfrac{3}{7})\cos 20° = 1.61 \text{ s} \qquad Ans.$$

The precession is retrograde, and the body cone is external to the space cone as shown in the illustration where the body-cone angle, from Eq. 7/29, is

$$\tan\beta = \frac{I}{I_0}\tan\theta = \frac{56/3}{32/3}(0.3640) = 0.6370 \qquad \beta = 32.50°$$

① Our theory is based on the assumption that $I_{xx} = I_{yy}$ = the moment of inertia about any axis through G perpendicular to the z-axis. Such is the case here, and students should prove it to their own satisfaction.

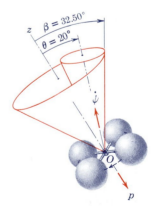

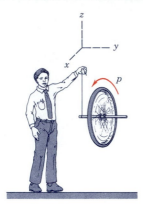

Problem 7/101

PROBLEMS

Introductory problems

7/101 A dynamics instructor demonstrates gyroscopic principles to his students. He suspends a rapidly spinning wheel with a string attached to one end of its horizontal axle. Describe the precession motion of the wheel.

   Ans. Precession is CCW as viewed from above

7/102 If the ship of Sample Problem 7/8 is on a straight course but its bow is falling as it drops into the trough of a wave, determine the direction of the gyroscopic moment exerted *by* the turbine rotor *on* the hull structure and its effect on the motion of the ship.

7/103 A car makes a turn to the right on a level road. Determine whether the normal reaction under the right rear wheel is increased or decreased as a result of the gyroscopic effect of the precessing wheels.

   Ans. Decreased

7/104 The two identical disks are rotating freely on the shaft, with angular velocities equal in magnitude and opposite in direction as shown. The shaft in turn is caused to rotate about the vertical axis in the sense indicated. Prove whether the shaft bends as in *A* or as in *B* because of gyroscopic action.

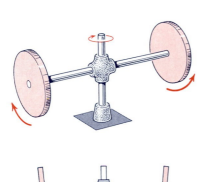

Problem 7/104

7/105 The special-purpose fan is mounted as shown. The motor armature, shaft, and blades have a combined mass of 2.2 kg with radius of gyration of 60 mm. The axial position b of the 0.8-kg block A can be adjusted. With the fan turned off, the unit is balanced about the x-axis when $b = 180$ mm. The motor and fan operate at 1725 rev/min in the direction shown. Determine the value of b that will produce a steady precession of 0.2 rad/s about the positive y-axis.

Ans. $b = 216$ mm

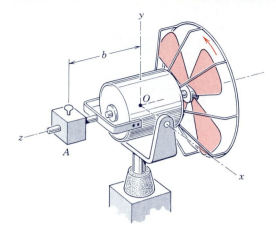

Problem 7/105

7/106 The blades and hub of the helicopter rotor weigh 140 lb and have a radius of gyration of 10 ft about the z-axis of rotation. With the rotor turning at 500 rev/min during a short interval following vertical lift-off, the helicopter tilts forward at the rate $\dot{\theta} = 10$ deg/sec in order to acquire forward velocity. Determine the gyroscopic moment M transmitted to the body of the helicopter by its rotor and indicate whether the helicopter tends to deflect clockwise or counterclockwise, as viewed by a passenger facing forward.

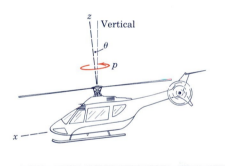

Problem 7/106

7/107 The 210-kg rotor of a turbojet aircraft engine has a radius of gyration of 220 mm and rotates counterclockwise at 18 000 rev/min when viewed from the front. If the aircraft is traveling at 1200 km/h and starts to execute an inside vertical loop of 3800-m radius, compute the gyroscopic moment M transmitted to the airframe. What correction to the controls does the pilot have to make in order to remain in the vertical plane?

Ans. $M = 1681$ N·m, Left rudder

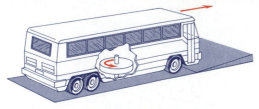

Problem 7/108

7/108 An experimental antipollution bus is powered by the kinetic energy stored in a large flywheel that spins at a high speed p in the direction indicated. As the bus encounters a short upward ramp, the front wheels rise, thus causing the flywheel to precess. What changes occur to the forces between the tires and the road during this sudden change?

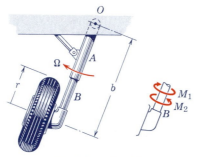

Problem 7/109

7/109 An airplane has just cleared the runway with a take-off speed v. Each of its freely spinning wheels has a mass m, with a radius of gyration k about its axle. As seen from the front of the airplane, the wheel precesses at the angular rate Ω as the landing strut is folded into the wing about its pivot O. As a result of the gyroscopic action, the supporting member A exerts a torsional moment M on B to prevent the tubular member from rotating in the sleeve at B. Determine M and identify whether it is in the sense of M_1 or M_2.

$$\text{Ans. } M = M_1 = mk^2\Omega\,\frac{v}{r}$$

Representative problems

7/110 A small air compressor for an aircraft cabin consists of the 3.50-kg turbine A that drives the 2.40-kg blower B at a speed of 20 000 rev/min. The shaft of the assembly is mounted transversely to the direction of flight and is viewed from the rear of the aircraft in the figure. The radii of gyration of A and B are 79.0 and 71.0 mm, respectively. Calculate the radial forces exerted on the shaft by the bearings at C and D if the aircraft executes a clockwise roll (rotation about the longitudinal flight axis) of 2 rad/s viewed from the rear of the aircraft. Neglect the small moments caused by the weights of the rotors. Draw a free-body diagram of the shaft as viewed from above and indicate the shape of its deflected centerline.

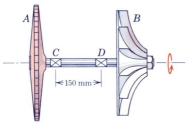

Problem 7/110

7/111 The figure shows a gyro mounted with vertical axis and used to stabilize a hospital ship against rolling. The motor A turns the pinion that precesses the gyro by rotating the large precession gear B and attached rotor assembly about a horizontal transverse axis in the ship. The rotor turns inside the housing at a clockwise speed of 960 rev/min as viewed from the top and has a mass of 80 Mg with radius of gyration of 1.45 m. Calculate the moment exerted on the hull structure by the gyro if the motor turns the precession gear B at the rate of 0.320 rad/s. In which of the two directions, (a) or (b), should the motor turn in order to counteract a roll of the ship to port?

Ans. $M = 5410$ kN·m, direction (b)

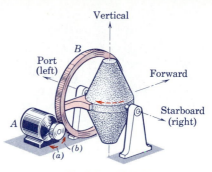

Problem 7/111

7/112 The electric motor has a total weight of 20 lb and is supported by the mounting brackets A and B attached to the rotating disk. The armature of the motor has a weight of 5 lb and a radius of gyration of 1.5 in. and turns counterclockwise at a speed of 1725 rev/min as viewed from A to B. The turntable revolves about its vertical axis at the constant rate of 48 rev/min in the direction shown. Determine the vertical components of the forces supported by the mounting brackets at A and B.

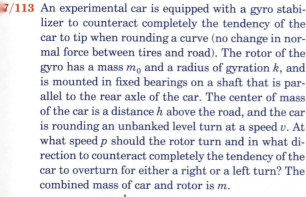

Problem 7/112

7/113 An experimental car is equipped with a gyro stabilizer to counteract completely the tendency of the car to tip when rounding a curve (no change in normal force between tires and road). The rotor of the gyro has a mass m_0 and a radius of gyration k, and is mounted in fixed bearings on a shaft that is parallel to the rear axle of the car. The center of mass of the car is a distance h above the road, and the car is rounding an unbanked level turn at a speed v. At what speed p should the rotor turn and in what direction to counteract completely the tendency of the car to overturn for either a right or a left turn? The combined mass of car and rotor is m.

Ans. $p = \dfrac{mvh}{m_0 k^2}$, opposite direction to car wheels

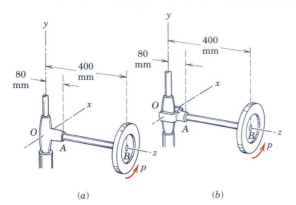

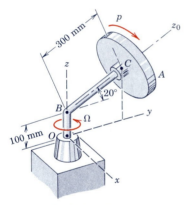

(a)   (b)

Problem 7/114

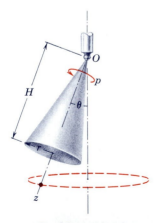

Problem 7/116

Problem 7/117

7/114 Each of the identical wheels has a mass of 4 kg and a radius of gyration $k_z = 120$ mm and is mounted on a horizontal shaft AB secured to the vertical shaft at O. In case (a), the horizontal shaft is fixed to a collar at O that is free to rotate about the vertical y-axis. In case (b), the shaft is secured by a yoke hinged about the x-axis to the collar. If the wheel has a large angular velocity $p = 3600$ rev/min about its z-axis in the position shown, determine any precession that occurs and the bending moment M_A in the shaft at A for each case. Neglect the small mass of the shaft and fitting at O.

7/115 If the wheel in case (a) of Prob. 7/114 is forced to precess about the vertical by a mechanical drive at the steady rate $\Omega = 2\mathbf{j}$ rad/s, determine the bending moment in the horizontal shaft at A. In the absence of friction, what torque M_O is applied to the collar at O to sustain this motion?

       Ans. $M_A = 30.9$ N·m, $M_O = 0$

7/116 The 5-kg disk and hub A have a radius of gyration of 85 mm about the z_0-axis and spin at the rate $p = 1250$ rev/min. Simultaneously, the assembly rotates about the vertical z-axis at the rate $\Omega = 400$ rev/min. Calculate the gyroscopic moment $\mathbf{M}$ exerted *on* the shaft at C *by* the disk and the bending moment M_A in the shaft at O. Neglect the mass of the shaft but otherwise account for all forces acting on it.

7/117 A solid right-circular cone of mass m, base radius $r = 100$ mm, and altitiude $H = 3r = 300$ mm is suspended freely at its vertex O and is spinning about its geometric axis at the rate $p = 2700$ rev/min. Determine the period for steady slow precession of the cone about the vertical. Does the result depend on the angle θ made by the cone axis with the vertical?   *Ans.* $\tau = 2.41$ s

7/118 The 4-oz top with radius of gyration about its spin axis of 0.62 in. is spinning at the rate $p = 3600$ rev/min in the sense shown, with its spin axis making an angle $\theta = 20°$ with the vertical. The distance from its tip O to its mass center G is $\bar{r} = 2.5$ in. Determine the precession Ω of the top and explain why θ gradually decreases as long as the spin rate remains large. An enlarged view of the contact of the tip is shown.

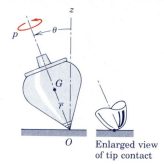

Problem 7/118

7/119 The 8-lb rotor with radius of gyration of 3 in. rotates on ball bearings at a speed of 3000 rev/min about its shaft OG. The shaft is free to pivot about the X-axis, as well as to rotate about the Z-axis. Calculate the vector Ω for precession about the Z-axis. Neglect the mass of shaft OG and compute the gyroscopic couple $\mathbf{M}$ exerted by the shaft on the rotor at G.

　　　Ans. $\Omega = -1.23\mathbf{K}$ rad/sec, $\mathbf{M} = 67.7\mathbf{i}$ lb-in.

7/120 The two solid cones with the same base and equal altitudes are spinning in space about their common axis at the rate p. For what ratio h/r will precession of their spin axis be impossible?

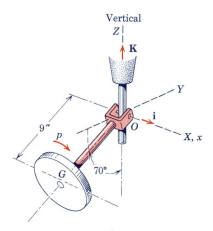

Problem 7/119

7/121 The two identical circular disks, each of mass m and radius r, are spinning as a rigid unit about their common axis. Determine the value of b for which no precessional motion can take place if the unit is free to move in space.　　　*Ans.* $b = r$

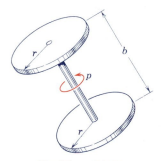

Problem 7/121

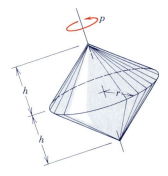

Problem 7/120

Problem 7/122

7/122 The thin ring is projected into the air with a spin velocity of 300 rev/min. If its geometric axis is observed to have a very slight precessional wobble, determine the frequency f of the wobble.

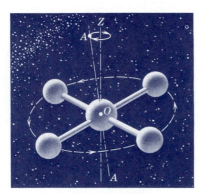

Problem 7/123

7/123 The primary structure of a proposed space station consists of five spherical shells connected by tubular spokes. The moment of inertia of the structure about its geometric axis A-A is twice as much as that about any axis through O normal to A-A. The station is designed to rotate about its geometric axis at the constant rate of 3 rev/min. If the spin axis A-A precesses about the Z-axis of fixed orientation and makes a very small angle with it, calculate the rate $\dot{\psi}$ at which the station wobbles. The mass center O has negligible acceleration.

       Ans. $\dot{\psi} = -6$ rev/min, retrograde precession

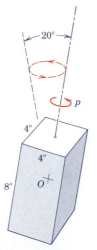

Problem 7/124

7/124 The rectangular bar is spinning in space about its longitudinal axis at the rate $p = 200$ rev/min. If its axis wobbles through a total angle of 20° as shown, calculate the period τ of the wobble.

7/125 A boy throws a thin circular disk (like a Frisbee) with a spin rate of 300 rev/min. The plane of the disk is seen to wobble through a total angle of 10°. Calculate the period τ of the wobble and indicate whether the precession is direct or retrograde.

Ans. $\tau = 0.0996$ s, retrograde

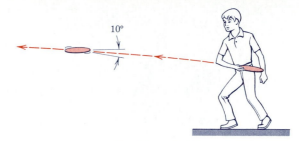

Problem 7/125

7/126 The figure shows a football in three common inflight configurations. Case (a) is a perfectly thrown spiral pass with a spin rate of 120 rev/min. Case (b) is a wobbly spiral pass again with a spin rate of 120 rev/min about its own axis, but with the axis wobbling through a total angle of 20°. Case (c) is an end-over-end place kick with a rotational rate of 120 rev/min. For each case, specify the values of p, θ, β, and ψ as defined in this article. The moment of inertia about the ball's long axis is 0.3 of that about the transverse axis of symmetry.

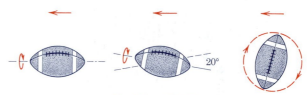

Problem 7/126

7/127 The spacecraft shown is symmetrical about its z-axis and has a radius of gyration of 720 mm about this axis. The radii of gyration about the x- and y-axes through the mass center are both equal to 540 mm. When moving in space, the z-axis is observed to generate a cone with a total vertex angle of 4° as it precesses about the axis of total angular momentum. If the spacecraft has a spin velocity $\dot{\phi}$ about its z-axis of 1.5 rad/s, compute the period τ of each full precession. Is the spin vector in the positive or negative z-direction?

Ans. $\tau = 1.831$ s
spin vector in negative z-direction

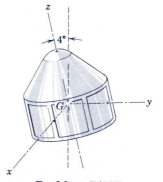

Problem 7/127

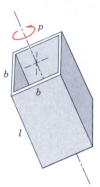

Problem 7/128

7/128 The open-ended thin-walled rectangular box of square cross section is rotating in space about its central longitudinal axis as shown. If the axis has a slight wobble, for what ratios l/b will the motion be direct or retrograde precession?

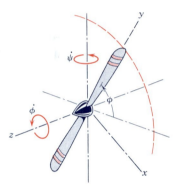

Problem 7/129

7/129 The two-bladed airplane propeller has a constant angular velocity $\dot{\phi}$ about its shaft z-axis. The airplane is turning to the pilot's left so that the airplane and propeller have a constant angular velocity $\dot{\psi}$ (precession) about a vertical axis. Determine the x-, y-, and z-components of the moment exerted on the propeller by its shaft as functions of the angular position ϕ of the propeller. The moments of inertia about the x-, y-, and z-axes are I_{xx}, 0, I_{zz}, respectively. (*Suggestion:* With axes x-y-z attached to the propeller, make use of Eqs. 7/20.)

$$Ans. \quad M_x = (I_{xx} + I_{zz})\dot{\psi}\dot{\phi} \sin \phi$$
$$M_y = (I_{zz} - I_{xx})\dot{\psi}\dot{\phi} \cos \phi$$
$$M_z = \tfrac{1}{2}I_{xx}\dot{\psi}^2 \sin 2\phi$$

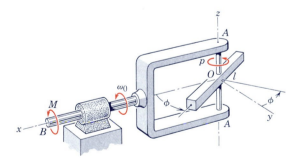

Problem 7/130

7/130 The uniform slender bar of mass m and length l is centrally mounted on the shaft A-A, about which it rotates with a constant speed $\dot{\phi} = p$. Simultaneously, the yoke is forced to rotate about the x-axis with a constant speed ω_0. As a function of ϕ, determine the magnitude of the torque M required to maintain the constant speed ω_0. (*Hint:* Apply Eq. 7/19 to obtain the x-component of M.)

7/131 Let the rotor of Fig. 7/18 have a spin velocity of 3600 rev/min and execute steady precession about the vertical Z-axis at the constant angle $\theta = 30°$. Furthermore, $\bar{r} = 200$ mm, $k = 150$ mm, and the moment of inertia about the x-axis is five times that about the spin axis. Calculate the precession $\dot{\psi} = \Omega$ using Eq. 7/25 and obtain a close approximation to the percentage error e in this value.

$$Ans. \quad \dot{\psi} = 0.231 \text{ rad/s}, \; e = 0.21\% \text{ too low}$$

7/132 The housing of the electric motor is freely pivoted about the horizontal x-axis which passes through the mass center G of the rotor. If the motor is turning at the constant rate $\dot{\phi} = p$, determine the angular acceleration $\ddot{\psi}$ which will result from the application of the moment M about the vertical shaft if $\dot{\gamma} = \dot{\psi} = 0$. The mass of the frame and housing is considered negligible compared with the mass m of the rotor. The radius of gyration of the rotor about the z-axis is k_z and that about the x-axis is k_x.

$$Ans. \quad \ddot{\psi} = \frac{M/m}{k_x{}^2 \cos^2\gamma + k_z{}^2 \sin^2\gamma}$$

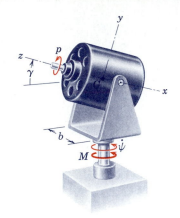

Problem 7/132

7/133 The earth-scanning satellite is in a circular orbit of period τ. The angular velocity of the satellite about its y- or pitch-axis is $\omega = 2\pi/\tau$, and the angular rates about the x- and z-axes are zero. Thus, the x-axis of the satellite always points to the center of the earth. The satellite has a reaction-wheel attitude-control system consisting of the three wheels shown, each of which may be variably torqued by its individual motor. The angular rate Ω_z of the z-wheel relative to the satellite is Ω_0 at time $t = 0$, and the x- and y-wheels are at rest relative to the satellite at $t = 0$. Determine the axial torques M_x, M_y, and M_z that must be exerted by the motors on the shafts of their respective wheels in order that the angular velocity $\boldsymbol{\omega}$ of the satellite will remain constant. The moment of inertia of each reaction wheel about its axis is I. The x and z reaction-wheel speeds are harmonic functions of the time with a period equal to that of the orbit. Plot the variations of the torques and the relative wheel speeds Ω_x, Ω_y, and Ω_z as functions of the time during one orbit period. (*Hint:* The torque to accelerate the x-wheel equals the reaction of the gyroscopic moment on the z-wheel, and vice versa.)

$$Ans. \quad M_x = -I\omega\Omega_0 \cos \omega t$$
$$M_y = 0$$
$$M_z = -I\omega\Omega_0 \sin \omega t$$

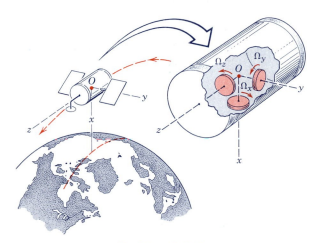

Problem 7/133

▶ **7/134** The steady-state precession of an axisymmetric body is described by Eq. 7/27, which is a quadratic equation in $\dot{\psi}$. Let $\Sigma M_x = M$ and determine the two roots $\dot{\psi}_1$ and $\dot{\psi}_2$ for large values of $\dot{\phi}$. First, express the solution in the general form $\dot{\psi} = [\ \ ](1 \pm \sqrt{1 - \{\ \ \}})$. Then expand the radical into a convergent series, retaining only the first two terms. Compare $\dot{\psi}_1 = \dot{\psi}_{(-)}$ with the precession derived for the rotor of Fig. 7/18 in the earlier part of the article. The root $\dot{\psi}_2 = \dot{\psi}_{(+)}$ represents a fast precession which occurs at a higher energy level difficult to achieve experimentally.

$$Ans. \quad \dot{\psi}_1 = \frac{M}{Ip \sin \theta}, \ \dot{\psi}_2 = \frac{Ip}{(I_0 - I) \cos \theta}$$

$$where \ p^2 > \frac{4}{I^2} M(I_0 - I) \cot \theta$$

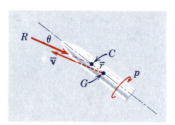

Problem 7/135

▶ **7/135** A projectile moving through the atmosphere with a velocity $\bar{\mathbf{v}}$ that makes a small angle θ with its geometric axis is subjected to a resultant aerodynamic force $\mathbf{R}$ essentially opposite in direction to $\bar{\mathbf{v}}$ as shown. If $\mathbf{R}$ passes through a point C slightly ahead of the mass center G, determine the expression for the minimum spin velocity p for which the projectile will be spin-stabilized with $\dot{\theta} = 0$. The moment of inertia about the spin axis is I and that about a transverse axis through G is I_0. (*Hint:* Determine M_x and substitute into Eq. 7/27. Express the result as a quadratic equation in $\dot{\psi}$ and determine the minimum value of p for which the expression under the radical is positive.)

$$Ans. \quad p_{min} = \frac{2}{I} \sqrt{R\bar{r}(I_0 - I) \cos \theta}$$

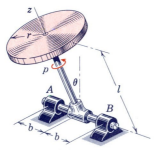

Problem 7/136

▶ **7/136** The solid circular disk of mass m and small thickness is spinning freely on its shaft at the rate p. If the assembly is released in the vertical position at $\theta = 0$ with $\dot{\theta} = 0$, determine the horizontal components of the forces A and B exerted by the respective bearings on the horizontal shaft as the position $\theta = \pi/2$ is passed. Neglect the mass of the two shafts compared with m and neglect all friction. Solve by using the appropriate moment equations.

$$Ans. \quad A_z = -\frac{m\dot{\theta}}{2}\left(\frac{r^2}{2b}p + l\dot{\theta}\right)$$

$$B_z = \frac{m\dot{\theta}}{2}\left(\frac{r^2}{2b}p - l\dot{\theta}\right)$$

$$where \ \dot{\theta} = 2\sqrt{\frac{2gl}{r^2 + 4l^2}}$$

▸ **7/137** Derive Eq. 7/24 by relating the forces to the accelerations for a differential element of the thin ring of mass m. The ring has a constant angular velocity p about the z-axis and is given an additional constant angular velocity Ω about the y-axis by the application of an external moment M (not shown). (*Hint:* The acceleration of the element in the z-direction is due to (*a*) the change in magnitude of its velocity component in this direction resulting from Ω and (*b*) the change in direction of its x-component of velocity.)

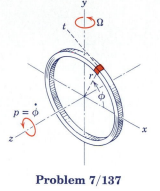

Problem 7/137

7/138 The elements of a gyrocompass are shown in the figure where the rotor, at north latitude γ on the surface of the earth, is mounted in a single gimbal ring which is free to rotate about the fixed vertical y-axis. The rotor axis is then able to rotate in the horizontal x-z plane, as measured by the angle β from the north direction. Assume that the gyro spins at the rate p and that it has mass moments of inertia about the spin axis and transverse axis through G of I and I_0, respectively. Show that the gyro axis oscillates about the north direction according to the equation $\ddot{\beta} + K^2\beta = 0$, where $K^2 = I\omega_0 p \cos \gamma / I_0$, and that the period of oscillation about the north direction for small values of β is $\tau = 2\pi\sqrt{I_0/(I\omega_0 p \cos \gamma)}$. (*Hint:* The components of the angular velocity Ω of the axes in terms of the angular velocity ω_0 of the earth are

$$\Omega_x = -\omega_0 \cos \gamma \sin \beta$$
$$\Omega_y = \omega_0 \sin \gamma + \dot{\beta}$$
$$\Omega_z = \omega_0 \cos \gamma \cos \beta$$

which may be used in the y-component of the moment equations, Eqs. 7/19, to determine β as a function of time. Note that the square of the angular velocity ω_0 of the earth is small and may be neglected compared with the product $\omega_0 p$.)

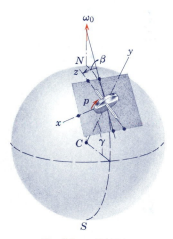

Problem 7/138

Vibration is an important subset of the discipline of dynamics. Bodies that are subjected to periodic disturbances can exhibit motion characterized by very large, even destructive, amplitudes. Shown here is an automobile whose suspension system is being excited by actuators placed beneath the tires. With such laboratory testing, a variety of road conditions and vehicle speeds can be simulated.

VIBRATION AND TIME RESPONSE

8

8/1 INTRODUCTION

An important and special class of problems in dynamics deals with the linear and angular motions of bodies that oscillate or respond to applied disturbances in the presence of restoring forces. A few examples of this class of dynamics problems are the response of an engineering structure to earthquakes, the vibration of an unbalanced rotating machine, the time response of the plucked string of a musical instrument, the wind-induced vibration of power lines, and the flutter of aircraft wings. In many cases, excessive vibration levels must be reduced due to material limitations or human factors.

In the analysis of every engineering problem, the system under scrutiny must be represented by a physical model. It is often permissible to represent a *continuous* or *distributed-parameter system* (one in which the mass and spring elements are continuously spread over space) by a *discrete* or *lumped-parameter model* (one in which the mass and spring elements are separate and concentrated). Such a modeling scheme is especially desirable when some portions of a continuous system are relatively massive in comparison to other portions. For example, the physical model of a ship propeller shaft is often assumed to be a massless but twistable rod with a disk rigidly attached to each end—one disk representing the turbine and the other representing the propeller. As a second example, we observe that the mass of springs may often be neglected in comparison to that of attached bodies. It should be noted that not every system is reducible to a discrete model. For example, the transverse vibration of a diving board after the departure of the diver is a somewhat difficult problem of distributed-parameter vibration. In this chapter, we shall begin the study of discrete systems, limiting our discussion to those whose configurations may be described with one displacement variable. Such systems are said to possess *one degree of freedom*. For a more detailed study that includes the treatment of two or more degrees of freedom and continuous systems, the student should consult one of the many textbooks devoted solely to the subject of vibrations.

Chapter 8 is divided into four subsequent sections: Article 8/2 treats the free vibration of particles and Art. 8/3 introduces the forced vibration of particles. Each of these two articles is subdivided into undamped- and damped-motion categories. In Art. 8/4 we discuss the vibration of rigid bodies. Finally, an energy approach to the solution of vibration problems is presented in Art. 8/5. The study of this chapter is greatly facilitated if one observes that the topic of vibrations is a direct application of the principles of kinetics as developed in Chapters 3 and 6. In particular, a complete free-body diagram *drawn for an arbitrary positive value of the displacement variable*, followed by application of the appropriate governing equations of dynamics, will yield the equation of motion. From this equation of motion, which is a second-order ordinary differential equation, one may obtain all information of interest, such as the motion frequency, period, or the motion itself as a function of time.

8/2   *FREE VIBRATION OF PARTICLES*

When a spring-mounted body is disturbed from its equilibrium position, its ensuing motion in the absence of any imposed external forces is termed *free vibration*. In every actual case of free vibration, there exists some retarding or damping force which tends to diminish the motion. Common damping forces are those due to mechanical and fluid friction. In part (*a*) we consider the ideal case where the damping forces are small enough to be neglected. In part (*b*) we treat the case where the damping is appreciable and must be accounted for.

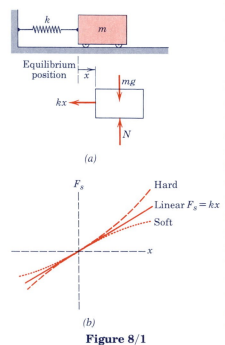

(a)

(b)

Figure 8/1

(a) Undamped free vibration.   We begin by considering the horizontal vibration of the simple frictionless spring-mass system of Fig. 8/1*a*. Note that the variable x denotes the displacement of the mass from the equilibrium position, which, for this system, is also the position of zero spring deflection. Figure 8/1*b* shows a plot of the force F_s necessary to deflect the spring versus the corresponding spring deflection for various types of springs. Although nonlinear hard and soft springs are useful in some applications, we will restrict our attention to the linear spring. Such a spring exerts a restoring force $-kx$ on the mass—that is, when the mass is displaced to the right, the spring force is to the left, and vice versa. We must be careful to distinguish between the forces of magnitude F_s that must be applied at both ends of the massless spring to cause tension or compression and the force $F = -kx$ of equal magnitude that the spring exerts on the mass. The constant of proportionality k is known as the *spring constant, modulus,* or *stiffness* and has the units N/m or lb/ft.

The equation of motion for the body of Fig. 8/1*a* is obtained by first drawing its free-body diagram. Applying Newton's second law in the form $\Sigma F_x = m\ddot{x}$ gives

$$-kx = m\ddot{x} \quad \text{or} \quad m\ddot{x} + kx = 0 \qquad (8/1)$$

The oscillation of a mass subjected to a linear restoring force as described by this equation is called *simple harmonic motion* and is characterized by acceleration that is proportional to the displacement but of opposite sign. Equation 8/1 is normally written as

$$\ddot{x} + \omega_n{}^2 x = 0 \qquad (8/2)$$

where

$$\omega_n = \sqrt{k/m} \qquad (8/3)$$

is a convenient substitution whose physical significance will be clarified shortly.

Because we anticipate an oscillatory motion, we look for a solution which gives x as a periodic function of time. Thus, a logical choice is

$$x = A \cos \omega_n t + B \sin \omega_n t \qquad (8/4)$$

or, alternatively,

$$x = C \sin (\omega_n t + \psi) \qquad (8/5)$$

Direct substitution of these expressions into Eq. 8/2 verifies that each expression is a valid solution to the equation of motion. The constants A and B, or C and ψ, are determined from knowledge of the initial displacement x_0 and initial velocity $\dot{x}_0$ of the mass. For example, if we work with the solution form of Eq. 8/4 and evaluate x and $\dot{x}$ at time $t = 0$, we obtain

$$x_0 = A \quad \text{and} \quad \dot{x}_0 = B\omega_n$$

Substitution of these values of A and B into Eq. 8/4 yields

$$x = x_0 \cos \omega_n t + \frac{\dot{x}_0}{\omega_n} \sin \omega_n t \qquad (8/6)$$

The constants C and ψ of Eq. 8/5 can be determined in terms of given initial conditions in a similar manner. Evaluation of Eq. 8/5 and its first time derivative at $t = 0$ gives

$$x_0 = C \sin \psi \quad \text{and} \quad \dot{x}_0 = C\omega_n \cos \psi$$

Solving for C and ψ yields

$$C = \sqrt{x_0{}^2 + (\dot{x}_0/\omega_n)^2} \qquad \psi = \tan^{-1}(x_0\omega_n/\dot{x}_0)$$

Substitution of these values into Eq. 8/5 gives

$$x = \sqrt{x_0{}^2 + (\dot{x}_0/\omega_n)^2} \sin [\omega_n t + \tan^{-1}(x_0\omega_n/\dot{x}_0)] \qquad (8/7)$$

Equations 8/6 and 8/7 represent two different mathematical expressions for the same time-dependent motion. We observe that $C = \sqrt{A^2 + B^2}$ and $\psi = \tan^{-1}(A/B)$.

The motion may be represented graphically, Fig. 8/2, where x is seen to be the projection onto a vertical axis of the rotating vector of length C. The vector rotates at the constant angular velocity $\omega_n = \sqrt{k/m}$ that is called the *natural circular frequency* and has the units radians per second. The number of complete cycles per unit

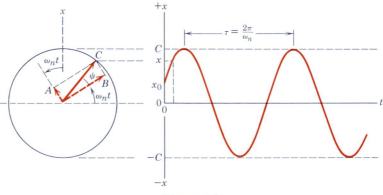

Figure 8/2

time is the *natural frequency* $f_n = \omega_n/2\pi$ and is expressed in hertz (1 hertz (Hz) = 1 cycle per second). The time required for one complete motion cycle (one rotation of the reference vector) is the *period* of the motion and is given by $\tau = 1/f_n = 2\pi/\omega_n$. We also see from the figure that x is the sum of the projections onto the vertical axis of two perpendicular vectors whose magnitudes are A and B and whose vector sum C is the *amplitude*. Vectors A, B, and C rotate together with the constant angular velocity ω_n. Thus, as we have already seen, $C = \sqrt{A^2 + B^2}$ and $\psi = \tan^{-1}(A/B)$.

As a further note on the free undamped vibration of particles, we see that, if the system of Fig. 8/1a is rotated 90° clockwise to obtain the system of Fig. 8/3 where the motion is vertical rather than horizontal, the equation of motion (and therefore all system properties) is unchanged if we continue to define x as the displacement from the equilibrium position. The equilibrium position now involves a nonzero spring deflection δ_{st}. From the free-body diagram of Fig. 8/3, Newton's second law gives

$$-k(\delta_{st} + x) + mg = m\ddot{x}$$

At the equilibrium position $x = 0$, the force sum must be zero, so that

$$-k\delta_{st} + mg = 0$$

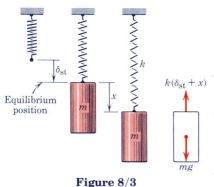

Figure 8/3

Thus, we see that the pair of forces $-k\delta_{st}$ and mg on the left side of the motion equation cancel, giving

$$m\ddot{x} + kx = 0$$

which is identical to Eq. 8/1. The lesson here is that by defining the displacement variable to be zero at equilibrium rather than at the position of zero spring deflection, we may ignore the equal and opposite forces associated with equilibrium.*

(b) Damped free vibration. Every mechanical system possesses some inherent degree of friction, which acts as a consumer of mechanical energy. Precise mathematical models of the dissipative friction forces are usually complex. The dashpot or viscous damper is a device intentionally added to systems for the purpose of limiting or retarding vibration. It consists of a cylinder filled with a viscous fluid and a piston with holes or other passages by which the fluid can flow from one side of the piston to the other. Simple dashpots arranged as shown schematically in Fig. 8/4a exert a force F_d whose magnitude is proportional to the velocity of the mass, as depicted in Fig. 8/4b. The constant of proportionality c is known as the *viscous damping coefficient* and has units of $N \cdot s/m$ or lb-sec/ft. The direction of the damping force as applied to the mass is opposite to that of the velocity $\dot{x}$. Hence, the force on the mass is $-c\dot{x}$.

Complex dashpots with internal flow-rate-dependent one-way valves can produce a different damping coefficient in extension than in compression; nonlinear characteristics are also possible. We shall restrict our attention to the simple linear dashpot.

The equation of motion for the body with damping is determined from the free-body diagram as shown in Fig. 8/4a. Newton's second law gives

$$-kx - c\dot{x} = m\ddot{x} \quad \text{or} \quad m\ddot{x} + c\dot{x} + kx = 0 \quad (8/8)$$

In addition to the variable substitution $\omega_n = \sqrt{k/m}$, it is convenient, for reasons that will shortly become evident, to introduce the combination of constants

$$\zeta = c/(2m\omega_n)$$

The quantity ζ (zeta) is called the *viscous damping factor* or *damping ratio* and is a measure of the severity of the damping. The student should verify that ζ is nondimensional. Equation 8/8 may now be written as

$$\ddot{x} + 2\zeta\omega_n\dot{x} + \omega_n^2 x = 0 \quad (8/9)$$

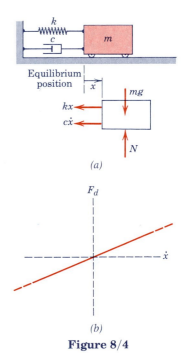

(a)

(b)

Figure 8/4

*For nonlinear systems, all forces, including the static forces associated with equilibrium, should be included in the analysis.

In order to solve the equation of motion, Eq. 8/9, we assume solutions of the form

$$x = Ae^{\lambda t}$$

Substitution into Eq. 8/9 yields the characteristic equation

$$\lambda^2 + 2\zeta\omega_n\lambda + \omega_n{}^2 = 0$$

whose roots are

$$\lambda_1 = \omega_n(-\zeta + \sqrt{\zeta^2 - 1}) \qquad \lambda_2 = \omega_n(-\zeta - \sqrt{\zeta^2 - 1})$$

By superposition, the general solution is

$$\begin{aligned} x &= A_1 e^{\lambda_1 t} + A_2 e^{\lambda_2 t} \\ &= A_1 e^{(-\zeta + \sqrt{\zeta^2 - 1})\omega_n t} + A_2 e^{(-\zeta - \sqrt{\zeta^2 - 1})\omega_n t} \end{aligned} \tag{8/10}$$

Since $0 \le \zeta \le \infty$, the radicand $(\zeta^2 - 1)$ may be positive, negative, or even zero, giving rise to the following three categories of damped motion:

I. $\zeta > 1$ (*overdamped*). The roots λ_1 and λ_2 are distinct real negative numbers. The motion as given by Eq. 8/10 decays so that x approaches zero for large values of time t. There is no oscillation and therefore no period associated with the motion.

II. $\zeta = 1$ (*critically damped*). The roots λ_1 and λ_2 are equal real negative numbers ($\lambda_1 = \lambda_2 = -\omega_n$) and the solution to the differential equation for the special case of equal roots is given by

$$x = (A_1 + A_2 t)e^{-\omega_n t}$$

Again, the motion decays with x approaching zero for large time, and the motion is nonperiodic. A critically damped system, when excited with an initial velocity or displacement (or both), will approach equilibrium faster than will an overdamped system. Figure 8/5 depicts an actual

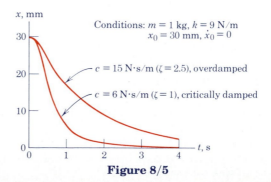

Figure 8/5

response for both an overdamped and a critically damped system to an initial displacement x_0 with no initial velocity $\dot{x}_0$.

III. $\zeta < 1$ (*underdamped*). Noting that the radicand $(\zeta^2 - 1)$ is negative and recalling that $e^{(a+b)} = e^a e^b$ enable us to rewrite Eq. 8/10 as

$$x = \{A_1 e^{i\sqrt{1-\zeta^2}\,\omega_n t} + A_2 e^{-i\sqrt{1-\zeta^2}\,\omega_n t}\}e^{-\zeta\omega_n t}$$

where $i = \sqrt{-1}$. It is convenient to let a new variable ω_d represent the combination $\omega_n\sqrt{1 - \zeta^2}$. Thus,

$$x = \{A_1 e^{i\omega_d t} + A_2 e^{-i\omega_d t}\}e^{-\zeta\omega_n t}$$

Use of the Euler formula $e^{\pm ix} = \cos x \pm i \sin x$ allows the previous equation to be written as

$$
\begin{aligned}
x &= \{A_1(\cos \omega_d t + i \sin \omega_d t) \\
&\quad + A_2(\cos \omega_d t - i \sin \omega_d t)\}e^{-\zeta\omega_n t} \\
&= \{(A_1 + A_2) \cos \omega_d t + i(A_1 - A_2) \sin \omega_d t\}e^{-\zeta\omega_n t} \\
&= \{A_3 \cos \omega_d t + A_4 \sin \omega_d t\}e^{-\zeta\omega_n t} \quad\quad (8/11)
\end{aligned}
$$

where $A_3 = (A_1 + A_2)$ and $A_4 = i(A_1 - A_2)$. We have shown with Eqs. 8/4 and 8/5 that the sum of two harmonics, such as those in the braces of Eq. 8/11, can be replaced by a single trigonometric function that involves a phase angle. Thus, Eq. 8/11 can be written as

$$x = \{C \sin (\omega_d t + \psi)\}e^{-\zeta\omega_n t}$$

or

$$x = Ce^{-\zeta\omega_n t} \sin (\omega_d t + \psi) \quad\quad (8/12)$$

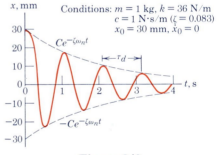

Figure 8/6

Conditions: $m = 1$ kg, $k = 36$ N/m
$c = 1$ N·s/m ($\zeta = 0.083$)
$x_0 = 30$ mm, $\dot{x}_0 = 0$

Equation 8/12 represents an exponentially decreasing harmonic function, as shown in Fig. 8/6 for specific numerical values. The frequency

$$\omega_d = \omega_n\sqrt{1 - \zeta^2}$$

is called the *damped natural frequency*. The *damped period* is given by $\tau_d = 2\pi/\omega_d = 2\pi/(\omega_n\sqrt{1 - \zeta^2})$.

It is important to note that the expressions developed for the constants C and ψ in terms of initial conditions for the case of no damping in part (a) are not valid for the case of damping in part (b). To find C and ψ if damping is present, one must begin anew, setting the general displacement expression of Eq. 8/12 and its first time derivative, both evaluated at time $t = 0$, equal to the initial displacement x_0 and initial velocity $\dot{x}_0$, respectively.

It is frequently desirable to experimentally determine the value of the damping ratio ζ for an underdamped system. The usual reason is that the value of the viscous damping coefficient c is not otherwise well known. The system is excited by initial conditions and a plot of the displacement x versus time t, such as that shown schematically in Fig. 8/7, is generated. Two successive amplitudes x_1 and x_2 are measured and their ratio

$$\frac{x_1}{x_2} = \frac{Ce^{-\zeta\omega_n t_1}}{Ce^{-\zeta\omega_n(t_1 + \tau_d)}} = e^{\zeta\omega_n \tau_d}$$

is formed. The *logarithmic decrement* δ is defined as

$$\delta = \ln\left(\frac{x_1}{x_2}\right) = \zeta\omega_n\tau_d = \zeta\omega_n\frac{2\pi}{\omega_n\sqrt{1 - \zeta^2}} = \frac{2\pi\zeta}{\sqrt{1 - \zeta^2}}$$

From this equation, we may solve for ζ and obtain

$$\zeta = \frac{\delta}{\sqrt{(2\pi)^2 + \delta^2}}$$

For a small damping ratio, $x_1 \cong x_2$ and $\delta << 1$, so that $\zeta \cong \delta/2\pi$. If x_1 and x_2 are so close in value that experimental distinction between them is impractical, the above analysis may be modified by using two observed amplitudes that are n cycles apart.

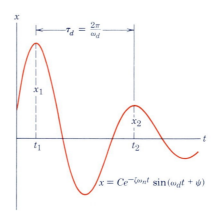

Figure 8/7

Sample Problem 8/1

A body weighing 25 lb is suspended from a spring of constant $k = 160$ lb/ft. At time $t = 0$, it has a downward velocity of 2 ft/sec as it passes through the position of static equilibrium. Determine

(a) the static spring deflection δ_{st}

(b) the natural frequency of the system in both rad/sec (ω_n) and cycles/sec (f_n)

(c) the system period τ

(d) the displacement x as a function of time, where x is measured from the position of static equilibrium

(e) the maximum velocity v_{max} attained by the mass

(f) the maximum acceleration a_{max} attained by the mass.

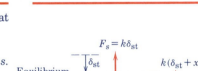

$k = 160$ lb/ft

$W = 25$ lb

Solution. (a) From the spring relationship $F_s = kx$, we see that at equilibrium,

① $$mg = k\delta_{st} \qquad \delta_{st} = \frac{mg}{k} = \frac{25}{160} = 0.1562 \text{ ft or } 1.875 \text{ in. } \textit{Ans.}$$

(b) $$\omega_n = \sqrt{\frac{k}{m}} = \sqrt{\frac{160}{25/32.2}} = 14.36 \text{ rad/sec} \qquad \textit{Ans.}$$

$$f_n = (14.36)\left(\frac{1}{2\pi}\right) = 2.28 \text{ cycles/sec} \qquad \textit{Ans.}$$

(c) $$\tau = \frac{1}{f_n} = \frac{1}{2.28} = 0.438 \text{ sec} \qquad \textit{Ans.}$$

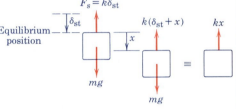

$F_s = k\delta_{st}$

Equilibrium position

δ_{st}

mg

$k(\delta_{st} + x)$ $\qquad$ kx

x

mg

② (d) From Eq. 8/6:

$$x = x_0 \cos \omega_n t + \frac{\dot{x}_0}{\omega_n} \sin \omega_n t$$

$$= (0) \cos 14.36t + \frac{2}{14.36} \sin 14.36t$$

$$= 0.1393 \sin 14.36t \qquad \textit{Ans.}$$

As an exercise, let us determine x from the alternative Eq. (8/7):

$$x = \sqrt{x_0^2 + (\dot{x}_0/\omega_n)^2} \sin\left[\omega_n t + \tan^{-1}\left(\frac{x_0 \omega_n}{\dot{x}_0}\right)\right]$$

$$= \sqrt{0^2 + \left(\frac{2}{14.36}\right)^2} \sin\left\{14.36t + \tan^{-1}\left[\frac{(0)(14.36)}{2}\right]\right\}$$

$$= 0.1393 \sin 14.36t$$

(e) The velocity is $\dot{x} = 14.36(0.1393) \cos 14.36t = 2 \cos 14.36t$. Since the cosine function cannot be greater than 1 or less than -1, the maximum velocity v_{max} is 2 ft/sec, which, in this case, is the initial velocity. $\qquad \textit{Ans.}$

(f) The acceleration is

$$\ddot{x} = -14.36(2) \sin 14.36t = -28.7 \sin 14.36t$$

The maximum acceleration a_{max} is 28.7 ft/sec². $\qquad \textit{Ans.}$

① The student should always exercise extreme caution in the matter of units. In the subject of vibrations, it is quite easy to commit errors due to mixing of feet and inches, cycles and radians, and other pairs that frequently enter the calculations.

② Recall that when we refer the motion to the position of static equilibrium, the equation of motion, and therefore its solution, for the present system is identical to that for the horizontally vibrating system.

Sample Problem 8/2

The 8-kg body is moved 0.2 m to the right of the equilibrium position and released from rest at time $t = 0$. Determine its displacement at time $t = 2$ s. The viscous damping coefficient c is 20 N·s/m, and the spring stiffness k is 32 N/m.

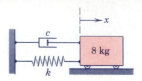

Solution. We must first determine whether the system is under-damped, critically damped, or overdamped. For that purpose, we compute the damping ratio ζ.

$$\omega_n = \sqrt{k/m} = \sqrt{32/8} = 2 \text{ rad/s} \qquad \zeta = \frac{c}{2m\omega_n} = \frac{20}{2(8)(2)} = 0.625$$

Since $\zeta < 1$, the system is underdamped. The damped natural frequency is $\omega_d = \omega_n\sqrt{1 - \zeta^2} = 2\sqrt{1 - (0.625)^2} = 1.561$ rad/s. The motion is given by Eq. 8/12 and is

$$x = Ce^{-\zeta\omega_n t} \sin(\omega_d t + \psi) = Ce^{-1.25t} \sin(1.561t + \psi)$$

The velocity is then

$$\dot{x} = -1.25Ce^{-1.25t} \sin(1.561t + \psi) + 1.561Ce^{-1.25t} \cos(1.561t + \psi)$$

Evaluating the displacement and velocity at time $t = 0$ gives

$$x_0 = C \sin\psi = 0.2 \qquad \dot{x}_0 = -1.25C \sin\psi + 1.561C \cos\psi = 0$$

Solving the two equations for C and ψ yields $C = 0.256$ m and $\psi = 0.896$ rad. Therefore, the displacement in meters is

$$x = 0.256e^{-1.25t} \sin(1.561t + 0.896)$$

① Evaluation for time $t = 2$ s gives $x_2 = -0.0162$ m. *Ans.*

① We note that the exponential factor $e^{-1.25t}$ is 0.082 at $t = 2$ s. Hence, $\zeta = 0.625$ represents severe damping, although the motion is still oscillatory.

Sample Problem 8/3

The two fixed counterrotating pulleys are driven at the same angular speed ω_0. A round bar is placed off center on the pulleys as shown. Determine the natural frequency of the resulting bar motion. The coefficient of kinetic friction between the bar and pulleys is μ_k.

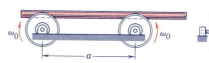

Solution. The free-body diagram of the bar is constructed for an arbitrary displacement x from the central position as shown. The governing equations are

$$[\Sigma F_x = m\ddot{x}] \qquad \mu_k N_A - \mu_k N_B = m\ddot{x}$$

$$[\Sigma F_y = 0] \qquad N_A + N_B - mg = 0$$

① $[\Sigma M_A = 0] \qquad aN_B - \left(\frac{a}{2} + x\right)mg = 0$

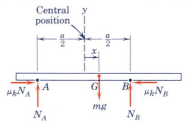

Eliminating N_A and N_B from the first equation yields

② $$\ddot{x} + \frac{2\mu_k g}{a}x = 0$$

We recognize the form of this equation as that of Eq. 8/2, so that the natural frequency in radians per second is $\omega_n = \sqrt{2\mu_k g/a}$ and the natural frequency in cycles per second is

$$f_n = \frac{1}{2\pi}\sqrt{2\mu_k g/a} \qquad \text{Ans.}$$

① Since the bar is slender and does not rotate, the use of a moment equilibrium equation is justified.

② We note that the angular speed ω_0 does not enter the equation of motion. The reason for this is our assumption that the kinetic friction force does not depend on the relative velocity at the contacting surface.

PROBLEMS

(Unless otherwise indicated, assume that all motion variables are referenced to the equilibrium position.)

Introductory problems—
undamped, free vibrations

8/1 When a 3-kg collar is placed upon the pan which is attached to the spring of unknown constant, the additional static deflection of the pan is observed to be 40 mm. Determine the spring constant k in N/m, lb/in., and lb/ft.

Ans. $k = 736$ N/m
$k = 4.20$ lb/in.
$k = 50.4$ lb/ft

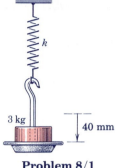

Problem 8/1

8/2 Determine the natural frequency of the spring-mass system in both rad/sec and cycles/sec (Hz).

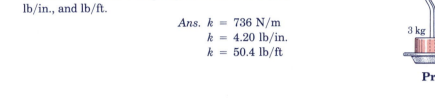

Problem 8/2

8/3 For the system of Prob. 8/2, determine the displacement x of the mass as a function of time if the mass is released from rest at time $t = 0$ from a position 2 in. to the right of the equilibrium position.

Ans. $x = 2 \cos 12t$ in.

8/4 For the system of Prob. 8/2, determine the displacement x of the mass as a function of time if the mass is released at time $t = 0$ from a position 2 in. to the left of the equilibrium position with an initial velocity of 7 in./sec to the right. Determine the amplitude C of the motion.

8/5 The vertical plunger has a mass of 2.5 kg and is supported by the two springs that are always in compression. Calculate the natural frequency f_n of vibration of the plunger if it is deflected from the equilibrium position and released. Friction in the guide is negligible.   Ans. $f_n = 7.40$ Hz

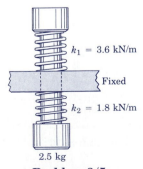

Problem 8/5

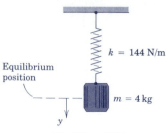

Problem 8/6

Equilibrium position

$k = 144$ N/m

$m = 4$ kg

y

8/6 For the spring-mass system shown, determine the static deflection δ_{st}, the system period τ, and the maximum velocity v_{max} which result if the cylinder is displaced 0.1 m downward from its equilibrium position and released.

8/7 The cylinder of the system of Prob. 8/6 is displaced 0.1 m downward from its equilibrium position and is released at time $t = 0$. Determine the displacement y and the velocity v when $t = 3$ s. What is the maximum acceleration?

$$Ans. \quad y = 0.0660 \text{ m}$$
$$v = 0.451 \text{ m/s}$$
$$a_{max} = 3.6 \text{ m/s}^2$$

Representative problems— undamped, free vibrations

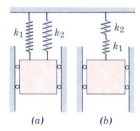

k_1 k_2 k_2

k_1

(a)   *(b)*

Problem 8/8

8/8 Replace the springs in each of the two cases shown by a single spring of stiffness k (equivalent spring stiffness) which will cause each mass to vibrate with its original frequency.

Electromagnet

Problem 8/9

8/9 An old car being moved by a magnetic crane pickup is dropped from a short distance above the ground. Neglect any damping effects of its worn-out shock absorbers and calculate the natural frequency f_n in cycles per second (Hz) of the vertical vibration which occurs after impact with the ground. Each of the four springs on the 1000-kg car has a constant of 17.5 kN/m. Because the center of mass is located midway between the axles and the car is level when dropped, there is no rotational motion. State any assumptions.   *Ans.* $f_n = 1.332$ Hz

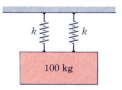

k k

100 kg

Problem 8/10

8/10 If the 100-kg mass has a downward velocity of 0.5 m/s as it passes through its equilibrium position, calculate the magnitude a_{max} of its maximum acceleration. Each of the two springs has a stiffness $k = 180$ kN/m.

8/11 With the assumption of no slipping, determine the mass m of the block that must be placed on the top of the 6-kg cart in order that the system period be 0.75 s. What is the minimum coefficient of static friction μ_s for which the block will not slip relative to the cart if the cart is displaced 50 mm from the equilibrium position and released?

Ans. $m = 2.55$ kg, $\mu_s = 0.358$

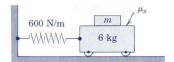

Problem 8/11

8/12 If both springs are unstretched when the mass is in the central position shown, determine the static deflection δ_{st} of the mass. What is the period of oscillatory motion about the position of static equilibrium?

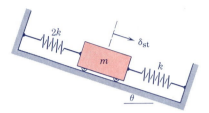

Problem 8/12

8/13 An energy-absorbing car bumper with its springs initially undeformed has an equivalent spring constant of 3000 lb/in. If the 2500-lb car approaches a massive wall with a speed of 5 mi/hr, determine (a) the velocity v of the car as a function of time during contact with the wall, where $t = 0$ is the beginning of the impact, and (b) the maximum deflection x_{max} of the bumper.

Ans. (a) $v = 88 \cos 21.5t$ in./sec
(b) $x_{max} = 4.09$ in.

Problem 8/13

8/14 A conventional spring scale registers the normal force which it exerts on the feet of the person being weighed. In the orbital environment aboard the space shuttle orbiter, such normal forces do not exist. Use your knowledge of vibration and explain how an astronaut might "weigh" himself or herself.

8/15 A 120-lb woman stands in the center of an end-supported board and causes a midspan deflection of 0.9 in. If she flexes her knees slightly in order to cause a vertical vibration, what is the frequency f_n of the motion? Assume elastic response of the board and neglect its relatively small mass.

Ans. $f_n = 3.30$ Hz

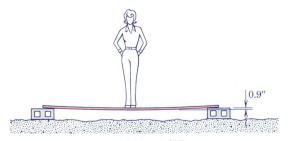

Problem 8/15

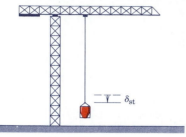

Problem 8/16

Problem 8/17

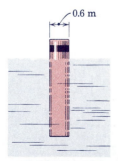

Problem 8/19

8/16 The large cement bucket suspended from the crane by an elastic cable has a mass of 4000 kg. When the bucket is disturbed, a vertical oscillation of period 0.5 s is observed. What is the static deflection δ_{st} of the bucket? Neglect the mass of the cable and assume that the crane is rigid for the inboard support position shown.

8/17 A small particle of mass m is attached to two highly tensioned wires as shown. Determine the system natural frequency ω_n for small vertical oscillations if the tension T in both wires is assumed to be constant. Is the calculation of the small static deflection of the particle necessary?

$$Ans. \quad \omega_n = \sqrt{\frac{2T}{ml}}$$

8/18 The particle of Prob. 8/17 is now considered to be supported from below by a smooth, horizontal surface and oscillates along a line midway between the supports. Rather than being in constant tension, assume that both wires are modeled as elastic bands. The tension is given as $T = k\Delta l$, where Δl is the change in length of each band as the mass is displaced from its equilibrium position. Derive the nonlinear equation of small motion for the system.

8/19 The cylindrical buoy floats in salt water (density 1030 kg/m³) and has a mass of 800 kg with a low center of mass to keep it stable in the upright position. Determine the frequency f_n of vertical oscillation of the buoy. Assume the water level remains undisturbed adjacent to the buoy.

$$Ans. \quad f_n = 0.301 \text{ Hz}$$

8/20 Shown in the figure is a model of a one-story building. The bar of mass m is supported by two light elastic upright columns whose upper and lower ends are fixed against rotation. For each column, if a force P and corresponding moment M were applied as shown in the right-hand part of the figure, the deflection δ would be given by $\delta = PL^3/12EI$, where L is the effective column length, E is Young's modulus, and I is the area moment of inertia of the column cross section with respect to its neutral axis. Determine the natural frequency of horizontal oscillation of the bar when the columns bend as shown in the figure.

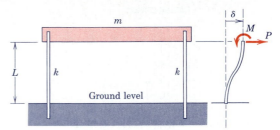

Problem 8/20

8/21 A 3-kg piece of putty is dropped 2 m onto the initially stationary 28-kg block, which is supported by four springs, each of which has a constant $k = 800$ N/m. Determine the displacement x as a function of time during the resulting vibration, where x is measured from the initial position of the block as shown.
Ans.
$$x = 9.20\,(1 - \cos 10.16t) + 59.7 \sin 10.16t \text{ mm}$$

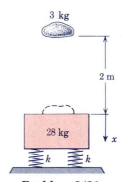

Problem 8/21

8/22 Calculate the frequency f_n of vertical oscillation of the 50-lb block when it is set in motion. Each spring has a stiffness of 6 lb/in. Neglect the mass of the pulleys.

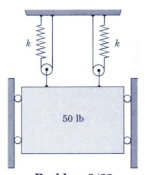

Problem 8/22

8/23 The weighing platform has a mass m and is connected to the spring of stiffness k by the system of levers shown. Derive the differential equation for small vertical oscillations of the platform and find the period τ. Designate y as the platform displacement from the equilibrium position and neglect the mass of the levers.

$$\text{Ans. } \ddot{y} + \frac{k}{m}\left(\frac{c}{b}\right)^4 y = 0, \ \tau = 2\pi \left(\frac{b}{c}\right)^2 \sqrt{\frac{m}{k}}$$

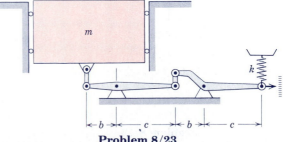

Problem 8/23

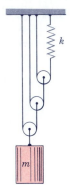

Problem 8/24

8/24 Determine the expression for the frequency f_n of vertical oscillation of the cylinder when it is set in vertical motion. Neglect the mass of the small pulleys.

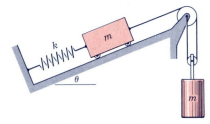

Problem 8/25

8/25 Calculate the natural frequency ω_n of the system shown in the figure. The mass and friction of the pulleys are negligible.

$$Ans. \ \omega_n = \sqrt{\frac{4k}{5m}}$$

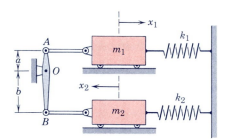

Problem 8/26

8/26 Derive the differential equation of motion for the system shown in terms of the variable x_1. The mass of the linkage is negligible. State the natural frequency $\omega_n{}'$ in rad/s for the case $k_1 = k_2 = k$ and $m_1 = m_2 = m$. Assume small oscillations throughout.

Introductory problems—damped, free vibrations

8/27 Determine the value of the damping ratio ζ for the simple spring-mass-dashpot system shown.

$$Ans. \ \zeta = 0.75$$

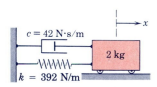

Problem 8/27

8/28 The period τ_d of damped linear oscillation for a certain 1-kg mass is 0.3 s. If the stiffness of the supporting linear spring is 800 N/m, calculate the damping coefficient c.

8/29 Viscous damping is added to an initially undamped spring-mass system. For what value of the damping ratio ζ will the damped natural frequency ω_d be equal to 90 percent of the natural frequency of the original undamped system?   *Ans.* $\zeta = 0.436$

8/30 The addition of damping to an undamped spring-mass system causes its period to increase by 25 percent. Determine the damping ratio ζ.

8/31 Determine the value of the viscous damping coefficient c for which the system shown is critically damped.   *Ans.* $c = 154.4$ lb-sec/ft

Problem 8/31

Representative problems—damped, free vibrations

8/32 Determine the value of the viscous damping coefficient c for which the system has a damping ratio of (*a*) 0.5 and (*b*) 1.5.

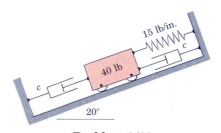

Problem 8/32

8/33 A linear harmonic oscillator having a mass of 1.10 kg is set into motion with viscous damping. If the frequency is 10 Hz and if two successive amplitudes a full cycle apart are measured to be 4.65 mm and 4.30 mm as shown, compute the viscous damping coefficient c.   *Ans.* $c = 1.721$ N·s/m

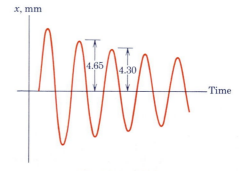

Problem 8/33

8/34 If the amplitude of the eighth cycle of a linear oscillator with viscous damping is sixteen times the amplitude of the twentieth cycle, calculate the damping ratio ζ.

8/35 The 2-kg mass of Prob. 8/27 is released from rest at a distance x_0 to the right of the equilibrium position. Determine the displacement x as a function of time t, where $t = 0$ is the time of release.

Ans. $x = x_0(\cos 9.26t + 1.134 \sin 9.26t)e^{-10.5t}$

8/36 A damped spring-mass system is released from rest from a positive initial displacement x_0. If the succeeding maximum positive displacement is $x_0/2$, determine the damping ratio ζ of the system.

8/37 If the body of mass m shown in Fig. 8/4 is critically damped, derive an expression for the displacement x measured from the equilibrium position at time t after the body is released from rest with a displacement x_0 from the equilibrium position.

$$Ans.\ x = x_0 \left(1 + \frac{c}{2m} t \right) e^{-(c/2m)t}$$

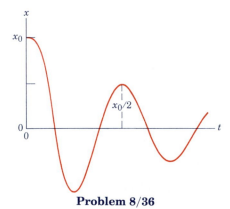

Problem 8/36

8/38 The system shown is released from rest from an initial position x_0. Determine the overshoot displacement x_1. Assume translational motion in the x-direction.

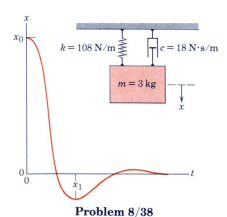

Problem 8/38

8/39 The signal from a vibration transducer records a decaying amplitude of a freely vibrating 4-kg mass. Calculate the indicated amplitude h, the damping coefficient c, and the stiffness k of the elastic support.

Ans. $h = 0.075$ mm
$c = 9.24$ N·s/m
$k = 115.0$ N/m

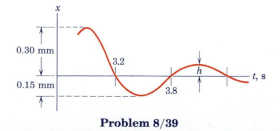

Problem 8/39

8/40 Develop an expression for the logarithmic decrement δ in terms of the displacements x_1 and x_{n+1} which are n cycles apart, as shown on the plot of displacement versus time.

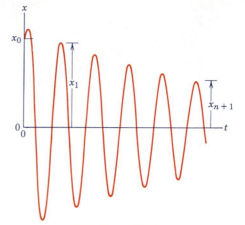

Problem 8/40

8/41 The mass of the system shown is released from rest at $x_0 = 6$ in. when $t = 0$. Determine the displacement x at $t = 0.5$ sec if (a) $c = 12$ lb-sec/ft and (b) $c = 18$ lb-sec/ft.   *Ans.* (a) $x = 4.42$ in.
                                           (b) $x = 4.72$ in.

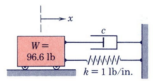

Problem 8/41

8/42 The small cannon fires a 10-lb cannonball with an absolute velocity of 800 ft/sec at 20° to the horizontal. The combined weight of the cannon and its cart is 1610 lb. If the recoil mechanism consists of the spring of constant $k = 150$ lb/in. and the damper with viscous coefficient $c = 600$ lb-sec/ft, determine the maximum recoil deflection x_{max} of the cannon unit.

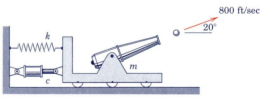

Problem 8/42

8/43 The mass of a given critically damped system is released at time $t = 0$ from the position $x_0 > 0$ with a negative initial velocity. Determine the critical value $(\dot{x}_0)_c$ of the initial velocity below which the mass will pass through the equilibrium position.
                       Ans. $(\dot{x}_0)_c = -\omega_n x_0$

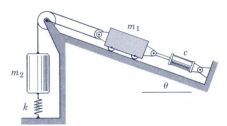

Problem 8/44

8/44 Determine the damping ratio ζ of the system depicted in the figure. The mass and friction of the pulleys are negligible, and the cable remains taut at all times.

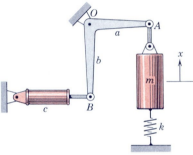

Problem 8/45

8/45 Develop the equation of motion in terms of the variable x for the system shown. Determine an expression for the damping ratio ζ in terms of the given system properties. Neglect the mass of the crank AB and assume small oscillations about the equilibrium position shown.

$$Ans. \quad \ddot{x} + \frac{b^2}{a^2} \frac{c}{m} \dot{x} + \frac{k}{m} x = 0$$

$$\zeta = \frac{1}{2} \frac{b^2}{a^2} \frac{c}{\sqrt{km}}$$

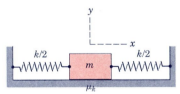

Problem 8/46

▶**8/46** Investigate the case of Coulomb damping for the block shown, where the coefficient of kinetic friction is μ_k and each spring has a stiffness $k/2$. The block is displaced a distance x_0 from the neutral position and released. Determine and solve the differential equation of motion. Plot the resulting vibration and indicate the rate r of decay of the amplitude with time.

$$Ans. \quad r = \frac{2\mu_k g}{\pi} \sqrt{\frac{m}{k}}$$

8/3 *FORCED VIBRATION OF PARTICLES*

Although there are many significant applications of free vibrations, the most important class of vibration problems is that where the motion is continuously excited by a disturbing force. The force may be externally applied or may be generated within the system by such means as unbalanced rotating parts. Forced vibrations may also be excited by the motion of the system foundation.

Various forms of forcing functions $F = F(t)$ and foundation displacements $x_B = x_B(t)$ are depicted in Fig. 8/8. The harmonic

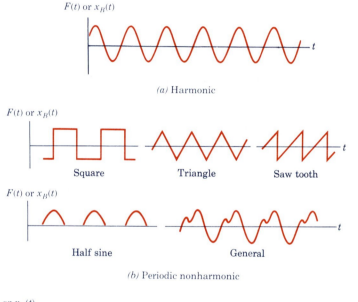

(a) Harmonic

Square   Triangle   Saw tooth

Half sine   General

(b) Periodic nonharmonic

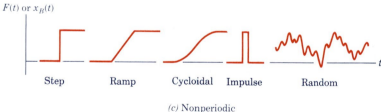

Step   Ramp   Cycloidal   Impulse   Random

(c) Nonperiodic

Figure 8/8

force shown in the *a*-part of the figure occurs frequently in engineering practice, and the understanding of the analysis associated with harmonic forces is a necessary first step in the study of more complex forms. For this reason, we will focus our attention on harmonic excitation.

We first consider the system of Fig. 8/9*a*, where the body is subjected to the external harmonic force $F = F_0 \sin \omega t$, in which F_0 is the force amplitude and ω is the driving frequency (in radians per

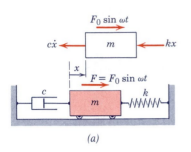

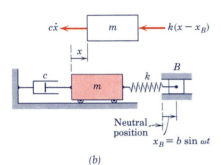

(a)

(b)

Figure 8/9

second). The student should take care to distinguish between $\omega_n = \sqrt{k/m}$, which is a property of the system, and ω, which is a property of the force applied to the system. We also note that for a force $F = F_0 \cos \omega t$, one merely substitutes $\cos \omega t$ for $\sin \omega t$ in the results about to be developed. From the free-body diagram of Fig. 8/9a, we may apply Newton's second law to obtain

$$-kx - c\dot{x} + F_0 \sin \omega t = m\ddot{x}$$

In standard form, with the same variable substitutions made in Art. 8/2, the equation of motion becomes

$$\ddot{x} + 2\zeta\omega_n\dot{x} + \omega_n^2 x = \frac{F_0 \sin \omega t}{m} \qquad (8/13)$$

In many cases, the excitation of the mass is due not to a directly applied force but to the movement of the base or foundation to which the mass is connected by springs or other compliant mountings. Examples of such applications are seismographs, vehicle suspensions, and structures shaken by earthquakes. Harmonic movement of the base is equivalent to the direct application of a harmonic force. To show this, consider the system of Fig. 8/9b where the spring is attached to the movable base. The free-body diagram shows the mass displaced a distance x from the neutral or equilibrium position it would have if the base were in its neutral position. The base, in turn, is assumed to have a harmonic movement $x_B = b \sin \omega t$. Note that the spring deflection is the difference between the inertial displacements of the mass and the base. From the free-body diagram, Newton's second law gives

$$-k(x - x_B) - c\dot{x} = m\ddot{x}$$

or

$$\ddot{x} + 2\zeta\omega_n\dot{x} + \omega_n^2 x = \frac{kb \sin \omega t}{m} \qquad (8/14)$$

We see immediately that Eq. 8/14 is exactly the same as our basic equation of motion, Eq. 8/13, in that F_0 is replaced by kb. Consequently, all the results about to be developed apply to either Eq. 8/13 or 8/14.

(a) Undamped forced vibration. First, we treat the case where damping is negligible ($c = 0$). Our basic equation of motion, Eq. 8/13, becomes

$$\ddot{x} + \omega_n^2 x = \frac{F_0}{m} \sin \omega t \qquad (8/15)$$

The complete solution to Eq. 8/15 is a sum of the complementary solution x_c, which is the general solution of Eq. 8/15 with the right side set to zero, and the particular solution x_p, which is *any* solution to the complete equation. Thus, $x = x_c + x_p$. We developed the complementary solution in section (*a*) of Art. 8/2. A particular solution is investigated by assuming that the response to the force should resemble the form of the force term. To that end, we assume

$$x_p = X \sin \omega t \qquad (8/16)$$

where X is the magnitude (in units of length) of the particular solution. Substituting this expression into Eq. 8/15 and solving for X yield

$$X = \frac{F_0/k}{1 - (\omega/\omega_n)^2} \qquad (8/17)$$

Thus, the particular solution becomes

$$x_p = \frac{F_0/k}{1 - (\omega/\omega_n)^2} \sin \omega t \qquad (8/18)$$

The complementary solution, known as the *transient solution*, is of no special interest here since, with time, it dies out with the small amount of damping that can never be completely eliminated. The particular solution x_p describes the continuing motion and is called the *steady-state solution*. Its period is $\tau = 2\pi/\omega$, the same as that of the forcing function. Of primary interest is the amplitude X of the motion. If we let δ_{st} stand for the magnitude of the static deflection of the mass under a static load F_0, then $\delta_{\text{st}} = F_0/k$, and we may form the ratio

$$M = \frac{X}{\delta_{\text{st}}} = \frac{1}{1 - (\omega/\omega_n)^2} \qquad (8/19)$$

The ratio M is known as the *amplitude ratio* or *magnification factor* and is a measure of the severity of the vibration. We especially note that M *approaches infinity* as ω approaches ω_n. Consequently, if the system possesses no damping and is excited by a harmonic force whose frequency ω approaches the natural frequency ω_n of the system, then M, and hence X, increase without limit. Physically, this means that the motion amplitude would reach the limits of the attached spring, which is a condition to be avoided. The value ω_n is known as the *resonant* or *critical frequency* of the system, and the condition of ω being close in value to ω_n with the resulting large displacement amplitude X is known as *resonance*. For $\omega < \omega_n$, the magnification factor M is positive, and the vibration is in phase with the force F. For $\omega > \omega_n$, the magnification factor is negative, and the vibration is 180° out of phase with F. Figure 8/10 shows a plot of the absolute value of M as a function of the driving-frequency ratio ω/ω_n.

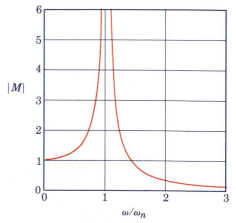

Figure 8/10

(b) Damped forced vibration. We now reintroduce damping in our expressions for forced vibration. Our basic differential equation of motion is

$$\ddot{x} + 2\zeta\omega_n\dot{x} + \omega_n^2 x = \frac{F_0 \sin \omega t}{m} \qquad [8/13]$$

As in section (a), the complete solution is the sum of the complementary solution x_c, which is the general solution of Eq. 8/13 with the right side equal to zero, and the particular solution x_p, which is *any* solution to the complete equation. We have already developed the complementary solution x_c in section (b) of Art. 8/2. By virtue of the influence of damping, we find that a single sine or cosine expression, such as we were able to use in section (a), is not sufficiently general for the particular solution in the case of damping. So we try

$$x_p = X_1 \cos \omega t + X_2 \sin \omega t \qquad \text{or} \qquad x_p = X \sin (\omega t - \phi)$$

Substitution of the latter expression into Eq. 8/13, matching coefficients of sin ωt and cos ωt, and solving the resulting two equations give

$$X = \frac{F_0/k}{\{[1 - (\omega/\omega_n)^2]^2 + [2\zeta\omega/\omega_n]^2\}^{1/2}} \qquad (8/20)$$

$$\phi = \tan^{-1}\left[\frac{2\zeta\omega/\omega_n}{1 - (\omega/\omega_n)^2}\right] \qquad (8/21)$$

The complete solution is now known, and for underdamped systems it can be written as

$$x = Ce^{-\zeta\omega_n t} \sin (\omega_d t + \psi) + X \sin (\omega t - \phi) \qquad (8/22)$$

Because the first term on the right side diminishes with time, it is known as the *transient solution*. The particular solution x_p is known as the *steady-state solution* and is the part of the solution in which we are primarily interested. All quantities on the right side of Eq. 8/22 are properties of the system and the applied force, except for C and ψ (which are determinable from initial conditions) and the running time variable t.

Near resonance the magnitude X of the steady-state solution is a strong function of the damping ratio ζ and the nondimensional frequency ratio ω/ω_n. It is again convenient to form the nondimensional ratio $M = X/(F_0/k)$, which is known as the *amplitude ratio* or *magnification factor*

$$M = \frac{1}{\{[1 - (\omega/\omega_n)^2]^2 + [2\zeta\omega/\omega_n]^2\}^{1/2}} \qquad (8/23)$$

An accurate plot of the magnification factor M versus the frequency ratio ω/ω_n for various values of the damping ratio ζ is shown

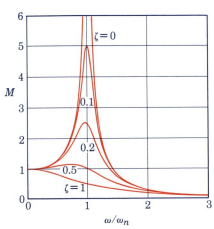

Figure 8/11

in Fig. 8/11. This figure reveals the most essential information pertinent to the forced vibration of a single-degree-of-freedom system under harmonic excitation. It is clear from the graph that, if a motion amplitude is excessive, two possible remedies would be to (*a*) increase the damping (to obtain a larger value of ζ) or (*b*) alter the driving frequency so that ω is farther from the resonant frequency ω_n. The addition of damping is most effective near resonance. Figure 8/11 also shows that, except for $\zeta = 0$, the magnification-factor curves do not actually peak at $\omega/\omega_n = 1$. The peak for any given value of ζ can be calculated by finding the maximum value of M from Eq. 8/23.

The phase angle ϕ, given by Eq. 8/21, can vary from 0 to π and represents the part of a cycle (and hence the time) by which the response x_p lags the forcing function F. Figure 8/12 shows how the phase angle ϕ varies with the frequency ratio for various values of the damping ratio ζ. Note that the value of ϕ at resonance is 90° for all values of ζ. To further illustrate the phase difference between the response and the forcing function, we show in Fig. 8/13 two examples of the variation of F and x_p with ωt. In the first example, $\omega < \omega_n$ and ϕ is taken to be $\pi/4$. In the second example, $\omega > \omega_n$ and ϕ is taken to be $3\pi/4$.

As a frequently encountered application of harmonic excitation, we consider vibration-measuring instruments such as seismometers and accelerometers. The elements of this class of instruments are

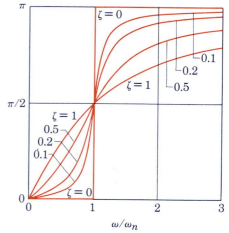

Figure 8/12

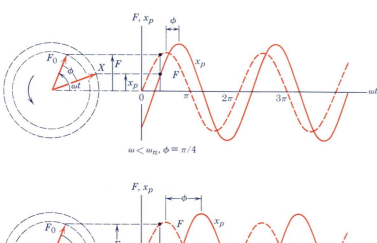
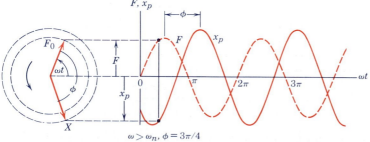

Figure 8/13

shown in Fig. 8/14*a*. We note that the entire system is subjected to the motion x_B of the frame. Letting x denote the position of the mass *relative* to the frame, we may apply Newton's second law and obtain

$$-c\dot{x} - kx = m\frac{d^2}{dt^2}(x + x_B) \quad \text{or} \quad \ddot{x} + \frac{c}{m}\dot{x} + \frac{k}{m}x = -\ddot{x}_B$$

where $(x + x_B)$ is the inertial displacement of the mass. If $x_B = b \sin \omega t$, then our equation of motion with the usual notation is

$$\ddot{x} + 2\zeta\omega_n\dot{x} + \omega_n{}^2x = b\omega^2 \sin \omega t$$

which is the same as Eq. 8/13 if $b\omega^2$ is substituted for F_0/m. Again, we are interested only in the steady-state solution x_p. Thus, from Eq. 8/20, we have

$$x_p = \frac{b(\omega/\omega_n)^2}{\{[1 - (\omega/\omega_n)^2]^2 + [2\zeta\omega/\omega_n]^2\}^{1/2}} \sin (\omega t - \phi)$$

If X represents the amplitude of the relative response x_p, then the nondimensional ratio X/b is

$$X/b = (\omega/\omega_n)^2M$$

where M is the magnification ratio of Eq. 8/23. A plot of X/b as a function of the driving-frequency ratio ω/ω_n is shown in the *b*-part of Fig. 8/14. The similarities and differences between the magnification ratios of Figs. 8/14(*b*) and 8/11 should be noted.

If the frequency ratio ω/ω_n is large, then $X/b \cong 1$ for all values of the damping ratio ζ. Under these conditions, the displacement of the mass relative to the frame is approximately the same as the absolute displacement of the frame, and the instrument acts as a *displacement meter*. To obtain a high value of ω/ω_n, we need a small

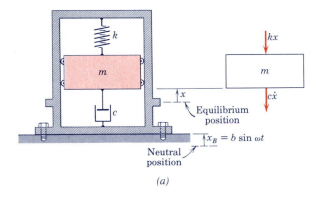

(*a*)

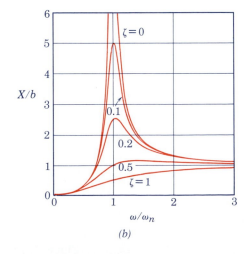

(*b*)

Figure 8/14

value of $\omega_n = \sqrt{k/m}$, which means a soft spring and a large mass. With such a combination, the mass will tend to stay inertially fixed. Displacement meters generally have very light damping.

On the other hand, if the frequency ratio ω/ω_n is small, then M approaches unity (see Fig. 8/11) and $X/b \cong (\omega/\omega_n)^2$ or $X \cong b(\omega/\omega_n)^2$. But $b\omega^2$ is the maximum acceleration of the frame. Hence, X is proportional to the maximum acceleration of the frame, and the instrument may be used as an *accelerometer*. The damping ratio is generally selected so that M approximates unity over the widest possible range of ω/ω_n. From Fig. 8/11, we see that a damping factor somewhere between $\zeta = 0.5$ and $\zeta = 1$ would meet this criterion.

An important analogy exists between electric circuits and mechanical spring-mass systems. Figure 8/15 shows a series circuit consisting of a voltage E that is a function of time, an inductance L, a capacitance C, and a resistance R. If we denote the charge by the symbol q, the equation that governs the charge is

$$L\ddot{q} + R\dot{q} + \frac{1}{C}q = E \qquad (8/24)$$

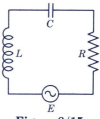

Figure 8/15

This equation has the same form as the equation for the mechanical system. Thus, by a simple interchange of symbols, the behavior of the electrical circuit may be used to predict the behavior of the mechanical system, or vice versa. The mechanical and electrical equivalents in the following table are worth noting:

MECHANICAL–ELECTRICAL EQUIVALENTS

MECHANICAL			ELECTRICAL		
QUANTITY	SYMBOL	SI UNIT	QUANTITY	SYMBOL	SI UNIT
Mass	m	kg	Inductance	L	H	henry
Spring stiffness	k	N/m	1/Capacitance	$1/C$	$1/F$	1/farad
Force	F	N	Voltage	E	V	volt
Velocity	$\dot{x}$	m/s	Current	I	A	ampere
Displacement	x	m	Charge	q	C	coulomb
Viscous damping constant	c	N·s/m	Resistance	R	Ω	ohm

Sample Problem 8/4

A 50-kg instrument is supported by four springs, each of stiffness 7500 N/m. If the instrument foundation undergoes harmonic motion given in meters by $x_B = 0.002 \cos 50t$, determine the amplitude of the steady-state motion of the instrument. Damping is negligible.

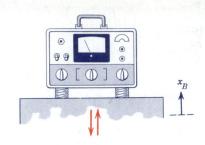

x_B

Solution. For harmonic oscillation of the base, we substitute kb for F_0 in our particular-solution results, so that, from Eq. 8/17, the steady-state amplitude becomes

① $$X = \frac{b}{1 - (\omega/\omega_n)^2}$$

The resonant frequency is $\omega_n = \sqrt{k/m} = \sqrt{4(7500)/50} = 24.5$ rad/s, and the impressed frequency $\omega = 50$ rad/s is given. Thus,

② $$X = \frac{0.002}{1 - (50/24.5)^2} = -6.32(10^{-4}) \text{ m} \quad \text{or} \quad -0.632 \text{ mm} \quad Ans.$$

Note that the frequency ratio ω/ω_n is approximately 2, so that the condition of resonance is avoided.

① Note that either $\sin 50t$ or $\cos 50t$ can be used for the forcing function with this same result.

② The minus sign indicates that the motion is 180° out of phase with the applied excitation.

Sample Problem 8/5

The spring attachment point B is given a horizontal motion $x_B = b \cos \omega t$. Determine the critical driving frequency ω_c for which the oscillations of the mass m tend to become excessively large. Neglect the friction and mass associated with the pulleys. The two springs have the same stiffness k.

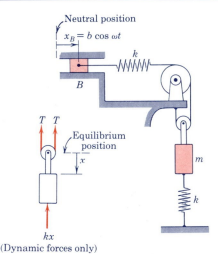

Neutral position

$x_B = b \cos \omega t$

k

B

$T \quad T$

Equilibrium position

x

m

k

kx
(Dynamic forces only)

Solution. The free-body diagram is drawn for arbitrary positive displacements x and x_B. The motion variable x is measured downward from the position of static equilibrium defined as that which exists when $x_B = 0$. The additional stretch in the upper spring, beyond that which

① exists at static equilibrium, is $2x - x_B$. Therefore, the *dynamic* spring
② force in the upper spring, and hence the *dynamic* tension T in the cable, is $k(2x - x_B)$. Summing forces in the x-direction gives

$[\Sigma F_x = m\ddot{x}] \qquad -2k(2x - x_B) - kx = m\ddot{x}$

which becomes

$$\ddot{x} + \frac{5k}{m}x = \frac{2kb \cos \omega t}{m}$$

The natural frequency of the system is $\omega_n = \sqrt{5k/m}$. Thus,

$$\omega_c = \omega_n = \sqrt{5k/m} \qquad Ans.$$

① If a review of the kinematics of constrained motion is necessary, see Art. 2/9.

② We learned from the discussion in Art. 8/2 that the equal and opposite forces associated with the position of static equilibrium may be omitted from the analysis. Our use of the terms *dynamic* spring force and *dynamic* tension stresses that only the force increments in addition to the static values need be considered.

Sample Problem 8/6

The 100-lb piston is supported by a spring of modulus $k = 200$ lb/in. A dashpot of damping coefficient $c = 85$ lb-sec/ft acts in parallel with the spring. A fluctuating pressure $p = 0.625 \sin 30t$ in lb/in.^{2} acts on the piston, whose top surface area is 80 in.^{2} Determine the steady-state displacement as a function of time and the maximum force transmitted to the base.

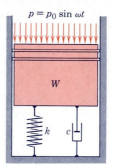

Solution. We begin by computing the system natural frequency and damping ratio:

$$\omega_n = \sqrt{\frac{k}{m}} = \sqrt{\frac{(200)(12)}{100/32.2}} = 27.8 \text{ rad/sec}$$

$$\zeta = \frac{c}{2m\omega_n} = \frac{85}{2\left(\dfrac{100}{32.2}\right)(27.8)} = 0.492 \text{ (underdamped)}$$

The steady-state amplitude, from Eq. 8/20, is

$$X = \frac{F_0/k}{\{[1 - (\omega/\omega_n)^2]^2 + [2\zeta\omega/\omega_n]^2\}^{1/2}}$$

$$= \frac{(0.625)(80)/[(200)(12)]}{\{[1 - (30/27.8)^2]^2 + [2(0.492)(30/27.8)]^2\}^{1/2}}$$

$$= 0.0194 \text{ ft}$$

The phase angle, from Eq. 8/21, is

$$\phi = \tan^{-1}\left[\frac{2\zeta\omega/\omega_n}{1 - (\omega/\omega_n)^2}\right]$$

$$= \tan^{-1}\left[\frac{2(0.492)(30/27.8)}{1 - (30/27.8)^2}\right]$$

$$= 1.724 \text{ rad}$$

The steady-state motion is then given by the second term on the right side of Eq. 8/22:

$$x_p = X \sin(\omega t - \phi) = 0.0194 \sin(30t - 1.724) \text{ ft} \qquad Ans.$$

The force F_{tr} transmitted to the base is the sum of the spring and damper forces, or

$$F_{\text{tr}} = kx_p + c\dot{x}_p = kX \sin(\omega t - \phi) + c\omega X \cos(\omega t - \phi)$$

The maximum value of F_{tr} is

$$(F_{\text{tr}})_{\max} = \sqrt{(kX)^2 + (c\omega X)^2} = X\sqrt{k^2 + c^2\omega^2}$$

$$= 0.0194\sqrt{[(200)(12)]^2 + (85)^2(30)^2}$$

$$= 67.9 \text{ lb} \qquad Ans.$$

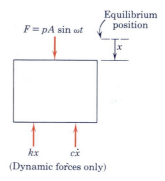

$F = pA \sin \omega t$

Equilibrium position

$kx \qquad c\dot{x}$
(Dynamic forces only)

① The student is encouraged to repeat these calculations with the damping coefficient c set to zero so as to observe the influence of the relatively large amount of damping present.

② Note that the argument of the inverse tangent expression for ϕ has a positive numerator and a negative denominator for the case at hand, thus placing ϕ in the second quadrant. Recall that the defined range of ϕ is $0 \leq \phi \leq \pi$.

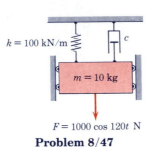

$k = 100$ kN/m c

$m = 10$ kg

$F = 1000 \cos 120t$ N

Problem 8/47

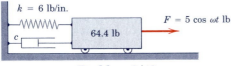

$k = 6$ lb/in.

$F = 5 \cos \omega t$ lb

64.4 lb

c

Problem 8/49

PROBLEMS

(Unless otherwise instructed, assume that the damping is light to moderate so that the amplitude of the forced response is a maximum at $\omega/\omega_n \cong 1$.)

Introductory problems

8/47 Determine the amplitude X of the steady-state motion of the 10-kg mass if (*a*) $c = 500$ N·s/m and (*b*) $c = 0$.   *Ans.* (*a*) $X = 1.344(10^{-2})$ m
(*b*) $X = 2.27(10^{-2})$ m

8/48 A viscously damped spring-mass system is excited by a harmonic force of constant amplitude F_0 but varying frequency ω. If the amplitude of the steady-state motion is observed to decrease by a factor of 8 as the frequency ratio ω/ω_n is varied from 1 to 2, determine the damping ratio ζ of the system.

8/49 The 64.4-lb cart is acted upon by the harmonic force shown in the figure. If $c = 0$, determine the range of the driving frequency ω for which the magnitude of the steady-state response is less than 3 in.
Ans. $\omega < 5.10$ rad/sec or $\omega > 6.78$ rad/sec

8/50 If the viscous damping coefficient of the damper in the system of Prob. 8/49 is $c = 2.4$ lb-sec/ft, determine the range of the driving frequency ω for which the magnitude of the steady-state response is less than 3 in. Justify your answer in comparison with that given for Prob. 8/49.

8/51 If the driving frequency for the system of Prob. 8/49 is $\omega = 6$ rad/sec, determine the required value of the damping coefficient c if the steady-state amplitude is not to exceed 3 in.
Ans. $c = 3.33$ lb-sec/ft

Representative problems

8/52 A viscously damped spring-mass system is forced harmonically at the undamped natural frequency ($\omega/\omega_n = 1$). If the damping ratio ζ is doubled from 0.1 to 0.2, compute the percentage reduction R_1 in the steady-state amplitude. Compare with the result R_2 of a similar calculation for the condition $\omega/\omega_n = 2$. Verify your results by inspecting Fig. 8/11.

8/53 Each 0.5-kg ball is attached to the end of the light elastic rod and deflects 4 mm when a 2-N force is statically applied to the ball. If the central collar is given a vertical harmonic movement with a frequency of 4 Hz and an amplitude of 3 mm, find the amplitude y_0 of vertical vibration of each ball.

Ans. $y_0 = 8.15$ mm

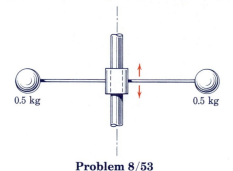

Problem 8/53

8/54 The motion of the outer cart B is given by $x_B = b \sin \omega t$. For what range of the driving frequency ω is the magnitude of the motion of the mass m relative to the cart less than $2b$?

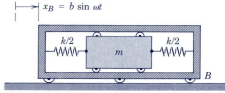

Problem 8/54

8/55 It was noted in the text that the maxima of the curves for the magnification factor M are not located at $\omega/\omega_n = 1$. Determine an expression in terms of the damping ratio ζ for the frequency ratio at which the maxima occur.

$$Ans. \quad \frac{\omega}{\omega_n} = \sqrt{1 - 2\zeta^2}$$

8/56 When the person stands in the center of the floor system shown, he causes a static deflection δ_{st} of the floor under his feet. If he walks (or runs quickly!) in the same area, how many steps per second would cause the floor to vibrate with the greatest vertical amplitude?

Problem 8/56

8/57 Attachment B is given a horizontal motion $x_B = b \cos \omega t$. Derive the equation of motion for the mass m and state the critical frequency ω_c for which the oscillations of the mass become excessively large.

Ans. $m\ddot{x} + c\dot{x} + (k_1 + k_2)x = k_2 b \cos \omega t$

$$\omega_c = \sqrt{\frac{k_1 + k_2}{m}}$$

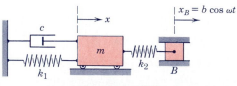

Problem 8/57

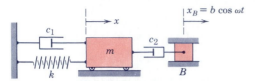

Problem 8/58

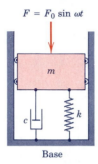

Base

Problem 8/59

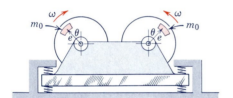

Problem 8/61

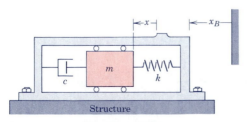

Structure

Problem 8/62

8/58 Attachment B is given a horizontal motion $x_B = b \cos \omega t$. Derive the equation of motion for the mass m and state the critical frequency ω_c for which the oscillations of the mass become excessively large. What is the damping ratio ζ for the system?

8/59 Derive an expression for the *transmission ratio* T for the system of the figure. This ratio is defined as the maximum force transmitted to the base divided by the amplitude F_0 of the forcing function. Express your answer in terms of ζ, ω, ω_n, and the magnification factor M.

$$Ans. \ T = M\sqrt{1 + \left(2\zeta \frac{\omega}{\omega_n}\right)^2}$$

8/60 Derive the equation of motion for the *inertial* displacement x_i of the mass of Fig. 8/14. Comment on, but do not carry out, the solution to the equation of motion.

8/61 A device to produce vibrations consists of the two counter-rotating wheels, each carrying an eccentric mass $m_0 = 1$ kg with a center of mass at a distance $e = 12$ mm from its axis of rotation. The wheels are synchronized so that the vertical positions of the unbalanced masses are always identical. The total mass of the device is 10 kg. Determine the two possible values of the equivalent spring constant k for the mounting which will permit the amplitude of the periodic force transmitted to the fixed mounting to be 1500 N due to the imbalance of the rotors at a speed of 1800 rev/min. Neglect damping.

$Ans. \ k = 823$ kN/m or 227 kN/m

8/62 The seismic instrument shown is attached to a structure which has a horizontal harmonic vibration at 3 Hz. The instrument has a mass $m = 0.5$ kg, a spring stiffness $k = 20$ N/m, and a viscous damping coefficient $c = 3$ N·s/m. If the maximum recorded value of x in its steady-state motion is $X = 2$ mm, determine the amplitude b of the horizontal movement x_B of the structure.

8/63 A device similar to that shown with Prob. 8/62 is to be used to measure the horizontal acceleration of the structure which is vibrating with a frequency of 5 Hz. The mass is $m = 0.008$ kg, the spring constant is $k = 150$ N/m, and the damping factor is $\zeta = 0.75$. If the amplitude of x is 4.0 mm, approximate the maximum acceleration a_{max} of the structure.

Ans. $a_{max} = 75.0$ m/s^2

8/64 The instrument shown has a mass of 43 kg and is spring-mounted to the horizontal base. If the amplitude of vertical vibration of the base is 0.10 mm, calculate the range of frequencies f_n of the base vibration which must be prohibited if the amplitude of vertical vibration of the instrument is not to exceed 0.15 mm. Each of the four identical springs has a stiffness of 7.2 kN/m.

Problem 8/64

8/65 The seismic instrument is mounted on a structure which has a vertical vibration with a frequency of 5 Hz and a double amplitude of 18 mm. The sensing element has a mass $m = 2$ kg, and the spring stiffness is $k = 1.5$ kN/m. The motion of the mass relative to the instrument base is recorded on a revolving drum and shows a double amplitude of 24 mm during the steady-state condition. Calculate the viscous damping constant c. *Ans.* $c = 44.6$ N·s/m

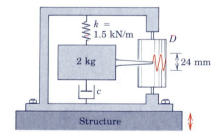

Problem 8/65

8/66 Derive and solve the equation of motion for the mass which is subjected to the suddenly applied force F that remains constant after application. The displacement and velocity of the mass are both zero at time $t = 0$. Plot x versus t for several motion cycles.

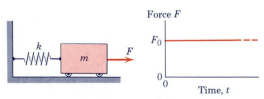

Problem 8/66

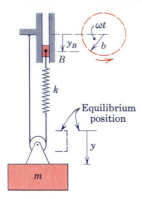

Problem 8/67

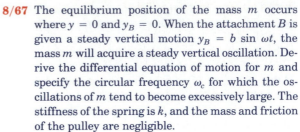

8/67 The equilibrium position of the mass m occurs where $y = 0$ and $y_B = 0$. When the attachment B is given a steady vertical motion $y_B = b \sin \omega t$, the mass m will acquire a steady vertical oscillation. Derive the differential equation of motion for m and specify the circular frequency ω_c for which the oscillations of m tend to become excessively large. The stiffness of the spring is k, and the mass and friction of the pulley are negligible.

$$Ans. \quad \ddot{y} + \frac{4k}{m} y = \frac{2kb}{m} \sin \omega t, \quad \omega_c = 2\sqrt{k/m}$$

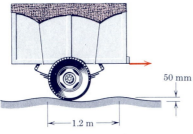

Problem 8/68

▶ **8/68** Determine the amplitude of vertical vibration of the spring-mounted trailer as it travels at a velocity of 25 km/h over the corduroy road whose contour may be expressed by a sine or cosine term. The mass of the trailer is 500 kg and that of the wheels alone may be neglected. During the loading each 75 kg added to the load caused the trailer to sag 3 mm on its springs. Assume that the wheels are in contact with the road at all times and neglect damping. At what critical speed v_c is the vibration of the trailer greatest?   *Ans.* $X = 14.75$ mm, $v_c = 15.23$ km/h

8/4 VIBRATION OF RIGID BODIES

The subject of planar rigid-body vibrations is entirely analogous to that of particle vibrations. In particle vibrations, the variable of interest is one of translation (x), while in rigid-body vibrations, the variable of primary concern is one of rotation (θ). Thus, the principles of rotational dynamics play a central role in the development of the equation of motion. We shall see that the equation of motion for rotational vibration of rigid bodies has a mathematical form identical to that developed in Arts. 8/2 and 8/3 for translational vibration of particles. As was the case with particles, it is convenient to draw the free-body diagram for an arbitrary positive value of the displacement variable, because a negative displacement value easily leads to sign errors in the equation of motion. The practice of measuring the displacement from the position of static equilibrium rather than from the position of zero spring deflection continues to simplify the formulation for linear systems because the equal and opposite forces and moments associated with the static equilibrium position cancel from the analysis.

Rather than individually treating the cases of (a) free vibration, undamped and damped, and (b) forced vibrations, undamped and damped, as was done with particles in Arts. 8/2 and 8/3, we shall go directly to the damped, forced problem.

As an illustrative example, consider the rotational vibration of the uniform slender bar of Fig. 8/16a. Figure 8/16b depicts the free-body diagram associated with the horizontal position of static equilibrium. Equating to zero the moment sum about O yields

$$-P\left(\frac{l}{2} + \frac{l}{6}\right) + mg\left(\frac{l}{6}\right) = 0 \qquad P = \frac{mg}{4}$$

where P is the magnitude of the static spring force.

Figure 8/16c depicts the free-body diagram associated with an arbitrary positive angular displacement θ. Using the rotational equation of motion $\Sigma M_O = I_O\ddot{\theta}$ as treated in Chapter 6, we write

$$(mg)\left(\frac{l}{6}\cos\theta\right) - \left(\frac{cl}{3}\dot{\theta}\cos\theta\right)\left(\frac{l}{3}\cos\theta\right) - \left(P + k\frac{2l}{3}\sin\theta\right)\left(\frac{2l}{3}\cos\theta\right)$$

$$+ (F_0\cos\omega t)\left(\frac{l}{3}\cos\theta\right) = \frac{1}{9}ml^2\ddot{\theta}$$

where $I_O = \bar{I} + md^2 = ml^2/12 + m(l/6)^2 = ml^2/9$ is obtained from the parallel-axis theorem for mass moments of inertia. For small angular deflections, the approximations $\sin\theta \cong \theta$ and $\cos\theta \cong 1$ may be used. With $P = mg/4$, the equation of motion, upon rearrangement and simplification, becomes

$$\ddot{\theta} + \frac{c}{m}\dot{\theta} + 4\frac{k}{m}\theta = \frac{(F_0 l/3)\cos\omega t}{ml^2/9} \qquad (8/25)$$

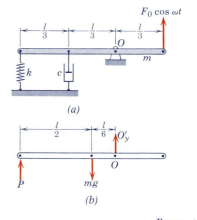

(a)

(b)

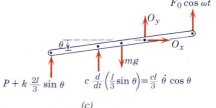

(c)

Figure 8/16

The right side has been left unsimplified in the form $M_0(\cos \omega t)/I_O$, where $M_0 = F_0 l/3$ is the magnitude of the moment about point O of the externally applied force. Note that the two equal and opposite moments associated with static equilibrium forces canceled on the left side of the equation of motion. Hence, it is not necessary to include the static-equilibrium forces and moments in the analysis.

At this point, we observe that Eq. 8/25 is identical in form to Eq. 8/13 for the linear case, so we may write

$$\ddot{\theta} + 2\zeta\omega_n\dot{\theta} + \omega_n{}^2\theta = \frac{M_0 \cos \omega t}{I_O} \qquad (8/26)$$

Thus, we may use all of the relations developed in Arts. 8/2 and 8/3 merely by replacing the linear quantities with their rotational counterparts. The following table shows the results of this procedure as applied to the rotating bar of Fig. 8/16:

LINEAR	ANGULAR (for current problem)
$\ddot{x} + \dfrac{c}{m}\dot{x} + \dfrac{k}{m}x = \dfrac{F_0 \cos \omega t}{m}$	$\ddot{\theta} + \dfrac{c}{m}\dot{\theta} + \dfrac{4k}{m}\theta = \dfrac{M_0 \cos \omega t}{I_O}$
$\omega_n = \sqrt{k/m}$	$\omega_n = \sqrt{4k/m} = 2\sqrt{k/m}$
$\zeta = \dfrac{c}{2m\omega_n} = \dfrac{c}{2\sqrt{km}}$	$\zeta = \dfrac{c}{2m\omega_n} = \dfrac{c}{4\sqrt{km}}$
$\omega_d = \omega_n\sqrt{1 - \zeta^2} = \dfrac{1}{2m}\sqrt{4km - c^2}$	$\omega_d = \omega_n\sqrt{1 - \zeta^2} = \dfrac{1}{2m}\sqrt{16km - c^2}$
$x_c = Ce^{-\zeta\omega_n t}\sin(\omega_d t + \psi)$	$\theta_c = Ce^{-\zeta\omega_n t}\sin(\omega_d t + \psi)$
$x_p = X\cos(\omega t - \phi)$	$\theta_p = \Theta\cos(\omega t - \phi)$
$X = M\left(\dfrac{F_0}{k}\right)$	$\Theta = M\left(\dfrac{M_0}{k_\theta}\right) = M\dfrac{F_0(l/3)}{\frac{4}{9}kl^2} = \dfrac{3F_0}{4kl}$

In the above table, the variable k_θ represents the equivalent torsional spring constant of the system of Fig. 8/16 and is determined by writing the restoring moment of the spring. For a small angle θ, this moment about O is

$$M_k = -[k(2l/3) \sin \theta][(2l/3) \cos \theta] \cong -(\tfrac{4}{9}kl^2)\theta$$

Thus, $k_\theta = \tfrac{4}{9}kl^2$. Note that M_0/k_θ is the static angular deflection that would be produced by a constant external moment M_0.

We conclude that there exists an exact analogy between the linear vibration of particles and the small angular vibration of rigid bodies. Furthermore, the utilization of this analogy can save the labor of complete rederivation of the governing relationships for a given problem of general rigid-body vibration.

Sample Problem 8/7

A simplified version of a pendulum used in impact tests is shown in the figure. Derive the equation of motion and determine the period for small oscillations about the pivot. The mass center G is located a distance $\bar{r} = 0.9$ m from O, and the radius of gyration about O is $k_O = 0.95$ m. The friction of the bearing is negligible.

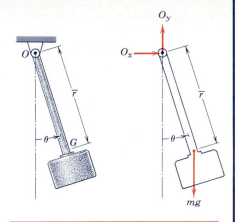

Solution. We draw the free-body diagram for an arbitrary, positive value of the angular-displacement variable θ, which is measured counterclockwise for the coordinate system chosen. Next we apply the governing equation of motion to obtain

① $[\Sigma M_O = I_O\ddot{\theta}]$ $\qquad -mg\bar{r}\sin\theta = mk_O^2\ddot{\theta}$

or $\qquad\qquad \ddot{\theta} + \dfrac{g\bar{r}}{k_O^2}\sin\theta = 0$ $\qquad$ *Ans.*

Note that the governing equation is independent of the mass. When θ is small, $\sin\theta \cong \theta$, and our equation of motion may be written as

$$\ddot{\theta} + \frac{g\bar{r}}{k_O^2}\theta = 0$$

② The frequency in cycles per second and the period in seconds are

$$f_n = \frac{1}{2\pi}\sqrt{\frac{g\bar{r}}{k_O^2}} \qquad \tau = \frac{1}{f_n} = 2\pi\sqrt{\frac{k_O^2}{g\bar{r}}} \qquad Ans.$$

For the given properties: $\qquad \tau = 2\pi\sqrt{\dfrac{(0.95)^2}{(9.81)(0.9)}} = 2.01$ s $\qquad$ *Ans.*

① With our choice of point O as the moment center, the bearing reactions O_x and O_y never enter the equation of motion.

② For large angles of oscillation, determining the period for the pendulum requires the evaluation of an elliptic integral.

Sample Problem 8/8

The uniform bar of mass m and length l is pivoted at its center. The spring of constant k at the left end is attached to a stationary surface, but the right-end spring, also of constant k, is attached to a support that undergoes a harmonic motion given by $y_B = b\sin\omega t$. Determine the driving frequency ω_c that causes resonance.

Solution. We use the moment equation of motion about the fixed point O to obtain

① $-\left(k\dfrac{l}{2}\sin\theta\right)\dfrac{l}{2}\cos\theta - k\left(\dfrac{l}{2}\sin\theta - y_B\right)\dfrac{l}{2}\cos\theta = \dfrac{1}{12}ml^2\ddot{\theta}$

Assuming small deflections and simplifying give us

$$\ddot{\theta} + \frac{6k}{m}\theta = \frac{6kb}{ml}\sin\omega t$$

② The natural frequency should be recognized from the now-familiar form of the equation to be

$$\omega_n = \sqrt{6k/m}$$

Hence, $\omega_c = \omega_n = \sqrt{6k/m}$ will result in resonance (as well as violation of our small-angle assumption!). $\qquad$ *Ans.*

① As previously, we consider only the changes in the forces due to a movement away from the equilibrium position.

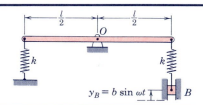

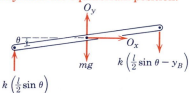

② The standard form here is $\ddot{\theta} + \omega_n^2\theta = \dfrac{M_0\sin\omega t}{I_O}$, where $M_0 = \dfrac{klb}{2}$ and $I_O = \frac{1}{12}ml^2$. The natural frequency ω_n of a system does not depend on the external disturbance.

Sample Problem 8/9

Derive the equation of motion for the homogeneous circular cylinder that rolls without slipping. If the cylinder mass is 50 kg, the cylinder radius 0.5 m, the spring constant 75 N/m, and the damping coefficient 10 N·s/m, determine

    (a) the undamped natural frequency
    (b) the damping ratio
    (c) the damped natural frequency
    (d) the period of the damped system.

In addition, determine x as a function of time if the cylinder is released from rest at the position $x = -0.2$ m when $t = 0$.

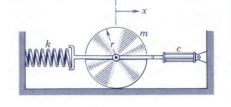

Solution. We have a choice of motion variables in that either x or ① the angular displacement θ of the cylinder may be used. Since the problem statement involves x, we draw the free-body diagram for an arbitrary, positive value of x and write the two motion equations for the cylinder as

② $[\Sigma F_x = m\ddot{x}]$ $-c\dot{x} - kx + F = m\ddot{x}$

    $[\Sigma M_G = \bar{I}\ddot{\theta}]$ $-Fr = \tfrac{1}{2}mr^2\ddot{\theta}$

The condition of rolling with no slip is $\ddot{x} = r\ddot{\theta}$. Substitution of this condition into the moment equation gives $F = -\tfrac{1}{2}m\ddot{x}$. Inserting this expression for the friction force into the force equation for the x-direction yields

$$-c\dot{x} - kx - \frac{1}{2}m\ddot{x} = m\ddot{x} \quad \text{or} \quad \ddot{x} + \frac{2}{3}\frac{c}{m}\dot{x} + \frac{2}{3}\frac{k}{m}x = 0$$

Comparing the above equation with that for the standard damped oscillator, Eq. 8/9, allows us to state directly

(a) $\omega_n{}^2 = \dfrac{2}{3}\dfrac{k}{m}$ $\omega_n = \sqrt{\dfrac{2}{3}\dfrac{k}{m}} = \sqrt{\dfrac{2}{3}\dfrac{75}{50}} = 1$ rad/s   *Ans.*

(b) $2\zeta\omega_n = \dfrac{2}{3}\dfrac{c}{m}$ $\zeta = \dfrac{1}{3}\dfrac{c}{m\omega_n} = \dfrac{10}{3(50)(1)} = 0.0667$ *Ans.*

Hence, the damped natural frequency and the damped period are

(c) $\omega_d = \omega_n\sqrt{1 - \zeta^2} = (1)\sqrt{1 - (0.0667)^2} = 0.998$ rad/s   *Ans.*

(d) $\tau_d = 2\pi/\omega_d = 2\pi/0.998 = 6.30$ s   *Ans.*

From Eq. 8/12, the underdamped solution to the equation of motion is

$$x = Ce^{-\zeta\omega_n t}\sin(\omega_d t + \psi) = Ce^{-(0.0667)(1)t}\sin(0.998t + \psi)$$

The velocity is $\dot{x} = -0.0667Ce^{-0.0667t}\sin(0.998t + \psi)$

$$+ 0.998Ce^{-0.0667t}\cos(0.998t + \psi)$$

At time $t = 0$, x and $\dot{x}$ become

$$x_0 = C\sin\psi = -0.2$$

$$\dot{x}_0 = -0.0667C\sin\psi + 0.998C\cos\psi = 0$$

The solution to the two equations in C and ψ gives

$$C = -0.2004 \text{ m} \qquad \psi = 1.504 \text{ rad}$$

Hence, the motion is given by

$$x = -0.2004e^{-0.0667t}\sin(0.998t + 1.504) \text{ m} \qquad \textit{Ans.}$$

① The angle θ is taken positive clockwise to be kinematically consistent with x.

② The friction force F may be assumed in either direction. We will find that the *actual* direction is to the right for $x > 0$ and to the left for $x < 0$; $F = 0$ when $x = 0$.

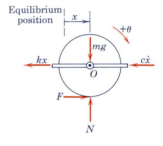

PROBLEMS

Introductory problems

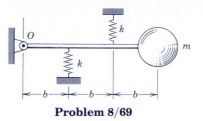

8/69 The light rod and attached sphere of mass m are at rest in the horizontal position shown. Determine the period τ for small oscillations in the vertical plane about the pivot O.

$$Ans. \ \ \tau = 6\pi\sqrt{\frac{m}{5k}}$$

Problem 8/69

8/70 The inverted and spring-loaded pendulum is shown in its equilibrium position. Determine the natural frequency f_n of small oscillations about the vertical, and state the minimum value of the spring constant k for which such oscillations will occur. The mass of the rod is negligible.

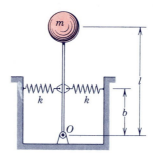

Problem 8/70

8/71 The thin square plate is suspended from a socket (not shown) that fits the small ball attachment at O. If the plate is made to swing about axis A-A, determine the period for small oscillations. Neglect the small offset, mass, and friction of the ball.

$$Ans. \ \ \tau = 2\pi\sqrt{\frac{2b}{3g}}$$

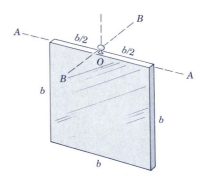

8/72 If the square plate of Prob. 8/71 is made to oscillate about axis B-B, determine the period of small oscillations and compare it with the answer for Prob. 8/71.

Problem 8/71

8/73 The rectangular frame is formed of a uniform slender rod and is suspended from a socket (not shown) that fits the small ball attachment at O. If the rectangle is made to swing about axis A-A, determine the natural frequency for small oscillations. Neglect the small offset, mass, and friction of the ball.

$$Ans. \ \ \omega_n = \frac{3}{2}\sqrt{\frac{g}{2b}}$$

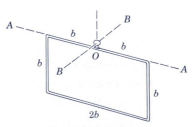

Problem 8/73

8/74 If the rectangular frame of Prob. 8/73 is made to oscillate about axis *B-B*, determine the natural frequency of small oscillations and compare it with the answer for Prob. 8/73.

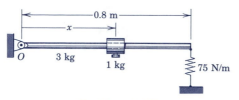

8/75 The mass of the uniform slender rod is 3 kg. Determine the position x for the 1-kg slider so that the system period is 0.9 s. Assume small oscillations about the horizontal equilibrium position shown.

Ans. $x = 0.587$ m

Problem 8/75

Representative problems

8/76 Determine the natural frequency f_n for small oscillations in the vertical plane about the bearing O for the semicircular disk of radius r.

Problem 8/76

8/77 The circular ring of radius r is suspended from a socket (not shown) that fits the small ball attachment at O. Determine the ratio R of the period of small oscillations about axis *B-B* to that about axis *A-A*. Neglect the small offset, mass, and friction of the ball.

Ans. $R = \dfrac{2}{\sqrt{3}}$

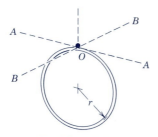

Problem 8/77

8/78 Determine the expression for the natural frequency f_n of small oscillations of the weighted arm about O. The stiffness of the spring is k and its length is adjusted so that the arm is in equilibrium in the horizontal position shown. Neglect the mass of the spring and arm compared with m.

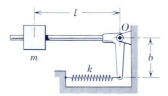

Problem 8/78

8/79 The mechanism shown oscillates in the vertical plane about the pivot O. The springs of equal stiffness k are both compressed in the equilibrium position $\theta = 0$. Determine an expression for the period τ of small oscillations about O. The mechanism has a mass m with mass center G, and the radius of gyration of the assembly about O is k_O.

$$Ans. \quad \tau = \frac{2\pi k_O}{\sqrt{\dfrac{2kb^2}{m} + g\bar{r}}}$$

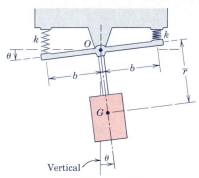

Problem 8/79

8/80 The flywheel is suspended from its center by a wire from a fixed support, and a period τ_1 is measured for torsional oscillation of the flywheel about the vertical axis. Two small weights, each of mass m, are next attached to the flywheel in opposite positions at a distance r from the center. This additional mass results in a slightly longer period τ_2. Write an expression for the moment of inertia I of the flywheel in terms of the measured quantities.

Problem 8/80

8/81 The uniform square plate is suspended in a horizontal plane by its four corner cables from fixed points A and B on a horizontal line a distance b above the plate. Determine an expression for the frequency f_n of small oscillations of the plate about the axis AB.

$$Ans. \quad f_n = \frac{1}{\pi}\sqrt{\frac{3g}{13b}}$$

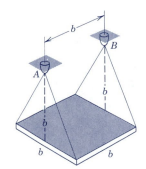

Problem 8/81

8/82 When the motor is slowly brought up to speed, a rather large vibratory oscillation of the entire motor about O-O occurs at a speed of 360 rev/min, which shows that this speed corresponds to the natural frequency of free oscillation of the motor. If the motor has a mass of 43 kg and a radius of gyration of 100 mm about O-O, determine the stiffness k of each of the four identical spring mounts.

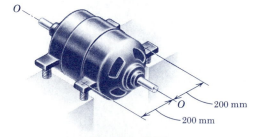

Problem 8/82

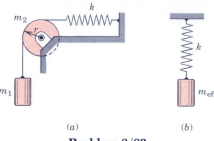

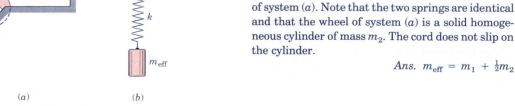

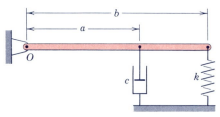

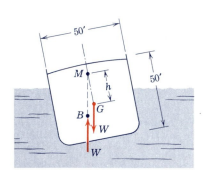

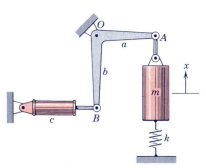

Problem 8/83

Problem 8/84

Problem 8/85

Problem 8/86

8/83 Determine the value m_{eff} of the mass of system (*b*) so that the frequency of system (*b*) is equal to that of system (*a*). Note that the two springs are identical and that the wheel of system (*a*) is a solid homogeneous cylinder of mass m_2. The cord does not slip on the cylinder.

> *Ans.* $m_{\text{eff}} = m_1 + \tfrac{1}{2}m_2$

8/84 The uniform rod of mass m is freely pivoted about point O. Assume small oscillations and determine an expression for the damping ratio ζ. For what value c_{cr} of the damping coefficient c will the system be critically damped?

8/85 The center of mass G of the ship may be assumed to be at the center of the equivalent 50-ft square section. The metacentric height h, determined by the intersection M of the ship's center line with the line of action of the total buoyancy force acting through the center of buoyancy B, is 3 ft. Determine the period τ of one complete roll of the ship if the amplitude is small and the resistance of the water is neglected. Neglect also the change in cross section of the ship at the bow and stern, and treat the loaded ship as a uniform solid block of square cross section.

> *Ans.* $\tau = 13.05$ sec

8/86 The system of Prob. 8/45 is repeated here. If the crank AB now has mass m_2 and a radius of gyration k_O about point O, determine expressions for the undamped natural frequency ω_n and the damping ratio ζ in terms of the given system properties. Assume small oscillations. The damping coefficient for the damper is c.

8/87 The spherical bob of the pendulum shown oscillates in a fluid, which exerts a drag force of form kv^2, where k is a constant and v is the speed of the bob. If the mass of the bob is m and the mass and drag of the rod and the friction in the bearing are negligible, derive the equation of motion of the pendulum for small oscillations and comment on the form of damping present.

$$\text{Ans. } \ddot{\theta} + \frac{kl}{m}|\dot{\theta}|\dot{\theta} + \frac{g}{l}\theta = 0$$

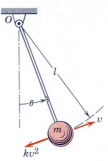

Problem 8/87

8/88 The cart B is given the harmonic displacement $x_B = b \sin \omega t$. Determine the steady-state amplitude Θ of the periodic oscillation of the uniform slender bar which is pinned to the cart at P. Assume small angles and neglect friction at the pivot. The torsional spring is undeformed when $\theta = 0$.

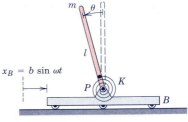

Problem 8/88

8/89 Two identical uniform bars are welded together at a right angle and are pivoted about a horizontal axis through point O as shown. Determine the critical driving frequency ω_c of the block B which will result in excessively large oscillations of the assembly. The mass of the welded assembly is m.

$$\text{Ans. } \omega_c = \sqrt{\frac{6}{5}\left(\frac{2k}{m} + \frac{g}{l}\right)}$$

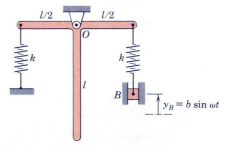

Problem 8/89

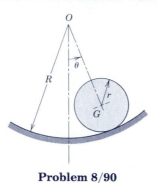

Problem 8/90

8/90 The uniform solid cylinder of mass m and radius r rolls without slipping during its oscillation on the circular surface of radius R. If the motion is confined to small amplitudes $\theta = \theta_0$, determine the period τ of the oscillations. Also determine the angular velocity ω of the cylinder as it crosses the vertical center line. (*Caution:* Do not confuse ω with $\dot{\theta}$ or with ω_n as used in the defining equations. Note also that θ is not the angular displacement of the cylinder.)

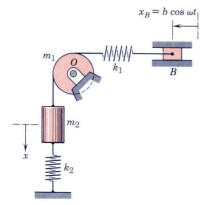

Problem 8/91

8/91 The homogeneous solid cylindrical pulley has mass m_1 and radius r. If the attachment at B undergoes the indicated harmonic displacement, determine the equation of motion of the system in terms of the variable x. The cord which connects mass m_2 to the upper spring does not slip on the pulley.

Ans. $(m_2 + \tfrac{1}{2}m_1)\ddot{x} + (k_1 + k_2)x = k_1 b \cos \omega t$

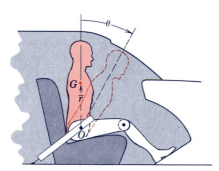

Problem 8/92

8/92 The segmented "dummy" of Prob. 6/118 is repeated here. The hip joint O is assumed to remain fixed to the car, and the torso above the hip is treated as a rigid body of mass m. The center of mass of the torso is at G and the radius of gyration of the torso about O is k_O. An internal torsional spring exerts a moment $M = K\theta$ on the upper torso, where K is the torsional spring constant and θ is the angular deflection from the initial vertical position. If the car is brought to a sudden stop with a constant deceleration a, derive the differential equation for the motion of the torso prior to its impact with the dashboard.

▶8/93   The elements of the "swing-axle" type of independent rear suspension for automobiles are depicted in the figure. The differential D is rigidly attached to the car frame. The half-axles are pivoted at their inboard ends (point O for the half-axle shown) and are rigidly attached to the wheels. Suspension elements not shown constrain the wheel motion to the plane of the figure. The weight of the wheel–tire assembly is $W = 100$ lb, and its mass moment of inertia about a diametral axis passing through its mass center G is 1 lb-ft-sec^2. The weight of the half-axle is negligible. The spring rate and shock-absorber damping coefficient are $k = 50$ lb/in. and $c = 200$ lb-sec/ft, respectively. If a static tire imbalance is present, as represented by the additional concentrated weight $w = 0.5$ lb as shown, determine the angular velocity ω which results in the suspension system being driven at its undamped natural frequency. What would be the corresponding vehicle speed v? Determine the damping ratio ζ. Assume small angular deflections and neglect gyroscopic effects and any car frame vibration. In order to avoid the complications associated with the varying normal force exerted by the road on the tire, treat the vehicle as being on a lift with the wheels hanging free.

Ans. $\omega = 10.24$ rad/sec, $v = 11.95$ ft/sec
$\zeta = 1.707$

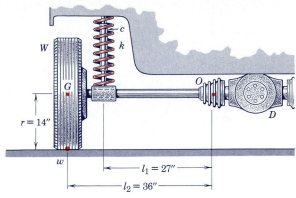

Problem 8/93

▶8/94   For the automobile suspension system of Prob. 8/93, determine the amplitude X of the vertical motion of point G if the angular velocity of the tire corresponds to (a) the undamped natural frequency of the system and (b) a vehicle speed of 55 mi/hr. Reconcile the two results.   *Ans.* (a) $X = 0.0198$ in.
(b) $X = 0.0614$ in.

8/5 ENERGY METHODS

In Arts. 8/2 through 8/4 we derived and solved the equations of motion for vibrating bodies by applying Newton's second law of motion to the body isolated with its free-body diagram. With this approach, we were able to account for the actions of all forces acting on the body, including frictional damping forces. There are many problems where the effect of damping is small and may be neglected, so that the total energy of the system is essentially conserved. For such systems, we find that the principle of conservation of energy may frequently be applied to considerable advantage in establishing the equation of motion and, when the motion is simple harmonic, in determining the frequency of vibration.

To illustrate this alternative approach, consider first the simple case of the body of mass m attached to the spring of stiffness k and vibrating in the vertical direction without damping, Fig. 8/17. As previously, we find it convenient to measure the motion variable x from the equilibrium position. With this datum, the total potential energy of the system, elastic plus gravitational, becomes

$$V = V_e + V_g = \tfrac{1}{2}k(x + \delta_{st})^2 - \tfrac{1}{2}k\delta_{st}^2 - mgx$$

where $\delta_{st} = mg/k$ is the initial static displacement. Substituting $k\delta_{st} = mg$ and simplifying give

$$V = \tfrac{1}{2}kx^2$$

Thus, the total energy of the system becomes

$$T + V = \tfrac{1}{2}m\dot{x}^2 + \tfrac{1}{2}kx^2$$

Because $T + V$ is constant for a conservative system, its time derivative is zero. Consequently,

$$\frac{d}{dt}(T + V) = m\dot{x}\ddot{x} + kx\dot{x} = 0$$

Canceling $\dot{x}$ gives us our basic differential equation of motion

$$m\ddot{x} + kx = 0$$

which is identical to Eq. 8/1 derived in Art. 8/2 for the same system of Fig. 8/3.

Conservation of energy may also be used to determine the period or frequency of vibration for a linear conservative system, without having to derive and solve the equation of motion. For a system that oscillates with simple harmonic motion about the equilibrium position, from which the displacement x is measured, the energy changes from maximum kinetic and zero potential at the equilibrium position $x = 0$ to zero kinetic and maximum potential at the position of maximum displacement $x = x_{max}$. Thus, we may write

$$T_{max} = V_{max}$$

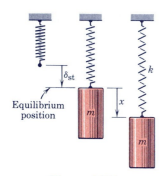

Figure 8/17

Equilibrium position

δ_{st}

k

x

m

m

The maximum kinetic energy is $\frac{1}{2}m(\dot{x}_{max})^2$, and the maximum potential energy is $\frac{1}{2}k(x_{max})^2$. For the harmonic oscillator of Fig. 8/17, we know that the displacement may be written $x = x_{max} \sin(\omega_n t + \psi)$ so that the maximum velocity is $\dot{x}_{max} = \omega_n x_{max}$. Thus, we may write

$$\tfrac{1}{2}m(\omega_n x_{max})^2 = \tfrac{1}{2}k(x_{max})^2$$

where x_{max} is the maximum displacement when the potential energy is a maximum. From this energy balance, we easily obtain

$$\omega_n = \sqrt{k/m}$$

This direct determination of the frequency may be used for any linear undamped vibration.

The main advantage of the energy approach for the free vibration of conservative systems is that it becomes unnecessary to dismember the system and account for all of the forces that act on each member. In Art. 3/7 of Chapter 3 and in Arts. 6/6 and 6/7 of Chapter 6, we learned for a system of interconnected bodies that an active-force diagram of the complete system enabled us to evaluate the work U' of the external active forces and to equate it to the change in the total mechanical energy $T + V$ of the system. So for a conservative mechanical system of interconnected parts with a single degree of freedom where $U' = 0$, we may obtain its equation of motion simply by setting the time derivative of its constant total mechanical energy to zero, giving

$$\frac{d}{dt}(T + V) = 0$$

Here $V = V_e + V_g$ is the sum of the elastic and gravitational potential energies of the system. Also, for an interconnected mechanical system as for a single body, the natural frequency of vibration is obtained by equating the expression for its maximum total kinetic energy to the expression for its maximum potential energy, where the potential energy is taken to be zero at the equilibrium position. This approach to the determination of natural frequency is valid only if it can be determined that the system vibrates with simple harmonic motion.

Sample Problem 8/10

The small sphere of mass m is mounted on the light rod pivoted at O and supported at end A by the vertical spring of stiffness k. End A is displaced a small distance y_0 below the horizontal equilibrium position and released. By the energy method, derive the differential equation of motion for small oscillations of the rod and determine the expression for its natural frequency ω_n of vibration. Damping is negligible.

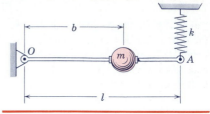

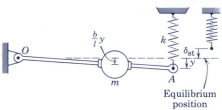

Equilibrium
position

Solution. With the displacement y of the end of the bar measured from the equilibrium position, the potential energy in the displaced position for small values of y becomes

① $$V = V_e + V_g = \frac{1}{2}k(y + \delta_{st})^2 - \frac{1}{2}k\delta_{st}^2 - mg\left(\frac{b}{l}y\right)$$

where δ_{st} is the static deflection of the spring at equilibrium. But the force in the spring in the equilibrium position, from a zero moment sum about O, is $(b/l)mg = k\delta_{st}$. Substituting this value in the expression for V and simplifying yield

② $$V = \frac{1}{2}ky^2$$

The kinetic energy in the displaced position is

$$T = \frac{1}{2}m\left(\frac{b}{l}\dot{y}\right)^2$$

where we see that the vertical displacement of m is $(b/l)y$. Thus, with the energy sum constant, its time derivative is zero, and we have

$$\frac{d}{dt}(T + V) = \frac{d}{dt}\left[\frac{1}{2}m\left(\frac{b}{l}\dot{y}\right)^2 + \frac{1}{2}ky^2\right] = 0$$

which yields

$$\ddot{y} + \frac{l^2}{b^2}\frac{k}{m}y = 0 \qquad\qquad Ans.$$

when $\dot{y}$ is canceled. By analogy with Eq. 8/2, we may write the motion frequency directly as

$$\omega_n = \frac{l}{b}\sqrt{k/m} \qquad\qquad Ans.$$

Alternatively, we can obtain the frequency by equating the maximum kinetic energy, which occurs at $y = 0$, to the maximum potential energy, which occurs at $y = y_0 = y_{max}$, where the deflection is a maximum. Thus,

$$T_{max} = V_{max} \qquad \text{gives} \qquad \frac{1}{2}m\left(\frac{b}{l}\dot{y}_{max}\right)^2 = \frac{1}{2}ky_{max}^2$$

Knowing that we have a harmonic oscillation, which can be expressed as $y = y_{max}\sin\omega_n t$, we have $\dot{y}_{max} = y_{max}\omega_n$. Substituting this relation into our energy balance gives us

$$\frac{1}{2}m\left(\frac{b}{l}y_{max}\omega_n\right)^2 = \frac{1}{2}ky_{max}^2 \qquad \text{so that} \qquad \omega_n = \frac{l}{b}\sqrt{k/m} \quad Ans.$$

as before.

① For large values of y, the circular motion of the end of the bar would cause our expression for the deflection of the spring to be in error.

② Here again, we note the simplicity of the expression for potential energy when the displacement is measured from the equilibrium position.

Sample Problem 8/11

Determine the natural frequency ω_n of vertical vibration of the 3-kg collar to which are attached the two uniform 1.2-kg links, which may be treated as slender bars. The stiffness of the spring, which is attached to both the collar and the foundation, is $k = 1.5$ kN/m, and the bars are both horizontal in the equilibrium position. A small roller on end B of each link permits end A to move with the collar. Frictional retardation is negligible.

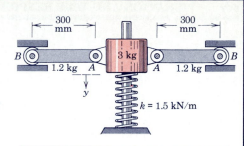

Solution. In the equilibrium position, the compression P in the spring equals the weight of the 3-kg collar, plus half the weight of each link or $P = 3(9.81) + 2(\frac{1}{2})(1.2)(9.81) = 41.20$ N. The corresponding static deflection of the spring is $\delta_{st} = P/k = 41.20/1.5(10^3) = 27.47(10^{-3})$ m. With the displacement variable y measured downward from the equilibrium position, which becomes the position of zero potential energy, the potential energy for each member in the displaced position is

$$(\text{Spring}) \quad V_e = \frac{1}{2} k(y + \delta_{st})^2 - \frac{1}{2} k\delta_{st}^2 = \frac{1}{2} ky^2 + k\delta_{st}y$$

$$= \frac{1}{2}(1.5)(10^3)y^2 + 1.5(10^3)(27.47)(10^{-3})y$$

$$= 750y^2 + 41.20y \text{ J}$$

$$(\text{Collar}) \quad V_g = -m_c gy = -3(9.81)y = -29.43y \text{ J}$$

① $\quad (\text{Each link}) \quad V_g = -m_l g\frac{y}{2} = -1.2(9.81)\frac{y}{2} = -5.89y \text{ J}$

The total potential energy of the system then becomes

② $\quad V = 750y^2 + 41.20y - 29.43y - 2(5.89)y = 750y^2 \text{ J}$

The maximum kinetic energy occurs at the equilibrium position, where the velocity $\dot{y}$ of the collar has its maximum value. For small motion of the collar, end B of each link has essentially no motion, so that ③ each link may be treated as though it were rotating about B as a fixed point with an angular velocity $\dot{y}/0.3$. Thus, the kinetic energy of each part is

$$(\text{Collar}) \quad T = \frac{1}{2} m_c \dot{y}^2 = \frac{3}{2} \dot{y}^2 \text{ J}$$

$$(\text{Each link}) \quad T = \frac{1}{2} I_B \omega^2 = \frac{1}{2}\left(\frac{1}{3} m_l l^2\right)(\dot{y}/l)^2 = \frac{1}{6} m_l \dot{y}^2$$

$$\frac{1}{6}(1.2)\dot{y}^2 = 0.2\dot{y}^2$$

Hence, the kinetic energy of the collar and both links is

$$T = \frac{3}{2} \dot{y}^2 + 2(0.2\dot{y}^2) = 1.9\dot{y}^2$$

With the harmonic motion expressed by $y = y_{max} \sin \omega_n t$, we have $\dot{y}_{max} = y_{max}\omega_n$, so that the energy balance $T_{max} = V_{max}$ with $\dot{y} = \dot{y}_{max}$ ④ becomes

⑤ $\quad 1.9(y_{max}\omega_n)^2 = 750y_{max}^2 \quad$ or $\quad \omega_n = \sqrt{750/1.9} = 19.87$ Hz $\quad$ *Ans.*

① Note that the mass center of each link moves down only half as far as the collar.

② We note again that measurement of the motion variable y from the equilibrium position results in the total potential energy being simply $V = \frac{1}{2}ky^2$.

③ Our knowledge of rigid-body kinematics is essential at this point.

④ To appreciate the advantage of the work-energy method for this and similar problems of interconnected systems, the student is encouraged to explore the steps required for solution by the force and moment equations of motion of the separate parts.

⑤ If the oscillations were large, the links could no longer be considered as bodies rotating about fixed points. In this event, we would find that the angular velocity of each link would equal $\dot{y}/\sqrt{0.09 - y^2}$, which would cause a nonlinear response no longer described by $y = y_{max} \sin \omega t$.

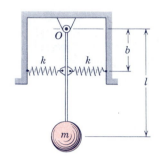

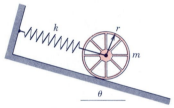

Problem 8/95

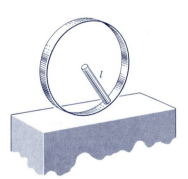

Problem 8/96

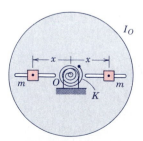

Problem 8/97

PROBLEMS

(Solve the following problems by the energy method of Art. 8/5.)

Introductory problems

8/95 Derive the differential equation of motion for the spring-loaded pendulum which consists of a particle and a light rod. Assume small oscillation.

$$Ans. \quad \ddot{\theta} + \left(\frac{2kb^2 + mgl}{ml^2}\right)\theta = 0$$

8/96 The spoked wheel of radius r, mass m, and centroidal radius of gyration $\bar{k}$ rolls without slipping on the incline. Determine the natural frequency of oscillation and explore the limiting cases of $\bar{k} = 0$ and $\bar{k} = r$.

8/97 A uniform rod of mass m and length l is welded at one end to the rim of a light circular hoop of radius l. The other end lies at the center of the hoop. Determine the period τ for small oscillations about the vertical position of the bar if the hoop rolls on the horizontal surface without slipping.

$$Ans. \quad \tau = 2\pi\sqrt{\frac{2l}{3g}}$$

8/98 The disk has mass moment of inertia I_O about O and is acted upon by a torsional spring of constant K. The position of the small sliders, each of which has mass m, is adjustable. Determine the value of x for which the system has a given period τ.

Problem 8/98

8/99 The system of Prob. 8/22 is repeated here. By the methods of this article, determine the period of vertical oscillation. Each spring has a stiffness of 6 lb/in.   *Ans.* $\tau = 0.326$ sec

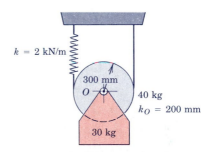

Problem 8/99

8/100 Determine the natural frequency ω_n of the system shown. The mass and friction of the pulleys are negligible.

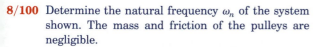

Problem 8/100

Representative problems

8/101 Calculate the frequency f_n of vertical oscillation of the system shown. The 40-kg pulley has a radius of gyration about its center O of 200 mm.
         Ans. $f_n = 1.519$ Hz

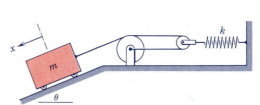

Problem 8/101

8/102 The homogeneous circular cylinder of Prob. 8/90, repeated here, rolls without slipping on the track of radius R. Determine the period τ for small oscillations.

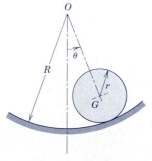

Problem 8/102

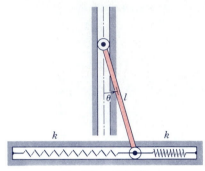

Problem 8/103

8/103 The ends of the uniform bar of mass m slide freely in the vertical and horizontal slots as shown. If the bar is in static equilibrium when $\theta = 0$, determine the natural frequency ω_n of small oscillations. What condition must be imposed on the spring constant k in order that oscillations take place?

$$Ans. \quad \omega_n = \sqrt{\frac{6k}{m} - \frac{3}{2}\frac{g}{l}}, \quad k > \frac{mg}{4l}$$

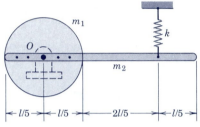

Problem 8/104

8/104 The uniform slender rod of length l and mass m_2 is secured to the uniform disk of radius $l/5$ and mass m_1. If the system is shown in its equilibrium position, determine the natural frequency ω_n and the maximum angular velocity ω for small oscillations of amplitude θ_0 about the pivot O.

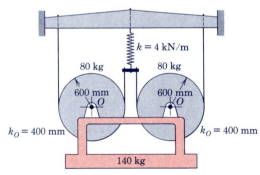

Problem 8/105

8/105 Determine the period τ of vertical oscillations for the system composed of the 140-kg frame and two 80-kg pulleys, each of which has a radius of gyration $k_O = 400$ mm. The flexible wires do not slip on the pulleys. $Ans. \quad \tau = 0.957$ s

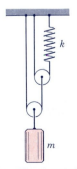

Problem 8/106

8/106 Determine the natural frequency f_n for vertical oscillation of the system shown in the figure.

8/107 Derive the natural frequency f_n of the system composed of two homogeneous circular cylinders, each of mass M, and the connecting link AB of mass m. Assume small oscillations.

$$Ans. \ f_n = \frac{1}{2\pi} \sqrt{\frac{mgr_0}{3Mr^2 + m(r - r_0)^2}}$$

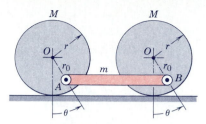

Problem 8/107

8/108 The thin homogeneous panel of mass m is hinged to swing freely about a fixed axis that makes an angle α with the vertical. Determine the period of small oscillations.

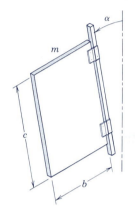

Problem 8/108

8/109 The 12-kg block is supported by the two 5-kg links with two torsion springs, each of constant $K = 500$ N·m/rad, arranged as shown. The springs are sufficiently stiff so that stable equilibrium is established in the position shown. Determine the natural frequency f_n for small oscillations about this equilibrium position. *Ans.* $f_n = 1.496$ Hz

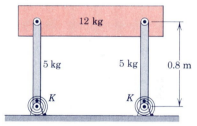

Problem 8/109

8/110 Each of the two slider blocks A has a mass m and is constrained to move in one of the smooth radial slots of the flywheel, which is driven at a constant angular speed ω. Each of the four springs has a stiffness k. Is it correct to state that the system composed of the flywheel, blocks, and springs possesses a constant energy? Explain your answer.

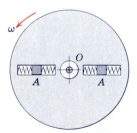

Problem 8/110

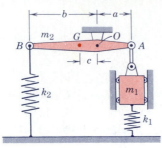

Problem 8/111

8/111 Determine the period τ of the system if the link *AB* is horizontal in the static equilibrium position shown. The radius of gyration about point *O* of the link *AB* is k_O and its center of mass is point *G*. Assume small oscillations.

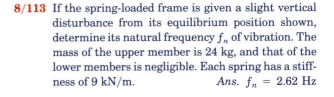

$$\textit{Ans.} \quad \tau = 2\pi \sqrt{\frac{m_1 + \dfrac{k_O^2}{a^2} m_2}{k_1 + \dfrac{b^2}{a^2} k_2}}$$

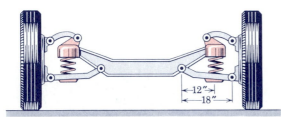

Problem 8/112

8/112 The front-end suspension of an automobile is shown. Each of the coil springs has a stiffness of 270 lb/in. If the weight of the front-end frame and equivalent portion of the body attached to the front end is 1800 lb, determine the natural frequency f_n of vertical oscillation of the frame and body in the absence of shock absorbers. (*Hint:* To relate the spring deflection to the deflection of the frame and body, consider the frame fixed and let the ground and wheels move vertically.)

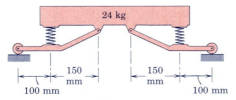

Problem 8/113

8/113 If the spring-loaded frame is given a slight vertical disturbance from its equilibrium position shown, determine its natural frequency f_n of vibration. The mass of the upper member is 24 kg, and that of the lower members is negligible. Each spring has a stiffness of 9 kN/m.   *Ans.* $f_n = 2.62$ Hz

Problem 8/114

8/114 The semicircular cylindrical shell of radius r with small but uniform wall thickness is set into small rocking oscillation on the horizontal surface. If no slipping occurs, determine the expression for the period τ of each complete oscillation.

▶**8/115** The concrete hopper has a total weight $W = 12,000$ lb and is being lowered at the rate of 4 ft/sec when the hoisting drum is suddenly stopped. Determine the additional downward deflection Δh of the hopper and the frequency f_n of the resulting vibration of the hopper for $h = 40$ ft. The length of the 1-in.-diameter steel cable from drum to hopper is $L = 50$ ft. The elastic modulus of steel is $E = 30(10^6)$ lb/in.² (Recall that $E = \sigma/\epsilon$ where the stress σ is force per unit area and the strain ϵ is the elastic elongation per unit length.)

Ans. $\Delta h = 1.811$ in., $f_n = 4.22$ Hz

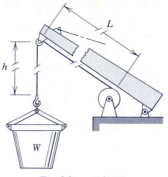

Problem 8/115

▶**8/116** A hole of radius $R/4$ is drilled through a cylinder of radius R to form a body of mass m as shown. If the body rolls on the horizontal surface without slipping, determine the period τ for small oscillations.

Ans. $\tau = 41.4\sqrt{\dfrac{R}{g}}$

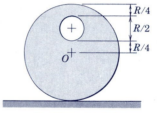

Problem 8/116

8/6 PROBLEM FORMULATION AND REVIEW

In studying the vibrations of particles and rigid bodies in Chapter 8, we have observed that the subject is simply a direct application of the fundamental principles of dynamics as derived in Chapters 3 and 6. In these previous chapters, we established the dynamic behavior of a body at an instant of time and also found the changes in motion resulting from finite intervals of displacement or time. Chapter 8, on the other hand, has dealt with the solution of the defining differential equations of motion, so that the linear or angular displacement can be expressed as a function of time. This information is of critical importance in many engineering problems.

Our study of the time response of particles has been divided into the two categories of free and forced motion, with the further subdivisions of negligible and significant damping. It has been seen that the damping ratio ζ is a convenient variable for determining the nature of unforced but viscously damped vibrations. The prime lesson associated with harmonic forcing is that driving a lightly damped system with a force whose frequency is near the natural frequency can cause motion of excessively large amplitude—a condition called resonance that usually must be carefully avoided.

In our study of rigid-body vibrations, we observed that the equation of motion has a form identical to that for particle vibrations. Whereas particle vibrations may be described completely by the equations governing linear motion, rigid-body vibrations usually require the equations of rotational dynamics.

In the final article of Chapter 8, we learned how the energy method can facilitate the determination of the natural frequency ω_n in free vibration problems where damping may be neglected. Here the total mechanical energy of the system is assumed to be constant. Setting its first time derivative to zero leads directly to the differential equation of motion for the system. The energy approach also permits the analysis of a conservative system of interconnected parts without dismembering the system. When mechanical energy dissipation cannot be neglected, we must use the laws of motion in terms of force, mass, and acceleration.

Throughout the chapter, we have restricted our attention to systems having one degree of freedom, where the position of the system can be specified by a single variable. One motion variable suffices for many engineering problems. If a system possesses n degrees of freedom, it has n natural frequencies. Thus, if a harmonic force is applied to such a system that is lightly damped, there are n driving frequencies that can cause motion of large amplitude. By a process called modal analysis, a complex system with n degrees of freedom can be reduced to n single-degree-of-freedom systems. For this reason, the thorough understanding of the material of this chapter is vital for the further study of vibrations.

REVIEW PROBLEMS

8/117 The 0.1-kg projectile is fired into the 10-kg block which is initially at rest with no force in the spring. The spring is attached at both ends. Calculate the maximum horizontal displacement X of the spring and the ensuing period of oscillation of the block and embedded projectile.

$$Ans. \ X = 0.287 \text{ m}, \ \tau = 0.365 \text{ s}$$

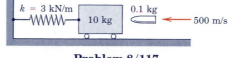

Problem 8/117

8/118 Determine the natural frequency f_n of vertical oscillations of the cylinder of mass m. The mass and friction of the stepped drum are negligible.

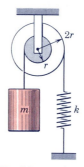

Problem 8/118

8/119 The uniform circular disk is suspended by a socket (not shown) which fits over the small ball attachment at O. Determine the period of small motion if the disk swings freely about (a) axis A-A and (b) axis B-B. Neglect the small offset, mass, and friction of the ball.

$$Ans. \ (a) \ \omega_n = 2\sqrt{\frac{g}{5r}}, \ (b) \ \omega_n = \sqrt{\frac{2g}{3r}}$$

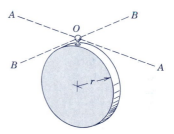

Problem 8/119

8/120 Determine the value of the damping coefficient c for which the system is critically damped if $k =$ 400 lb/in. and $W = 200$ lb.

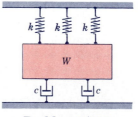

Problem 8/120

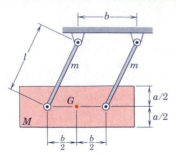

Problem 8/121

8/121 The block of mass M is suspended by the two uniform slender rods each of mass m. Determine the natural frequency ω_n of small oscillations for the system shown.

$$\text{Ans.} \quad \omega_n = \sqrt{\frac{g}{l}\left(\frac{M + m}{M + \frac{2}{3}m}\right)}$$

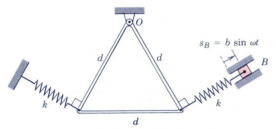

Problem 8/122

8/122 The triangular frame is constructed of uniform slender rod and pivots about a horizontal axis through point O. Determine the critical driving frequency ω_c of the block B which will result in oscillations of the assembly which would tend to become excessively large. The total mass of the frame is m.

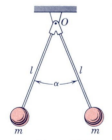

Problem 8/123

8/123 Determine the period τ for small oscillations of the assembly composed of two light bars and two particles, each of mass m. Investigate your expression as the angle α approaches values of 0 and 180°.

$$\text{Ans.} \quad \tau = 2\pi\sqrt{\frac{l}{g\cos\dfrac{\alpha}{2}}}$$

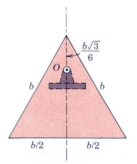

Problem 8/124

8/124 The uniform triangular plate pivots freely about a horizontal axis through point O. Determine the natural frequency of small oscillations.

8/125 Calculate the damping ratio ζ of the system shown if the weight and radius of gyration of the stepped cylinder are $W = 20$ lb and $\bar{k} = 5.5$ in., the spring constant is $k = 15$ lb/in., and the damping coefficient of the hydraulic cylinder is $c = 2$ lb-sec/ft. The cylinder rolls without slipping on the radius $r = 6$ in. and the spring can support tension as well as compression.　　　　*Ans.* $\zeta = 0.070$

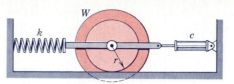

Problem 8/125

8/126 Determine the largest amplitude x_0 for which the uniform circular disk will roll without slipping on the horizontal surface.

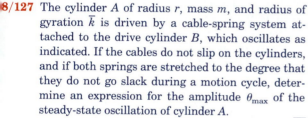

Problem 8/126

8/127 The cylinder A of radius r, mass m, and radius of gyration $\bar{k}$ is driven by a cable-spring system attached to the drive cylinder B, which oscillates as indicated. If the cables do not slip on the cylinders, and if both springs are stretched to the degree that they do not go slack during a motion cycle, determine an expression for the amplitude θ_{max} of the steady-state oscillation of cylinder A.

Ans. $\theta_{max} = \phi_0 \dfrac{r_0/r}{1 - (\omega/\omega_n)^2}$ where $\omega_n = \dfrac{r}{\bar{k}} \sqrt{\dfrac{2k}{m}}$

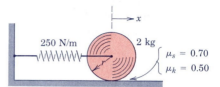

Problem 8/127

8/128 An experimental engine weighing 480 lb is mounted on a test stand with spring mounts at A and B, each with a stiffness of 600 lb/in. The radius of gyration of the engine about its mass center G is 4.60 in. With the motor not running, calculate the natural frequency $(f_n)_y$ of vertical vibration and $(f_n)_\theta$ of rotation about G. If vertical motion is suppressed and a slight rotational imbalance occurs, at what speed N should the engine not be run?

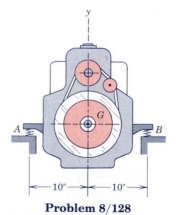

Problem 8/128

Problem 8/129

Problem 8/130

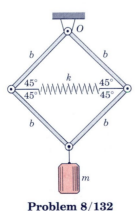

Problem 8/132

8/129 The uniform bar of mass M and length l has a small roller of mass m with negligibie bearing friction at each end. Determine the period τ of the system for small oscillations on the curved track.

$$Ans. \quad \tau = 2\pi \sqrt{\frac{2mR^2 + MR^2 - \frac{1}{6}Ml^2}{g\sqrt{4R^2 - l^2}\,(m + M/2)}}$$

8/130 A 60-g bullet is fired with a velocity of 300 m/s at the 5-kg block mounted on a stiff but light cantilever beam. The bullet is embedded in the block, which is then observed to vibrate with a frequency of 4 Hz. Compute the maximum displacement A in the vibration and find the damping constant c in N·s/m if the ratio of two amplitudes ten full cycles apart is 0.6. Neglect any energy loss during the first quarter cycle.

▶**8/131** A 200-kg machine rests on four floor mounts each of which has an effective spring constant $k = 250$ kN/m and an effective viscous damping coefficient $c = 1000$ N·s/m. The floor is known to vibrate vertically with a frequency of 24 Hz. What would be the effect on the amplitude of the absolute machine oscillation if the mounts were replaced with new ones which have the same effective spring constant but twice the effective damping coefficient?
                Ans. Amplitude increases by 28.9%!

▶**8/132** The system composed of four light rigid rods and the cylinder of mass m is shown in its equilibrium position. Determine the natural frequency ω_n of small-amplitude vertical oscillations of the cylinder. (*Hint:* Retain only first-order terms.)

$$Ans. \quad \omega_n = \sqrt{\frac{k}{m} + \frac{g}{b}\sqrt{2}}$$

Computer-oriented problems

8/133 The 2-kg mass is released from rest with an initial displacement $x = 15$ mm from its equilibrium position at time $t = 0$. Successive positive amplitudes are observed to diminish by a factor of 4. Plot the displacement x of the mass in terms of time t for $0 < t < 1$ s.

Ans.
$$x = (15 \cos 19.53t + 3.31 \sin 19.53t)e^{-4.31t} \text{ mm}$$

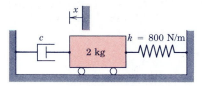

Problem 8/133

8/134 The mass of a critically damped system having a natural frequency $\omega_n = 4$ rad/s is released from rest at an initial displacement x_0. Determine the time t required for the mass to reach the position $x = 0.1x_0$.

8/135 The mass of the system shown is released with the initial conditions $x_0 = 0.1$ m and $\dot{x}_0 = -5$ m/s at $t = 0$. Plot the response of the system and determine the time(s) (if any) at which $x = -0.05$ m.

Ans. $t_1 = 0.0544$ s, $t_2 = 0.442$ s

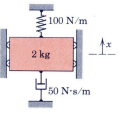

Problem 8/135

8/136 Plot the response x of the 50-lb body over the time interval $0 \le t \le 1$ second. Determine the maximum and minimum values of x and their respective times. The initial conditions are $x_0 = 0$ and $\dot{x}_0 = 6$ ft/sec.

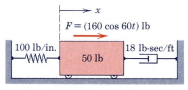

Problem 8/136

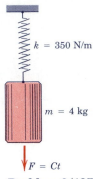

$k = 350$ N/m

$m = 4$ kg

$F = Ct$

Problem 8/137

*8/137 The 4-kg mass is suspended by the spring of stiffness $k = 350$ N/m and is initially at rest in the equilibrium position. If a downward force $F = Ct$ is applied to the body and reaches a value of 40 N when $t = 1$ s, derive the differential equation of motion, obtain its solution, and plot the displacement y in millimeters as a function of time during the first second. Damping is negligible.

Ans. $y = 114.3(t - 0.1069 \sin 9.35t)$ mm

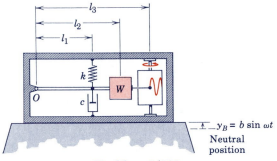

l_3

l_2

l_1

k

W

O

c

$y_B = b \sin \omega t$

Neutral
position

Problem 8/138

*8/138 Shown in the figure are the elements of a displacement meter used to study the motion $y_B = b \sin \omega t$ of the base. The motion of the mass relative to the frame is recorded on the rotating drum. If $l_1 = 1.2$ ft, $l_2 = 1.6$ ft, $l_3 = 2$ ft, $W = 2$ lb, $c = 0.1$ lb-sec/ft, and $\omega = 10$ rad/sec, determine the range of the spring constant k over which the magnitude of the recorded relative displacement is less than $1.5b$. It is assumed that the ratio ω/ω_n must remain greater than unity.

AREA MOMENTS OF INERTIA

A

See Appendix A of *Vol. 1 Statics* for a treatment of the theory and calculation of area moments of inertia. Because this quantity plays an important role in the design of structures, especially those dealt with in statics, we present only a brief definition in this *Dynamics* volume so that the student can appreciate the basic differences between area and mass moments of inertia.

The moments of inertia of a plane area A about the x- and y-axes in its plane and about the z-axis normal to its plane, Fig. A/1, are defined by

$$I_x = \int y^2 \, dA \qquad I_y = \int x^2 \, dA \qquad I_z = \int r^2 \, dA$$

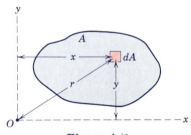

Figure A/1

where dA is the differential element of area and $r^2 = x^2 + y^2$. Clearly, the polar moment of inertia I_z equals the sum $I_x + I_y$ of the rectangular moments of inertia. For thin flat plates, the area moment of inertia is useful in the calculation of the mass moment of inertia, as explained in Appendix B.

The area moment of inertia is a measure of the distribution of area about the axis in question and, for that axis, is a constant property of the area. The dimensions of area moment of inertia are (distance)^4 expressed in m^4 or mm^4 in SI units and ft^4 or in.^{4} in U.S. customary units. In contrast, mass moment of inertia is a measure of the distribution of mass about the axis in question, and its dimensions are (mass)(distance)^2 which are expressed in kg·m^2 in SI units and in lb-ft-sec^2 or lb-in.-sec^2 in U.S. customary units.

MASS MOMENTS OF INERTIA

B

B/1 MASS MOMENTS OF INERTIA ABOUT AN AXIS

The rotational equation of motion about an axis normal to the plane of motion for a rigid body in plane motion contains an integral that depends on the distribution of mass with respect to the moment axis. This integral occurs whenever a rigid body has an angular acceleration about its axis of rotation.

Consider a body of mass m, Fig. B/1, rotating about an axis O-O with an angular acceleration α. All particles of the body move in parallel planes that are normal to the rotation axis O-O. We may choose any one of the planes as the plane of motion, although the one containing the center of mass is usually the one so designated. An element of mass dm has a component of acceleration tangent to its circular path equal to $r\alpha$, and by Newton's second law of motion the resultant tangential force on this element equals $r\alpha\,dm$. The moment of this force about the axis O-O is $r^2\alpha\,dm$, and the sum of the moments of these forces for all elements is $\int r^2\alpha\,dm$. For a rigid body, α is the same for all radial lines in the body and we may take it outside the integral sign. The remaining integral is known as the moment of inertia I of the mass m about the axis O-O and is

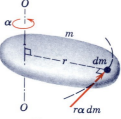

Figure B/1

$$ I = \int r^2\,dm \qquad \text{(B/1)} $$

This integral represents an important property of a body and is involved in the analysis of any body that has rotational acceleration about a given axis. Just as the mass m of a body is a measure of the resistance to translational acceleration, the moment of inertia I is a measure of resistance to rotational acceleration of the body.

The moment-of-inertia integral may be expressed alternatively as

$$I = \Sigma r_i^2 m_i \tag{B/1a}$$

where r_i is the radial distance from the inertia axis to the representative particle of mass m_i and where the summation is taken over all particles of the body.

If the density ρ is constant throughout the body, the moment of inertia becomes

$$I = \rho \int r^2 \, dV$$

where dV is the element of volume. In this case, the integral by itself defines a purely geometrical property of the body. When the density is not constant but is expressed as a function of the coordinates of the body, it must be left within the integral sign and its effect accounted for in the integration process.

In general, the coordinates that best fit the boundaries of the body should be used in the integration. It is particularly important that we make a good choice of the element of volume dV. To simplify the integration, an element of lowest possible order should be chosen, and the correct expression for the moment of inertia of the element about the axis involved should be used. For example, in finding the moment of inertia of a solid right-circular cone about its central axis, we may choose an element in the form of a circular slice of infinitesimal thickness, Fig. B/2a. The differential moment

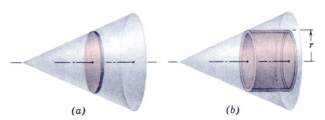

(a)   (b)

Figure B/2

of inertia for this element is the expression for the moment of inertia of a circular cylinder of infinitesimal altitude about its central axis. (This expression will be obtained in Sample Problem B/1.) Alternatively, we could choose an element in the form of a cylindrical shell of infinitesimal thickness as shown in Fig. B/2b. Since all of the mass of the element is at the same distance r from the inertia axis, the differential moment of inertia for this element is merely $r^2 \, dm$ where dm is the differential mass of the elemental shell.

From the definition of mass moment of inertia, its dimensions are (mass)(distance)^2 and are expressed in the units kg·m^2 in SI units and in lb-ft-sec^2 in U.S. customary units.

(a) Radius of gyration. The radius of gyration k of a mass m about an axis for which the moment of inertia is I is defined as

$$k = \sqrt{\frac{I}{m}} \quad \text{or} \quad I = k^2 m \qquad \textbf{(B/2)}$$

Thus, k is a measure of the distribution of mass of a given body about the axis in question, and its definition is analogous to the definition of the radius of gyration for area moments of inertia. If all the mass m of a body could be concentrated at a distance k from the axis, the moment of inertia would be unchanged. The moment of inertia of a body about a particular axis is frequently indicated by specifying the mass of the body and the radius of gyration of the body about the axis. The moment of inertia is then calculated from Eq. B/2.

(b) Transfer of axes. If the moment of inertia of a body is known about a centroidal axis, it may be determined easily about any parallel axis. To prove this statement, consider the two parallel axes in Fig. B/3, one being a centroidal axis through the mass center G and the other a parallel axis through some other point C. The radial distances from the two axes to any element of mass dm are r_0 and r, and the separation of the axes is d. Substituting the law of cosines $r^2 = r_0{}^2 + d^2 + 2r_0 d \cos \theta$ into the definition for the moment of inertia about the noncentroidal axis through C gives

$$I = \int r^2 \, dm = \int (r_0{}^2 + d^2 + 2r_0 d \cos \theta) \, dm$$
$$= \int r_0{}^2 \, dm + d^2 \int dm + 2d \int u \, dm$$

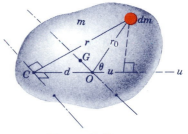

Figure B/3

The first integral is the moment of inertia $\bar{I}$ about the mass-center axis, the second term is md^2, and the third integral equals zero, since the u-coordinate of the mass center with respect to the axis through G is zero. Thus, the parallel-axis theorem is

$$I = \bar{I} + md^2 \qquad \textbf{(B/3)}$$

It must be remembered that the transfer cannot be made unless one axis passes through the center of mass and unless the axes are parallel. When the expressions for the radii of gyration are substituted in Eq. B/3, there results

$$k^2 = \bar{k}^2 + d^2 \qquad \textbf{(B/3a)}$$

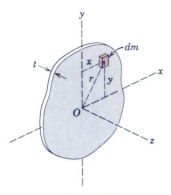

Figure B/4

which is the parallel-axis theorem for obtaining the radius of gyration k about an axis a distance d from a parallel centroidal axis for which the radius of gyration is $\bar{k}$.

For plane-motion problems where rotation occurs about an axis normal to the plane of motion, a single subscript for I is sufficient to designate the inertia axis. Thus, if the plate of Fig. B/4 has plane motion in the x-y plane, the moment of inertia of the plate about the z-axis through O is designated I_O. For three-dimensional motion, however, where components of rotation may occur about more than one axis, we use a double subscript to preserve notational symmetry with product-of-inertia terms, which are described in Art. B/2. Thus, the moments of inertia about the x-, y-, and z-axes are labeled I_{xx}, I_{yy}, and I_{zz}, respectively, and from Fig. B/5 we see that they become

$$
\begin{aligned}
I_{xx} &= \int r_x^2 \, dm = \int (y^2 + z^2) \, dm \\
I_{yy} &= \int r_y^2 \, dm = \int (z^2 + x^2) \, dm \\
I_{zz} &= \int r_z^2 \, dm = \int (x^2 + y^2) \, dm
\end{aligned}
\qquad \textbf{(B/4)}
$$

These integrals are cited in Eqs. 7/10 of Art. 7/7 on angular momentum in three-dimensional rotation.

The similarity between the defining expressions for mass moments of inertia and area moments of inertia is easily observed. An exact relationship between the two moment-of-inertia expressions exists in the case of flat plates. Consider the flat plate of uniform thickness in Fig. B/4. If the constant thickness is t and the density is ρ, the mass moment of inertia I_{zz} of the plate about the z-axis normal to the plate is

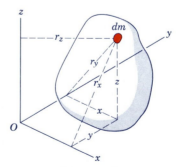

Figure B/5

$$I_{zz} = \int r^2 \, dm = \rho t \int r^2 \, dA = \rho t I_z \qquad (B/5)$$

Thus, the mass moment of inertia about the z-axis equals the mass per unit area ρt times the polar moment of inertia I_z of the plate area about the z-axis. If t is small compared with the dimensions of the plate in its plane, the mass moments of inertia I_{xx} and I_{yy} of the plate about the x- and y-axes are closely approximated by

$$I_{xx} = \int y^2 \, dm = \rho t \int y^2 \, dA = \rho t I_x$$
$$I_{yy} = \int x^2 \, dm = \rho t \int x^2 \, dA = \rho t I_y \qquad (B/6)$$

Hence, the mass moments of inertia equal the mass per unit area ρt times the corresponding area moments of inertia. The double subscripts for mass moments of inertia distinguish these quantities from area moments of inertia.

Inasmuch as $I_z = I_x + I_y$ for area moments of inertia, we have

$$I_{zz} = I_{xx} + I_{yy} \qquad (B/7)$$

which holds *only* for a thin flat plate. This restriction is observed from Eqs. B/6, which do not hold true unless the thickness t or z-coordinate of the element is negligible compared with the distance of the element from the corresponding x- or y-axis. Equation B/7 is very useful when dealing with a differential mass element taken as a flat slice of differential thickness, say, dz. In this case, Eq. B/7 holds exactly and becomes

$$dI_{zz} = dI_{xx} + dI_{yy} \qquad (B/7a)$$

(c) Composite bodies.   The defining integral, Eq. B/1, involves the square of the distance from the axis to the element and so is always positive. Thus, as in the case of area moments of inertia, the mass moment of inertia of a composite body is the sum of the moments of inertia of the individual parts about the same axis. It is often convenient to consider a composite body as defined by positive volumes and negative volumes. The moment of inertia of a negative element, such as the material removed to form a hole, must be considered a negative quantity.

Thorough familiarity with the calculation of mass moments of inertia for rigid bodies is an absolute necessity for the study of the dynamics of rotation.

A summary of some of the more useful formulas for mass moments of inertia of various masses of common shape is given in Table D/4, Appendix D.

Sample Problem   B/1

Determine the moment of inertia and radius of gyration of a homogeneous right-circular cylinder of mass m and radius r about its central axis O-O.

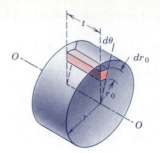

Solution.   An element of mass in cylindrical coordinates is $dm = \rho\, dV = \rho t r_0\, dr_0\, d\theta$, where ρ is the density of the cylinder. The moment ① of inertia about the axis of the cylinder is

$$I = \int r_0{}^2\, dm = \rho t \int_0^{2\pi}\!\!\int_0^r r_0{}^3\, dr_0\, d\theta = \rho t\, \frac{\pi r^4}{2} = \tfrac{1}{2}mr^2 \qquad Ans.$$

②

The radius of gyration is

$$k = \sqrt{\frac{I}{m}} = \frac{r}{\sqrt{2}} \qquad\qquad Ans.$$

① If we had started with a cylindrical shell of radius r_0 and axial length t as our mass element dm, then $dI = r_0{}^2\, dm$ directly. The student should evaluate the integral.

② The result $I = \tfrac{1}{2}mr^2$ applies *only* to a solid homogeneous circular cylinder and cannot be used for any other wheel of circular periphery.

Sample Problem   B/2

Determine the moment of inertia and radius of gyration of a homogeneous solid sphere of mass m and radius r about a diameter.

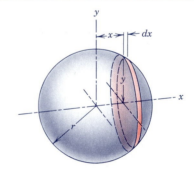

Solution.   A circular slice of radius y and thickness dx is chosen as the volume element. From the results of Sample Problem B/1, the moment of inertia about the x-axis of the elemental cylinder is

①

$$dI_{xx} = \tfrac{1}{2}(dm)y^2 = \tfrac{1}{2}(\pi\rho y^2\, dx)y^2 = \frac{\pi\rho}{2}(r^2 - x^2)^2\, dx$$

where ρ is the constant density of the sphere. The total moment of inertia about the x-axis is

$$I_{xx} = \frac{\pi\rho}{2}\int_{-r}^{r}(r^2 - x^2)^2\, dx = \tfrac{8}{15}\pi\rho r^5 = \tfrac{2}{5}mr^2 \qquad Ans.$$

The radius of gyration is

$$k = \sqrt{\frac{I}{m}} = \sqrt{\frac{2}{5}}\, r \qquad\qquad Ans.$$

① Here is an example where we utilize a previous result to express the moment of inertia of the chosen element, which in this case is a right-circular cylinder of differential axial length dx. It would be foolish to start with a third-order element, such as $\rho\, dx\, dy\, dz$, when we can easily solve the problem with a first-order element.

Sample Problem B/3

Determine the moments of inertia of the homogeneous rectangular parallelepiped of mass m about the centroidal x_0- and z-axes and about the x-axis through one end.

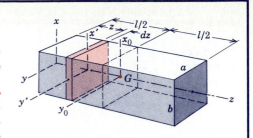

Solution. A transverse slice of thickness dz is selected as the element of volume. The moment of inertia of this slice of infinitesimal thickness equals the moment of inertia of the area of the section times the mass per unit area $\rho\, dz$. Thus, the moment of inertia of the transverse slice about the y'-axis is

$$dI_{y'y'} = (\rho\, dz)(\tfrac{1}{12}ab^3)$$

and that about the x'-axis is

$$dI_{x'x'} = (\rho\, dz)(\tfrac{1}{12}a^3b)$$

① Refer to Eqs. B/6 and recall the expression for the area moment of inertia of a rectangle about an axis through its center parallel to its base.

As long as the element is a plate of differential thickness, the principle given by Eq. B/7a may be applied to give

$$dI_{zz} = dI_{x'x'} + dI_{y'y'} = (\rho\, dz)\frac{ab}{12}(a^2 + b^2)$$

These expressions may now be integrated to obtain the desired results.

The moment of inertia about the z-axis is

$$I_{zz} = \int dI_{zz} = \frac{\rho ab}{12}(a^2 + b^2)\int_0^l dz = \tfrac{1}{12}m(a^2 + b^2) \qquad Ans.$$

where m is the mass of the block. By interchange of symbols, the moment of inertia about the x_0-axis is

$$I_{x_0x_0} = \tfrac{1}{12}m(a^2 + l^2) \qquad Ans.$$

The moment of inertia about the x-axis may be found by the parallel-axis theorem, Eq. B/3. Thus,

$$I_{xx} = I_{x_0x_0} + m\left(\frac{l}{2}\right)^2 = \tfrac{1}{12}m(a^2 + 4l^2) \qquad Ans.$$

This last result may be obtained by expressing the moment of inertia of the elemental slice about the x-axis and integrating the expression over the length of the bar. Again, by the parallel-axis theorem

$$dI_{xx} = dI_{x'x'} + z^2\, dm = (\rho\, dz)(\tfrac{1}{12}a^3b) + z^2\rho ab\, dz = \rho ab\left(\frac{a^2}{12} + z^2\right)dz$$

Integrating gives the result obtained previously:

$$I_{xx} = \rho ab \int_0^l \left(\frac{a^2}{12} + z^2\right)dz = \frac{\rho abl}{3}\left(l^2 + \frac{a^2}{4}\right) = \tfrac{1}{12}m(a^2 + 4l^2)$$

The expression for I_{xx} may be simplified for a long prismatic bar or slender rod whose transverse dimensions are small compared with the length. In this case, a^2 may be neglected compared with $4l^2$, and the moment of inertia of such a slender bar about an axis through one end normal to the bar becomes $I = \tfrac{1}{3}ml^2$. By the same approximation, the moment of inertia about a centroidal axis normal to the bar is $I = \tfrac{1}{12}ml^2$.

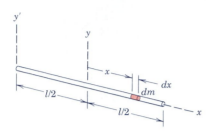

Problem B/1

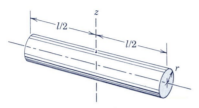

Problem B/2

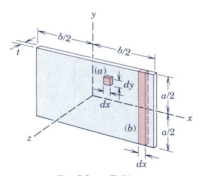

Problem B/3

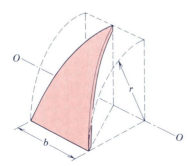

Problem B/4

PROBLEMS

Introductory problems

B/1 Use the mass element $dm = \rho\, dx$, where ρ is the mass per unit length, and determine the mass moments of inertia I_{yy} and $I_{y'y'}$ of the homogeneous slender rod of mass m and length l.

Ans. $I_{yy} = \frac{1}{12}ml^2,\ I_{y'y'} = \frac{1}{3}ml^2$

B/2 Every "slender" bar has a finite radius r. Refer to Table D/4 and derive an expression for the percentage error e that results if one neglects the radius r of a homogeneous solid cylindrical bar of length l when calculating its moment of inertia I_{zz}. Evaluate your expression for $\dfrac{r}{l} = 0.01, 0.1,$ and 0.5.

B/3 In order to better appreciate the greater ease of integration with lower-order elements, determine the mass moment of inertia I_{xx} of the homogeneous thin plate by using the square element (a) and then by using the rectangular element (b). The mass of the plate is m. Then by inspection state I_{yy}, and finally, determine I_{zz}.

Ans. $I_{xx} = \frac{1}{12}ma^2$
$I_{yy} = \frac{1}{12}mb^2$
$I_{zz} = \frac{1}{12}m(a^2 + b^2)$

B/4 The wedge-shaped segment of mass m is cut from a uniform cylindrical shell of radius r and with a thickness that is small compared with r. By inspection, write the moment of inertia of the segment about the central axis *O-O*.

B/5 The moment of inertia of a solid homogeneous sphere of radius r about any noncentroidal axis x may be obtained approximately by multiplying the mass of the sphere by the square of the distance d between the x-axis and the parallel centroidal axis. What percentage error e results if (*a*) $d = 2r$ and (*b*) $d = 10r$?

Ans. (*a*) $e = -9.09\%$, (*b*) $e = -0.398\%$

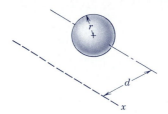

Problem B/5

B/6 Determine the radius of gyration of the thin triangular plate of mass m about the x-x axis through its base.

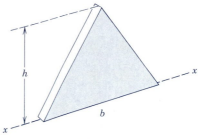

Problem B/6

B/7 Determine the moment of inertia of the thin equilateral triangular plate of mass m about the z-z axis normal to the plate through its mass center G. Solve by using the results of Prob. B/6, the relations developed for thin flat plates, and the tranfer-of-axis theorem.

Ans. $I_{zz} = \frac{1}{12}mb^2$

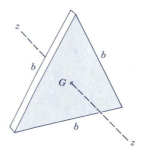

Problem B/7

B/8 A pattern is made from thin steel plate that has a mass of 31.3 kg per square meter of area. The area moments of inertia of the face of the plate are $I_x = 25.2(10^6)$ mm^4 and $I_y = 76.6(10^6)$ mm^4. Calculate the mass moment of inertia of the plate about the z-axis.

Problem B/8

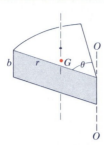

Problem B/9

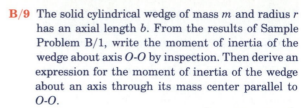

B/9 The solid cylindrical wedge of mass m and radius r has an axial length b. From the results of Sample Problem B/1, write the moment of inertia of the wedge about axis O-O by inspection. Then derive an expression for the moment of inertia of the wedge about an axis through its mass center parallel to O-O.

$$\text{Ans. } I_{OO} = \tfrac{1}{2}mr^2, \ I_{GG} = mr^2\left(\frac{1}{2} - \frac{16}{9}\frac{\sin^2 \theta/2}{\theta^2}\right)$$

Problem B/10

B/10 Determine I_{xx} for the cylinder with a centered circular hole. The mass of the body is m.

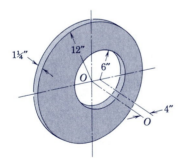

Problem B/11

B/11 Calculate the radius of gyration about axis O-O for the steel disk with the hole.   *Ans.* $k_O = 9.20$ in.

Representative problems

B/12 The semicircular disk weighs 5 lb, and its small thickness may be neglected compared with its 10-in. radius. Compute the moment of inertia of the disk about the x-, y-, y'-, and z-axes.

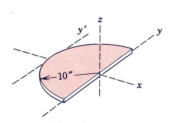

Problem B/12

B/13 A badminton racket is constructed of uniform slender rods bent into the shape shown. Neglect the strings and the wooden handle and estimate the mass moment of inertia about the y-axis through O, which is the location of the player's hand. The mass per unit length of the rod material is ρ.

$$Ans. \ I_{yy} = \left(\tfrac{43}{192} + \tfrac{83}{128}\pi\right)\rho L^3$$

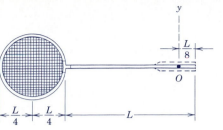

Problem B/13

B/14 Calculate the moment of inertia of the steel control wheel, shown in section, about its central axis. There are eight spokes, each of which has a constant cross-sectional area of 200 mm². What percent n of the total moment of inertia is contributed by the outer rim?

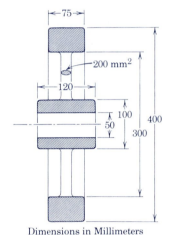

Dimensions in Millimeters

Problem B/14

B/15 The clock pendulum consists of the slender rod of length l and mass m and the bob of mass $7m$. Neglect the effects of the radius of the bob and determine I_O in terms of the bob position x. Calculate the ratio R of I_O evaluated for $x = \tfrac{3}{4}l$ to I_O evaluated for $x = l$.

$$Ans. \ I_O = m(7x^2 + \tfrac{1}{3}l^2), \ R = 0.582$$

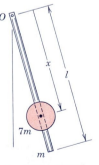

Problem B/15

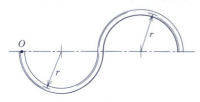

Problem B/16

B/16 A slender rod is bent into the shape of the two semi-circles and has a total mass m. Determine the moment of inertia of the rod about an axis through O perpendicular to the plane of the bent rod. Obtain your result without using the centroidal relation for a semicircular arc.

Problem B/17

B/17 The uniform circular cylinder has mass m, radius r, and length l. Derive the expression for its moment of inertia about the end axis x-x.

$$Ans.\ I_{xx} = m\left(\frac{r^2}{4} + \frac{l^2}{3}\right)$$

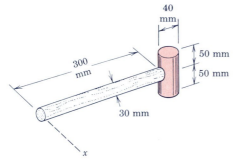

Problem B/18

B/18 Determine the moment of inertia of the mallet about the x-axis. The density of the wooden handle is 800 kg/m³ and that of the soft-metal head is 9000 kg/m³. The longitudinal axis of the cylindrical head is normal to the x-axis.

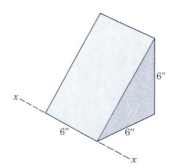

Problem B/19

B/19 A 6-in. steel cube is cut along its diagonal plane. Calculate the moment of inertia of the resulting prism about the edge x-x.   *Ans.* $I_{xx} = 1.898$ lb-in.-sec²

B/20 Calculate the moment of inertia of the solid steel semicylinder about the *x-x* axis and about the parallel x_0-x_0 axis. (See Table D/1 for the density of steel.)

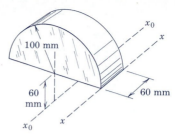

Problem B/20

B/21 Determine the moment of inertia about the tangent *x-x* axis for the full ring of mass m_1 and the half-ring of mass m_2.

$$\text{Ans. } I_{xx} = \tfrac{3}{2}m_1 r^2, \; I_{xx} = m_2 r^2 \left(\frac{3}{2} - \frac{4}{\pi} \right)$$

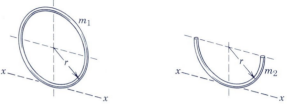

Problem B/21

B/22 Determine the moment of inertia of the half-ring of mass *m* about its diametral axis *a-a* and about axis *b-b* through the midpoint of the arc normal to the plane of the ring. The radius of the circular cross section is small compared with *r*.

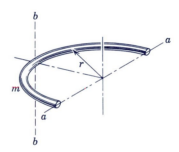

Problem B/22

B/23 The radius of the solid of revolution is proportional to the square of the *x*-coordinate. If the mass of the body is *m*, determine I_{xx}. $\text{Ans. } I_{xx} = \tfrac{5}{18}mr^2$

Problem B/23

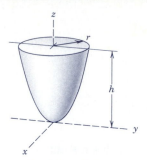

Problem B/24

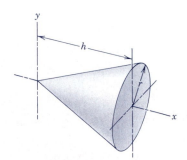

Problem B/26

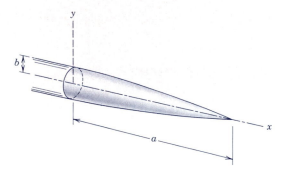

Problem B/27

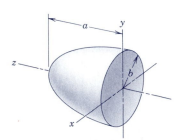

Problem B/28

B/24 Determine the radius of gyration about the *z*-axis of the paraboloid of revolution shown. The mass of the homogeneous body is *m*.

B/25 Determine the moment of inertia about the *y*-axis for the paraboloid of revolution of Prob. B/24.

$$Ans. \ I_{yy} = \tfrac{1}{2}m\left(h^2 + \frac{r^2}{3} \right)$$

B/26 Calculate the moment of inertia of the homogeneous right-circular cone of mass *m*, base radius *r*, and altitude *h* about the cone axis *x* and about the *y*-axis through its vertex.

B/27 A cylindrical model rocket has the solid nose cone shown in the figure. The boundary in the first quadrant of the *x-y* plane is given by the parabola $y = b\left(1 - \frac{x^2}{a^2} \right)$. If the mass of the homogeneous nose cone is *m*, determine its moment of inertia about the *x*-axis. $Ans. \ I_{xx} = \tfrac{8}{21}mb^2$

B/28 Determine the moment of inertia about the *x*-axis of the homogeneous solid semiellipsoid of revolution having mass *m*.

B/29 Determine the mass moment of inertia about the x-axis of the thin elliptical plate of mass m.

$$Ans. \quad I_{xx} = \tfrac{1}{4}mb^2$$

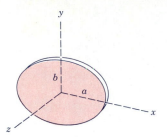

Problem B/29

B/30 A homogeneous solid of mass m is formed by revolving the 45° right triangle about the z-axis. Determine the radius of gyration of the solid about the z-axis.

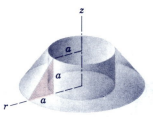

Problem B/30

B/31 Determine the mass moment of inertia about the x-axis of the solid spherical segment of mass m.

$$Ans. \quad I_{xx} = \tfrac{53}{200}mR^2$$

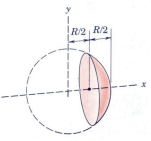

Problem B/31

B/32 Determine the moment of inertia about the x-axis of the portion of the homogeneous sphere shown. The mass of the sphere portion is m.

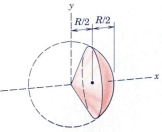

Problem B/32

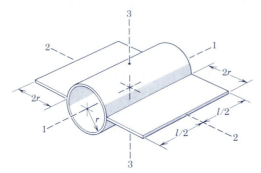

Problem B/33

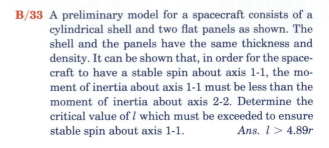

B/33 A preliminary model for a spacecraft consists of a cylindrical shell and two flat panels as shown. The shell and the panels have the same thickness and density. It can be shown that, in order for the spacecraft to have a stable spin about axis 1-1, the moment of inertia about axis 1-1 must be less than the moment of inertia about axis 2-2. Determine the critical value of l which must be exceeded to ensure stable spin about axis 1-1.   *Ans.* $l > 4.89r$

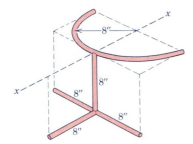

Problem B/34

B/34 The welded assembly shown is made from steel rod that weighs 0.667 lb per foot of length. Calculate the moment of inertia of the assembly about the x-x axis.

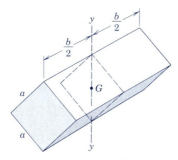

Problem B/35

B/35 The homogeneous bar of mass m and length b has a square cross section. Determine the moment of inertia of the bar about the y-y axis through the mass center G of the bar and along a diagonal of the central square section.   *Ans.* $I_{yy} = \frac{1}{12}m(a^2 + b^2)$

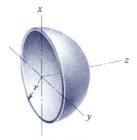

Problem B/36

B/36 Determine the moments of inertia of the half-spherical shell with respect to the x- and z-axes. The mass of the shell is m, and its thickness is negligible compared with the radius r.

B/37 Determine the moment of inertia about the generating axis of a complete ring of mass m having a circular section (torus) with the dimensions shown in the sectional view.   *Ans.* $I = m(R^2 + \frac{3}{4}a^2)$

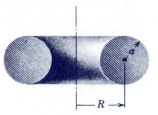

Problem B/37

B/38 Determine the moment of inertia, about the generating axis, of the hollow circular tube of mass m obtained by revolving the thin ring shown in the sectional view completely around the generating axis.

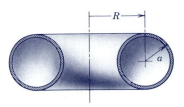

Problem B/38

B/39 Determine I_{xx} for the cone frustum, which has base radii r_1 and r_2 and mass m.

$$\text{Ans. } I = \tfrac{3}{10}m\,\frac{r_2{}^5 - r_1{}^5}{r_2{}^3 - r_1{}^3}$$

▶**B/40** Determine the mass moment of inertia about the centroidal $\bar{y}$-axis of the cone frustum of Prob. B/39.

$$\text{Ans. } \bar{I}_{yy} = \frac{m}{2P_1P_2}\left\{\tfrac{3}{10}[P_2 + 4h^2](r_2{}^5 - r_1{}^5)\right.$$
$$\left. - \tfrac{9}{8}h^2\,\frac{r_2{}^4 - r_1{}^4}{P_1}\right\}$$

where $P_1 = r_2{}^3 - r_1{}^3$, $P_2 = (r_2 - r_1)^2$

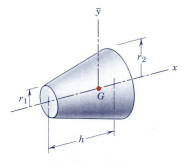

Problem B/39

▶**B/41** The cube with semicircular grooves in two opposite faces is cast of lead. Calculate the moment of inertia of the solid about the a-a axis.
     Ans. $I_{aa} = 0.367$ kg·m^2

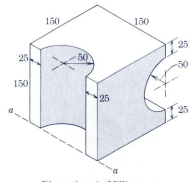

Dimensions in Millimeters

Problem B/41

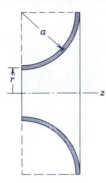

Problem B/42

▶**B/42** A shell of mass m is obtained by revolving the circular section about the z-axis. If the thickness of the shell is small compared with a and if $r = a/3$, determine the radius of gyration of the shell about the z-axis.   *Ans.* $k_z = 0.890a$

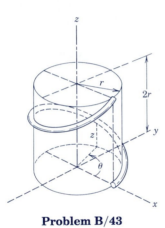

Problem B/43

▶**B/43** A slender rod of mass ρ per unit length is wrapped around the circular cylinder in the form of a helix for one complete turn as shown. The z-coordinate is proportional to θ. Determine the moments of inertia about the z- and x-axes.

Ans. $I_{zz} = 2\rho r^3\sqrt{1 + \pi^2}$, $I_{xx} = \frac{11}{3}\rho r^3\sqrt{1 + \pi^2}$

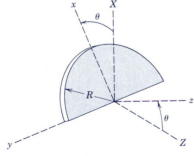

Problem B/44

▶**B/44** By direct integration, determine the moment of inertia about the Z-axis of the thin semicircular disk of mass m and radius R inclined an angle θ from the X-y plane.   *Ans.* $I_{ZZ} = \frac{1}{4}mR^2(1 + \cos^2 \theta)$

B/2 *PRODUCTS OF INERTIA*

For problems in the rotation of three-dimensional rigid bodies, the expression for angular momentum contains, in addition to the moment-of-inertia terms, *product-of-inertia* terms defined as

$$
\begin{aligned}
I_{xy} &= I_{yx} = \int xy\, dm \\[6pt]
I_{xz} &= I_{zx} = \int xz\, dm \\[6pt]
I_{yz} &= I_{zy} = \int yz\, dm
\end{aligned}
\qquad \textbf{(B/8)}
$$

These expressions were cited in Eqs. 7/10 in the expansion of the expression for angular momentum, Eq. 7/9.

The calculation of products of inertia involves the same basic procedure that we have followed in calculating moments of inertia and in evaluating other volume integrals as far as the choice of element and the limits of integration are concerned. The only special precaution we need to observe is to be doubly watchful of the algebraic signs in the expressions. Whereas moments of inertia are always positive, products of inertia may be either positive or negative. The units of products of inertia are the same as those of moments of inertia.

We have seen that the calculation of moments of inertia is often simplified by using the transfer-of-axis theorem. A similar theorem exists for transferring products of inertia, and we prove it easily as follows. In Fig. B/6 is shown the *x-y* view of a rigid body with parallel axes x_0-y_0 passing through the mass center G and located from the *x-y* axes by the distances d_x and d_y. The product of inertia about the *x-y* axes by definition is

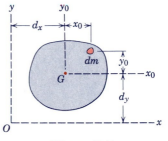

Figure B/6

$$
\begin{aligned}
I_{xy} &= \int xy\, dm = \int (x_0 + d_x)(y_0 + d_y)\, dm \\[6pt]
&= \int x_0 y_0\, dm + d_x d_y \int dm + d_x \int y_0\, dm + d_y \int x_0\, dm \\[6pt]
&= I_{x_0 y_0} + m d_x d_y
\end{aligned}
$$

The last two integrals vanish since the first moments of mass about the mass center are necessarily zero. Similar relations exist for the remaining two product-of-inertia terms. Dropping the zero subscripts and using the bar to designate the mass-center quantity give

$$
\begin{aligned}
I_{xy} &= \bar{I}_{xy} + m d_x d_y \\[4pt]
I_{xz} &= \bar{I}_{xz} + m d_x d_z \\[4pt]
I_{yz} &= \bar{I}_{yz} + m d_y d_z
\end{aligned}
\qquad \textbf{(B/9)}
$$

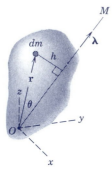

Figure B/7

These transfer-of-axis relations are valid *only* for transfer to or from *parallel axes* through the *mass center*.

With the aid of the product-of-inertia terms, we can calculate the moment of inertia of a rigid body about any prescribed axis through the origin of coordinates. For the rigid body of Fig. B/7, let it be required to determine the moment of inertia about axis *OM*. The direction cosines of *OM* are l, m, n, and a unit vector $\boldsymbol{\lambda}$ along *OM* may be written $\boldsymbol{\lambda} = l\mathbf{i} + m\mathbf{j} + n\mathbf{k}$. The moment of inertia about *OM* is

$$I_M = \int h^2\, dm = \int (\mathbf{r} \times \boldsymbol{\lambda})\cdot(\mathbf{r} \times \boldsymbol{\lambda})\, dm$$

where $|\mathbf{r} \times \boldsymbol{\lambda}| = r \sin\theta = h$. The cross product is

$$(\mathbf{r} \times \boldsymbol{\lambda}) = (yn - zm)\mathbf{i} + (zl - xn)\mathbf{j} + (xm - yl)\mathbf{k}$$

and, after we collect terms, the dot-product expansion gives

$$(\mathbf{r} \times \boldsymbol{\lambda})\cdot(\mathbf{r} \times \boldsymbol{\lambda}) = h^2 = (y^2 + z^2)l^2 + (x^2 + z^2)m^2 + (x^2 + y^2)n^2$$
$$- 2xylm - 2xzln - 2yzmn$$

Thus, with the substitution of the expressions of Eqs. B/4 and B/8, we have

$$\boxed{I_M = I_{xx}l^2 + I_{yy}m^2 + I_{zz}n^2 - 2I_{xy}lm - 2I_{xz}ln - 2I_{yz}mn} \quad \textbf{(B/10)}$$

This expression gives the moment of inertia about any axis *OM* in terms of the direction cosines of the axis and the moments and products of inertia about the coordinate directions.

(a) Principal axes of inertia.
Expansion of the angular momentum expression, Eq. 7/11, for a rigid body with attached axes yields the array

$$\begin{bmatrix} I_{xx} & -I_{xy} & -I_{xz} \\ -I_{yx} & I_{yy} & -I_{yz} \\ -I_{zx} & -I_{zy} & I_{zz} \end{bmatrix}$$

which is known as the *inertia matrix* or *inertia tensor*. If we examine the moment- and product-of-inertia terms for all possible orientations of the axes with respect to the body for a given origin, we will find in the general case one unique orientation *x-y-z* for which the product-of-inertia terms vanish and the array takes the diagonalized form

$$\begin{bmatrix} I_{xx} & 0 & 0 \\ 0 & I_{yy} & 0 \\ 0 & 0 & I_{zz} \end{bmatrix}$$

Axes *x-y-z* are called the *principal axes of inertia*, and I_{xx}, I_{yy}, and I_{zz} are called the *principal moments of inertia* and represent the maximum, minimum, and intermediate values of the moments of inertia for the particular origin chosen.

It may be shown* that for any given orientation of axes *x-y-z* the solution of the determinant equation

$$\begin{vmatrix} I_{xx} - I & -I_{xy} & -I_{xz} \\ -I_{yx} & I_{yy} - I & -I_{yz} \\ -I_{zx} & -I_{zy} & I_{zz} - I \end{vmatrix} = 0 \qquad (B/11)$$

for *I* yields three roots I_1, I_2, and I_3 of the resulting cubic equation which are the three principal moments of inertia. Also, the direction cosines *l*, *m*, and *n* of a principal inertia axis are given by

$$(I_{xx} - I)l - I_{xy}m - I_{xz}n = 0$$

$$-I_{yx}l + (I_{yy} - I)m - I_{yz}n = 0 \qquad (B/12)$$

$$-I_{zx}l - I_{zy}m + (I_{zz} - I)n = 0$$

These equations along with $l^2 + m^2 + n^2 = 1$ will enable a solution for the direction cosines to be made for each of the three *I*'s separately.

To assist with the visualization of these conclusions, consider the rectangular block, Fig. B/8, with an arbitrary orientation with

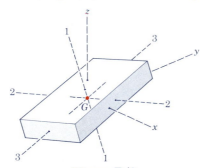

Figure B/8

respect to the *x-y-z* axes. For simplicity, the mass center *G* is located at the origin of the coordinates. If the moments and products of inertia for the block about the *x-y-z* axes are known, then solution of Eq. B/11 would give the three roots, I_1, I_2, and I_3, which are the principal moments of inertia. Solution of Eq. B/12 for each of the three *I*'s, in turn, along with $l^2 + m^2 + n^2 = 1$ would give the direction cosines *l*, *m*, and *n* for each respective principal axis. From the proportions of the block as drawn, we see that I_1 is the maximum moment of inertia, I_2 is the intermediate value, and I_3 is the minimum moment of inertia.

*See, for example, the senior author's *Dynamics, SI Version*, 1975, John Wiley & Sons, Art. 41.

Sample Problem B/4

The bent plate has a uniform thickness t that is negligible compared with its other dimensions. The density of the plate material is ρ. Determine the products of inertia of the plate with respect to the axes as chosen.

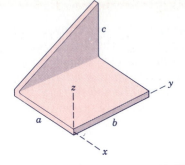

Solution. Each of the two parts is analyzed separately.

① *Rectangular part.* In the separate view of this part, we introduce parallel axes x_0-y_0 through the mass center G and use the transfer-of-axis theorem. By symmetry, we see that $\bar{I}_{xy} = I_{x_0 y_0} = 0$ so that

$$[I_{xy} = \bar{I}_{xy} + md_x d_y] \qquad I_{xy} = 0 + \rho tab\left(-\frac{a}{2}\right)\left(\frac{b}{2}\right) = -\frac{1}{4}\rho t a^2 b^2$$

Since the z-coordinate of all elements of the plate is zero, it follows that $I_{xz} = I_{yz} = 0$.

Triangular part. In the separate view of this part, we locate the mass center G and construct x_0-, y_0-, and z_0-axes through G. Since the x_0-coordinate of all elements is zero, it follows that $\bar{I}_{xy} = I_{x_0 y_0} = 0$ and $\bar{I}_{xz} = I_{x_0 z_0} = 0$. The transfer-of-axis theorems then give us

$$[I_{xy} = \bar{I}_{xy} + md_x d_y] \qquad I_{xy} = 0 + \rho t\frac{b}{2}c(-a)\left(\frac{2b}{3}\right) = -\frac{1}{3}\rho t a b^2 c$$

$$[I_{xz} = \bar{I}_{xz} + md_x d_z] \qquad I_{xz} = 0 + \rho t\frac{b}{2}c(-a)\left(\frac{c}{3}\right) = -\frac{1}{6}\rho t a b c^2$$

We obtain I_{yz} by direct integration, noting that the distance a of the plane of the triangle from the y-z plane in no way affects the y- and z-coordinates. With the mass element $dm = \rho t\, dy\, dz$, we have

$$②\left[I_{yz} = \int yz\, dm\right] \qquad I_{yz} = \rho t\int_0^b \int_0^{cy/b} yz\, dz\, dy = \rho t\int_0^b y\left[\frac{z^2}{2}\right]_0^{cy/b} dy$$

$$= \frac{\rho t c^2}{2b^2}\int_0^b y^3\, dy = \frac{1}{8}\rho t b^2 c^2$$

Adding the expressions for the two parts gives

$$I_{xy} = -\tfrac{1}{4}\rho t a^2 b^2 - \tfrac{1}{3}\rho t a b^2 c = -\tfrac{1}{12}\rho t a b^2(3a + 4c) \qquad \text{Ans.}$$

$$I_{xz} = 0 \qquad -\tfrac{1}{6}\rho t a b c^2 = -\tfrac{1}{6}\rho t a b c^2 \qquad \text{Ans.}$$

$$I_{yz} = 0 \qquad +\tfrac{1}{8}\rho t b^2 c^2 = +\tfrac{1}{8}\rho t b^2 c^2 \qquad \text{Ans.}$$

① We must be careful to preserve the same sense of the coordinates. Thus, plus x_0 and y_0 must agree with plus x and y.

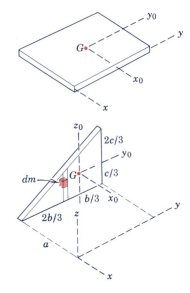

② We choose to integrate with respect to z first, where the upper limit is the variable height $z = cy/b$. If we were to integrate first with respect to y, the limits of the first integral would be from the variable $y = bz/c$ to b.

Sample Problem   B/5

The angle bracket is made from aluminum plate with a mass of 13.45 kg per square meter. Calculate the principal moments of inertia about the origin O and the direction cosines of the principal axes of inertia. The thickness of the plate is small compared with the other dimensions.

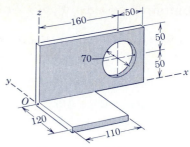

Dimensions in Millimeters

Solution.   The masses of the three parts are

$$m_1 = 13.45(0.21)(0.1) = 0.282 \text{ kg}$$

①
$$m_2 = -13.45\pi(0.035)^2 = -0.0518 \text{ kg}$$

$$m_3 = 13.45(0.12)(0.11) = 0.1775 \text{ kg}$$

① Note that the mass of the hole is treated as a negative number.

Part 1

$$I_{xx} = \tfrac{1}{3}mb^2 = \tfrac{1}{3}(0.282)(0.1)^2 = 9.42(10^{-4}) \text{ kg} \cdot \text{m}^2$$

②
$$I_{yy} = \tfrac{1}{3}m(a^2 + b^2) = \tfrac{1}{3}(0.282)[(0.21)^2 + (0.1)^2] = 50.9(10^{-4}) \text{ kg} \cdot \text{m}^2$$

$$I_{zz} = \tfrac{1}{3}ma^2 = \tfrac{1}{3}(0.282)(0.21)^2 = 41.5(10^{-4}) \text{ kg} \cdot \text{m}^2$$

$$I_{xy} = 0 \qquad I_{yz} = 0$$

$$I_{xz} = 0 + m\,\frac{a}{2}\frac{b}{2} = 0.282(0.105)(0.05) = 14.83(10^{-4}) \text{ kg} \cdot \text{m}^2$$

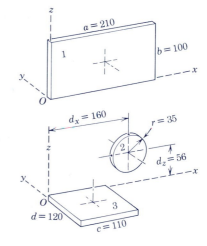

Part 2

$$I_{xx} = \tfrac{1}{4}mr^2 + md_z{}^2 = -0.0518\left[\frac{(0.035)^2}{4} + (0.050)^2\right]$$

$$= -1.45(10^{-4}) \text{ kg} \cdot \text{m}^2$$

$$I_{yy} = \tfrac{1}{2}mr^2 + m(d_x{}^2 + d_z{}^2)$$

$$= -0.0518\left[\frac{(0.035)^2}{2} + (0.16)^2 + (0.05)^2\right]$$

$$= -14.86(10^{-4}) \text{ kg} \cdot \text{m}^2$$

② You can easily derive this formula. Also check Table D/4.

$$I_{zz} = \tfrac{1}{4}mr^2 + md_x{}^2 = -0.0518\left[\frac{(0.035)^2}{4} + (0.16)^2\right]$$

$$= -13.41(10^{-4}) \text{ kg} \cdot \text{m}^2$$

$$I_{xy} = 0 \qquad I_{yz} = 0$$

$$I_{xz} = 0 + md_xd_z = -0.0518(0.16)(0.05) = -4.14(10^{-4}) \text{ kg} \cdot \text{m}^2$$

Sample Problem   B/5 (Continued)

Part 3

$$I_{xx} = \tfrac{1}{3}md^2 = \tfrac{1}{3}(0.1775)(0.12)^2 = 8.52(10^{-4}) \text{ kg·m}^2$$

$$I_{yy} = \tfrac{1}{3}mc^2 = \tfrac{1}{3}(0.1775)(0.11)^2 = 7.16(10^{-4}) \text{ kg·m}^2$$

$$I_{zz} = \tfrac{1}{3}m(c^2 + d^2) = \tfrac{1}{3}(0.1775)[(0.11)^2 + (0.12)^2]$$

$$= 15.68(10^{-4}) \text{ kg·m}^2$$

$$I_{xy} = m\,\frac{c}{2}\left(\frac{-d}{2}\right) = 0.1775(0.055)(-0.06) = -5.86(10^{-4}) \text{ kg·m}^2$$

$$I_{yz} = 0 \qquad I_{xz} = 0$$

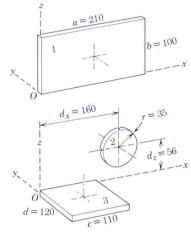

Totals

$$I_{xx} = 16.48(10^{-4}) \text{ kg·m}^2 \qquad I_{xy} = -5.86(10^{-4}) \text{ kg·m}^2$$

$$I_{yy} = 43.23(10^{-4}) \text{ kg·m}^2 \qquad I_{yz} = 0$$

$$I_{zz} = 43.79(10^{-4}) \text{ kg·m}^2 \qquad I_{xz} = 10.69(10^{-4}) \text{ kg·m}^2$$

Substitution into Eq. B/11, expansion of the determinant, and simplification yield

$$I^3 - 103.51(10^{-4})I^2 + 3179(10^{-8})I - 24\,769(10^{-12}) = 0$$

(Figure repeated)

③ Solution of this cubic equation yields the following roots that are the principal moments of inertia

$$I_1 = 48.4(10^{-4}) \text{ kg·m}^2$$

$$I_2 = 11.82(10^{-4}) \text{ kg·m}^2$$

$$I_3 = 43.3(10^{-4}) \text{ kg·m}^2$$

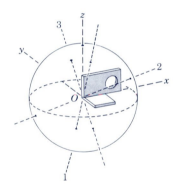

The direction cosines of each principal axis are obtained by substituting each root, in turn, into Eq. B/12 and using $l^2 + m^2 + n^2 = 1$. The results are

$$l_1 = 0.363 \qquad l_2 = 0.934 \qquad l_3 = 0.0183$$

$$m_1 = 0.409 \qquad m_2 = -0.174 \qquad m_3 = 0.895$$

$$n_1 = -0.837 \qquad n_2 = -0.312 \qquad n_3 = -0.445$$

The bottom figure shows a pictorial view of the bracket and the orientation of its principal axes of inertia.

③ A computer program for the solution of a cubic equation may be used, or an algebraic solution using the formula cited in item 4 of Art. C/4, Appendix C, may be employed.

PROBLEMS

Introductory problems

B/45 Determine the products of inertia about the coordinate axes for the unit that consists of four small particles, each of mass m, connected by the light but rigid slender rods.

Ans. $I_{xy} = -2ml^2$, $I_{xz} = -4ml^2$, $I_{yz} = 0$

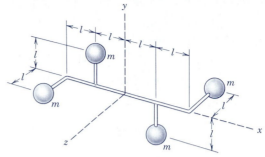

Problem B/45

B/46 Determine the products of inertia of the uniform slender rod of mass m about the coordinate axes shown.

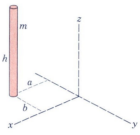

Problem B/46

B/47 Determine the products of inertia about the coordinate axes for the thin square plate with two circular holes. The mass of the plate material per unit area is ρ.

Ans. $I_{xy} = \dfrac{\rho \pi b^4}{512}$, $I_{xz} = I_{yz} = 0$

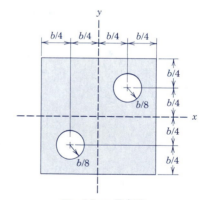

Problem B/47

B/48 The uniform rectangular block weighs 50 lb. Calculate its products of inertia about the coordinate axes shown.

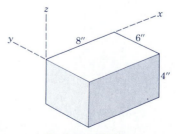

Problem B/48

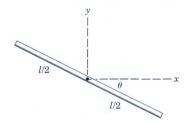

Problem B/49

B/49 Determine the product of inertia I_{xy} for the slender rod of mass m.   *Ans.* $I_{xy} = -\frac{1}{24}ml^2 \sin 2\theta$

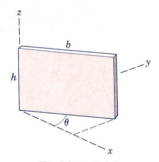

Problem B/50

B/50 Determine the products of inertia of the solid homogeneous half-cylinder of mass m for the axes shown.

Representative problems

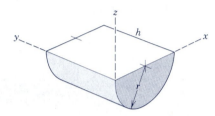

Problem B/51

B/51 Determine the three products of inertia with respect to the given axes for the uniform rectangular plate of mass m.   *Ans.* $I_{xy} = \frac{1}{6}mb^2 \sin 2\theta$

$$I_{xz} = \frac{1}{4}mbh \cos \theta$$
$$I_{yz} = \frac{1}{4}mbh \sin \theta$$

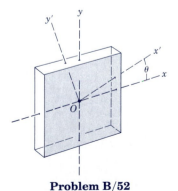

Problem B/52

B/52 Without calculation, show that $I_{x'y'} = 0$ for x'-y' axes through the center of the square block and parallel to its face.

B/53 The semicircular disk of mass m and radius R, inclined at an angle θ from the X-y plane, of Prob. B/44 is repeated here. By the methods of this article, determine the moment of inertia about the Z-axis.

$$\text{Ans. } I_{ZZ} = \tfrac{1}{4}mR^2(1 + \cos^2 \theta)$$

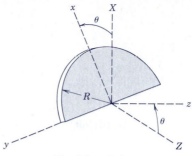

Problem B/53

B/54 Prove that the moment of inertia of the rigid assembly of three identical balls, each of mass m and radius r, has the same value for all axes through O. Neglect the mass of the connecting rods.

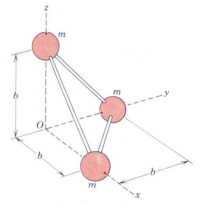

Problem B/54

B/55 For the slender rod of mass m bent into the configuration shown, determine its products of inertia I_{xy}, I_{xz}, and I_{yz}.

$$\text{Ans. } I_{xy} = \frac{mb^2}{4\sqrt{2}}, \; I_{xz} = -\frac{1}{12}mb^2, \; I_{yz} = -\frac{mb^2}{4\sqrt{2}}$$

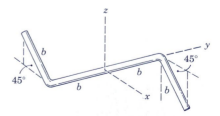

Problem B/55

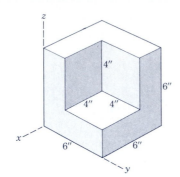

Problem B/56

B/56 The aluminum casting consists of a 6-in. cube with a 4-in. cubical recess. Calculate the products of inertia of the casting about the axes shown.

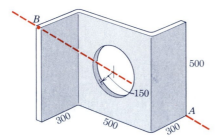

Dimensions in Millimeters
Problem B/57

B/57 The steel plate with two right-angle bends and a central hole has a thickness of 15 mm. Calculate its moment of inertia about the diagonal axis through the corners A and B. *Ans.* $I_{AB} = 2.58$ kg·m^2

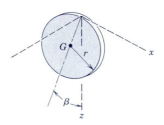

Problem B/58

B/58 The plane of the thin circular disk of mass m and radius r makes an angle β with the x-z plane. Determine the product of inertia of the disk with respect to the y-z plane.

Computer-oriented problems

*B/59 Each of the spheres of mass m has a diameter that is small compared with the dimension b. Neglect the mass of the connecting struts and determine the principal moments of inertia of the assembly with respect to the coordinates shown. Determine also the direction cosines of the axis of maximum moment of inertia.

Ans. $I_1 = 7.525mb^2,\ l_1 = 0.521$
$I_2 = 6.631mb^2,\ m_1 = -0.756$
$I_3 = 1.844mb^2,\ n_1 = 0.397$

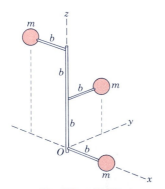

Problem B/59

*B/60 Determine the principal moments of inertia for the bent rod, which has a mass ρ per unit length.

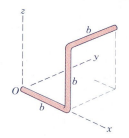

Problem B/60

*B/61 The curved rod of Prob. B/16 is shown here with axis *A-A* through the mass center *G* in the plane of the rod. Determine the moment of inertia *I* of the rod about axis *A-A* in terms of θ and plot *I* versus θ from $\theta = 0$ to a value that discloses both I_{min} and I_{max}. Check your plotted values of I_{min} and I_{max} with the values calculated from Eqs. A/11 of *Vol. 1 Statics.*   *Ans.* $I_{min} = 0.1905mr^2$ at $\theta = 25.9°$
$$I_{max} = 1.809mr^2 \text{ at } \theta = 115.9°$$

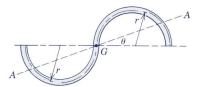

Problem B/61

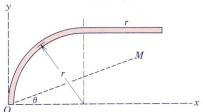

B/62 Determine the moment of inertia *I* about axis *OM* for the uniform slender rod bent into the shape shown. Plot *I* versus θ from $\theta = 0$ to $\theta = 90°$ and determine the minimum value of *I* and the angle α that its axis makes with the *x*-direction. (*Note:* Because the analysis does not involve the *z*-coordinate, the expressions developed for area moments of inertia, Eqs. A/9, A/10, and A/11 in Appendix A of *Vol. 1 Statics*, may be utilized for this problem in place of the three-dimensional relations of Appendix B.) The rod has a mass ρ per unit length.

Problem B/62

B/63 The thin plate has a mass ρ per unit area and is formed into the shape shown. Determine the principal moments of inertia of the plate about axes through *O*.   *Ans.* $I_1 = 3.779\rho b^4$
$$I_2 = 0.612\rho b^4$$
$$I_3 = 3.609\rho b^4$$

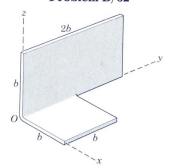

Problem B/63

B/64 The slender rod has a mass ρ per unit length and is formed into the shape shown. Determine the principal moments of inertia about axes through *O* and calculate the direction cosines of the axis of minimum moment of inertia.

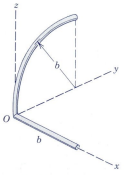

Problem B/64

SELECTED TOPICS OF MATHEMATICS

C

C/1 INTRODUCTION

Appendix C contains an abbreviated summary and reminder of selected topics in basic mathematics which find frequent use in mechanics. The relationships are cited without proof. The student of mechanics will have frequent occasion to use many of these relations, and he or she will be handicapped if they are not well in hand. Other topics not listed will also be needed from time to time.

As the reader reviews and applies mathematics, he or she should bear in mind that mechanics is an applied science descriptive of real bodies and actual motions. Therefore, the geometric and physical interpretation of the applicable mathematics should be kept clearly in mind during the development of theory and the formulation and solution of problems.

C/2 PLANE GEOMETRY

1. When two intersecting lines are, respectively, perpendicular to two other lines, the angles formed by each pair are equal.

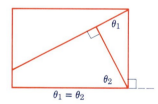

$$\theta_1 = \theta_2$$

2. Similar triangles

$$\frac{x}{b} = \frac{h - y}{h}$$

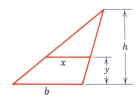

3. Any triangle

$$\text{Area} = \tfrac{1}{2}bh$$

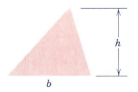

4. Circle

Circumference $= 2\pi r$
Area $= \pi r^2$
Arc length $s = r\theta$
Sector area $= \tfrac{1}{2}r^2\theta$

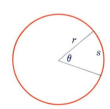

5. Every triangle inscribed within a semicircle is a right triangle.

$$\theta_1 + \theta_2 = \pi/2$$

6. Angles of a triangle

$$\theta_1 + \theta_2 + \theta_3 = 180°$$
$$\theta_4 = \theta_1 + \theta_2$$

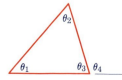

C/3 SOLID GEOMETRY

1. Sphere

   Volume $= \frac{4}{3}\pi r^3$
   Surface area $= 4\pi r^2$

3. Right-circular cone

   Volume $= \frac{1}{3}\pi r^2 h$
   Lateral area $= \pi r L$
   $L = \sqrt{r^2 + h^2}$

2. Spherical wedge

   Volume $= \frac{2}{3}r^3\theta$

4. Any pyramid or cone

   Volume $= \frac{1}{3}Bh$
   where B = area of base

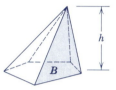

C/4 ALGEBRA

1. Quadratic equation

   $ax^2 + bx + c = 0$

   $x = \dfrac{-b \pm \sqrt{b^2 - 4ac}}{2a}$, $b^2 \geq 4ac$ for real roots

2. Logarithms

   $b^x = y$, $x = \log_b y$

   Natural logarithms

   $b = e = 2.718\ 282$
   $e^x = y$, $x = \log_e y = \ln y$
   $\log (ab) = \log a + \log b$
   $\log (a/b) = \log a - \log b$
   $\log (1/n) = -\log n$
   $\log a^n = n \log a$
   $\log 1 = 0$
   $\log_{10} x = 0.4343 \ln x$

3. Determinants
   2nd order

   $\begin{vmatrix} a_1 & b_1 \\ a_2 & b_2 \end{vmatrix} = a_1 b_2 - a_2 b_1$

   3rd order

   $\begin{vmatrix} a_1 & b_1 & c_1 \\ a_2 & b_2 & c_2 \\ a_3 & b_3 & c_3 \end{vmatrix} = \begin{aligned} &+a_1 b_2 c_3 + a_2 b_3 c_1 + a_3 b_1 c_2 \\ &-a_3 b_2 c_1 - a_2 b_1 c_3 - a_1 b_3 c_2 \end{aligned}$

4. Cubic equation

   $x^3 = Ax + B$

   Let $p = A/3$, $q = B/2$.

   Case I: $q^2 - p^3$ negative (three roots real and distinct)

   $$\cos u = q/(p\sqrt{p}),\ 0 < u < 180°$$
   $$x_1 = 2\sqrt{p}\, \cos (u/3)$$
   $$x_2 = 2\sqrt{p}\, \cos (u/3 + 120°)$$
   $$x_3 = 2\sqrt{p}\, \cos (u/3 + 240°)$$

   Case II: $q^2 - p^3$ positive (one root real, two roots imaginary)

   $$x_1 = (q + \sqrt{q^2 - p^3})^{1/3} + (q - \sqrt{q^2 - p^3})^{1/3}$$

   Case III: $q^2 - p^3 = 0$ (three roots real, two roots equal)

   $$x_1 = 2q^{1/3},\ x_2 = x_3 = -q^{1/3}$$

   For general cubic equation

   $$x^3 + ax^2 + bx + c = 0$$

   Substitute $x = x_0 - a/3$ and get $x_0^3 = Ax_0 + B$. Then proceed as above to find values of x_0 from which $x = x_0 - a/3$.

C/5 ANALYTIC GEOMETRY

1. Straight line

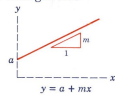

$$y = a + mx$$

$$\frac{x}{a} + \frac{y}{b} = 1$$

2. Circle

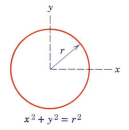

$$x^2 + y^2 = r^2$$

$$(x - a)^2 + (y - b)^2 = r^2$$

3. Parabola

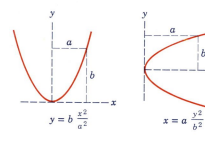

$$y = b\,\frac{x^2}{a^2}$$

$$x = a\,\frac{y^2}{b^2}$$

4. Ellipse

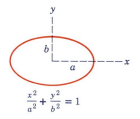

$$\frac{x^2}{a^2} + \frac{y^2}{b^2} = 1$$

5. Hyperbola

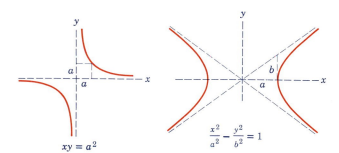

$$xy = a^2$$

$$\frac{x^2}{a^2} - \frac{y^2}{b^2} = 1$$

C/6 TRIGONOMETRY

1. Definitions

$\sin \theta = a/c \quad \csc \theta = c/a$
$\cos \theta = b/c \quad \sec \theta = c/b$
$\tan \theta = a/b \quad \cot \theta = b/a$

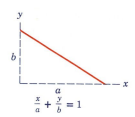

2. Signs in the four quadrants

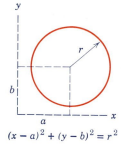

	I	II	III	IV
$\sin \theta$	+	+	−	−
$\cos \theta$	+	−	−	+
$\tan \theta$	+	−	+	−
$\csc \theta$	+	+	−	−
$\sec \theta$	+	−	−	+
$\cot \theta$	+	−	+	−

3. Miscellaneous relations

$\sin^2 \theta + \cos^2 \theta = 1$
$1 + \tan^2 \theta = \sec^2 \theta$
$1 + \cot^2 \theta = \csc^2 \theta$

$\sin \dfrac{\theta}{2} = \sqrt{\tfrac{1}{2}(1 - \cos \theta)}$

$\cos \dfrac{\theta}{2} = \sqrt{\tfrac{1}{2}(1 + \cos \theta)}$

$\sin 2\theta = 2 \sin \theta \cos \theta$
$\cos 2\theta = \cos^2 \theta - \sin^2 \theta$
$\sin (a \pm b) = \sin a \cos b \pm \cos a \sin b$
$\cos (a \pm b) = \cos a \cos b \mp \sin a \sin b$

4. Law of sines

$\dfrac{a}{b} = \dfrac{\sin A}{\sin B}$

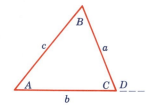

5. Law of cosines

$c^2 = a^2 + b^2 - 2ab \cos C$
$c^2 = a^2 + b^2 + 2ab \cos D$

C/7 *VECTOR OPERATIONS*

1. *Notation.*   Vector quantities are printed in boldface type, and scalar quantities appear in lightface italic type. Thus, the vector quantity **V** has a scalar magnitude V. In longhand work vector quantities should always be consistently indicated by a symbol such as $\underline{V}$ or $\vec{V}$ to distinguish them from scalar quantities.

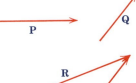

2. *Addition*

   Triangle addition $\mathbf{P} + \mathbf{Q} = \mathbf{R}$
   Parallelogram addition $\mathbf{P} + \mathbf{Q} = \mathbf{R}$
   Commutative law $\mathbf{P} + \mathbf{Q} = \mathbf{Q} + \mathbf{P}$
   Associative law $\mathbf{P} + (\mathbf{Q} + \mathbf{R}) = (\mathbf{P} + \mathbf{Q}) + \mathbf{R}$

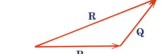

3. *Subtraction*

$$\mathbf{P} - \mathbf{Q} = \mathbf{P} + (-\mathbf{Q})$$

4. *Unit vectors* **i, j, k**

$$\mathbf{V} = V_x \mathbf{i} + V_y \mathbf{j} + V_z \mathbf{k}$$

where $|\mathbf{V}| = V = \sqrt{V_x^2 + V_y^2 + V_z^2}$

5. *Direction cosines* l, m, n are the cosines of the angles between **V** and the x-, y-, z-axes. Thus,

$$l = V_x/V \qquad m = V_y/V \qquad n = V_z/V$$

so that $\mathbf{V} = V(l\mathbf{i} + m\mathbf{j} + n\mathbf{k})$

and $l^2 + m^2 + n^2 = 1$

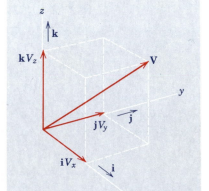

6. ***Dot or scalar product***

$$\mathbf{P} \cdot \mathbf{Q} = PQ \cos \theta$$

This product may be viewed as the magnitude of **P** multiplied by the component $Q \cos \theta$ of **Q** in the direction of **P**, or as the magnitude of **Q** multiplied by the component $P \cos \theta$ of **P** in the direction of **Q**.

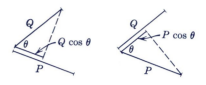

Commutative law $\quad \mathbf{P} \cdot \mathbf{Q} = \mathbf{Q} \cdot \mathbf{P}$

From the definition of the dot product

$$\mathbf{i} \cdot \mathbf{i} = \mathbf{j} \cdot \mathbf{j} = \mathbf{k} \cdot \mathbf{k} = 1$$

$$\mathbf{i} \cdot \mathbf{j} = \mathbf{j} \cdot \mathbf{i} = \mathbf{i} \cdot \mathbf{k} = \mathbf{k} \cdot \mathbf{i} = \mathbf{j} \cdot \mathbf{k} = \mathbf{k} \cdot \mathbf{j} = 0$$

$$\mathbf{P} \cdot \mathbf{Q} = (P_x \mathbf{i} + P_y \mathbf{j} + P_z \mathbf{k}) \cdot (Q_x \mathbf{i} + Q_y \mathbf{j} + Q_z \mathbf{k})$$
$$= P_x Q_x + P_y Q_y + P_z Q_z$$

$$\mathbf{P} \cdot \mathbf{P} = P_x{}^2 + P_y{}^2 + P_z{}^2$$

It follows from the definition of the dot product that two vectors **P** and **Q** are perpendicular when their dot product vanishes, $\mathbf{P} \cdot \mathbf{Q} = 0$.

The angle θ between two vectors $\mathbf{P}_1$ and $\mathbf{P}_2$ may be found from their dot product expression $\mathbf{P}_1 \cdot \mathbf{P}_2 = P_1 P_2 \cos \theta$, which gives

$$\cos \theta = \frac{\mathbf{P}_1 \cdot \mathbf{P}_2}{P_1 P_2} = \frac{P_{1_x} P_{2_x} + P_{1_y} P_{2_y} + P_{1_z} P_{2_z}}{P_1 P_2} = l_1 l_2 + m_1 m_2 + n_1 n_2$$

where l, m, n stand for the respective direction cosines of the vectors. It is also observed that two vectors are perpendicular to each other when their direction cosines obey the relation $l_1 l_2 + m_1 m_2 + n_1 n_2 = 0$.

Distributive law $\quad \mathbf{P} \cdot (\mathbf{Q} + \mathbf{R}) = \mathbf{P} \cdot \mathbf{Q} + \mathbf{P} \cdot \mathbf{R}$

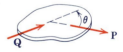

7. ***Cross or vector product.*** The cross product $\mathbf{P} \times \mathbf{Q}$ of the two vectors **P** and **Q** is defined as a vector with a magnitude

$$|\mathbf{P} \times \mathbf{Q}| = PQ \sin \theta$$

and a direction specified by the right-hand rule as shown. Reversing the vector order and using the right-hand rule give $\mathbf{Q} \times \mathbf{P} = -\mathbf{P} \times \mathbf{Q}$.

Distributive law $\quad \mathbf{P} \times (\mathbf{Q} + \mathbf{R}) = \mathbf{P} \times \mathbf{Q} + \mathbf{P} \times \mathbf{R}$

From the definition of the cross product, using a *right-handed coordinate system*, we get

$$\mathbf{i} \times \mathbf{j} = \mathbf{k} \qquad \mathbf{j} \times \mathbf{k} = \mathbf{i} \qquad \mathbf{k} \times \mathbf{i} = \mathbf{j}$$

$$\mathbf{j} \times \mathbf{i} = -\mathbf{k} \qquad \mathbf{k} \times \mathbf{j} = -\mathbf{i} \qquad \mathbf{i} \times \mathbf{k} = -\mathbf{j}$$

$$\mathbf{i} \times \mathbf{i} = \mathbf{j} \times \mathbf{j} = \mathbf{k} \times \mathbf{k} = 0$$

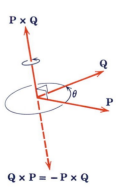

With the aid of these identities and the distributive law, the vector product may be written

$$\mathbf{P} \times \mathbf{Q} = (P_x\mathbf{i} + P_y\mathbf{j} + P_z\mathbf{k}) \times (Q_x\mathbf{i} + Q_y\mathbf{j} + Q_z\mathbf{k})$$
$$= (P_yQ_z - P_zQ_y)\mathbf{i} + (P_zQ_x - P_xQ_z)\mathbf{j} + (P_xQ_y - P_yQ_x)\mathbf{k}$$

The cross product may also be expressed by the determinant

$$\mathbf{P} \times \mathbf{Q} = \begin{vmatrix} \mathbf{i} & \mathbf{j} & \mathbf{k} \\ P_x & P_y & P_z \\ Q_x & Q_y & Q_z \end{vmatrix}$$

8. *Additional relations*

Triple scalar product $(\mathbf{P} \times \mathbf{Q}) \cdot \mathbf{R} = \mathbf{R} \cdot (\mathbf{P} \times \mathbf{Q})$. The dot and cross may be interchanged as long as the order of the vectors is maintained. Parentheses are unnecessary since $\mathbf{P} \times (\mathbf{Q} \cdot \mathbf{R})$ is meaningless because a vector $\mathbf{P}$ cannot be crossed into a scalar $\mathbf{Q} \cdot \mathbf{R}$. Thus, the expression may be written

$$\mathbf{P} \times \mathbf{Q} \cdot \mathbf{R} = \mathbf{P} \cdot \mathbf{Q} \times \mathbf{R}$$

The triple scalar product has the determinant expansion

$$\mathbf{P} \times \mathbf{Q} \cdot \mathbf{R} = \begin{vmatrix} P_x & P_y & P_z \\ Q_x & Q_y & Q_z \\ R_x & R_y & R_z \end{vmatrix}$$

Triple vector product $(\mathbf{P} \times \mathbf{Q}) \times \mathbf{R} = -\mathbf{R} \times (\mathbf{P} \times \mathbf{Q})$
$$= \mathbf{R} \times (\mathbf{Q} \times \mathbf{P})$$

Here the parentheses must be used since an expression $\mathbf{P} \times \mathbf{Q} \times \mathbf{R}$ would be ambiguous because it would not identify the vector to be crossed. It may be shown that the triple vector product is equivalent to

$$(\mathbf{P} \times \mathbf{Q}) \times \mathbf{R} = \mathbf{R} \cdot \mathbf{P}\mathbf{Q} - \mathbf{R} \cdot \mathbf{Q}\mathbf{P}$$
or
$$\mathbf{P} \times (\mathbf{Q} \times \mathbf{R}) = \mathbf{P} \cdot \mathbf{R}\mathbf{Q} - \mathbf{P} \cdot \mathbf{Q}\mathbf{R}$$

The first term in the first expression, for example, is the dot product $\mathbf{R} \cdot \mathbf{P}$, a scalar, multiplied by the vector $\mathbf{Q}$.

9. *Derivatives of vectors* obey the same rules as they do for scalars.

$$\frac{d\mathbf{P}}{dt} = \dot{\mathbf{P}} = \dot{P}_x\mathbf{i} + \dot{P}_y\mathbf{j} + \dot{P}_z\mathbf{k}$$

$$\frac{d(\mathbf{P}u)}{dt} = \mathbf{P}\dot{u} + \dot{\mathbf{P}}u$$

$$\frac{d(\mathbf{P} \cdot \mathbf{Q})}{dt} = \mathbf{P} \cdot \dot{\mathbf{Q}} + \dot{\mathbf{P}} \cdot \mathbf{Q}$$

$$\frac{d(\mathbf{P} \times \mathbf{Q})}{dt} = \mathbf{P} \times \dot{\mathbf{Q}} + \dot{\mathbf{P}} \times \mathbf{Q}$$

10. ***Integration of vectors.*** If $\mathbf{V}$ is a function of x, y, and z and an element of volume is $d\tau = dx\,dy\,dz$, the integral of $\mathbf{V}$ over the volume may be written as the vector sum of the three integrals of its components. Thus,

$$\int \mathbf{V}\,d\tau = \mathbf{i} \int V_x\,d\tau + \mathbf{j} \int V_y\,d\tau + \mathbf{k} \int V_z\,d\tau$$

C/8 SERIES

(Expression in brackets following series indicates range of convergence.)

$$(1 \pm x)^n = 1 \pm nx + \frac{n(n-1)}{2!}x^2 \pm \frac{n(n-1)(n-2)}{3!}x^3 + \cdots \quad [x^2 < 1]$$

$$\sin x = x - \frac{x^3}{3!} + \frac{x^5}{5!} - \frac{x^7}{7!} + \cdots \qquad\qquad [x^2 < \infty]$$

$$\cos x = 1 - \frac{x^2}{2!} + \frac{x^4}{4!} - \frac{x^6}{6!} + \cdots \qquad\qquad [x^2 < \infty]$$

$$\sinh x = \frac{e^x - e^{-x}}{2} = x + \frac{x^3}{3!} + \frac{x^5}{5!} + \frac{x^7}{7!} + \cdots \qquad [x^2 < \infty]$$

$$\cosh x = \frac{e^x + e^{-x}}{2} = 1 + \frac{x^2}{2!} + \frac{x^4}{4!} + \frac{x^6}{6!} + \cdots \qquad [x^2 < \infty]$$

$$f(x) = \frac{a_0}{2} + \sum_{n=1}^{\infty} a_n \cos \frac{n\pi x}{l} + \sum_{n=1}^{\infty} b_n \sin \frac{n\pi x}{l}$$

where $a_n = \dfrac{1}{l} \displaystyle\int_{-l}^{l} f(x) \cos \frac{n\pi x}{l}\,dx,$ $\qquad b_n = \dfrac{1}{l} \displaystyle\int_{-l}^{l} f(x) \sin \frac{n\pi x}{l}\,dx$

$$[\text{Fourier expansion for } -l < x < l]$$

C/9 DERIVATIVES

$$\frac{dx^n}{dx} = nx^{n-1}, \qquad \frac{d(uv)}{dx} = u\frac{dv}{dx} + v\frac{du}{dx}, \qquad \frac{d\left(\dfrac{u}{v}\right)}{dx} = \frac{v\dfrac{du}{dx} - u\dfrac{dv}{dx}}{v^2}$$

$$\lim_{\Delta x \to 0} \sin \Delta x = \sin dx = \tan dx = dx$$

$$\lim_{\Delta x \to 0} \cos \Delta x = \cos dx = 1$$

$$\frac{d \sin x}{dx} = \cos x, \qquad \frac{d \cos x}{dx} = -\sin x, \qquad \frac{d \tan x}{dx} = \sec^2 x$$

$$\frac{d \sinh x}{dx} = \cosh x, \qquad \frac{d \cosh x}{dx} = \sinh x, \qquad \frac{d \tanh x}{dx} = \operatorname{sech}^2 x$$

10 INTEGRALS

$$\int x^n \, dx = \frac{x^{n-1}}{n+1}$$

$$\int \frac{dx}{x} = \ln x$$

$$\int \sqrt{a+bx} \, dx = \frac{2}{3b} \sqrt{(a+bx)^3}$$

$$\int x\sqrt{a+bx} \, dx = \frac{2}{15b^2} (3bx - 2a)\sqrt{(a+bx)^3}$$

$$\int x^2\sqrt{a+bx} \, dx = \frac{2}{105b^3} (8a^2 - 12abx + 15b^2x^2)\sqrt{(a+bx)^3}$$

$$\int \frac{dx}{\sqrt{a+bx}} = \frac{2\sqrt{a+bx}}{b}$$

$$\int \frac{\sqrt{a+x}}{\sqrt{b-x}} \, dx = -\sqrt{a+x}\sqrt{b-x} + (a+b) \sin^{-1}\sqrt{\frac{a+x}{a+b}}$$

$$\int \frac{x \, dx}{a+bx} = \frac{1}{b^2} [a + bx - a \ln (a+bx)]$$

$$\int \frac{x \, dx}{(a+bx)^n} = \frac{(a+bx)^{1-n}}{b^2} \left(\frac{a+bx}{2-n} - \frac{a}{1-n} \right)$$

$$\int \frac{dx}{a+bx^2} = \frac{1}{\sqrt{ab}} \tan^{-1} \frac{x\sqrt{ab}}{a} \qquad \text{or} \qquad \frac{1}{\sqrt{-ab}} \tanh^{-1} \frac{x\sqrt{-ab}}{a}$$

$$\int \frac{x \, dx}{a+bx^2} = \frac{1}{2b} \ln (a+bx^2)$$

$$\int \sqrt{x^2 \pm a^2} \, dx = \tfrac{1}{2}[x\sqrt{x^2 \pm a^2} \pm a^2 \ln (x + \sqrt{x^2 \pm a^2})]$$

$$\int \sqrt{a^2 - x^2} \, dx = \tfrac{1}{2} \left(x\sqrt{a^2 - x^2} + a^2 \sin^{-1} \frac{x}{a} \right)$$

$$\int x\sqrt{a^2 - x^2} \, dx = -\tfrac{1}{3}\sqrt{(a^2 - x^2)^3}$$

$$\int x^2\sqrt{a^2 - x^2} \, dx = -\frac{x}{4} \sqrt{(a^2 - x^2)^3} + \frac{a^2}{8} \left(x\sqrt{a^2 - x^2} + a^2 \sin^{-1} \frac{x}{a} \right)$$

$$\int x^3\sqrt{a^2 - x^2} \, dx = -\tfrac{1}{5}(x^2 + \tfrac{2}{3}a^2)\sqrt{(a^2 - x^2)^3}$$

$$\int \frac{dx}{\sqrt{a+bx+cx^2}} = \frac{1}{\sqrt{c}} \ln \left(\sqrt{a+bx+cx^2} + x\sqrt{c} + \frac{b}{2\sqrt{c}} \right) \qquad \text{or} \qquad \frac{-1}{\sqrt{-c}} \sin^{-1} \left(\frac{b+2cx}{\sqrt{b^2 - 4ac}} \right)$$

$$\int \frac{dx}{\sqrt{x^2 \pm a^2}} = \ln (x + \sqrt{x^2 \pm a^2})$$

$$\int \frac{dx}{\sqrt{a^2 - x^2}} = \sin^{-1} \frac{x}{a}$$

$$\int \frac{x\,dx}{\sqrt{x^2 - a^2}} = \sqrt{x^2 - a^2}$$

$$\int \frac{x\,dx}{\sqrt{a^2 \pm x^2}} = \pm\sqrt{a^2 \pm x^2}$$

$$\int x\sqrt{x^2 \pm a^2}\,dx = \tfrac{1}{3}\sqrt{(x^2 \pm a^2)^3}$$

$$\int x^2\sqrt{x^2 \pm a^2}\,dx = \frac{x}{4}\sqrt{(x^2 \pm a^2)^3} \mp \frac{a^2}{8}\,x\sqrt{x^2 \pm a^2} - \frac{a^4}{8}\ln\,(x + \sqrt{x^2 \pm a^2})$$

$$\int \sin x\,dx = -\cos x$$

$$\int \cos x\,dx = \sin x$$

$$\int \sec x\,dx = \frac{1}{2}\ln\frac{1 + \sin x}{1 - \sin x}$$

$$\int \sin^2 x\,dx = \frac{x}{2} - \frac{\sin 2x}{4}$$

$$\int \cos^2 x\,dx = \frac{x}{2} + \frac{\sin 2x}{4}$$

$$\int \sin x \cos x\,dx = \frac{\sin^2 x}{2}$$

$$\int \sinh x\,dx = \cosh x$$

$$\int \cosh x\,dx = \sinh x$$

$$\int \tanh x\,dx = \ln \cosh x$$

$$\int \ln x\,dx = x \ln x - x$$

$$\int e^{ax}\,dx = \frac{e^{ax}}{a}$$

$$\int x e^{ax}\,dx = \frac{e^{ax}}{a^2}(ax - 1)$$

$$\int e^{ax} \sin px\,dx = \frac{e^{ax}(a \sin px - p \cos px)}{a^2 + p^2}$$

$$\int e^{ax} \cos px\,dx = \frac{e^{ax}(a \cos px + p \sin px)}{a^2 + p^2}$$

$$\int e^{ax} \sin^2 x \, dx = \frac{e^{ax}}{4 + a^2} \left(a \sin^2 x - \sin 2x + \frac{2}{a} \right)$$

$$\int e^{ax} \cos^2 x \, dx = \frac{e^{ax}}{4 + a^2} \left(a \cos^2 x + \sin 2x + \frac{2}{a} \right)$$

$$\int e^{ax} \sin x \cos x \, dx = \frac{e^{ax}}{4 + a^2} \left(\frac{a}{2} \sin 2x - \cos 2x \right)$$

$$\int \sin^3 x \, dx = -\frac{\cos x}{3} (2 + \sin^2 x)$$

$$\int \cos^3 x \, dx = \frac{\sin x}{3} (2 + \cos^2 x)$$

$$\int \cos^5 x \, dx = \sin x - \tfrac{2}{3} \sin^3 x + \tfrac{1}{5} \sin^5 x$$

$$\int x \sin x \, dx = \sin x - x \cos x$$

$$\int x \cos x \, dx = \cos x + x \sin x$$

$$\int x^2 \sin x \, dx = 2x \sin x - (x^2 - 2) \cos x$$

$$\int x^2 \cos x \, dx = 2x \cos x + (x^2 - 2) \sin x$$

Radius of curvature
$$\begin{cases} \rho_{xy} = \dfrac{\left[1 + \left(\dfrac{dy}{dx} \right)^2 \right]^{3/2}}{\dfrac{d^2y}{dx^2}} \\[20pt] \rho_{r\theta} = \dfrac{\left[r^2 + \left(\dfrac{dr}{d\theta} \right)^2 \right]^{3/2}}{r^2 + 2 \left(\dfrac{dr}{d\theta} \right)^2 - r \dfrac{d^2r}{d\theta^2}} \end{cases}$$

C/11 NEWTON'S METHOD FOR SOLVING INTRACTABLE EQUATIONS

Frequently, the application of the fundamental principles of mechanics leads to an algebraic or transcendental equation which is not solvable (or easily solvable) in closed form. In such cases, an iterative technique, such as Newton's method, can be a powerful tool for obtaining a good estimate to the root or roots of the equation.

Let us place the equation to be solved in the form $f(x) = 0$. The a-part of the accompanying figure depicts an arbitrary function $f(x)$ for values of x in the vicinity of the desired root x_r. Note that x_r is merely the value of x at which the function crosses the x-axis. Suppose that we have available (perhaps via a hand-drawn plot) a rough estimate x_1 of this root. Provided that x_1 does not closely correspond to a maximum or minimum value of the function $f(x)$, we may obtain a better estimate of the root x_r by projecting the tangent to $f(x)$ at

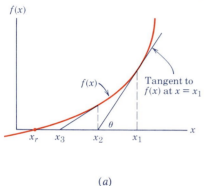

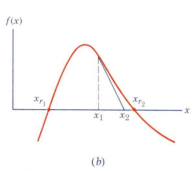

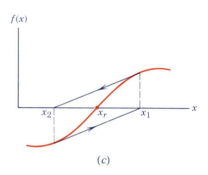

(a)   (b)   (c)

x_1 such that it intersects the x-axis at x_2. From the geometry of the figure, we may write

$$\tan \theta = f'(x_1) = \frac{f(x_1)}{x_1 - x_2}$$

where $f'(x_1)$ denotes the derivative of $f(x)$ with respect to x evaluated at $x = x_1$. Solving the above equation for x_2 results in

$$x_2 = x_1 - \frac{f(x_1)}{f'(x_1)}$$

The term $-f(x_1)/f'(x_1)$ is the correction to the initial root estimate x_1. Once x_2 is calculated, we may repeat the process to obtain x_3, and so forth.

Thus, we generalize the above equation to

$$x_{k+1} = x_k - \frac{f(x_k)}{f'(x_k)}$$

where

$$x_{k+1} = \text{the } (k + 1)\text{th estimate to the desired root } x_r$$

$$x_k = \text{the } k\text{th estimate to the desired root } x_r$$

$$f(x_k) = \text{the function } f(x) \text{ evaluated at } x = x_k$$

$$f'(x_k) = \text{the function derivative evaluated at } x = x_k$$

This equation is repetitively applied until $f(x_{k+1})$ is sufficiently close to zero and $x_{k+1} \cong x_k$. The student should verify that the equation is valid for all possible sign combinations of x_k, $f(x_k)$, and $f'(x_k)$.

Several cautionary notes are in order:

1. Clearly, $f'(x_k)$ must not be zero or close to zero. This would mean, as restricted above, that x_k exactly or approximately corresponds to a minimum or maximum of $f(x)$. If the slope $f'(x_k)$ is zero, then the slope projection never intersects the x-axis. If the slope $f'(x_k)$ is small, then the correction to x_k may be so large that x_{k+1} is a worse root estimate than x_k. For this reason, experienced engineers usually limit the size of the correction term; that is, if the absolute value of $f(x_k)/f'(x_k)$ is larger than a preselected maximum value, the maximum value is used.

2. If there are several roots of the equation $f(x) = 0$, we must be in the vicinity of the desired root x_r in order that the algorithm actually converges to that root. The b-part of the figure depicts the condition that the initial estimate x_1 will result in convergence to x_{r_2} rather than x_{r_1}.

3. Oscillation from one side of the root to the other can occur if, for example, the function is antisymmetric about a root which is an inflection point. The use of one-half of the correction will usually prevent this behavior, which is depicted in the c-part of the accompanying figure.

Example: Beginning with an initial estimate of $x_1 = 5$, estimate the single root of the equation $e^x - 10 \cos x - 100 = 0$.

The table below summarizes the application of Newton's method to the given equation. The iterative process was terminated when the absolute value of the correction $-f(x_k)/f'(x_k)$ became less than 10^{-6}.

k	x_k	$f(x_k)$	$f'(x_k)$	$x_{k+1} - x_k = -\dfrac{f(x_k)}{f'(x_k)}$
1	5.000 000	45.576 537	138.823 916	−0.328 305
2	4.671 695	7.285 610	96.887 065	−0.075 197
3	4.596 498	0.292 886	89.203 650	−0.003 283
4	4.593 215	0.000 527	88.882 536	−0.000 006
5	4.593 209	$-2(10^{-8})$	88.881 956	$2.25\,(10^{-10})$

C/12 *SELECTED TECHNIQUES FOR NUMERICAL INTEGRATION*

1. Area determination.   Consider the problem of determining the shaded area under the curve $y = f(x)$ from $x = a$ to $x = b$ as depicted in the *a*-part of the figure and suppose that analytical integration is not feasible. The function may be known in tabular form from experimental measurements or it may be known in analytical form. The function is taken to be continuous within the interval $a < x < b$. We may divide the area into n vertical strips, each of width $\Delta x = (b - a)/n$, and then add the areas of all strips to obtain $A = \int y \, dx$. A representative strip of area A_i is shown with darker shading in the figure. Three useful numerical approximations are cited. In each case the greater the number of strips, the more accurate becomes the approximation geometrically. As a general rule, one can begin with a relatively small number of strips and increase the number until the resulting changes in the area approximation no longer improve the desired accuracy.

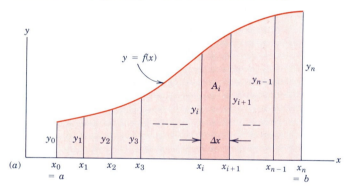

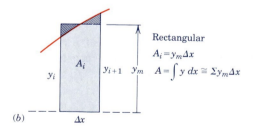

I. *Rectangular* [Figure (*b*)]   The areas of the strips are taken to be rectangles, as shown by the representative strip whose height y_m is chosen visually so that the small cross-hatched areas are as nearly equal as possible. Thus, we form the sum Σy_m of the effective heights and multiply by Δx. For a function known in analytical form, a value for y_m equal to that of the function at the midpoint $x_i + \Delta x/2$ may be calculated and used in the summation.

II. *Trapezoidal* [Figure (c)]   The areas of the strips are taken to be trapezoids, as shown by the representative strip. The area A_i is the average height $(y_i + y_{i+1})/2$ times Δx. Adding the areas gives the area approximation as tabulated. For the example with the curvature shown, clearly the approximation will be on the low side. For the reverse curvature, the approximation will be on the high side.

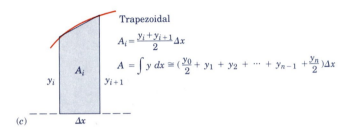

Trapezoidal

$$A_i = \frac{y_i + y_{i+1}}{2} \Delta x$$

$$A = \int y \, dx \cong \left(\frac{y_0}{2} + y_1 + y_2 + \cdots + y_{n-1} + \frac{y_n}{2}\right) \Delta x$$

(c)

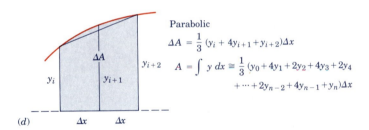

Parabolic

$$\Delta A = \frac{1}{3}(y_i + 4y_{i+1} + y_{i+2}) \Delta x$$

$$A = \int y \, dx \cong \frac{1}{3}(y_0 + 4y_1 + 2y_2 + 4y_3 + 2y_4$$
$$+ \cdots + 2y_{n-2} + 4y_{n-1} + y_n) \Delta x$$

(d)

III. *Parabolic* [Figure (d)]   The area between the chord and the curve (neglected in the trapezoidal solution) may be accounted for by approximating the function by a parabola passing through the points defined by three successive values of y. This area may be calculated from the geometry of the parabola and added to the trapezoidal area of the pair of strips to give the area ΔA of the pair as cited. Adding all of the ΔA's produces the tabulation shown, which is known as Simpson's rule. To use Simpson's rule, the number n of strips must be even.

Example: Determine the area under the curve $y = x\sqrt{1 + x^2}$ from $x = 0$ to $x = 2$. (An integrable function is chosen here so that the three approximations can be compared with the exact value, which is $A = \int_0^2 x\sqrt{1 + x^2}\, dx = \frac{1}{3}(1 + x^2)^{3/2}\big|_0^2 = \frac{1}{3}(5\sqrt{5} - 1) = 3.393\,447$.)

NUMBER OF	AREA APPROXIMATIONS		
SUBINTERVALS	RECTANGULAR	TRAPEZOIDAL	PARABOLIC
4	3.361 704	3.456 731	3.392 214
10	3.388 399	3.403 536	3.393 420
50	3.393 245	3.393 850	3.393 447
100	3.393 396	3.393 547	3.393 447
1000	3.393 446	3.393 448	3.393 447
2500	3.393 447	3.393 447	3.393 447

Note that the worst approximation error is less than 2 percent, even with only four strips.

2. Integration of first-order ordinary differential equations. The application of the fundamental principles of mechanics frequently results in differential relationships. Let us consider the first-order form $\frac{dy}{dt} = f(t)$, where the function $f(t)$ may not be readily integrable or may be known only in tabular form. We may numerically integrate by means of a simple slope-projection technique, known as Euler integration, which is illustrated in the figure.

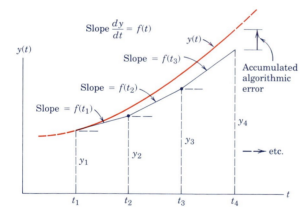

Beginning at t_1, at which the value y_1 is known, we project the slope over a horizontal subinterval or step $(t_2 - t_1)$ and see that $y_2 = y_1 + f(t_1)(t_2 - t_1)$. At t_2, the process may be repeated beginning at y_2, and so forth until the desired value of t is reached. Hence, the general expression is

$$y_{k+1} = y_k + f(t_k)(t_{k+1} - t_k)$$

If y versus t were linear, i.e., if $f(t)$ were constant, the method would be exact, and there would be no need for a numerical approach in that case. Changes in the slope over the subinterval introduce error. For the case shown in the figure, the estimate y_2 is clearly less than the true value of the function $y(t)$ at t_2. More accurate integration techniques (such as Runge–Kutta methods) take into account changes in the slope over the subinterval and thus provide better results.

As with the area-determination techniques, experience is helpful in the selection of a subinterval or step size when dealing with analytical functions. As a rough rule, one begins with a relatively large step size and then steadily decreases the step size until the corresponding changes in the integrated result are much smaller than the desired accuracy. A step size which is too small, however, can result in increased error due to a very large number of computer operations. This type of error is generally known as "round-off error", while the error which results from a large step size is known as algorithm error.

Example: For the differential equation $\dfrac{dy}{dt} = 5t$ with the initial condition $y = 2$ when $t = 0$, determine the value of y for $t = 4$.

Application of the Euler integration technique yields the following results:

NUMBER OF SUBINTERVALS	STEP SIZE	y at $t = 4$	PERCENT ERROR
10	0.4	38	9.5
100	0.04	41.6	0.95
500	0.008	41.92	0.19
1000	0.004	41.96	0.10

This simple example may be integrated analytically. The result is $y = 42$ (exactly).

USEFUL TABLES

D

TABLE D/1   PHYSICAL PROPERTIES

Density (kg/m³) *and specific weight* (lb/ft³)

	kg/m³	lb/ft³		kg/m³	lb/ft³
Air*	1.2062	0.07530			
Aluminum	2 690	168	Iron (cast)	7 210	450
Concrete (av.)	2 400	150	Lead	11 370	710
Copper	8 910	556	Mercury	13 570	847
Earth (wet, av.)	1 760	110	Oil (av.)	900	56
(dry, av.)	1 280	80	Steel	7 830	489
Glass	2 590	162	Titanium	3 080	192
Gold	19 300	1205	Water (fresh)	1 000	62.4
Ice	900	56	(salt)	1 030	64
* At 20°C (68°F) and atmospheric			Wood (soft pine)	480	30
pressure			(hard oak)	800	50

Coefficients of friction

(The coefficients in the following table represent typical values under normal working conditions. Actual coefficients for a given situation will depend on the exact nature of the contacting surfaces. A variation of 25 to 100 percent or more from these values could be expected in an actual application, depending on prevailing conditions of cleanliness, surface finish, pressure, lubrication, and velocity.)

	TYPICAL VALUES OF COEFFICIENT OF FRICTION	
CONTACTING SURFACE	STATIC, μ_s	KINETIC, μ_k
Steel on steel (dry)	0.6	0.4
Steel on steel (greasy)	0.1	0.05
Teflon on steel	0.04	0.04
Steel on babbitt (dry)	0.4	0.3
Steel on babbitt (greasy)	0.1	0.07
Brass on steel (dry)	0.5	0.4
Brake lining on cast iron	0.4	0.3
Rubber tires on smooth pavement (dry)	0.9	0.8
Wire rope on iron pulley (dry)	0.2	0.15
Hemp rope on metal	0.3	0.2
Metal on ice		0.02

TABLE D/2   SOLAR SYSTEM CONSTANTS

Universal gravitational constant	$G = 6.673(10^{-11})$ m^3/(kg·s^2)
	$= 3.439(10^{-8})$ ft^4/(lbf-s^4)
Mass of Earth	$m_e = 5.976(10^{24})$ kg
	$= 4.095(10^{23})$ lbf-s^2/ft
Period of Earth's rotation (1 sidereal day)	$= 23$ h 56 min 4 s
	$= 23.9344$ h
Angular velocity of Earth	$\omega = 0.7292(10^{-4})$ rad/s
Mean angular velocity of Earth–Sun line	$\omega' = 0.1991(10^{-6})$ rad/s
Mean velocity of Earth's center about Sun	$= 107\ 200$ km/h
	$= 66,610$ mi/h

BODY	MEAN DISTANCE TO SUN km (mi)	ECCENTRICITY OF ORBIT e	PERIOD OF ORBIT solar days	MEAN DIAMETER km (mi)	MASS RELATIVE TO EARTH	SURFACE GRAVITATIONAL ACCELERATION m/s^2 (ft/s^2)	ESCAPE VELOCITY km/s (mi/s)
Sun	—	—	—	1 392 000 (865 000)	333 000	274 (898)	616 (383)
Moon	384 398* (238 854)*	0.055	27.32	3 476 (2 160)	0.0123	1.62 (5.32)	2.37 (1.47)
Mercury	57.3×10^6 (35.6×10^6)	0.206	87.97	5 000 (3 100)	0.054	3.47 (11.4)	4.17 (2.59)
Venus	108×10^6 (67.2×10^6)	0.0068	224.70	12 400 (7 700)	0.815	8.44 (27.7)	10.24 (6.36)
Earth	149.6×10^6 (92.96×10^6)	0.0167	365.26	12 742† (7 918)†	1.000	9.821‡ (32.22)‡	11.18 (6.95)
Mars	227.9×10^6 (141.6×10^6)	0.093	686.98	6 788 (4 218)	0.107	3.73 (12.3)	5.03 (3.13)

* Mean distance to Earth (center-to-center)
† Diameter of sphere of equal volume, based on a spheroid Earth with a polar diameter of 12 714 km (7900 mi) and an equatorial diameter of 12 756 km (7926 mi)
‡ For nonrotating spherical Earth, equivalent to absolute value at sea level and latitude 37.5°

TABLE D/3   PROPERTIES OF PLANE FIGURES

FIGURE	CENTROID	AREA MOMENTS OF INERTIA
Arc Segment	$\bar{r} = \dfrac{r \sin \alpha}{\alpha}$	—
Quarter and Semicircular Arcs	$\bar{y} = \dfrac{2r}{\pi}$	—
Circular Area	—	$I_x = I_y = \dfrac{\pi r^4}{4}$ $I_z = \dfrac{\pi r^4}{2}$
Semicircular Area	$\bar{y} = \dfrac{4r}{3\pi}$	$I_x = I_y = \dfrac{\pi r^4}{8}$ $\bar{I}_x = \left(\dfrac{\pi}{8} - \dfrac{8}{9\pi}\right) r^4$ $I_z = \dfrac{\pi r^4}{4}$
Quarter-Circular Area	$\bar{x} = \bar{y} = \dfrac{4r}{3\pi}$	$I_x = I_y = \dfrac{\pi r^4}{16}$ $\bar{I}_x = \bar{I}_y = \left(\dfrac{\pi}{16} - \dfrac{4}{9\pi}\right) r^4$ $I_z = \dfrac{\pi r^4}{8}$
Area of Circular Sector	$\bar{x} = \dfrac{2}{3} \dfrac{r \sin \alpha}{\alpha}$	$I_x = \dfrac{r^4}{4} \left(\alpha - \tfrac{1}{2} \sin 2\alpha\right)$ $I_y = \dfrac{r^4}{4} \left(\alpha + \tfrac{1}{2} \sin 2\alpha\right)$ $I_z = \tfrac{1}{2} r^4 \alpha$

TABLE D/3   PROPERTIES OF PLANE FIGURES   *Continued*

FIGURE	CENTROID	AREA MOMENTS OF INERTIA
Rectangular Area 	—	$I_x = \dfrac{bh^3}{3}$ $\bar{I}_x = \dfrac{bh^3}{12}$ $\bar{I}_z = \dfrac{bh}{12}(b^2 + h^2)$
Triangular Area 	$\bar{x} = \dfrac{a+b}{3}$ $\bar{y} = \dfrac{h}{3}$	$I_x = \dfrac{bh^3}{12}$ $\bar{I}_x = \dfrac{bh^3}{36}$ $I_{x_1} = \dfrac{bh^3}{4}$
Area of Elliptical Quadrant 	$\bar{x} = \dfrac{4a}{3\pi}$ $\bar{y} = \dfrac{4b}{3\pi}$	$I_x = \dfrac{\pi ab^3}{16}, \quad \bar{I}_x = \left(\dfrac{\pi}{16} - \dfrac{4}{9\pi}\right)ab^3$ $I_y = \dfrac{\pi a^3 b}{16}, \quad \bar{I}_y = \left(\dfrac{\pi}{16} - \dfrac{4}{9\pi}\right)a^3 b$ $I_z = \dfrac{\pi ab}{16}(a^2 + b^2)$
Subparabolic Area $y = kx^2 = \dfrac{b}{a^2}x^2$ Area $A = \dfrac{ab}{3}$	$\bar{x} = \dfrac{3a}{4}$ $\bar{y} = \dfrac{3b}{10}$	$I_x = \dfrac{ab^3}{21}$ $I_y = \dfrac{a^3 b}{5}$ $I_z = ab\left(\dfrac{a^2}{5} + \dfrac{b^2}{21}\right)$
Parabolic Area $y = kx^2 = \dfrac{b}{a^2}x^2$ Area $A = \dfrac{2ab}{3}$	$\bar{x} = \dfrac{3a}{8}$ $\bar{y} = \dfrac{3b}{5}$	$I_x = \dfrac{2ab^3}{7}$ $I_y = \dfrac{2a^3 b}{15}$ $I_z = 2ab\left(\dfrac{a^2}{15} + \dfrac{b^2}{7}\right)$

TABLE D/4   PROPERTIES OF HOMOGENEOUS SOLIDS

(m = mass of body shown)

BODY	MASS CENTER	MASS MOMENTS OF INERTIA
Circular Cylindrical Shell	—	$I_{xx} = \frac{1}{2}mr^2 + \frac{1}{12}ml^2$ $I_{x_1x_1} = \frac{1}{2}mr^2 + \frac{1}{3}ml^2$ $I_{zz} = mr^2$
Half Cylindrical Shell	$\bar{x} = \dfrac{2r}{\pi}$	$I_{xx} = I_{yy}$ $\quad = \frac{1}{2}mr^2 + \frac{1}{12}ml^2$ $I_{x_1x_1} = I_{y_1y_1}$ $\quad = \frac{1}{2}mr^2 + \frac{1}{3}ml^2$ $I_{zz} = mr^2$ $\bar{I}_{zz} = \left(1 - \dfrac{4}{\pi^2}\right)mr^2$
Circular Cylinder	—	$I_{xx} = \frac{1}{4}mr^2 + \frac{1}{12}ml^2$ $I_{x_1x_1} = \frac{1}{4}mr^2 + \frac{1}{3}ml^2$ $I_{zz} = \frac{1}{2}mr^2$
Semicylinder	$\bar{x} = \dfrac{4r}{3\pi}$	$I_{xx} = I_{yy}$ $\quad = \frac{1}{4}mr^2 + \frac{1}{12}ml^2$ $I_{x_1x_1} = I_{y_1y_1}$ $\quad = \frac{1}{4}mr^2 + \frac{1}{3}ml^2$ $I_{zz} = \frac{1}{2}mr^2$ $\bar{I}_{zz} = \left(\dfrac{1}{2} - \dfrac{16}{9\pi^2}\right)mr^2$
Rectangular Parallelepiped	—	$I_{xx} = \frac{1}{12}m(a^2 + l^2)$ $I_{yy} = \frac{1}{12}m(b^2 + l^2)$ $I_{zz} = \frac{1}{12}m(a^2 + b^2)$ $I_{y_1y_1} = \frac{1}{12}mb^2 + \frac{1}{3}ml^2$ $I_{y_2y_2} = \frac{1}{3}m(b^2 + l^2)$

TABLE D/4   PROPERTIES OF HOMOGENEOUS SOLIDS   *Continued*

(m = mass of body shown)

BODY	MASS CENTER	MASS MOMENTS OF INERTIA
Spherical Shell	—	$I_{zz} = \frac{2}{3}mr^2$
Hemispherical Shell	$\bar{x} = \dfrac{r}{2}$	$I_{xx} = I_{yy} = I_{zz} = \frac{2}{3}mr^2$ $\bar{I}_{yy} = \bar{I}_{zz} = \frac{5}{12}mr^2$
Sphere	—	$I_{zz} = \frac{2}{5}mr^2$
Hemisphere	$\bar{x} = \dfrac{3r}{8}$	$I_{xx} = I_{yy} = I_{zz} = \frac{2}{5}mr^2$ $\bar{I}_{yy} = \bar{I}_{zz} = \frac{83}{320}mr^2$
Uniform Slender Rod	—	$I_{yy} = \frac{1}{12}ml^2$ $I_{y_1 y_1} = \frac{1}{3}ml^2$

TABLE D/4   PROPERTIES OF HOMOGENEOUS SOLIDS   *Continued*

(m = mass of body shown)

BODY	MASS CENTER	MASS MOMENTS OF INERTIA
Quarter-Circular Rod	$\bar{x} = \bar{y}$ $= \dfrac{2r}{\pi}$	$I_{xx} = I_{yy} = \frac{1}{2}mr^2$ $I_{zz} = mr^2$
Elliptical Cylinder	—	$I_{xx} = \frac{1}{4}ma^2 + \frac{1}{12}ml^2$ $I_{yy} = \frac{1}{4}mb^2 + \frac{1}{12}ml^2$ $I_{zz} = \frac{1}{4}m(a^2 + b^2)$ $I_{y_1 y_1} = \frac{1}{4}mb^2 + \frac{1}{3}ml^2$
Conical Shell	$\bar{z} = \dfrac{2h}{3}$	$I_{yy} = \frac{1}{4}mr^2 + \frac{1}{2}mh^2$ $I_{y_1 y_1} = \frac{1}{4}mr^2 + \frac{1}{6}mh^2$ $I_{zz} = \frac{1}{2}mr^2$ $\bar{I}_{yy} = \frac{1}{4}mr^2 + \frac{1}{18}mh^2$
Half Conical Shell	$\bar{x} = \dfrac{4r}{3\pi}$ $\bar{z} = \dfrac{2h}{3}$	$I_{xx} = I_{yy}$ $\qquad = \frac{1}{4}mr^2 + \frac{1}{2}mh^2$ $I_{x_1 x_1} = I_{y_1 y_1}$ $\qquad = \frac{1}{4}mr^2 + \frac{1}{6}mh^2$ $I_{zz} = \frac{1}{2}mr^2$ $\bar{I}_{zz} = \left(\dfrac{1}{2} - \dfrac{16}{9\pi^2} \right) mr^2$
Right-Circular Cone	$\bar{z} = \dfrac{3h}{4}$	$I_{yy} = \frac{3}{20}mr^2 + \frac{3}{5}mh^2$ $I_{y_1 y_1} = \frac{3}{20}mr^2 + \frac{1}{10}mh^2$ $I_{zz} = \frac{3}{10}mr^2$ $\bar{I}_{yy} = \frac{3}{20}mr^2 + \frac{3}{80}mh^2$

TABLE D/4   PROPERTIES OF HOMOGENEOUS SOLIDS   *Continued*

(m = mass of body shown)

BODY	MASS CENTER	MASS MOMENTS OF INERTIA
Half Cone	$\bar{x} = \dfrac{r}{\pi}$ $\bar{z} = \dfrac{3h}{4}$	$I_{xx} = I_{yy}$ $\quad = \frac{3}{20}mr^2 + \frac{3}{5}mh^2$ $I_{x_1x_1} = I_{y_1y_1}$ $\quad = \frac{3}{20}mr^2 + \frac{1}{10}mh^2$ $I_{zz} = \frac{3}{10}mr^2$ $\bar{I}_{zz} = \left(\dfrac{3}{10} - \dfrac{1}{\pi^2}\right)mr^2$
Semiellipsoid $\dfrac{x^2}{a^2} + \dfrac{y^2}{b^2} + \dfrac{z^2}{c^2} = 1$	$\bar{z} = \dfrac{3c}{8}$	$I_{xx} = \frac{1}{5}m(b^2 + c^2)$ $I_{yy} = \frac{1}{5}m(a^2 + c^2)$ $I_{zz} = \frac{1}{5}m(a^2 + b^2)$ $\bar{I}_{xx} = \frac{1}{5}m(b^2 + \frac{19}{64}c^2)$ $\bar{I}_{yy} = \frac{1}{5}m(a^2 + \frac{19}{64}c^2)$
Elliptic Paraboloid $\dfrac{x^2}{a^2} + \dfrac{y^2}{b^2} = \dfrac{z}{c}$	$\bar{z} = \dfrac{2c}{3}$	$I_{xx} = \frac{1}{6}mb^2 + \frac{1}{2}mc^2$ $I_{yy} = \frac{1}{6}ma^2 + \frac{1}{2}mc^2$ $I_{zz} = \frac{1}{6}m(a^2 + b^2)$ $\bar{I}_{xx} = \frac{1}{6}m(b^2 + \frac{1}{3}c^2)$ $\bar{I}_{yy} = \frac{1}{6}m(a^2 + \frac{1}{3}c^2)$
Rectangular Tetrahedron	$\bar{x} = \dfrac{a}{4}$ $\bar{y} = \dfrac{b}{4}$ $\bar{z} = \dfrac{c}{4}$	$I_{xx} = \frac{1}{10}m(b^2 + c^2)$ $I_{yy} = \frac{1}{10}m(a^2 + c^2)$ $I_{zz} = \frac{1}{10}m(a^2 + b^2)$ $\bar{I}_{xx} = \frac{3}{80}m(b^2 + c^2)$ $\bar{I}_{yy} = \frac{3}{80}m(a^2 + c^2)$ $\bar{I}_{zz} = \frac{3}{80}m(a^2 + b^2)$
Half Torus	$\bar{x} = \dfrac{a^2 + 4R^2}{2\pi R}$	$I_{xx} = I_{yy} = \frac{1}{2}mR^2 + \frac{5}{8}ma^2$ $I_{zz} = mR^2 + \frac{3}{4}ma^2$

PHOTO CREDITS

Chapter Openers

Chapter 1: Courtesy McDonnell-Douglas Corp.

Chapter 2: PSSC Physics, 2e, 1965, D. C. Heath & Co. with Education Development Center.

Chapter 3: Hank Morgan/Rainbow.

Chapter 4: NASA.

Chapter 5: Bob Daemmrich/The Image Works.

Chapter 6: Courtesy Lufthansa Airlines.

Chapter 7: Runk/Schoenberger/Grant Heilman Photography.

Chapter 8: Courtesy MTS Systems.

INDEX

CONVERSION CHARTS BETWEEN SI AND U.S. CUSTOMARY UNITS

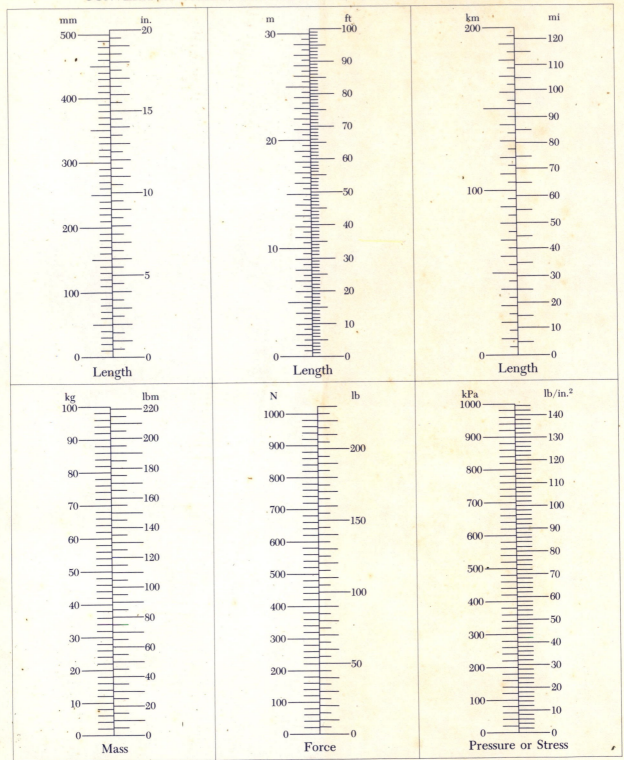